Michel Marie Deza • Elena Deza

Encyclopedia of Distances

Third Edition

Springer

Michel Marie Deza
Ecole Normale Supérieure
Paris, France

Elena Deza
Moscow State Pedagogical University
Moscow, Russia

ISBN 978-3-662-51868-7 ISBN 978-3-662-44342-2 (eBook)
DOI 10.1007/978-3-662-44342-2
Springer Heidelberg New York Dordrecht London

Mathematics Subject Classification (2010): 46-XX, 51-XX, 12-XX, 15-XX, 60-XX, 52-XX, 92-XX, 70-XX, 86-XX, 91-XX, 94-XX

Printed on acid-free paper

Springer is part of Springer Science+Business Media (www.springer.com)

Encyclopedia of Distances

In 1906, Maurice Fréchet submitted his outstanding thesis *Sur quelques points du calcul functionnel* introducing (within a systematic study of functional operations) the notion of metric space (*E-espace*, *E* from *écart*, i.e., gap).

Also, in 1914, Felix Hausdorff published his famous *Grundzüge der Mengenlehre* where the theory of topological and metric spaces (*metrische Räume*) was created.

Let this Encyclopedia be our homage to the memory of these great mathematicians and their lives of dignity through the hard times of the first half of the twentieth century.

Maurice Fréchet (1878–1973)

Felix Hausdorff (1868–1942)

Preface

Since the publication of the second edition in 2012, several people have again given us their valued feedback, and have thus contributed to the publication of this third edition. We are thankful to them for their input.

In the latest edition, new items from very active research areas in the use of distances and metrics such as geometry, graph theory, probability theory, and analysis have been added. We have kept the structure, but have revised many topics, simplifying, shortening, and updating them, especially in Chaps. 23–25 and 27–29.

Among the new topics included are, for example, polyhedral metric spaces, nearness matrix problems, distances between belief assignments, distance-related animal settings, diamond-cutting distances, natural units of length, Heidegger's de-severance distance, and brain distances in Chaps. 9, 12, 14, 23, 24, 27, 28, and 29, respectively.

We would also like to thank the team at Springer for their very efficient and friendly assistance.

Paris, France
Moscow, Russia
May 2014

Michel Marie Deza
Elena Deza

Preface to the Second Edition

The preparation of the second edition of Encyclopedia of Distances has presented a welcome opportunity to improve the first edition published in 2009 by updating and streamlining many sections, and by adding new items (especially in Chaps. 1, 15, 18, 23, 25, 27–29), increasing the book's size by about 70 pages. This new edition preserves, except for Chaps. 18, 23, 25 and 28, the structure of the first edition.

The first large conference with a scope matching that of this Encyclopedia is MDA 2012, the International Conference "Mathematics of Distances and Applications", held in July 2012 in Varna, Bulgaria; cf. [DPM12].

Preface to the First Edition

Encyclopedia of Distances is the result of re-writing and extending of our Dictionary of Distances published in 2006 (and put online http://www.sciencedirect.com/science/book/9780444520876) by Elsevier. About one-third of the items are new, and majority of the remaining ones are upgraded.

We were motivated by the growing intensity of research on metric spaces and, especially, in distance design for applications. Even if we do not address the practical questions arising during the selection of a "good" distance function, just a sheer listing of the main available distances should be useful for the distance design community.

This Encyclopedia is the first one treating fully the general notion of distance. This broad scope is useful *per se*, but it also limited our options for referencing. We give an original reference for many definitions but only when it was not too difficult to do so. On the other hand, citing somebody who well developed the notion but was not the original author may induce problems. However, with our data (usually, author name(s) and year), a reader can easily search sources using the Internet.

We found many cases when authors developed very similar distances in different contexts and, clearly, were unaware of it. Such connections are indicated by a simple "cf." in both definitions, without going into priority issues explicitly.

Concerning the style, we tried to make it a mixture of resource and coffee-table book, with maximal independence of its parts and many cross-references.

Preface to *Dictionary of Distances*, 2006

The concept of *distance* is a basic one in the whole human experience. In everyday life it usually means some degree of closeness of two physical objects or ideas, i.e., length, time interval, gap, rank difference, coolness or remoteness, while the term *metric* is often used as a standard for a measurement.

But here we consider, except for the last two chapters, the mathematical meaning of those terms which is an abstraction of measurement. The mathematical notions of *distance metric* (i.e., a function $d(x, y)$ from $X \times X$ to the set of real numbers satisfying to $d(x, y) \geq 0$ with equality only for $x = y$, $d(x, y) = d(y, x)$, and $d(x, y) \leq d(x, z) + d(z, y)$) and of *metric space* (X, d) were originated a century ago by M. Fréchet (1906) and F. Hausdorff (1914) as a special case of an infinite topological space. The *triangle inequality* above appears already in Euclid. The infinite metric spaces are usually seen as a generalization of the metric $|x - y|$ on the real numbers. Their main classes are the measurable spaces (add measure) and Banach spaces (add norm and completeness).

However, starting from K. Menger (who, in 1928, introduced metric spaces in Geometry) and L.M. Blumenthal (1953), an explosion of interest in both finite and infinite metric spaces occurred. Another trend: many mathematical theories, in the process of their generalization, settled on the level of metric space. It is an ongoing process, for example, for Riemannian geometry, Real Analysis, Approximation Theory.

Distance metrics and distances have become now an essential tool in many areas of Mathematics and its applications including Geometry, Probability, Statistics, Coding/Graph Theory, Clustering, Data Analysis, Pattern Recognition, Networks, Engineering, Computer Graphics/Vision, Astronomy, Cosmology, Molecular Biology, and many other areas of science. Devising the most suitable distance metrics and similarities, in order to quantify the proximity between objects, has become a standard task for many researchers. Especially intense ongoing search for such distances occurs, for example, in Computational Biology, Image Analysis, Speech Recognition, and Information Retrieval.

Often the same distance metric appears independently in several different areas; for example, the edit distance between words, the evolutionary distance in Biology, the Levenstein metric in Coding Theory, and the Hamming+Gap or shuffle-Hamming distance.

This body of knowledge has become too big and disparate to operate within. The number of worldwide web entries offered by Google on the topics "distance", "metric space" and "distance metric" is about 216, 3 and 9 million, respectively, not to mention all the printed information outside the Web, or the vast "invisible Web" of searchable databases. About 15,000 books on Amazon.com contains "distance" in their titles. However, this huge information on distances is too scattered: the works evaluating distance from some list usually treat very specific areas and are hardly accessible for nonexperts.

Therefore many researchers, including us, keep and cherish a collection of distances for use in their areas of science. In view of the growing general need for an accessible interdisciplinary source for a vast multitude of researchers, we have expanded our private collection into this Dictionary. Some additional material was reworked from various encyclopedias, especially Encyclopedia of Mathematics [EM98], MathWorld [Weis99], PlanetMath [PM] and Wikipedia [WFE]. However, the majority of distances are extracted directly from specialist literature.

Besides distances themselves, we collected here many distance-related notions (especially in Chap. 1) and paradigms, enabling people from applications to get those (arcane for nonspecialists) research tools, in ready-to-use fashion. This and the appearance of some distances in different contexts can be a source of new research.

In the time when over-specialization and terminology fences isolate researchers, this Dictionary tries to be "centripetal" and "ecumenical", providing some access and altitude of vision but without taking the route of scientific vulgarization. This attempted balance defined the structure and style of the Dictionary.

This reference book is a specialized encyclopedic dictionary organized by subject area. It is divided into 29 chapters grouped into 7 parts of about the same length. The titles of the parts are purposely approximative: they just allow a reader to figure out her/his area of interest and competence. For example, Parts II, III and IV, V require some culture in, respectively, pure and applied Mathematics. Part VII can be read by a layman.

The chapters are thematic lists, by areas of Mathematics or applications, which can be read independently. When necessary, a chapter or a section starts with a short introduction: a field trip with the main concepts. Besides these introductions, the main properties and uses of distances are given, within items, only exceptionally. We also tried, when it was easy, to trace distances to their originator(s), but the proposed extensive bibliography has a less general ambition: just to provide convenient sources for a quick search.

Each chapter consists of items ordered in a way that hints of connections between them. All item titles and (with majuscules only for proper nouns) selected key terms can be traced in the large Subject Index; they are boldfaced unless the meaning is clear from the context. So, the definitions are easy to locate, by subject, in chapters and/or, by alphabetic order, in the Subject Index.

The introductions and definitions are reader-friendly and maximally independent each from another; still they are interconnected, in the 3D HTML manner, by hyperlink-like boldfaced references to similar definitions.

Many nice curiosities appear in this "Who is Who" of distances. Examples of such sundry terms are: ubiquitous Euclidean distance ("as-the-crow-flies"), flower-shop metric (shortest way between two points, visiting a "flower-shop" point first), knight-move metric on a chessboard, Gordian distance of knots, Earth Mover's distance, biotope distance, Procrustes distance, lift metric, Post Office metric, Internet hop metric, WWW hyperlink quasi-metric, Moscow metric, dogkeeper distance.

Besides abstract distances, the distances having physical meaning appear also (especially in Part VI); they range from 1.6×10^{-35} m (Planck length) to 8.8×10^{26} m (the estimated size of the observable Universe, about 5.4×10^{61} Planck lengths).

The number of distance metrics is infinite, and therefore our Dictionary cannot enumerate all of them. But we were inspired by several successful thematic dictionaries on other infinite lists; for example, on Numbers, Integer Sequences, Inequalities, Random Processes, and by atlases of Functions, Groups, Fullerenes, etc. On the other hand, the large scope often forced us to switch to the mode of laconic tutorial.

The target audience consists of all researchers working on some measuring schemes and, to a certain degree, students and a part of the general public interested in science.

We tried to address all scientific uses of the notion of distance. But some distances did not made it to this Dictionary due to space limitations (being too specific and/or complex) or our oversight. In general, the size/interdisciplinarity cut-off, i.e., decision where to stop, was our main headache. We would be grateful to the readers who will send us their favorite distances missed here.

Contents

Part I
Mathematics of Distances

Chapter 1
General Definitions

In this core Chapter, the main metrics and metric-related notions are given.

1.1 Basic Definitions

- **Distance**
  A **distance space** (X, d) is a set X (*carrier*) equipped with a **distance** d.
  A function $d : X \times X \to \mathbb{R}$ is called a **distance** (or **dissimilarity**) on X if, for all $x, y \in X$, it holds:

  1. $d(x, y) \geq 0$ (*nonnegativity*);
  2. $d(x, y) = d(y, x)$ (*symmetry*);
  3. $d(x, x) = 0$ (*reflexivity*).

  In Topology, a distance with $d(x, y) = 0$ implying $x = y$ is called a **symmetric**.
  For any distance d, the function D_1 defined for $x \neq y$ by $D_1(x, y) = d(x, y) + c$, where $c = \max_{x,y,z \in X}(d(x, y) - d(x, z) - d(y, z))$, and $D(x, x) = 0$, is a **metric**. Also, $D_2(x, y) = d(x, y)^c$ is a metric for sufficiently small $c \geq 0$.
  The function $D_3(x, y) = \inf \sum_i d(z_i, z_{i+1})$, where the infimum is taken over all sequences $x = z_0, \dots, z_{n+1} = y$, is the **path semimetric** of the complete weighted graph on X, where, for any $x, y \in X$, the weight of edge xy is $d(x, y)$.
- **Similarity**
  Let X be a set. A function $s : X \times X \to \mathbb{R}$ is called a **similarity** on X if s is nonnegative, symmetric and the inequality

  $$s(x, y) \leq s(x, x)$$

  holds for all $x, y \in X$, with equality if and only if $x = y$.

M.M. Deza, E. Deza, *Encyclopedia of Distances*,
DOI 10.1007/978-3-662-44342-2_1

The main transforms used to obtain a distance (dissimilarity) d from a similarity s bounded by 1 from above are: $d = 1 - s$, $d = \frac{1-s}{s}$, $d = \sqrt{1-s}$, $d = \sqrt{2(1-s^2)}$, $d = \arccos s$, $d = -\ln s$ (cf. Chap. 4).

- **Semimetric**
  Let X be a set. A function $d : X \times X \to \mathbb{R}$ is called a **semimetric** on X if d is nonnegative, symmetric, *reflexive* ($d(x,x) = 0$ for $x \in X$) and it holds

$$d(x,y) \le d(x,z) + d(z,y)$$

  for all $x, y, z \in X$ (**triangle inequality** or, sometimes, *triangular inequality*).
  In Topology, it is called a **pseudo-metric** (or, rarely, **semidistance**), while the term *semimetric* is sometimes used for a **symmetric** (a distance $d(x,y)$ with $d(x,y) = 0$ only if $x = y$); cf. **symmetrizable space** in Chap. 2.
  For a semimetric d, the triangle inequality is equivalent, for each fixed $n \ge 4$ and all $x, y, z_1, \dots, z_{n-2} \in X$, to the following *n-gon inequality*

$$d(x,y) \le d(x,z_1) + d(z_1,z_2) + \cdots + d(z_{n-2},y).$$

  Equivalent *rectangle inequality* is $|d(x,y) - d(z_1,z_2)| \le d(x,z_1) + +d(y,z_2)$.
  For a semimetric d on X, define an equivalence relation, called **metric identification**, by $x \sim y$ if $d(x,y) = 0$; equivalent points are equidistant from all other points. Let $[x]$ denote the equivalence class containing x; then $D([x],[y]) = d(x,y)$ is a **metric** on the set $\{[x] : x \in X\}$ of equivalence classes.
- **Metric**
  Let X be a set. A function $d : X \times X \to \mathbb{R}$ is called a **metric** on X if, for all $x, y, z \in X$, it holds:

  1. $d(x,y) \ge 0$ (*nonnegativity*);
  2. $d(x,y) = 0$ if and only if $x = y$ (*identity of indiscernibles*);
  3. $d(x,y) = d(y,x)$ (*symmetry*);
  4. $d(x,y) \le d(x,z) + d(z,y)$ (**triangle inequality**).

  In fact, the above condition 1. follows from above 3. and 4.
  If 2. is dropped, then d is called (Bukatin, 2002) **relaxed semimetric**. If 2. is weakened to "$d(x,x) = d(x,y) = d(y,y)$ implies $x = y$", then d is called **relaxed metric**. A **partial metric** is a **partial semimetric**, which is a relaxed metric.
  If above 2. is weakened to "$d(x,y) = 0$ implies $x = y$", then d is called (Amini-Harandi, 2012) **metric-like function**. Any **partial metric** is metric-like.
- **Metric space**
  A **metric space** (X,d) is a set X equipped with a metric d.
  It is called a **metric frame** (or *metric scheme*, *integral*) if d is integer-valued.
  A **pointed metric space** (or *rooted metric space*) (X,d,x_0) is a metric space (X,d) with a selected base point $x_0 \in X$.
- **Extended metric**
  An **extended metric** is a generalization of the notion of metric: the value ∞ is allowed for a metric d.

- **Quasi-distance**
  Let X be a set. A function $d : X \times X \to \mathbb{R}$ is called a **quasi-distance** on X if d is nonnegative, and $d(x,x) = 0$ holds for all $x \in X$. It is also called a **premetric** or **prametric** in Topology and a *divergence* in Probability.
  If a quasi-distance d satisfies the **strong triangle inequality** $d(x,y) \le d(x,z) + d(y,z)$, then (Lindenbaum, 1926) it is symmetric and so, a semimetric. A **quasi-semimetric** d is a semimetric if and only if (Weiss, 2012) it satisfies the **full triangle inequality** $|d(x,z) - d(z,y)| \le d(x,z) \le d(x,z) + d(z,y)$.
  The distance/metric notions are usually named as weakenings or modifications of the fundamental notion of **metric**, using various prefixes and modifiers. But, perhaps, extended (i.e., the value ∞ is allowed) quasi-distance or (as suggested in Lawvere, 2002) quasi-semimetric should be used as the basic term.
- **Quasi-semimetric**
  A function $d : X \times X \to \mathbb{R}$ is called a **quasi-semimetric** (or **hemimetric**, *ostensible metric*) on X if $d(x,x) = 0$, $d(x,y) \ge 0$ and the **oriented triangle inequality**

  $$d(x,y) \le d(x,z) + d(z,y)$$

  hold for all $x, y, z \in X$. The set X can be partially ordered by the *specialization order*: $x \preceq y$ if and only if $d(x,y) = 0$.
  A **weak quasi-metric** is a quasi-semimetric d on X with *weak symmetry*, i.e., for all $x, y \in X$ the equality $d(x,y) = 0$ implies $d(y,x) = 0$.
  An **Albert quasi-metric** is a quasi-semimetric d on X with *weak definiteness*, i.e., for all $x, y \in X$ the equality $d(x,y) = d(y,x) = 0$ implies $x = y$.
  A **weightable quasi-semimetric** is a quasi-semimetric d on X with *relaxed symmetry*, i.e., for all $x, y, z \in X$

  $$d(x,y) + d(y,z) + d(z,x) = d(x,z) + d(z,y) + d(y,x),$$

  holds or, equivalently, there exists a weight function $w(x) \in \mathbb{R}$ on X with $d(x,y) - d(y,x) = w(y) - w(x)$ for all $x, y \in X$ (i.e., $d(x,y) + \frac{1}{2}(w(x) - w(y))$ is a semimetric). If d is a weightable quasi-semimetric, then $d(x,y) + w(x)$ is a **partial semimetric** (moreover, a **partial metric** if d is an Albert quasi-metric).
- **Partial metric**
  Let X be a set. A nonnegative symmetric function $p : X \times X \to \mathbb{R}$ is called a **partial metric** [Matt92] if, for all $x, y, z \in X$, it holds:

  1. $p(x,x) \le p(x,y)$, i.e., every **self-distance** (or *extent*) $p(x,x)$ is *small*;
  2. $x = y$ if $p(x,x) = p(x,y) = p(y,y) = 0$ (T_0 *separation axiom*);
  3. $p(x,y) \le p(x,z) + p(z,y) - p(z,z)$ (**sharp triangle inequality**).

  If the above separation axiom is dropped, the function p is called a **partial semimetric**. The nonnegative function p is a partial semimetric if and only if $p(x,y) - p(x,x)$ is a **weightable quasi-semimetric** with $w(x) = p(x,x)$.

If the 1st above condition is also dropped, the function p is called (Heckmann, 1999) a **weak partial semimetric**. The nonnegative function p is a weak partial semimetric if and only if $2p(x,y) - p(x,x) - p(y,y)$ is a semimetric.
Sometimes, the term *partial metric* is used when a metric $d(x,y)$ is defined only on a subset of the set of all pairs x, y of points.

- **Protometric**
A function $p : X \times X \to \mathbb{R}$ is called a **protometric** if, for all (equivalently, for all different) $x, y, z \in X$, the **sharp triangle inequality** holds:

$$p(x,y) \le p(x,z) + p(z,y) - p(z,z).$$

For finite X, the matrix $((p(x,y)))$ is (Burkard et al., 1996) *weak Monge array*.
A **strong protometric** is a protometric p with $p(x,x) = 0$ for all $x \in X$. Such a protometric is exactly a quasi-semimetric, but with the condition $p(x,y) \ge 0$ (for any $x, y \in X$) being relaxed to $p(x,y) + p(y,x) \ge 0$.
A **partial semimetric** is a **symmetric protometric** (i.e., $p(x,y) = p(y,x)$) with $p(x,y) \ge p(x,x) \ge 0$ for all $x, y \in X$. An example of a nonpositive symmetric protometric is given by $p(x,y) = -(x.y)_{x_0} = \frac{1}{2}(d(x,y) - d(x,x_0) - d(y,y_0))$, where (X,d) is a metric space with a fixed base point $x_0 \in X$; see **Gromov product similarity** $(x.y)_{x_0}$ and, in Chap. 4, **Farris transform metric** $C - (x.y)_{x_0}$.
A 0-**protometric** is a protometric p for which all sharp triangle inequalities (equivalently, all inequalities $p(x,y) + p(y,x) \ge p(x,x) + p(y,y)$ implied by them) hold as equalities. For any $u \in X$, denote by A'_u, A''_u the 0-protometrics p with $p(x,y) = 1_{x=u}, 1_{y=u}$, respectively. The protometrics on X form a flat convex cone in which the 0-protometrics form the largest linear space. For finite $|X|$, a basis of this space is given by all but one A'_u, A''_u (since $\sum_u A'_u = \sum_u A''_u$) and, for the flat subcone of all symmetric 0-protometrics on X, by all $A'_u + A''_u$.
A **weighted protometric** on X is a protometric with a point-weight function $w : X \to \mathbb{R}$. The mappings $p(x,y) = \frac{1}{2}(d(x,y) + w(x) + w(y))$ and $d(x,y) = 2p(x,y) - p(x,x) - p(y,y)$, $w(x) = p(x,x)$ establish a bijection between the weighted strong protometrics (d,w) and the protometrics p on X, as well as between the weighted semimetrics and the symmetric protometrics. For example, a weighted semimetric (d,w) with $w(x) = -d(x,x_0)$ corresponds to a protometric $-(x.y)_{x_0}$. For finite $|X|$, the above mappings amount to the representation

$$2p = d + \sum_{u \in X} p(u,u)(A'_u + A''_u).$$

- **Quasi-metric**
A function $d : X \times X \to \mathbb{R}$ is called a **quasi-metric** (or *asymmetric metric*, *directed metric*) on X if $d(x,y) \ge 0$ holds for all $x, y \in X$ with equality if and only if $x = y$, and for all $x, y, z \in X$ the **oriented triangle inequality**

$$d(x, y) \le d(x, z) + d(z, y)$$

holds. A *quasi-metric space* (X, d) is a set X equipped with a quasi-metric d. For any quasi-metric d, the functions $\max\{d(x, y), d(y, x)\}$ (called sometimes *bi-distance*), $\min\{d(x, y), d(y, x)\}$,
$\frac{1}{2}(d^p(x, y) + d^p(y, x))^{\frac{1}{p}}$ with given $p \ge 1$ are **metric generating**; cf. Chap. 4.
A **non-Archimedean quasi-metric** d is a quasi-distance on X which, for all $x, y, z \in X$, satisfies the following strengthened oriented triangle inequality:

$$d(x, y) \le \max\{d(x, z), d(z, y)\}.$$

- **Directed-metric**
  Let X be a set. A function $d : X \times X \to \mathbb{R}$ is called (Jegede, 2005) a **directed-metric** on X if, for all $x, y, z \in X$, it holds that $d(x, y) = -d(y, x)$ and

  $$|d(x, y)| \le |d(x, z)| + |d(z, y)|.$$

  Cf. **displacement** in Chap. 24 and **rigid motion of metric space**.
- **Coarse-path metric**
  Let X be a set. A metric d on X is called a **coarse-path metric** if, for a fixed $C \ge 0$ and for every pair of points $x, y \in X$, there exists a sequence $x = x_0, x_1, \dots, x_t = y$ for which $d(x_{i-1}, x_i) \le C$ for $i = 1, \dots, t$, and it holds

  $$d(x, y) \ge d(x_0, x_1) + d(x_1, x_2) + \cdots + d(x_{t-1}, x_t) - C.$$

- **Near-metric**
  Let X be a set. A distance d on X is called a **near-metric** (or C-*near-metric*) if $d(x, y) > 0$ for $x \ne y$ and the C-*relaxed triangle inequality*

  $$d(x, y) \le C(d(x, z) + d(z, y))$$

  holds for all $x, y, z \in X$ and some constant $C \ge 1$.
  If $d(x, y) > 0$ for $x \ne y$ and the C-*asymmetric triangle inequality* $d(x, y) \le d(x, z) + Cd(z, y)$ holds, d is a $\frac{C+1}{2}$-near-metric.
  A C**-inframetric** is a C-near-metric, while a C-near-metric is a $2C$-inframetric.
  Some recent papers use the term *quasi-triangle inequality* for the above inequality and so, *quasi-metric* for the notion of near-metric.
  The **power transform** (cf. Chap. 4) $(d(x, y))^\alpha$ of any near-metric is a near-metric for any $\alpha > 0$. Also, any near-metric d admits a **bi-Lipschitz mapping** on $(D(x, y))^\alpha$ for some semimetric D on the same set and a positive number α.
  A near-metric d on X is called a **Hölder near-metric** if the inequality

  $$|d(x, y) - d(x, z)| \le \beta d(y, z)^\alpha (d(x, y) + d(x, z))^{1-\alpha}$$

holds for some $\beta > 0$, $0 < \alpha \le 1$ and all points $x, y, z \in X$. Cf. **Hölder mapping**.

- **Weak ultrametric**

  A **weak ultrametric** (or **C-inframetric**, *C-pseudo-distance*) d is a distance on X such that $d(x, y) > 0$ for $x \neq y$ and the *C-inframetric inequality*

  $$d(x, y) \le C \max\{d(x, z), d(z, y)\}$$

  holds for all $x, y, z \in X$ and some constant $C \ge 1$.

  The term **pseudo-distance** is also used, in some applications, for any of a **pseudo-metric**, a **quasi-distance**, a **near-metric**, a distance which can be infinite, a distance with an error, etc. Another unsettled term is **weak metric**: it is used for both a **near-metric** and a **quasi-semimetric**.

- **Ultrametric**

  An **ultrametric** (or *non-Archimedean metric*) is (Krasner, 1944) a metric d on X which satisfies, for all $x, y, z \in X$, the following strengthened version of the triangle inequality (Hausdorff, 1934), called the **ultrametric inequality**:

  $$d(x, y) \le \max\{d(x, z), d(z, y)\}$$

  An ultrametric space is also called an *isosceles space* since at least two of $d(x, y), d(z, y), d(x, z)$ are equal. An ultrametric on a set V has at most $|V|$ different values.

  A metric d is an ultrametric if and only if its **power transform** (see Chap. 4) d^α is a metric for any real positive number α. Any ultrametric satisfies the **four-point inequality**. A metric d is an ultrametric if and only if it is a **Farris transform metric** (cf. Chap. 4) of a **four-point inequality metric**.

- **Robinsonian distance**

  A distance d on X is called a **Robinsonian distance** (or *monotone distance*) if there exists a total order $\preceq$ on X *compatible* with it, i.e., for $x, y, w, z \in X$,

  $$x \preceq y \preceq w \preceq z \text{ implies } d(y, w) \le d(x, z),$$

  or, equivalently, for $x, y, z \in X$,

  $$x \preceq y \preceq z \text{ implies } d(x, y) \le \max\{d(x, z), d(z, y)\}.$$

  Any **ultrametric** is a Robinsonian distance.

- **Four-point inequality metric**

  A metric d on X is a **four-point inequality metric** (or **additive metric**) if it satisfies the following strengthened version of the triangle inequality called the **four-point inequality** (Buneman, 1974): for all $x, y, z, u \in X$

  $$d(x, y) + d(z, u) \le \max\{d(x, z) + d(y, u), d(x, u) + d(y, z)\}$$

holds. Equivalently, among the three sums $d(x,y)+d(z,u)$, $d(x,z)+d(y,u)$, $d(x,u)+d(y,z)$ the two largest sums are equal.
A metric satisfies the four-point inequality if and only if it is a **tree-like metric**. Any metric, satisfying the four-point inequality, is a **Ptolemaic metric** and an L_1-*metric*. Cf. L_p**-metric** in Chap. 5.
A **bush metric** is a metric for which all four-point inequalities are equalities, i.e., $d(x,y)+d(u,z)=d(x,u)+d(y,z)$ holds for any $u,x,y,z \in X$.

- **Relaxed four-point inequality metric**
A metric d on X satisfies the **relaxed four-point inequality** if, for all $x,y,z,u \in X$, among the three sums

$$d(x,y)+d(z,u), d(x,z)+d(y,u), d(x,u)+d(y,z)$$

at least two (not necessarily the two largest) are equal. A metric satisfies this inequality if and only if it is a **relaxed tree-like metric**.

- **Ptolemaic metric**
A **Ptolemaic metric** d is a metric on X which satisfies the **Ptolemaic inequality**

$$d(x,y)d(u,z) \le d(x,u)d(y,z)+d(x,z)d(y,u)$$

for all $x,y,u,z \in X$. A classical result, attributed to Ptolemy, says that this inequality holds in the Euclidean plane, with equality if and only if the points x,y,u,z lie on a circle in that order.
A *Ptolemaic space* is a *normed vector space* $(V, ||.||)$ such that its norm metric $||x-y||$ is a Ptolemaic metric. A normed vector space is a Ptolemaic space if and only if it is an **inner product space** (cf. Chap. 5); so, a **Minkowskian metric** (cf. Chap. 6) is Euclidean if and only if it is Ptolemaic.
The *involution space* $(X\backslash z, d_z)$, where $d_z(x,y)=\frac{d(x,y)}{d(x,z)d(y,z)}$, is a metric space, for any $z \in X$, if and only if d is Ptolemaic [FoSc06].
For any metric d, the metric $\sqrt{d}$ is Ptolemaic [FoSc06].

- **δ-Hyperbolic metric**
Given a number $\delta \ge 0$, a metric d on a set X is called **δ-hyperbolic** if it satisfies the following *Gromov δ-hyperbolic inequality* (another weakening of the **four-point inequality**): for all $x,y,z,u \in X$, it holds that

$$d(x,y)+d(z,u) \le 2\delta + \max\{d(x,z)+d(y,u), d(x,u)+d(y,z)\}.$$

A metric space (X,d) is δ-hyperbolic if and only if for all $x_0,x,y,z \in X$ it holds

$$(x.y)_{x_0} \ge \min\{(x.z)_{x_0}, (y.z)_{x_0}\} - \delta,$$

where $(x.y)_{x_0} = \frac{1}{2}(d(x_0, x) + d(x_0, y) - d(x, y))$ is the **Gromov product** of the points x and y of X with respect to the base point $x_0 \in X$.
A metric space (X, d) is 0-hyperbolic exactly when d satisfies the **four-point inequality**. Every bounded metric space of diameter D is D-hyperbolic. The n-dimensional *hyperbolic space* is ln 3-hyperbolic.
Every δ-hyperbolic metric space is isometrically embeddable into a **geodesic metric space** (Bonk and Schramm, 2000).

- **Gromov product similarity**
Given a metric space (X, d) with a fixed point $x_0 \in X$, the **Gromov product similarity** (or *Gromov product, covariance, overlap function*) $(.)_{x_0}$ is a similarity on X defined by

$$(x.y)_{x_0} = \frac{1}{2}(d(x, x_0) + d(y, x_0) - d(x, y)).$$

The triangle inequality for d implies $(x.y)_{x_0} \geq (x.z)_{x_0} + (y.z)_{x_0} - (z.z)_{x_0}$ (**covariance triangle inequality**), i.e., **sharp triangle inequality** for **protometric** $-(x.y)_{x_0}$.
If (X, d) is a tree, then $(x.y)_{x_0} = d(x_0, [x, y])$. If (X, d) is a **measure semimetric space**, i.e., $d(x, y) = \mu(x \triangle y)$ for a Borel measure μ on X, then $(x.y)_\emptyset = \mu(x \cap y)$. If d is a **distance of negative type**, i.e., $d(x, y) = d_E^2(x, y)$ for a subset X of a Euclidean space $\mathbb{E}^n$, then $(x.y)_0$ is the usual *inner product* on $\mathbb{E}^n$.
Cf. **Farris transform metric** $d_{x_0}(x, y) = C - (x.y)_{x_0}$ in Chap. 4.

- **Cross-difference**
Given a metric space (X, d) and quadruple (x, y, z, w) of its points, the **cross-difference** is the real number *cd* defined by

$$cd(x, y, z, w) = d(x, y) + d(z, w) - d(x, z) - d(y, w).$$

In terms of the **Gromov product similarity**, for all $x, y, z, w, p \in X$, it holds

$$\frac{1}{2}cd(x, y, z, w) = -(x.y)_p - (z.w)_p + (x.z)_p + (y.w)_p;$$

in particular, it becomes $(x.y)_p$ if $y = w = p$.
If $x \neq z$ and $y \neq w$, the **cross-ratio** is the positive number defined by

$$cr((x, y, z, w), d) = \frac{d(x, y)d(z, w)}{d(x, z)d(y, w)}.$$

- **$2k$-gonal distance**
A $2k$**-gonal distance** d is a distance on X which satisfies, for all distinct elements $x_1, \ldots, x_n \in X$, the $2k$**-gonal inequality**

$$\sum_{1\le i<j\le n} b_i b_j d(x_i, x_j) \le 0$$

for all $b \in \mathbb{Z}^n$ with $\sum_{i=1}^n b_i = 0$ and $\sum_{i=1}^n |b_i| = 2k$.

- **Distance of negative type**
  A **distance of negative type** d is a distance on X which is $2k$**-gonal** for any $k \ge 1$, i.e., satisfies the **negative type inequality**

$$\sum_{1\le i<j\le n} b_i b_j d(x_i, x_j) \le 0$$

  for all $b \in \mathbb{Z}^n$ with $\sum_{i=1}^n b_i = 0$, and for all distinct elements $x_1, \dots, x_n \in X$.
  A distance can be of negative type without being a semimetric. Cayley proved that a metric d is an L_2**-metric** if and only if d^2 is a distance of negative type.
- $(2k+1)$**-gonal distance**
  A $(2k+1)$**-gonal distance** d is a distance on X which satisfies, for all distinct elements $x_1, \dots, x_n \in X$, the $(2k+1)$**-gonal inequality**

$$\sum_{1\le i<j\le n} b_i b_j d(x_i, x_j) \le 0$$

  for all $b \in \mathbb{Z}^n$ with $\sum_{i=1}^n b_i = 1$ and $\sum_{i=1}^n |b_i| = 2k+1$.
  The $(2k+1)$-gonal inequality with $k = 1$ is the usual triangle inequality. The $(2k+1)$-gonal inequality implies the $2k$**-gonal inequality**.
- **Hypermetric**
  A **hypermetric** d is a distance on X which is $(2k+1)$**-gonal** for any $k \ge 1$, i.e., satisfies the **hypermetric inequality** (Deza, 1960)

$$\sum_{1\le i<j\le n} b_i b_j d(x_i, x_j) \le 0$$

  for all $b \in \mathbb{Z}^n$ with $\sum_{i=1}^n b_i = 1$, and for all distinct elements $x_1, \dots, x_n \in X$.
  Any hypermetric is a semimetric, a **distance of negative type** and, moreover, it can be isometrically embedded into some n-sphere $\mathbb{S}^n$ with squared Euclidean distance. Any L_1*-metric* (cf. L_p**-metric** in Chap. 5) is a hypermetric.
- P**-metric**
  A P**-metric** d is a metric on X with values in $[0, 1]$ which satisfies the **correlation triangle inequality**

$$d(x,y) \le d(x,z) + d(z,y) - d(x,z)d(z,y).$$

  The equivalent inequality $1 - d(x,y) \ge (1 - d(x,z))(1 - d(z,y))$ expresses that the probability, say, to reach x from y via z is either equal to

$(1 - d(x, z))(1 - d(z, y))$ (independence of reaching z from x and y from z), or greater than it (positive correlation). A metric is a P-metric if and only if it is a **Schoenberg transform metric** (cf. Chap. 4).

1.2 Main Distance-Related Notions

- **Metric ball**
  Given a metric space (X, d), the **metric ball** (or *closed metric ball*) with center $x_0 \in X$ and radius $r > 0$ is defined by $\overline{B}(x_0, r) = \{x \in X : d(x_0, x) \leq r\}$, and the **open metric ball** with center $x_0 \in X$ and radius $r > 0$ is defined by $B(x_0, r) = \{x \in X : d(x_0, x) < r\}$. The closed ball is a subset of the closure of the open ball; it is a proper subset for, say, the **discrete metric** on X.
  The **metric sphere** with center $x_0 \in X$ and radius $r > 0$ is defined by $S(x_0, r) = \{x \in X : d(x_0, x) = r\}$.
  For the **norm metric** on an n-dimensional *normed vector space* $(V, ||.||)$, the metric ball $\overline{B}^n = \{x \in V : ||x|| \leq 1\}$ is called the *unit ball*, and the set $S^{n-1} = \{x \in V : ||x|| = 1\}$ is called the *unit sphere*. In a two-dimensional vector space, a metric ball (closed or open) is called a **metric disk** (closed or open, respectively).
- **Metric hull**
  Given a metric space (X, d), let M be a **bounded** subset of X.
  The **metric hull** $H(M)$ of M is the intersection of all metric balls containing M. The set of *surface points* $S(M)$ of M is the set of all $x \in H(M)$ such that x lies on the sphere of one of the metric balls containing M.
- **Distance-invariant metric space**
  A metric space (X, d) is **distance-invariant** if all **metric balls** $\overline{B}(x_0, r) = \{x \in X : d(x_0, x) \leq r\}$ of the same radius have the same number of elements.
  Then the **growth rate of a metric space** (X, d) is the function $f(n) = |\overline{B}(x, n)|$. (X, d) is a *metric space of polynomial growth* if there are some positive constants k, C such that $f(n) \leq Cn^k$ for all $n \geq 0$. Cf. **graph of polynomial growth**, including the group case, in Chap. 15.
  For a **metrically discrete metric space** (X, d) (i.e., with $a = \inf_{x,y \in X, x \neq y} d(x, y) > 0$), its *growth rate* was defined also (Gordon–Linial–Rabinovich, 1998) by

$$\max_{x \in X, r \geq 2} \frac{\log |\overline{B}(x, ar)|}{\log r}.$$

- **Ahlfors q-regular metric space**
  A metric space (X, d) endowed with a Borel measure μ is called **Ahlfors q-regular** if there exists a constant $C \geq 1$ such that for every ball in (X, d) with radius $r < diam(X, d)$ it holds

$$C^{-1}r^q \le \mu(\overline{B}(x_0, r)) \le Cr^q.$$

If such an (X, d) is locally compact, then the **Hausdorff q-measure** can be taken as μ and q is the **Hausdorff dimension**. For two disjoint **continua** (nonempty **connected compact metric subspaces**) C_1, C_2 of such space (X, d), let Γ be the set of rectifiable curves connecting C_1 to C_2. The *q-modulus* between C_1 and C_2 is $M_q(C_1, C_2) = \inf\{\int_X \rho^q : \inf_{\gamma\in\Gamma} \int_\gamma \rho \ge 1\}$, where $\rho : X \to \mathbb{R}_{>0}$ is any density function on X; cf. the **modulus metric** in Chap. 6.
The *relative distance* between C_1 and C_2 is $\delta(C_1, C_2) = \frac{\inf\{d(p_1,p_2):p_1\in C_1,p_2\in C_2\}}{\min\{diam(C_1),diam(C_2)\}}$.
(X, d) is a **q-Loewner space** if there are increasing functions $f, g : [0, \infty) \to [0, \infty)$ such that for all C_1, C_2 it holds $f(\delta(C_1, C_2)) \le M_q(C_1, C_2) \le g(\delta(C_1, C_2))$.

- **Connected metric space**
  A metric space (X, d) is called **connected** if it cannot be partitioned into two nonempty **open** sets. Cf. **connected space** in Chap. 2.
  The maximal connected subspaces of a metric space are called its *connected components*. A **totally disconnected metric space** is a space in which all connected subsets are $\emptyset$ and one-point sets.
  A **path-connected metric space** is a connected metric space such that any two its points can be joined by an **arc** (cf. **metric curve**).
- **Cantor connected metric space**
  A metric space (X, d) is called **Cantor** (or *pre-*) *connected* if, for any two its points x, y and any $\epsilon > 0$, there exists an *ϵ-chain* joining them, i.e., a sequence of points $x = z_0, z_1, \dots, z_{n-1}, z_n = y$ such that $d(z_k, z_{k+1}) \le \epsilon$ for every $0 \le k \le n$. A metric space (X, d) is Cantor connected if and only if it cannot be partitioned into two *remote parts* A and B, i.e., such that $\inf\{d(x, y) : x \in A, y \in B\} > 0$.
  The maximal Cantor connected subspaces of a metric space are called its *Cantor connected components*. A **totally Cantor disconnected metric** is the metric of a metric space in which all Cantor connected components are one-point sets.
- **Indivisible metric space**
  A metric space (X, d) is called **indivisible** if it cannot be partitioned into two parts, neither of which contains an isometric copy of (X, d). Any indivisible metric space with $|X| \ge 2$ is infinite, bounded and **totally Cantor disconnected** (Delhomme–Laflamme–Pouzet–Sauer, 2007).
  A metric space (X, d) is called an **oscillation stable metric space** (Nguyen Van Thé, 2006) if, given any $\epsilon > 0$ and any partition of X into finitely many pieces, the **ϵ-neighborhood** of one of the pieces includes an isometric copy of (X, d).
- **Closed subset of metric space**
  Given a subset M of a metric space (X, d), a point $x \in X$ is called a *limit* (or *accumulation*) *point of M* if any **open metric ball** $B(x, r) = \{y \in X : d(x, y) < r\}$ contains a point $x' \in M$ with $x' \ne x$. The *boundary* $\vartheta(M)$

of M is the set of all its limit points. The *closure of* M, denoted by $cl(M)$, is $M \cup \vartheta(M)$, and M is called **closed subset**, if $M = cl(M)$, and **dense subset**, if $X = cl(M)$.
Every point of M which is not its limit point, is called an *isolated point*. The *interior* $int(M)$ of M is the set of all its isolated points, and the *exterior* $ext(M)$ of M is $int(X \backslash M)$. A subset M is called *nowhere dense* if $int(cl(M)) = \emptyset$.
A subset M is called *topologically discrete* (cf. **metrically discrete metric space**) if $int(M) = M$ and *dense-in-itself* if $int(M) = \emptyset$. A dense-in-itself subset is called *perfect* (cf. **perfect metric space**) if it is closed. The subsets *Irr* (irrational numbers) and $\mathbb{Q}$ (rational numbers) of $\mathbb{R}$ are dense, dense-in-itself but not perfect. The set $\mathbb{Q} \cap [0, 1]$ is dense-in-itself but not dense in $\mathbb{R}$.

- **Open subset of metric space**
  A subset M of a metric space (X, d) is called *open* if, given any point $x \in M$, the **open metric ball** $B(x, r) = \{y \in X : d(x, y) < r\}$ is contained in M for some number $r > 0$. The family of open subsets of a metric space forms a natural topology on it. A **closed subset** is the complement of an open subset.
  An open subset is called *clopen*, if it is closed, and a *domain* if it is **connected**.
  A *door space* is a metric (in general, topological) space in which every subset is either open or closed.
- **Metric topology**
  A **metric topology** is a *topology* induced by a metric; cf. **equivalent metrics**. More exactly, the **metric topology** on a metric space (X, d) is the set of all *open sets* of X, i.e., arbitrary unions of (finitely or infinitely many) open metric balls $B(x, r) = \{y \in X : d(x, y) < r\}$, $x \in X$, $r \in \mathbb{R}$, $r > 0$.
  A topological space which can arise in this way from a metric space is called a **metrizable space** (cf. Chap. 2). **Metrization theorems** are theorems which give sufficient conditions for a topological space to be metrizable.
  On the other hand, the adjective *metric* in several important mathematical terms indicates connection to a measure, rather than distance, for example, *metric Number Theory, metric Theory of Functions, metric transitivity.*
- **Equivalent metrics**
  Two metrics d_1 and d_2 on a set X are called **equivalent** if they define the same *topology* on X, i.e., if, for every point $x_0 \in X$, every open metric ball with center at x_0 defined with respect to d_1, contains an open metric ball with the same center but defined with respect to d_2, and conversely.
  Two metrics d_1 and d_2 are equivalent if and only if, for every $\epsilon > 0$ and every $x \in X$, there exists $\delta > 0$ such that $d_1(x, y) \leq \delta$ implies $d_2(x, y) \leq \epsilon$ and, conversely, $d_2(x, y) \leq \delta$ implies $d_1(x, y) \leq \epsilon$.
  All metrics on a finite set are equivalent; they generate the *discrete topology*.
- **Metric betweenness**
  The **metric betweenness** of a metric space (X, d) is (Menger, 1928) the set of all ordered triples (x, y, z) such that x, y, z are (not necessarily distinct) points of X for which the **triangle equality** $d(x, y) + d(y, z) = d(x, z)$ holds.

- **Closed metric interval**
Given two different points $x, y \in X$ of a metric space (X, d), the **closed metric interval** between them (or *line induced by*) them is the set of the points z, for which the **triangle equality** (or **metric betweenness** (x, z, y)) holds:

$$I(x, y) = \{z \in X : d(x, y) = d(x, z) + d(z, y)\}.$$

Cf. **inner product space** (Chap. 5) and **cutpoint additive metric** (Chap. 15).
- **Underlying graph of a metric space**
The **underlying graph** (or *neighborhood graph*) of a metric space (X, d) is a graph with the vertex-set X and xy being an edge if $I(x, y) = \{x, y\}$, i.e., there is no third point $z \in X$, for which $d(x, y) = d(x, z) + d(z, y)$.
- **Distance monotone metric space**
A metric space (X, d) is called **distance monotone** if for any its **closed metric interval** $I(x, y)$ and $u \in X \setminus I(x, y)$, there exists $z \in I(x, xy)$ with $d(u, z) > d(x, y)$.
- **Metric triangle**
Three distinct points $x, y, z \in X$ of a metric space (X, d) form a **metric triangle** if the **closed metric intervals** $I(x, y), I(y, z)$ and $I(z, x)$ intersect only in the common endpoints.
- **Metric space having collinearity**
A metric space (X, d) has **collinearity** if for any $\epsilon > 0$ each of its infinite subsets contains distinct ϵ-*collinear* (i.e., with $d(x, y) + d(y, z) - d(x, z) \leq \epsilon$) points x, y, z.
- **Modular metric space**
A metric space (X, d) is called **modular** if, for any three different points $x, y, z \in X$, there exists a point $u \in I(x, y) \cap I(y, z) \cap I(z, x)$. This should not be confused with **modular distance** in Chap. 10 and **modulus metric** in Chap. 6.
- **Median metric space**
A metric space (X, d) is called a **median metric space** if, for any three points $x, y, z \in X$, there exists a unique point $u \in I(x, y) \cap I(y, z) \cap I(z, x)$.
Any median metric space is an L_1-*metric*; cf. L_p**-metric** in Chap. 5 and **median graph** in Chap. 15.
A metric space (X, d) is called an **antimedian metric space** if, for any three points $x, y, z \in X$, there exists a unique point $u \in X$ maximizing $d(x, u) + d(y, u) + d(z, u)$.
- **Metric quadrangle**
Four different points $x, y, z, u \in X$ of a metric space (X, d) form a **metric quadrangle** if $x, z \in I(y, u)$ and $y, u \in I(x, z)$; then $d(x, y) = d(z, u)$ and $d(x, u) = d(y, z)$.
A metric space (X, d) is called *weakly spherical* if any three different points $x, y, z \in X$ with $y \in I(x, z)$, form a metric quadrangle with some point $u \in X$.
- **Metric curve**
A **metric curve** (or, simply, *curve*) γ in a metric space (X, d) is a continuous mapping $\gamma : I \to X$ from an interval I of $\mathbb{R}$ into X. A curve is called an **arc**

(or **path**, *simple curve*) if it is injective. A curve $\gamma : [a, b] \to X$ is called a *Jordan curve* (or *simple closed curve*) if it does not cross itself, and $\gamma(a) = \gamma(b)$.
The **length of a curve** $\gamma : [a, b] \to X$ is the number $l(\gamma)$ defined by

$$l(\gamma) = \sup\{\sum_{1 \leq i \leq n} d(\gamma(t_i), \gamma(t_{i-1})) : n \in \mathbb{N}, a = t_0 < t_1 < \cdots < t_n = b\}.$$

A *rectifiable curve* is a curve with a finite length. A metric space (X, d), where every two points can be joined by a rectifiable curve, is called a **quasi-convex metric space** (or, specifically, C**-quasi-convex metric space**) if there exists a constant $C \geq 1$ such that every pair $x, y \in X$ can be joined by a rectifiable curve of length at most $Cd(x, y)$. If $C = 1$, then this length is equal to $d(x, y)$, i.e., (X, d) is a **geodesic metric space** (cf. Chap. 6).
In a quasi-convex metric space (X, d), the infimum of the lengths of all rectifiable curves, connecting $x, y \in X$ is called the **internal metric**.
The metric d on X is called the **intrinsic metric** (and then (X, d) is called a **length space**) if it coincides with the internal metric of (X, d).
If, moreover, any pair x, y of points can be joined by a curve of length $d(x, y)$, the metric d is called **strictly intrinsic**, and the length space (X, d) is a geodesic metric space. Hopf-Rinow, 1931, showed that any complete locally compact length space is geodesic and **proper**. The **punctured plane** $(\mathbb{R}^2 \setminus \{0\}, ||x - y||_2)$ is **locally compact** and **path-connected** but not geodesic: the distance between $(-1, 0)$ and $(1, 0)$ is 2 but there is no geodesic realizing this distance.
The **metric derivative** of a metric curve $\gamma : [a, b] \to X$ at a limit point t is

$$\lim_{s \to 0} \frac{d(\gamma(t + s), \gamma(t))}{|s|},$$

if it exists. It is the rate of change, with respect to t, of the length of the curve at almost every point, i.e., a generalization of the notion of *speed* to metric spaces.

- **Geodesic**
Given a metric space (X, d), a **geodesic** is a locally shortest **metric curve**, i.e., it is a locally isometric embedding of $\mathbb{R}$ into X; cf. Chap. 6.
A subset S of X is called a **geodesic segment** (or **metric segment**, *shortest path*, *minimizing geodesic*) between two distinct points x and y in X, if there exists a *segment* (closed interval) [a,b] on the real line $\mathbb{R}$ and an isometric embedding $\gamma : [a, b] \to X$, such that $\gamma[a, b] = S$, $\gamma(a) = x$ and $\gamma(b) = y$.
A **metric straight line** is a geodesic which is minimal between any two of its points; it is an isometric embedding of the whole of $\mathbb{R}$ into X. A **metric ray** and **metric great circle** are isometric embeddings of, respectively, the half-line $\mathbb{R}_{\geq 0}$ and a circle $S^1(0, r)$ into X.
A **geodesic metric space** (cf. Chap. 6) is a metric space in which any two points are joined by a geodesic segment. If, moreover, the geodesic is unique, the space is called *totally geodesic* (or *uniquely geodesic*).

A geodesic metric space (X, d) is called *geodesically complete* if every geodesic is a subarc of a metric straight line. If (X, d) is **complete**, then it is geodesically complete. The **punctured plane** $(\mathbb{R}^2 \setminus \{0\}, ||x - y||_2)$ is not geodesically complete: any geodesic going to 0 is not a subarc of a metric straight line.

- **Length spectrum**
  Given a metric space (X, d), a *closed geodesic* is a map $\gamma : \mathbb{S}^1 \to X$ which is locally minimizing around every point of $\mathbb{S}^1$.
  If (X, d) is a compact **length space**, its **length spectrum** is the collection of lengths of closed geodesics. Each length is counted with *multiplicity* equal to the number of distinct *free homotopy* classes that contain a closed geodesic of such length. The **minimal length spectrum** is the set of lengths of closed geodesics which are the shortest in their free homotopy class. Cf. the **distance list**.
- **Systole of metric space**
  Given a compact metric space (X, d), its **systole** $sys(X, d)$ is the length of the shortest noncontractible loop in X; such a loop is a closed geodesic. So, $sys(X, d) = 0$ exactly if (X, d) is **simply connected**. Cf. **connected space** in Chap. 2.
  If (X, d) is a graph with path metric, then its systole is referred to as the *girth*. If (X, d) is a closed surface, then its *systolic ratio* is the ratio $SR = \frac{sys^2(X,d)}{area(X,d)}$.
  Some tight upper bounds of SR for every metric on a surface are: $\frac{2}{\sqrt{3}} = \gamma_2$ (*Hermite constant* in 2D) for 2-torus (Loewner, 1949), $\frac{\pi}{2}$ for the real projective plane (Pu, 1952) and $\frac{\pi}{\sqrt{8}}$ for the Klein bottle (Bavard, 1986). Tight asymptotic bounds for a surface S of large genus g are $\frac{4}{9}\frac{\log^2 g}{\pi g} \leq SR(S) \leq \frac{\log^2 g}{\pi g}$ (Katz et al., 2007).
- **Shankar–Sormani radii**
  Given a **geodesic metric space** (X, d), Shankar and Sormani, 2009, defined its **unique injectivity radius** $Uirad(X)$ as the supremum over all $r \geq 0$ such that any two points at distance at most r are joined by a unique geodesic, and its **minimal radius** $Mrad(X)$ as $\inf_{p \in X} d(p, MinCut(p))$.
  Here the *minimal cut locus of p* $MinCut(p)$ is the set of points $q \in X$ for which there is a geodesic γ running from p to q such that γ extends past q but is not minimizing from p to any point past q. If (X, d) is a Riemannian space, then the distance function from p is a smooth function except at p itself and the cut locus. Cf. **medial axis and skeleton** in Chap. 21.
  It holds $Uirad(X) \leq Mrad(X)$ with equality if (X, d) is a Riemannian space in which case it is the **injectivity radius**. It holds $Uirad(X) = \infty$ for a flat disk but $Mrad(X) < \infty$ if (X, d) is compact and at least one geodesic is extendible.
- **Geodesic convexity**
  Given a **geodesic metric space** (X, d) and a subset $M \subset X$, the set M is called **geodesically convex** (or *convex*) if, for any two points of M, there exists a geodesic segment connecting them which lies entirely in M; the space is **strongly convex** if such a segment is unique and no other geodesic connecting those points

lies entirely in M. The space is called **locally convex** if such a segment exists for any two sufficiently close points in M.
For a given point $x \in M$, the **radius of convexity** is $r_x = \sup\{r \geq 0 : B(x,r) \subset M\}$, where the **metric ball** $B(x,r)$ is convex. The point x is called the *center of mass* of points $y_1, \ldots, y_k \in M$ if it minimizes the function $\sum_i d(x, y_i)^2$ (cf. **Fréchet mean**); such point is unique if $d(y_i, y_j) < r_x$ for all $1 \leq i < j \leq k$.
The **injectivity radius** of the set M is the supremum over all $r \geq 0$ such that any two points in M at distance $\leq r$ are joined by unique geodesic segment which lies in M. The **Hawaiian Earring** is a compact complete metric space consisting of a set of circles of radius $\frac{1}{i}$ for each $i \in \mathbb{N}$ all joined at a common point; its injectivity radius is 0. It is **path-connected** but not **simply connected**.
The set $M \subset X$ is called a **totally convex metric subspace** of (X, d) if, for any two points of M, any geodesic segment connecting them lies entirely in M.

- **Busemann convexity**
A **geodesic metric space** (X, d) is called **Busemann convex** (or *Busemann space, nonpositively curved in the sense of Busemann*) if, for any three points $x, y, z \in X$ and *midpoints* $m(x, z)$ and $m(y, z)$ (i.e., $d(x, m(x, z)) = d(m(x, z), z) = \frac{1}{2}d(x, z)$ and $d(y, m(y, z)) = d(m(y, z), z) = \frac{1}{2}d(y, z)$), there holds
$$d(m(x,z), m(y,z)) \leq \frac{1}{2}d(x, y).$$
The *flat Euclidean strip* $\{(x, y) \in \mathbb{R}^2 : 0 < x < 1\}$ is **Gromov hyperbolic metric space** (Chap. 6) but not Busemann convex one. In a complete Busemann convex metric space any two points are joined by a unique geodesic segment.
A locally geodesic metric space (X, d) is called **Busemann locally convex** if the above inequality holds locally. Any **locally CAT(0)** metric space is Busemann locally convex.
- **Menger convexity**
A metric space (X, d) is called **Menger convex** if, for any different points $x, y \in X$, there exists a third point $z \in X$ for which $d(x, y) = d(x, z) + d(z, y)$, i.e., $|I(x, y)| > 2$ holds for the **closed metric interval** $I(x, y) = \{z \in X : (x, y) = d(x, z) + d(z, y)\}$. It is called **strictly Menger convex** if such a z is unique for all $x, y \in X$.
Geodesic convexity implies Menger convexity. The converse holds for **complete** metric spaces.
A subset $M \subset X$ is called (Menger, 1928) a *d-convex set* (or *interval-convex set*) if $I(x, y) \subset M$ for any different points $x, y \in M$. A function $f : M \to \mathbb{R}$ defined on a d-convex set $M \subset X$ is a **d-convex function** if for any $z \in I(x, y) \subset M$
$$f(z) \leq \frac{d(y,z)}{d(x,y)} f(x) + \frac{d(x,z)}{d(x,y)} f(y).$$

A subset $M \subset X$ is a *gated set* if for every $x \in X$ there exists a unique $x' \in M$, the *gate*, such that $d(x, y) = d(x, x') + d(x', y)$ for $y \in M$. Any such set is d-convex.

- **Midpoint convexity**
  A metric space (X, d) is called **midpoint convex** (or **having midpoints**, *admitting a midpoint map*) if, for any different points $x, y \in X$, there exists a third point $m(x, y) \in X$ for which $d(x, m(x, y)) = d(m(x, y), y) = \frac{1}{2}d(x, y)$. Such a point $m(x, y)$ is called a *midpoint* and the map $m : X \times X \to X$ is called a *midpoint map* (cf. **midset**); this map is unique if $m(x, y)$ is unique for all $x, y \in X$.
  For example, the geometric mean $\sqrt{xy}$ is the midpoint map for the metric space $(\mathbb{R}_{>0}, d(x, y) = |\log x - \log y|)$.
  A **complete** metric space is **geodesic** if and only if it is midpoint convex.
  A metric space (X, d) is said to have **approximate midpoints** if, for any points $x, y \in X$ and any $\epsilon > 0$, there exists an *ϵ-midpoint*, i.e., a point $z \in X$ such that $d(x, z) \leq \frac{1}{2}d(x, y) + \epsilon \geq d(z, y)$.
- **Ball convexity**
  A **midpoint convex** metric space (X, d) is called **ball convex** if

$$d(m(x, y), z) \leq \max\{d(x, z), d(y, z)\}$$

  for all $x, y, z \in X$ and any midpoint map $m(x, y)$.
  Ball convexity implies that all metric balls are **totally convex** and, in the case of a **geodesic** metric space, vice versa. Ball convexity implies also the uniqueness of a midpoint map (geodesics in the case of **complete** metric space).
  The metric space $(\mathbb{R}^2, d(x, y) = \sum_{i=1}^{2} \sqrt{|x_i - y_i|})$ is not ball convex.
- **Distance convexity**
  A **midpoint convex** metric space (X, d) is called **distance convex** if

$$d(m(x, y), z) \leq \frac{1}{2}(d(x, z) + d(y, z)).$$

  A **geodesic metric space** is distance convex if and only if the restriction of the distance function $d(x, \cdot)$, $x \in X$, to every geodesic segment is a convex function. Distance convexity implies **ball convexity** and, in the case of **Busemann convex** metric space, vice versa.
- **Metric convexity**
  A metric space (X, d) is called **metrically convex** if, for any different points $x, y \in X$ and any $\lambda \in (0, 1)$, there exists a third point $z = z(x, y, \lambda) \in X$ for which $d(x, y) = d(x, z) + d(z, y)$ and $d(x, z) = \lambda d(x, y)$.
  The space is called **strictly metrically convex** if such a point $z(x, y, \lambda)$ is unique for all $x, y \in X$ and any $\lambda \in (0, 1)$.
  A metric space (X, d) is called **strongly metrically convex** if, for any different points $x, y \in X$ and any $\lambda_1, \lambda_2 \in (0, 1)$, there exists a third point $z = z(x, y, \lambda) \in X$ for which $d(z(x, y, \lambda_1), z(x, y, \lambda_2)) = |\lambda_1 - \lambda_2| d(x, y)$.

Metric convexity implies **Menger convexity**, and every Menger convex **complete** metric space is strongly metrically convex.
A metric space (X, d) is called **nearly convex** (Mandelkern, 1983) if, for any different points $x, y \in X$ and any $\lambda, \mu > 0$ such that $d(x, y) < \lambda + \mu$, there exists a third point $z \in X$ for which $d(x,z) < \lambda$ and $d(z, y) < \mu$, i.e., $z \in B(x, \lambda) \cap B(y, \mu)$. Metric convexity implies near convexity.

- **Takahashi convexity**
A metric space (X, d) is called **Takahashi convex** if, for any different points $x, y \in X$ and any $\lambda \in (0, 1)$, there exists a third point $z = z(x, y, \lambda) \in X$ such that $d(z(x, y, \lambda), u) \leq \lambda d(x, u) + (1 - \lambda)d(y, u)$ for all $u \in X$. Any convex subset of a normed space is a Takahashi convex metric space with $z(x, y, \lambda) = \lambda x + (1 - \lambda)y$.
A set $M \subset X$ is *Takahashi convex* if $z(x, y, \lambda) \in M$ for all $x, y \in X$ and any $\lambda \in [0, 1]$. In a Takahashi convex metric space, all metric balls, open metric balls, and arbitrary intersections of Takahashi convex subsets are all Takahashi convex.
- **Hyperconvexity**
A metric space (X, d) is called **hyperconvex** (Aronszajn–Panitchpakdi, 1956) if it is **metrically convex** and its metric balls have the *infinite Helly property*, i.e., any family of mutually intersecting closed balls in X has nonempty intersection. A metric space (X, d) is hyperconvex if and only if it is an **injective metric space**.
The spaces l^n_∞, l^∞_∞ and l^2_1 are hyperconvex but l^∞_2 is not.
- **Distance matrix**
Given a finite metric space $(X = \{x_1, \cdots, x_n\}, d)$, its **distance matrix** is the symmetric $n \times n$ matrix $((d_{ij}))$, where $d_{ij} = d(x_i, x_j)$ for any $1 \leq i, j \leq n$.
The probability that a symmetric $n \times n$ matrix, whose diagonal elements are zeros and all other elements are uniformly random real numbers, is a distance matrix is (Mascioni, 2005) $\frac{1}{2}$, $\frac{17}{120}$ for $n = 3, 4$, respectively.
- **Distance product of matrices**
Given $n \times n$ matrices $A = ((a_{ij}))$ and $B = ((b_{ij}))$, their **distance** (or *min-plus*) **product** is the $n \times n$ matrix $C = ((c_{ij}))$ with $c_{ij} = \min^n_{k=1}(a_{ik} + b_{kj})$.
It is the usual matrix multiplication in the *tropical semiring* $(\mathbb{R} \cup \{\infty\}, \min, +)$ (Chap. 18). If A is the matrix of weights of an edge-weighted complete graph K_n, then its *direct power* A^n is the (shortest path) **distance matrix** of this graph.
- **Distance list**
Given a metric space (X, d), its **distance set** and **distance list** are the set and the *multiset* (i.e., multiplicities are counted) and of all pairwise distances.
Two subsets $A, B \subset X$ are said to be *homometric sets* if they have the same distance list. Cf. **homometric structures** in Chap. 24.
- **Degree of distance near-equality**
Given a finite metric space (X, d) with $|X| = n \geq 3$, let $f = \min |\frac{d(x,y)}{d(a,b)} - 1|$ (**degree of distance near-equality**) and $f' = \min |\frac{d(x,y)}{d(x,b)} - 1|$, where the minimum is over different 2-subsets $\{x, y\}, \{a, b\}$ of X and, respectively, over

different $x, y, b \in X$. [Ophi14] proved $f \leq \frac{9 \log n}{n^2}$ and $f' \leq \frac{3}{n}$, while $f \geq \frac{\log n}{20n^2}$ and $f' \geq \frac{1}{2n}$ for some (X, d).

- **Semimetric cone**
  The **semimetric cone** MET_n is the polyhedral cone in $\mathbb{R}^{\binom{n}{2}}$ of all **distance matrices** of semimetrics on the set $V_n = \{1, \dots, n\}$. Vershik, 2004, considers MET_∞, i.e., the weakly closed convex cone of infinite distance matrices of semimetrics on $\mathbb{N}$.
  The cone of n-point **weightable quasi-semimetrics** is a projection along an extreme ray of the semimetric cone Met_{n+1} (Grishukhin–Deza–Deza, 2011).
  The **metric fan** is a canonical decomposition MF_n of MET_n into subcones whose faces belong to the fan, and the intersection of any two of them is their common boundary. Two semimetrics $d, d' \in MET_n$ lie in the same cone of the metric fan if the subdivisions $\delta_d, \delta_{d'}$ of the polyhedron $\delta(n, 2) = conv\{e_i + e_j : 1 \leq i < j \leq n\} \subset \mathbb{R}^n$ are equal. Here a subpolytope P of $\delta(n, 2)$ is a cell of the subdivision δ_d if there exists $y \in \mathbb{R}^n$ satisfying $y_i + y_j = d_{ij}$ if $e_i + e_j$ is a vertex of P, and $y_i + y_j > d_{ij}$, otherwise. The complex of bounded faces of the polyhedron dual to δ_d is the **tight span** of the semimetric d.
- **Cayley–Menger matrix**
  Given a finite metric space $(X = \{x_1, \cdots, x_n\}, d)$, its **Cayley–Menger matrix** is the symmetric $(n + 1) \times (n + 1)$ matrix

$$CM(X, d) = \begin{pmatrix} 0 & e \\ e^T & D \end{pmatrix},$$

  where $D = ((d^2(x_i, x_j)))$ and e is the n-vector all components of which are 1. The determinant of $CM(X, d)$ is called the *Cayley–Menger determinant*. If (X, d) is a metric subspace of the Euclidean space $\mathbb{E}^{n-1}$, then $CM(X, d)$ is $(-1)^n 2^{n-1}((n-1)!)^2$ times the squared $(n-1)$*-dimensional volume* of the convex hull of X in $\mathbb{R}^{n-1}$.
- **Gram matrix**
  Given elements $v_1, \dots, v_k$ of a Euclidean space, their **Gram matrix** is the symmetric $k \times k$ matrix VV^T, where $V = ((v_{ij}))$, of pairwise *inner products* of $v_1, \dots, v_k$:

$$G(v_1, \dots, v_k) = ((\langle v_i, v_j \rangle)).$$

  It holds $G(v_1, \dots, v_k) = \frac{1}{2}((d_E^2(v_0, v_i) + d_E^2(v_0, v_j) - d_E^2(v_i, v_j)))$, i.e., the inner product $\langle , \rangle$ is the **Gromov product similarity** of the **squared Euclidean distance** d_E^2. A $k \times k$ matrix $((d_E^2(v_i, v_j)))$ is called *Euclidean distance matrix* (or *EDM*). It defines a **distance of negative type** on $\{1, \dots, k\}$; all such matrices form the (nonpolyhedral) closed convex cone of all such distances.
  The determinant of a Gram matrix is called the *Gram determinant*; it is equal to the square of the k*-dimensional volume* of the *parallelotope* constructed on $v_1, \dots, v_k$.

A symmetric $k \times k$ real matrix M is said to be *positive-semidefinite* (PSD) if $xMx^T \geq 0$ for any nonzero $x \in \mathbb{R}^k$ and *positive-definite* (PD) if $xMx^T > 0$. A matrix is PSD if and only if it is a Gram matrix; it is PD if and only the vectors $v_1, \dots, v_k$ are linearly independent. In Statistics, the *covariance matrices* and *correlation matrices* are exactly PSD and PD ones, respectively.

- **Midset**
  Given a metric space (X, d) and distinct $y, z \in X$, the **midset** (or *bisector*) of points y and z is the set $M = \{x \in X : d(x, y) = d(x, z)\}$ of *midpoints* x.
  A metric space is said to have the *n-point midset property* if, for every pair of its points, the midset has exactly n points. The one-point midset property means uniqueness of the *midpoint map*. Cf. **midpoint convexity**.
- **Distance k-sector**
  Given a metric space (X, d) and disjoint subsets $Y, Z \subset X$, the *bisector* of Y and Z is the set $M = \{x \in X : \inf_{y \in Y} d(x, y) = \inf_{z \in Z} d(x, z)\}$.
  The **distance k-sector** of Y and Z is the sequence $M_1, \dots, M_{k-1}$ of subsets of X such that M_i, for any $1 \leq i \leq k - 1$, is the bisector of sets M_{i-1} and M_{i+1}, where $Y = M_0$ and $Z = M_k$. Asano–Matousek–Tokuyama, 2006, considered the distance k-sector on the Euclidean plane $(\mathbb{R}^2, l_2)$; for compact sets Y and Z, the sets $M_1, \dots, M_{k-1}$ are curves partitioning the plane into k parts.
- **Metric basis**
  Given a metric space (X, d) and a subset $M \subset X$, for any point $x \in X$, its *metric M-representation* is the set $\{(m, d(x, m)) : m \in M\}$ of its *metric M-coordinates* $(m, d(x, m))$. The set M is called (Blumenthal, 1953) a **metric basis** (or *resolving set, locating set, set of uniqueness, metric generator*) if distinct points $x \in X$ have distinct M-representations. A vertex-subset M of a connected graph is (Okamoto et al., 2009) a *local metric basis* if adjacent vertices have distinct M-representations.
  The *resolving number* of a finite (X, d) is (Chartrand–Poisson–Zhang, 2000) minimum k such that any k-subset of X is a metric basis.
  The vertices of a non degenerate simplex form a metric basis of $\mathbb{E}^n$, but l_1- and l_∞-metrics on $\mathbb{R}^n$, $n > 1$, have no finite metric basis.
  The **distance similarity** is (Saenpholphat–Zhang, 2003) an equivalence relation on X defined by $x \sim y$ if $d(z, x) = d(z, y)$ for any $z \in X \setminus \{x, y\}$. Any metric basis contains all or all but one elements from each equivalence class.

1.3 Metric Numerical Invariants

- **Resolving dimension**
  Given a metric space (X, d), its **resolving dimension** (or **location number** (Slater, 1975), *metric dimension* (Harary–Melter, 1976)) is the minimum cardinality of its **metric basis**. The **upper resolving dimension** of (X, d) is the maximum cardinality of its metric basis not containing another metric basis

as a proper subset. *Adjacency dimension* of (X, d) is the metric dimension of $(X, \min(2, d))$.

A **metric independence number** of (X, d) is (Currie–Oellermann, 2001) the maximum cardinality I of a collection of pairs of points of X, such that for any two, (say, (x, y) and (x', y')) of them there is no point $z \in X$ with $d(z, x) \neq d(z, y)$ and $d(z, x') \neq d(z, y')$. A function $f : X \to [0, 1]$ is a *resolving function* of (X, d) if $\sum_{z \in X : d(x,z) \neq d(y,z)} f(z) \geq 1$ for any distinct $x, y \in X$. The *fractional resolving dimension* of (X, d) is $F = \min \sum_{x \in X} g(x)$, where the minimum is taken over resolving functions f such that any function f' with f', f is not resolving.

The *partition dimension* of (X, d) is (Chartrand–Salevi–Zhang, 1998) the minimum cardinality P of its *resolving partition*, i.e., a partition $X = \cup_{1 \leq i \leq k} S_i$ such that no two points have, for $1 \leq i \leq k$, the same minimal distances to the set S_i. Related *locating a robber* game on a graph $G = (V, E)$ was considered in 2012 by Seager and by Carraher et al.: *cop win* on G if every sequence $r = r_1, \dots, r_n$ of robber's steps ($r_i \in V$ and $d_{\text{path}}(r_i, r_{i+1}) \leq 1$) is uniquely identified by a sequence $d(r_1, c_1), \dots, d(r_n, c_n)$ of cop's distance queries for some $c_1, \dots, c_n \in V$.

- **Metric dimension**

  For a metric space (X, d) and a number $\epsilon > 0$, let C_ϵ be the minimal size of an **ϵ-net** of (X, d), i.e., a subset $M \subset X$ with $\cup_{x \in M} B(x, \epsilon) = X$. The number

$$dim(X, d) = \lim_{\epsilon \to 0} \frac{\ln C_\epsilon}{-\ln \epsilon}$$

  (if it exists) is called the **metric dimension** (or **Minkowski–Bouligand dimension, box-counting dimension**) of X. If the limit above does not exist, then the following notions of dimension are considered:

  1. $\underline{dim}(X, d) = \underline{\lim}_{\epsilon \to 0} \frac{\ln C_\epsilon}{-\ln \epsilon}$ called the **lower Minkowski dimension** (or *lower dimension, lower box dimension, Pontryagin–Snirelman dimension*);
  2. $\overline{dim}(X, d) = \overline{\lim}_{\epsilon \to 0} \frac{\ln C_\epsilon}{-\ln \epsilon}$ called the **Kolmogorov–Tikhomirov dimension** (or *upper dimension, entropy dimension, upper box dimension*).

  See below examples of other, less prominent, notions of *metric dimension*.

  1. The (equilateral) *metric dimension* of a metric space is the maximum cardinality of its *equidistant* subset, i.e., such that any two of its distinct points are at the same distance. For a normed space, this dimension is equal to the maximum number of translates of its unit ball that touch pairwise.
  2. For any $c > 1$, the (normed space) *metric dimension* $dim_c(X)$ of a finite metric space (X, d) is the least dimension of a real *normed space* $(V, ||.||)$ such that there is an embedding $f : X \to V$ with $\frac{1}{c} d(x, y) \leq ||f(x) - f(y)|| \leq d(x, y)$.

3. The (Euclidean) *metric dimension* of a finite metric space (X, d) is the least dimension n of a Euclidean space $\mathbb{E}^n$ such that $(X, f(d))$ is its metric subspace, where the minimum is taken over all continuous monotone increasing functions $f(t)$ of $t \geq 0$.
4. The *dimensionality* of a metric space is $\frac{\mu^2}{2\sigma^2}$, where μ and σ^2 are the mean and variance of its histogram of distance values; this notion is used in Information Retrieval for proximity searching.
   The term *dimensionality* is also used for the minimal dimension, if it is finite, of Euclidean space in which a given metric space embeds isometrically.

- **Hausdorff dimension**
  Given a metric space (X, d) and $p, q > 0$, let $H_p^q = \inf \sum_{i=1}^{\infty} (diam(A_i))^p$, where the infimum is taken over all countable coverings $\{A_i\}$ with diameter of A_i less than q. The **Hausdorff q-measure** of X is the **metric outer measure** defined by

$$H^p = \lim_{q \to 0} H_p^q.$$

  The **Hausdorff dimension** (or **fractal dimension**) of (X, d) is defined by

$$dim_{Haus}(X, d) = \inf\{p \geq 0 : H^p(X) = 0\}.$$

  Any countable metric space has $dim_{Haus} = 0$, $dim_{Haus}(\mathbb{E}^n) = n$, and any $X \subset \mathbb{E}^n$ with *Int* $X \neq \emptyset$ has $dim_{Haus} = \underline{dim}$. For any **totally bounded** (X, d), it holds

$$dim_{top} \leq dim_{Haus} \leq \underline{dim} \leq dim \leq \overline{dim}.$$

- **Rough dimension**
  Given a metric space (X, d), its *rough n-volume* $Vol_n X$ is $\overline{\lim}_{\epsilon \to 0} \epsilon^n \beta_X(\epsilon)$, where $\epsilon > 0$ and $\beta_X(\epsilon) = \max |Y|$ for $Y \subseteq X$ with $d(a, b) \geq \epsilon$ if $a \in Y, b \in Y \setminus \{a\}$; $\beta_X(\epsilon) = \infty$ is permitted. The **rough dimension** is defined [BBI01] by

$$dim_{rough}(X, d) = \sup\{n : Vol_n X = \infty\} \text{ or, equivalently, } = \inf\{n : Vol_n X = 0\}.$$

  The space (X, d) can be not locally compact. It holds $dim_{Haus} \leq dim_{rough}$.
- **Packing dimension**
  Given a metric space (X, d) and $p, q > 0$, let $P_p^q = \sup \sum_{i=1}^{\infty} (diam(B_i))^p$, where the supremum is taken over all countable packings (by disjoint balls) $\{B_i\}$ with the diameter of B_i less than q.
  The **packing q-pre-measure** is $P_0^p = \lim_{q \to 0} P_p^q$. The **packing q-measure** is the **metric outer measure** which is the infimum of packing q-pre-measures of countable coverings of X. The **packing dimension** of (X, d) is defined by

$$dim_{pack}(X, d) = \inf\{p \geq 0 : P^p(X) = 0\}.$$

- **Topological dimension**
  For any compact metric space (X, d) its **topological dimension** (or **Lebesgue covering dimension**) is defined by
  $$\dim_{top}(X, d) = \inf_{d'}\{\dim_{Haus}(X, d')\},$$
  where d' is any metric on X **equivalent** to d. So, it holds $\dim_{top} \leq \dim_{Haus}$. A **fractal** (cf. Chap. 18) is a metric space for which this inequality is strict.
  This dimension does not exceed also the **Assouad–Nagata dimension** of (X, d).
  In general, the **topological dimension** of a topological space X is the smallest integer n such that, for any finite open covering of X, there exists a finite open refinement of it with no point of X belonging to more than $n + 1$ elements.
  The *geometric dimension* is (Kleiner, 1999) $\sup dim_{top}(Y, d)$ over compact $Y \subset X$.
- **Doubling dimension**
  The **doubling dimension** ($dim_{doubl}(X, d)$) of a metric space (X, d) is the smallest integer n (or ∞ if such an n does not exist) such that every metric ball (or, say, a set of finite diameter) can be covered by a family of at most 2^n metric balls (respectively, sets) of half the diameter.
  If (X, d) has finite doubling dimension, then d is called a **doubling metric** and the smallest integer m such that every metric ball can be covered by a family of at most m metric balls of half the diameter is called *doubling constant*.
- **Assouad–Nagata dimension**
  The **Assouad–Nagata dimension** $dim_{AN}(X, d)$ of a metric space (X, d) is the smallest integer n (or ∞ if such an n does not exist) for which there exists a constant $C > 0$ such that, for all $s > 0$, there exists a covering of X by its subsets of diameter $\leq Cs$ with every subset of X of diameter $\leq s$ meeting $\leq n + 1$ elements of covering. It holds (LeDonne–Rajala, 2014) $dim_{AN} \leq dim_{doubl}$; but $dim_{AN} = 1$, while $dim_{doubl} = \infty$, holds (Lang–Schlichenmaier, 2014) for some **real trees** (X, d).
  Replacing "for all $s > 0$" in the above definition by "for $s > 0$ sufficiently large" or by "for $s > 0$ sufficiently small", gives the *microscopic mi-*$dim_{AN}(X, d)$ and *macroscopic ma-*$dim_{AN}(X, d)$ Assouad–Nagata dimensions, respectively. Then (Brodskiy et al., 2006) *mi-*$dim_{AN}(X, d) = dim_{AN}(X, \min\{d, 1\})$ and *ma-*$dim_{AN}(X, d) = dim_{AN}(X, \max\{d, 1\})$ (here $\max\{d(x, y), 1\}$ means 0 for $x = y$).
  The Assouad–Nagata dimension is preserved (Lang–Schlichenmaier, 2004) under **quasi-symmetric mapping** but, in general, not under **quasi-isometry**.
- **Vol'berg–Konyagin dimension**
  The **Vol'berg–Konyagin dimension** of a metric space (X, d) is the smallest constant $C > 1$ (or ∞ if such a C does not exist) for which X carries a *doubling measure*, i.e., a Borel measure μ such that, for all $x \in X$ and $r > 0$, it holds that
  $$\mu(\overline{B}(x, 2r)) \leq C\mu(\overline{B}(x, r)).$$

A metric space (X, d) carries a doubling measure if and only if d is a **doubling metric**, and any complete doubling metric carries a doubling measure.
The **Karger–Ruhl constant** of a metric space (X, d) is the smallest $c > 1$ (or ∞ if such a c does not exist) such that for all $x \in X$ and $r > 0$ it holds

$$|\overline{B}(x, 2r)| \leq c|\overline{B}(x, r)|.$$

If c is finite, then the **doubling dimension** of (X, d) is at most $4c$.

- **Hyperbolic dimension**
  A metric space (X, d) is called an (R, N)-*large-scale doubling* if there exists a number $R > 0$ and integer $N > 0$ such that every ball of radius $r \geq R$ in (X, d) can be covered by N balls of radius $\frac{r}{2}$.
  The **hyperbolic dimension** $hypdim(X, d)$ of a metric space (X, d) (Buyalo–Schroeder, 2004) is the smallest integer n such that for every $r > 0$ there are $R > 0$, an integer $N > 0$ and a covering of X with the following properties:

  1. Every ball of radius r meets at most $n + 1$ elements of the covering;
  2. The covering is an (R, N)-large-scale doubling, and any finite union of its elements is an (R', N)-large-scale doubling for some $R' > 0$.

  The hyperbolic dimension is 0 if (X, d) is a large-scale doubling, and it is n if (X, d) is n-dimensional hyperbolic space.
  Also, $hypdim(X, d) \leq asdim(X, d)$ since the **asymptotic dimension** $asdim(X, d)$ corresponds to the case $N = 1$ in the definition of $hypdim(X, d)$.
  The hyperbolic dimension is preserved under a **quasi-isometry**.
- **Asymptotic dimension**
  The **asymptotic dimension** $asdim(X, d)$ of a metric space (X, d) (Gromov, 1993) is the smallest integer n such that, for every $r > 0$, there exists a constant $D = D(r)$ and a covering of X by its subsets of diameter at most D such that every ball of radius r meets at most $n + 1$ elements of the covering.
  The asymptotic dimension is preserved under a **quasi-isometry**.
- **Width dimension**
  Let (X, d) be a compact metric space. For a given number $\epsilon > 0$, the **width dimension** $Widim_{\epsilon}(X, d)$ of (X, d) is (Gromov, 1999) the minimum integer n such that there exists an n-dimensional polyhedron P and a continuous map $f : X \to P$ (called an ϵ-*embedding*) with $diam(f^{-1}(y)) \leq \epsilon$ for all $y \in P$.
  The width dimension is a *macroscopic dimension at the scale* $\geq\epsilon$ of (X, d), because its limit for $\epsilon \to 0$ is the **topological dimension** of (X, d).
- **Godsil–McKay dimension**
  We say that a metric space (X, d) has **Godsil–McKay dimension** $n \geq 0$ if there exists an element $x_0 \in X$ and two positive constants c and C such that the inequality $ck^n \leq |\{x \in X : d(x, x_0) \leq k\}| \leq Ck^n$ holds for every integer $k \geq 0$.

This notion was introduced in [GoMc80] for the **path metric** of a countable locally finite graph. They proved that, if the group $\mathbb{Z}^n$ acts faithfully and with a finite number of orbits on the vertices of the graph, then this dimension is n.

- **Metric outer measure**

  A *σ-algebra over X* is any nonempty collection Σ of subsets of X, including X itself, that is closed under complementation and countable unions of its members. Given a σ-algebra Σ over X, a *measure* on (X, Σ) is a function $\mu : \Sigma \to [0, \infty]$ with the following properties:

  1. $\mu(\emptyset) = 0$;
  2. For any sequence $\{A_i\}$ of pairwise disjoint subsets of X, $\mu(\sum_i A_i) = \sum_i \mu(A_i)$ (*countable σ-additivity*).

  The triple (X, Σ, μ) is called a *measure space*. If $M \subset A \in \Sigma$ and $\mu(A) = 0$ implies $M \in \Sigma$, then (X, Σ, μ) is called a *complete measure space*. A measure μ with $\mu(X) = 1$ is called a *probability measure*.

  If X is a *topological space* (see Chap. 2), then the σ-algebra over X, consisting of all *open* and *closed sets* of X, is called the *Borel σ-algebra* of X, (X, Σ) is called a *Borel space*, and a measure on Σ is called a *Borel measure*. So, any metric space (X, d) admits a Borel measure coming from its **metric topology**, where the *open set* is an arbitrary union of open **metric d-balls**.

  An *outer measure* on X is a function $\nu : P(X) \to [0, \infty]$ (where $P(X)$ is the set of all subsets of X) with the following properties:

  1. $\nu(\emptyset) = 0$;
  2. For any subsets $A, B \subset X$, $A \subset B$ implies $\nu(A) \leq \nu(B)$ (*monotonicity*);
  3. For any sequence $\{A_i\}$ of subsets of X, $\nu(\sum_i A_i) \leq \sum_i \nu(A_i)$ (*countable subadditivity*).

  A subset $M \subset X$ is called *ν-measurable* if $\nu(A) = \nu(A \cup M) + \nu(A \setminus M)$ for any $A \subset X$. The set Σ' of all ν-measurable sets forms a σ-algebra over X, and (X, Σ', ν) is a complete measure space.

  A **metric outer measure** is an outer measure ν defined on the subsets of a given metric space (X, d) such that $\nu(A \cup B) = \nu(A) + \nu(B)$ for every pair of nonempty subsets $A, B \subset X$ with positive **set-set distance** $\inf_{a \in A, b \in B} d(a, b)$. An example is **Hausdorff q-measure**; cf. **Hausdorff dimension**.

- **Length of metric space**

  The **Fremlin length** of a metric space (X, d) is its **Hausdorff 1-measure** $H^1(X)$.

  The **Hejcman length** $lng(M)$ of a subset $M \subset X$ of a metric space (X, d) is $\sup\{lng(M') : M' \subset M, |M'| < \infty\}$. Here $lng(\emptyset) = 0$ and, for a finite subset $M' \subset X$, $lng(M') = \min \sum_{i=1}^{n} d(x_{i-1}, x_i)$ over all sequences $x_0, \dots, x_n$ such that $\{x_i : i = 0, 1, \dots, n\} = M'$.

The **Schechtman length** of a finite metric space (X,d) is $\inf\sqrt{\sum_{i=1}^{n}a_i^2}$ over all sequences $a_1,\dots,a_n$ of positive numbers such that there exists a sequence $X_0,\dots,X_n$ of partitions of X with following properties:

1. $X_0=\{X\}$ and $X_n=\{\{x\}:x\in X\}$;
2. X_i refines X_{i-1} for $i=1,\dots,n$;
3. For $i=1,\dots,n$ and $B,C\subset A\in X_{i-1}$ with $B,C\in X_i$, there exists a one-to-one map f from B onto C such that $d(x,f(x))\le a_i$ for all $x\in B$.

- **Volume of finite metric space**
  Given a metric space (X,d) with $|X|=k<\infty$, its **volume** (Feige, 2000) is the maximal $(k-1)$-dimensional volume of the simplex with vertices $\{f(x):x\in X\}$ over all **metric mappings** $f:(X,d)\to(\mathbb{R}^{k-1},l_2)$. The volume coincides with the metric for $k=2$. It is monotonically increasing and continuous in the metric d.
- **Rank of metric space**
  The **Minkowski rank of a metric space** (X,d) is the maximal dimension of a normed vector space $(V,||.||)$ such that there is an isometry $(V,||.||)\to(X,d)$.
  The **Euclidean rank of a metric space** (X,d) is the maximal dimension of a *flat* in it, that is of a Euclidean space $\mathbb{E}^n$ such that there is an isometric embedding $\mathbb{E}^n\to(X,d)$.
  The **quasi-Euclidean rank of a metric space** (X,d) is the maximal dimension of a *quasi-flat* in it, i.e., of an Euclidean space $\mathbb{E}^n$ admitting a **quasi-isometry** $\mathbb{E}^n\to(X,d)$. Every **Gromov hyperbolic metric space** has this rank 1.
- **Roundness of metric space**
  The **roundness of a metric space** (X,d) is the supremum of all q such that

$$d(x_1,x_2)^q+d(y_1,y_2)^q\le d(x_1,y_1)^q+d(x_1,y_2)^q+d(x_2,y_1)^q+d(x_2,y_2)^q$$

  for any four points $x_1,x_2,y_1,y_2\in X$.
  Every metric space has roundness ≥ 1; it is ≤ 2 if the space has **approximate midpoints**. The roundness of L_p**-space** is p if $1\le p\le 2$.
  The *generalized roundness of a metric space* (X,d) is (Enflo, 1969) the supremum of all q such that, for any $2k\ge 4$ points $x_i,y_i\in X$ with $1\le i\le k$,

$$\sum_{1\le i<j\le k}(d(x_i,x_j)^q+d(y_i,y_j)^q)\le\sum_{1\le i,j\le k}d(x_i,y_j)^q.$$

  Lennard–Tonge–Weston, 1997, have shown that the generalized roundness is the supremum of q such that d is of q**-negative type**, i.e., d^q is of **negative type**.
  Every **CAT(0) space** (cf. Chap. 6) has roundness 2, but some of them have generalized roundness 0 (Lafont–Prassidis, 2006).

- **Type of metric space**
The **Enflo type** of a metric space (X, d) is p if there exists a constant $1 \leq C < \infty$ such that, for every $n \in \mathbb{N}$ and every function $f : \{-1, 1\}^n \to X$, $\sum_{\epsilon \in \{-1,1\}^n} d^p(f(\epsilon), f(-\epsilon))$ is at most
$C^p \sum_{j=1}^{n} \sum_{\epsilon \in \{-1,1\}^n} d^p(f(\epsilon_1, \dots, \epsilon_{j-1}, \epsilon_j, \epsilon_{j+1}, \dots, \epsilon_n), f(\epsilon_1, \dots, \epsilon_{j-1}, -\epsilon_j, \epsilon_{j+1}, \dots, \epsilon_n))$.
A Banach space $(V, ||.||)$ of Enflo type p has *Rademacher type* p, i.e., for every $x_1, \dots, x_n \in V$, it holds
$$\sum_{\epsilon \in \{-1,1\}^n} ||\sum_{j=1}^{n} \epsilon_j x_j||^p \leq C^p \sum_{j=1}^{n} ||x_j||^p.$$
Given a metric space (X, d), a *symmetric Markov chain on* X is a Markov chain $\{Z_l\}_{l=0}^{\infty}$ on a state space $\{x_1, \dots, x_m\} \subset X$ with a symmetrical transition $m \times m$ matrix $((a_{ij}))$, such that $P(Z_{l+1} = x_j : Z_l = x_i) = a_{ij}$ and $P(Z_0 = x_i) = \frac{1}{m}$ for all integers $1 \leq i, j \leq m$ and $l \geq 0$. A metric space (X, d) has **Markov type** p (Ball, 1992) if $\sup_T M_p(X, T) < \infty$ where $M_p(X, T)$ is the smallest constant $C > 0$ such that the inequality
$$\mathbb{E}d^p(Z_T, Z_0) \leq TC^p \mathbb{E}d^p(Z_1, Z_0)$$
holds for every symmetric Markov chain $\{Z_l\}_{l=0}^{\infty}$ on X holds, in terms of expected value (mean) $\mathbb{E}[X] = \sum_x xp(x)$ of the discrete random variable X.
A metric space of Markov type p has Enflo type p.
- **Strength of metric space**
Given a finite metric space (X, d) with s different nonzero values of $d_{ij} = d(i, j)$, its **strength** is the largest number t such that, for any integers $p, q \geq 0$ with $p + q \leq t$, there is a polynomial $f_{pq}(s)$ of degree at most $\min\{p, q\}$ such that $((d_{ij}^{2p}))((d_{ij}^{2q})) = ((f_{pq}(d_{ij}^2)))$.
- **Rendez-vous number**
Given a metric space (X, d), its **Rendez-vous number** (or *Gross number, magic number*) is a positive real number $r(X, d)$ (if it exists) defined by the property that for each integer n and all (not necessarily distinct) $x_1, \dots, x_n \in X$ there exists a point $x \in X$ such that
$$r(X, d) = \frac{1}{n} \sum_{i=1}^{n} d(x_i, x).$$
If the number $r(X, d)$ exists, then it is said that (X, d) has the **average distance property**. Every compact connected metric space has this property. The *unit ball* $\{x \in V : ||x|| \leq 1\}$ of a **Banach space** $(V, ||.||)$ has the rendez-vous number 1.
- **Wiener-like distance indices**
Given a finite subset M of a metric space (X, d) and a parameter q, the **Wiener polynomial** of M (as defined by Hosoya, 1988, for the graphic metric d_{path}) is

$$W(M;q) = \frac{1}{2} \sum_{x,y \in M: x \neq y} q^{d(x,y)}.$$

It is a *generating function* for the **distance distribution** (cf. Chap. 16) of M, i.e., the coefficient of q^i in $W(M;q)$ is the number $|\{\{x,y\} \in M \times M : d(x,y) = i\}|$.

In the main case when M is the vertex-set V of a connected graph $G = (V, E)$ and d is the **path metric** of G, the number $W(M;1) = \frac{1}{2}\sum_{x,y \in M} d(x,y)$ is called the **Wiener index** of G. This notion is originated (Wiener, 1947) and applied, together with its many analogs, in Chemistry; cf. **chemical distance** in Chap. 24.

The *hyper-Wiener index* is $\sum_{x,y \in M}(d(x,y) + d(x,y)^2)$. The *reverse-Wiener index* is $\frac{1}{2}\sum_{x,y \in M}(D - d(x,y))$, where D is the diameter of M. The *complementary reciprocal Wiener index* is $\frac{1}{2}\sum_{x,y \in M}(1 + D - d(x,y))^{-1}$. The *Harary index* is $\sum_{x,y \in M}(d(x,y))^{-1}$. The *Szeged index* and the *vertex PI index* are $\sum_{e \in E} n_x(e)n_y(e)$ and $\sum_{e \in E}(n_x(e) + n_y(e))$, where $e = (xy)$ and $n_x(e) = |\{z \in V : d(x,z) < d(y,z)\}|$.

Two studied *edge-Wiener indices* of G are the Wiener index of its *line graph* and $\sum_{(xy),(x'y') \in E} \max\{d(x,x'), d(x,y'), d(y,x'), d(y,y')\}$.

The *Gutman–Schultz index*, **degree distance** (Dobrynin–Kochetova, 1994), *reciprocal degree distance* and *terminal Wiener index* are:

$$\sum_{x,y \in M} r_x r_y d(x,y), \ \sum_{x,y \in M} d(x,y)(r_x + r_y),$$

$$\times \sum_{x,y \in M} \frac{1}{d(x,y)}(r_x + r_y), \quad \sum_{x,y \in \{z \in M : r_z = 1\}} d(x,y),$$

where r_z is the degree of the vertex $z \in M$. The **eccentric distance sum** (Gupta et al., 2002) is $\sum_{y \in M}(\max\{d(x,y) : x \in M\}d_y$, where d_y is $\sum_{x \in M} d(x,y)$. The *Balaban index* is $\frac{|E|}{c+1}\sum_{(yz) \in E}(\sqrt{d_y d_z})^{-1}$, where c is the number of primitive cycles.

Above indices are called (corresponding) *Kirchhoff indices* if d the **resistance metric** (cf. Chap. 15) of G.

The *average distance* of M is the number $\frac{1}{|M|(|M|-1)}\sum_{x,y \in M} d(x,y)$. In general, for a quasi-metric space (X,d), the numbers $\sum_{x,y \in M} d(x,y)$ and $\frac{1}{|M|(|M|-1)}\sum_{x,y \in M, x \neq y}\frac{1}{d(x,y)}$ are called, respectively, the *transmission* and *global efficiency* of M.

- **Distance polynomial**

  Given an ordered finite subset M of a metric space (X,d), let D be the **distance matrix** of M. The **distance polynomial** of M is the *characteristic polynomial* of D, i.e., the determinant $det(D - \lambda I)$.

Usually, D is the distance matrix (of path metric) of a graph. Sometimes, the distance polynomial is defined as $det(\lambda I - D)$ or $(-1)^n det(D - \lambda I)$.
The roots of the distance polynomial constitute the **distance spectrum** (or *D-spectrum* of *D-eigenvalues*) of M. Let $\rho_{\max}$ and $\rho_{\min}$ be the largest and the smallest roots; then $\rho_{\max}$ and $\rho_{\max} - \rho_{\min}$ are called (distance spectral) *radius* and *spread* of M. The **distance degree** of $x \in M$ is $\sum_{y \in M} d(x, y)$. The **distance energy** of M is the sum of the absolute values of its D-eigenvalues. It is $2\rho_{\max}$ if (as, for example, for the path metric of a tree) exactly one D-eigenvalue is positive.

- ***s*-energy**
  Given a finite subset M of a metric space (X, d) and a number $s > 0$, the ***s*-energy** and **0-energy** of M are, respectively, the numbers

$$\sum_{x,y \in M, x \neq y} \frac{1}{d^s(x, y)} \text{ and } \sum_{x,y \in M, x \neq y} \log \frac{1}{d(x, y)} = -\log \prod_{x,y \in M, x \neq y} d(x, y).$$

  The (unnormalized) ***s*-moment** of M is the number $\sum_{x,y \in M} d^s(x, y)$.
  The *discrete Riesz s-energy* is the s-energy for Euclidean distance d. In general, let μ be a finite Borel probability measure on (X, d). Then $U_s^{\mu}(x) = \int \frac{\mu(dy)}{d(x,y)^s}$ is the (abstract) *s-potential* at a point $x \in X$. The *Newton gravitational potential* is the case $(X, d) = (\mathbb{R}^3, |x - y|)$, $s = 1$, for the mass distribution μ.
  The *s-energy* of μ is $E_s^{\mu} = \int U_s^{\mu}(x)\mu(dx) = \int\int \frac{\mu(dx)\mu(dy)}{d(x,y)^s}$, and the *s-capacity* of (X, d) is $(\inf_{\mu} E_s^{\mu})^{-1}$. Cf. the **metric capacity**.

- **Fréchet mean**
  Given a metric space (X, d) and a number $s > 0$, the *Fréchet function* is $F_s(x) = \mathbb{E}[d^s(x, y)]$. For a finite subset M of X, this expected value is the mean $F_s(x) = \sum_{y \in M} w(y) d^s(x, y)$, where $w(y)$ is a weight function on M.
  The points, minimizing $F_1(x)$ and $F_2(x)$, are called the **Fréchet median** (or *weighted geometric median*) and **Fréchet mean** (or *Karcher mean*), respectively. If $(X, d) = (\mathbb{R}^n, ||x - y||_2)$ and the weights are equal, these points are called the *geometric median* (or *Fermat–Weber point*, *1-median*) and the *centroid* (or *geometric center*, *barycenter*), respectively.
  The *k-median* and *k-mean* of M are the k-sets C minimizing, respectively, the sums $\sum_{y \in M} \min_{c \in C} d(y, c) = \sum_{y \in M} d(y, C)$ and $\sum_{y \in M} d^2(y, C)$.
  Let (X, d) be the metric space $(\mathbb{R}_{>0}, |f(x) - f(y)|)$, where $f : \mathbb{R}_{>0} \to \mathbb{R}$ is a given injective and continuous function. Then the Fréchet mean of $M \subset \mathbb{R}_{>0}$ is the *f-mean* (or *Kolmogorov mean*, *quasi-arithmetic mean*) $f^{-1}(\frac{\sum_{x \in M} f(x)}{|M|})$. It is the arithmetic, geometric, harmonic, and power mean if $f = x, \log(x), \frac{1}{x}$, and $f = x^p$ (for a given $p \neq 0$), respectively. The cases $p \to +\infty, p \to -\infty$ correspond to maximum and minimum, while $p = 2, = 1, \to 0, \to -1$ correspond to the quadratic, arithmetic, geometric and harmonic mean.
  Given a *completely monotonic* (i.e., $(-1)^k f^{(k)} \geq 0$ for any k) function $f \in \mathbb{C}^{\infty}$, the ***f*-potential energy** of a finite subset M of (X, d) is

$\sum_{x,y \in M, x \neq y} f(d^2(x,y))$. The set M is called (Cohn–Kumar, 2007) *universally optimal* if it minimizes, among sets $M' \subset X$ with $|M'| = |M|$, the f-potential energy for any such f. Among universally optimal subsets of $(\mathbb{S}^{n-1}, ||x-y||_2)$, there are the vertex-sets of a polygon, simplex, cross-polytope, icosahedron, 600-cell, E_8 root system.

- **Distance-weighted mean**
In Statistics, the **distance-weighted mean** between given data points $x_1, \dots, x_n$ is defined (Dodonov–Dodonova, 2011) by

$$\frac{\sum_{1 \le i \le n} w_i x_i}{\sum_{1 \le i \le n} w_i} \quad \text{with } w_i = \frac{n-1}{\sum_{1 \le j \le n} |x_i - x_j|}.$$

The case $w_i = 1$ for all i corresponds to the arithmetic mean.

- **Inverse distance weighting**
In Numerical Analysis, *multivariate* (or *spatial*) interpolation is interpolation on functions of more than one variable. **Inverse distance weighting** is a method (Shepard, 1968) for multivariate interpolation. Let $x_1, \dots, x_n$ be interpolating points (i.e., samples $u_i = u(x_i)$ are known), x be an interpolated (unknown) point and $d(x, x_i)$ be a given distance. A general form of interpolated value $u(x)$ is

$$u(x) = \frac{\sum_{1 \le i \le n} w_i(x) u_i}{\sum_{1 \le i \le n} w_i(x)}, \text{ with } w_i(x) = \frac{1}{(d(x, x_i))^p},$$

where $p > 0$ (usually $p = 2$) is a fixed *power parameter*.

- **Transfinite diameter**
The *n-th diameter* $D_n(M)$ and the *n-th Chebyshev constant* $C_n(M)$ of a set $M \subseteq X$ in a metric space (X, d) are defined (Fekete, 1923, for the complex plane $\mathbb{C}$) as

$$D_n(M) = \sup_{x_1, \dots, x_n \in M} \prod_{i \neq j} d(x_i, x_j)^{\frac{1}{n(n-1)}} \text{ and}$$

$$C_n(M) = \inf_{x \in X} \sup_{x_1, \dots, x_n \in M} \prod_{j=1}^{n} d(x, x_j)^{\frac{1}{n}}.$$

The number $\log D_n(M)$ (the supremum of the average distance) is called the *n-extent* of M. The numbers $D_n(M), C_n(M)$ come from the geometric mean averaging; they also come as the limit case $s \to 0$ of the *s-moment* $\sum_{i \neq j} d(x_i, x_j)^s$ averaging.
The **transfinite diameter** (or ∞*-th diameter*) and the ∞*-th Chebyshev constant* $C_\infty(M)$ of M are defined as

$$D_\infty(M) = \lim_{n \to \infty} D_n(M) \text{ and } C_\infty(M) = \lim_{n \to \infty} C_n(M);$$

these limits existing since $\{D_n(M)\}$ and $\{C_n(M)\}$ are nonincreasing sequences of nonnegative real numbers. Define $D_\infty(\emptyset) = 0$.

The transfinite diameter of a compact subset of $\mathbb{C}$ is its **conformal radius** at infinity (cf. Chap. 6); for a segment in $\mathbb{C}$, it is $\frac{1}{4}$ of its length.

- **Metric diameter**
The **metric diameter** (or **diameter**, *width*) $diam(M)$ of a set $M \subseteq X$ in a metric space (X, d) is defined by

$$\sup_{x,y \in M} d(x, y).$$

The *diameter graph* of M has, as vertices, all points $x \in M$ with $d(x, y) = diam(M)$ for some $y \in M$; it has, as edges, all pairs of its vertices at distance $diam(M)$ in (X, d). (X, d) is called a **diametrical metric space** if any $x \in X$ has the *antipode*, i.e., a unique $x' \in X$ such that the **closed metric interval** $I(x, x')$ is X.
The *furthest neighbor digraph* of M is a directed graph on M, where xy is an arc (called a *furthest neighbor pair*) whenever y is at maximal distance from x.
In a metric space endowed with a measure, one says that the *isodiametric inequality* holds if the metric balls maximize the measure among all sets with given diameter. It holds for the volume in Euclidean space but not, for example, for the **Heisenberg metric** on the *Heisenberg group* (cf. Chap. 10).
The k**-ameter** (Grove–Markvorsen, 1992) is $\sup_{K \subseteq X : |K| = k} \frac{1}{2} \sum_{x,y \in K} d(x, y)$, and the k**-diameter** (Chung–Delorme–Sole, 1999) is $\sup_{K \subseteq X : |K| = k} \inf_{x,y \in K : x \neq y} d(x, y)$.
Given a property $P \subseteq X \times X$ of a pair (K, K') of subsets of a finite metric space (X, d), the **conditional diameter** (called P-*diameter* in Balbuena et al., 1996) is $\max_{(K,K') \in P} \min_{(x,y) \in K \times K'} d(x, y)$. It is $diam(X, d)$ if $P = \{(K, K') \in X \times X : |K| = |K'| = 1\}$. When (X, d) models an interconnection network, the P-diameter corresponds to the maximum delay of the messages interchanged between any pair of clusters of nodes, K and K', satisfying a given property P of interest.

- **Metric spread**
A subset M of a metric space (X, d) is called **Delone set** (or *separated* ϵ**-net**, (A, a)*-Delone set*) if it is **bounded** (with a finite **diameter** $A = \sup_{x,y \in M} d(x, y)$) and **metrically discrete** (with a *separation* $a = \inf_{x,y \in M, x \neq y} d(x, y) > 0$).
The **metric spread** (or *distance ratio, normalized diameter*) of M is the ratio $\frac{A}{a}$.
The **aspect ratio** (or *axial ratio*) of a shape is the ratio of its longer and shorter dimensions, say, the length and diameter of a rod, major and minor axes of a torus or width and height of a rectangle (image, display, pixel, etc.). The *Feret ratio* is the reciprocal of the aspect ratio; cf. **shape parameters** in Chap. 21.
In Physics, the *aspect ratio* is the ratio of height-to-length scale characteristics. *Dynamic range* DNR is the ratio between the largest and smallest possible values of a quantity, such as in sound or light signals; cf. **SNR distance** in Chap. 21.

In the Theory of Approximation and Interpolation, the separation a and the **covering radius** (or *mesh norm*) $c = \sup_{y \in X} \inf_{x \in M} d(x, y)$ of M are used to measure the stability and error of the approximation. The *mesh ratio* of M is $\frac{c}{a}$.

- **Eccentricity**
Given a bounded metric space (X, d), the **eccentricity** (or *Koenig number*) of a point $x \in X$ is the number $e(x) = \max_{y \in X} d(x, y)$.
The numbers $D = \max_{x \in X} e(x)$ and $r = \min_{x \in X} e(x)$ are called the **diameter** and the **radius** of (X, d), respectively. The point $z \in X$ is called *central* if $e(z) = r$, *peripheral* if $e(z) = D$, and *pseudo-peripheral* if for each point x with $d(z, x) = e(z)$ it holds that $e(z) = e(x)$. For finite $|X|$, the *average eccentricity* is $\frac{1}{|X|} \sum_{x \in X} e(x)$, and the *contour* of (X, d) is the set of points $x \in X$ such that no *neighbor* (closest point) of x has an eccentricity greater than x.
The **eccentric digraph** (Buckley, 2001) of (X, d) has, as vertices, all points $x \in X$ and, as arcs, all ordered pairs (x, y) of points with $d(x, y) = e(y)$. The **eccentric graph** (Akyiama–Ando–Avis, 1976) of (X, d) has, as vertices, all points $x \in X$ and, as edges, all pairs (x, y) of points at distance $\min\{e(x), e(y)\}$. The **super-eccentric graph** (Iqbalunnisa–Janairaman–Srinivasan, 1989) of (X, d) has, as vertices, all points $x \in X$ and, as edges, all pairs (x, y) of points at distance no less than the radius of (X, d). The **radial graph** (Kathiresan–Marimuthu, 2009) of (X, d) has, as vertices, all points $x \in X$ and, as edges, all pairs (x, y) of points at distance equal to the radius of (X, d).
The sets $\{x \in X : e(x) \leq e(z)$ for any $z \in X\}$, $\{x \in X : e(x) \geq e(z)$ for any $z \in X\}$ and $\{x \in X : \sum_{y \in X} d(x, y) \leq \sum_{y \in X} d(z, y)$ for any $z \in X\}$ are called, respectively, the **metric center** (or *eccentricity center*, *center*), **metric antimedian** (or *periphery*) and the **metric median** (or *distance center*) of (X, d).
- **Radii of metric space**
Given a bounded metric space (X, d) and a set $M \subseteq X$ of **diameter** D, its **metric radius** (or **radius**) Mr, **covering radius** (or **directed Hausdorff distance** from X to M) Cr and **remoteness** (or **Chebyshev radius**) Re are the numbers $\inf_{x \in M} \sup_{y \in M} d(x, y)$, $\sup_{x \in X} \inf_{y \in M} d(x, y)$ and $\inf_{x \in X} \sup_{y \in M} d(x, y)$., respectively. It holds that $\frac{D}{2} \leq Re \leq Mr \leq D$ with $Mr = \frac{D}{2}$ in any **injective metric space**. Somemimes, $\frac{D}{2}$ is called the *radius*.
For $m > 0$, a **minimax distance design of size** m is an m-subset of X having smallest covering radius. This radius is called the *m-point mesh norm* of (X, d).
The **packing radius** Pr of M is the number $\sup\{r : \inf_{x, y \in M, x \neq y} d(x, y) > 2r\}$.
For $m > 0$, a **maximum distance design of size** m is an m-subset of X having largest packing radius. This radius is the *m-point best packing distance* on (X, d).
- **ϵ-Net**
Given a metric space (X, d), a subset $M \subset X$, and a number $\epsilon > 0$, the **ϵ-neighborhood** of M is the set $M^{\epsilon} = \cup_{x \in M} B(x, \epsilon)$.

The set M is called an **ϵ-net** (or *ϵ-covering*, *ϵ-approximation*) of (X, d) if $M^\epsilon = X$, i.e., the **covering radius** of M is at most ϵ.
Let C_ϵ denote the *ϵ-covering number*, i.e., the smallest size of an ϵ-net in (X, d). The number $\lg_2 C_\epsilon$ is called (Kolmogorov–Tikhomirov, 1959) the **metric entropy** (or *ϵ-entropy*) of (X, d). It holds $P_\epsilon \leq C_\epsilon \leq P_{\frac{\epsilon}{2}}$, where P_ϵ denote the *ϵ-packing number* of (X, d), i.e., $\sup\{|M| : M \subset X, \overline{B}(x, \epsilon) \cap \overline{B}(y, \epsilon) = \emptyset$ for any $x, y \in M, x \neq y\}$. The number $\lg_2 P_\epsilon$ is called the **metric capacity** (or *ϵ-capacity*) of (X, d).

- **Steiner ratio**
Given a metric space (X, d) and a finite subset $V \subset X$, let $G = (V, E)$ be the complete weighted graph on V with edge-weights $d(x, y)$ for all $x, y \in V$.
Given a tree T, its *weight* is the sum $d(T)$ of its edge-weights. A *spanning tree* of V is a subset of $|V| - 1$ edges forming a tree on V. Let $MSpT_V$ be a *minimum spanning tree* of V, i.e., a spanning tree with the minimal weight $d(MSpT_V)$.
A *Steiner tree* of V is a tree on Y, $V \subset Y \subset X$, connecting vertices from V; elements of $Y \setminus V$ are called *Steiner points*. Let $StMT_V$ be a *minimum Steiner tree* of V, i.e., a Steiner tree with the minimal weight $d(StMT_V) = \inf_{Y \subset X : V \subset Y} d(MSpT_Y)$. This weight is called the **Steiner diversity** of V. It is the **Steiner distance of set** V (cf. Chap. 15) if (X, d) is graphic metric space.
The **Steiner ratio** $St(X, d)$ of the metric space (X, d) is defined by
$$\inf_{V \subset X} \frac{d(StMT_V)}{d(MSpT_V)}.$$
Cf. **arc routing problems** in Chap. 15.
- **Diversity**
Given a set X, a function f from its finite subsets to $\mathbb{R}_{\geq 0}$ is called (Bryant–Tupper, 2012) *diversity on* X if $f(A) = 0$ for all $A \subset X$ with $|A| \leq 1$ and
$$f(A \cup B) + f(B \cup C) \geq f(A \cup C) \text{ for all } A, B, C \subset X \text{ with } B \neq \emptyset.$$
The **induced diversity metric** $d(x, y)$ is $f(\{x, y\})$. For any diversity $f(A)$ with induced metric space (X, d), it holds $f_{diam}(A) \leq f(A) \leq f_S(A) \leq (|A|-1) f_{diam}(A)$, where the **diameter diversity** $f_{diam}(A)$ is $\max_{x,y \in A} d(x, y) = diam(A)$ and the **Steiner diversity** $f_S(A)$ is the minimum weight of a Steiner tree connecting elements of A. Also, the *Traveling Salesman diversity* is the minimum of $\frac{1}{2}(d(a_1, a_2) + d(a_2, a_3) + \cdots + d(a_{|A|}, a_1))$ over all orderings $a_1, a_2, \ldots, a_{|A|}$ of A.
- **Chromatic numbers of metric space**
Given a metric space (X, d) and a set D of positive real numbers, the **D-chromatic number** of (X, d) is the standard *chromatic number* of its **D-distance graph**, i.e., the graph (X, E) with the vertex-set X and the edge-set $E = \{xy : d(x, y) \in D\}$ (Chap. 15). Usually, (X, d) is an **l_p-space** and $D = \{1\}$ (**Benda–Perles chromatic number**) or $D = [1 - \epsilon, 1 + \epsilon]$.

For a metric space (X,d), the **polychromatic number** is the minimum number of colors needed to color all the points $x \in X$ so that, for each color class C_i, there is a distance d_i such that no two points of C_i are at distance d_i.
For a metric space (X,d), the **packing chromatic number** is the minimum number of colors needed to color all the points $x \in X$ so that, for each color class C_i, no two distinct points of C_i are at distance at most i.
For any integer $t > 0$, the **t-distance chromatic number** of a metric space (X,d) is the minimum number of colors needed to color all the points $x \in X$ so that any two points whose distance is $\leq t$ have distinct colors. Cf. **k-distance chromatic number** in Chap. 15.
For any integer $t > 0$, the **t-th Babai number** of a metric space (X,d) is the minimum number of colors needed to color all the points in X so that, for any set D of positive distances with $|D| \leq t$, any two points $x, y \in X$ with $d(x,y) \in D$ have distinct colors.

- **Congruence order of metric space**
A metric space (X,d) has **congruence order** n if every finite metric space which is not **isometrically embeddable** in (X,d) has a subspace with at most n points which is not isometrically embeddable in (X,d). For example, the congruence order of l_2^n is $n+3$ (Menger, 1928); it is 4 for the **path metric** of a tree.

1.4 Main Mappings of Metric Spaces

- **Distance function**
In Topology, the term *distance function* is often used for **distance**. But, in general, a **distance function** (or *ray function*) is a continuous function on a metric space (X,d) (usually, on a Euclidean space $\mathbb{E}^n$) $f : X \rightarrow \mathbb{R}_{\geq 0}$ which is *homogeneous*, i.e., $f(tx) = tf(x)$ for all $t \geq 0$ and all $x \in X$.
Such function f is called *positive* if $f(x) > 0$ for all $x \neq 0$, *symmetric* if $f(x) = f(-x)$, *convex* if $f(tx + (1-t)y) \leq tf(x) + (1-t)f(y)$ for any $0 < t < 1$ and $x \neq y$, and *strictly convex* if this inequality is strict.
If $X = \mathbb{E}^n$, the set $S_f = \{x \in \mathbb{R}^n : f(x) < 1\}$ is *star body*, i.e., $x \in S_f$ implies $[0,x] \subset S_f$. Any star body S corresponds to a unique distance function $g(x) = \inf_{tx \in S, t>0} \frac{1}{t}$, and $S = S_g$. The star body is bounded if f is positive, symmetric about the origin if f is symmetric, convex if f is convex, and *strictly convex* (i.e., the boundary ∂B does not contain a segment) if f is strictly convex.
For a quadratic distance function of the form $f_A = xAx^T$, where A is a real matrix and $x \in \mathbb{R}^n$, the matrix A is *positive-definite* (i.e., the **Gram matrix** $VV^T = ((\langle v_i, v_j \rangle))$ of n linearly independent vectors $v_i = (v_{i1}, \dots, v_{in})$) if and only if f_A is symmetric and strictly convex function. The *homogeneous minimum* of f_A is

$$\min(f_A) = \inf_{x \in \mathbb{Z}^n \setminus \{0\}} f_A(x) = \inf_{x \in L \setminus \{0\}} \sum_{1 \leq i \leq n} x_i^2,$$

where $L = \{\sum x_i v_i : x_i \in \mathbb{Z}\}$ is a *lattice*, i.e., a discrete subgroup of $\mathbb{R}^n$ spanning it. The *Hermite constant* γ_n, a central notion in Geometry of Numbers, is the supremum, over all positive-definite $(n \times n)$-matrices, of $\min(f_A) \det(A)^{\frac{1}{n}}$. It is known only for $2 \le n \le 8$ and $n = 24$; cf. **systole of metric space**.

- **Convex distance function**
  Given a compact convex region $B \subset \mathbb{R}^n$ containing the origin O in its interior, the **convex distance function** (or **Minkowski distance function**, *Minkowski seminorm, gauge*) is the function $||P||_B$ whose value at a point $P \in \mathbb{R}^n$ is the *distance ratio* $\frac{OP}{OQ}$, where $Q \in B$ is the furthest from O point on the ray OP.
  Then $d_B(x, y) = ||x - y||_B$ is the quasi-metric on $\mathbb{R}^n$ defined, for $x \neq y$, by
  $$\inf\{\alpha > 0 : y - x \in \alpha B\},$$
  and $B = \{x \in \mathbb{R}^n : d_B(0, x) \le 1\}$ with equality only for $x \in \partial B$.
  The function $||P||_B$ is called a **polyhedral distance function** if B is a n-polytope, *simplicial distance function* if it is a n-simplex, and so on.
  If B is centrally-symmetric with respect to the origin, then d_B is a **Minkowskian metric** (cf. Chap. 6) whose unit ball is B. This is the l_1-metric if B is the n-cross-polytope and the l_∞-metric if B is the n-cube.
- **Funk distance**
  Let B be an nonempty open convex subset of $\mathbb{R}^n$. For any $x, y \in B$, denote by $R(x, y)$ the ray from x through y. The **Funk distance** (Funk, 1929) on B is the quasi-semimetric defined, for any $x, y \in B$, as 0 if the boundary $\partial(B)$ and $R(x, y)$ are disjoint, and, otherwise, i.e., if $R(x, y) \cap \partial B = \{z\}$, by
  $$\ln \frac{||x - z||_2}{||y - z||_2}.$$
  The **Hilbert projective metric** in Chap. 6 is a symmetrization of this distance.
- **Metric projection**
  Given a metric space (X, d) and a subset $M \subset X$, an element $u_0 \in M$ is called an **element of best approximation** to a given element $x \in X$ if $d(x, u_0) = \inf_{u \in M} d(x, u)$, i.e., if $d(x, u_0)$ is the **point-set distance** $d(x, M)$.
  A **metric projection** (or *operator of best approximation, nearest point map*) is a multivalued mapping associating to each element $x \in X$ the set of elements of best approximation from the set M (cf. **distance map**).
  A **Chebyshev set** in a metric space (X, d) is a subset $M \subset X$ containing a unique element of best approximation for every $x \in X$.
  A subset $M \subset X$ is called a **semi-Chebyshev set** if the number of such elements is at most one, and a **proximinal set** if this number is at least one.
  The **Chebyshev radius** (or **remoteness**) of the set M is $\inf_{x \in X} \sup_{y \in M} d(x, y)$, and a **Chebyshev center** of M is an element $x_0 \in X$ realizing this infimum. Sometimes (say, for a finite graphic metric space), $\inf_{x \in X} \sum_{y \in M} d(x, y)$ and $\sup_{x \in X} \sum_{y \in M} d(x, y)$ are called *proximity* and *remoteness* of M.

- **Distance map**
Given a metric space (X, d) and a subset $M \subset X$, the **distance map** is a function $f_M : X \to \mathbb{R}_{\geq 0}$, where $f_M(x) = \inf_{u \in M} d(x, u)$ is the **point-set distance** $d(x, M)$ (cf. **metric projection**).
If the boundary $B(M)$ of the set M is defined, then the **signed distance function** g_M is defined by $g_M(x) = -\inf_{u \in B(M)} d(x, u)$ for $x \in M$, and $g_M(x) = \inf_{u \in B(M)} d(x, u)$, otherwise. If M is a (closed orientable) n-**manifold** (Chap. 2), then g_M is the solution of the *eikonal equation* $|\nabla g| = 1$ for its *gradient* ∇.
If $X = \mathbb{R}^n$ and, for every $x \in X$, there is unique element $u(x)$ with $d(x, M) = d(x, u(x))$ (i.e., M is a **Chebyshev set**), then $||x - u(x)||$ is called a **vector distance function**.
Distance maps are used in Robot Motion (M being the set of obstacle points) and, especially, in Image Processing (M being the set of all or only boundary pixels of the image). For $X = \mathbb{R}^2$, the graph $\{(x, f_M(x)) : x \in X\}$ of $d(x, M)$ is called the *Voronoi surface* of M.
- **Isometry**
Given metric spaces (X, d_X) and (Y, d_Y), a function $f : X \to Y$ is called an **isometric embedding** of X into Y if it is injective and the equality $d_Y(f(x), f(y)) = d_X(x, y)$ holds for all $x, y \in X$.
An **isometry** (or *congruence mapping*) is a bijective isometric embedding. Two metric spaces are called **isometric** (or *isometrically isomorphic*) if there exists an isometry between them.
A property of metric spaces which is invariant with respect to isometries (completeness, boundedness, etc.) is called a **metric property** (or *metric invariant*).
A **path isometry** (or *arcwise isometry*) is a mapping from X into Y (not necessarily bijective) preserving lengths of curves.
- **Rigid motion of metric space**
A **rigid motion** (or, simply, **motion**) of a metric space (X, d) is an **isometry** of (X, d) onto itself.
For a motion f, the **displacement function** $d_f(x)$ is $d(x, f(x))$. The motion f is called *semisimple* if $\inf_{x \in X} d_f(x) = d(x_0, f(x_0))$ for some $x_0 \in X$, and *parabolic*, otherwise. A semisimple motion is called *elliptic* if $\inf_{x \in X} d_f(x) = 0$, and *axial* (or *hyperbolic*), otherwise. A motion is called a *Clifford translation* if the displacement function $d_f(x)$ is a constant for all $x \in X$.
- **Symmetric metric space**
A metric space (X, d) is called **symmetric** if, for any point $p \in X$, there exists a *symmetry* relative to that point, i.e., a **motion** f_p of this metric space such that $f_p(f_p(x)) = x$ for all $x \in X$, and p is an isolated fixed point of f_p.
- **Homogeneous metric space**
A metric space is called **homogeneous** (or *point-homogeneous*) if, for any two points of it, there exists a motion mapping one of the points to the other.
In general, a *homogeneous space* is a set together with a given transitive group of *symmetries*. Moss, 1992, defined similar *distance-homogeneous distanced graph*.

A metric space is called **ultrahomogeneous space** (or *highly transitive*) if any isometry between two of its finite subspaces extends to the whole space.
A metric space (X, d) is called (Grünbaum–Kelly) **metrically homogeneous metric space** if $\{d(x, z) : z \in X\} = \{d(y, z) : z \in X\}$ for any $x, y \in X$.

- **Flat space**
  A **flat space** is any metric space with **local isometry** to some $\mathbb{E}^n$, i.e., each point has a neighborhood isometric to an open set in $\mathbb{E}^n$. A space is *locally Euclidean* if every point has a neighborhood homeomorphic to an open subset in $\mathbb{E}^n$.
- **Dilation of metric space**
  Given a metric space (X, d), its **dilation** (or r**-dilation**) is a mapping $f : X \to X$ with $d(f(x), f(y)) = rd(x, y)$ for some $r > 0$ and any $x \in X$.
- **Wobbling of metric space**
  Given a metric space (X, d), its **wobbling** (or r**-wobbling**) is a mapping $f : X \to X$ with $d(x, f(x)) < r$ for some $r > 0$ and any $x \in X$.
- **Paradoxical metric space**
  Given a metric space (X, d) and an equivalence relation on the subsets of X, the space (X, d) is called **paradoxical** if X can be decomposed into two disjoint sets M_1, M_2 so that M_1, M_2 and X are pairwise equivalent.
  Deuber, Simonovitz and Sós, 1995, introduced this idea for *wobbling equivalent* subsets $M_1, M_2 \subset X$, i.e., there is a bijective r**-wobbling** $f : M_1 \to M_2$. For example, $(\mathbb{R}^2, l_2)$ is paradoxical for wobbling but not for isometry equivalence.
- **Metric cone**
  A **pointed metric space** (X, d, x_0) is called a **metric cone**, if it is isometric to $(\lambda X, d, x_0)$ for all $\lambda > 0$. A **metric cone structure** on (X, d, x_0) is a (pointwise) continuous family f_t ($t \in \mathbb{R}_{>0}$) of **dilations** of X, leaving the point x_0 invariant, such that $d(f_t(x), f_t(y)) = td(x, y)$ for all x, y and $f_t \circ f_s = f_{ts}$. A Banach space has such a structure for the dilations $f_t(x) = tx$ ($t \in \mathbb{R}_{>0}$). The *Euclidean cone over a metric space* (cf. **cone over metric space** in Chap. 9) is another example.
  The **tangent metric cone** over a metric space (X, d) at a point x_0 is (for all dilations $tX = (X, td)$) the closure of $\cup_{t>0} tX$, i.e., of $\lim_{t\to\infty} tX$ taken in the pointed Gromov–Hausdorff topology (cf. **Gromov–Hausdorff metric**).
  The **asymptotic metric cone** over (X, d) is its tangent metric cone "at infinity", i.e., $\cap_{t>0} tX = \lim_{t\to 0} tX$. Cf. **boundary of metric space** in Chap. 6.
  The term *metric cone* was also used by Bronshtein, 1998, for a convex cone C equipped with a complete metric compatible with its operations of addition (continuous on $C \times C$) and multiplication (continuous on $C \times \mathbb{R}_{\geq 0}$). by all $\lambda \geq 0$.
- **Metric fibration**
  Given a **complete** metric space (X, d), two subsets M_1 and M_2 of X are called *equidistant* if for each $x \in M_1$ there exists $y \in M_2$ with $d(x, y)$ being equal to the **Hausdorff metric** between the sets M_1 and M_2. A **metric fibration** of (X, d) is a partition $\mathcal{F}$ of X into isometric mutually equidistant closed sets.

The quotient metric space $X/\mathcal{F}$ inherits a natural metric for which the **distance map** is a **submetry**.

- **Homeomorphic metric spaces**
  Two metric spaces (X, d_X) and (Y, d_Y) are called **homeomorphic** (or *topologically isomorphic*) if there exists a *homeomorphism* from X to Y, i.e., a bijective function $f : X \to Y$ such that f and f^{-1} are *continuous* (the preimage of every open set in Y is open in X).
  Two metric spaces (X, d_X) and (Y, d_Y) are called *uniformly isomorphic* if there exists a bijective function $f : X \to Y$ such that f and f^{-1} are *uniformly continuous*. A function g is *uniformly continuous* if, for any $\epsilon > 0$, there exists $\delta > 0$ such that, for any $x, y \in X$, the inequality $d_X(x, y) < \delta$ implies that $d_Y(g(x), f(y)) < \epsilon$; a continuous function is uniformly continuous if X is compact.
- **Möbius mapping**
  Given distinct points x, y, z, w of a metric space (X, d), their **cross-ratio** is

$$cr((x, y, z, w), d) = \frac{d(x, y)d(z, w)}{d(x, z)d(y, w)} > 0.$$

  Given metric spaces (X, d_X) and (Y, d_Y), a **homeomorphism** $f : X \to Y$ is called a **Möbius mapping** if, for every distinct points $x, y, z, w \in X$, it holds

$$cr((x, y, z, w), d_X) = cr((f(x), f(y), f(z), f(w)), d_Y).$$

  A homeomorphism $f : X \to Y$ is called a **quasi-Möbius mapping** (Väisälä, 1984) if there exists a homeomorphism $\tau : [0, \infty) \to [0, \infty)$ such that, for every quadruple x, y, z, w of distinct points of X, it holds

$$cr((f(x), f(y), f(z), f(w)), d_Y) \leq \tau(cr((x, y, z, w), d_X)).$$

  A metric space (X, d) is called *metrically dense* (or *μ-dense* for given $\mu > 1$, Aseev–Trotsenko, 1987) if for any $x, y \in X$, there exists a sequence $\{z_i, i \in \mathbb{Z}\}$ with $z_i \to x$ as $i \to -\infty$, $z_i \to y$ as $i \to \infty$, and $\log cr((x, z_i, z_{i+1}, y), d) \leq \log \mu$ for all $i \in \mathbb{Z}$. The space (X, d) is μ-dense if and only if (Tukia-Väisälä, 1980), for any $x, y \in X$, there exists $z \in X$ with $\frac{d(x,y)}{6\mu} \leq d(x, z) \leq \frac{d(x,y)}{4}$.
- **Quasi-symmetric mapping**
  Given metric spaces (X, d_X) and (Y, d_Y), a **homeomorphism** $f : X \to Y$ is called a **quasi-symmetric mapping** (Tukia–Väisälä, 1980) if there is a homeomorphism $\tau : [0, \infty) \to [0, \infty)$ such that, for every triple (x, y, z) of distinct points of X,

$$\frac{d_Y(f(x), f(y))}{d_Y(f(x), f(z))} \leq \tau \frac{d_X(x, y)}{d_X(x, z)}.$$

Quasi-symmetric mappings are **quasi-Möbius**, and quasi-Möbius mappings between bounded metric spaces are quasi-symmetric. In the case $f : \mathbb{R}^n \to \mathbb{R}^n$, quasi-symmetric mappings are exactly the same as **quasi-conformal mappings**.

- **Conformal metric mapping**
  Given metric spaces (X, d_X) and (Y, d_Y) which are domains in $\mathbb{R}^n$, a **homeomorphism** $f : X \to Y$ is called a **conformal metric mapping** if, for any nonisolated point $x \in X$, the limit $\lim_{y \to x} \frac{d_Y(f(x), f(y))}{d(x,y)}$ exists, is finite and positive.
  A homeomorphism $f : X \to Y$ is called a **quasi-conformal mapping** (or, specifically, *C-quasi-conformal mapping*) if there exists a constant C such that

$$\limsup_{r \to 0} \frac{\max\{d_Y(f(x), f(y)) : d_X(x, y) \le r\}}{\min\{d_Y(f(x), f(y)) : d_X(x, y) \ge r\}} \le C$$

  for each $x \in X$. The smallest such constant C is called the **conformal dilation**.
  The **conformal dimension** of a metric space (X, d) (Pansu, 1989) is the infimum of the **Hausdorff dimension** over all quasi-conformal mappings of (X, d) into some metric space. For the middle-third Cantor set on $[0, 1]$, it is 0 but, for any of its quasi-conformal images, it is positive.
- **Hölder mapping**
  Let $c, \alpha \ge 0$ be constants. Given metric spaces (X, d_X) and (Y, d_Y), a function $f : X \to Y$ is called the **Hölder mapping** (or *α-Hölder mapping* if the constant α should be mentioned) if for all $x, y \in X$

$$d_Y(f(x), f(y)) \le c(d_X(x, y))^\alpha.$$

  A 1-Hölder mapping is a **Lipschitz mapping**; 0-Hölder mapping means that the metric d_Y is bounded.
- **Lipschitz mapping**
  Let c be a positive constant. Given metric spaces (X, d_X) and (Y, d_Y), a function $f : X \to Y$ is called a **Lipschitz** (or **Lipschitz continuous**, *c-Lipschitz* if the constant c should be mentioned) **mapping** if for all $x, y \in X$ it holds

$$d_Y(f(x), f(y)) \le c d_X(x, y).$$

  A c-Lipschitz mapping is called a **metric mapping** if $c = 1$, and is called a **contraction** if $c < 1$.
- **Bi-Lipschitz mapping**
  Given metric spaces (X, d_X), (Y, d_Y) and a constant $c > 1$, a function $f : X \to Y$ is called a **bi-Lipschitz mapping** (or *c-bi-Lipschitz mapping*, **c-embedding**) if there exists a number $r > 0$ such that for any $x, y \in X$ it holds

$$r d_X(x, y) \le d_Y(f(x), f(y)) \le c r d_X(x, y).$$

  Every bi-Lipschitz mapping is a **quasi-symmetric mapping**.

The smallest c for which f is a c-bi-Lipschitz mapping is called the **distortion** of f. Bourgain, 1985, proved that every k-point metric space c-embeds into a Euclidean space with distortion $O(\ln k)$. Gromov's *distortion for curves* is the maximum ratio of arc length to chord length.
Two metrics d_1 and d_2 on X are called **bi-Lipschitz equivalent metrics** if there are positive constants c and C such that $Cd_1(x, y) \leq d_2(x, y) \leq Cd_1(x, y)$ for all $x, y \in X$, i.e., the identity mapping is a bi-Lipschitz mapping from (X, d_1) into (X, d_2). Bi-Lipschitz equivalent metrics are **equivalent**, i.e., generate the same topology but, for example, equivalent L_1-metric and L_2-metric (cf. L_p-**metric** in Chap. 5) on $\mathbb{R}$ are not bi-Lipschitz equivalent.
A bi-Lipschitz mapping $f : X \to Y$ is a c-**isomorphism** $f : X \to f(X)$.

- **c-Isomorphism of metric spaces**
Given two metric spaces (X, d_X) and (Y, d_Y), the *Lipschitz norm* $||.||_{Lip}$ on the set of all injective mappings $f : X \to Y$ is defined by

$$||f||_{Lip} = \sup_{x,y \in X, x \neq y} \frac{d_Y(f(x), f(y))}{d_X(x, y)}.$$

Two metric spaces X and Y are called c-**isomorphic** if there exists an injective mapping $f : X \to Y$ such that $||f||_{Lip}||f^{-1}||_{Lip} \leq c$.

- **Metric Ramsey number**
For a given class $\mathcal{M}$ of metric spaces (usually, l_p-*spaces*), an integer $n \geq 1$, and a real number $c \geq 1$, the **metric Ramsey number** (or c-*metric Ramsey number*) $R_{\mathcal{M}}(c, n)$ is the largest integer m such that every n-point metric space has a subspace of cardinality m that c-embeds into a member of $\mathcal{M}$ (see [BLMN05]).
The *Ramsey number* R_n is the minimal number of vertices of a complete graph such that any edge-coloring with n colors produces a monochromatic triangle. The following metric analog of R_n was considered in [Masc04]: the least number of points a finite metric space must contain in order to contain an equilateral triangle, i.e., to have **equilateral metric dimension** greater than two.

- **Uniform metric mapping**
Given metric spaces (X, d_X) and (Y, d_Y), a function $f : X \to Y$ is called a **uniform metric mapping** if there are two nondecreasing functions g_1 and g_2 from $\mathbb{R}_{\geq 0}$ to itself with $\lim_{r \to \infty} g_i(r) = \infty$ for $i = 1, 2$, such that the inequality

$$g_1(d_X(x, y)) \leq d_Y(f(x), f(y)) \leq g_2(d_X(x, y))$$

holds for all $x, y \in X$. A **bi-Lipschitz mapping** is a uniform metric mapping with linear functions g_1, g_2.

- **Metric compression**
Given metric spaces (X, d_X) (unbounded) and (Y, d_Y), a function $f : X \to Y$ is a *large scale Lipschitz mapping* if, for some $c > 0, D \geq 0$ and all $x, y \in X$,

$$d_Y(f(x), f(y)) \leq cd_X(x, y) + D.$$

The *compression* of such a mapping f is $\rho_f(r) = \inf_{d_X(x,y) \geq r} d_Y(f(x), f(y))$. The **metric compression** of (X, d_X) in (Y, d_Y) is defined by

$$R(X, Y) = \sup_f \{ \underline{\lim}_{r \to \infty} \frac{\log \max\{\rho_f(r), 1\}}{\log r} \},$$

where the supremum is over all large scale Lipschitz mappings f.

In the main interesting case—when (Y, d_Y) is a Hilbert space and (X, d_X) is a (finitely generated discrete) group with **word metric**—$R(X, Y) = 0$ if there is no (Guentner–Kaminker, 2004) **uniform metric mapping** $(X, d_X) \to (Y, d_Y)$, and $R(X, Y) = 1$ for free groups, even if there is no **quasi-isometry**. Arzhantzeva–Guba–Sapir, 2006, found groups with $\frac{1}{2} \leq R(X, Y) \leq \frac{3}{4}$.

- **Quasi-isometry**

  Given metric spaces (X, d_X) and (Y, d_Y), a function $f : X \to Y$ is called a **quasi-isometry** (or (C, c)**-quasi-isometry**) if it holds

  $$C^{-1} d_X(x, y) - c \leq d_Y(f(x), f(y)) \leq C d_X(x, y) + c,$$

  for some $C \geq 1, c \geq 0$, and $Y = \cup_{x \in X} B_{d_Y}(f(x), c)$, i.e., for every point $y \in Y$, there exists $x \in X$ such that $d_Y(y, f(x)) < \frac{c}{2}$. Quasi-isometry is an equivalence relation on metric spaces; it is a bi-Lipschitz equivalence up to small distances. Quasi-isometry means that metric spaces contain bi-Lipschitz equivalent **Delone sets**.

  A quasi-isometry with $C = 1$ is called a **coarse isometry** (or *rough isometry, almost isometry*). Cf. **quasi-Euclidean rank of a metric space**.

- **Coarse embedding**

  Given metric spaces (X, d_X) and (Y, d_Y), a function $f : X \to Y$ is called a **coarse embedding** if there exist nondecreasing functions $\rho_1, \rho_2 : [0, \infty) \to [0, \infty)$ with $\rho_1(d_X(x, x')) \leq d_Y(f(x), f(x')) \leq \rho_2(d_X(x, x'))$ if $x, x' \in X$ and $\lim_{t \to \infty} \rho_1(t) = +\infty$.

  Metrics d_1, d_2 on X are called **coarsely equivalent metrics** if there exist nondecreasing functions $f, g : [0, \infty) \to [0, \infty)$ such that $d_1 \leq f(d_2), d_2 \leq g(d_1)$.

- **Metrically regular mapping**

  Let (X, d_X) and (Y, d_Y) be metric spaces, and let F be a set-valued mapping from X to Y, having *inverse* F^{-1}, i.e., with $x \in F^{-1}(y)$ if and only if $y \in F(x)$. The mapping F is said to be **metrically regular at** $\overline{x}$ **for** $\overline{y}$ (Dontchev–Lewis–Rockafeller, 2002) if there exists $c > 0$ such that it holds

  $$d_X(x, F^{-1}(y)) \leq c d_Y(y, F(x))$$

  for all (x, y) close to $(\overline{x}, \overline{y})$. Here $d(z, A) = \inf_{a \in A} d(z, a)$ and $d(z, \emptyset) = +\infty$.

- **Contraction**
  Given metric spaces (X, d_X) and (Y, d_Y), a function $f : X \to Y$ is called a **contraction** if the inequality

$$d_Y(f(x), f(y)) \le c d_X(x, y)$$

  holds for all $x, y \in X$ and some real number c, $0 \le c < 1$.
  Every contraction is a **contractive mapping**, and it is uniformly continuous. *Banach fixed point theorem* (or *contraction principle*): every contraction from a **complete** metric space into itself has a unique fixed point.
- **Contractive mapping**
  Given metric spaces (X, d_X) and (Y, d_Y), a function $f : X \to Y$ is called a **contractive** (or *strictly short*, *distance-decreasing*) **mapping** if

$$d_Y(f(x), f(y)) < d_X(x, y)$$

  holds for all different $x, y \in X$. A function $f : X \to Y$ is called a **noncontractive mapping** (or *dominating mapping*) if for all $x, y \in X$ it holds

$$d_Y(f(x), f(y)) \ge d_X(x, y).$$

  Every noncontractive bijection from a **totally bounded** metric space onto itself is an **isometry**.
- **Short mapping**
  Given metric spaces (X, d_X) and (Y, d_Y), a function $f : X \to Y$ is called a **short** (or *1-Lipschitz*, *nonexpansive*, *distance-noninreasing*, *metric*) **mapping** (or *semicontraction*) if for all $x, y \in X$ it holds

$$d_Y(f(x), f(y)) \le d_X(x, y).$$

  A **submetry** is a short mapping such that the image of any metric ball is a metric ball of the same radius.
  The set of short mappings $f : X \to Y$ for bounded metric spaces (X, d_X) and (Y, d_Y) is a metric space under the **uniform metric** $\sup\{d_Y(f(x), g(x)) : x \in X\}$.
  Two subsets A and B of a metric space (X, d) are called (Gowers, 2000) **similar** if there exist short mappings $f : A \to X$, $g : B \to X$ and a small $\epsilon > 0$ such that every point of A is within ϵ of some point of B, every point of B is within ϵ of some point of A, and $|d(x, g(f(x))) - d(y, f(g(y)))| \le \epsilon$ for any $x \in A, y \in B$.
- **Category of metric spaces**
  A *category* Ψ consists of a class $Ob(\Psi)$ of *objects* and a class $Mor(\Psi)$ of *morphisms* (or *arrows*) satisfying the following conditions

  1. To each ordered pair of objects A, B is associated a set $\Psi(A, B)$ of morphisms, and each morphism belongs to only one set $\Psi(A, B)$;
  2. The composition $f \cdot g$ of two morphisms $f : A \to B, g : C \to D$ is defined if $B = C$ in which case it belongs to $\Psi(A, D)$, and it is associative;

3. Each set $\Psi(A, A)$ contains, as an *identity*, a morphism id_A such that $f \cdot id_A = f$ and $id_A \cdot g = g$ for any morphisms $f : X \to A$ and $g : A \to Y$.

The **category of metric spaces**, denoted by *Met* (see [Isbe64]), is a category which has metric spaces as objects and **short mappings** as morphisms. A unique **injective envelope** exists in this category for every one of its objects; it can be identified with its **tight span**. In *Met*, the *monomorphisms* are injective short mappings, and *isomorphisms* are **isometries.** *Met* is a subcategory of the category which has metric spaces as objects and **Lipschitz mappings** as morphisms.
Cf. **metric 1-space** on the objects of a category in Chap. 3.

- **Injective metric space**
A metric space (X, d) is called **injective** if, for every isometric embedding $f : X \to X'$ of (X, d) into another metric space (X', d'), there exists a **short mapping** f' from X' into X with $f' \cdot f = id_X$, i.e., X is a *retract* of X'. Equivalently, X is an *absolute retract*, i.e., a retract of every metric space into which it embeds isometrically. A metric space (X, d) is injective if and only if it is **hyperconvex**. Examples of such metric spaces are l_1^2-space, l_∞^n-space, any **real tree** and the **tight span** of a metric space.
- **Injective envelope**
The **injective envelope** (introduced first in [Isbe64] as *injective hull*) is a generalization of **Cauchy completion**. Given a metric space (X, d), it can be embedded isometrically into an **injective metric space** $(\hat{X}, \hat{d})$; given any such isometric embedding $f : X \to \hat{X}$, there exists a unique smallest injective subspace $(\overline{X}, \overline{d})$ of $(\hat{X}, \hat{d})$ containing $f(X)$ which is called the **injective envelope** of X. It is isometrically identified with the **tight span** of (X, d).
A metric space coincides with its injective envelope if and only if it is injective.
- **Tight extension**
An extension (X', d') of a metric space (X, d) is called a **tight extension** if, for every semimetric d'' on X' satisfying the conditions $d''(x_1, x_2) = d(x_1, x_2)$ for all $x_1, x_2 \in X$, and $d''(y_1, y_2) \le d'(y_1, y_2)$ for any $y_1, y_2 \in X'$, one has $d''(y_1, y_2) = d'(y_1, y_2)$ for all $y_1, y_2 \in X'$.
The **tight span** is the *universal tight extension* of X, i.e., it contains, up to isometries, every tight extension of X, and it has no proper tight extension itself.
- **Tight span**
Given a metric space (X, d) of finite diameter, consider the set $\mathbb{R}^X = \{f : X \to \mathbb{R}\}$. The **tight span** $T(X, d)$ of (X, d) is defined as the set $T(X, d) = \{f \in \mathbb{R}^X : f(x) = \sup_{y \in X}(d(x, y) - f(y))$ for all $x \in X\}$, endowed with the metric induced on $T(X, d)$ by the *sup norm* $||f|| = \sup_{x \in X} |f(x)|$.
The set X can be identified with the set $\{h_x \in T(X, d) : h_x(y) = d(y, x)\}$ or, equivalently, with the set $T^0(X, d) = \{f \in T(X, d) : 0 \in f(X)\}$. The **injective envelope** $(\overline{X}, \overline{d})$ of X is isometrically identified with the tight span $T(X, d)$ by

$$\overline{X} \to T(X, d),\ \overline{x} \to h_{\overline{x}} \in T(X, d) : h_{\overline{x}}(y) = \overline{d}(f(y), \overline{x}).$$

The tight span $T(X,d)$ of a finite metric space is the metric space $(T(X), D(f,g) = \max|f(x) - g(x)|)$, where $T(X)$ is the set of functions $f : X \to \mathbb{R}$ such that for any $x, y \in X$, $f(x) + f(y) \geq d(x,y)$ and, for each $x \in X$, there exists $y \in X$ with $f(x) + f(y) = d(x,y)$. The mapping of any x into the function $f_x(y) = d(x,y)$ gives an isometric embedding of (X,d) into $T(X,d)$. For example, if $X = \{x_1, x_2\}$, then $T(X,d)$ is the interval of length $d(x_1, x_2)$.
The tight span of a metric space (X,d) of finite diameter can be considered as a polytopal complex of bounded faces of the polyhedron

$$\{y \in \mathbb{R}^n_{\geq 0} : y_i + y_j \geq d(x_i, x_j) \text{ for } 1 \leq i < j \leq n\}$$

if, for example, $X = \{x_1, \dots, x_n\}$. The dimension of this complex is called (Dress, 1984) dimension of (X,d).

- **Real tree**
A metric space (X,d) is called (Tits, 1977) a **real tree** (or **$\mathbb{R}$-tree**) if, for all $x, y \in X$, there exists a unique **arc** from x to y, and this arc is a **geodesic segment**. So, an $\mathbb{R}$-tree is a (uniquely) arcwise connected metric space in which each arc is isometric to a subarc of $\mathbb{R}$. $\mathbb{R}$-tree is not related to a **metric tree** in Chap. 17.
A metric space (X,d) is a real tree if and only if it is **path-connected** and Gromov 0-**hyperbolic** (i.e., satisfies the **four-point inequality**). The plane $\mathbb{R}^2$ with the **Paris metric** or **lift metric** (cf. Chap. 19) are examples of an $\mathbb{R}$-tree.
Real trees are exactly **tree-like** metric spaces which are **geodesic**; they are **injective** metric spaces among tree-like spaces. Tree-like metric spaces are by definition metric subspaces of real trees.
If (X,d) is a finite metric space, then the **tight span** $T(X,d)$ is a real tree and can be viewed as an edge-weighted graph-theoretical tree.
A metric space is a complete real tree if and only if it is **hyperconvex** and any two points are joined by a **metric segment**.
A geodesic metric space (X,d) is called (Druţu–Sapir, 2005) *tree-graded with respect to* a collection $\mathcal{P}$ of connected proper subsets with $|P \cap P'| \leq 1$ for any distinct $P, P' \in \mathcal{P}$, if every its simple loop composed of three geodesics is contained in one $P \in \mathcal{P}$. $\mathbb{R}$-trees are tree-graded with respect to the empty set.

1.5 General Distances

- **Discrete metric**
Given a set X, the **discrete metric** (or **trivial metric**, **sorting distance**, **drastic distance**, **Dirac distance**, *overlap*) is a metric on X, defined by $d(x,y) = 1$ for all distinct $x, y \in X$ and $d(x,x) = 0$. Cf. the much more general notion of a (metrically or topologically) **discrete metric space**.

- **Indiscrete semimetric**
  Given a set X, the **indiscrete semimetric** d is a semimetric on X defined by $d(x, y) = 0$ for all $x, y \in X$.
- **Equidistant metric**
  Given a set X and a positive real number t, the **equidistant metric** d is a metric on X defined by $d(x, y) = t$ for all distinct $x, y \in X$ (and $d(x, x) = 0$).
- $(1, 2) - B$**-metric**
  Given a set X, the $(1, 2) - B$**-metric** d is a metric on X such that, for any $x \in X$, the number of points $y \in X$ with $d(x, y) = 1$ is at most B, and all other distances are equal to 2. The $(1, 2) - B$**-metric** is the **truncated metric** of a graph with maximal vertex degree B.
- **Permutation metric**
  Given a finite set X, a metric d on it is called a **permutation metric** (or *linear arrangement metric*) if there exists a bijection $\omega : X \to \{1, \dots, |X|\}$ such that

$$d(x, y) = |\omega(x) - \omega(y)|$$

  holds for for all $x, y \in X$. Even–Naor–Rao–Schieber, 2000, defined a more general **spreading metric**, i.e., any metric d on $\{1, \dots, n\}$ such that $\sum_{y \in M} d(x, y) \geq \frac{|M|(|M|+2)}{4}$ for any $1 \leq x \leq n$ and $M \subseteq \{1, \dots, n\} \setminus \{x\}$ with $|M| \geq 2$.
- **Induced metric**
  Given a metric space (X, d) and a subset $X' \subset X$, an **induced metric** (or **submetric**) is the restriction d' of d to X'. A metric space (X', d') is called a **metric subspace** of (X, d), and (X, d) is called a **metric extension** of (X', d').
- **Katětov mapping**
  Given a metric space (X, d), the mapping $f : X \to \mathbb{R}$ is a **Katětov mapping** if

$$|f(x) - f(y)| \leq d(x, y) \leq f(x) + f(y)$$

  for any $x, y \in X$, i.e., setting $d(x, z) = f(x)$ defines a one-point **metric extension** $(X \cup \{z\}, d)$ of (X, d).
  The set $E(X)$ of Katětov mappings on X is a complete metric space with metric $D(f, g) = \sup_{x \in X} |f(x) - g(x)|$; (X, d) embeds isometrically in it via the *Kuratowski mapping* $x \to d(x, .)$, with unique extension of each isometry of X to one of $E(X)$.
- **Dominating metric**
  Given metrics d and d_1 on a set X, d_1 **dominates** d if $d_1(x, y) \geq d(x, y)$ for all $x, y \in X$. Cf. **noncontractive mapping** (or *dominating mapping*).
- **Barbilian semimetric**
  Given sets X and P, the function $f : P \times X \to \mathbb{R}_{>0}$ is called an *influence* (of P over X) if for any $x, y \in X$ the ratio $g_{xy}(p) = \frac{f(p,x)}{f(p,y)}$ has a maximum when $p \in P$.

The **Barbilian semimetric** is defined on the set X by

$$\ln \frac{\max_{p \in P} g_{xy}(p)}{\min_{p \in P} g_{xy}(p)}$$

for any $x, y \in X$. Barbilian, 1959, proved that the above function is well defined (moreover, $\min_{p \in P} g_{xy}(p) = \frac{1}{\max_{p \in P} g_{yx}(p)}$) and is a semimetric. Also, it is a metric if the influence f is *effective*, i.e., there is no pair $x, y \in X$ such that $g_{xy}(p)$ is constant for all $p \in P$. Cf. a special case **Barbilian metric** in Chap. 6.

- **Metric transform**
A **metric transform** is a distance obtained as a function of a given metric (cf. Chap. 4).
- **Complete metric**
Given a metric space (X, d), a sequence $\{x_n\}$, $x_n \in X$, is said to have *convergence to* $x^* \in X$ if $\lim_{n\to\infty} d(x_n, x^*) = 0$, i.e., for any $\epsilon > 0$, there exists $n_0 \in \mathbb{N}$ such that $d(x_n, x^*) < \epsilon$ for any $n > n_0$. Any sequence converges to at most one limit in X; it is not so, in general, if d is a semimetric.
A sequence $\{x_n\}_n$, $x_n \in X$, is called a *Cauchy sequence* if, for any $\epsilon > 0$, there exists $n_0 \in \mathbb{N}$ such that $d(x_n, x_m) < \epsilon$ for any $m, n > n_0$.
A metric space (X, d) is called a **complete metric space** if every *Cauchy sequence* in it converges. In this case the metric d is called a **complete metric**. An example of an incomplete metric space is $(\mathbb{N}, d(m, n) = \frac{|m-n|}{mn})$.
- **Cauchy completion**
Given a metric space (X, d), its **Cauchy completion** is a metric space (X^*, d^*) on the set X^* of all equivalence classes of *Cauchy sequences*, where the sequence $\{x_n\}_n$ is called *equivalent to* $\{y_n\}_n$ if $\lim_{n\to\infty} d(x_n, y_n) = 0$. The metric d^* is defined by

$$d^*(x^*, y^*) = \lim_{n\to\infty} d(x_n, y_n),$$

for any $x^*, y^* \in X^*$, where $\{z_n\}_n$ is any element in the equivalence class z^*.
The Cauchy completion (X^*, d^*) is a unique, up to isometry, **complete** metric space, into which the metric space (X, d) embeds as a dense metric subspace.
The Cauchy completion of the metric space $(\mathbb{Q}, |x - y|)$ of rational numbers is the *real line* $(\mathbb{R}, |x - y|)$. A **Banach space** is the Cauchy completion of a *normed vector space* $(V, ||.||)$ with the **norm metric** $||x - y||$. A **Hilbert space** corresponds to the case an *inner product norm* $||x|| = \sqrt{\langle x, x\rangle}$.
- **Perfect metric space**
A complete metric space (X, d) is called **perfect** if every point $x \in X$ is a *limit point*, i.e., $|B(x, r) = \{y \in X : d(x, y) < r\}| > 1$ holds for any $r > 0$.
A topological space is a **Cantor space** (i.e., *homeomorphic* to the *Cantor set* with the natural metric $|x - y|$) if and only if it is nonempty, perfect, **totally disconnected**, compact and metrizable. The totally disconnected countable metric space $(\mathbb{Q}, |x - y|)$ of rational numbers also consists only of limit points but it is not complete and not **locally compact**.

Every proper metric ball of radius r in a metric space has diameter at most $2r$. Given a number $0 < c \leq 1$, a metric space is called a c**-uniformly perfect metric space** if this diameter is at least $2cr$. Cf. the **radii of metric space**.

- **Metrically discrete metric space**
  A metric space (X, d) is called **metrically** (or *uniformly*) **discrete** if there exists a number $r > 0$ such that $B(x, r) = \{y \in X : d(x, y) < r\} = \{x\}$ for every $x \in X$.
  (X, d) is a **topologically discrete metric space** (or a *discrete metric space*) if the underlying topological space is **discrete**, i.e., each point $x \in X$ is an *isolated point*: there exists a number $r(x) > 0$ such that $B(x, r(x)) = \{x\}$. For $X = \{\frac{1}{n} : n = 1, 2, 3, \ldots\}$, the metric space $(X, |x - y|)$ is topologically but not metrically discrete. Cf. **translation discrete metric** in Chap. 10.
  Alternatively, a metric space (X, d) is called *discrete* if any of the following holds:

  1. (Burdyuk–Burdyuk 1991) it has a proper *isolated subset*, i.e., $M \subset X$ with $\inf\{d(x, y) : x \in M, y \notin M\} > 0$ (any such space admits a unique decomposition into *continuous*, i.e., nondiscrete, components);
  2. (Lebedeva–Sergienko–Soltan, 1984) for any distinct points $x, y \in X$, there exists a point z of the **closed metric interval** $I(x, y)$ with $I(x, z) = \{x, z\}$;
  3. a stronger property holds: for any two distinct points $x, y \in X$, every sequence of points $z_1, z_2, \ldots$ with $z_k \in I(x, y)$ but $z_{k+1} \in I(x, z_k) \setminus \{z_k\}$ for $k = 1, 2, \ldots$ is a finite sequence.

- **Locally finite metric space**
  Let (X, d) be a **metrically discrete metric space**. Then it is called **locally finite** if for every $x \in X$ and every $r \geq 0$, the ball $|B(x, r)|$ is finite.
  If, moreover, $|B(x, r)| \leq C(r)$ for some number $C(r)$ depending only on r, then (X, d) is said to have *bounded geometry*.
- **Bounded metric space**
  A metric (moreover, a distance) d on a set X is called **bounded** if there exists a constant $C > 0$ such that $d(x, y) \leq C$ for any $x, y \in X$.
  For example, given a metric d on X, the metric D on X, defined by $D(x, y) = \frac{d(x,y)}{1+d(x,y)}$, is bounded with $C = 1$.
  A metric space (X, d) with a bounded metric d is called a **bounded metric space**.
- **Totally bounded metric space**
  A metric space (X, d) is called **totally bounded** if, for every $\epsilon > 0$, there exists a finite ϵ**-net**, i.e., a finite subset $M \subset X$ with the **point-set distance** $d(x, M) < \epsilon$ for any $x \in X$ (cf. **totally bounded space** in Chap. 2).
  Every totally bounded metric space is **bounded** and **separable**. A metric space is totally bounded if and only if its **Cauchy completion** is **compact**.
- **Separable metric space**
  A metric space (X, d) is called **separable** if it contains a countable **dense subset** M, i.e., a subset with which all its elements can be approached: X is the *closure* $cl(M)$ (M together with all its limit points).

A metric space is separable if and only if it is **second-countable** (cf. Chap. 2).

- **Compact metric space**
  A **compact metric space** (or **metric compactum**) is a metric space in which every sequence has a *Cauchy subsequence*, and those subsequences are convergent. A metric space is compact if and only if it is **totally bounded** and **complete**. Every bounded and closed subset of a Euclidean space is compact. Every finite metric space is compact. Every compact metric space is **second-countable**.
  A **continuum** is a nonempty **connected** metric compactum.
- **Proper metric space**
  A metric space is called **proper** (or *finitely compact, having the Heine–Borel property*) if every its closed metric ball is compact. Any such space is **complete**.
- **UC metric space**
  A metric space is called a **UC metric space** (or *Atsuji space*) if any continuous function from it into an arbitrary metric space is *uniformly continuous*.
  Every such space is **complete**. Every **metric compactum** is a UC metric space.
- **Metric measure space**
  A **metric measure space** (or *mm-space*, *metric triple*) is a triple (X, d, μ), where (X, d) is a *Polish* (i.e., complete separable; cf. Chap. 2) *metric space* and (X, Σ, μ) is a *probability measure space* ($\mu(X) = 1$) with Σ being a *Borel σ-algebra* of all open and closed sets of the **metric topology** (cf. Chap. 2) induced by the metric d on X. Cf. **metric outer measure**.
- **Norm metric**
  Given a *normed vector space* $(V, ||.||)$, the **norm metric** on V is defined by

$$||x - y||.$$

  The metric space $(V, ||x - y||)$ is called a **Banach space** if it is **complete**. Examples of norm metrics are l_p**-** and L_p**-metrics**, in particular, the **Euclidean metric**.
  Any metric space (X, d) admits an isometric embedding into a Banach space B such that its convex hull is dense in B (cf. **Monge–Kantorovich metric** in Chap. 14); (X, d) is a **linearly rigid metric space** if such an embedding is unique up to isometry. A metric space isometrically embeds into the unit sphere of a Banach space if and only if its diameter is at most 2.
- **Path metric**
  Given a connected graph $G = (V, E)$, its **path metric** (or *graphic metric*) d_{path} is a metric on V defined as the length (i.e., the number of edges) of a shortest path connecting two given vertices x and y from V (cf. Chap. 15).
- **Editing metric**
  Given a finite set X and a finite set $\mathcal{O}$ of (unary) *editing operations* on X, the **editing metric** on X is the **path metric** of the graph with the vertex-set X and xy being an edge if y can be obtained from x by one of the operations from $\mathcal{O}$.

- **Gallery metric**
A *chamber system* is a set X (its elements are called *chambers*) equipped with n equivalence relations $\sim_i$, $1 \le i \le n$. A *gallery* is a sequence of chambers $x_1, \dots, x_m$ such that $x_i \sim_j x_{i+1}$ for every i and some j depending on i.
The **gallery metric** is an **extended metric** on X which is the length of the shortest gallery connecting x and $y \in X$ (and is equal to ∞ if there is no connecting gallery). The gallery metric is the (extended) **path metric** of the graph with the vertex-set X and xy being an edge if $x \sim_i y$ for some $1 \le i \le n$.
- **Metric on incidence structure**
An *incidence structure* (P, L, I) consists of three sets: points P, lines L and flags $I \subset P \times L$, where a point $p \in P$ is said to be *incident* with a line $l \in L$ if $(p, l) \in I$.
If, moreover, for any pair of distinct points, there is at most one line incident with both of them, then the collinearity graph is a graph whose vertices are the points with two vertices being adjacent if they determine a line.
The **metric on incidence structure** is the **path metric** of this graph.
- **Riemannian metric**
Given a connected n-dimensional smooth *manifold* M^n (cf. Chaps. 2, 7), its **Riemannian metric** is a collection of positive-definite symmetric bilinear forms $((g_{ij}))$ on the tangent spaces of M^n which varies smoothly from point to point.
The length of a curve γ on M^n is expressed as $\int_\gamma \sqrt{\sum_{i,j} g_{ij} dx_i dx_j}$, and the **intrinsic metric** on M^n, also called the **Riemannian distance**, is the infimum of lengths of curves connecting any two given points $x, y \in M^n$. Cf. Chap. 7.
- **Linearly additive metric**
A **linearly additive** (or *additive on lines*) **metric** is a continuous metric d on $\mathbb{R}^n$ which, for any points x, y, z lying in that order on a common line, satisfies

$$d(x, z) = d(x, y) + d(y, z).$$

Hilbert's 4th problem asked in 1900 to classify such metrics; it is solved only for dimension $n = 2$ [Amba76]. Cf. **projective metric** in Chap. 6.
Every **norm metric** on $\mathbb{R}^n$ is; linearly additive. Every linearly additive metric on $\mathbb{R}^2$ is a **hypermetric**.
- **Hamming metric**
The **Hamming metric** d_H (called sometimes *Dalal distance* in Semantics) is a metric on $\mathbb{R}^n$ defined (Hamming, 1950) by

$$|\{i : 1 \le i \le n, x_i \ne y_i\}|.$$

On binary vectors $x, y \in \{0, 1\}^n$ the Hamming metric and the l_1*-metric* (cf. L_p**-metric** in Chap. 5) coincide; they are equal to $|I(x) \Delta I(y)| = |I(x) \backslash I(y)| + |I(y) \setminus I(x)|$, where $I(z) = \{1 \le t \le n : z_i = 1\}$.
In fact, $\max\{|I(x) \setminus I(y)|, |I(y) \setminus I(x)|\}$ is also a metric.

- **Lee metric**
Given $m, n \in \mathbb{N}$, $m \geq 2$, the **Lee metric** d_{Lee} is a metric on $\mathbb{Z}_m^n = \{0, 1, \ldots, m-1\}^n$ defined (Lee, 1958) by

$$\sum_{1 \leq i \leq n} \min\{|x_i - y_i|, m - |x_i - y_i|\}.$$

The metric space (Z_m^n, d_{Lee}) is a discrete analog of the *elliptic space*.
The Lee metric coincides with the Hamming metric d_H if $m = 2$ or $m = 3$. The metric spaces (Z_4^n, d_{Lee}) and $Z_2^{2n}, d_H)$ are isometric. Lee and Hamming metrics are applied for phase and orthogonal modulation, respectively.
Cf. **absolute summation distance** and **generalized Lee metric** in Chap. 16.
- **Enomoto–Katona metric**
Given a finite set X and an integer k, $2k \leq |X|$, the **Enomoto–Katona metric** (2001) is the distance between unordered pairs (X_1, X_2) and (Y_1, Y_2) of disjoint k-subsets of X defined by

$$\min\{|X_1 \setminus Y_1| + |X_2 \setminus Y_2|, |X_1 \setminus Y_2| + |X_2 \setminus Y_1|\}.$$

Cf. **Earth Mover's distance**, **transportation distance** in Chaps. 21 and 14.
- **Symmetric difference metric**
Given a *measure space* $(\Omega, \mathcal{A}, \mu)$, the **symmetric difference** (or *measure*) *semimetric* on the set $\mathcal{A}_\mu = \{A \in \mathcal{A} : \mu(A) < \infty\}$ is defined by

$$od_\Delta(A, B) = \mu(A \Delta B),$$

where $A \Delta B = (A \cup B) \backslash (A \cap B)$ is the *symmetric difference* of A and $B \in \mathcal{A}_\mu$. The value $d_\Delta(A, B) = 0$ if and only if $\mu(A \Delta B) = 0$, i.e., A and B are equal *almost everywhere*. Identifying two sets $A, B \in \mathcal{A}_\mu$ if $\mu(A \Delta B) = 0$, we obtain the **symmetric difference metric** (or **Fréchet–Nikodym–Aronszyan distance**, **measure metric**).
If μ is the *cardinality measure*, i.e., $\mu(A) = |A|$, then $d_\Delta(A, B) = |A \Delta B| = |A \setminus B| + |B \setminus A|$. In this case $|A \Delta B| = 0$ if and only if $A = B$.
The metrics $d_{\max}(A, B) = \max(|A \backslash B|, |B \backslash A|)$ and $1 - \frac{|A \cap B|}{\max(|A|, |B|)}$ (its normalised version) are special cases of **Zelinka distance** and **Bunke–Shearer metric** in Chap. 15. For each $p \geq 1$, the p**-difference metric** (Noradam–Nyblom, 2014) is $d_p(A, B) = (|A \setminus B|^p + |B \setminus A|^p)^{\frac{1}{p}}$; so, $d_1 = d_\Delta$ and $\lim_{p \to \infty} d_p = d_{\max}$.
The **Johnson distance** between k-sets A and B is $\frac{|A \Delta B|}{2} = k - |A \cap B|$.
The *symmetric difference metric between ordered* q*-partitions* $A = (A_1, \ldots, A_q)$ and $B = (B_1, \ldots, B_q)$ is $\sum_{i=1}^q |A_i \Delta B_i|$. Cf. **metrics between partitions** in Chap. 10.

- **Steinhaus distance**
Given a *measure space* $(\Omega, \mathcal{A}, \mu)$, the **Steinhaus distance** d_{St} is a semimetric on the set $\mathcal{A}_\mu = \{A \in \mathcal{A} : \mu(A) < \infty\}$ defined as 0 if $\mu(A) = \mu(B) = 0$, and by

$$\frac{\mu(A\triangle B)}{\mu(A \cup B)} = 1 - \frac{\mu(A \cap B)}{\mu(A \cup B)}$$

if $\mu(A \cup B) > 0$. It becomes a metric on the set of equivalence classes of elements from $\mathcal{A}_\mu$; here $A, B \in \mathcal{A}_\mu$ are called *equivalent* if $\mu(A\triangle B) = 0$.
The **biotope** (or **Tanimoto**, **Jaccard**) **distance** $\frac{|A\triangle B|}{|A\cup B|}$ is the special case of Steinhaus distance obtained for the *cardinality measure* $\mu(A) = |A|$ for finite sets.
Cf. also the **generalized biotope transform metric** in Chap. 4.
- **Fréchet metric**
Let (X, d) be a metric space. Consider a set $\mathcal{F}$ of all continuous mappings $f : A \to X, g : B \to X, \dots$, where $A, B, \dots$ are subsets of $\mathbb{R}^n$, homeomorphic to $[0, 1]^n$ for a fixed dimension $n \in \mathbb{N}$.
The *Fréchet semimetric* d_F is a semimetric on $\mathcal{F}$ defined by

$$\inf_{\sigma} \sup_{x\in A} d(f(x), g(\sigma(x))),$$

where the infimum is taken over all orientation preserving homeomorphisms $\sigma : A \to B$. It becomes the **Fréchet metric** on the set of equivalence classes $f^* = \{g : d_F(g, f) = 0\}$. Cf. the **Fréchet surface metric** in Chap. 8.
- **Hausdorff metric**
Given a metric space (X, d), the **Hausdorff metric** (or *two-sided Hausdorff distance*) is a metric on the family $\mathcal{F}$ of nonempty compact subsets of X defined by

$$d_{Haus} = \max\{d_{dHaus}(A, B), d_{dHaus}(B, A)\},$$

where $d_{dHaus}(A, B) = \max_{x\in A} \min_{y\in B} d(x, y)$ is the **directed Hausdorff distance** (or *one-sided Hausdorff distance*) from A to B. The metric space $(\mathcal{F}, d_{Haus})$ is called **hyperspace of metric space** (X, d); cf. **hyperspace** in Chap. 2.
In other words, $d_{Haus}(A, B)$ is the minimal number ϵ (called also the **Blaschke distance**) such that a closed **ϵ-neighborhood** of A contains B and a closed ϵ-neighborhood of B contains A. Then $d_{Haus}(A, B)$ is equal to

$$\sup_{x\in X} |d(x, A) - d(x, B)|,$$

where $d(x, A) = \min_{y\in A} d(x, y)$ is the **point-set distance**.

If the above definition is extended for noncompact closed subsets A and B of X, then $d_{Haus}(A, B)$ can be infinite, i.e., it becomes an **extended metric**.
For not necessarily closed subsets A and B of X, the **Hausdorff semimetric** between them is defined as the Hausdorff metric between their closures. If X is finite, d_{Haus} is a metric on the class of all subsets of X.

- **L_p-Hausdorff distance**
Given a finite metric space (X, d), the **L_p-Hausdorff distance** [Badd92] between two subsets A and B of X is defined by

$$(\sum_{x \in X} |d(x, A) - d(x, B)|^p)^{\frac{1}{p}},$$

where $d(x, A)$ is the **point-set distance**. The usual **Hausdorff metric** corresponds to the case $p = \infty$.

- **Generalized G-Hausdorff metric**
Given a group $(G, \cdot, e)$ acting on a metric space (X, d), the **generalized G-Hausdorff metric** between two closed bounded subsets A and B of X is

$$\min_{g_1, g_2 \in G} d_{Haus}(g_1(A), g_2(B)),$$

where d_{Haus} is the **Hausdorff metric**. If $d(g(x), g(y)) = d(x, y)$ for any $g \in G$ (i.e., if the metric d is *left-invariant* with respect of G), then above metric is equal to $\min_{g \in G} d_{Haus}(A, g(B))$.

- **Gromov–Hausdorff metric**
The **Gromov–Hausdorff metric** is a metric on the set of all *isometry classes* of compact metric spaces defined by

$$\inf d_{Haus}(f(X), g(Y))$$

for any two classes X^* and Y^* with the representatives X and Y, respectively, where d_{Haus} is the **Hausdorff metric**, and the minimum is taken over all metric spaces M and all isometric embeddings $f : X \to M$, $g : Y \to M$. The corresponding metric space is called the *Gromov–Hausdorff space*.
The **Hausdorff–Lipschitz distance** between isometry classes of compact metric spaces X and Y is defined by

$$\inf\{d_{GH}(X, X_1) + d_L(X_1, Y_1) + d_{GH}(Y, Y_1)\},$$

where d_{GH} is the Gromov–Hausdorff metric, d_L is the **Lipschitz metric**, and the minimum is taken over all (isometry classes of compact) metric spaces X_1, Y_1.

- **Kadets distance**
The *gap* (or *opening*) between two closed subspaces X and Y of a Banach space $(V, ||.||)$ is defined by

$$gap(X, Y) = \max\{\delta(X, Y), \delta(Y, X)\},$$

where $\delta(X, Y) = \sup\{\inf_{y \in Y} ||x - y|| : x \in X, ||x|| = 1\}$ (cf. **gap distance** in Chap. 12 and **gap metric** in Chap. 18).
The **Kadets distance** between two Banach spaces V and W is a semimetric defined (Kadets, 1975) by

$$\inf_{Z, f, g} gap(\overline{B}_{f(V)}, \overline{B}_{g(W)}),$$

where the infimum is taken over all Banach spaces Z and all linear isometric embeddings $f : V \to Z$ and $g : W \to Z$; here $\overline{B}_{f(V)}$ and $\overline{B}_{g(W)}$ are the closed unit balls of Banach spaces $f(V)$ and $g(W)$, respectively.
The nonlinear analog of the Kadets distance is the following **Gromov–Hausdorff distance between Banach spaces** U and W:

$$\inf_{Z, f, g} d_{Haus}(f(\overline{B}_V), g(\overline{B}_W)),$$

where the infimum is taken over all metric spaces Z and all isometric embeddings $f : V \to Z$ and $g : W \to Z$; here d_{Haus} is the **Hausdorff metric**.
The **Kadets path distance** between Banach spaces V and W is defined (Ostrovskii, 2000) as the infimum of the length (with respect to the Kadets distance) of all curves joining V and W (and is equal to ∞ if there is no such curve).

- **Banach–Mazur distance**
The **Banach–Mazur distance** d_{BM} between two Banach spaces V and W is

$$\ln \inf_{T} ||T|| \cdot ||T^{-1}||,$$

where the infimum is taken over all isomorphisms $T : V \to W$.
It can also be written as $\ln d(V, W)$, where the number $d(V, W)$ is the smallest positive $d \geq 1$ such that $\overline{B}^n_W \subset T(\overline{B}^n_V) \subset d\,\overline{B}^n_W$ for some linear invertible transformation $T : V \to W$. Here $\overline{B}^n_V = \{x \in V : ||x||_V \leq 1\}$ and $\overline{B}^n_W = \{x \in W ; ||x||_W \leq 1\}$ are the *unit balls* of the normed spaces $(V, ||.||_V)$ and $(W, ||.||_W)$, respectively.
One has $d_{BM}(V, W) = 0$ if and only if V and W are *isometric*, and d_{BM} becomes a metric on the set X^n of all equivalence classes of n-dimensional normed spaces, where $V \sim W$ if they are isometric. The pair (X^n, d_{BM}) is a compact metric space which is called the **Banach–Mazur compactum**.
The **modified Banach–Mazur distance** (Glushkin, 1963, and Khrabrov, 2001) is

$$\inf\{||T||_{X \to Y} : |detT| = 1\} \cdot \inf\{||T||_{Y \to X} : |detT| = 1\}.$$

The **weak Banach–Mazur distance** (Tomczak–Jaegermann, 1984) is

$$\max\{\overline{\gamma}_Y(id_X), \overline{\gamma}_X(id_Y)\},$$

where id is the identity map and, for an operator $U : X \to Y$, $\overline{\gamma}_Z(U)$ denotes $\inf \sum ||W_k|| ||V_k||$. Here the infimum is taken over all representations $U = \sum W_k V_k$ for $W_k : X \to Z$ and $V_k : Z \to Y$. This distance never exceeds the corresponding Banach–Mazur distance.

- **Lipschitz distance**
Given $\alpha \geq 0$ and two metric spaces (X, d_X), (Y, d_Y), the *α-Hölder norm* $||.||_{Hol}$ on the set of all injective functions $f : X \to Y$ is defined by

$$||f||_{Hol} = \sup_{x,y \in X, x \neq y} \frac{d_Y(f(x), f(y))}{d_X(x,y)^\alpha}.$$

The *Lipschitz norm* $||.||_{Lip}$ is the case $\alpha = 1$ of $||.||_{Hol}$.
The **Lipschitz distance** between metric spaces (X, d_X) and (Y, d_Y) is defined by

$$\ln \inf_f ||f||_{Lip} \cdot ||f^{-1}||_{Lip},$$

where the infimum is taken over all bijective functions $f : X \to Y$. Equivalently, it is the infimum of numbers $\ln a$ such that there exists a bijective **bi-Lipschitz mapping** between (X, d_X) and (Y, d_Y) with constants $\exp(-a)$, $\exp(a)$.
It becomes a metric—**Lipschitz metric**—on the set of all isometry classes of compact metric spaces. Cf. **Hausdorff–Lipschitz distance**.
This distance is an analog to the **Banach–Mazur distance** and, in the case of finite-dimensional real Banach spaces, coincides with it.
It also coincides with the **Hilbert projective metric** on *nonnegative* projective spaces, obtained by starting with $\mathbb{R}^n_{>0}$ and identifying any point x with cx, $c > 0$.

- **Lipschitz distance between measures**
Given a compact metric space (X, d), the *Lipschitz seminorm* $||.||_{Lip}$ on the set of all functions $f : X \to \mathbb{R}$ is defined by $||f||_{Lip} = \sup_{x,y \in X, x \neq y} \frac{|f(x)-f(y)|}{d(x,y)}$.
The **Lipschitz distance between measures** μ and ν on X is defined by

$$\sup_{||f||_{Lip} \leq 1} \int f d(\mu - \nu).$$

It is the **transportation distance** (Chap. 14) if μ, ν are probability measures. Let a such measure $m_x(.)$ be attached to any $x \in X$; for distinct x, y the *coarse Ricci curvature along* (xy) is defined (Ollivier, 2009) as $\kappa(x, y) = 1 - \frac{W_1(m_x, m_y)}{d(x,y)}$. *Ollivier's curvature* generalizes the *Ricci curvature* in Riemannian space (cf. Chap. 7).

- **Barycentric metric space**
Given a metric space (X, d), let $(B(X), ||\mu - \nu||_{TV})$ be the metric space, where $B(X)$ is the set of all regular Borel probability measures on X with bounded support, and $||\mu - \nu||_{TV}$ is the **variational distance** $\int_X |p(\mu) - p(\nu)| d\lambda$ (cf. Chap. 14). Here $p(\mu)$ and $p(\nu)$ are the density functions of measures μ and ν, respectively, with respect to the σ-finite measure $\frac{\mu+\nu}{2}$.

A metric space (X, d) is **barycentric** if there exists a constant $\beta > 0$ and a surjection $f : B(X) \to X$ such that for any measures $\mu, \nu \in B(X)$ it holds the inequality

$$d(f(\mu), f(\nu)) \leq \beta diam(supp(\mu + \nu))||\mu - \nu||_{TV}.$$

Any Banach space $(X, d = ||x - y||)$ is a barycentric metric space with the smallest β being 1 and the map $f(\mu)$ being the usual *center of mass* $\int_X x d\mu(x)$. Any **Hadamard** (i.e., a complete **CAT(0) space**, cf. Chap. 6, is barycentric with the smallest β being 1 and the map $f(\mu)$ being the unique minimizer of the function $g(y) = \int_X d^2(x, y) d\mu(x)$ on X.

- **Point-set distance**

  Given a metric space (X, d), the **point-set distance** $d(x, A)$ between a point $x \in X$ and a subset A of X is defined as

  $$\inf_{y \in A} d(x, y).$$

  For any $x, y \in X$ and for any nonempty subset A of X, we have the following version of the triangle inequality: $d(x, A) \leq d(x, y) + d(y, A)$ (cf. **distance map**).

  For a given point-measure $\mu(x)$ on X and a *penalty function* p, an **optimal quantizer** is a set $B \subset X$ such that $\int p(d(x, B)) d\mu(x)$ is as small as possible.

- **Set-set distance**

  Given a metric space (X, d), the **set-set distance** between two subsets A and B of X is defined by

  $$d_{ss}(A, B) = \inf_{x \in A, y \in B} d(x, y).$$

  This distance can be 0 even for disjoint sets, for example, for the intervals $(1, 2)$, $(2, 3)$ on $\mathbb{R}$. The sets A and B are *positively separated* if $d_{ss}(A, B) > 0$. A constructive *appartness space* is a generalization of this relation on subsets of X.

  The **spanning distance** between A and B is $\sup_{x \in A, y \in B} d(x, y)$.

  In Data Analysis, (cf. Chap. 17) the set-set and spanning distances between clusters are called the **single** and **complete linkage**, respectively.

- **Matching distance**

  Given a metric space (X, d), the **matching distance** (or *multiset-multiset distance*) between two multisets A and B in X is defined by

  $$\inf_{\phi} \max_{x \in A} d(x, \phi(x)),$$

  where ϕ runs over all bijections between A and B, as multisets.

The matching distance is not related to the **perfect matching distance** in Chap. 15 and to the **nonlinear elastic matching distance** in Chap. 21.

- **Metrics between multisets**
  A *multiset* (or *bag*) drawn from a set S is a mapping $m : S \to \mathbb{Z}_{\geq 0}$, where $m(x)$ represents the "multiplicity" of $x \in S$. The *dimensionality*, *cardinality* and *height* of multiset m is $|S|$, $|m| = \sum_{x \in S} m(x)$ and $\max_{x \in S} m(x)$, respectively. Multisets are good models for multi-attribute objects such as, say, all symbols in a string, all words in a document, etc.
  A multiset m is finite if S and all $m(x)$ are finite; the *complement* of a finite multiset m is the multiset $\overline{m} : S \to \mathbb{Z}_{\geq 0}$, where $\overline{m}(x) = \max_{y \in S} m(y) - m(x)$. Given two multisets m_1 and m_2, denote by $m_1 \cup m_2$, $m_1 \cap m_2$, $m_1 \backslash m_2$ and $m_1 \Delta m_2$ the multisets on S defined , for any $x \in S$, by $m_1 \cup m_2(x) = \max\{m_1(x), m_2(x)\}$, $m_1 \cap m_2(x) = \min\{m_1(x), m_2(x)\}$, $m_1 \backslash m_2(x) = \max\{0, m_1(x) - m_2(x)\}$ and $m_1 \Delta m_2(x) = |m_1(x) - m_2(x)|$, respectively. Also, $m_1 \subseteq m_2$ denotes that $m_1(x) \leq m_2(x)$ for all $x \in S$.
  The *measure* $\mu(m)$ of a multiset m is a linear combination $\mu(m) = \sum_{x \in S} \lambda(x) m(x)$ with $\lambda(x) \geq 0$. In particular, $|m|$ is the *counting measure*.
  For any measure $\mu(m) \in \mathbb{R}_{\geq 0}$, Miyamoto, 1990, and Petrovsky, 2003, proposed several **semimetrics between multisets** m_1 and m_2 including $d_1(m_1, m_2) = \mu(m_1 \Delta m_2)$ and $d_2(m_1, m_2) = \frac{\mu(m_1 \Delta m_2)}{\mu(m_1 \cup m_2)}$ (with $d_2(\emptyset, \emptyset) = 0$ by definition). Cf. **symmetric difference metric** and **Steinhaus distance**.
  Among examples of other metrics between multisets are **matching distance**, **metric space of roots** in Chap. 12, **μ-metric** in Chap. 15 and, in Chap. 11, **bag distance** $\max\{|m_1 \backslash m_2|, |m_2 \backslash m_1|\}$ and **q-gram similarity**.
- **Metrics between fuzzy sets**
  A *fuzzy subset* of a set S is a mapping $\mu : S \to [0, 1]$, where $\mu(x)$ represents the "degree of membership" of $x \in S$. It is an ordinary (*crisp*) if all $\mu(x)$ are 0 or 1. Fuzzy sets are good models for *gray scale images* (cf. **gray scale images distances** in Chap. 21), random objects and objects with nonsharp boundaries.
  Bhutani–Rosenfeld, 2003, introduced the following two metrics between two fuzzy subsets μ and ν of a finite set S. The **diff-dissimilarity** is a metric (a fuzzy generalization of **Hamming metric**), defined by

$$d(\mu, \nu) = \sum_{x \in S} |\mu(x) - \nu(x)|.$$

  The **perm-dissimilarity** is a semimetric defined by

$$\min\{d(\mu, p(\nu))\},$$

  where the minimum is taken over all permutations p of S.
  The **Chaudhuri–Rosenfeld metric** (1996) between two fuzzy sets μ and ν *with crisp points* (i.e., the sets $\{x \in S : \mu(x) = 1\}$ and $\{x \in S : \nu(x) = 1\}$ are nonempty) is an **extended metric**, defined the **Hausdorff metric** d_{Haus} by

$$\int_0^1 2td_{Haus}(\{x \in S : \mu(x) \geq t\}, \{x \in S : \nu(x) \geq t\})dt.$$

A *fuzzy number* is a fuzzy subset μ of the real line $\mathbb{R}$, such that the *level set* (or t-*cut*) $A_\mu(t) = \{x \in \mathbb{R} : \mu(x) \geq t\}$ is convex for every $t \in [0, 1]$. The *sendograph* of a fuzzy set μ is the set $send(\mu) = \{(x, t) \in S \times [0, 1] : \mu(x) > 0, \ \mu(x) \geq t\}$.
The **sendograph metric** (Kloeden, 1980) between two fuzzy numbers μ, ν with crisp points and compact sendographs is the **Hausdorff metric**

$$\max\{\sup_{a=(x,t)\in send(\mu)} d(a, send(\nu)), \sup_{b=(x',t')\in send(\nu)} d(b, send(\mu))\},$$

where $d(a, b) = d((x, t), (x', t'))$ is a **box metric** (cf. Chap. 4) $\max\{|x - x'|, |t - t'|\}$.
The **Klement–Puri–Ralesku metric** (1988) between fuzzy numbers μ, ν is

$$\int_0^1 d_{Haus}(A_\mu(t), A_\nu(t))dt,$$

where $d_{Haus}(A_\mu(t), A_\nu(t))$ is the **Hausdorff metric**

$$\max\{\sup_{x\in A_\mu(t)} \inf_{y\in A_\nu(t)} |x - y|, \sup_{x\in A_\nu(t)} \inf_{x\in A_\mu(t)} |x - y|\}.$$

Several other Hausdorff-like metrics on some families of fuzzy sets were proposed by Boxer in 1997, Fan in 1998 and Brass in 2002; Brass also argued the nonexistence of a "good" such metric.
If q is a quasi-metric on $[0, 1]$ and S is a finite set, then $Q(\mu, \nu) = \sup_{x\in S} q(\mu(x), \nu(x))$ is a quasi-metric on fuzzy subsets of S.
Cf. **fuzzy Hamming distance** in Chap. 11 and, in Chap. 23, **fuzzy set distance** and **fuzzy polynucleotide metric**. Cf. **fuzzy metric spaces** in Chap. 3 for fuzzy-valued generalizations of metrics and, for example, [Bloc99] for a survey.

- **Metrics between intuitionistic fuzzy sets**
  An *intuitionistic fuzzy subset* of a set S is (Atanassov, 1999) an ordered pair of mappings $\mu, \nu :\rightarrow [0, 1]$ with $0 \leq \mu(x) + \nu(x) \leq 1$ for all $x \in S$, representing the "degree of membership" and the "degree of nonmembership" of $x \in S$, respectively. It is an ordinary *fuzzy subset* if $\mu(x) + \nu(x) = 1$ for all $x \in S$.
  Given two intuitionistic fuzzy subsets $(\mu(x), \nu(x))$ and $(\mu'(x), \nu'(x))$ of a finite set $S = \{x_1, \ldots, x_n\}$, their **Atanassov distances** (1999) are:

$$\frac{1}{2}\sum_{i=1}^{n}(|\mu(x_i) - \mu'(x_i)| + |\nu(x_i) - \nu'(x_i)|) \text{ (Hamming distance)}$$

and, in general, for any given numbers $p \geq 1$ and $0 \leq q \leq 1$, the distance

$$(\sum_{i=1}^{n}(1-q)(\mu(x_i)-\mu'(x_i))^p + q(\nu(x_i)-\nu'(x_i)^p)^{\frac{1}{p}}.$$

Their **Grzegorzewski distances** (2004) are:

$$\sum_{i=1}^{n} \max\{|\mu(x_i)-\mu'(x_i)|, |\nu(x_i)-\nu'(x_i)|\} \text{ (Hamming distance)},$$

$$\sqrt{\sum_{i=1}^{n} \max\{(\mu(x_i)-\mu'(x_i))^2, (\nu(x_i)-\nu'(x_i))^2\}} \text{ (Euclidean distance)}.$$

The normalized versions (dividing the above sums by n) were also proposed. Szmidt–Kacprzyk, 1997, proposed a modification of the above, adding $\pi(x) - \pi'(x)$, where $\pi(x)$ is the third mapping $1-\mu(x)-\nu(x)$.
An *interval-valued fuzzy subset* of a set S is a mapping $\mu :\rightarrow [I]$, where $[I]$ is the set of closed intervals $[a^-, a^+] \subseteq [0,1]$. Let $\mu(x) = [\mu^-(x), \mu^+(x)]$, where $0 \leq \mu^-(x) \leq \mu^+(x) \leq 1$ and an interval-valued fuzzy subset is an ordered pair of mappings μ^- and μ^+. This notion is close to the above intuitionistic one; so, above distance can easily be adapted. For example, $\sum_{i=1}^{n} \max\{|\mu^-(x_i) - \mu'^-(x_i)|, |\mu^+(x_i) - \mu'^+(x_i)|\}$ is a Hamming distance between interval-valued fuzzy subsets (μ^-, μ^+) and (μ'^-, μ'^+).

- **Polynomial metric space**
  Let (X, d) be a metric space with a finite diameter D and a finite normalized measure μ_X. Let the Hilbert space $L_2(X, d)$ of complex-valued functions decompose into a countable (when X is infinite) or a finite (with $D+1$ members when X is finite) direct sum of mutually orthogonal subspaces $L_2(X, d) = V_0 \oplus V_1 \oplus \dots$.
  Then (X, d) is a **polynomial metric space** if there exists an ordering of the spaces $V_0, V_1, \dots$ such that, for $i = 0, 1, \dots$, there exist *zonal spherical functions*, i.e., real polynomials $Q_i(t)$ of degree i such that

$$Q_i(t(d(x,y))) = \frac{1}{r_i} \sum_{j=1}^{r_i} v_{ij}(x)\overline{v_{ij}(y)}$$

  for all $x, y \in X$, where r_i is the dimension of V_i, $\{v_{ii}(x) : 1 \leq j \leq r_i\}$ is an orthonormal basis of V_i, and $t(d)$ is a continuous decreasing real function such that $t(0) = 1$ and $t(D) = -1$. The zonal spherical functions constitute an orthogonal system of polynomials with respect to some weight $w(t)$.
  The finite polynomial metric spaces are also called *(P and Q)-polynomial association schemes*; cf. **distance-regular graph** in Chap. 15. The infinite

polynomial metric spaces are the *compact connected two-point homogeneous spaces*. Wang, 1952, classified them as the Euclidean unit spheres, the real, complex, quaternionic projective spaces or the Cayley projective line and plane.

- **Universal metric space**
A metric space (U, d) is called **universal** for a collection $\mathcal{M}$ of metric spaces if any metric space (M, d_M) from $\mathcal{M}$ is *isometrically embeddable* in (U, d), i.e., there exists a mapping $f : M \to U$ which satisfies $d_M(x, y) = d(f(x), f(y))$ for any $x, y \in M$. Some examples follow.
Every separable metric space (X, d) isometrically embeds (Fréchet, 1909) in (a nonseparable) **Banach space** l_∞^∞. In fact, $d(x, y) = \sup_i |d(x, a_i) - d(y, a_i)|$, where $(a_1, \dots, a_i, \dots)$ is a dense countable subset of X.
Every metric space isometrically embeds (Kuratowski, 1935) in the **Banach space** $L^\infty(X)$ of bounded functions $f : X \to \mathbb{R}$ with the norm $\sup_{x \in X} |f(x)|$.
The **Urysohn space** is a **homogeneous** complete separable space which is the universal metric space for all separable metric spaces. The **Hilbert cube** (Chap. 4) is the universal space for the class of metric spaces with a countable base.
The **graphic** metric space of the **random graph** (Rado, 1964; the vertex-set consists of all prime numbers $p \equiv 1 (\mod 4)$ with pq being an edge if p is a quadratic residue modulo q) is the universal metric space for any finite or countable metric space with distances 0, 1 and 2 only. It is a discrete analog of the Urysohn space.
There exists a metric d on $\mathbb{R}$, inducing the usual (interval) topology, such that $(\mathbb{R}, d)$ is a universal metric space for all finite metric spaces (Holsztynski, 1978).
The Banach space l_∞^n is a universal metric space for all metric spaces (X, d) with $|X| \leq n + 2$ (Wolfe, 1967). The Euclidean space $\mathbb{E}^n$ is a universal metric space for all ultrametric spaces (X, d) with $|X| \leq n+1$; the space of all finite functions $f(t) : \mathbb{R}_{\geq 0} \to \mathbb{R}$ equipped with the metric $d(f, g) = \sup\{t : f(t) \neq g(t)\}$ is a universal metric space for all ultrametric spaces (Lemin–Lemin, 1996).
The universality can be defined also for mappings, other than isometric embeddings, of metric spaces, say, a bi-Lipschitz embedding, etc. For example, any compact metric space is a continuous image of the **Cantor set** with the natural metric $|x - y|$ inherited from $\mathbb{R}$, and any complete separable metric space is a continuous image of the space of irrational numbers.
- **Constructive metric space**
A **constructive metric space** is a pair (X, d), where X is a set of constructive objects (say, words over an alphabet), and d is an algorithm converting any pair of elements of X into a constructive real number $d(x, y)$ such that d is a metric on X.
- **Computable metric space**
Let $\{x_n\}_{n \in \mathbb{N}}$ be a sequence of elements from a given *Polish* (i.e., complete separable) *metric space* (X, d) such that the set $\{x_n : n \in \mathbb{N}\}$ is dense in (X, d). Let $\mathcal{N}(m, n, k)$ be the *Cantor tuple function* of a triple $(n, m, k) \in \mathbb{N}^3$, and let $\{q_k\}_{k \in \mathbb{N}}$ be a fixed total standard numbering of the set $\mathbb{Q}$ of rational numbers.

The triple $(X, d, \{x_n\}_{n \in \mathbb{N}})$ is called an *effective* (or *semicomputable*) *metric space* [Hemm02] if the set $\{\mathcal{N}(n, m, k) : d(x_m, x_n) < q_k\}$ is *recursively enumerable*, i.e., there is an algorithm that enumerates the members of this set. If, moreover, the set $\{\mathcal{N}(n, m, k) : d(s_m, s_m) > q_k\}$ is recursively enumerable, then this triple is called (Lacombe, 1951) **computable metric space**, (or **recursive metric space**). In other words, the map $d \circ (q \times q) : \mathbb{N}^2 \to \mathbb{R}$ is a computable (double) sequence of real numbers, i.e., is recursive over $\mathbb{R}$.

Chapter 2
Topological Spaces

A *topological space* (X, τ) is a set X with a *topology* τ, i.e., a collection of subsets of X with the following properties:

1. $X \in \tau, \emptyset \in \tau$;
2. If $A, B \in \tau$, then $A \cap B \in \tau$;
3. For any collection $\{A_\alpha\}_\alpha$, if all $A_\alpha \in \tau$, then $\cup_\alpha A_\alpha \in \tau$.

The sets in τ are called *open sets*, and their complements are called *closed sets*. A *base* of the topology τ is a collection of open sets such that every open set is a union of sets in the base. The coarsest topology has two open sets, the empty set and X, and is called the *trivial topology* (or *indiscrete topology*). The finest topology contains all subsets as open sets, and is called the *discrete topology*.

In a metric space (X, d) define the *open ball* as the set $B(x, r) = \{y \in X : d(x, y) < r\}$, where $x \in X$ (the *center* of the ball), and $r \in \mathbb{R}, r > 0$ (the *radius* of the ball). A subset of X which is the union of (finitely or infinitely many) open balls, is called an *open set*. Equivalently, a subset U of X is called *open* if, given any point $x \in U$, there exists a real number $\epsilon > 0$ such that, for any point $y \in X$ with $d(x, y) < \epsilon$, $y \in U$.

Any metric space is a topological space, the topology (**metric topology**, *topology induced by the metric* d) being the set of all open sets. The metric topology is always T_4 (see below a list of topological spaces). A topological space which can arise in this way from a metric space, is called a **metrizable space**.

A *quasi-pseudo-metric topology* is a topology on X induced similarly by a quasi-semimetric d on X, using the set of open d-balls $B(x, r)$ as the base. In particular, *quasi-metric topology* and *pseudo-metric topology* are the terms used for the case of, respectively, quasi-metric and semimetric d. In general, those topologies are not T_0.

Given a topological space (X, τ), a *neighborhood* of a point $x \in X$ is a set containing an open set which in turn contains x. The *closure* of a subset of a topological space is the smallest closed set which contains it. An *open cover* of X is a collection $\mathcal{L}$ of open sets, the union of which is X; its *subcover* is a cover $\mathcal{K}$

M.M. Deza, E. Deza, *Encyclopedia of Distances*,
DOI 10.1007/978-3-662-44342-2_2

such that every member of $\mathcal{K}$ is a member of $\mathcal{L}$; its *refinement* is a cover $\mathcal{K}$, where every member of $\mathcal{K}$ is a subset of some member of $\mathcal{L}$. A collection of subsets of X is called *locally finite* if every point of X has a neighborhood which meets only finitely many of these subsets.

A subset $A \subset X$ is called *dense* if $X = cl(A)$, i.e., it consists of A and its *limit points*; cf. **closed subset of metric space** in Chap. 1. The *density* of a topological space is the least cardinality of its dense subset. A *local base* of a point $x \in X$ is a collection $\mathcal{U}$ of neighborhoods of x such that every neighborhood of x contains some member of $\mathcal{U}$.

A function from one topological space to another is called *continuous* if the preimage of every open set is open. Roughly, given $x \in X$, all points close to x map to points close to $f(x)$. A function f from one metric space (X, d_X) to another metric space (Y, d_Y) is *continuous* at the point $c \in X$ if, for any positive real number ϵ, there exists a positive real number δ such that all $x \in X$ satisfying $d_X(x, c) < \delta$ will also satisfy $d_Y(f(x), f(c)) < \epsilon$; the function is continuous on an interval I if it is continuous at any point of I.

The following classes of topological spaces (up to T_4) include any metric space.

- **T_0-space**
  A **T_0-space** (or *Kolmogorov space*) is a topological space in which every two distinct points are *topologically distinguishable*, i.e., have different neighborhoods.
- **T_1-space**
  A **T_1-space** (or *accessible space*) is a topological space in which every two distinct points are *separated*, i.e., each does not belong to other's closure. T_1-spaces are always T_0.
- **T_2-space**
  A **T_2-space** (or **Hausdorff space**) is a topological space in which every two distinct points are *separated by neighborhoods*, i.e., have disjoint neighborhoods. T_2-spaces are always T_1.
  A space is T_2 if and only if it is both T_0 and *pre-regular*, i.e., any two *topologically distinguishable* points are separated by neighborhoods.
- **Regular space**
  A **regular space** is a topological space in which every neighborhood of a point contains a closed neighborhood of the same point.
- **T_3-space**
  A **T_3-space** (or *Vietoris space*, *regular Hausdorff space*) is a topological space which is T_1 and **regular**.
- **Completely regular space**
  A **completely regular space** (or *Tychonoff space*) is a **Hausdorff space** (X, τ) in which any closed set A and any $x \notin A$ are *functionally separated*.
  Two subsets A and B of X are *functionally separated* if there is a continuous function $f : X \to [0, 1]$ such that $f(A) = 0$ and $f(B) = 1$.
- **Perfectly normal space**
  A **perfectly normal space** is a topological space (X, τ) in which any two disjoint closed subsets of X are *functionally separated*.

- **Normal space**
  A **normal space** is a topological space in which, for any two disjoint closed sets A and B, there exist two disjoint open sets U and V such that $A \subset U$, and $B \subset V$.
- **T_4-space**
  A **T_4-space** (or *Tietze space, normal Hausdorff space*) is a topological space which is T_1 and **normal**. Any metric space is a perfectly normal T_4-space.
- **Completely normal space**
  A **completely normal space** is a topological space in which any two separated sets have disjoint neighborhoods. It also called a *hereditarily normal space* since it is exactly one in which every subspace with subspace topology is a normal space.
  Sets A and B are *separated* in X if each is disjoint from the other's closure.
- **Monotonically normal space**
  A **monotonically normal space** is a **completely normal space** in which any two *separated* subsets A and B are *strongly separated*, i.e., there exist open sets U and V with $A \subset U$, $B \subset V$ and $Cl(U) \cap Cl(V) = \emptyset$.
- **T_5-space**
  A **T_5-space** (or *completely normal Hausdorff space*) is a topological space which is **completely normal** and T_1. T_5-spaces are always T_4.
- **T_6-space**
  A **T_6-space** (or *perfectly normal Hausdorff space*) is a topological space which is T_1 and **perfectly normal**. T_6-spaces are always T_5.
- **Moore space**
  A **Moore space** is a **regular space** with a *development.*
  A *development* is a sequence $\{\mathcal{U}_n\}_n$ of open covers such that, for every $x \in X$ and every open set A containing x, there exists n such that $St(x, \mathcal{U}_n) = \cup\{U \in \mathcal{U}_n : x \in U\} \subset A$, i.e., $\{St(x, \mathcal{U}_n)\}_n$ is a *neighborhood base* at x.
- **Polish space**
  A **separable space** is a topological space which has a countable dense subset.
  A **Polish space** is a separable space which can be equipped with a complete metric. A *Lusin space* is a topological space such that some weaker topology makes it into a Polish space; every Polish space is Lusin. A *Souslin space* is a continuous image of a Polish space; every Lusin space is Suslin.
- **Lindelöf space**
  A **Lindelöf space** is a topological space in which every open cover has a countable subcover.
  An *L-space* is a hereditarily Lindelöf space which is not hereditarily separable.
- **First-countable space**
  A topological space is called **first-countable** if every point has a countable local base. Any metric space is first-countable.
- **Second-countable space**
  A topological space is called **second-countable** if its topology has a countable base. Such space is **quasi-metrizable** and, if and only if it is a T_3-space, **metrizable**.

Second-countable spaces are **first-countable**, **separable** and **Lindelöf**. The properties **second-countable**, **separable** and **Lindelöf** are equivalent for metric spaces.
The Euclidean space $\mathbb{E}^n$ with its usual topology is second-countable.

- **Baire space**
  A **Baire space** is a topological space in which every intersection of countably many dense open sets is dense. Every complete metric space is a Baire space. Every locally compact T_2-space (hence, every n-**manifold**) is a Baire space.
- **Alexandrov space**
  An **Alexandrov space** is a topological space in which every intersection of arbitrarily many open sets is open.
  A topological space is called a P**-space** if every G_δ*-set* (i.e., the intersection of countably many open sets) is open.
  A topological space (X, τ) is called a Q**-space** if every subset $A \subset X$ is a G_δ-set.
- **Connected space**
  A topological space (X, τ) is called **connected** if it is not the union of a pair of disjoint nonempty open sets. In this case the set X is called a *connected set*.
  A connected topological space (X, τ) is called *unicoherent* if the intersection $A \cap B$ is connected for any closed connected sets A, B with $A \cup B = X$.
  A topological space (X, τ) is called **locally connected** if every point $x \in X$ has a local base consisting of connected sets.
  A topological space (X, τ) is called **path-connected** (or 0*-connected*) if for every points $x, y \in X$ there is a *path* γ from x to y, i.e., a continuous function $\gamma : [0, 1] \to X$ with $\gamma(x) = 0, \gamma(y) = 1$.
  A topological space (X, τ) is called **simply connected** (or 1*-connected*) if it consists of one piece, and has no circle-shaped "holes" or "handles" or, equivalently, if every continuous curve of X is *contractible*, i.e., can be reduced to one of its points by a *continuous deformation*.
  A topological space (X, τ) is called **hyperconnected** (or *irreducible*) if X cannot be written as the union of two proper closed sets.
- **Sober space**
  A topological space (X, τ) is called **sober** if every **hyperconnected** closed subset of X is the closure of exactly one point of X. Any sober space is a T_0-space. Any T_2-space is a sober T_1-space but some sober T_1-spaces are not T_2.
- **Paracompact space**
  A topological space is called **paracompact** if every open cover of it has an open locally finite refinement. Every **metrizable** space is paracompact.
- **Totally bounded space**
  A topological space (X, τ) is called **totally bounded** (or *pre-compact*) if it can be covered by finitely many subsets of any fixed cardinality.
  A metric space (X, d) is a **totally bounded metric space** if, for every real number $r > 0$, there exist finitely many open balls of radius r, whose union is equal to X.

- **Compact space**
  A topological space (X, τ) is called **compact** if every open cover of X has a finite subcover.
  Compact spaces are always **Lindelöf**, **totally bounded**, and **paracompact**. A metric space is compact if and only if it is **complete** and **totally bounded**. A subset of a Euclidean space $\mathbb{E}^n$ is compact if and only if it is closed and bounded.
  There exist a number of topological properties which are equivalent to compactness in metric spaces, but are nonequivalent in general topological spaces. Thus, a metric space is compact if and only if it is a *sequentially compact space* (every sequence has a convergent subsequence), or a *countably compact space* (every countable open cover has a finite subcover), or a *pseudo-compact space* (every real-valued continuous function on the space is bounded), or a *weakly countably compact space* (i.e., every infinite subset has an accumulation point).
  Sometimes, a compact **connected** T_2**-space** is called *continuum*; cf. **continuum** in Chap. 1.
- **Locally compact space**
  A topological space is called **locally compact** if every point has a local base consisting of compact neighborhoods. The Euclidean spaces $\mathbb{E}^n$ and the spaces $\mathbb{Q}_p$ of *p-adic numbers* are locally compact.
  A topological space (X, τ) is called a *k-space* if, for any compact set $Y \subset X$ and $A \subset X$, the set A is closed whenever $A \cap Y$ is closed. The k-spaces are precisely quotient images of locally compact spaces.
- **Locally convex space**
  A **topological vector space** is a real (complex) vector space V which is a T_2**-space** with continuous vector addition and scalar multiplication. It is a **uniform space** (cf. Chap. 3).
  A **locally convex space** is a topological vector space whose topology has a base, where each member is a *convex balanced absorbent* set. A subset A of V is called *convex* if, for all $x, y \in A$ and all $t \in [0, 1]$, the point $tx + (1 - t)y \in A$, i.e., every point on the *line segment* connecting x and y belongs to A. A subset A is *balanced* if it contains the line segment between x and $-x$ for every $x \in A$; A is *absorbent* if, for every $x \in V$, there exist $t > 0$ such that $tx \in A$.
  The locally convex spaces are precisely vector spaces with topology induced by a family $\{||.||_\alpha\}$ of seminorms such that $x = 0$ if $||x||_\alpha = 0$ for every α.
  Any metric space $(V, ||x - y||)$ on a real (complex) vector space V with a **norm metric** $||x - y||$ is a locally convex space; each point of V has a local base consisting of convex sets. Every L_p with $0 < p < 1$ is an example of a vector space which is not locally convex.
- n**-manifold**
  Broadly, a *manifold* is a topological space locally homeomorphic to a **topological vector space** over the reals.
  But usually, a **topological manifold** is a **second-countable** T_2**-space** that is locally homeomorphic to Euclidean space. An n**-manifold** is a topological manifold such that every point has a neighborhood homeomorphic to $\mathbb{E}^n$.

- **Fréchet space**
A **Fréchet space** is a **locally convex space** (V, τ) which is complete as a **uniform space** and whose topology is defined using a countable set of seminorms $||.||_1, \dots, ||.||_n, \dots$, i.e., a subset $U \subset V$ is *open in* (V, τ) if, for every $u \in U$, there exist $\epsilon > 0$ and $N \geq 1$ with $\{v \in V : ||u - v||_i < \epsilon \text{ if } i \leq N\} \subset U$.
A Fréchet space is precisely a locally convex **F-space** (cf. Chap. 5). Its topology can be induced by a **translation invariant metric** (Chap. 5) and it is a complete and **metrizable space** with respect to this topology. But this topology may be induced by many such metrics. Every **Banach space** is a Fréchet space.
- **Countably-normed space**
A **countably-normed space** is a **locally convex space** (V, τ) whose topology is defined using a countable set of *compatible norms* $||.||_1, \dots, ||.||_n, \dots$. It means that, if a sequence $\{x_n\}_n$ of elements of V that is fundamental in the norms $||.||_i$ and $||.||_j$ converges to zero in one of these norms, then it also converges in the other. A countably-normed space is a **metrizable space**, and its metric can be defined by
$$\sum_{n=1}^{\infty} \frac{1}{2^n} \frac{||x - y||_n}{1 + ||x - y||_n}.$$
- **Metrizable space**
A topological space (T, τ) is called **metrizable** if it is homeomorphic to a metric space, i.e., X admits a metric d such that the set of open d-balls $\{B(x, r) : r > 0\}$ forms a neighborhood base at each point $x \in X$. If, moreover, (X, d) is a complete metric space for one of such metrics d, then (X, d) is a *completely metrizable* (or *topologically complete*) space.
Metrizable spaces are always **paracompact** T_2**-spaces** (hence, **normal** and **completely regular**), and **first-countable**.
A topological space is called **locally metrizable** if every point in it has a metrizable neighborhood.
A topological space (X, τ) is called **submetrizable** if there exists a metrizable topology τ' on X which is coarser than τ.
A topological space (X, τ) is called **proto-metrizable** if it is paracompact and has an *orthobase*, i.e., a base $\mathcal{B}$ such that, for $\mathcal{B}' \subset \mathcal{B}$, either $\cap\mathcal{B}'$ is open, or $\mathcal{B}'$ is a local base at the unique point in $\cap\mathcal{B}'$. It is not related to the **protometric** in Chap. 1.
Some examples of other direct generalizations of metrizable spaces follow.
A **sequential space** is a quotient image of a metrizable space.
Morita's M**-space** is a topological space (X, τ) from which there exists a continuous map f onto a metrizable topological space (Y, τ') such that f is closed and $f^{-1}(y)$ is countably compact for each $y \in Y$.
Ceder's M_1**-space** is a topological space (X, τ) having a σ-closure-preserving base (metrizable spaces have σ-locally finite bases).

Okuyama's **σ-space** is a topological space (X, τ) having a σ-locally finite *net*, i.e., a collection $\mathcal{U}$ of subsets of X such that, given of a point $x \in U$ with U open, there exists $U' \in \mathcal{U}$ with $x \in U' \subset U$ (a base is a net consisting of open sets). Every compact subset of a σ-space is metrizable.

Michael's **cosmic space** is a topological space (X, τ) having a countable net (equivalently, a Lindelöf σ-space). It is exactly a continuous image of a separable metric space. A **T_2-space** is called **analytic** if it is a continuous image of a complete separable metric space; it is called a **Lusin space** if, moreover, the image is one-to-one.

- **Quasi-metrizable space**

A topological space (X, τ) is called a **quasi-metrizable space** if X admits a quasi-metric d such that the set of open d-balls $\{B(x, r) : r > 0\}$ forms a neighborhood base at each point $x \in X$.

A more general **γ-space** is a topological space admitting a **γ-metric** d (i.e., a function $d : X \times X \to \mathbb{R}_{\geq 0}$ with $d(x, z_n) \to 0$ whenever $d(x, y_n) \to 0$ and $d(y_n, z_n) \to 0$) such that the set of open *forward* d-balls $\{B(x, r) : r > 0\}$ forms a neighborhood base at each point $x \in X$.

The *Sorgenfrey line* is the topological space $(\mathbb{R}, \tau)$ defined by the base $\{[a, b) : a, b \in \mathbb{R}, a < b\}$. It is not metrizable but it is a first-countable separable and paracompact **T_5-space**; neither it is second-countable, nor locally compact or locally connected. However, the Sorgenfrey line is quasi-metrizable by the **Sorgenfrey quasi-metric** (cf. Chap. 12) defined as $y - x$ if $y \geq x$, and 1, otherwise.

- **Symmetrizable space**

A topological space (X, τ) is called **symmetrizable** (and τ is called the **distance topology**) if there is a **symmetric** d on X (i.e., a distance $d : X \times X \to \mathbb{R}_{\geq 0}$ with $d(x, y) = 0$ implying $x = y$) such that a subset $U \subset X$ is open if and only if, for each $x \in U$, there exists $\epsilon > 0$ with $B(x, \epsilon) = \{y \in X : d(x, y) < \epsilon\} \subset U$. In other words, a subset $H \subset X$ is closed if and only if $d(x, H) = \inf_y\{d(x, y) : y \in H\} > 0$ for each $x \in X \backslash U$. A symmetrizable space is **metrizable** if and only if it is a Morita's **M-space**.

In Topology, the term **semimetrizable space** refers to a topological space (X, τ) admitting a symmetric d such that, for each $x \in X$, the family $\{B(x, \epsilon) : \epsilon > 0\}$ of balls forms a (not necessarily open) neighborhood base at x. In other words, a point x is in the closure of a set H if and only if $d(x, H) = 0$.

A topological space is semimetrizable if and only if it is symmetrizable and **first-countable**. Also, a symmetrizable space is semimetrizable if and only if it is a *Fréchet-Urysohn space* (or *E-space*), i.e., for any subset A and for any point x of its closure, there is a sequence in A converging to x.

- **Hyperspace**

A **hyperspace** of a topological space (X, τ) is a topological space on the set $CL(X)$ of all nonempty closed (or, moreover, compact) subsets of X. The topology of a hyperspace of X is called a *hypertopology*. Examples of such a *hit-and-miss topology* are the *Vietoris topology*, and the *Fell topology*. Examples of such a *weak hyperspace topology* are the *Hausdorff metric topology*, and the *Wijsman topology*.

- **Discrete topological space**
  A topological space (X, τ) is **discrete** if τ is the *discrete topology* (the finest topology on X), i.e., containing all subsets of X as open sets. Equivalently, it does not contain any *limit point*, i.e., it consists only of *isolated points*.
- **Indiscrete topological space**
  A topological space (X, τ) is **indiscrete** if τ is the *indiscrete topology* (the coarsest topology on X), i.e., having only two open sets, $\emptyset$ and X.
  It can be considered as the semimetric space (X, d) with the **indiscrete semimetric**: $d(x, y) = 0$ for any $x, y \in X$.
- **Extended topology**
  Consider a set X and a map $cl : P(X) \to P(X)$, where $P(X)$ is the set of all subsets of X. The set $cl(A)$ (for $A \subset X$), its dual set $int(A) = X \backslash cl(X \backslash A)$ and the map $N : X \to P(X)$ with $N(x) = \{A \subset X : x \in int(A)\}$ are called the *closure*, *interior* and *neighborhood* map, respectively.
  So, $x \in cl(A)$ is equivalent to $X \backslash A \in P(X) \backslash N(x)$. A subset $A \subset X$ is *closed* if $A = cl(A)$ and *open* if $A = int(A)$. Consider the following possible properties of cl; they are meant to hold for all $A, B \in P(X)$.

  1. $cl(\emptyset) = \emptyset$;
  2. $A \subseteq B$ implies $cl(A) \subseteq cl(B)$ (*isotony*);
  3. $A \subseteq cl(A)$(*enlarging*);
  4. $cl(A \cup B) = cl(A) \cup cl(B)$ (*linearity*, and, in fact, 4 implies 2);
  5. $cl(cl(A)) = cl(A)$ (*idempotency*).

  The pair (X, cl) satisfying 1 is called an **extended topology** if 2 holds, a **Brissaud space** (Brissaud, 1974) if 3 holds, a **neighborhood space** (Hammer, 1964) if 2 and 3 hold, a **Smyth space** (Smyth, 1995) if 4 holds, a **pre-topology** (Čech, 1966) if 3 and 4 hold, and a **closure space** (Soltan, 1984) if 2, 3 and 5 hold.
  (X, cl) is the usual topology, in closure terms, if 1, 3, 4 and 5 hold.

Chapter 3
Generalizations of Metric Spaces

Some immediate generalizations of the notion of metric, for example, **quasi-metric**, **near-metric**, **extended metric**, were defined in Chap. 1. Here we give some generalizations in the direction of Topology, Probability, Algebra, etc.

3.1 m-Tuple Generalizations of Metrics

In the definition of a metric, for every *two points* there is a *unique associated number*. Here we group some generalizations of metrics in which *several points* or *several numbers* are considered instead.

- **m-Hemimetric**
  Let X be a nonempty set. A function $d : X^{m+1} \to \mathbb{R}_{\geq 0}$ is called a **m-hemimetric** (Deza–Rosenberg, 2000) if it have the following properties:

  1. d is *totally symmetric*, i.e., satisfies $d(x_1, \dots, x_{m+1}) = d(x_{\pi(1)}, \dots, x_{\pi(m+1)})$ for all $x_1, \dots, x_{m+1} \in X$ and for any permutation π of $\{1, \dots, m+1\}$;
  2. $d(x_1, \dots, x_{m+1}) = 0$ if $x_1, \dots, x_{m+1}$ are not pairwise distinct;
  3. for all $x_1, \dots, x_{m+2} \in X$, d satisfies the **m-simplex inequality**:

  $$d(x_1, \dots, x_{m+1}) \leq \sum_{i=1}^{m+1} d(x_1, \dots, x_{i-1}, x_{i+1}, \dots, x_{m+2}).$$

  Cf. unrelated **hemimetric** (i.e., a quasi-semimetric) in Chap. 1.
  If in above 3. $d(x_1, \dots, x_{m+1})$ is replaced by $sd(x_1, \dots, x_{m+1})$ for some $s, 0 < s \leq 1$, then d is called **(m, s)-super-metric** [DeDu03]. $(m, 1)$- and $(1, s)$-super-metrics are exactly m-hemimetric and $\frac{1}{s}$-*near-semimetric*; cf. **near-metric** in Chap. 1.

M.M. Deza, E. Deza, *Encyclopedia of Distances*,
DOI 10.1007/978-3-662-44342-2_3

If above 3. is dropped, d is called m-**dissimilarity.** 1-dissimilarity and 1-hemimetric are exactly a distance and a semimetric.

- **2-Metric**
A m-**hemimetric** with $m = 2$ satisfies 2-**simplex** (or *tetrahedron*) **inequality**

$$d(x_1, x_2, x_3) \leq d(x_4, x_2, x_3) + d(x_1, x_4, x_3) + d(x_1, x_2, x_4).$$

A 2-**metric** (Gähler, 1963 and 1966) is a 2-**hemimetric** d in which, for any distinct points x_1, x_2, there is a point x_3 with $d(x_1, x_2, x_3) > 0$. The area of the triangle spanned by x_1, x_2, x_3 on $\mathbb{R}^2$ or $\mathbb{S}^2$ is a 2-metric.
A *D-space* (Dhage, 1992) is an 2-**hemimetric space** (X, d) in which the condition "$d(x_1, x_2, x_3) = 0$ if two of x_1, x_2, x_3 are equal" is replaced by "$d(x_1, x_2, x_3) = 0$ if and only if $x_1 = x_2 = x_3$." Mustafa and Sims, 2003, showed that D-spaces are not suitable for topological constructions. In 2006, they defined instead a function, let us call it *MS* – 2-**metric**, $D : X^3 \to \mathbb{R}_{\geq 0}$ which satisfies

1. $D(x_1, x_2, x_3) = 0$ if $x_1 = x_2 = x_3$;
2. $D(x_1, x_1, x_2) > 0$ whenever $x_1 \neq x_2$;
3. $D(x_1, x_2, x_3) \geq D(x_1, x_1, x_2)$ whenever $x_3 \neq x_2$;
4. D is a *totally symmetric* function of its three variables, and
5. $D(x_1, x_2, x_3) \leq D(x_1, x_4, x_4) + D(x_4, x_2, x_3)$ for all $x_1, x_2, x_3, x_4 \in X$.

The perimeter of the triangle spanned by x_1, x_2, x_3 on $\mathbb{R}^2$ is a *MS* – 2-metric. If d is a metric, then $\frac{1}{2}(d(x_1, x_2) + d(x_2, x_3) + d(x_1, x_3))$ and $\max(d(x_1, x_2), d(x_2, x_3), d(x_1, x_3))$ are *MS* – 2-metrics. If D is a *MS* – 2-metric, then $D(x_1, x_2, x_2) + D(x_1, x_1, x_2)$ is a metric. If (X, D) is a *MS* – 2-metric space, the open *D-ball with center* x_0 *and radius* r is $B_D(x_0, r) = \{x_1 \in X : D(x_0, x_1, x_1) < r\}$.

- **Multidistance**
Let X be a set and let $\mathcal{X}$ denote $\cup_{m=1}^{\infty} X^m$. A function $d : \mathcal{X} \to [0, \infty]$ is called a **multidistance** (Martin–Major, 2009) if it have the following properties for all m and all $x_1, \ldots, x_m, y \in X$:

1. $d(x_1, \ldots, x_m) = d(x_{\pi(1)}, \ldots, x_{\pi(m)})$ for any permutation π of $\{1, \ldots, m\}$;
2. $d(x_1, \ldots, x_m) = 0$ if and only if $x_1 = \cdots = x_m$;
3. $d(x_1, \ldots, x_m) \leq \sum_{i=1}^{m} d(x_i, y)$.

A multidistance is *regular*, if, moreover, $d(x_1, \ldots, x_m) \leq d(x_1, \ldots, x_m, y)$ holds.

- **Multimetric**
A **multimetric space** is the union of some metric spaces $(X_i, d_i), i \in J$.
In the case $X_i = X, i \in J$, the **multimetric** is defined as the sequence-valued map $d(x, y) = (d_i), i \in J$, from $X \times X$ to $R_{\geq 0}^{|J|}$.
Cf. **bimetric theory of gravity** in Chap. 24 and (in the item **meter-related terms**) *multimetric crystallography* in Chap. 27. Also, Jörnsten, 2007, consider *Clustering* (cf. Chap. 17) under several distance metrics simultaneously.

- **Metric 1-space**
  A *category* Ψ consists (Eilenberg and MacLane, 1945) of a set $Ob(\Psi)$ of *objects*, a set $Mor(\Psi)$ of *morphisms* (or *arrows*)) and a set-valued map associating a set $\Psi(x, y)$ of arrows to each ordered pair of objects x, y, so that each arrow belongs to only one set $\Psi(x, y)$. An element of $\Psi(x, y)$ is also denoted by $f : x \to y$. Moreover, the composition $f \cdot g \in \Psi(x, z)$ of two arrows $f : x \to y, g : y \to z$ is defined, and it is associative. Finally, each set $\Psi(x, x)$ contains an *identity arrow* id_x such that $f \cdot id_x = f$ and $id_x \cdot g = g$ for any arrows $f : y \to x$ and $g : x \to z$. Cf. **category of metric spaces** in Chap. 1.
  Weiss defined in [Weis12] a **metric 1-space** as a category Ψ together with a weight-function $w : \Psi(x, y) \to \mathbb{R}_{\geq 0} \cup \{\infty\}$ on arrows, which satisfies

  1. $w(id_x) = 0$ holds for each object $x \in Ob(\Psi)$ (*reflexivity*).
  2. $|w(g) - w(f)| \leq w(g \cdot f) \leq w(g) + w(f)$ holds for any objects x, y, z and arrows $f : x \to y, g : y \to z$ (*full triangle inequality*).

  Any set X produces an *indiscrete category* I_X, in which $Ob(I_X) = X$ and $|I_X(x, y)| = 1$ for all $x, y \in X$. Any metric space (X, d) produces a metric 1-space on I_X by defining $w(f) = d(x, y)$, and it is unique metric 1-space on I_X. But, in general, the function w on arrows can be seen as a multivalued function on $Ob \times Ob$.
  [Weis12] also outlined a **metric** m**-space** as a kind of an m**-hemimetric** on an m*-category* consisting of i-dimensional cells, $0 \leq i \leq m$ (objects, arrows, ...) and a associative-like composition rule for the cells with matching boundaries.

3.2 Indefinite Metrics

- **Indefinite metric**
  An **indefinite metric** (or G*-metric*) on a real (complex) vector space V is a *bilinear* (in the complex case, *sesquilinear*) *form* G on V, i.e., a function $G : V \times V \to \mathbb{R}$ ($\mathbb{C}$), such that, for any $x, y, z \in V$ and for any scalars α, β, we have the following properties: $G(\alpha x + \beta y, z) = \alpha G(x, z) + \beta G(y, z)$, and $G(x, \alpha y + \beta z) = \overline{\alpha} G(x, y) + \overline{\beta} G(x, z)$, where $\overline{\alpha} = \overline{a + bi} = a - bi$ denotes the *complex conjugation*.
  If a positive-definite form G is symmetric, then it is an *inner product* on V, and one can use it to canonically introduce a *norm* and the corresponding **norm metric** on V. In the case of a general form G, there is neither a norm, nor a metric canonically related to G, and the term **indefinite metric** only recalls the close relation of such forms with certain metrics in vector spaces (cf. Chaps. 7 and 26).
  The pair (V, G) is called a *space with an indefinite metric*. A finite-dimensional space with an indefinite metric is called a *bilinear metric space*. A **Hilbert space** H, endowed with a continuous G-metric, is called a *Hilbert space with an indefinite metric*. The most important example of such space is a J*-space*; cf. J**-metric**.

A subspace L in a space (V, G) with an indefinite metric is called a *positive subspace*, *negative subspace*, or *neutral subspace*, depending on whether $G(x, x) > 0$, $G(x, x) < 0$, or $G(x, x) = 0$ for all $x \in L$.

- **Hermitian G-metric**
  A **Hermitian G-metric** is an **indefinite metric** G^H on a complex vector space V such that, for all $x, y \in V$, we have the equality

$$G^H(x, y) = \overline{G^H(y, x)},$$

  where $\overline{\alpha} = \overline{a + bi} = a - bi$ denotes the *complex conjugation*.

- **Regular G-metric**
  A **regular G-metric** is a continuous **indefinite metric** G on a **Hilbert space** H over $\mathbb{C}$, generated by an invertible *Hermitian operator* T by the formula

$$G(x, y) = \langle T(x), y \rangle,$$

  where $\langle , \rangle$ is the *inner product* on H.
  A *Hermitian operator* on a Hilbert space H is a *linear operator* T on H defined on a *domain* $D(T)$ of H such that $\langle T(x), y \rangle = \langle x, T(y) \rangle$ for any $x, y \in D(T)$. A bounded Hermitian operator is either defined on the whole of H, or can be so extended by continuity, and then $T = T^*$. On a finite-dimensional space a Hermitian operator can be described by a *Hermitian matrix* $((a_{ij})) = ((\overline{a}_{ji}))$.

- **J-metric**
  A **J-metric** is a continuous **indefinite metric** G on a **Hilbert space** H over $\mathbb{C}$ defined by a certain *Hermitian involution* J on H by the formula

$$G(x, y) = \langle J(x), y \rangle,$$

  where $\langle , \rangle$ is the *inner product* on H.
  An *involution* is a mapping H onto H whose square is the *identity mapping*. The involution J may be represented as $J = P_+ - P_-$, where P_+ and P_- are orthogonal projections in H, and $P_+ + P_- = H$. The *rank of indefiniteness* of the J-metric is defined as $\min\{\dim P_+, \dim P_-\}$.
  The space (H, G) is called a *J-space*. A J-space with finite rank of indefiniteness is called a *Pontryagin space*.

3.3 Topological Generalizations

- **Metametric space**
  A **metametric space** (Väisälä, 2003) is a pair (X, d), where X is a set, and d is a nonnegative symmetric function $d : X \times X \to \mathbb{R}$ such that $d(x, y) = 0$ implies $x = y$ and triangle inequality $d(x, y) \leq d(x, z) + d(z, y)$ holds for all $x, y, z \in X$.

A metametric space is metrizable: the metametric d defines the same topology as the metric d' defined by $d'(x,x) = 0$ and $d'(x,y) = d(x,y)$ if $x \neq y$. A metametric d induces a Hausdorff topology with the usual definition of a *ball* $B(x_0, r) = \{x \in X : d(x_0, x) < r\}$. Any **partial metric** (cf. Chap. 1) is a metametric.

- **Resemblance**
  Let X be a set. A function $d : X \times X \to \mathbb{R}$ is called (Batagelj-Bren, 1993) a **resemblance** on X if d is *symmetric* and if, for all $x, y \in X$, either $d(x,x) \leq d(x,y)$ (in which case d is called a **forward resemblance**), or $d(x,x) \geq d(x,y)$ (in which case d is called a **backward resemblance**).
  Every resemblance d induces a *strict partial order* $\prec$ on the set of all unordered pairs of elements of X by defining $\{x,y\} \prec \{u,v\}$ if and only if $d(x,y) < d(u,v)$.
  For any backward resemblance d, the forward resemblance $-d$ induces the same partial order.
- **w-distance**
  Given a metric space (X,d), a **w-distance** on X (Kada–Suzuki–Takahashi, 1996) is a nonnegative function $p : X \times X \to \mathbb{R}$ which satisfies the following conditions:

  1. $p(x,z) \leq p(x,y) + p(y,z)$ for all $x, y, z \in X$;
  2. for any $x \in X$, the function $p(x,.) : X \to \mathbb{R}$ is *lower semicontinuous*, i.e., if a sequence $\{y_n\}_n$ in X converges to $y \in X$, then $p(x,y) \leq \underline{\lim}_{n\to\infty} p(x, y_n)$;
  3. for any $\epsilon > 0$, there exists $\delta > 0$ such that $p(z,x) \leq \delta$ and $p(z,y) \leq \delta$ imply $d(x,y) \leq \epsilon$, for each $x, y, z \in X$.

- **τ-Distance space**
  A **τ-distance space** is a pair (X, f), where X is a topological space and f is an Aamri-Moutawakil's *τ-distance* on X, i.e., a nonnegative function $f : X \times X \to \mathbb{R}$ such that, for any $x \in X$ and any neighborhood U of x, there exists $\epsilon > 0$ with $\{y \in X : f(x,y) < \epsilon\} \subset U$.
  Any distance space (X,d) is a τ-distance space for the topology τ_f defined as follows: $A \in \tau_f$ if, for any $x \in X$, there exists $\epsilon > 0$ with $\{y \in X : f(x,y) < \epsilon\} \subset A$. However, there exist nonmetrizable τ-distance spaces. A τ-distance $f(x,y)$ need be neither symmetric, nor vanishing for $x = y$; for example, $e^{|x-y|}$ is a τ-distance on $X = \mathbb{R}$ with usual topology.
- **Proximity space**
  A **proximity space** (Efremovich, 1936) is a set X with a binary relation δ on the *power set* $P(X)$ of all of its subsets which satisfies the following conditions:

  1. $A\delta B$ if and only if $B\delta A$ (*symmetry*);
  2. $A\delta(B \cup C)$ if and only if $A\delta B$ or $A\delta C$ (*additivity*);
  3. $A\delta A$ if and only if $A \neq \emptyset$ (*reflexivity*).

  The relation δ defines a **proximity** (or *proximity structure*) on X. If $A\delta B$ fails, the sets A and B are called *remote sets*.

Every metric space (X, d) is a proximity space: define $A\delta B$ if and only if $d(A, B) = \inf_{x\in A, y\in B} d(x, y) = 0$.
Every proximity on X induces a (**completely regular**) topology on X by defining the *closure operator* $cl : P(X) \to P(X)$ on the set of all subsets of X as $cl(A) = \{x \in X : \{x\}\delta A\}$.

- **Uniform space**
A **uniform space** is a topological space (with additional structure) providing a generalization of metric space, based on **set-set distance**.
A **uniform space** (Weil, 1937) is a set X with an *uniformity* (or *uniform structure*) $\mathcal{U}$, i.e., a nonempty collection of subsets of $X \times X$, called *entourages*, with the following properties:

    1. Every subset of $X \times X$ which contains a set of $\mathcal{U}$ belongs to $\mathcal{U}$;
    2. Every finite intersection of sets of $\mathcal{U}$ belongs to $\mathcal{U}$;
    3. Every set $V \in \mathcal{U}$ contains the *diagonal*, i.e., the set $\{(x, x) : x \in X\} \subset X \times X$;
    4. If V belongs to $\mathcal{U}$, then the set $\{(y, x) : (x, y) \in V\}$ belongs to $\mathcal{U}$;
    5. If V belongs to $\mathcal{U}$, then there exists $V' \in \mathcal{U}$ such that $(x, z) \in V$ whenever $(x, y), (y, z) \in V'$.

    Every metric space (X, d) is a uniform space. An entourage in (X, d) is a subset of $X \times X$ which contains the set $V_\epsilon = \{(x, y) \in X \times X : d(x, y) < \epsilon\}$ for some positive real number ϵ. Other basic example of uniform space are *topological groups*.
Every uniform space $(X, \mathcal{U})$ generates a topology consisting of all sets $A \subset X$ such that, for any $x \in A$, there is a set $V \in \mathcal{U}$ with $\{y : (x, y) \in V\} \subset A$.
Every uniformity induces a **proximity** σ where $A\sigma B$ if and only if $A \times B$ has nonempty intersection with any entourage.
A topological space admits a uniform structure inducing its topology if only if the topology is **completely regular** (cf. Chap. 2) and, also, if only if it is a *gauge space*, i.e., the topology is defined by a $\geq$*-filter* of semimetrics.
- **Nearness space**
A **nearness space** (Herrich, 1974) is a set X with a *nearness structure*, i.e., a nonempty collection $\mathcal{U}$ of families of subsets of X, called *near families*, with the following properties:

    1. Each family refining a near family is near;
    2. Every family with nonempty intersection is near;
    3. V is near if $\{cl(A) : A \in V\}$ is near, where $cl(A)$ is $\{x \in X : \{\{x\}, A\} \in \mathcal{U}\}$;
    4. $\emptyset$ is near, while the set of all subsets of X is not;
    5. If $\{A \cup B : A \in \mathcal{F}_1, B \in \mathcal{F}_2\}$ is near family, then so is $\mathcal{F}_1$ or $\mathcal{F}_2$.

    The **uniform spaces** are precisely **paracompact** nearness spaces.
- **Approach space**
An **approach space** is a topological space providing a generalization of metric space, based on **point-set distance**.

An **approach space** (Lowen, 1989) is a pair (X, D), where X is a set and D is a **point-set distance**, i.e., a function $X \times P(X) \to [0, \infty]$ (where $P(X)$ is the set of all subsets of X) satisfying, for all $x \in X$ and all $A, B \in P(X)$, the following conditions:

1. $D(x, \{x\}) = 0$;
2. $D(x, \{\emptyset\}) = \infty$;
3. $D(x, A \cup B) = \min\{D(x, A), D(x, B)\}$;
4. $D(x, A) \leq D(x, A^{\epsilon}) + \epsilon$ for any $\epsilon \in [0, \infty]$, where $A^{\epsilon} = \{x : D(x, A) \leq \epsilon\}$ is the "ϵ-ball" with center x.

Every metric space (X, d) (moreover, any extended quasi-semimetric space) is an approach space with $D(x, A)$ being the usual point-set distance $\min_{y \in A} d(x, y)$.
Given a **locally compact separable** metric space (X, d) and the family $\mathcal{F}$ of its nonempty closed subsets, the **Baddeley–Molchanov distance function** gives a tool for another generalization. It is a function $D : X \times \mathcal{F} \to \mathbb{R}$ which is lower semicontinuous with respect to its first argument, measurable with respect to the second, and satisfies the following two conditions: $F = \{x \in X : D(x, F) \leq 0\}$ for $F \in \mathcal{F}$, and $D(x, F_1) \geq D(x, F_2)$ for $x \in X$, whenever $F_1, F_2 \in \mathcal{F}$ and $F_1 \subset F_2$.
The additional conditions $D(x, \{y\}) = D(y, \{x\})$, and $D(x, F) \leq D(x, \{y\}) + D(y, F)$ for all $x, y \in X$ and every $F \in \mathcal{F}$, provide analogs of symmetry and the triangle inequality. The case $D(x, F) = d(x, F)$ corresponds to the usual point-set distance for the metric space (X, d); the case $D(x, F) = d(x, F)$ for $x \in X \backslash F$ and $D(x, F) = -d(x, X \backslash F)$ for $x \in X$ corresponds to the **signed distance function** in Chap. 1.

- **Metric bornology**
  Given a topological space X, a *bornology* of X is any family $\mathcal{A}$ of proper subsets A of X such that the following conditions hold:

  1. $\cup_{A \in \mathcal{A}} A = X$;
  2. $\mathcal{A}$ is an *ideal*, i.e., contains all subsets and finite unions of its members.
     The family $\mathcal{A}$ is a **metric bornology** [Beer99] if, moreover
  3. $\mathcal{A}$ contains a countable base;
  4. For any $A \in \mathcal{A}$ there exists $A' \in \mathcal{A}$ such that the closure of A coincides with the interior of A'.

  The metric bornology is called *trivial* if $\mathcal{A}$ is the set $P(X)$ of all subsets of X; such a metric bornology corresponds to the family of bounded sets of some bounded metric. For any noncompact **metrizable** topological space X, there exists an unbounded metric compatible with this topology. A nontrivial metric bornology on such a space X corresponds to the family of bounded subsets with respect to some such unbounded metric. A noncompact metrizable topological space X admits uncountably many nontrivial metric bornologies.

3.4 Beyond Numbers

- **Probabilistic metric space**
A notion of **probabilistic metric space** is a generalization of the notion of metric space (see, for example, [ScSk83]) in two ways: distances become probability distributions, and the sum in the triangle inequality becomes a **triangle operation**.
Formally, let A be the set of all *probability distribution functions*, whose support lies in $[0, \infty]$. For any $a \in [0, \infty]$ define *step functions* $\epsilon_a \in A$ by $\epsilon_a(x) = 1$ if $x > a$ or $x = \infty$, and $\epsilon_a(x) = 0$, otherwise. The functions in A are ordered by defining $F \leq G$ to mean $F(x) \leq G(x)$ for all $x \geq 0$; the minimal element is ϵ_0.
A commutative and associative operation τ on A is called a **triangle function** if $\tau(F, \epsilon_0) = F$ for any $F \in A$ and $\tau(E, F) \leq \tau(G, H)$ whenever $E \leq G$, $F \leq H$. The semigroup (A, τ) generalizes the group $(\mathbb{R}, +)$.
A **probabilistic metric space** is a triple (X, D, τ), where X is a set, D is a function $X \times X \to A$, and τ is a triangle function, such that for any $p, q, r \in X$

  1. $D(p, q) = \epsilon_0$ if and only if $p = q$;
  2. $D(p, q) = D(q, p)$;
  3. $D(p, r) \geq \tau(D(p, q), D(q, r))$.

  For any metric space (X, d) and any triangle function τ, such that $\tau(\epsilon_a, \epsilon_b) \geq \epsilon_{a+b}$ for all $a, b \geq 0$, the triple $(X, D = \epsilon_{d(x,y)}, \tau)$ is a probabilistic metric space.
For any $x \geq 0$, the value $D(p, q)$ at x can be interpreted as "the probability that the distance between p and q is less than x"; this was approach of Menger, who proposed in 1942 the original version, *statistical metric space*, of this notion.
A probabilistic metric space is called a *Wald space* if the triangle function is a convolution, i.e., of the form $\tau_x(E, F) = \int_{\mathbb{R}} E(x - t)dF(t)$.
A probabilistic metric space is called a **generalized Menger space** if the triangle function has form $\tau_x(E, F) = \sup_{u+v=x} T(E(u), F(v))$ for a *t-norm* T, i.e., such a commutative and associative operation on $[0, 1]$ that $T(a, 1) = a$, $T(0, 0) = 0$ and $T(c, d) \geq T(a, b)$ whenever $c \geq a, d \geq b$.
- **Fuzzy metric spaces**
A *fuzzy subset* of a set S is a mapping $\mu : S \to [0, 1]$, where $\mu(x)$ represents the "degree of membership" of $x \in S$.
A *continuous t-norm* is a binary commutative and associative continuous operation T on $[0, 1]$, such that $T(a, 1) = a$ and $T(c, d) \geq T(a, b)$ whenever $c \geq a, d \geq b$.
A **KM fuzzy metric space** (Kramosil–Michalek, 1975) is a pair $(X, (\mu, T))$, where X is a nonempty set and a *fuzzy metric* (μ, T) is a pair comprising a continuous t-norm T and a fuzzy set $\mu : X^2 \times \mathbb{R}_{\geq 0} \to [0, 1]$, such that, for $x, y, z \in X$ and $s, t \geq 0$, the following conditions hold:

  1. $\mu(x, y, 0) = 0$;
  2. $\mu(x, y, t) = 1$ if and only if $x = y, t > 0$;
  3. $\mu(x, y, t) = \mu(y, x, t)$;

4. $T(\mu(x,y,t),\mu(y,z,s)) \leq \mu(x,z,t+s)$;
5. the function $\mu(x,y,\cdot) : \mathbb{R}_{\geq 0} \to [0,1]$ is left continuous.

A KM fuzzy metric space is called also a **fuzzy Menger space** since by defining $D_t(p,q) = \mu(p,q,t)$ one gets a **generalized Menger space**. The following modification of the above notion, using a stronger form of metric fuzziness, it a generalized Menger space with $D_t(p,q)$ positive and continuous on $\mathbb{R}_{>0}$ for all p,q.

A **GV fuzzy metric space** (George–Veeramani, 1994) is a pair $(X,(\mu,T))$, where X is a nonempty set, and a *fuzzy metric* (μ,T) is a pair comprising a continuous t-norm T and a fuzzy set $\mu : X^2 \times \mathbb{R}_{>0} \to [0,1]$, such that for $x,y,z \in X$ and $s,t > 0$

1. $\mu(x,y,t) > 0$;
2. $\mu(x,y,t) = 1$ if and only if $x = y$;
3. $\mu(x,y,t) = \mu(y,x,t)$;
4. $T(\mu(x,y,t),\mu(y,z,s)) \leq \mu(x,z,t+s)$;
5. the function $\mu(x,y,\cdot) : \mathbb{R}_{>0} \to [0,1]$ is continuous.

An example of a GV fuzzy metric space comes from any metric space (X,d) by defining $T(a,b) = b - ab$ and $\mu(x,y,t) = \frac{t}{t+d(x,y)}$. Conversely, any GV fuzzy metric space (and also any KM fuzzy metric space) generates a metrizable topology. Most GV fuzzy metrics are *strong*, i.e., $T(\mu(x,y,t),\mu(y,z,t)) \leq \mu(x,z,t)$ holds.

A *fuzzy number* is a fuzzy set $\mu : \mathbb{R} \to [0,1]$ which is *normal* ($\{x \in \mathbb{R} : \mu(x) = 1\} \neq \emptyset$), *convex* ($\mu(tx + (1-t)y) \geq \min\{\mu(x),\mu(y)\}$ for every $x,y \in \mathbb{R}$ and $t \in [0,1]$) and *upper semicontinuous* (at each point x_0, the values $\mu(x)$ for x near x_0 are either close to $\mu(x_0)$ or less than $\mu(x_0)$). Denote the set of all fuzzy numbers which are *nonnegative*, i.e., $\mu(x) = 0$ for all $x < 0$, by G. The additive and multiplicative identities of fuzzy numbers are denoted by $\tilde{0}$ and $\tilde{1}$, respectively. The *level set* $[\mu]_t = \{x : \mu(x) \geq t\}$ of a fuzzy number μ is a closed interval.

Given a nonempty set X and a mapping $d : X^2 \to G$, let the mappings $L, R : [0,1]^2 \to [0,1]$ be symmetric and nondecreasing in both arguments and satisfy $L(0,0) = 0$, $R(1,1) = 1$. For all $x,y \in X$ and $t \in (0,1]$, let $[d(x,y)]_t = [\lambda_t(x,y),\rho_t(x,y)]$.

A **KS fuzzy metric space** (Kaleva–Seikkala, 1984) is a quadruple (X,d,L,R) with *fuzzy metric* d, if for all $x,y,z \in X$

1. $d(x,y) = \tilde{0}$ if and only if $x = y$;
2. $d(x,y) = d(y,x)$;
3. $d(x,y)(s+t) \geq L(d(x,z)(s),d(z,y)(t))$ whenever $s \leq \lambda_1(x,z)$, $t \leq \lambda_1(z,y)$, and $s+t \leq \lambda_1(x,y)$;
4. $d(x,y)(s+t) \leq R(d(x,z)(s),d(z,y)(t))$ whenever $s \geq \lambda_1(x,z)$, $t \geq \lambda_1(z,y)$, and $s+t \geq \lambda_1(x,y)$.

The following functions are some frequently used choices for L and R:

$$\max\{a+b-1,0\}, ab, \min\{a,b\}, \max\{a,b\}, a+b-ab, \min\{a+b,1\}.$$

Several other notions of **fuzzy metric space** were proposed, including those by Erceg, 1979, Deng, 1982, and Voxman, 1998, Xu–Li, 2001, Tran–Duckstein, 2002, Chakraborty–Chakraborty, 2006. Cf. also **metrics between fuzzy sets**, **fuzzy Hamming distance**, **gray-scale image distances** and **fuzzy polynucleotide metric** in Chaps. 1, 11, 21 and 23, respectively.

- **Interval-valued metric space**
Let $I(\mathbb{R}_{\geq 0})$ denote the set of closed intervals of $\mathbb{R}_{\geq 0}$.
An **interval-valued metric space** (Coppola–Pacelli, 2006) is a pair $((X,\leq),\Delta)$, where $(X,\leq)$ is a partially ordered set and Δ is an interval-valued mapping $\Delta : X\times X \to I(\mathbb{R}_{\geq 0})$, such that for every $x,y,z\in X$

  1. $\Delta(x,x) \star [0,1] = \Delta(x,x)$;
  2. $\Delta(x,y) = \Delta(y,x)$;
  3. $\Delta(x,y) - \Delta(z,z) \preceq \Delta(x,z) + \Delta(z,y)$;
  4. $\Delta(x,y) - \Delta(x,y) \preceq \Delta(x,x) + \Delta(y,y)$;
  5. $x \leq x'$ and $y \leq y'$ imply $\Delta(x,y) \subseteq \Delta(x',y')$;
  6. $\Delta(x,y) = 0$ if and only if $x = y$ and x,y are *atoms* (minimal elements of $(X,\leq)$).

  Here the following *interval arithmetic* rules hold: $[u,v] \preceq [u',v']$ if and only if $u \leq u'$,
$[u,v]+[u',v'] = [u+u', v+v']$, $[u,v]-[u',v'] = [u-u', v-v']$,
$[u,v] \star [u',v'] = [\min\{uu',uv',vu',vv'\}, \max\{uu',uv',vu',vv'\}]$ and
$\frac{[u,v]}{[u',v']} = [\min\{\frac{u}{u'},\frac{u}{v'},\frac{v}{u'},\frac{v}{v'}\}, \max\{\frac{u}{u'},\frac{u}{v'},\frac{v}{u'},\frac{v}{v'}\}]$ when $0 \notin [u',v']$.
The addition and multiplication operations are commutative, associative and *subdistributive*: it holds $X \star (Y+Z) \subset (X\star Y + X\star Z)$.
Cf. **metric between intervals** in Chap. 10.
The usual metric spaces coincide with above spaces in which all $x\in X$ are atoms.
- **Direction distance**
Given a normed real vector space $(V,||.||)$, for any $x \in V\setminus\{0\}$, denote by $[x]$ the *direction* (ray) $\{\lambda x : \lambda > 0\}$ and by x_0 the point $\frac{x}{||x||}$. An *oriented angle* is an ordered pair $([x],[y])$ of directions. The **direction distance** from x to y is defined (Busch–Ruch, 1992) as the family of distances $||\alpha x_0 - \beta y_0||$ with $\alpha,\beta\in\mathbb{R}_{>0}$.
The **mixing distance** is defined as the restriction of the direction distance to pairs of directions in the cone $\{\lambda v : v\in V, \lambda > 0\}$. In fact, authors introduced these distances on some special normed spaces used in Quantum Mechanics.
- **Generalized metric**
Let X be a set. Let $(V,+,\leq)$ be an *ordered semigroup* (not necessarily commutative) with a least element θ and with $x \leq y, x_1 \leq y_1$ implying $x+x_1 \leq y+y_1$. Let $(V,+)$ be also endowed with an order-preserving *involution* x^* (i.e., $(x^*)^* = x$), which is operation-reversing, i.e., $(x+y)^* = y^* + x^*$.

A function $d : X \times X \to G$ is called (Li–Wang–Pouzet, 1987) a **generalized metric** over $(V, +, \leq)$ if the following conditions hold:

1. $d(x, y) = \theta$ if and only if $x = y$;
2. $d(x, y) \leq d(x, z) + d(z, y)$ for all $x, y \in X$;
3. $d^*(x, y) = d(y, x)$.

- **Cone metric**
Let C be a *proper cone* in a real Banach space W, i.e., C is closed, $C \neq \emptyset$, the interior of C is not equal to $\{\theta\}$ (where θ is the zero vector in W) and

1. if $x, y \in C$ and $a, b \in \mathbb{R}_{\geq 0}$, then $ax + by \in C$;
2. if $x \in C$ and $-x \in C$, then $x = 0$.

Define a partial ordering $(W, \leq)$ on W by letting $x \leq y$ if $y - x \in C$. The following variation of **generalized metric** and **partially ordered distance** was defined in Huang–Zhang, 2007, and, partially, in Rzepecki, 1980. Given a set X, a **cone metric** is a mapping $d : X \times X \to (W, \leq)$ such that

1. $\theta \leq d(x, y)$ with equality if and only if $x = y$;
2. $d(x, y) = d(y, x)$ for all $x, y \in X$;
3. $d(x, y) \leq d(x, z) + d(z, y)$ for all $x, y \in X$;

The pair (X, d) is called a **cone metric space**.

- **W-distance on building**
Let X be a set, and let $(W, \cdot, 1)$ be a group. A W*-distance* on X is a W-valued map $\sigma : X \times X \to W$ having the following properties:

1. $\sigma(x, y) = 1$ if and only if $x = y$;
2. $\sigma(y, x) = (\sigma(x, y))^{-1}$.

A natural W-distance on W is $\sigma(x, y) = x^{-1}y$.
A *Coxeter group* is a group $(W, \cdot, 1)$ generated by the elements

$$\{w_1, \dots, w_n : (w_i w_j)^{m_{ij}} = 1, 1 \leq i, j \leq n\}.$$

Here $M = ((m_{ij}))$ is a *Coxeter matrix*, i.e., an arbitrary symmetric $n \times n$ matrix with $m_{ii} = 1$, and the other values are positive integers or ∞. The *length* $l(x)$ of $x \in W$ is the smallest number of generators $w_1, \dots, w_n$ needed to represent x.
Let X be a set, let $(W, \cdot, 1)$ be a Coxeter group and let $\sigma(x, y)$ be a W-distance on X. The pair (X, σ) is called (Tits, 1981) a *building* over $(W, \cdot, 1)$ if it holds

1. the relation $\sim_i$ defined by $x \sim_i y$ if $\sigma(x, y) = 1$ or w_i, is an equivalence relation;
2. given $x \in X$ and an equivalence class C of $\sim_i$, there exists a unique $y \in C$ such that $\sigma(x, y)$ is *shortest* (i.e., of smallest length), and $\sigma(x, y') = \sigma(x, y)w_i$ for any $y' \in C, y' \neq y$.

The **gallery distance on building** d is a usual metric on X defined by $l(d(x, y))$.

The distance d is the **path metric** in the graph with the vertex-set X and xy being an edge if $\sigma(x, y) = w_i$ for some $1 \leq i \leq n$. The gallery distance on building is a special case of a **gallery metric** (of *chamber system* X).

- **Boolean metric space**
  A *Boolean algebra* (or *Boolean lattice*) is a *distributive lattice* $(B, \vee, \wedge)$ admitting a least element 0 and greatest element 1 such that every $x \in B$ has a *complement* $\overline{x}$ with $x \vee \overline{x} = 1$ and $x \wedge \overline{x} = 0$.
  Let X be a set, and let $(B, \vee, \wedge)$ be a Boolean algebra. The pair (X, d) is called a **Boolean metric space** over B if the function $d : X \times X \to B$ has the following properties:

  1. $d(x, y) = 0$ if and only if $x = y$;
  2. $d(x, y) \leq d(x, z) \vee d(z, y)$ for all $x, y, z \in X$.

- **Space over algebra**
  A **space over algebra** is a metric space with a differential-geometric structure, whose points can be provided with coordinates from some *algebra* (usually, an associative algebra with identity).
  A *module* over an algebra is a generalization of a vector space over a field, and its definition can be obtained from the definition of a vector space by replacing the field by an associative algebra with identity. An *affine space over an algebra* is a similar generalization of an *affine space* over a field. In affine spaces over algebras one can specify a Hermitian metric, while in the case of commutative algebras even a quadratic metric can be given. To do this one defines in a unital module a *scalar product* $\langle x, y\rangle$, in the first case with the property $\langle x, y\rangle = J(\langle y, x\rangle)$, where J is an *involution* of the algebra, and in the second case with the property $\langle y, x\rangle = \langle x, y\rangle$.
  The n-dimensional *projective space over an algebra* is defined as the variety of one-dimensional submodules of an $(n + 1)$-dimensional unital module over this algebra. The introduction of a *scalar product* $\langle x, y\rangle$ in a unital module makes it possible to define a Hermitian metric in a projective space constructed by means of this module or, in the case of a commutative algebra, quadratic elliptic and hyperbolic metrics. The metric invariant of the points of these spaces is the **cross-ratio** $W = \langle x, x\rangle^{-1}\langle x, y\rangle\langle y, y\rangle^{-1}\langle y, x\rangle$. If W is a real number, then $w = \arccos \sqrt{W}$ is called the **distance** between x and y **in the space over algebra**.
- **Partially ordered distance**
  Let X be a set. Let $(G, \leq)$ be a *partially ordered set* with a least element g_0. A **partially ordered distance** is a function $d : X \times X \to G$ such that, for any $x, y \in X$, $d(x, y) = g_0$ if and only if $x = y$.
  A **generalized ultrametric** (Priess-Crampe and Ribenboim, 1993) is a symmetric (i.e., $d(x, y) = d(y, x)$) partially ordered distance, such that $d(z, x) \leq g$ and $d(z, y) \leq g$ imply $d(x, y) \leq g$ for any $x, y, z \in X$ and $g \in G$.

Suppose that $G' = G\backslash\{g_0\} \neq \emptyset$ and, for any $g_1, g_2 \in G'$, there exists $g_3 \in G'$ such that $g_3 \leq g_1$ and $g_3 \leq g_2$. Consider the following possible properties:

1. For any $g_1 \in G'$, there exists $g_2 \in G'$ such that, for any $x, y \in X$, from $d(x, y) \leq g_2$ it follows that $d(y, x) \leq g_1$;
2. For any $g_1 \in G'$, there exist $g_2, g_3 \in G'$ such that, for any $x, y, z \in X$, from $d(x, y) \leq g_2$ and $d(y, z) \leq g_3$ it follows that $d(x, z) \leq g_1$;
3. For any $g_1 \in G'$, there exists $g_2 \in G'$ such that, for any $x, y, z \in X$, from $d(x, y) \leq g_2$ and $d(y, z) \leq g_2$ it follows that $d(y, x) \leq g_1$;
4. G' has no first element;
5. $d(x, y) = d(y, x)$ for any $x, y \in X$;
6. For any $g_1 \in G'$, there exists $g_2 \in G'$ such that, for any $x, y, z \in X$, from $d(x, y) <^* g_2$ and $d(y, z) <^* g_2$ it follows that $d(x, z) <^* g_1$; here $p <^* q$ means that either $p < q$, or p is not comparable to q;
7. The order relation $<$ is a total ordering of G.

In terms of above properties, d is called: the **Appert partially ordered distance** if 1 and 2 hold; the **Golmez partially ordered distance of first type** if 4, 5, and 6 hold; the **Golmez partially ordered distance of second type** if 3, 4, and 5 hold; the **Kurepa–Fréchet distance** if 3, 4, 5, and 7 hold.
The case $G = \mathbb{R}_{\geq 0}$ of the Kurepa–Fréchet distance corresponds to the **Fréchet V-space**, i.e., a pair (X, d), where X is a set, and $d(x, y)$ is a symmetric function $d : X \times X \to \mathbb{R}_{\geq 0}$ (**voisinage of two points** x and y) such that $d(x, y) = 0$ if and only if $x = y$, and there exists a function $f : \mathbb{R} \to \mathbb{R}_{\geq 0}$ with $\lim_{t \to 0} f(t) = 0$ and $\max\{d(x, y), d(y, z)\} \leq r$ implying $d(x, z) \leq f(r)$ for all $x, y, z \in X$ and all $r > 0$. The general case was considered in Kurepa, 1934, and rediscovered in Fréchet, 1946.

- **Distance from measurement**
  Distance from measurement is an analog of distance on domains in Computer Science; it was developed in [Mart00].
  A *po* (partially ordered set) $(D, \preceq)$ is called *dcpo* (directed-complete po) if every *directed subset* $S \subset D$ (i.e., $S \neq \emptyset$ and any pair $x, y \in S$ is *bounded*: there is $z \in S$ with $x, y \preceq z$) has a *supremum* $\sqcup S$, i.e., the least of such upper bounds z. For $x, y \in D$, y is an *approximation* of x if, for all directed subsets $S \subset D$, $x \preceq \sqcup S$ implies $y \preceq s$ for some $s \in S$. A dcpo $(D, \preceq)$ is *continuous* if for all $x \in D$ the set of all approximations of x is directed and x is its supremum. A *domain* is a continuous dcpo $(D, \preceq)$ such that for all $x, y \in D$ there is $z \in D$ with $z \preceq x, y$. A *Scott domain* is a domain with least element, in which any bounded pair has a supremum.
  A subset U of a dcpo $(D, \preceq)$ is *Alexandrov open* if, for any $x \in U$ and $y \in D$, $x \preceq y$ implies $y \in U$; it is *Scott open* if also, for any directed subset $S \subset D$, $\sqcup S \in U$ implies $S \cap U \neq \emptyset$. The set of Scott open sets form the *Scott topology*; it is a T_0**-space** (Chap. 2) with generalized metrization by a **partial metric** (Chap. 1).

A *measurement* is a mapping $\mu : D \to \mathbb{R}_{\geq 0}$ between dcpo $(D, \preceq)$ and dcpo $(\mathbb{R}_{\geq 0}, \preceq)$, where $\mathbb{R}_{\geq 0}$ is ordered as $x \preceq y$ if $y \leq x$, such that

1. $x \preceq y$ implies $\mu(x) \preceq \mu(y)$;
2. $\mu(\sqcup S) = \sqcup(\{\mu(s) : s \in S\})$ for every directed subset $S \subset D$;
3. For all $x \in D$ with $\mu(x) = 0$ and all sequences $(x_n), n \to \infty$, of approximations of x with $\lim_{n\to\infty} \mu(x_n) = \mu(x)$, one has $\sqcup(\cup_{n=1}^{\infty}\{x_n\}) = x$.

Given a measurement μ, the **distance from measurement** is a mapping $d : D \times D \to \mathbb{R}_{\geq 0}$ given by

$$d(x, y) = \inf\{\mu(z) : z \text{ approximates } x, y\} = \inf\{\mu(z) : z \preceq x, y\}.$$

One has $d(x, x) \preceq \mu(x)$. The function $d(x, y)$ is a metric on the set $\{x \in D : \mu(x) = 0\}$ if μ satisfies the following **measurement triangle inequality**: for all bounded pairs $x, y \in D$, there is an element $z \preceq x, y$ such that $\mu(z) \leq \mu(x) + \mu(y)$.
Waszkiewicz, 2001, found topological connections between topologies coming from a distance from measurement and from a **partial metric** defined in Chap. 1.

Chapter 4
Metric Transforms

There are many ways to obtain new distances (metrics) from given distances (metrics). Metric transforms give new distances as a functions of given metrics (or given distances) on the same set X. A metric so obtained is called a **transform metric**. We give some important examples of transform metrics in Sect. 4.1.

Given a metric space (X, d), one can construct a new metric on an extension of X; similarly, given a collection of metrics on sets $X_1, \dots, X_n$, one can obtain a new metric on an extension of $X_1, \dots, X_n$. Examples of such operations are given in Sect. 4.2. There are many distances on other structures connected with X, say, on the set of all subsets of X. The main distances of this kind are considered in Sect. 4.3.

4.1 Metrics on the Same Set

- **Metric transform**

  A **metric transform** is a distance on a set X, obtained as a function of given metrics (or given distances) on X.

  In particular, given a continuous monotone increasing function $f(x)$ of $x \geq 0$ with $f(0) = 0$, called the *scale*, and a distance space (X, d), one obtains another distance space (X, d_f), called a **scale metric transform** of X, defining $d_f(x, y) = f(d(x, y))$. For every finite distance space (X, d), there exists a scale f, such that (X, d_f) is a metric subspace of a Euclidean space $\mathbb{E}^n$.

  If (X, d) is a metric space and f is a continuous differentiable strictly increasing scale with $f(0) = 0$ and nonincreasing f', then (X, d_f) is a metric space (cf. **functional transform metric**).

  The metric d is an **ultrametric** if and only if $f(d)$ is a metric for every nondecreasing function $f : \mathbb{R}_{\geq 0} \to \mathbb{R}_{\geq 0}$.

M.M. Deza, E. Deza, *Encyclopedia of Distances*,
DOI 10.1007/978-3-662-44342-2_4

- **Transform metric**
  A **transform metric** is a metric on a set X which is a **metric transform**, i.e., is obtained as a function of a given metric (or given metrics) on X. In particular, transform metrics can be obtained from a given metric d (or given metrics d_1 and d_2) on X by any of the following operations (here $t > 0$):

  1. $td(x, y)$ (**t-scaled metric**, or **dilated metric**, **similar metric**);
  2. $\min\{t, d(x, y)\}$ (**t-truncated metric**);
  3. $\max\{t, d(x, y)\}$ for $x \neq y$ (**t-uniformly discrete metric**);
  4. $d(x, y) + t$ for $x \neq y$ (**t-translated metric**);
  5. $\frac{kd(x,y)}{1+d(x,y)}$ (this metric has diameter less than k);
  6. $d^p(x, y) = \frac{2d(x,y)}{d(x,p)+d(y,p)+d(x,y)}$, where p is an fixed element of X (**biotope transform metric**, or **p-smoothing distance** on $X \setminus \{p\}$);
  7. $\max\{d_1(x, y), d_2(x, y)\}$;
  8. $\alpha d_1(x, y) + \beta d_2(x, y)$, where $\alpha, \beta > 0$ (cf. **semimetric cone** in Chap. 1).

- **Generalized biotope transform metric**
  For a given metric d on a set X and a closed set $M \subset X$, the **generalized biotope transform metric** d^M on X is defined by

$$d^M(x, y) = \frac{2d(x, y)}{d(x, y) + \inf_{z\in M}(d(x, z) + d(y, z))}.$$

  In fact, $d^M(x, y)$ and its **1-truncation** $\min\{1, d^M(x, y)\}$ are both metrics.
  The **biotope transform metric** is $d^M(x, y)$ with $|M| = 1$. The **Steinhaus distance** from Chap. 1 is the case $d(x, y) = \mu(x \Delta y)$ with $p \neq \emptyset$ and the **biotope distance** from Chap. 23 is its subcase $d(x, y) = \mu(x \Delta y) = |x \Delta y|$.
- **Metric-preserving function**
  A function $f : \mathbb{R}_{\geq 0} \to \mathbb{R}_{\geq 0}$ with $f^{-1}(0) = \{0\}$ is a **metric-preserving function** if, for each metric space (X, d), the **metric transform**

$$d_f(x, y) = f(d(x, y))$$

  is a metric on X; cf. [Cora99]. In this case d_f is called a **functional transform metric**. For example, αd ($\alpha > 0$), $d^\alpha (0 < \alpha \leq 1)$, $\ln(1 + d)$, $\operatorname{arcsinh} d$, $\operatorname{arccosh}(1 + d)$, and $\frac{d}{1+d}$ are functional transform metrics.
  The superposition, sum and maximum of two metric-preserving functions are metric-preserving. If f is *subadditive*, i.e. $f(x + y) \leq f(x) + f(y)$ for all $x, y \geq 0$, and nondecreasing, then it is metric-preserving. But, for example, the function $f(x) = \frac{x+2}{x+1}$, for $x > 0$, and $f(0) = 0$, is decreasing and metric-preserving. If f is metric-preserving, then it is subadditive.
  If f is *concave*, i.e., $f(\frac{x+y}{2}) \geq \frac{f(x)+f(y)}{2}$ for all $x, y \geq 0$, then it is metric-preserving. In particular, a twice differentiable function $f : \mathbb{R}_{\geq 0} \to \mathbb{R}_{\geq 0}$ such that $f(0) = 0$, $f'(x) > 0$ for all $x \geq 0$, and $f''(x) \leq 0$ for all $x \geq 0$, is metric-preserving.

The function f is **strongly metric-preserving function** if d and $f(d(x,y))$ are **equivalent metrics** on X, for each metric space (X,d). A metric-preserving function is strongly metric-preserving if and only if it is continuous at 0.

- **Metric aggregating function**
  A function $f:\mathbb{R}^2_{\geq 0}\to\mathbb{R}_{\geq 0}$ with $f(a,b)=0$ if and only if $a=b=0$, is said to be **metric** (respectively, **quasi-metric**) **aggregating function** if the function $d_f : X\times X\to\mathbb{R}_{\geq 0}$ is a metric for every pair of metric spaces (respectively, a quasi-metric for every pair of quasi-metric spaces) (X_1,d_1) and (X_2,d_2), where $X=X_1\times X_2$ and, for all $(x,z),(y,w)\in X$, it holds

$$d_f((x,z),(y,w)) = f(d_1(x,z),d_2(y,w)).$$

  Borsík–Doboš, 1981, proved that a function f is metric aggregating if and only if, for all $a,b,c,a',b',c'\geq 0$ with $|a-b|\leq c\leq a+b$ and $|a'-b'|\leq c'\leq a'+b'$, it holds

$$|f(a,a')-f(b,b')|\leq f(c,c')\leq f(a,a')+f(b,b').$$

  Cf. **spin triangle inequality** in Chap. 15.
  Major–Valero, 2008, proved that a function f is quasi-metric aggregating if and only if it holds $f(a,a')\leq f(b,c')+f(c,b')$ for all $a,b,c,a',b',c'\geq 0$ such that $a\leq b+c$ and $a'\leq b'+c'$; so, any quasi-metric aggregating function is metric aggregating.
- **Metric generating function**
  A symmetric function $f:\mathbb{R}^2_{\geq 0}\to\mathbb{R}_{\geq 0}$ with $f(a,b)=0$ if and only if $a=b=0$, is said to be **metric generating** if the function defined by

$$d_f(x,y) = f(d(x,y),d(y,x))$$

  for all $x,y\in X$ is a metric on X for every quasi-metric space (X,d).
  Martin–Major–Valero, 2013, proved that a function f is metric generating if and only if it holds $f(a,a')\leq f(b,c')+f(c,b')$ for all a,b,c,a',b',c' such that $a\leq b+c, b\leq a+b', c\leq c'+a$ and $a'\leq b'+c', b'\leq a'+b, c'\leq c+a'$.
- **Power transform metric**
  Let $0<\alpha\leq 1$. Given a metric space (X,d), the **power** (or *α-snowflake*) **transform metric** is a **functional transform metric** on X defined by

$$(d(x,y))^{\alpha}.$$

  The distance $d(x,y)=(\sum_1^n|x_i-y_i|^p)^{\frac{1}{p}}$ with $0<p=\alpha<1$ is not a metric on $\mathbb{R}^n$, but its power transform $(d(x,y)^{\alpha})$ is a metric.
  For a given metric d on X and any $\alpha>1$, the function d^{α} is, in general, only a distance on X. It is a metric, for any positive α, if and only if d is an **ultrametric**.

A metric d is a **doubling metric** if and only if (Assouad, 1983) the power transform metric d^{α} admits a **bi-Lipschitz embedding** in some Euclidean space for every $0 < \alpha < 1$ (cf. Chap. 1 for definitions).

- **Quadrance**
  A distance which is a squared distance d^2 is called a **quadrance**.
  Rational trigonometry is the proposal (Wildberger, 2007) to use as its fundamental units, quadrance and *spread* (square of sine of angle), instead of distance and angle.
  It makes some problems easier to computers: solvable with only addition, subtraction, multiplication, and division, while avoiding square roots, sine, and cosine functions. Also, such trigonometry can be done over any field.
- **Schoenberg transform metric**
  Let $\lambda > 0$. Given a metric space (X, d), the **Schoenberg transform metric** is a **functional transform metric** on X defined by

$$1 - e^{-\lambda d(x,y)}.$$

  The Schoenberg transform metrics are exactly P**-metrics** (cf. Chap. 1).
- **Pullback metric**
  Given two metric spaces (X, d_X), (Y, d_Y) and an injective mapping $g : X \to Y$, the **pullback metric** (of (Y, d_Y) by g) on X is defined by

$$d_Y(g(x), g(y)).$$

  If $(X, d_X) = (Y, d_Y)$, then the pullback metric is called a g**-transform metric**.
- **Internal metric**
  Given a metric space (X, d) in which every pair of points x, y is joined by a *rectifiable curve*, the **internal metric** (or **inner metric**, induced **intrinsic metric**, **interior metric**) D is a **transform metric** on X, obtained from d as the infimum of the lengths of all rectifiable curves connecting two given points x and $y \in X$.
  The metric d is called an **intrinsic metric** (or **length metric** if it coincides with its internal metric. Cf. Chap. 6 and **metric curve** in Chap. 1.
- **Farris transform metric**
  Given a metric space (X, d) and a point $z \in X$, the **Farris transform** is a metric transform D_z on $X \backslash \{z\}$ defined by $D_z(x, x) = 0$ and, for different $x, y \in X \backslash \{z\}$, by

$$D_z(x, y) = C - (x.y)_z,$$

  where C is a positive constant, and $(x.y)_z = \frac{1}{2}(d(x, z) + d(y, z) - d(x, y))$ is the **Gromov product** (cf. Chap. 1). It is a metric if $C \geq \max_{x \in X \backslash \{z\}} d(x, z)$; in fact, there exists a number $C_0 \in (\max_{x,y \in X \backslash \{z\}, x \neq y}(x.y)_z, \max_{x \in X \backslash \{z\}} d(x, z)]$ such that it is a metric if and only if $C \geq C_0$. The Farris transform is an **ultrametric** if and only if d satisfies the **four-point inequality**. In Phylogenetics, where it was applied first, the term *Farris transform* is used for the function $d(x, y) - d(x,z) - d(y,z)$.

- **Involution transform metric**
Given a metric space (X, d) and a point $z \in X$, the **involution transform metric** is a metric transform d_z on $X \backslash \{z\}$ defined by

$$d_z(x, y) = \frac{d(x, y)}{d(x,z)d(y,z)}.$$

It is a metric for any $z \in X$, if and only if d is a **Ptolemaic metric** [FoSc06].

4.2 Metrics on Set Extensions

- **Extension distances**
If d is a metric on $V_n = \{1, \dots, n\}$, and $\alpha \in \mathbb{R}, \alpha > 0$, then the following extension distances (see, for example, [DeLa97]) are used.
The **gate extension distance** $gat = gat_\alpha^d$ is a metric on $V_{n+1} = \{1, \dots, n+1\}$ defined by the following conditions:

1. $gat(1, n+1) = \alpha$;
2. $gat(i, n+1) = \alpha + d(1, i)$ if $2 \le i \le n$;
3. $gat(i, j) = d(i, j)$ if $1 \le i < j \le n$.

The distance gat_0^d is called the **gate 0-extension** or, simply, **0-extension** of d.
If $\alpha \ge \max_{2 \le i \le n} d(1, i)$, then the **antipodal extension distance** $ant = ant_\alpha^d$ is a distance on V_{n+1} defined by the following conditions:

1. $ant(1, n+1) = \alpha$;
2. $ant(i, n+1) = \alpha - d(1, i)$ if $2 \le i \le n$;
3. $ant(i, j) = d(i, j)$ if $1 \le i < j \le n$.

If $\alpha \ge \max_{1 \le i,j \le n} d(i, j)$, then the **full antipodal extension distance** $Ant = Ant_\alpha^d$ is a distance on $V_{2n} = \{1, \dots, 2n\}$ defined by the following conditions:

1. $Ant(i, n+i) = \alpha$ if $1 \le i \le n$;
2. $Ant(i, n+j) = \alpha - d(i, j)$ if $1 \le i \ne j \le n$;
3. $Ant(i, j) = d(i, j)$ if $1 \le i \ne j \le n$;
4. $Ant(n+i, n+j) = d(i, j)$ if $1 \le i \ne j \le n$.

It is obtained by applying the antipodal extension operation iteratively n times, starting from d.
The **spherical extension distance** $sph = sph_\alpha^d$ is a metric on V_{n+1} defined by the following conditions:

1. $sph(i, n+1) = \alpha$ if $1 \le i \le n$;
2. $sph(i, j) = d(i, j)$ if $1 \le i < j \le n$.

- **1-sum distance**
Let d_1 be a distance on a set X_1, let d_2 be a distance on a set X_2, and suppose that $X_1 \cap X_2 = \{x_0\}$. The **1-sum distance** of d_1 and d_2 is the distance d on $X_1 \cup X_2$ defined by the following conditions:

$$d(x,y) = \begin{cases} d_1(x,y), & \text{if} \quad x,y \in X_1, \\ d_2(x,y), & \text{if} \quad x,y \in X_2, \\ d(x,x_0)+d(x_0,y), & \text{if} \ x \in X_1, y \in X_2. \end{cases}$$

In Graph Theory, the 1-sum distance is a **path metric**, corresponding to the clique 1-sum operation for graphs.
- **Disjoint union metric**
Given a family (X_t, d_t), $t \in T$, of metric spaces, the **disjoint union metric** is an **extended metric** on the set $\bigcup_t X_t \times \{t\}$ defined by

$$d((x,t_1),(y,t_2)) = d_t(x,y)$$

for $t_1 = t_2$, and $d((x,t_1),(y,t_2)) = \infty$, otherwise.
- **Metric bouquet**
Given a family (X_t, d_t), $t \in T$, of metric spaces with marked points x_t, the **metric bouquet** is obtained from their **disjoint union** by gluing all points x_t together.
- **Product metric**
Given finite or countable number n of metric spaces (X_1,d_1), (X_2,d_2), $\dots$, (X_n,d_n), the **product metric** is a metric on the *Cartesian product* $X_1 \times X_2 \times \dots \times X_n = \{x = (x_1,x_2,\dots,x_n) : x_1 \in X_1,\dots,x_n \in X_n\}$ defined as a function of $d_1,\dots,d_n$. The simplest finite product metrics are defined by

1. $\sum_{i=1}^{n} d_i(x_i,y_i)$;
2. $(\sum_{i=1}^{n} d_i^p(x_i,y_i))^{\frac{1}{p}}$, $1 < p < \infty$;
3. $\max_{1 \le i \le n} d_i(x_i,y_i)$;
4. $\sum_{i=1}^{n} \frac{1}{2^i} \frac{d_i(x_i,y_i)}{1+d_i(x_i,y_i)}$.

The last metric is **bounded** and can be extended to the product of countably many metric spaces.
If $X_1 = \dots = X_n = \mathbb{R}$, and $d_1 = \dots = d_n = d$, where $d(x,y) = |x-y|$ is the **natural metric** on $\mathbb{R}$, all product metrics above induce the Euclidean topology on the n-dimensional space $\mathbb{R}^n$. They do not coincide with the Euclidean metric on $\mathbb{R}^n$, but they are equivalent to it. In particular, the set $\mathbb{R}^n$ with the Euclidean metric can be considered as the Cartesian product $\mathbb{R} \times \dots \times \mathbb{R}$ of n copies of the *real line* $(\mathbb{R}, d)$ with the product metric defined by $\sqrt{\sum_{i=1}^{n} d^2(x_i,y_i)}$.
- **Box metric**
Let (X,d) be a metric space and I the unit interval of $\mathbb{R}$. The **box metric** is the **product metric** d' on the Cartesian product $X \times I$ defined by

$$d'((x_1,t_1),(x_2,t_2)) = \max(d(x_1,x_2),|t_1-t_2|).$$

Cf. unrelated **bounded box metric** in Chap. 18.

- **Fréchet product metric**
  Let (X,d) be a metric space with a **bounded** metric d. Let $X^\infty = X \times \cdots \times X \cdots = \{x = (x_1,\dots,x_n,\dots) : x_1 \in X_1,\dots,x_n \in X_n,\dots\}$ be the *countable Cartesian product space* of X.
  The **Fréchet product metric** is a **product metric** on X^∞ defined by

$$\sum_{n=1}^{\infty} A_n d(x_n,y_n),$$

  where $\sum_{n=1}^{\infty} A_n$ is any convergent series of positive terms. Usually, $A_n = \frac{1}{2^n}$ is used.
  A metric (sometimes called the *Fréchet metric*) on the set of all sequences $\{x_n\}_n$ of real (complex) numbers, defined by

$$\sum_{n=1}^{\infty} A_n \frac{|x_n-y_n|}{1+|x_n-y_n|},$$

  where $\sum_{n=1}^{\infty} A_n$ is any convergent series of positive terms, is a Fréchet product metric of countably many copies of $\mathbb{R}$ ($\mathbb{C}$). Usually, $A_n = \frac{1}{n!}$ or $A_n = \frac{1}{2^n}$ are used.
- **Hilbert cube metric**
  The *Hilbert cube* $I^{\aleph_0}$ is the *Cartesian product* of countable many copies of the interval [0, 1], equipped with the metric

$$\sum_{i=1}^{\infty} 2^{-i}|x_i-y_i|$$

  (cf. **Fréchet infinite metric product**). It also can be identified up to homeomorphisms with the compact metric space formed by all sequences $\{x_n\}_n$ of real numbers such that $0 \le x_n \le \frac{1}{n}$, where the metric is defined as $\sqrt{\sum_{n=1}^{\infty}(x_n-y_n)^2}$.
  The Cartesian products $[0,1]^\tau$ and $\{0,1\}^\tau$, where τ is an arbitrary cardinal number, are called a *Tikhonov cube* and *Cantor cube*, respectively.
- **Hamming cube**
  Given integers $n \ge 1$ and $q \ge 2$, the *Hamming space* $H(n,q)$ is the set of all n-tuples over an alphabet of size q (say, the *Cartesian product* of n copies of the set $\{0,1,\dots,q-1\}$), equipped with the **Hamming metric** (cf. Chap. 1), i.e., the distance between two n-tuples is the number of coordinates where they differ. The **Hamming cube** is the Hamming space $H(n,2)$.

The *infinite Hamming cube* $H(\infty, 2)$ is the set of all infinite strings over the alphabet $\{0, 1\}$ containing only finitely many 1's, equipped with the Hamming metric.

The *Fibonacci cube* $F(n)$ is the set of all n-tuples over $\{0, 1\}$ that contain no two consecutive 1's, equipped with the Hamming metric; it is a **partial cube** (cf. Chap. 15), i.e., an isometric subgraph of $H(n, 2)$. The *Lucas cube* $L(n)$ is obtained from $F(n)$ by removing n-tuples that start and end with 1.

- **Cameron–Tarzi cube**

  Given integers $n \geq 1$ and $q \geq 2$, the *normalized Hamming space* $H_n(q)$ is the set of all n-tuples over an alphabet of size q, equipped with the **Hamming metric** divided by n. Clearly, there are isometric embeddings

$$H_1(q) \to H_2(q) \to H_4(q) \to H_8(q) \to \dots$$

  Let $H(q)$ denote the **Cauchy completion** (cf. Chap. 1) of the union (denote it by $H_\omega(q)$) of all metric spaces $H_n(q)$ with $n \geq 1$. This metric space was introduced in [CaTa08]. Call $H(2)$ the **Cameron–Tarzi cube**.

  It is shown in [CaTa08] that $H_\omega(2)$ is the **word metric** space (cf. Chap. 10) of the *countable Nim group*, i.e., the elementary Abelian 2-group of all natural numbers under bitwise addition modulo 2 of the number expressions in base 2. The Cameron–Tarzi cube is also the word metric space of an Abelian group.

- **Rubik cube**

  There is a bijection between legal positions of the *Rubik* $3 \times 3 \times 3$ *cube* and elements of the subgroup G of the group Sym_{48} (of all permutations of $6(9-1)$ movable facets) generated by the 6 face rotations. The number of possible positions attainable by the cube is $|G| \approx 43 \times 10^{18}$.

  The maximum number of face turns needed to solve any instance of the Rubik cube is the diameter (maximal **word metric**), 20, of the *Cayley graph* of G.

- **Warped product metric**

  Let (X, d_X) and (Y, d_Y) be two complete **length spaces** (cf. Chap. 6), and let $f : X \to \mathbb{R}$ be a positive continuous function. Given a curve $\gamma : [a, b] \to X \times Y$, consider its projections $\gamma_1 : [a, b] \to X$ and $\gamma_2 : [a, b] \to Y$ to X and Y, and define the length of γ by the formula $\int_a^b \sqrt{|\gamma_1'|^2(t) + f^2(\gamma_1(t))|\gamma_2'|^2(t)}dt$.

  The **warped product metric** is a metric on $X \times Y$, defined as the infimum of lengths of all rectifiable curves connecting two given points in $X \times Y$ (see [BBI01]).

4.3 Metrics on Other Sets

Given a metric space (X, d), one can construct several distances between some subsets of X. The main such distances are: the **point-set distance** $d(x, A) = \inf_{y \in A} d(x, y)$ between a point $x \in X$ and a subset $A \subset X$, the **set-set distance**

$\inf_{x\in A, y\in B} d(x, y)$ between two subsets A and B of X, and the **Hausdorff metric** between compact subsets of X which are considered in Chap. 1. In this section we list some other distances of this kind.

- **Line-line distance**
The **line-line distance** (or *vertical distance between lines*) is the **set-set distance** in $\mathbb{E}^3$ between two *skew* lines, i.e., two straight lines that do not lie in a plane. It is the length of the segment of their common perpendicular whose endpoints lie on the lines. For l_1 and l_2 with equations l_1: $x = p + qt$, $t \in \mathbb{R}$, and l_2: $x = r + st$, $t \in \mathbb{R}$, the distance is given by

$$\frac{|\langle r - p, q \times s\rangle|}{||q \times s||_2},$$

where $\times$ is the *cross product* on $\mathbb{E}^3$, $\langle , \rangle$ is the *inner product* on $\mathbb{E}^3$, and $||.||_2$ is the Euclidean norm. For $x = (q_1, q_2, q_3)$, $s = (s_1, s_2, s_3)$, one has $q \times s = (q_2s_3 - q_3s_2, q_3s_1 - q_1s_3, q_1s_2 - q_2s_1)$.
- **Point-line distance**
The **point-line distance** is the **point-set distance** between a point and a line. In $\mathbb{E}^2$, the distance between a point $P = (x_1, y_1)$ and a line l: $ax + by + c = 0$ (in Cartesian coordinates) is the **perpendicular distance** given by

$$\frac{|ax_1 + by_1 + c|}{\sqrt{a^2 + b^2}}.$$

In $\mathbb{E}^3$, the distance between a point P and a line l: $x = p + qt$, $t \in \mathbb{R}$ (in vector formulation) is given by

$$\frac{||q \times (p - P)||_2}{||q||_2},$$

where $\times$ is the *cross product* on $\mathbb{E}^3$, and $||.||_2$ is the Euclidean norm.
- **Point-plane distance**
The **point-plane distance** is the **point-set distance** in $\mathbb{E}^3$ between a point $P = (x_1, y_1, z_1)$ and a plane α: $ax + by + cz + d = 0$ given by

$$\frac{|ax_1 + by_1 + cz_1 + d|}{\sqrt{a^2 + b^2 + c^2}}.$$

- **Prime number distance**
The **prime number distance** is the **point-set distance** in $(\mathbb{N}, |n - m|)$ between a number $n \in \mathbb{N}$ and the set of prime numbers $P \subset \mathbb{N}$. It is the absolute difference between n and the nearest prime number.

- **Distance up to nearest integer**
  The **distance up to nearest integer** is the **point-set distance** in $(\mathbb{R}, |x-y|)$ between a number $x \in \mathbb{R}$ and the set of integers $\mathbb{Z} \subset \mathbb{R}$, i.e., $\min_{n \in \mathbb{Z}} |x-n|$.
- **Busemann metric of sets**
  Given a metric space (X, d), the **Busemann metric of sets** (see [Buse55]) is a metric on the set of all nonempty closed subsets of X defined by

$$\sup_{x \in X} |d(x, A) - d(x, B)| e^{-d(p,x)},$$

  where $p \in X$ is fixed, and $d(x, A) = \min_{y \in A} d(x, y)$ is the **point-set distance**. Instead of the weighting factor $e^{-d(p,x)}$, one can take any distance transform function which decreases fast enough (cf. L_p**-Hausdorff distance** in Chap. 1, and the list of variations of the **Hausdorff metric** in Chap. 21).
- **Quotient semimetric**
  Given an **extended metric space** (X, d) (i.e., a possibly infinite metric) and an equivalence relation $\sim$ on X, the **quotient semimetric** is a semimetric on the set $\overline{X} = X/\sim$ of equivalence classes defined, for any $\overline{x}, \overline{y} \in \overline{X}$, by

$$\overline{d}(\overline{x}, \overline{y}) = \inf_{m \in \mathbb{N}} \sum_{i=1}^{m} d(x_i, y_i),$$

  where the infimum is taken over all sequences $x_1, y_1, x_2, y_2, \dots, x_m, y_m$ with $x_1 \in \overline{x}$, $y_m \in \overline{y}$, and $y_i \sim x_{i+1}$ for $i = 1, 2, \dots, m-1$. One has $\overline{d}(\overline{x}, \overline{y}) \leq d(x, y)$ for all $x, y \in X$, and $\overline{d}$ is the biggest semimetric on $\overline{X}$ with this property.

Chapter 5
Metrics on Normed Structures

In this chapter we consider a special class of metrics defined on some *normed structures*, as the norm of the difference between two given elements. This structure can be a group (with a *group norm*), a vector space (with a *vector norm* or, simply, a *norm*), a vector lattice (with a *Riesz norm*), a field (with a *valuation*), etc.

Any norm is *subadditive*, i.e., triangle inequality $||x + y|| \leq ||x|| + ||y||$ holds. A norm is *submultiplicative* if **multiplicative triangle inequality** $||xy|| \leq ||x||||y||$ holds.

- **Group norm metric**
  A **group norm metric** is a metric on a *group* $(G, +, 0)$ defined by

$$||x + (-y)|| = ||x - y||,$$

  where $||.||$ is a *group norm* on G, i.e., a function $||.|| : G \to \mathbb{R}$ such that, for all $x, y \in G$, we have the following properties:

  1. $||x|| \geq 0$, with $||x|| = 0$ if and only if $x = 0$;
  2. $||x|| = || - x||$;
  3. $||x + y|| \leq ||x|| + ||y||$ (triangle inequality).

  Any group norm metric d is **right-invariant**, i.e., $d(x, y) = d(x + z, y + z)$ for any $x, y, z \in G$. Conversely, any right-invariant (as well as any **left-invariant**, and, in particular, any **bi-invariant**) metric d on G is a group norm metric, since one can define a group norm on G by $||x|| = d(x, 0)$.
- ***F*-norm metric**
  A *vector space* (or *linear space*) over a *field* $\mathbb{F}$ is a set V equipped with operations of *vector addition* $+ : V \times V \to V$ and *scalar multiplication* $\cdot : \mathbb{F} \times V \to V$ such that $(V, +, 0)$ forms an *Abelian group* (where $0 \in V$ is the *zero vector*), and, for all *vectors* $x, y \in V$ and any *scalars* $a, b \in \mathbb{F}$, we have the following properties: $1 \cdot x = x$ (where 1 is the multiplicative unit of $\mathbb{F}$), $(ab) \cdot x = a \cdot (b \cdot x)$, $(a + b) \cdot x = a \cdot x + b \cdot x$, and $a \cdot (x + y) = a \cdot x + a \cdot y$.

M.M. Deza, E. Deza, *Encyclopedia of Distances*,
DOI 10.1007/978-3-662-44342-2_5

A vector space over the field $\mathbb{R}$ of real numbers is called a *real vector space*. A vector space over the field $\mathbb{C}$ of complex numbers is called *complex vector space*.

A F**-norm metric** is a metric on a real (complex) vector space V defined by

$$||x - y||_F,$$

where $||.||_F$ is an F-*norm* on V, i.e., a function $||.||_F : V \to \mathbb{R}$ such that, for all $x, y \in V$ and for any scalar a with $|a| = 1$, we have the following properties:

1. $||x||_F \geq 0$, with $||x||_F = 0$ if and only if $x = 0$;
2. $||ax||_F \leq ||x||_F$ if $|a| \leq 1$;
3. $\lim_{a \to 0} ||ax||_F = 0$;
4. $||x + y||_F \leq ||x||_F + ||y||_F$ (triangle inequality).

An F-norm is called *p-homogeneous* if $||ax||_F = |a|^p||x||_F$ for any scalar a.

Any F-norm metric d is a **translation invariant metric**, i.e., $d(x, y) = d(x + z, y + z)$ for all $x, y, z \in V$. Conversely, if d is a translation invariant metric on V, then $||x||_F = d(x, 0)$ is an F-norm on V.

- F^***-metric**

  An F^***-metric** is an F**-norm metric** $||x - y||_F$ on a real (complex) vector space V such that the operations of scalar multiplication and vector addition are continuous with respect to $||.||_F$. Thus $||.||_F$ is a function $||.||_F : V \to \mathbb{R}$ such that, for all $x, y, x_n \in V$ and for all scalars a, a_n, we have the following properties:

  1. $||x||_F \geq 0$, with $||x||_F = 0$ if and only if $x = 0$;
  2. $||ax||_F = ||x||_F$ for all a with $|a| = 1$;
  3. $||x + y||_F \leq ||x||_F + ||y||_F$;
  4. $||a_n x||_F \to 0$ if $a_n \to 0$;
  5. $||ax_n||_F \to 0$ if $x_n \to 0$;
  6. $||a_n x_n||_F \to 0$ if $a_n \to 0$, $x_n \to 0$.

  The metric space $(V, ||x - y||_F)$ with an F^*-metric is called a nF^***-space**. Equivalently, an F^*-space is a metric space (V, d) with a **translation invariant metric** d such that the operation of scalar multiplication and vector addition are continuous with respect to this metric.

  A **complete** F^*-space is called an F**-space**. A **locally convex** F-space is known as a **Fréchet space** (cf. Chap. 2) in Functional Analysis.

  A **modular space** is an F^*-space $(V, ||.||_F)$ in which the F-norm $||.||_F$ is defined by

  $$||x||_F = \inf\{\lambda > 0 : \rho\left(\frac{x}{\lambda}\right) < \lambda\},$$

  and ρ is a *metrizing modular* on V, i.e., a function $\rho : V \to [0, \infty]$ such that, for all $x, y, x_n \in V$ and for all scalars a, a_n, we have the following properties:

1. $\rho(x) = 0$ if and only if $x = 0$;
2. $\rho(ax) = \rho(x)$ implies $|a| = 1$;
3. $\rho(ax + by) \leq \rho(x) + \rho(y)$ implies $a, b \geq 0, a + b = 1$;
4. $\rho(a_n x) \to 0$ if $a_n \to 0$ and $\rho(x) < \infty$;
5. $\rho(ax_n) \to 0$ if $\rho(x_n) \to 0$ (*metrizing property*);
6. For any $x \in V$, there exists $k > 0$ such that $\rho(kx) < \infty$.

- **Norm metric**
  A **norm metric** is a metric on a real (complex) vector space V defined by

$$||x - y||,$$

where $||.||$ is a *norm* on V, i.e., a function $||.|| : V \to \mathbb{R}$ such that, for all $x, y \in V$ and for any scalar a, we have the following properties:

1. $||x|| \geq 0$, with $||x|| = 0$ if and only if $x = 0$;
2. $||ax|| = |a|||x||$;
3. $||x + y|| \leq ||x|| + ||y||$ (triangle inequality).

Therefore, a norm $||.||$ is a 1-*homogeneous* F-*norm*. The vector space $(V, ||.||)$ is called a *normed vector space* or, simply, *normed space*.
Any metric space can be embedded isometrically in some normed vector space as a closed linearly independent subset. Every finite-dimensional normed space is **complete**, and all norms on it are equivalent.
In general, the norm $||.||$ is equivalent (Maligranda, 2008) to the norm

$$||x||_{u,p} = (||x + ||x|| \cdot u||^p + ||x - ||x|| \cdot u||^p)^{\frac{1}{p}},$$

introduced, for any $u \in V$ and $p \geq 1$, by Odell and Schlumprecht, 1998.
The **norm-angular distance** between x and y is defined (Clarkson, 1936) by

$$d(x, y) = ||\frac{x}{||x||} - \frac{y}{||y||}||.$$

The following sharpening of the triangle inequality (Maligranda, 2003) holds:

$$\frac{||x - y|| - |||x|| - ||y|||}{\min\{||x||, ||y||\}} \leq d(x, y) \leq \frac{||x - y|| + |||x|| - ||y|||}{\max\{||x||, ||y||\}}, \text{ i.e.,}$$

$$(2 - d(x, -y)) \min\{||x||, ||y||\} \leq ||x|| + ||y|| - ||x + y||$$
$$\leq (2 - d(x, -y)) \max\{||x||, ||y||\}.$$

Dragomir, 2004, call $|\int_a^b f(x)dx| \leq \int_a^b |f(x)|dx$ *continuous triangle inequality*.

- **Reverse triangle inequality**
  The triangle inequality $||x + y|| \leq ||x|| + ||y||$ in a normed space $(V, ||.||)$ is equivalent to the following inequality, for any $x_1, \ldots, x_n \in V$ with $n \geq 2$:

$$||\sum_{i=1}^{n} x_i|| \leq \sum_{i=1}^{n} ||x_i||.$$

If in the normed space $(V, ||.||)$, for some $C \geq 1$ one has

$$C||\sum_{i=1}^{n} x_i|| \geq \sum_{i=1}^{n} ||x_i||,$$

then this inequality is called the **reverse triangle inequality**.
This term is used, sometimes, also for the **inverse triangle inequality** (cf. **kinematic metric** in Chap. 26) and for the eventual inequality $Cd(x,z) \geq d(x,y) + d(y,z)$ with $C \geq 1$ in a metric space (X, d).
The triangle inequality $||x + y|| \leq ||x|| + ||y||$, for any $x, y \in V$, in a normed space $(V, ||.||)$ is, for any number $q > 1$, equivalent (Belbachir, Mirzavaziri and Moslenian, 2005) to the following inequality:

$$||x + y||^q \leq 2^{q-1}(||x||^q + ||y||^q).$$

The *parallelogram inequality* $||x + y||^2 \leq 2(||x||^2 + ||y||^2)$ is the case $q = 2$ of above.
Given a number q, $0 < q \leq 1$, the norm is called *q-subadditive* if $||x + y||^q \leq ||x||^q + ||y||^q$ holds for $x, y \in V$.

- **Seminorm semimetric**
A **seminorm semimetric** on a real (complex) vector space V is defined by

$$||x - y||,$$

where $||.||$ is a *seminorm* (or *pseudo-norm*) on V, i.e., a function $||.|| : V \rightarrow \mathbb{R}$ such that, for all $x, y \in V$ and for any scalar a, we have the following properties:

1. $||x|| \geq 0$, with $||0|| = 0$;
2. $||ax|| = |a|||x||$;
3. $||x + y|| \leq ||x|| + ||y||$ (triangle inequality).

The vector space $(V, ||.||)$ is called a *seminormed vector space*. Many *normed vector spaces*, in particular, **Banach spaces**, are defined as the quotient space by the subspace of elements of seminorm zero.
A *quasi-normed space* is a vector space V, on which a *quasi-norm* is given. A *quasi-norm* on V is a nonnegative function $||.|| : V \rightarrow \mathbb{R}$ which satisfies the same axioms as a norm, except for the triangle inequality which is replaced by the weaker requirement: there exists a constant $C > 0$ such that, for all $x, y \in V$, the following **C-triangle inequality** (cf. **near-metric** in Chap. 1) holds:

$$||x + y|| \leq C(||x|| + ||y||)$$

An example of a quasi-normed space, that is not normed, is the *Lebesgue space* $L_p(\Omega)$ with $0 < p < 1$ in which a quasi-norm is defined by

$$||f|| = \left(\int_{\Omega} |f(x)|^p dx\right)^{1/p}, f \in L_p(\Omega).$$

- **Banach space**

  A **Banach space** (or *B-space*) is a **complete** metric space $(V, ||x - y||)$ on a vector space V with a norm metric $||x - y||$. Equivalently, it is the complete *normed space* $(V, ||.||)$. In this case, the norm $||.||$ on V is called the *Banach norm*. Some examples of Banach spaces are:

  1. l_p^n*-spaces*, l_p^∞*-spaces*, $1 \le p \le \infty$, $n \in \mathbb{N}$;
  2. The space C of convergent numerical sequences with the norm $||x|| = \sup_n |x_n|$;
  3. The space C_0 of numerical sequences which converge to zero with the norm $||x|| = \max_n |x_n|$;
  4. The space $C_{[a,b]}^p$, $1 \le p \le \infty$, of continuous functions on $[a, b]$ with the L_p*-norm* $||f||_p = (\int_a^b |f(t)|^p dt)^{\frac{1}{p}}$;
  5. The space C_K of continuous functions on a compactum K with the norm $||f|| = \max_{t \in K} |f(t)|$;
  6. The space $(C_{[a,b]})^n$ of functions on $[a, b]$ with continuous derivatives up to and including the order n with the norm $||f||_n = \sum_{k=0}^n \max_{a \le t \le b} |f^{(k)}(t)|$;
  7. The space $C^n[I^m]$ of all functions defined in an m-dimensional cube that are continuously differentiable up to and including the order n with the norm of uniform boundedness in all derivatives of order at most n;
  8. The space $M_{[a,b]}$ of bounded measurable functions on $[a, b]$ with the norm

  $$||f|| = ess \sup_{a \le t \le b} |f(t)| = \inf_{e, \mu(e)=0} \sup_{t \in [a,b] \setminus e} |f(t)|;$$

  9. The space $A(\Delta)$ of functions analytic in the open *unit disk* $\Delta = \{z \in \mathbb{C} : |z| < 1\}$ and continuous in the closed disk $\overline{\Delta}$ with the norm $||f|| = \max_{z \in \overline{\Delta}} |f(z)|$;
  10. The **Lebesgue spaces** $L_p(\Omega)$, $1 \le p \le \infty$;
  11. The *Sobolev spaces* $W^{k,p}(\Omega)$, $\Omega \subset \mathbb{R}^n$, $1 \le p \le \infty$, of functions f on Ω such that f and its derivatives, up to some order k, have a finite L_p*-norm*, with the norm $||f||_{k,p} = \sum_{i=0}^k ||f^{(i)}||_p$;
  12. The *Bohr space AP* of almost periodic functions with the norm

  $$||f|| = \sup_{-\infty < t < +\infty} |f(t)|.$$

  A finite-dimensional real Banach space is called a *Minkowskian space*. A norm metric of a Minkowskian space is called a **Minkowskian metric** (cf. Chap. 6). In particular, any l_p**-metric** is a Minkowskian metric.

All n-dimensional Banach spaces are pairwise isomorphic; the set of such spaces becomes compact if one introduces the **Banach–Mazur distance** by $d_{BM}(V, W) = \ln \inf_T ||T||\cdot||T^{-1}||$, where the infimum is taken over all operators which realize an isomorphism $T : V \to W$.

- **l_p-metric**
The **l_p-metric** d_{l_p}, $1 \leq p \leq \infty$, is a norm metric on $\mathbb{R}^n$ (or on $\mathbb{C}^n$), defined by

$$||x - y||_p,$$

where the l_p-*norm* $||.||_p$ is defined by

$$||x||_p = (\sum_{i=1}^{n} |x_i|^p)^{\frac{1}{p}}.$$

For $p = \infty$, we obtain $||x||_\infty = \lim_{p\to\infty} \sqrt[p]{\sum_{i=1}^{n} |x_i|^p} = \max_{1\leq i\leq n} |x_i|$. The metric space $(\mathbb{R}^n, d_{l_p})$ is abbreviated as l_p^n and is called l_p^n-*space*.
The **l_p-metric**, $1 \leq p \leq \infty$, on the set of all sequences $x = \{x_n\}_{n=1}^{\infty}$ of real (complex) numbers, for which the sum $\sum_{i=1}^{\infty} |x_i|^p$ (for $p = \infty$, the sum $\sum_{i=1}^{\infty} |x_i|$) is finite, is

$$(\sum_{i=1}^{\infty} |x_i - y_i|^p)^{\frac{1}{p}}.$$

For $p = \infty$, we obtain $\max_{i\geq 1} |x_i - y_i|$. This metric space is abbreviated as l_p^∞ and is called l_p^∞-*space*.
Most important are l_1-, l_2- and l_∞-metrics. Among l_p-metrics, only l_1- and l_∞-metrics are **crystalline metrics**, i.e., metrics having polygonal unit balls. On $\mathbb{R}$ all l_p-metrics coincide with the **natural metric** (cf. Chap. 12) $|x - y|$.
The l_2-norm $||(x_1, x_2)||_2 = \sqrt{x_1^2 + x_2^2}$ on $\mathbb{R}^2$ is also called *Pythagorean addition* of the numbers x_1 and x_2. Under this commutative operation, $\mathbb{R}$ form a semigroup, and $\mathbb{R}_{\geq 0}$ form a *monoid* (semigroup with identity, 0).

- **Euclidean metric**
The **Euclidean metric** (or **Pythagorean distance**, **as-the-crow-flies distance**, **beeline distance**) d_E is the metric on $\mathbb{R}^n$ defined by

$$||x - y||_2 = \sqrt{(x_1 - y_1)^2 + \cdots + (x_n - y_n)^2}.$$

It is the ordinary **l_2-metric** on $\mathbb{R}^n$. The metric space $(\mathbb{R}^n, d_E)$ is abbreviated as $\mathbb{E}^n$ and is called **Euclidean space** "Euclidean space" stands for the case $n = 3$, as opposed, for $n = 2$, to *Euclidean plane* and, for $n = 1$, *Euclidean* (or *real*) *line*.
In fact, $\mathbb{E}^n$ is an **inner product space** (and even a **Hilbert space**), i.e., $d_E(x, y) = ||x - y||_2 = \sqrt{\langle x - y, x - y\rangle}$, where $\langle x, y\rangle$ is the *inner product*

on $\mathbb{R}^n$ which is given in the Cartesian coordinate system by $\langle x, y\rangle = \sum_{i=1}^{n} x_i y_i$. In a standard coordinate system one has $\langle x, y\rangle = \sum_{i,j} g_{ij} x_i y_j$, where $g_{ij} = \langle e_i, e_j\rangle$, and the **metric tensor** $((g_{ij}))$ (cf. Chap. 7) is a positive-definite symmetric $n \times n$ matrix.

In general, a Euclidean space is defined as a space, the properties of which are described by the axioms of *Euclidean Geometry*.

- **Norm transform metric**
  A **norm transform metric** is a metric $d(x, y)$ on a vector space $(V, ||.||)$, which is a function of $||x||$ and $||y|$. Usually, $V = \mathbb{R}^n$ and, moreover, $\mathbb{E}^n = (\mathbb{R}^n, ||.||_2)$. Some examples are **$(p, q)$-relative metric**, **$M$-relative metric** and, from Chap. 19, the **British Rail metric** $||x|| + ||y||$ for $x \neq y$, (and equal to 0, otherwise), the **radar screen metric** $\min\{1, ||x - y||\}$ and $\max\{1, ||x - y||\}$ for $x \neq y$. Cf. **t-truncated** and **t-uniformly discrete** metrics in Chap. 4.
- **(p, q)-relative metric**
  Let $0 < q \leq 1$, and $p \geq \max\{1 - q, \frac{2-q}{3}\}$. Let $(V, ||.||)$ be a *Ptolemaic space*, i.e., the norm metric $||x - y||$ is a **Ptolemaic metric** (cf. Chap. 1).
  The **(p, q)-relative metric** on $(V, ||.||)$ is defined, for x or $y \neq 0$, by

$$\frac{||x - y||}{(\frac{1}{2}(||x||^p + ||y||^p))^{\frac{q}{p}}}$$

  (and equal to 0, otherwise). In the case of $p = \infty$, it has the form

$$\frac{||x - y||}{(\max\{||x||, ||y||\})^q}.$$

  $(p, 1)$-, $(\infty, 1)$- and the original $(1, 1)$-relative metric on $\mathbb{E}^n$ are called **p-relative** (or **Klamkin–Meir metric**), **relative metric** and **Schattschneider metric**.
- **M-relative metric**
  Let $f : [0, \infty) \to (0, \infty)$ be a convex increasing function such that $\frac{f(x)}{x}$ is decreasing for $x > 0$. Let $(V, ||.||)$ be a *Ptolemaic space*, i.e., $||x - y||$ is a **Ptolemaic metric**.
  The **M-relative metric** on $(V, ||.||)$ is defined by

$$\frac{||x - y||}{f(||x||) \cdot f(||y||)}.$$

- **Unitary metric**
  The **unitary** (or *complex Euclidean*) **metric** is the **l_2-metric** on $\mathbb{C}^n$ defined by

$$||x - y||_2 = \sqrt{|x_1 - y_1|^2 + \cdots + |x_n - y_n|^2}.$$

For $n = 1$, it is the **complex modulus metric** $|x - y| = \sqrt{(x-y)\overline{(x-y)}}$ on the *Wessel–Argand plane* (cf. Chap. 12).

- **L_p-metric**
An **L_p-metric** d_{L_p}, $1 \le p \le \infty$, is a norm metric on $L_p(\Omega, \mathcal{A}, \mu)$ defined by

$$||f - g||_p$$

for any $f, g \in L_p(\Omega, \mathcal{A}, \mu)$. The metric space $(L_p(\Omega, \mathcal{A}, \mu), d_{L_p})$ is called the **L_p-space** (or **Lebesgue space**).
Here Ω is a set, and $\mathcal{A}$ is n *σ-algebra* of subsets of Ω, i.e., a collection of subsets of Ω satisfying the following properties:

1. $\Omega \in \mathcal{A}$;
2. If $A \in \mathcal{A}$, then $\Omega \backslash A \in \mathcal{A}$;
3. If $A = \cup_{i=1}^{\infty} A_i$ with $A_i \in \mathcal{A}$, then $A \in \mathcal{A}$.

A function $\mu : \mathcal{A} \to \mathbb{R}_{\ge 0}$ is called a *measure* on $\mathcal{A}$ if it is *additive*, i.e., $\mu(\cup_{i \ge 1} A_i) = \sum_{i \ge 1} \mu(A_i)$ for all pairwise disjoint sets $A_i \in \mathcal{A}$, and satisfies $\mu(\emptyset) = 0$. A *measure space* is a triple $(\Omega, \mathcal{A}, \mu)$.
Given a function $f : \Omega \to \mathbb{R}(\mathbb{C})$, its *$L_p$-norm* is defined by

$$||f||_p = \left(\int_{\Omega} |f(\omega)|^p \mu(d\omega) \right)^{\frac{1}{p}}.$$

Let $L_p(\Omega, \mathcal{A}, \mu) = L_p(\Omega)$ denote the set of all functions $f : \Omega \to \mathbb{R}\ (\mathbb{C})$ such that $||f||_p < \infty$. Strictly speaking, $L_p(\Omega, \mathcal{A}, \mu)$ consists of equivalence classes of functions, where two functions are *equivalent* if they are equal *almost everywhere*, i.e., the set on which they differ has measure zero. The set $L_\infty(\Omega, \mathcal{A}, \mu)$ is the set of equivalence classes of measurable functions $f : \Omega \to \mathbb{R}\ (\mathbb{C})$ whose absolute values are bounded almost everywhere.
The most classical example of an L_p-metric is d_{L_p} on the set $L_p(\Omega, \mathcal{A}, \mu)$, where Ω is the open interval $(0, 1)$, $\mathcal{A}$ is the *Borel σ-algebra* on $(0, 1)$, and μ is the *Lebesgue measure*. This metric space is abbreviated by $L_p(0, 1)$ and is called *$L_p(0, 1)$-space*.
In the same way, one can define the L_p-metric on the set $C_{[a,b]}$ of all real (complex) continuous functions on $[a, b]$: $d_{L_p}(f, g) = (\int_a^b |f(x) - g(x)|^p dx)^{\frac{1}{p}}$. For $p = \infty$, $d_{L_\infty}(f, g) = \max_{a \le x \le b} |f(x) - g(x)|$. This metric space is abbreviated by $C_{[a,b]}^p$ and is called *$C_{[a,b]}^p$-space*.
If $\Omega = \mathbb{N}$, $\mathcal{A} = 2^{\Omega}$ is the collection of all subsets of Ω, and μ is the *cardinality measure* (i.e., $\mu(A) = |A|$ if A is a finite subset of Ω, and $\mu(A) = \infty$, otherwise), then the metric space $(L_p(\Omega, 2^{\Omega}, |.|), d_{L_p})$ coincides with the space l_p^{∞}.
If $\Omega = V_n$ is a set of cardinality n, $\mathcal{A} = 2^{V_n}$, and μ is the cardinality measure, then the metric space $(L_p(V_n, 2^{V_n}, |.|), d_{L_p})$ coincides with the space l_p^n.

- **Dual metrics**
The l_p**-metric** and the l_q**-metric**, $1 < p, q < \infty$, are called **dual** if $1/p + 1/q = 1$.
In general, when dealing with a *normed vector space* $(V, ||.||_V)$, one is interested in the *continuous* linear functionals from V into the base field ($\mathbb{R}$ or $\mathbb{C}$). These functionals form a **Banach space** $(V', ||.||_{V'})$, called the *continuous dual* of V. The norm $||.||_{V'}$ on V' is defined by $||T||_{V'} = \sup_{||x||_V \le 1} |T(x)|$.
The continuous dual for the metric space l_p^n (l_p^∞) is l_q^n (l_q^∞, respectively). The continuous dual of l_1^n (l_1^∞) is l_∞^n (l_∞^∞, respectively). The continuous duals of the Banach spaces C (consisting of all convergent sequences, with l_∞**-metric**) and C_0 (consisting of the sequences converging to zero, with l_∞**-metric**) are both naturally identified with l_1^∞.
- **Inner product space**
An **inner product space** (or *pre-Hilbert space*) is a metric space $(V, ||x - y||)$ on a real (complex) vector space V with an *inner product* $\langle x, y\rangle$ such that the norm metric $||x - y||$ is constructed using the *inner product norm* $||x|| = \sqrt{\langle x, x\rangle}$.
An *inner product* $\langle , \rangle$ on a real (complex) vector space V is a *symmetric bilinear* (in the complex case, *sesquilinear*) form on V, i.e., a function $\langle , \rangle : V \times V \longrightarrow \mathbb{R}$ ($\mathbb{C}$) such that, for all $x, y, z \in V$ and for all scalars α, β, we have the following properties:

1. $\langle x, x\rangle \ge 0$, with $\langle x, x\rangle = 0$ if and only if $x = 0$;
2. $\langle x, y\rangle = \overline{\langle y, x\rangle}$, where the bar denotes *complex conjugation*;
3. $\langle \alpha x + \beta y, z\rangle = \alpha\langle x, z\rangle + \beta\langle y, z\rangle$.

For a complex vector space, an inner product is called also a *Hermitian inner product*, and the corresponding metric space is called a *Hermitian inner product space*.
A norm $||.||$ in a *normed space* $(V, ||.||)$ is generated by an inner product if and only if, for all $x, y \in V$, we have: $||x + y||^2 + ||x - y||^2 = 2(||x||^2 + ||y||^2)$.
In an inner product space, the **triangle equality** (Chap. 1) $||x - y|| = ||x|| + ||y||$, for $x, y \neq 0$, holds if and only if $\frac{x}{||x||} = \frac{y}{||y||}$, i.e., $x - y \in [x, y]$.
- **Hilbert space**
A **Hilbert space** is an **inner product space** which, as a metric space, is **complete**. More precisely, a Hilbert space is a complete metric space $(H, ||x - y||)$ on a real (complex) vector space H with an *inner product* $\langle , \rangle$ such that the norm metric $||x - y||$ is constructed using the *inner product norm* $||x|| = \sqrt{\langle x, x\rangle}$.
Any Hilbert space is a **Banach space**.
An example of a Hilbert space is the set of all sequences $x = \{x_n\}_n$ of real (complex) numbers such that $\sum_{i=1}^\infty |x_i|^2$ converges, with the **Hilbert metric** defined by

$$(\sum_{i=1}^{\infty}(x_i - y_i)^2)^{\frac{1}{2}}.$$

Other examples of Hilbert spaces are any L_2**-space**, and any finite-dimensional inner product space. In particular, any Euclidean space is a Hilbert space.
A direct product of two **Hilbert spaces** is called a *Liouville space* (or *line space*, *extended Hilbert space*).
Given an infinite cardinal number τ and a set A of the cardinality τ, let $\mathbb{R}_a$, $a \in A$, be the copies of $\mathbb{R}$. Let $H(A) = \{\{x_a\} \in \prod_{a \in A} \mathbb{R}_a : \sum_a x_a^2 < \infty\}$; then $H(A)$ with the metric defined for $\{x_a\}, \{y_a\} \in H(A)$ as

$$(\sum_{a \in A}(x_a - y_a)^2)^{\frac{1}{2}},$$

is called the **generalized Hilbert space** of weight τ.

- **Erdös space**
  The **Erdös space** (or *rational Hilbert space*) is the metric subspace of l_2 consisting of all vectors in l_2 with only rational coordinates. It has topological dimension 1 and is not complete. Erdös space is **homeomorphic** to its countable infinite power, and every nonempty open subset of it is homeomorphic to whole space.
  The **complete Erdös space** (or *irrational Hilbert space*) is the complete metric subspace of l_2 consisting of all vectors in l_2 the coordinates of which are all irrational.
- **Riesz norm metric**
  A *Riesz space* (or *vector lattice*) is a partially ordered vector space $(V_{Ri}, \preceq)$ in which the following conditions hold:

  1. The vector space structure and the partial order structure are compatible, i.e., from $x \preceq y$ it follows that $x + z \preceq y + z$, and from $x \succ 0$, $a \in \mathbb{R}$, $a > 0$ it follows that $ax \succ 0$;
  2. For any two elements $x, y \in V_{Ri}$, there exist the *join* $x \vee y \in V_{Ri}$ and *meet* $x \wedge y \in V_{Ri}$ (cf. Chap. 10).

  The **Riesz norm metric** is a norm metric on V_{Ri} defined by

  $$||x - y||_{Ri},$$

  where $||.||_{Ri}$ is a *Riesz norm* on V_{Ri}, i.e., a norm such that, for any $x, y \in V_{Ri}$, the inequality $|x| \preceq |y|$, where $|x| = (-x) \vee (x)$, implies $||x||_{Ri} \leq ||y||_{Ri}$.
  The space $(V_{Ri}, ||.||_{Ri})$ is called a *normed Riesz space*. In the case of completeness, it is called a *Banach lattice*.
- **Banach–Mazur compactum**
  The **Banach–Mazur distance** d_{BM} between two n-dimensional *normed spaces* $(V, ||.||_V)$ and $(W, ||.||_W)$ is defined by

  $$\ln \inf_T ||T|| \cdot ||T^{-1}||,$$

where the infimum is taken over all isomorphisms $T : V \to W$. It is a metric on the set X^n of all equivalence classes of n-dimensional normed spaces, where $V \sim W$ if and only if they are *isometric*. Then the pair (X^n, d_{BM}) is a compact metric space which is called the **Banach–Mazur compactum**.

- **Quotient metric**
  Given a *normed space* $(V, ||.||_V)$ with a norm $||.||_V$ and a closed subspace W of V, let $(V/W, ||.||_{V/W})$ be the normed space of cosets $x+W = \{x+w : w \in W\}$, $x \in V$, with the *quotient norm* $||x + W||_{V/W} = \inf_{w\in W} ||x + w||_V$.
  The **quotient metric** is a norm metric on V/W defined by

$$||(x + W) - (y + W)||_{V/W}.$$

- **Tensor norm metric**
  Given *normed spaces* $(V, ||.||_V)$ and $(W, ||.||_W)$, a norm $||.||_\otimes$ on the *tensor product* $V \otimes W$ is called *tensor norm* (or *cross norm*) if $||x \otimes y||_\otimes = ||x||_V ||y||_W$ for all *decomposable* tensors $x \otimes y$.
  The **tensor product metric** is a norm metric on $V \otimes W$ defined by

$$||z - t||_\otimes.$$

  For any $z \in V \otimes W$, $z = \sum_j x_j \otimes y_j$, $x_j \in V$, $y_j \in W$, the *projective norm* (or π-*norm*) of z is defined by $||z||_{pr} = \inf \sum_j ||x_j||_V ||y_j||_W$, where the infimum is taken over all representations of z as a sum of decomposable vectors. It is the largest tensor norm on $V \otimes W$.
- **Valuation metric**
  A **valuation metric** is a metric on a *field* $\mathbb{F}$ defined by

$$||x - y||,$$

  where $||.||$ is a *valuation* on $\mathbb{F}$, i.e., a function $||.|| : \mathbb{F} \to \mathbb{R}$ such that, for all $x, y \in \mathbb{F}$, we have the following properties:

  1. $||x|| \geq 0$, with $||x|| = 0$ if and only if $x = 0$;
  2. $||xy|| = ||x|| \, ||y||$,
  3. $||x + y|| \leq ||x|| + ||y||$ (triangle inequality).

  If $||x + y|| \leq \max\{||x||, ||y||\}$, the valuation $||.||$ is called *non-Archimedean*. In this case, the valuation metric is an **ultrametric**. The simplest valuation is the *trivial valuation* $||.||_{tr}$: $||0||_{tr} = 0$, and $||x||_{tr} = 1$ for $x \in \mathbb{F}\backslash\{0\}$. It is non-Archimedean.
  There are different definitions of valuation in Mathematics. Thus, the function $\nu : \mathbb{F} \to \mathbb{R} \cup \{\infty\}$ is called a *valuation* if $\nu(x) \geq 0$, $\nu(0) = \infty$, $\nu(xy) = \nu(x) + \nu(y)$, and $\nu(x + y) \geq \min\{\nu(x), \nu(y)\}$ for all $x, y \in \mathbb{F}$. The valuation $||.||$ can be obtained from the function ν by the formula $||x|| = \alpha^{\nu(x)}$ for some fixed $0 < \alpha < 1$ (cf. p**-adic metric** in Chap. 12).

The *Kürschák valuation* $|.|_{Krs}$ is a function $|.|_{Krs} : \mathbb{F} \to \mathbb{R}$ such that $|x|_{Krs} \geq 0$, $|x|_{Krs} = 0$ if and only if $x = 0$, $|xy|_{Krs} = |x|_{Krs}|y|_{Krs}$, and $|x + y|_{Krs} \leq C \max\{|x|_{Krs}, |y|_{Krs}\}$ for all $x, y \in \mathbb{F}$ and for some positive constant C, called the *constant of valuation*. If $C \leq 2$, one obtains the ordinary valuation $||.||$ which is non-Archimedean if $C \leq 1$. In general, any $|.|_{Krs}$ is *equivalent* to some $||.||$, i.e., $|.|^p_{Krs} = ||.||$ for some $p > 0$.
Finally, given an *ordered group* $(G, \cdot, e, \leq)$ equipped with zero, the *Krull valuation* is a function $|.| : \mathbb{F} \to G$ such that $|x| = 0$ if and only if $x = 0$, $|xy| = |x||y|$, and $|x + y| \leq \max\{|x|, |y|\}$ for any $x, y \in \mathbb{F}$. It is a generalization of the definition of non-Archimedean valuation $||.||$ (cf. **generalized metric** in Chap. 3).

- **Power series metric**
Let $\mathbb{F}$ be an arbitrary algebraic field, and let $\mathbb{F}\langle x^{-1}\rangle$ be the field of power series of the form $w = \alpha_{-m}x^m + \cdots + \alpha_0 + \alpha_1 x^{-1} + \ldots$, $\alpha_i \in \mathbb{F}$. Given $l > 1$, a *non-Archimedean valuation* $||.||$ on $\mathbb{F}\langle x^{-1}\rangle$ is defined by

$$||w|| = \begin{cases} l^m, & \text{if } w \neq 0, \\ 0, & \text{if } w = 0. \end{cases}$$

The **power series metric** is the **valuation metric** $||w - v||$ on $\mathbb{F}\langle x^{-1}\rangle$.

Part II
Geometry and Distances

Chapter 6
Distances in Geometry

Geometry arose as the field of knowledge dealing with spatial relationships. It was one of the two fields of pre-modern Mathematics, the other being the study of numbers.

Earliest known evidence of abstract representation—ochre rocks marked with cross hatches and lines to create a consistent complex geometric motif, dated about 75,000 BC—were found in Blombos Cave, South Africa. In modern times, geometric concepts have been generalized to a high level of abstraction and complexity.

6.1 Geodesic Geometry

In Mathematics, the notion of "geodesic" is a generalization of the notion of "straight line" to curved spaces. This term is taken from *Geodesy*, the science of measuring the size and shape of the Earth.

Given a metric space (X, d), a **metric curve** γ is a continuous function $\gamma : I \rightarrow X$, where I is an *interval* (i.e., nonempty connected subset) of $\mathbb{R}$. If γ is r times continuously differentiable, it is called a *regular curve* of class C^r; if $r = \infty$, γ is called a *smooth curve*.

In general, a curve may cross itself. A curve is called a *simple curve* (or *arc*, *path*) if it does not cross itself, i.e., if it is injective. A curve $\gamma : [a, b] \rightarrow X$ is called a *Jordan curve* (or *simple closed curve*) if it does not cross itself, and $\gamma(a) = \gamma(b)$.

The *length* (which may be equal to ∞) $l(\gamma)$ of a curve $\gamma : [a, b] \rightarrow X$ is defined by $\sup \sum_{i=1}^{n} d(\gamma(t_{i-1}), \gamma(t_i))$, where the supremum is taken over all finite decompositions $a = t_0 < t_1 < \ldots < t_n = b$, $n \in \mathbb{N}$, of $[a, b]$.

A curve with finite length is called *rectifiable*. For each regular curve $\gamma : [a, b] \rightarrow X$ define the *natural parameter* s of γ by $s = s(t) = l(\gamma|_{[a,t]})$, where $l(\gamma|_{[a,t]})$ is the length of the part of γ corresponding to the interval $[a, t]$. A curve with this *natural*

M.M. Deza, E. Deza, *Encyclopedia of Distances*,
DOI 10.1007/978-3-662-44342-2_6

parametrization $\gamma = \gamma(s)$ is called **of unit speed**, (or *parametrized by arc length, normalized*); in this parametrization, for any $t_1, t_2 \in I$, one has $l(\gamma|_{[t_1,t_2]}) = |t_2 - t_1|$, and $l(\gamma) = |b - a|$.

The length of any curve $\gamma : [a, b] \to X$ is at least the distance between its endpoints: $l(\gamma) \geq d(\gamma(a), \gamma(b))$. The curve γ, for which $l(\gamma) = d(\gamma(a), \gamma(b))$, is called the **geodesic segment** (or *shortest path*) from $x = \gamma(a)$ to $y = \gamma(b)$, and denoted by $[x, y]$.

Thus, a geodesic segment is a shortest join of its endpoints; it is an isometric embedding of $[a, b]$ in X. In general, geodesic segments need not exist, unless the segment consists of one point only. A geodesic segment joining two points need not be unique.

A **geodesic** (cf. Chap. 1) is a curve which extends indefinitely in both directions and behaves locally like a segment, i.e., is everywhere locally a distance minimizer.

More exactly, a curve $\gamma : \mathbb{R} \to X$, given in the natural parametrization, is called a *geodesic* if, for any $t \in \mathbb{R}$, there exists a *neighborhood* U of t such that, for any $t_1, t_2 \in U$, we have $d(\gamma(t_1), \gamma(t_2)) = |t_1 - t_2|$. Thus, any geodesic is a locally isometric embedding of the whole of $\mathbb{R}$ in X.

A geodesic is called a **metric straight line** if the equality $d(\gamma(t_1), \gamma(t_2)) = |t_1 - t_2|$ holds for all $t_1, t_2 \in \mathbb{R}$. Such a geodesic is an isometric embedding of the whole real line $\mathbb{R}$ in X. A geodesic is called a **metric great circle** if it is an isometric embedding of a circle $S^1(0, r)$ in X. In general, geodesics need not exist.

- **Geodesic metric space**
  A metric space (X, d) is called **geodesic** if any two points in X can be joined by a **geodesic segment**, i.e., for any two points $x, y \in X$, there is an isometry from the segment $[0, d(x, y)]$ into X. Examples of geodesic spaces are complete *Riemannian spaces*, **Banach spaces** and **metric graphs** from Chap. 15.
  A metric space (X, d) is called a **locally geodesic metric space** if any two sufficiently close points in X can be joined by a geodesic segment; it is called D**-geodesic** if any two points at distance $< D$ can be joined by a geodesic segment.
- **Geodesic distance**
  The **geodesic distance** (or **shortest path distance**) is the length of a **geodesic segment** (i.e., a *shortest path*) between two points.
- **Intrinsic metric**
  Given a metric space (X, d) in which every two points are joined by a rectifiable curve, the **internal metric** (cf. Chap. 4) D on X is defined as the infimum of the lengths of all rectifiable curves, connecting two given points $x, y \in X$.
  The metric d on X is called the **intrinsic metric** (or **length metric**) if it coincides with its internal metric D. A metric space with the intrinsic metric is called a **length space** (or **path metric space**, *inner metric space*, *intrinsic space*).
  If, moreover, any pair x, y of points can be joined by a curve of length $d(x, y)$, the intrinsic metric d is called *strictly intrinsic*, and the length space (X, d) is a **geodesic metric space** (or *shortest path metric space*).

A complete metric space (X, d) is a length space if and only if it is having **approximate midpoints**, i.e., for any points $x, y \in X$ and for any $\epsilon > 0$, there exists a third point $z \in X$ with $d(x, z), d(y, z) \leq \frac{1}{2}d(x, y)+\epsilon$. A complete metric space (X, d) is a **geodesic metric space** if and only if it is having **midpoints**. Any complete locally compact length space is a **proper** geodesic metric space.

- **G-space**

A G-**space** (or **space of geodesics**) is a metric space (X, d) with the geometry characterized by the fact that extensions of geodesics, defined as locally shortest lines, are unique. Such geometry is a generalization of *Hilbert Geometry* (see [Buse55]).

More exactly, a G-space (X, d) is defined by the following conditions:

1. It is **proper** (or *finitely compact*), i.e., all metric balls are compact;
2. It is **Menger-convex**, i.e., for any different $x, y \in X$, there exists a third point $z \in X, z \neq x, y$, such that $d(x, z) + d(z, y) = d(x, y)$;
3. It is *locally extendable*, i.e., for any $a \in X$, there exists $r > 0$ such that, for any distinct points x, y in the ball $B(a, r)$, there exists z distinct from x and y such that $d(x, y) + d(y, z) = d(x, z)$;
4. It is *uniquely extendable*, i.e., if in 3 above two points z_1 and z_2 were found, so that $d(y, z_1) = d(y, z_2)$, then $z_1 = z_2$.

The existence of geodesic segments is ensured by finite compactness and Menger-convexity: any two points of a finitely compact Menger-convex set X can be joined by a geodesic segment in X. The existence of geodesics is ensured by the axiom of local prolongation: if a finitely compact Menger-convex set X is locally extendable, then there exists a geodesic containing a given segment. Finally, the uniqueness of prolongation ensures the assumption of Differential Geometry that a *line element* determines a geodesic uniquely.

All *Riemannian* and *Finsler spaces* are G-spaces. A 1D G-space is a metric straight line or metric great circle. Any 2D G-space is a **topological manifold** (Chap. 2).

Every G-space is a **chord metric space**, i.e., a metric space with a set distinguished geodesic segments such that any two points are joined by a unique such segment (see [BuPh87]).

- **Desarguesian space**

A **Desarguesian space** is a G-**space** (X, d) in which the role of geodesics is played by ordinary straight lines. Thus, X may be topologically mapped into a *projective space* $\mathbb{R}P^n$ so that each geodesic of X is mapped into a straight line of $\mathbb{R}P^n$.

Any X mapped into $\mathbb{R}P^n$ must either cover all of $\mathbb{R}P^n$ and, in such a case, the geodesics of X are all metric great circles of the same length, or X may be considered as an open convex subset of an affine space A^n.

A space (X, d) of geodesics is a Desarguesian space if and only if the following conditions hold:

1. The geodesic passing through two different points is unique;
2. For dimension $n = 2$, both the direct and the converse *Desargues theorems* are valid and, for dimension $n > 2$, any three points in X lie in one plane.

Among *Riemannian spaces*, the only Desarguesian spaces are Euclidean, *hyperbolic*, and *elliptic* spaces. An example of the non-Riemannian Desarguesian space is the *Minkowskian space* which can be regarded as the prototype of all non-Riemannian spaces, including *Finsler spaces*.

- **G-space of elliptic type**
  A **G-space of elliptic type** is a **G-space** in which the geodesic through two points is unique, and all geodesics are the metric great circles of the same length. Every G-space such that there is unique geodesic through each given pair of points is either a G-space of elliptic type, or a **straight G-space**.
- **Straight G-space**
  A **straight G-space** is a **G-space** in which extension of a geodesic is possible globally, so that any segment of the geodesic remains a shortest path. In other words, for any two points $x, y \in X$, there is a unique geodesic segment joining x to y, and a unique metric straight line containing x and y.
  Any geodesic in a straight G-space is a metric straight line, and is uniquely determined by any two of its points. Any such 2D space is homeomorphic to the plane.
  All simply connected *Riemannian spaces* of nonpositive curvature (including Euclidean and *hyperbolic* spaces), *Hilbert geometries*, and Teichmüller spaces of compact Riemann surfaces of genus $g > 1$ (when metrized by the **Teichmüller metric**) are straight G-spaces.
- **Gromov hyperbolic metric space**
  A metric space (X, d) is called **Gromov hyperbolic** if it is **geodesic** and **δ-hyperbolic** for some $\delta \geq 0$.
  An important class of such spaces are the *hyperbolic groups*, i.e., finitely generated groups whose **word metric** is Gromov hyperbolic. A metric space is a **real tree** exactly when it is 0-hyperbolic.
  Every bounded metric space X is (diam(X,d))-hyperbolic. A normed vector space is Gromov hyperbolic if and only it has dimension 1. Any complete simply connected *Riemannian space* of *sectional curvature* $k \leq -a^2 < 0$ is $\frac{\ln 3}{a}$-hyperbolic. Every **CAT(κ)** space with $\kappa < 0$ is Gromov hyperbolic.
- **CAT(κ) space**
  Let (X, d) be a metric space. Let M^2 be a simply connected 2D *Riemannian manifold* (cf. Chap. 7) of *constant curvature* κ, i.e., the 2-sphere S^2_κ with $\kappa > 0$, the Euclidean plane $\mathbb{E}^2$ with $\kappa = 0$, or the hyperbolic plane H^2_κ with $\kappa < 0$. Let D_κ denote the *diameter* of M^2, i.e., $D_\kappa = \frac{\pi}{\sqrt{\kappa}}$ if $\kappa > 0$, and $D_\kappa = \infty$ if $\kappa \leq 0$.
  A *triangle* T in X consists of three points in X together with three *geodesic segments* joining them pairwise; the segments are called the *sides of the triangle*. For a triangle $T \subset X$, a *comparison triangle* for T in M^2 is a triangle $T' \subset M^2$ together with a map f_T which sends each side of T isometrically onto a side

of T'. A triangle T is said (Gromov, 1987) to satisfy the **CAT(κ) inequality** (for Cartan, Alexandrov and Toponogov) if, for every $x, y \in T$, we have

$$d(x, y) \leq d_{M^2}(f_T(x), f_T(y)),$$

where f_T is the map associated to a comparison triangle for T in M^2. So, the geodesic triangle T is at least as "thin" as its comparison triangle in M^2.

The metric space (X, d) is a **CAT(κ) space** if it is D_κ**-geodesic** (i.e., any two points at distance $< D_\kappa$ can be joined by a geodesic segment), and all triangles T with perimeter $< 2D_\kappa$ satisfy the CAT(κ) inequality.

Every CAT(κ_1) space is a CAT(κ_2) space if $\kappa_1 < \kappa_2$. Every **real tree** is a CAT($-\infty$) *space*, i.e., is a CAT(κ_1) space for all $\kappa \in \mathbb{R}$.

A **locally CAT(κ) space** (called **metric space with curvature** $\leq\kappa$ in Alexandrov, 1951) is a metric space (X, d) in which every point $p \in X$ has a neighborhood U such that any two points $x, y \in U$ are connected by a geodesic segment, and the CAT(κ) inequality holds for any $x, y, z \in U$. A Riemannian manifold is locally CAT(κ) if and only if its *sectional curvature* is at most κ.

A **metric space with curvature** $\geq\kappa$ is (Alexandrov, 1951) a metric space (X, d) in which every $p \in X$ has a neighborhood U such that any $x, y \in U$ are connected by a geodesic segment, and the *reverse CAT(κ) inequality*

$$d(x, y) \geq d_{M^2}(f_T(x), f_T(y))$$

holds for any $x, y, z \in U$, where f_T is the map associated to a comparison triangle for T in M^2. It is a **generalized Riemannian space** (cf. Chap. 7).

Above two definitions differ only by the sign of $d(x, y) - d_{M^2}(f_T(x), f_T(y))$. In the case $\kappa = 0$, the above spaces are called **nonpositively curved** and **nonnegatively curved** metric spaces, respectively. For complete metric spaces, they differ also (Bruhat–Tits, 1972) by the sign ($\leq$0 or $\geq$0, respectively) of

$$F(x, y, z) = 4d^2(z, m(x, y)) - (d^2(z, x) + d^2(z, y) - d^2(x, y)),$$

where x, y, z are any three points and $m(x, y)$ is the midpoint of the **metric interval** $I(x, y)$. A complete CAT(0) space is called **Hadamard space**.

The inequality $F(x, y, z) \leq 0$ for all $x, y, z \in X$, characterizing Hadamard spaces, is called *semiparallelogram inequality*, because the usual vector parallelogram law $||u-v||^2+||u+v||^2 = 2||u||^2+2||v||^2$, characterizing norms induced by inner products, is equivalent to the equality $F(x, y, z) = 0$. A normed space is an Hadamard space if and only if it is a **Hilbert space**.

Every two points in an Hadamard space are connected by a unique geodesic (and hence unique shortest path), while in a general CAT(0) space, they are connected by a unique geodesic segment, and the distance is a convex function.

Foertsch–Lytchack–Schroeder, 2007, proved that a metric space is CAT(0) if and only if it is **Busemann convex** and **Ptolemaic**; cf. Chap. 1. Euclidean spaces, hyperbolic spaces, and trees are CAT(0) spaces.

- **δ-Bolic metric space**
  Given a number $\delta > 0$, a metric space (X, d) is called **δ-bolic** (Kasparov–Skandalis, 1994, simplified by Bucher–Karlsson, 2002) if for any $x, y, z \in X$ and some function $m : X \times X \to X$, it holds
  $$2d(z, m(x, y)) \leq \sqrt{2d^2(z, x) + 2d^2(z, y) - d^2(x, y)} + \frac{4}{3}\delta.$$
  A **δ-hyperbolic space** with **approximate δ-midpoints** (Chap. 1) is $\frac{3\delta}{2}$-bolic. Every CAT(0)-space is δ-bolic for any $\delta > 0$; for complete spaces the converse holds as well. An l_p-metric space of dimension > 1 is δ-bolic for any $\delta > 0$ only if $p = 2$.
- **Boundary of metric space**
  There are many notions of the **boundary ∂X of a metric space** (X, d). We give below some of the most general among them. Usually, if (X, d) is locally compact, $X \cup \partial X$ is its *compactification*.

  1. **Ideal boundary** (or *boundary at ∞*). Given a geodesic metric space (X, d), let γ^1 and γ^2 be two **metric rays**, i.e., geodesics with isometry of $\mathbb{R}_{\geq 0}$ into X. These rays are called *equivalent* if the **Hausdorff distance** between them (associated with the metric d) is finite, i.e., if $\sup_{t \geq 0} d(\gamma^1(t), \gamma^2(t)) < \infty$. The **ideal boundary** of (X, d) is the set $\partial_\infty X$ of equivalence classes γ_∞ of all metric rays. Cf. **asymptotic metric cone** (Chap. 1).
     If (X, d) is a complete CAT(0) space, then the **Tits metric** (or *asymptotic angle of divergence*) on $\partial_\infty X$ is defined by $2 \arcsin\left(\frac{\rho}{2}\right)$ for all $\gamma^1_\infty, \gamma^2_\infty \in \partial_\infty X$, where $\rho = \lim_{t \to \infty} \frac{1}{t} d(\gamma^1(t), \gamma^2(t))$. The set $\partial_\infty X$ equipped with the Tits metric is called the **Tits boundary** of X.
     If (X, d, x_0) is a pointed complete CAT(-1) space, then the **Bourdon metric** (or **visual distance**) on $\partial_\infty X$ is defined, for any distinct $x, y \in \partial_\infty X$, by $e^{-(x.y)}$, where $(x.y)$ denotes the **Gromov product** $(x.y)_{x_0}$.
     The **visual sphere of (X, d) at a point** $x_0 \in X$ is the set of equivalence classes of all metric rays emanating from x_0.
  2. **Gromov boundary**. Given a **pointed metric space** (X, d, x_0) (i.e., one with a selected base point $x_0 \in X$), the **Gromov boundary** of it (as generalized by Buckley and Kokkendorff, 2005, from the case of the Gromov hyperbolic space) is the set $\partial_G X$ of equivalence classes of *Gromov sequences*.
     A sequence $x = \{x_n\}_n$ in X is a *Gromov sequence* if the Gromov product $(x_i.x_j)_{x_0} \to \infty$ as $i, j \to \infty$. Two Gromov sequences x and y are *equivalent* if there is a finite chain of Gromov sequences x^k, $0 \leq k \leq k'$, such that $x = x^0$, $y = x^{k'}$, and $\lim_{i,j \to \infty} \inf(x^{k-1}_i.x^k_j) = \infty$ for $0 \leq k \leq k'$.
     In a **proper** geodesic Gromov hyperbolic space (X, d), the visual sphere does not depends on the base point x_0 and is naturally isomorphic to its *Gromov boundary* $\partial_G X$ which can be identified with $\partial_\infty X$.
  3. **g-Boundary**. Denote by $\overline{X_d}$ the metric completion of (X, d) and, viewing X as a subset of $\overline{X_d}$, denote by ∂X_d the difference $\overline{X_d} \backslash X$. Let (X, l, x_0) be a

pointed unbounded **length space**, i.e., its metric coincides with the **internal metric** l of (X,d). Given a measurable function $g : \mathbb{R}_{\geq 0} \to \mathbb{R}_{\geq 0}$, the g-**boundary** of (X,d,x_0) (as generalized by Buckley–Kokkendorff, 2005, from *spherical* and *Floyd boundaries*) is $\partial_g X = \partial X_\sigma \backslash \partial X_l$, where $\sigma(x,y) = \inf \int_\gamma g(z)dl(z)$ for all $x,y \in X$ (here the infimum is taken over all metric segments $\gamma = [x,y]$).

4. **Hotchkiss boundary**. Given a pointed proper **Busemann convex** metric space (X,d,x_0), the **Hotchkiss boundary** of it is the set $\partial_H(X,x_0)$ of isometries $f : \mathbb{R}_{\geq 0} \to X$ with $f(0) = x_0$. The boundaries $\partial_H^{x_0} X$ and $\partial_H^{x_1} X$ are homeomorphic for distinct $x_0, x_1 \in X$. In a Gromov hyperbolic space, $\partial_H^{x_0} X$ is homeomorphic to the Gromov boundary $\partial_G X$.
5. **Metric boundary**. Given a pointed metric space (X,d,x_0) and an unbounded subset S of $\mathbb{R}_{\geq 0}$, a ray $\gamma : S \to X$ is called a *weakly geodesic ray* if, for every $x \in X$ and every $\epsilon > 0$, there is an integer N such that $|d(\gamma(t),\gamma(0))-t| < \epsilon$, and $|d(\gamma(t),x) - d(\gamma(s),x) - (t-s)| < \epsilon$ for all $s,t \in T$ with $s,t \geq N$.
Let $\mathcal{G}(X,d)$ be the *commutative unital* C^**-algebra* with the norm $||.||_\infty$, generated by the (bounded, continuous) functions which vanish at infinity, the constant functions, and the functions of the form $g_y(x) = d(x,x_0) - d(x,y)$; cf. **Rieffel metric space** in Chap. 7 for definitions.
The Rieffel's **metric boundary** $\partial_R X$ of (X,d) is the difference $\overline{X}^d \backslash X$, where $\overline{X}^d$ is the *metric compactification* of (X,d), i.e., the maximum ideal space (the set of *pure states*) of this C^*-algebra.
For a proper metric space (X,d) (cf. Chap. 1) with a countable base, the boundary $\partial_R X$ consists of the limits $\lim_{t\to\infty} f(\gamma(t))$ for every weakly geodesic ray γ and every function f from the above C^*-algebra (Rieffel, 2002).

- **Projectively flat metric space**
A metric space, in which geodesics are defined, is called **projectively flat** if it locally admits a *geodesic mapping* (or *projective mapping*), i.e., a mapping preserving geodesics into an Euclidean space. Cf. Euclidean **rank of metric space** in Chap. 1; similar terms are: *affinely flat*, *conformally flat*, etc.
A Riemannian space is projectively flat if and only if it has constant (sectional) curvature. Cf. **flat metric** in Chap. 8.

6.2 Projective Geometry

Projective Geometry is a branch of Geometry dealing with the properties and invariants of geometric figures under *projection*. Affine Geometry, Metric Geometry and Euclidean Geometry are subsets of Projective Geometry of increasing complexity. The main invariants of Projective, Affine, Metric, Euclidean Geometry are, respectively, **cross-ratio**, parallelism (and relative distances), angles (and relative distances), absolute distances.

An n-dimensional *projective space* $\mathbb{F}P^n$ is the space of one-dimensional vector subspaces of a given $(n+1)$-dimensional vector space V over a field $\mathbb{F}$. The basic construction is to form the set of equivalence classes of nonzero vectors in V under the relation of scalar proportionality. This idea goes back to mathematical descriptions of *perspective*.

The use of a basis of V allows the introduction of *homogeneous coordinates* of a point in $\mathbb{F}P^n$ which are usually written as $(x_1 : x_2 : \dots : x_n : x_{n+1})$—a vector of length $n+1$, other than $(0 : 0 : 0 : \dots : 0)$. Two sets of coordinates that are proportional denote the same point of the projective space. Any point of projective space which can be represented as $(x_1 : x_2 : \dots : x_n : 0)$ is called a *point at infinity*. The part of a projective space $\mathbb{F}P^n$ not "at infinity", i.e., the set of points of the projective space which can be represented as $(x_1 : x_2 : \dots : x_n : 1)$, is an n-dimensional *affine space* A^n.

The notation $\mathbb{R}P^n$ denotes the *real projective space* of dimension n, i.e., the space of 1D vector subspaces of $\mathbb{R}^{n+1}$. The notation $\mathbb{C}P^n$ denotes the *complex projective space* of dimension n. The projective space $\mathbb{R}P^n$ carries a natural structure of a compact smooth n-**manifold**. It can be viewed as the space of lines through the zero element 0 of $\mathbb{R}^{n+1}$ (i.e., as a *ray space*). It can be viewed also as the set $\mathbb{R}^n$, considered as an *affine space*, together with its points at infinity. Also it can be seen as the set of points of an n-sphere in $\mathbb{R}^{n+1}$ with identified diametrically-opposite points.

The projective points, projective straight lines, projective planes, ..., projective hyperplanes of $\mathbb{F}P^n$ are one-, two-, three-, ..., n-dimensional subspaces of V, respectively. Any two projective straight lines in a projective plane have one and only one common point. A *projective transformation* (or *collineation*, *projectivity*) is a bijection of a projective space onto itself, preserving collinearity (the property of points to be on one line) in both directions. Any projective transformation is a composition of a pair of *perspective projections*. Projective transformations do not preserve sizes or angles but do preserve *type* (that is, points remain points, and lines remain lines), *incidence* (that is, whether a point lies on a line), and **cross-ratio** (cf. Chap. 1).

Here, given four collinear points $x, y, z, t \in \mathbb{F}P^n$, their **cross-ratio** (x, y, z, t) is $\frac{(x-z)(y-t)}{(y-z)(x-t)}$, where $\frac{x-z}{x-t}$ denotes the ratio $\frac{f(x)-f(z)}{f(x)-f(t)}$ for some affine bijection f from the straight line $l_{x,y}$ through the points x and y onto $\mathbb{F}$.

Given four projective straight lines l_x, l_y, l_z, l_t, containing points x, y, z, t, respectively, and passing through a given point, their **cross-ratio** (l_x, l_y, l_z, l_t) is $\frac{\sin(l_x,l_z)\sin(l_y,l_t)}{\sin(l_y,l_z)\sin(l_x,l_t)}$, coincides with (x, y, z, t). The cross-ratio (x, y, z, t) of four complex numbers x, y, z, t is $\frac{(x-z)(y-t)}{(y-z)(x-t)}$. It is real if and only if the four numbers are either collinear or concyclic.

- **Projective metric**
  Given a convex subset D of a projective space $\mathbb{R}P^n$, the **projective metric** d is a metric on D such that shortest paths with respect to this metric are parts

of or entire projective straight lines. It is assumed that the following conditions hold:

1. D does not belong to a hyperplane;
2. For any three noncollinear points $x, y, z \in D$, the triangle inequality holds in the strict sense: $d(x, y) < d(x, z) + d(z, y)$;
3. If x, y are different points in D, then the intersection of the straight line $l_{x,y}$ through x and y with D is either all of $l_{x,y}$, and forms a **metric great circle**, or is obtained from $l_{x,y}$ by discarding some segment (which can be reduced to a point), and forms a **metric straight line**.

The metric space (D, d) is called a **projective metric space**. The problem of determining all projective metrics on $\mathbb{R}^n$ (called **linearly additive metrics** in Chap. 1) is the *4th problem of Hilbert*; it has been solved only for $n = 2$. In fact, given a smooth measure on the space of hyperplanes in $\mathbb{R}P^n$, define the distance between any two points $x, y \in \mathbb{R}P^n$ as one-half the measure of all hyperplanes intersecting the line segment joining x and y. The obtained metric is projective; it is the *Busemann's construction* of projective metrics. [Amba76] proved that all projective metrics on $\mathbb{R}^2$ can be obtained by this construction.

In a projective metric space there cannot simultaneously be both types of straight lines: they are either all metric straight lines, or they are all metric great circles of the same length (*Hamel's theorem*). Spaces of the first kind are called *open*. They coincide with subspaces of an affine space; the geometry of open projective metric spaces is a *Hilbert Geometry*. *Hyperbolic Geometry* is a Hilbert Geometry in which there exist reflections at all straight lines.

Thus, the set D has Hyperbolic Geometry if and only if it is the interior of an ellipsoid. The geometry of open projective metric spaces whose subsets coincide with all of affine space, is a *Minkowski Geometry*. *Euclidean Geometry* is a Hilbert Geometry and a Minkowski Geometry, simultaneously. Spaces of the second kind are called *closed*; they coincide with the whole of $\mathbb{R}P^n$. *Elliptic Geometry* is the geometry of a projective metric space of the second kind.

- **Strip projective metric**

  The **strip projective metric** [BuKe53] is a **projective metric** on the strip $St = \{x \in \mathbb{R}^2 : -\pi/2 < x_2 < \pi/2\}$ defined by

$$\sqrt{(x_1 - y_1)^2 + (x_2 - y_2)^2} + |\tan x_2 - \tan y_2|.$$

  The Euclidean metric $\sqrt{(x_1 - y_1)^2 + (x_2 - y_2)^2}$ is not a projective metric on *St*.

- **Half-plane projective metric**

  The **half-plane projective metric** [BuKe53] is a **projective metric** on $\mathbb{R}^2_+ = \{x \in \mathbb{R}^2 : x_2 > 0\}$ defined by

$$\sqrt{(x_1 - y_1)^2 + (x_2 - y_2)^2} + \left|\frac{1}{x_2} - \frac{1}{y_2}\right|.$$

- **Hilbert projective metric**
  Given a set H, the **Hilbert projective metric** h is a **complete projective metric** on H. It means that H contains, together with two arbitrary distinct points x and y, also the points z and t for which $h(x,z) + h(z,y) = h(x,y)$, $h(x,y) + h(y,t) = h(x,t)$, and that H is homeomorphic to a convex set in an n-dimensional affine space A^n, the geodesics in H being mapped to straight lines of A^n.
  The metric space (H,h) is called the *Hilbert projective space*, and the geometry of a Hilbert projective space is called *Hilbert Geometry*.
  Formally, let D be a nonempty convex open set in A^n with the boundary ∂D not containing two proper coplanar but noncollinear segments (ordinarily the boundary of D is a strictly convex closed curve, and D is its interior). Let $x, y \in D$ be located on a straight line which intersects ∂D at z and t, z is on the side of y, and t is on the side of x. Then the **Hilbert projective metric** h on D is the symmetrization of the **Funk distance** (cf. Chap. 1):

$$h(x,y) = \frac{1}{2}\left(\ln \frac{x-z}{y-z} + \ln \frac{x-t}{y-t}\right) = \frac{1}{2}\ln(x,y,z,t),$$

  where (x,y,z,t) is the **cross-ratio** of x,y,z,t.
  The metric space (D,h) is a **straight G-space**. If D is an ellipsoid, then h is the **hyperbolic metric**, and defines *Hyperbolic Geometry* on D. On the *unit disk* $\Delta = \{z \in \mathbb{C} : |z| < 1\}$ the metric h coincides with the **Cayley–Klein–Hilbert metric**. If $n = 1$, the metric h makes D isometric to the Euclidean line.
  If ∂D contains coplanar but noncollinear segments, a projective metric on D can be given by $h(x,y) + d(x,y)$, where d is any **Minkowskian metric**.
- **Minkowskian metric**
  The **Minkowskian metric** (or **Minkowski–Hölder distance**) is the **norm metric** of a finite-dimensional real **Banach space**.
  Formally, let $\mathbb{R}^n$ be an n-dimensional real vector space, let K be a *symmetric convex body* in $\mathbb{R}^n$, i.e., an open neighborhood of the origin which is bounded, convex, and *symmetric* ($x \in K$ if and only if $-x \in K$). Then the **Minkowski distance function** $||x||_K : \mathbb{R}^n \to [0,\infty)$, defined as $\inf\{\alpha > 0 : \frac{x}{\alpha} \in \partial K\}$ (cf. Chap. 1), is a *norm* on $\mathbb{R}^n$, and the Minkowskian metric m_K on $\mathbb{R}^n$ is defined by

$$m_K(x,y) = ||x-y||_K.$$

  The metric space $(\mathbb{R}^n, m)$ is called *Minkowskian space*; its geometry is *Minkowski Geometry*. It can be seen as an affine space A^n with a metric m in which the *unit ball* is the body K. For a strictly convex symmetric body the Minkowskian metric is a **projective metric**, and $(\mathbb{R}^n, m)$ is a **G-straight space**. A Minkowski Geometry is Euclidean if and only if its *unit sphere* is an ellipsoid.
  The Minkowskian metric m is proportional to the Euclidean metric d_E on every given line l, i.e., $m(x,y) = \phi(l)d_E(x,y)$. Thus, the Minkowskian metric can be considered as a metric which is defined in the whole affine space A^n and has the

property that the *affine ratio* $\frac{ac}{ab}$ of any three collinear points a, b, c (cf. Sect. 6.3) is equal to their *distance ratio* $\frac{m(a,c)}{m(a,b)}$.
Given a convex body C in a Minkowskian space with unit ball K, the *Minkowskian thickness* and *Minkowskian diameter* of C are (Averkov, 2003):

$$\sup\{\alpha > 0 : \alpha K \subset C - C\} \text{ and } \inf\{\alpha > 0 : C - C \subset \alpha K\}.$$

- **C-distance**
Given a convex body $C \subset \mathbb{E}^n$, the **C-distance** (or *relative distance*; Lassak, 1991) is a distance on $\mathbb{E}^n$ defined, for any $x, y \in \mathbb{E}^n$, by

$$d_C(x, y) = 2\frac{d_E(x, y)}{d_E(x', y')},$$

where $x'y'$ is the longest chord of C parallel to the segment xy. C-distance is not related to C-metric in Chap. 10 and to **rotating C-metric** in Chap. 26.
The unit ball of the normed space with the norm $||x|| = d_C(x, 0)$ is $\frac{1}{2}(C - C)$. For every $r \in [-1, 1]$, it holds $d_C(x, y) = d_{rC+(1-r)(-C)}(x, y)$.
- **Busemann metric**
The **Busemann metric** [Buse55] is a metric on the real n-dimensional projective space $\mathbb{R}P^n$ defined by

$$\min\left\{\sum_{i=1}^{n+1}\left|\frac{x_i}{||x||} - \frac{y_i}{||y||}\right|, \sum_{i=1}^{n+1}\left|\frac{x_i}{||x||} + \frac{y_i}{||y||}\right|\right\}$$

for any $x = (x_1 : \ldots : x_{n+1}), y = (y_1 : \ldots : y_{n+1}) \in \mathbb{R}P^n$, where $||x|| = \sqrt{\sum_{i=1}^{n+1} x_1^2}$.
- **Flag metric**
Given an n-dimensional *projective space* $\mathbb{F}P^n$, the **flag metric** d is a metric on $\mathbb{F}P^n$ defined by a *flag*, i.e., an *absolute* consisting of a collection of m-planes α_m, $m = 0, \ldots, n-1$, with α_{i-1} belonging to α_i for all $i \in \{1, \ldots, n-1\}$. The metric space $(\mathbb{F}P^n, d)$ is abbreviated by F^n and is called a *flag space*.
If one chooses an affine coordinate system $(x_i)_i$ in a space F^n, so that the vectors of the lines passing through the $(n-m-1)$-plane α_{n-m-1} are defined by the condition $x_1 = \ldots x_m = 0$, then the flag metric $d(x, y)$ between the points $x = (x_1, \ldots, x_n)$ and $y = (y_1, \ldots, y_n)$ is defined by

$$d(x, y) = |x_1 - y_1|, \text{ if } x_1 \neq y_1,\ d(x, y) = |x_2 - y_2|, \text{ if } x_1 = y_1, x_2 \neq y_2, \ldots$$
$$\ldots, d(x, y) = |x_k - y_k|, \text{ if } x_1 = y_1, \ldots, x_{k-1} = y_{k-1}, x_k \neq y_k, \ldots .$$

- **Projective determination of a metric**
The **projective determination of a metric** is an introduction, in subsets of a projective space, of a metric such that these subsets become isomorphic to a Euclidean, *hyperbolic*, or *elliptic space*.
To obtain a *Euclidean determination of a metric* in $\mathbb{R}P^n$, one should distinguish in this space an $(n-1)$-dimensional hyperplane π, called the *hyperplane at infinity*, and define $\mathbb{E}^n$ as the subset of the projective space obtained by removing from it this hyperplane π. In terms of homogeneous coordinates, π consists of all points $(x_1 : \ldots : x_n : 0)$, and $\mathbb{E}^n$ consists of all points $(x_1 : \ldots : x_n : x_{n+1})$ with $x_{n+1} \neq 0$. Hence, it can be written as $\mathbb{E}^n = \{x \in \mathbb{R}P^n : x = (x_1 : \ldots : x_n : 1)\}$. The Euclidean metric d_E on $\mathbb{E}^n$ is defined by
$$\sqrt{\langle x-y, x-y\rangle},$$
where, for any $x = (x_1 : \ldots : x_n : 1), y = (y_1 : \ldots : y_n : 1) \in \mathbb{E}^n$, one has $\langle x, y\rangle = \sum_{i=1}^{n} x_i y_i$.
To obtain a *hyperbolic determination of a metric* in $\mathbb{R}P^n$, a set D of interior points of a real oval hypersurface Ω of order two in $\mathbb{R}P^n$ is considered. The **hyperbolic metric** d_{hyp} on D is defined by
$$\frac{r}{2}|\ln(x, y, z, t)|,$$
where z and t are the points of intersection of the straight line $l_{x,y}$ through the points x and y with Ω, (x, y, z, t) is the **cross-ratio** of the points x, y, z, t, and $r > 0$ is a fixed constant. If, for any $x = (x_1 : \ldots : x_{n+1}), y = (y_1 : \ldots : y_{n+1}) \in \mathbb{R}P^n$, the *scalar product* $\langle x, y\rangle = -x_1y_1 + \sum_{i=2}^{n+1} x_i y_i$ is defined, the hyperbolic metric on the set $D = \{x \in \mathbb{R}P^n : \langle x, x\rangle < 0\}$ can be written, for a fixed constant $r > 0$, as
$$r \operatorname{arccosh} \frac{|\langle x, y\rangle|}{\sqrt{\langle x, x\rangle\langle y, y\rangle}},$$
where arccosh denotes the nonnegative values of the inverse hyperbolic cosine.
To obtain an *elliptic determination of a metric* in $\mathbb{R}P^n$, one should consider, for any $x = (x_1 : \ldots : x_{n+1}), y = (y_1 : \ldots : y_{n+1}) \in \mathbb{R}P^n$, the *inner product* $\langle x, y\rangle = \sum_{i=1}^{n+1} x_i y_i$. The **elliptic metric** d_{ell} on $\mathbb{R}P^n$ is defined now by
$$r \arccos \frac{|\langle x, y\rangle|}{\sqrt{\langle x, x\rangle\langle y, y\rangle}},$$
where $r > 0$ is a fixed constant, and arccos is the inverse cosine in $[0, \pi]$.
In all the considered cases, some hypersurfaces of the second-order remain invariant under given **motions**, i.e., projective transformations preserving a given metric. These hypersurfaces are called *absolutes*. In the case of a Euclidean

determination of a metric, the absolute is an imaginary $(n-2)$-dimensional oval surface of order two, in fact, the degenerate absolute $x_1^2+\cdots+x_n^2=0, x_{n+1}=0$. In the case of a hyperbolic determination of a metric, the absolute is a real $(n-1)$-dimensional oval hypersurface of order two, in the simplest case, the absolute $-x_1^2+x_2^2+\cdots+x_{n+1}^2=0$. In the case of an elliptic determination of a metric, the absolute is an imaginary $(n-1)$-dimensional oval hypersurface of order two, in fact, the absolute $x_1^2+\cdots+x_{n+1}^2=0$.

6.3 Affine Geometry

An n-dimensional *affine space* over a field $\mathbb{F}$ is a set A^n (the elements of which are called *points* of the affine space) to which corresponds an n-dimensional vector space V over $\mathbb{F}$ (called the *space associated to* A^n) such that, for any $a \in A^n$, $A = a + V = \{a + v : v \in V\}$. In the other words, if $a = (a_1, \dots, a_n)$ and $b = (b_1, \dots, b_n) \in A^n$, then the vector $\overrightarrow{ab} = (b_1 - a_1, \dots, b_n - a_n)$ belongs to V.

In an affine space, one can add a vector to a point to get another point, and subtract points to get vectors, but one cannot add points, since there is no origin. Given points $a, b, c, d \in A^n$ such that $c \neq d$, and the vectors $\overrightarrow{ab}$ and $\overrightarrow{cd}$ are collinear, the scalar λ, defined by $\overrightarrow{ab} = \lambda\overrightarrow{cd}$, is called the *affine ratio* of ab and cd, and is denoted by $\frac{ab}{cd}$.

An *affine transformation* (or *affinity*) is a bijection of A^n onto itself which preserves *collinearity* and *ratios of distances* In this sense, *affine* indicates a special class of *projective transformations* that do not move any objects from the affine space to the plane at infinity or conversely. Any affine transformation is a composition of *rotations*, *translations*, *dilations*, and *shears*. The set of all affine transformations of A^n forms a group $Aff(A^n)$, called the *general affine group* of A^n. Each element $f \in Aff(A)$ can be given by a formula $f(a) = b$, $b_i = \sum_{j=1}^n p_{ij}a_j + c_j$, where $((p_{ij}))$ is an invertible matrix.

The subgroup of $Aff(A^n)$, consisting of affine transformations with $det\,((p_{ij})) = 1$, is called the *equi-affine group* of A^n. An *equi-affine space* is an affine space with the equi-affine group of transformations. The fundamental invariants of an equi-affine space are volumes of parallelepipeds. In an *equi-affine plane* A^2, any two vectors v_1, v_2 have an invariant $|v_1 \times v_2|$ (the modulus of their cross product)—the surface area of the parallelogram constructed on v_1 and v_2.

Given a nonrectilinear curve $\gamma = \gamma(t)$, its *affine parameter* (or *equi-affine arc length*) is an invariant $s = \int_{t_0}^{t} |\gamma' \times \gamma''|^{1/3} dt$. The invariant $k = \frac{d^2\gamma}{ds^2} \times \frac{d^3\gamma}{ds^3}$ is called the *equi-affine curvature* of γ. Passing to the general affine group, two more invariants of the curve are considered: the *affine arc length* $\sigma = \int k^{1/2} ds$, and the *affine curvature* $\mathrm{k} = \frac{1}{k^{3/2}} \frac{dk}{ds}$.

For A^n, $n > 2$, the *affine parameter* (or *equi-affine arc length*) of a curve $\gamma = \gamma(t)$ is defined by $s = \int_{t_0}^{t} |(\gamma', \gamma'', \dots, \gamma^{(n)})|^{\frac{2}{n(n+1)}} dt$, where the invariant

$(v_1, \dots, v_n)$ is the (oriented) volume spanned by the vectors $v_1, \dots, v_n$ which is equal to the determinant of the $n \times n$ matrix whose i-th column is the vector v_i.

- **Affine distance**
  Given an *affine plane* A^2, a *line element* (a, l_a) of A^2 consists of a point $a \in A^2$ together with a straight line $l_a \subset A^2$ passing through a.
  The **affine distance** is a distance on the set of all line elements of A^2 defined by

  $$2f^{1/3},$$

  where, for a given line elements (a, l_a) and (b, l_b), f is the surface area of the triangle abc if c is the point of intersection of the straight lines l_a and l_b. The affine distance between (a, l_a) and (b, l_b) can be interpreted as the affine length of the arc ab of a parabola such that l_a and l_b are tangent to the parabola at a and b, respectively.
- **Affine pseudo-distance**
  Let A^2 be an *equi-affine plane*, and let $\gamma = \gamma(s)$ be a curve in A^2 defined as a function of the *affine parameter* s. The **affine pseudo-distance** dp_{aff} for A^2 is

  $$dp_{aff}(a, b) = \left|\overrightarrow{ab} \times \frac{d\gamma}{ds}\right|,$$

  i.e., it is equal to the surface area of the parallelogram constructed on the vectors $\overrightarrow{ab}$ and $\frac{d\gamma}{ds}$, where b is an arbitrary point in A^2, a is a point on γ, and $\frac{d\gamma}{ds}$ is the tangent vector to the curve γ at the point a.
  Similarly, the **affine pseudo-distance** for an *equi-affine space* A^3 is defined as

  $$\left|(\overrightarrow{ab}, \frac{d\gamma}{ds}, \frac{d^2\gamma}{ds^2})\right|,$$

  where $\gamma = \gamma(s)$ is a curve in A^3, defined as a function of the *affine parameter* s, $b \in A^3$, a is a point of γ, and the vectors $\frac{d\gamma}{ds}, \frac{d^2\gamma}{ds^2}$ are obtained at the point a.
  For A^n, $n > 3$, we have $dp_{aff}(a, b) = |(\overrightarrow{ab}, \frac{d\gamma}{ds}, \dots, \frac{d^{n-1}\gamma}{ds^{n-1}})|$. For an arbitrary parametrization $\gamma = \gamma(t)$, one obtains $dp_{aff}(a, b) = |(\overrightarrow{ab}, \gamma', \dots, \gamma^{(n-1)})||(\gamma', \dots, \gamma^{(n-1)})|^{\frac{1-n}{1+n}}$.
- **Affine metric**
  The **affine metric** is a metric on a *nondevelopable surface* $r = r(u_1, u_2)$ in an *equi-affine space* A^3, given by its **metric tensor** $((g_{ij}))$:

  $$g_{ij} = \frac{a_{ij}}{|det((a_{ij}))|^{1/4}},$$

  where $a_{ij} = (\partial_1 r, \partial_2 r, \partial_{ij} r)$, $i, j \in \{1, 2\}$.

6.4 Non-Euclidean Geometry

The term *non-Euclidean Geometry* describes both *Hyperbolic Geometry* (or *Lobachevsky-Bolyai-Gauss Geometry*) and *Elliptic Geometry* which are contrasted with *Euclidean Geometry* (or *Parabolic Geometry*). The essential difference between Euclidean and non-Euclidean Geometry is the nature of parallel lines. In Euclidean Geometry, if we start with a line l and a point a, which is not on l, then there is only one line through a that is parallel to l. In Hyperbolic Geometry there are infinitely many lines through a parallel to l. In Elliptic Geometry, parallel lines do not exist. The *Spherical Geometry* is also "non-Euclidean", but it fails the axiom that any two points determine exactly one line.

- **Spherical metric**
  Let $S^n(0,r) = \{x \in \mathbb{R}^{n+1} : \sum_{i=1}^{n+1} x_i^2 = r^2\}$ be the sphere in $\mathbb{R}^{n+1}$ with the center 0 and the radius $r > 0$.
  The **spherical metric** (or *great circle metric*) is a metric on $S^n(0,r)$ defined by

$$d_{sph} = r \arccos\left(\frac{|\sum_{i=1}^{n+1} x_i y_i|}{r^2}\right),$$

  where arccos is the inverse cosine in $[0, \pi]$. It is the length of the *great circle* arc, passing through x and y. In terms of the standard *inner product* $\langle x, y\rangle = \sum_{i=1}^{n+1} x_i y_i$ on $\mathbb{R}^{n+1}$, the spherical metric can be written as $r \arccos \frac{|\langle x,y\rangle|}{\sqrt{\langle x,x\rangle\langle y,y\rangle}}$.
  The metric space $(S^n(0,r), d_{sph})$ is called *n-dimensional spherical space*. It is a space of curvature $1/r^2$, and r is the radius of curvature. It is a model of n-dimensional *Spherical Geometry*. The great circles of the sphere are its geodesics and all geodesics are closed and of the same length. (See, for example, [Blum70].)
- **Elliptic metric**
  Let $\mathbb{R}P^n$ be the real n-dimensional projective space. The **elliptic metric** d_{ell} is a metric on $\mathbb{R}P^n$ defined by

$$r \arccos \frac{|\langle x, y\rangle|}{\sqrt{\langle x, x\rangle\langle y, y\rangle}},$$

  where, for any $x = (x_1 : \ldots : x_{n+1})$ and $y = (y_1 : \ldots : y_{n+1}) \in \mathbb{R}P^n$, one has $\langle x, y\rangle = \sum_{i=1}^{n+1} x_i y_i$, $r > 0$ is a constant and arccos is the inverse cosine in $[0, \pi]$.
  The metric space $(\mathbb{R}P^n, d_{ell})$ is called *n-dimensional elliptic space*. It is a model of n-dimensional *Elliptic Geometry*. It is the space of curvature $1/r^2$, and r is the radius of curvature. As $r \to \infty$, the metric formulas of Elliptic Geometry yield formulas of Euclidean Geometry (or become meaningless).

If $\mathbb{R}P^n$ is viewed as the set $E^n(0,r)$, obtained from the sphere $S^n(0,r) = \{x \in \mathbb{R}^{n+1} : \sum_{i=1}^{n+1} x_i^2 = r^2\}$ in $\mathbb{R}^{n+1}$ with center 0 and radius r by identifying diametrically-opposite points, then the elliptic metric on $E^n(0,r)$ can be written as $d_{sph}(x,y)$ if $d_{sph}(x,y) \le \frac{\pi}{2}r$, and as $\pi r - d_{sph}(x,y)$ if $d_{sph}(x,y) > \frac{\pi}{2}r$, where d_{sph} is the **spherical metric** on $S^n(0,r)$. Thus, no two points of $E^n(0,r)$ have distance exceeding $\frac{\pi}{2}r$. The elliptic space $(E^2(0,r), d_{ell})$ is called the *Poincaré sphere*.

If $\mathbb{R}P^n$ is viewed as the set E^n of lines through the zero element 0 in $\mathbb{R}^{n+1}$, then the elliptic metric on E^n is defined as the angle between the corresponding subspaces.

An n-dimensional elliptic space is a *Riemannian space* of constant positive curvature. It is the only such space which is topologically equivalent to a projective space. (See, for example, [Blum70, Buse55].)

- **Hermitian elliptic metric**

  Let $\mathbb{C}P^n$ be the n-dimensional complex projective space. The **Hermitian elliptic metric** d_{ell}^H (see, for example, [Buse55]) is a metric on $\mathbb{C}P^n$ defined by

$$r \arccos \frac{|\langle x, y\rangle|}{\sqrt{\langle x, x\rangle\langle y, y\rangle}},$$

  where, for any $x = (x_1 : \ldots : x_{n+1})$ and $y = (y_1 : \ldots : y_{n+1}) \in \mathbb{C}P^n$, one has $\langle x, y\rangle = \sum_{i=1}^{n+1} \overline{x}_i y_i$, $r > 0$ is a constant and arccos is the inverse cosine in $[0, \pi]$.

  The metric space $(\mathbb{C}P^n, d_{ell}^H)$ is called n-dimensional *Hermitian elliptic space* (cf. **Fubini–Study metric** in Chap. 7).

- **Elliptic plane metric**

  The **elliptic plane metric** is the **elliptic metric** on the *elliptic plane* $\mathbb{R}P^2$.

  If $\mathbb{R}P^2$ is viewed as the *Poincaré sphere* (i.e., a sphere in $\mathbb{R}^3$ with identified diametrically-opposite points) of diameter 1 tangent to the extended complex plane $\overline{\mathbb{C}} = \mathbb{C} \cup \{\infty\}$ at the point $z = 0$, then, under the stereographic projection from the "north pole" $(0, 0, 1)$, $\overline{\mathbb{C}}$ with identified points z and $-\frac{1}{\overline{z}}$ is a model of the elliptic plane.

  The elliptic plane metric d_{ell} on it is defined by its *line element* $ds^2 = \frac{|dz|^2}{(1+|z|^2)^2}$.

- **Pseudo-elliptic distance**

  The **pseudo-elliptic distance** (or *elliptic pseudo-distance*) dp_{ell} is defined, on the extended complex plane $\overline{\mathbb{C}} = \mathbb{C} \cup \{\infty\}$ with identified points z and $-\frac{1}{\overline{z}}$, by

$$\left|\frac{z - u}{1 + \overline{z}u}\right|.$$

  In fact, $dp_{ell}(z, u) = \tan d_{ell}(z, u)$, where d_{ell} is the **elliptic plane metric**.

- **Hyperbolic metric**
Let $\mathbb{R}P^n$ be the n-dimensional real projective space. Let, for any $x = (x_1 : \ldots : x_{n+1})$, $y = (y_1 : \ldots : y_{n+1}) \in \mathbb{R}P^n$, their *scalar product* $\langle x, y\rangle$ be $-x_1y_1 + \sum_{i=2}^{n+1} x_i y_i$.
The **hyperbolic metric** d_{hyp} is a metric on the set $H^n=\{x \in \mathbb{R}P^n : \langle x, x\rangle<0\}$ defined, for a fixed constant $r > 0$, by
$$r\operatorname{arccosh}\frac{|\langle x, y\rangle|}{\sqrt{\langle x, x\rangle\langle y, y\rangle}},$$
where arccosh denotes the nonnegative values of the inverse hyperbolic cosine.
In this construction, the points of H^n can be viewed as the one-spaces of the *pseudo-Euclidean space* $\mathbb{R}^{n,1}$ inside the cone $C = \{x \in \mathbb{R}^{n,1} : \langle x, x\rangle = 0\}$.
The metric space (H^n, d_{hyp}) is called *n-dimensional hyperbolic space*. It is a model of n-dimensional *Hyperbolic Geometry*. It is the space of curvature $-1/r^2$, and r is the radius of curvature. Replacement of r by ir transforms all metric formulas of Hyperbolic Geometry into the corresponding formulas of Elliptic Geometry. As $r \to \infty$, both systems yield formulas of Euclidean Geometry (or become meaningless).
If H^n is viewed as the set $\{x \in \mathbb{R}^n : \sum_{i=1}^{n} x_i^2 < K\}$, where $K > 1$ is any fixed constant, the hyperbolic metric can be written as
$$\frac{r}{2}\ln\frac{1+\sqrt{1-\gamma(x, y)}}{1-\sqrt{1-\gamma(x, y)}},$$
where $\gamma(x, y) = \frac{(K-\sum_{i=1}^{n} x_i^2)(K-\sum_{i=1}^{n} y_i^2)}{(K-\sum_{i=1}^{n} x_i y_i)^2}$, and $r > 0$ is a number with $\tanh\frac{1}{r} = \frac{1}{\sqrt{K}}$.
If H^n is viewed as a submanifold of the $(n + 1)$-dimensional *pseudo-Euclidean space* $\mathbb{R}^{n,1}$ with the scalar product $\langle x, y\rangle = -x_1y_1 + \sum_{i=2}^{n+1} x_i y_i$ (in fact, as the top sheet $\{x \in \mathbb{R}^{n,1} : \langle x, x\rangle = -1, x_1 > 0\}$ of the two-sheeted *hyperboloid of revolution*), then the hyperbolic metric on H^n is induced from the **pseudo-Riemannian metric** on $R^{n,1}$ (cf. **Lorentz metric** in Chap. 26).
An n-dimensional hyperbolic space is a *Riemannian space* of constant negative curvature. It is the only such space which is **complete** and topologically equivalent to an Euclidean space. (See, for example, [Blum70, Buse55].)
- **Hermitian hyperbolic metric**
Let $\mathbb{C}P^n$ be the n-dimensional complex projective space. Let, for any $x = (x_1 : \ldots : x_{n+1})$, $y = (y_1 : \ldots : y_{n+1}) \in \mathbb{C}P^n$, their *scalar product* $\langle x, y\rangle$ be $-\overline{x}_1y_1 + \sum_{i=2}^{n+1} \overline{x}_i y_i$.
The **Hermitian hyperbolic metric** d_{hyp}^H (see, for example, [Buse55]) is a metric on the set $\mathbb{C}H^n = \{x \in \mathbb{C}P^n : \langle x, x\rangle < 0\}$ defined, for a fixed constant $r > 0$, by

$$\operatorname{arccosh}\frac{|\langle x,y\rangle|}{\sqrt{\langle x,x\rangle\langle y,y\rangle}},$$

where arccosh denotes the nonnegative values of the inverse hyperbolic cosine. The metric space $(\mathbb{C}H^n, d^H_{hyp})$ is called n-dimensional *Hermitian hyperbolic space*.

- **Poincaré metric**
  The **Poincaré metric** d_P is the **hyperbolic metric** for the *Poincaré disk model* of Hyperbolic Geometry. In this model the *unit disk* $\Delta = \{z \in \mathbb{C} : |z| < 1\}$ is called the *hyperbolic plane*, every point of Δ is called a *hyperbolic point*, circular arcs (and diameters) in Δ which are orthogonal to the *absolute* $\Omega = \{z \in \mathbb{C} : |z| = 1\}$ are called *hyperbolic straight lines*. Every point of Ω is called an *ideal point*.
  The angular measurements in this model are the same as in Hyperbolic Geometry, i.e., it is a *conformal model*. There is a one-to-one correspondence between segments and acute angles. The Poincaré metric on Δ is defined by its *line element*

$$ds^2 = \frac{|dz|^2}{(1-|z|^2)^2} = \frac{dz_1^2 + dz_2^2}{(1-z_1^2-z_2^2)^2}.$$

  The distance d_P between two points z and u of Δ can be written as

$$\frac{1}{2}\ln\frac{|1-z\overline{u}|+|z-u|}{|1-z\overline{u}|-|z-u|} = \operatorname{arctanh}\frac{|z-u|}{|1-z\overline{u}|}.$$

  In terms of **cross-ratio**, it is equal to

$$\frac{1}{2}\ln(z,u,z^*,u^*) = \frac{1}{2}\ln\frac{(z^*-z)(u^*-u)}{(z^*-u)(u^*-z)},$$

  where z^* and u^* are the points of intersection of the hyperbolic straight line passing through z and u with Ω, z^* on the side of u, and u^* on the side of z.
  The **multiplicative distance function** on the segments zu of Δ is defined (Hartshorne, 2003) by $\mu(zu) = (z,u,z^*,u^*)^{-1}$; it allows the definition of trigonometric functions in the absence of continuity.
  In the conformal *Poincaré half-plane model* of Hyperbolic Geometry the *hyperbolic plane* is the upper half-plane $H^2 = \{z \in \mathbb{C} : z_2 > 0\}$, and the *hyperbolic lines* are semicircles and half-lines which are orthogonal to the real axis. The *absolute* (i.e., the set of *ideal points*) is the real axis together with the point at infinity.
  The *line element* of the **Poincaré metric** on H^2 is given by

$$ds^2 = \frac{|dz|^2}{(\Im z)^2} = \frac{dz_1^2 + dz_2^2}{z_2^2}.$$

The distance between two points z, u can be written as

$$\frac{1}{2}\ln\frac{|z-\overline{u}|+|z-u|}{|z-\overline{u}|-|z-u|} = \operatorname{arctanh}\frac{|z-u|}{|z-\overline{u}|} = \frac{1}{2}\ln(z,u,z^*,u^*) = \frac{1}{2}\ln\frac{(z^*-z)(u^*-u)}{(z^*-u)(u^*-z)},$$

where z^* is the ideal point of the half-line emanating from z and passing through u, and u^* is the ideal point of the half-line emanating from u and passing through z.

In general, the **hyperbolic metric** in any *domain* $D \subset \mathbb{C}$ with at least three boundary points is defined as the preimage of the Poincaré metric in Δ under a *conformal mapping* $f : D \to \Delta$. Its *line element* has the form

$$ds^2 = \frac{|f'(z)|^2|dz|^2}{(1-|f(z)|^2)^2}.$$

The distance between two points z and u in D can be written as

$$\frac{1}{2}\ln\frac{|1-f(z)\overline{f(u)}|+|f(z)-f(u)|}{|1-f(z)\overline{f(u)}|-|f(z)-f(u)|}.$$

- **Pseudo-hyperbolic distance**
  The **pseudo-hyperbolic distance** (or **Gleason distance**, *hyperbolic pseudo-distance*) dp_{hyp} is a metric on the *unit disk* $\Delta = \{z \in \mathbb{C} : |z| < 1\}$, defined by

$$\left|\frac{z-u}{1-\overline{z}u}\right|.$$

  In fact, $dp_{hyp}(z,u) = \tanh d_P(z,u)$, where d_P is the **Poincaré metric** on Δ.
- **Cayley–Klein–Hilbert metric**
  The **Cayley–Klein–Hilbert metric** d_{CKH} is the **hyperbolic metric** for the *Klein model* (or *projective disk model*, for Hyperbolic Geometry. In this model the *hyperbolic plane* is realized as the *unit disk* $\Delta = \{z \in \mathbb{C} : |z| < 1\}$, and the *hyperbolic straight lines* are realized as the chords of Δ. Every point of the *absolute* $\Omega = \{z \in \mathbb{C} : |z| = 1\}$ is called an *ideal point*. This model is not conformal: the angular measurements are distorted. The **Cayley–Klein–Hilbert metric** on Δ is given by its **metric tensor** $((g_{ij}))$, $i, j = 1, 2$:

$$g_{11} = \frac{r^2(1-z_2^2)}{(1-z_1^2-z_2^2)^2},\quad g_{12} = \frac{r^2 z_1 z_2}{(1-z_1^2-z_2^2)^2},\quad g_{22} = \frac{r^2(1-z_1^2)}{(1-z_1^2-z_2^2)^2},$$

  where r is any positive constant. The distance between points z and u in Δ is

$$r \operatorname{arccosh}\left(\frac{1-z_1u_1-z_2u_2}{\sqrt{1-z_1^2-z_2^2}\sqrt{1-u_1^2-u_2^2}}\right),$$

where arccosh denotes the nonnegative values of the inverse hyperbolic cosine.

- **Weierstrass metric**
  Given a real n-dimensional **inner product space** $(\mathbb{R}^n, \langle , \rangle)$, $n \geq 2$, the **Weierstrass metric** d_W is a metric on $\mathbb{R}^n$ defined by

$$\operatorname{arccosh}(\sqrt{1 + \langle x, x\rangle}\sqrt{1 + \langle y, y\rangle} - \langle x, y\rangle),$$

  where arccosh denotes the nonnegative values of the inverse hyperbolic cosine. Here, $(x, \sqrt{1 + \langle x, x\rangle}) \in \mathbb{R}^n \oplus \mathbb{R}$ are the *Weierstrass coordinates* of $x \in \mathbb{R}^n$, and the metric space $(\mathbb{R}^n, d_W)$ can be seen as the *Weierstrass model* of Hyperbolic Geometry.
  The **Cayley–Klein–Hilbert metric** $d_{CKH}(x, y) = \operatorname{arccosh}\frac{1-\langle x,y\rangle}{\sqrt{1-\langle x,x\rangle}\sqrt{1-\langle y,y\rangle}}$ on the open ball $B^n = \{x \in \mathbb{R}^n : \langle x, x\rangle < 1\}$ can be obtained from d_W by $d_{GKH}(x, y) = d_W(\mu(x), \mu(y))$, where $\mu : \mathbb{R}^n \to B^n$ is the *Weierstrass mapping*: $\mu(x) = \frac{x}{\sqrt{1-\langle x,x\rangle}}$.
- **Harnack metric**
  Given a *domain* $D \subset \mathbb{R}^n$, $n \geq 2$, the **Harnack metric** is a metric on D defined by

$$\sup_f |\log \frac{f(x)}{f(y)}|,$$

  where the supremum is taken over all positive functions which are harmonic on D.
- **Quasi-hyperbolic metric**
  Given a *domain* $D \subset \mathbb{R}^n$, $n \geq 2$, the **quasi-hyperbolic metric** on D is defined by

$$\inf_{\gamma\in\Gamma} \int_\gamma \frac{|dz|}{\rho(z)},$$

  where the infimum is taken over the set Γ of all rectifiable curves connecting x and y in D, $\rho(z) = \inf_{u\in\partial D} ||z - u||_2$ is the distance between z and the boundary ∂D of D, and $||.||_2$ is the Euclidean norm on $\mathbb{R}^n$.
  This metric is **Gromov hyperbolic** if the domain D is *uniform*, i.e., there exist constants C, C' such that each pair of points $x, y \in D$ can be joined by a rectifiable curve $\gamma = \gamma(x, y) \in D$ of length $l(\gamma)$ at most $C|x - y|$, and $\min\{l(\gamma(x, z)), l(\gamma(z, y))\} \leq C'd(z, \partial D)$ holds for all $z \in \gamma$. Also, the quasi-hyperbolic metric is the **inner metric** (cf. Chap. 4) of the **Vuorinen metric**.
  For $n = 2$, one can define the **hyperbolic metric** on D by

$$\inf_{\gamma\in\Gamma} \int_\gamma \frac{2|f'(z)|}{1 - |f(z)|^2}|dz|,$$

where $f : D \to \Delta$ is any conformal mapping of D onto the *unit disk* $\Delta = \{z \in \mathbb{C} : |z| < 1\}$. For $n \geq 3$, it is defined only for the half-hyperplane H^n and for the *open unit ball* B^n as the infimum over all $\gamma \in \Gamma$ of the integrals $\int_\gamma \frac{|dz|}{z_n}$ and $\int_\gamma \frac{2|dz|}{1-||z||_2^2}$.

- **Apollonian metric**
Let $D \subset \mathbb{R}^n$, $D \neq \mathbb{R}^n$, be a *domain* such that its complement is not contained in a hyperplane or a sphere. The **Apollonian metric** (or **Barbilian metric**, [Barb35]) on D is defined (denoting the boundary of D by ∂D) by the **cross-ratio** as

$$\sup_{a,b \in \partial D} \ln \frac{||a-x||_2||b-y||_2}{||a-y||_2||b-x||_2}.$$

This metric is **Gromov hyperbolic**.

- **Half-Apollonian metric**
Given a *domain* $D \subset \mathbb{R}^n$, $D \neq \mathbb{R}^n$, the **half-Apollonian metric** η_D (Hăsto and Lindén, 2004) on D is defined (denoting the boundary of D by ∂D) by

$$\sup_{a \in \partial D} \left| \ln \frac{||a-y||_2}{||a-x||_2} \right|.$$

This metric is **Gromov hyperbolic** only if the domain is $\mathbb{R}^n \backslash \{x\}$.

- **Gehring metric**
Given a *domain* $D \subset \mathbb{R}^n$, $D \neq \mathbb{R}^n$, the **Gehring metric** $\tilde{j}_D$ (Gehring, 1982) is a metric on D, defined by

$$\frac{1}{2} \ln \left(\left(1 + \frac{||x-y||_2}{\rho(x)}\right) \left(1 + \frac{||x-y||_2}{\rho(y)}\right) \right),$$

where $\rho(x) = \inf_{u \in \partial D} ||x-u||_2$ is the distance between x and the boundary of D. This metric is **Gromov hyperbolic**.

- **Vuorinen metric**
Given a *domain* $D \subset \mathbb{R}^n$, $D \neq \mathbb{R}^n$, the **Vuorinen** (or *distance ratio*, j_D-) **metric**; Vuorinen, 1988) is a metric on D defined by

$$\ln \left(1 + \frac{||x-y||_2}{\min\{\rho(x), \rho(y)\}}\right),$$

where $\rho(x) = \inf_{u \in \partial D} ||x-u||_2$ is the distance between x and the boundary of D. This metric is **Gromov hyperbolic** only if the domain is $\mathbb{R}^n \backslash \{x\}$.

- **Ferrand metric**
Given a *domain* $D \subset \mathbb{R}^n$, $D \neq \mathbb{R}^n$, the **Ferrand metric** σ_D (Ferrand, 1987) is a metric on D defined by

$$\inf_{\gamma \in \Gamma} \int_\gamma \sup_{a,b \in \partial D} \frac{||a-b||_2}{||z-a||_2||z-b||_2} |dz|,$$

where the infimum is taken over the set Γ of all rectifiable curves connecting x and y in D, ∂D is the boundary of D, and $||.||_2$ is the Euclidean norm on $\mathbb{R}^n$. This metric is the **inner metric** (cf. Chap. 4) of the **Möbius metric**.

- **Möbius metric**
Given a *domain* $D \subset \mathbb{R}^n$, $D \neq \mathbb{R}^n$, the **Möbius** (or *absolute ratio*, δ_D-) **metric**; Siettenranta, 1999) is a metric on D defined by

$$\sup_{a,b \in \partial D} \ln\left(1 + \frac{||a-x||_2||b-y||_2}{||a-b||_2||x-y||_2}\right).$$

This metric is **Gromov hyperbolic**.

- **Modulus metric**
Let $D \subset \mathbb{R}^n$, $D \neq \mathbb{R}^n$, be a *domain*. The *conformal modulus* of a family Γ of locally rectifiable curves in D is $M(\Gamma) = \inf_\rho \int_{\mathbb{R}^n} \rho^n dm$, where m is the n-dimensional Lebesgue measure, and ρ is any Borel-measurable function with $\int_\gamma \rho ds \geq 1$ and $\rho \geq 0$ for each $\gamma \in \Gamma$. Cf. general *modulus* in **extremal metric**, Chap. 8.
Let $\Delta(E, F; D)$ denote the family of all closed nonconstant curves in D joining E and F. The **modulus metric** μ_D (Gál, 1960) is a metric on D, defined by

$$\inf_{C_{xy}} M(\Delta(C_{xy}, \partial D; D)),$$

where C_{xy} is a compact connected set such that for some $\gamma : [0,1] \to D$, it holds $C_{xy} = \gamma([0,1])$ and $\gamma(0) = x$, $\gamma(1) = y$.
The **Ferrand second metric** λ^*_D (Ferrand, 1997) is a metric on D, defined by

$$\left(\inf_{C_x, C_y} M(\Delta(C_x, C_y; D))\right)^{\frac{1}{1-n}},$$

where C_z $(z = x, y)$ is a compact connected set such that, for some $\gamma_z : [0,1] \to D$, it holds $C_z = \gamma([0,1))$, $z \in |\gamma_z|$ and $\gamma_z(t) \to \partial D$ as $t \to 1$.
Above two metrics are **Gromov hyperbolic** if D is the open ball $B^n = \{x \in \mathbb{R}^n : \langle x, x \rangle < 1\}$ or a simply connected domain in $\mathbb{R}^2$.

- **Conformal radius**
Let $D \subset \mathbb{C}$, $D \neq \mathbb{C}$, be a simply connected *domain* and let $z \in D$, $z \neq \infty$. The **conformal** (or *harmonic*) **radius** is defined by

$$rad(z, D) = (f'(z))^{-1},$$

where $f : D \to \Delta$ is the *uniformizing map*, i.e., the unique conformal mapping onto the unit disk with $f(z) = 0 \in \Delta$ and $f'(z) > 0$.
The Euclidean distance between z and the boundary ∂D of D (i.e., the radius of the largest disk inscribed in D) lies in the segment $[\frac{rad(z,D)}{4}, rad(z, D)]$.

If D is compact, define $rad(\infty, D)$ as $\lim_{z\to\infty}\frac{f(z)}{z}$, where $f : (\mathbb{C}\backslash\Delta) \to (\mathbb{C}\backslash D)$ is the unique conformal mapping with $f(\infty) = \infty$ and positive above limit. This radius is the **transfinite diameter** from Chap. 1.

- **Parabolic distance**
  The **parabolic distance** is a metric on $\mathbb{R}^{n+1}$, considered as $\mathbb{R}^n \times \mathbb{R}$ defined by

$$\sqrt{(x_1 - y_1)^2 + \cdots + (x_n - y_n)^2 + |t_x - t_y|^{1/m}}, m \in \mathbb{N},$$

  for any $x = (x_1, \ldots, x_n, t_x)$, $y = (y_1, \ldots, y_n, t_y) \in \mathbb{R}^n \times \mathbb{R}$.
  The space $\mathbb{R}^n \times \mathbb{R}$ can be interpreted as multidimensional *space-time*.
  Usually, the value $m = 2$ is applied. There exist some variants of the parabolic distance, for example, the parabolic distance

$$\sup\{|x_1 - y_1|, |x_2 - y_2|^{1/2}\}$$

  on $\mathbb{R}^2$ (cf. also **Rickman's rug metric** in Chap. 19), or the **half-space parabolic distance** on $\mathbb{R}^3_+ = \{x \in \mathbb{R}^3 : x_1 \geq 0\}$ defined by

$$\frac{|x_1 - y_1| + |x_2 - y_2|}{\sqrt{x_1} + \sqrt{x_2} + \sqrt{|x_2 - y_2|}} + \sqrt{|x_3 - y_3|}.$$

Chapter 7
Riemannian and Hermitian Metrics

Riemannian Geometry is a multidimensional generalization of the intrinsic geometry of 2D surfaces in the Euclidean space $\mathbb{E}^3$. It studies *real smooth manifolds* equipped with **Riemannian metrics**, i.e., collections of positive-definite symmetric bilinear forms $((g_{ij}))$ on their tangent spaces which vary smoothly from point to point. The geometry of such (*Riemannian*) manifolds is based on the *line element* $ds^2 = \sum_{i,j} g_{ij}dx_i dx_j$. This gives, in particular, local notions of angle, length of curve, and volume.

From these notions some other global quantities can be derived, by integrating local contributions. Thus, the value ds is interpreted as the length of the vector $(dx_1, \dots, dx_n)$, and it is called the **infinitesimal distance**. The arc length of a curve γ is expressed by $\int_\gamma \sqrt{\sum_{i,j} g_{ij}dx_i dx_j}$, and then the **intrinsic metric** on a Riemannian manifold is defined as the infimum of lengths of curves joining two given points of the manifold.

Therefore, a Riemannian metric is not an ordinary metric, but it induces an ordinary metric, in fact, the intrinsic metric, called **Riemannian distance**, on any connected Riemannian manifold. A Riemannian metric is an infinitesimal form of the corresponding Riemannian distance.

As particular special cases of Riemannian Geometry, there occur *Euclidean Geometry* as well as the two standard types, *Elliptic Geometry* and *Hyperbolic Geometry*, of *non-Euclidean Geometry*. If the bilinear forms $((g_{ij}))$ are nondegenerate but indefinite, one obtains *pseudo-Riemannian Geometry*. In the case of dimension four (and *signature* $(1, 3)$) it is the main object of the General Theory of Relativity.

If $ds = F(x_1, \dots, x_n, dx_1, \dots, dx_n)$, where F is a real positive-definite convex function which cannot be given as the square root of a symmetric bilinear form (as in the Riemannian case), one obtains the *Finsler Geometry* generalizing Riemannian Geometry.

M.M. Deza, E. Deza, *Encyclopedia of Distances*,
DOI 10.1007/978-3-662-44342-2_7

Hermitian Geometry studies *complex manifolds* equipped with **Hermitian metrics**, i.e., collections of positive-definite symmetric *sesquilinear forms* (or $\frac{3}{2}$*-linear forms*) since they are linear in one argument and *antilinear* in the other) on their tangent spaces, which vary smoothly from point to point. It is a complex analog of Riemannian Geometry.

A special class of Hermitian metrics form **Kähler metrics** which have a closed fundamental form w. A generalization of Hermitian metrics give **complex Finsler metrics** which cannot be written as a bilinear symmetric positive-definite sesqulinear form.

7.1 Riemannian Metrics and Generalizations

A *real n-manifold* M^n *with boundary* is (cf. Chap. 2) a **Hausdorff space** in which every point has an open neighborhood homeomorphic to either an open subset of $\mathbb{E}^n$, or an open subset of the closed half of $\mathbb{E}^n$. The set of points which have an open neighborhood homeomorphic to $\mathbb{E}^n$ is called the *interior* (of the manifold); it is always nonempty.

The complement of the interior is called the *boundary* (of the manifold); it is an $(n-1)$-dimensional manifold. If it is empty, one obtains a *real n-manifold without boundary*. Such manifold is called *closed* if it is compact, and *open*, otherwise.

An open set of M^n together with a homeomorphism between the open set and an open set of $\mathbb{E}^n$ is called a *coordinate chart*. A collection of charts which cover M^n is an *atlas* on M^n. The homeomorphisms of two overlapping charts provide a transition mapping from a subset of $\mathbb{E}^n$ to some other subset of $\mathbb{E}^n$.

If all these mappings are continuously differentiable, then M^n is a *differentiable manifold*. If they are k times (infinitely often) continuously differentiable, then the manifold is a C^k *manifold* (respectively, a *smooth manifold*, or C^∞ *manifold*).

An atlas of a manifold is called *oriented* if the Jacobians of the coordinate transformations between any two charts are positive at every point. An *orientable manifold* is a manifold admitting an oriented atlas.

Manifolds inherit many local properties of the Euclidean space: they are locally path-connected, locally compact, and locally metrizable. Every smooth Riemannian manifold embeds isometrically (Nash, 1956) in some finite-dimensional Euclidean space.

Associated with every point on a differentiable manifold is a *tangent space* and its dual, a *cotangent space*. Formally, let M^n be a C^k manifold, $k \geq 1$, and p a point of M^n. Fix a chart $\varphi : U \to \mathbb{E}^n$, where U is an open subset of M^n containing p. Suppose that two curves $\gamma^1 : (-1,1) \to M^n$ and $\gamma^2 : (-1,1) \to M^n$ with $\gamma^1(0) = \gamma^2(0) = p$ are given such that $\varphi \cdot \gamma^1$ and $\varphi \cdot \gamma^2$ are both differentiable at 0.

Then γ^1 and γ^2 are called *tangent at* 0 if $(\varphi \cdot \gamma^1)'(0) = (\varphi \cdot \gamma^2)'(0)$. If the functions $\varphi \cdot \gamma^i : (-1,1) \to \mathbb{E}^n$, $i = 1, 2$, are given by n real-valued component functions $(\varphi \cdot \gamma^i)_1(t), \dots, (\varphi \cdot \gamma^i)_n(t)$, the condition above means that their Jacobians

$\left(\frac{d(\varphi\cdot\gamma^i)_1(t)}{dt}, \ldots, \frac{d(\varphi\cdot\gamma^i)_n(t)}{dt}\right)$ coincide at 0. This is an equivalence relation, and the equivalence class $\gamma'(0)$ of the curve γ is called a *tangent vector* of M^n at p.

The *tangent space* $T_p(M^n)$ of M^n at p is defined as the set of all tangent vectors at p. The function $(d\varphi)_p : T_p(M^n) \to \mathbb{E}^n$ defined by $(d\varphi)_p(\gamma'(0)) = (\varphi \cdot \gamma)'(0)$, is bijective and can be used to transfer the vector space operations from $\mathbb{E}^n$ over to $T_p(M^n)$.

All the tangent spaces $T_p(M^n)$, $p \in M^n$, when "glued together", form the *tangent bundle* $T(M^n)$ of M^n. Any element of $T(M^n)$ is a pair (p, v), where $v \in T_p(M^n)$.

If for an open neighborhood U of p the function $\varphi : U \to \mathbb{R}^n$ is a coordinate chart, then the preimage V of U in $T(M^n)$ admits a mapping $\psi : V \to \mathbb{R}^n \times \mathbb{R}^n$ defined by $\psi(p, v) = (\varphi(p), d\varphi(p))$. It defines the structure of a smooth $2n$-dimensional manifold on $T(M^n)$. The *cotangent bundle* $T^*(M^n)$ of M^n is obtained in similar manner using cotangent spaces $T_p^*(M^n)$, $p \in M^n$.

A *vector field* on a manifold M^n is a *section* of its tangent bundle $T(M^n)$, i.e., a smooth function $f : M^n \to T(M^n)$ which assigns to every point $p \in M^n$ a vector $v \in T_p(M^n)$.

A *connection* (or *covariant derivative*) is a way of specifying a derivative of a vector field along another vector field on a manifold.

Formally, the covariant derivative ∇ of a vector u (defined at a point $p \in M^n$) in the direction of the vector v (defined at the same point p) is a rule that defines a third vector at p, called $\nabla_v u$ which has the properties of a derivative. A Riemannian metric uniquely defines a special covariant derivative called the *Levi-Civita connection*. It is the torsion-free connection ∇ of the tangent bundle, preserving the given Riemannian metric.

The *Riemann curvature tensor* R is the standard way to express the *curvature* of *Riemannian manifolds*. The Riemann curvature tensor can be given in terms of the Levi-Civita connection ∇ by the following formula:

$$R(u, v)w = \nabla_u \nabla_v w - \nabla_v \nabla_u w - \nabla_{[u,v]} w,$$

where $R(u, v)$ is a linear transformation of the tangent space of the manifold M^n; it is linear in each argument. If $u = \frac{\partial}{\partial x_i}$ and $v = \frac{\partial}{\partial x_j}$ are coordinate vector fields, then $[u, v] = 0$, and the formula simplifies to $R(u, v)w = \nabla_u \nabla_v w - \nabla_v \nabla_u w$, i.e., the curvature tensor measures anti-commutativity of the covariant derivative. The linear transformation $w \to R(u, v)w$ is also called the *curvature transformation*.

The *Ricci curvature tensor* (or *Ricci curvature*) *Ric* is obtained as the trace of the full curvature tensor R. It can be thought of as a Laplacian of the Riemannian metric tensor in the case of Riemannian manifolds. Ricci curvature is a linear operator on the tangent space at a point. Given an orthonormal basis $(e_i)_i$ in the tangent space $T_p(M^n)$, we have

$$Ric(u) = \sum_i R(u, e_i)e_i.$$

The value of $Ric(u)$ does not depend on the choice of an orthonormal basis. Starting with dimension four, the Ricci curvature does not describe the curvature tensor completely.

The *Ricci scalar* (or *scalar curvature*) Sc of a Riemannian manifold M^n is the full trace of the curvature tensor; given an orthonormal basis $(e_i)_i$ at $p \in M^n$, we have

$$Sc = \sum_{i,j} \langle R(e_i, e_j)e_j, e_i \rangle = \sum_i \langle Ric(e_i), e_i \rangle.$$

The *sectional curvature* $K(\sigma)$ of a Riemannian manifold M^n is defined as the *Gauss curvature* of an σ-*section* at a point $p \in M^n$, where a σ-*section* is a locally-defined piece of surface which has the 2-plane σ as a tangent plane at p, obtained from geodesics which start at p in the directions of the image of σ under the exponential mapping.

- **Metric tensor**

  The **metric** (or *basic, fundamental*) **tensor** is a symmetric tensor of rank 2, that is used to measure distances and angles in a real n-dimensional differentiable manifold M^n. Once a local coordinate system $(x_i)_i$ is chosen, the metric tensor appears as a real symmetric $n \times n$ matrix $((g_{ij}))$.

  The assignment of a metric tensor on M^n introduces a *scalar product* (i.e., symmetric bilinear, but in general not positive-definite, form) $\langle , \rangle_p$ on the tangent space $T_p(M^n)$ at any $p \in M^n$ defined by

  $$\langle x, y \rangle_p = g_p(x, y) = \sum_{i,j} g_{ij}(p) x_i y_j,$$

  where $x = (x_1, \dots, x_n)$, $y = (y_1, \dots, y_n) \in T_p(M^n)$. The collection of all these scalar products is called the **metric** g with the metric tensor $((g_{ij}))$. The length ds of the vector $(dx_1, \dots, dx_n)$ is expressed by the quadratic differential form

  $$ds^2 = \sum_{i,j} g_{ij} dx_i dx_j,$$

  which is called the *line element* (or *first fundamental form*) of the metric g.

  The *length* of a curve γ is expressed by the formula $\int_\gamma \sqrt{\sum_{i,j} g_{ij} dx_i dx_j}$. In general it may be real, purely imaginary, or zero (an *isotropic curve*).

  Let p, q and r be the numbers of positive, negative and zero *eigenvalues* of the matrix $((g_{ij}))$ of the metric g; so, $p + q + r = n$. The **metric signature** (or, simply, *signature*) of g is the pair (p, q). A **nondegenerated metric** (i.e., one with $r = 0$) is Riemannian or pseudo-Riemannian if its signature is *positive-definite* ($q = 0$) or *indefinite* ($pq > 0$), respectively.

  The **nonmetricity tensor** is the *covariant derivative* of a metric tensor. It is 0 for **Riemannian metrics** but can be $\neq 0$ for **pseudo-Riemannian** ones.

- **Nondegenerate metric**
  A **nondegenerate metric** is a metric g with the metric tensor $((g_{ij}))$, for which the *metric discriminant* $\det((g_{ij})) \neq 0$. All Riemannian and pseudo-Riemannian metrics are nondegenerate.
  A **degenerate metric** is a metric g with $\det((g_{ij})) = 0$ (cf. **semi-Riemannian metric** and **semi-pseudo-Riemannian metric**). A manifold with a degenerate metric is called an *isotropic manifold.*
- **Diagonal metric**
  A **diagonal metric** is a metric g with a metric tensor $((g_{ij}))$ which is zero for $i \neq j$. The Euclidean metric is a diagonal metric, as its metric tensor has the form $g_{ii} = 1, g_{ij} = 0$ for $i \neq j$.
- **Riemannian metric**
  Consider a real n-dimensional differentiable manifold M^n in which each tangent space is equipped with an *inner product* (i.e., a symmetric positive-definite bilinear form) which varies smoothly from point to point.
  A **Riemannian metric** on M^n is a collection of inner products $\langle , \rangle_p$ on the tangent spaces $T_p(M^n)$, one for each $p \in M^n$.
  Every inner product $\langle , \rangle_p$ is completely defined by inner products $\langle e_i, e_j \rangle_p = g_{ij}(p)$ of elements $e_1, \dots, e_n$ of a standard basis in $\mathbb{E}^n$, i.e., by the real symmetric and positive-definite $n \times n$ matrix $((g_{ij})) = ((g_{ij}(p)))$, called a **metric tensor**. In fact, $\langle x, y \rangle_p = \sum_{i,j} g_{ij}(p) x_i y_j$, where $x = (x_1, \dots, x_n)$ and $y = (y_1, \dots, y_n) \in T_p(M^n)$. The smooth function g completely determines the Riemannian metric.
  A Riemannian metric on M^n is not an ordinary metric on M^n. However, for a connected manifold M^n, every Riemannian metric on M^n induces an ordinary metric on M^n, in fact, the **intrinsic metric** of M^n,
  For any points $p, q \in M^n$ the **Riemannian distance** between them is defined as

$$\inf_{\gamma} \int_0^1 \langle \frac{d\gamma}{dt}, \frac{d\gamma}{dt} \rangle^{\frac{1}{2}} dt = \inf_{\gamma} \int_0^1 \sqrt{\sum_{i,j} g_{ij} \frac{dx_i}{dt} \frac{dx_j}{dt}} dt,$$

  where the infimum is over all rectifiable curves $\gamma : [0, 1] \to M^n$, connecting p and q.
  A *Riemannian manifold* (or *Riemannian space*) is a real n-dimensional differentiable manifold M^n equipped with a Riemannian metric. The theory of Riemannian spaces is called *Riemannian Geometry*. The simplest examples of Riemannian spaces are Euclidean spaces, *hyperbolic spaces*, and *elliptic spaces*.
- **Conformal metric**
  A *conformal structure on a vector space* V is a class of pairwise-homothetic Euclidean metrics on V. Any Euclidean metric d_E on V defines a conformal structure $\{\lambda d_E : \lambda > 0\}$.

A *conformal structure on a manifold* is a field of conformal structures on the tangent spaces or, equivalently, a class of *conformally equivalent Riemannian metrics*. Two Riemannian metrics g and h on a smooth manifold M^n are called *conformally equivalent* if $g = f \cdot h$ for some positive function f on M^n, called a *conformal factor*.

A **conformal metric** is a Riemannian metric that represents the conformal structure. Cf. **conformally invariant metric** in Chap. 8.

- **Conformal space**

  The **conformal space** (or *inversive space*) is the Euclidean space $\mathbb{E}^n$ extended by an ideal point (at infinity). Under *conformal* transformations, i.e., continuous transformations preserving local angles, the ideal point can be taken to be an ordinary point. Therefore, in a conformal space a sphere is indistinguishable from a plane: a plane is a sphere passing through the ideal point.

  Conformal spaces are considered in *Conformal* (or *angle-preserving, Möbius*) *Geometry* in which properties of figures are studied that are invariant under conformal transformations. It is the set of transformations that map spheres into spheres, i.e., generated by the Euclidean transformations together with *inversions* which in coordinate form are conjugate to $x_i \to \frac{r^2 x_i}{\sum_j x_j^2}$, where r is the radius of the inversion. An inversion in a sphere becomes an everywhere well defined automorphism of period two. Any angle inverts into an equal angle.

  The 2D conformal space is the *Riemann sphere*, on which the conformal transformations are given by the *Möbius transformations* $z \to \frac{az+b}{cz+d}$, $ad-bc \neq 0$.

  In general, a **conformal mapping** between two Riemannian manifolds is a diffeomorphism between them such that the pulled back metric is *conformally equivalent* to the original one. A *conformal Euclidean space* is a *Riemannian space* admitting a conformal mapping onto an Euclidean space.

  In the General Theory of Relativity, conformal transformations are considered on the *Minkowski space* $\mathbb{R}^{1,3}$ extended by two ideal points.

- **Space of constant curvature**

  A **space of constant curvature** is a *Riemannian space* M^n for which the sectional curvature $K(\sigma)$ is constant in all 2D directions σ.

  A *space form* is a connected complete space of constant curvature k. Examples of a *flat space form*, i.e., with $k = 0$, are the Euclidean space and flat torus. The sphere and hyperbolic space are space forms with $k > 0$ and $k < 0$, respectively.

- **Generalized Riemannian space**

  A **generalized Riemannian space** is a metric space with the **intrinsic metric**, subject to certain restrictions on the curvature. Such spaces include *spaces of bounded curvature*, *Riemannian spaces*, etc. They are defined and investigated on the basis of their metric alone, without coordinates.

  A *space of bounded curvature* ($\leq k$ and $\geq k'$) is defined by the condition: for any sequence of *geodesic triangles* T_n contracting to a point, we have

$$k \geq \overline{\lim} \frac{\overline{\delta}(T_n)}{\sigma(T_n^0)} \geq \underline{\lim} \frac{\overline{\delta}(T_n)}{\sigma(T_n^0)} \geq k',$$

where a *geodesic triangle* $T = xyz$ is the triplet of geodesic segments $[x, y]$, $[y, z]$, $[z, x]$ (the sides of T) connecting in pairs three different points x, y, z, $\delta(T) = \alpha + \beta + \gamma - \pi$ is the *excess* of the geodesic triangle T, and $\sigma(T^0)$ is the area of a Euclidean triangle T^0 with the sides of the same lengths. The **intrinsic metric** on the space of bounded curvature is called a **metric of bounded curvature**.

Such a space turns out to be Riemannian under two additional conditions: local compactness of the space (this ensures the condition of local existence of geodesics), and local extendability of geodesics. If in this case $k = k'$, it is a Riemannian space of constant curvature k (cf. **space of geodesics** in Chap. 6).

A space of curvature $\leq k$ is defined by the condition $\overline{\lim}\frac{\overline{\delta}(T_n)}{\sigma(T_n^0)} \leq k$. In such a space any point has a neighborhood in which the sum $\alpha + \beta + \gamma$ of the angles of a geodesic triangle T does not exceed the sum $\alpha_k + \beta_k + \gamma_k$ of the angles of a triangle T^k with sides of the same lengths in a space of constant curvature k. The intrinsic metric of such space is called a k**-concave metric**.

A space of curvature $\geq k$ is defined by the condition $\underline{\lim}\frac{\overline{\delta}(T_n)}{\sigma(T_n^0)} \geq k$. In such a space any point has a neighborhood in which $\alpha + \beta + \gamma \geq \alpha_k + \beta_k + \gamma_k$ for triangles T and T^k. The intrinsic metric of such space is called a K**-concave metric**.

An *Alexandrov metric space* is a generalized Riemannian space with upper, lower or integral curvature bounds. Cf. a **CAT**(κ_1) **space** in Chap. 6.

- **Complete Riemannian metric**

  A Riemannian metric g on a manifold M^n is called **complete** if M^n forms a complete metric space with respect to g.

  Any Riemannian metric on a compact manifold is complete.

- **Ricci-flat metric**

  A **Ricci-flat metric** is a Riemannian metric with vanished Ricci curvature tensor. A *Ricci-flat manifold* is a Riemannian manifold equipped with a Ricci-flat metric. Ricci-flat manifolds represent vacuum solutions to the *Einstein field equation*, and are special cases of *Kähler–Einstein manifolds*. Important Ricci-flat manifolds are *Calabi–Yau manifolds*, and *hyper-Kähler manifolds*.

- **Osserman metric**

  An **Osserman metric** is a Riemannian metric for which the Riemannian curvature tensor R is *Osserman*, i.e., the eigenvalues of the *Jacobi operator* $\mathcal{J}(x) : y \to R(y, x)x$ are constant on the *unit sphere* S^{n-1} in $\mathbb{E}^n$ (they are independent of the unit vectors x).

- G**-invariant Riemannian metric**

  Given a *Lie group* $(G, \cdot, id)$ of transformations, a Riemannian metric g on a differentiable manifold M^n is called G**-invariant**, if it does not change under any $x \in G$. The group $(G, \cdot, id)$ is called the *group of motions* (or *group of isometries*) of the Riemannian space (M^n, g). Cf. G**-invariant metric** in Chap. 10.

- **Ivanov–Petrova metric**
Let R be the Riemannian curvature tensor of a Riemannian manifold M^n, and let $\{x, y\}$ be an orthogonal basis for an oriented 2-plane π in the tangent space $T_p(M^n)$ at a point p of M^n.
The **Ivanov–Petrova metric** is a Riemannian metric on M^n for which the eigenvalues of the antisymmetric curvature operator $\mathcal{R}(\pi) = R(x, y)$ [IvSt95] depend only on the point p of a Riemannian manifold M^n, but not upon the plane π.
- **Zoll metric**
A **Zoll metric** is a Riemannian metric on a smooth manifold M^n whose geodesics are all simple closed curves of an equal length. A 2D sphere S^2 admits many such metrics, besides the obvious metrics of constant curvature. In terms of cylindrical coordinates (z, θ) ($z \in [-1, 1]$, $\theta \in [0, 2\pi]$), the *line element*

$$ds^2 = \frac{(1 + f(z))^2}{1 - z^2} dz^2 + (1 - z^2) d\theta^2$$

defines a Zoll metric on S^2 for any smooth odd function $f : [-1, 1] \to (-1, 1)$ which vanishes at the endpoints of the interval.
- **Berger metric**
The **Berger metric** is a Riemannian metric on the *Berger sphere* (i.e., the three-sphere S^3 squashed in one direction) defined by the *line element*

$$ds^2 = d\theta^2 + \sin^2\theta d\phi^2 + \cos^2\alpha(d\psi + \cos\theta d\phi)^2,$$

where α is a constant, and θ, ϕ, ψ are *Euler angles*.
- **Cycloidal metric**
The **cycloidal metric** is a Riemannian metric on the half-plane $\mathbb{R}^2_+ = \{x \in \mathbb{R}^2 : x_2 > 0\}$ defined by the *line element*

$$ds^2 = \frac{dx_1^2 + dx_2^2}{2x_2}.$$

It is called *cycloidal* because its geodesics are cycloid curves. The corresponding distance $d(x, y)$ between two points $x, y \in \mathbb{R}^2_+$ is equivalent to the distance

$$\rho(x, y) = \frac{|x_1 - y_1| + |x_2 - y_2|}{\sqrt{x_1} + \sqrt{x_2} + \sqrt{|x_2 - y_2|}}$$

in the sense that $d \leq C\rho$, and $\rho \leq Cd$ for some positive constant C.

- **Klein metric**
The **Klein metric** is a Riemannian metric on the *open unit ball* $B^n = \{x \in \mathbb{R}^n : ||x||_2 < 1\}$ in $\mathbb{R}^n$ defined by

$$\frac{\sqrt{||y||_2^2 - (||x||_2^2||y||_2^2 - \langle x, y\rangle^2)}}{1 - ||x||_2^2}$$

for any $x \in B^n$ and $y \in T_x(B^n)$, where $||.||_2$ is the Euclidean norm on $\mathbb{R}^n$, and $\langle , \rangle$ is the ordinary *inner product* on $\mathbb{R}^n$.
The Klein metric is the hyperbolic case $a = -1$ of the general form

$$\frac{\sqrt{(1 + a||x||^2)||y||^2 - a\langle x, y\rangle^2}}{1 + a||x||^2},$$

while $a = 0, 1$ correspond to the Euclidean and spherical cases.
- **Carnot–Carathéodory metric**
A *distribution* (or *polarization*) on a manifold M^n is a subbundle of the tangent bundle $T(M^n)$ of M^n. Given a distribution $H(M^n)$, a vector field in $H(M^n)$ is called *horizontal*. A curve γ on M^n is called *horizontal* (or *distinguished*, *admissible*) with respect to $H(M^n)$ if $\gamma'(t) \in H_{\gamma(t)}(M^n)$ for any t.
A distribution $H(M^n)$ is called *completely nonintegrable* if the Lie brackets of $H(M^n)$, i.e., $[\cdots, [H(M^n), H(M^n)]]$, span the tangent bundle $T(M^n)$, i.e., for all $p \in M^n$ any tangent vector v from $T_p(M^n)$ can be presented as a linear combination of vectors of the following types: u, $[u, w]$, $[u, [w, t]]$, $[u, [w, [t, s]]], \dots \in T_p(M^n)$, where all vector fields $u, w, t, s, \dots$ are horizontal.
The **Carnot–Carathéodory metric** (or **CC metric**, **sub-Riemannian metric**, *control metric*) is a metric on a manifold M^n with a completely nonintegrable horizontal distribution $H(M^n)$ defined as the section g_C of positive-definite *scalar products* on $H(M^n)$. The distance $d_C(p, q)$ between any points $p, q \in M^n$ is defined as the infimum of the g_C-lengths of the horizontal curves joining p and q.
A *sub-Riemannian manifold* (or *polarized manifold*) is a manifold M^n equipped with a Carnot–Carathéodory metric. It is a generalization of a Riemannian manifold. Roughly, in order to measure distances in a sub-Riemannian manifold, one is allowed to go only along curves tangent to horizontal spaces.
- **Pseudo-Riemannian metric**
Consider a real n-dimensional differentiable manifold M^n in which every tangent space $T_p(M^n)$, $p \in M^n$, is equipped with a *scalar product* which varies smoothly from point to point and is nondegenerate, but indefinite.
A **pseudo-Riemannian metric** on M^n is a collection of scalar products $\langle , \rangle_p$ on the tangent spaces $T_p(M^n)$, $p \in M^n$, one for each $p \in M^n$.
Every scalar product $\langle , \rangle_p$ is completely defined by scalar products $\langle e_i, e_j\rangle_p = g_{ij}(p)$ of elements $e_1, \dots, e_n$ of a standard basis in $\mathbb{E}^n$, i.e., by the real symmetric indefinite $n \times n$ matrix $((g_{ij})) = ((g_{ij}(p)))$, called a **metric tensor**

(cf. **Riemannian metric** in which case this tensor is not only nondegenerate but, moreover, positive-definite).

In fact, $\langle x, y\rangle_p = \sum_{i,j} g_{ij}(p)x_i y_j$, where $x = (x_1, \dots, x_n)$ and $y = (y_1, \dots, y_n) \in T_p(M^n)$. The smooth function g determines the pseudo-Riemannian metric.

The length ds of the vector $(dx_1, \dots, dx_n)$ is given by the quadratic differential form

$$ds^2 = \sum_{i,j} g_{ij} dx_i dx_j .$$

The length of a curve $\gamma : [0, 1] \to M^n$ is expressed by the formula

$$\int_\gamma \sqrt{\sum_{i,j} g_{ij} dx_i dx_j} = \int_0^1 \sqrt{\sum_{i,j} g_{ij} \frac{dx_i}{dt}\frac{dx_j}{dt}} dt.$$

In general it may be real, purely imaginary or zero (an *isotropic curve*).

A pseudo-Riemannian metric on M^n is a metric with a fixed, but indefinite signature (p, q), $p + q = n$. A pseudo-Riemannian metric is nondegenerate, i.e., its metric discriminant $\det((g_{ij})) \neq 0$. Therefore, it is a **nondegenerate indefinite metric**.

A *pseudo-Riemannian manifold* (or *pseudo-Riemannian space*) is a real n-dimensional differentiable manifold M^n equipped with a pseudo-Riemannian metric. The theory of pseudo-Riemannian spaces is called *Pseudo-Riemannian Geometry*.

- **Pseudo-Euclidean distance**

  The model space of a **pseudo-Riemannian space** of signature (p, q) is the *pseudo-Euclidean space* $\mathbb{R}^{p,q}$, $p + q = n$ which is a real n-dimensional vector space $\mathbb{R}^n$ equipped with the metric tensor $((g_{ij}))$ of signature (p, q) defined, for $i \neq j$, by $g_{11} = \dots = g_{pp} = 1$, $g_{p+1,p+1} = \dots = g_{nn} = -1$, $g_{ij} = 0$.

  The *line element* of the corresponding metric is given by

$$ds^2 = dx_1^2 + \dots + dx_p^2 - dx_{p+1}^2 - \dots - dx_n^2.$$

  The **pseudo-Euclidean distance** of signature $(p, q = n - p)$ on $\mathbb{R}^n$ is defined by

$$d_{pE}^2(x, y) = D(x, y) = \sum_{i=1}^{p}(x_i - y_i)^2 - \sum_{i=p+1}^{n}(x_i - y_i)^2.$$

  Such a pseudo-Euclidean space can be seen as $\mathbb{R}^p \times i\mathbb{R}^q$, where $i = \sqrt{-1}$.

  The pseudo-Euclidean space with $(p, q) = (1, 3)$ is used as flat space-time model of Special Relativity; cf. **Minkowski metric** in Chap. 26.

The points correspond to *events*; the line spanned by x and y is *space-like* if $D(x, y) > 0$ and *time-like* if $D(x, y) < 0$. If $D(x, y) > 0$, then $\sqrt{D(x, y)}$ is Euclidean distance and if $D(x, y) < 0$, then $\sqrt{|D(x, y)|}$ is the lifetime of a particle (from x to y).

The pseudo-Euclidean distance of signature $(p, q = n - p)$ is the case $A = \text{diag}(a_i)$ with $a_i = 1$ for $1 \leq i \leq p$ and $a_i = -1$ for $p + 1 \leq i \leq n$, of the **weighted Euclidean distance** $\sqrt{\sum_{1 \leq i \leq n} a_i (x_i - y_i)^2}$ in Chap. 17.

- **Blaschke metric**
  The **Blaschke metric** on a nondegenerate hypersurface is a pseudo-Riemannian metric, associated to the affine normal of the immersion $\phi : M^n \to \mathbb{R}^{n+1}$, where M^n is an n-dimensional manifold, and $\mathbb{R}^{n+1}$ is considered as an affine space.
- **Semi-Riemannian metric**
  A **semi-Riemannian metric** on a real n-dimensional differentiable manifold M^n is a degenerate Riemannian metric, i.e., a collection of positive-semidefinite *scalar products* $\langle x, y \rangle_p = \sum_{i,j} g_{ij}(p) x_i y_j$ on the tangent spaces $T_p(M^n)$, $p \in M^n$; the metric discriminant $det((g_{ij})) = 0$.
  A *semi-Riemannian manifold* (or *semi-Riemannian space*) is a real n-dimensional differentiable manifold M^n equipped with a semi-Riemannian metric.
  The model space of a semi-Riemannian manifold is the *semi-Euclidean space* R^n_d, $d \geq 1$ (sometimes denoted also by $\mathbb{R}^n_{n-d}$), i.e., a real n-dimensional vector space $\mathbb{R}^n$ equipped with a semi-Riemannian metric.
  It means that there exists a scalar product of vectors such that, relative to a suitably chosen basis, the scalar product $\langle x, x \rangle$ has the form $\langle x, x \rangle = \sum_{i=1}^{n-d} x_i^2$.
  The number $d \geq 1$ is called the *defect* (or *deficiency*) of the space.
- **Grushin metric**
  The **Grushin metric** is a semi-Riemannian metric on $\mathbb{R}^2$ defined by the *line element*

$$ds^2 = dx_1^2 + \frac{dx_2^2}{x_1^2}.$$

- **Agmon distance**
  The **Agmon metric** attached to an energy E and a potential V is defined as

$$ds^2 = \max\{0, V(x) - E_0(h)\} dx^2,$$

where dx^2 is the standard metric on $\mathbb{R}^d$. Then the **Agmon distance** on $\mathbb{R}^d$ is the corresponding Riemannian distance defined, for any $x, y \in \mathbb{R}^d$, by

$$\inf_{\gamma}\{\int_0^1 \sqrt{\max\{V(\gamma(s)) - E_0(h), 0\}} \cdot |\gamma'(s)| ds : \gamma(0) = x, \gamma(1) = y, \gamma \in C^1\}.$$

- **Semi-pseudo-Riemannian metric**
A **semi-pseudo-Riemannian metric** on a real n-dimensional differentiable manifold M^n is a degenerate pseudo-Riemannian metric, i.e., a collection of degenerate indefinite *scalar products* $\langle x, y\rangle_p = \sum_{i,j} g_{ij}(p)x_i y_j$ on the tangent spaces $T_p(M^n)$, $p \in M^n$; the metric discriminant $det((g_{ij})) = 0$. In fact, a semi-pseudo-Riemannian metric is a **degenerate indefinite metric**.
A *semi-pseudo-Riemannian manifold* (or *semi-pseudo-Riemannian space*) is a real n-dimensional differentiable manifold M^n equipped with a semi-pseudo-Riemannian metric. The model space of such manifold is the *semi-pseudo-Euclidean space* $\mathbb{R}^n_{\substack{l_1,\dots,l_r\\ m_1,\dots,m_{r-1}}}$, i.e., a vector space $\mathbb{R}^n$ equipped with a semi-pseudo-Riemannian metric.
It means that there exist r scalar products $\langle x, y\rangle_a = \sum \epsilon_{i_a} x_{i_a} y_{i_a}$, where $a = 1, \dots r$, $0 = m_0 < m_1 < \dots < m_r = n$, $i_a = m_{a-1} + 1, \dots m_a$, $\epsilon_{i_a} = \pm 1$, and -1 occurs l_a times among the numbers ϵ_{i_a}. The product $\langle x, y\rangle_a$ is defined for those vectors for which all coordinates $x_i, i \leq m_{a-1}$ or $i > m_a + 1$ are zero.
The first scalar square of an arbitrary vector x is a degenerate quadratic form $\langle x, x\rangle_1 = -\sum_{i=1}^{l_1} x_i^2 + \sum_{j=l_1+1}^{n-d} x_j^2$. The number $l_1 \geq 0$ is called the *index*, and the number $d = n - m_1$ is called the *defect* of the space. If $l_1 = \dots = l_r = 0$, we obtain a *semi-Euclidean space*. The spaces $\mathbb{R}^n_{m}$ and $\mathbb{R}^n_{\substack{k,l\\ m}}$ are called *quasi-Euclidean spaces*.
The *semi-pseudo-non-Euclidean space* $\mathbb{S}^n_{\substack{l_1,\dots,l_r\\ m_1,\dots,m_{r-1}}}$ is a hypersphere in $\mathbb{R}^{n+1}_{\substack{l_1,\dots,l_r\\ m_1,\dots,m_{r-1}}}$ with identified antipodal points. It is called *semielliptic* (or *semi-non-Euclidean*) *space* if $l_1 = \dots = l_r = 0$ and a *semihyperbolic space* if there exist $l_i \neq 0$.
- **Finsler metric**
Consider a real n-dimensional differentiable manifold M^n in which every tangent space $T_p(M^n)$, $p \in M^n$, is equipped with a *Banach norm* $||.||$ such that the Banach norm as a function of position is smooth, and the matrix $((g_{ij}))$,

$$g_{ij} = g_{ij}(p, x) = \frac{1}{2}\frac{\partial^2 ||x||^2}{\partial x_i \partial x_j},$$

is positive-definite for any $p \in M^n$ and any $x \in T_p(M^n)$.
A **Finsler metric** on M^n is a collection of Banach norms $||.||$ on the tangent spaces $T_p(M^n)$, one for each $p \in M^n$. Its *line element* has the form

$$ds^2 = \sum_{i,j} g_{ij} dx_i dx_j.$$

The Finsler metric can be given by *fundamental function*, i.e., a real positive-definite convex function $F(p, x)$ of $p \in M^n$ and $x \in T_p(M^n)$ acting at the point p. $F(p, x)$ is positively homogeneous of degree one in x: $F(p, \lambda x) = \lambda F(p, x)$ for every $\lambda > 0$. Then $F(p, x)$ is the length of the vector x.

The *Finsler metric tensor* has the form $((g_{ij})) = ((\frac{1}{2}\frac{\partial^2 F^2(p,x)}{\partial x_i \partial x_j}))$. The length of a curve $\gamma : [0,1] \to M^n$ is given by $\int_0^1 F(p, \frac{dp}{dt})dt$. For each fixed p the Finsler metric tensor is Riemannian in the variables x.

The Finsler metric is a generalization of the Riemannian metric, where the general definition of the length $||x||$ of a vector $x \in T_p(M^n)$ is not necessarily given in the form of the square root of a symmetric bilinear form as in the Riemannian case.

A *Finsler manifold* (or *Finsler space*) is a real differentiable n-manifold M^n equipped with a Finsler metric. The theory of such spaces is *Finsler Geometry*. The difference between a Riemannian space and a Finsler space is that the former behaves locally like a Euclidean space, and the latter locally like a *Minkowskian space* or, analytically, the difference is that to an ellipsoid in the Riemannian case there corresponds an arbitrary convex surface which has the origin as the center.

A **pseudo-Finsler metric** F is defined by weakening the definition of a Finsler metric): $((g_{ij}))$ should be nondegenerate and of constant signature (not necessarily positive-definite) and hence F could be negative. The pseudo-Finsler metric is a generalization of the pseudo-Riemannian metric.

- **(α, β)-metric**

  Let $\alpha(x,y) = \sqrt{\alpha_{ij}(x)y^i y^j}$ be a Riemannian metric and $\beta(x,y) = b_i(x)y^i$ be a 1-form on a n-dimensional manifold M^n. Let $s = \frac{\beta}{\alpha}$ and $\phi(s)$ is an C^∞-positive function on some symmetric interval $(-r, r)$ with $r > \frac{\beta}{\alpha}$ for all (x, y) in the tangent bundle $TM = \cup_{x \in M} T_x(M^n)$ of the tangent spaces $T_x(M^n)$. Then $F = \alpha\phi(s)$ is a Finsler metric (Matsumoto, 1972) called an **(α, β)-metric**. The main examples of (α, β)-metrics follow.

  The **Kropina metric** is the case $\phi(s) = \frac{1}{s}$, i.e., $F = \frac{\alpha^2}{\beta}$.

  The **generalized Kropina metric** is the case $\phi(s) = s^m$, i.e., $F = \beta^m \alpha^{1-m}$.

  The **Randers metric** (1941) is the case $\phi(s) = 1 + s$, i.e., $F = \alpha + \beta$.

  The **Matsumoto slope metric** is the case $\phi(s) = \frac{1}{1-s}$, i.e., $F = \frac{\alpha^2}{\alpha - \beta}$.

  The **Riemann-type** (α, β)-**metric** is the case $\phi(s) = \sqrt{1+s^2}$, i.e., $F = \alpha^2 + \beta^2$.

  Park and Lee, 1998, considered the case $\phi(s) = 1 + s^2$, i.e., $F = \alpha + \frac{\beta^2}{\alpha}$.

- **Shen metric**

  Given a vector $a \in \mathbb{R}^n$, $||a||_2 < 1$, the **Shen metric** (2003) is a Finsler metric on the *open unit ball* $B^n = \{x \in \mathbb{R}^n : ||x||_2 < 1\}$ in $\mathbb{R}^n$ defined by

$$\frac{\sqrt{||y||_2^2 - (||x||_2^2||y||_2^2 - \langle x, y\rangle^2)} + \langle x, y\rangle}{1 - ||x||_2^2} + \frac{\langle a, y\rangle}{1 + \langle a, x\rangle}$$

  for any $x \in B^n$ and $y \in T_x(B^n)$, where $||.||_2$ is the Euclidean norm on $\mathbb{R}^n$, and $\langle , \rangle$ is the ordinary *inner product* on $\mathbb{R}^n$. It is a **Randers metric** and a **projective metric**. Cf. **Klein metric** and **Berwald metric**.

- **Berwald metric**
  The **Berwald metric** (1929) is a Finsler metric F_{Be} on the *open unit ball* $B^n = \{x \in \mathbb{R}^n : ||x||_2 < 1\}$ in $\mathbb{R}^n$ defined, for any $x \in B^n$ and $y \in T_x(B^n)$, by

$$\frac{\left(\sqrt{||y||_2^2 - (||x||_2^2||y||_2^2 - \langle x, y\rangle^2)} + \langle x, y\rangle\right)^2}{(1 - ||x||_2^2)^2\sqrt{||y||_2^2 - (||x||_2^2||y||_2^2 - \langle x, y\rangle^2)}},$$

  where $||.||_2$ is the Euclidean norm on $\mathbb{R}^n$, and $\langle , \rangle$ is the *inner product* on $\mathbb{R}^n$. It is a **projective metric** and an **(α, β)-metric** with $\phi(s) = (1+s)^2$, i.e., $F = \frac{(\alpha+\beta)^2}{\alpha}$. The Riemannian metrics are special Berwald metrics. Every Berwald metric is affinely equivalent to a Riemannian metric.
  In general, every Finsler metric on a manifold M^n induces a *spray* (second-order homogeneous ordinary differential equation) $y_i \frac{\partial}{\partial x_i} - 2G^i \frac{\partial}{\partial y_i}$ which determines the geodesics. A Finsler metric is a Berwald metric if the spray coefficients $G^i = G^i(x, y)$ are quadratic in $y \in T_x(M^n)$ at any point $x \in M^n$, i.e., $G^i = \frac{1}{2}\Gamma^i_{jk}(x)y^j y^k$.
  A Finsler metric is a more general **Landsberg metric** $\Gamma^i_{jk} = \frac{1}{2}\partial_{y^j}\partial_{y^k}(\Gamma^i_{jk}(x)y^j y^k)$. The Landsberg metric is the one for which the *Landsberg curvature* (the covariant derivative of the *Cartan torsion along a geodesic*) is zero.
- **Douglas metric**
  A **Douglas metric** a Finsler metric for which the *spray coefficients* $G^i = G^i(x, y)$ have the following form:

$$G^i = \frac{1}{2}\Gamma^i_{jk}(x)y_i y_k + P(x, y)y_i.$$

  Every Finsler metric which is projectively equivalent to a **Berwald metric** is a Douglas metric. Every **Berwald metric** is a Douglas metric. Every known Douglas metric is (locally) projectively equivalent to a Berwald metric.
- **Bryant metric**
  Let α be an angle with $|\alpha| < \frac{\pi}{2}$. Let, for any $x, y \in \mathbb{R}^n$, $A = ||y||_2^4 \sin^2 2\alpha + \left(||y||_2^2 \cos 2\alpha + ||x||_2^2||y||_2^2 - \langle x, y\rangle^2\right)^2$, $B = ||y||_2^2 \cos 2\alpha + ||x||_2^2||y||_2^2 - \langle x, y\rangle^2$, $C = \langle x, y\rangle \sin 2\alpha$, $D = ||x||_2^4 + 2||x||_2^2 \cos 2\alpha + 1$. Then we get a Finsler metric

$$\sqrt{\frac{\sqrt{A} + B}{2D} + \left(\frac{C}{D}\right)^2} + \frac{C}{D}.$$

  On the 2D *unit sphere* S^2, it is the **Bryant metric** (1996).

- **m-th root pseudo-Finsler metric**
  An **m-th root pseudo-Finsler metric** is (Shimada, 1979) a **pseudo-Finsler metric** defined (with $a_{i_1 \dots i_m}$ symmetric in all its indices) by

$$F(x, y) = (a_{i_1 \dots i_m}(x) y^{i_1 \dots i_m})^{\frac{1}{m}}.$$

  For $m = 2$, it is a pseudo-Riemannian metric. The 3rd and 4th root pseudo-Finsler metrics are called *cubic metric* and *quartic metric*, respectively.
- **Antonelli–Shimada metric**
  The **Antonelli–Shimada metric** (or *ecological Finsler metric*) is an **m-th root pseudo-Finsler metric** defined, via linearly independent 1-forms a^i, by

$$F(x, y) = \left(\sum_{i=1}^{n} (a^i)^m \right)^{\frac{1}{m}}.$$

  The **Uchijo metric** is defined, for a constant k, by

$$F(x, y) = \left(\sum_{i=1}^{n} (a^i)^2 \right)^{\frac{1}{2}} + ka^1.$$

- **Berwald–Moör metric**
  The **Berwald–Moör metric** is an **m-th root pseudo-Finsler metric**, defined by

$$F(x, y) = (y^1 \dots y^n)^{\frac{1}{n}}.$$

  More general **Asanov metric** is defined, via linearly independent 1-forms a^i, by

$$F(x, y) = (a^1 \dots a^n)^{\frac{1}{n}}.$$

  The Berwald–Moör metrics with $n = 4$ and $n = 6$ are applied in Relativity Theory and Diffusion Imaging, respectively. The pseudo-Finsler spaces which are locally isomorphic to the 4th root Berwald–Moör metric, are expected to be more general and productive space-time models than usual pseudo-Riemannian spaces, which are locally isomorphic to the Minkowski metric.
- **Kawaguchi metric**
  The **Kawaguchi metric** is a metric on a smooth n-dimensional manifold M^n, given by the arc element ds of a regular curve $x = x(t)$, $t \in [t_0, t_1]$ via the formula

$$ds = F(x, \frac{dx}{dt}, \dots, \frac{d^k x}{dt^k})dt,$$

where the *metric function* F satisfies Zermelo's conditions: $\sum_{s=1}^{k} sx^{(s)} F_{(s)i} = F$, $\sum_{s=r}^{k} \binom{s}{k} x^{(s-r+1)i} F_{(s)i} = 0$, $x^{(s)i} = \frac{d^s x^i}{dt^s}$, $F_{(s)i} = \frac{\partial F}{\partial x^{(s)i}}$, and $r = 2, \dots, k$. These conditions ensure that the arc element ds is independent of the parametrization of the curve $x = x(t)$.
A *Kawaguchi manifold* (or *Kawaguchi space*) is a smooth manifold equipped with a Kawaguchi metric. It is a generalization of a *Finsler manifold*.

- **Lagrange metric**
Consider a real n-dimensional manifold M^n. A set of symmetric nondegenerated matrices $((g_{ij}(p,x)))$ define a **generalized Lagrange metric** on M^n if a change of coordinates $(p,x) \to (q,y)$, such that $q_i = q_i(p_1, \dots, p_n)$, $y_i = (\partial_j q_i)x_j$ and rank $(\partial_j q_i) = n$, implies $g_{ij}(p,x) = (\partial_i q_i)(\partial_j q_j) g_{ij}(q,y)$.
A generalized Lagrange metric is called a **Lagrange metric** if there exists a *Lagrangian*, i.e., a smooth function $L(p,x)$ such that it holds

$$g_{ij}(p,x) = \frac{1}{2} \frac{\partial^2 L(p,x)}{\partial x_i \partial x_j}.$$

Every Finsler metric is a Lagrange metric with $L = F^2$.

- **DeWitt supermetric**
The **DeWitt supermetric** (or **Wheeler–DeWitt supermetric**) $G = ((G_{ijkl}))$ calculates distances between metrics on a given manifold, and it is a generalization of a Riemannian (or pseudo-Riemannian) metric $g = ((g_{ij}))$.
For example, for a given connected smooth 3-dimensional manifold M^3, consider the space $\mathcal{M}(M^3)$ of all Riemannian (or pseudo-Riemannian) metrics on M^3. Identifying points of $\mathcal{M}(M^3)$ that are related by a diffeomorphism of M^3, one obtains the space $Geom(M^3)$ of 3-*geometries* (of fixed topology), points of which are the classes of diffeomorphically equivalent metrics. The space $Geom(M^3)$ is called a *superspace*. It plays an important role in several formulations of Quantum Gravity.
A **supermetric**, i.e., a "metric on metrics", is a metric on $\mathcal{M}(M^3)$ (or on $Geom(M^3)$) which is used for measuring distances between metrics on M^3 (or between their equivalence classes). Given $g = ((g_{ij})) \in \mathcal{M}(M^3)$, we obtain

$$||\delta g||^2 = \int_{M^3} d^3x G^{ijkl}(x) \delta g_{ij}(x) \delta g_{kl}(x),$$

where G^{ijkl} is the inverse of the **DeWitt supermetric**

$$G_{ijkl} = \frac{1}{2\sqrt{det((g_{ij}))}} (g_{ik}g_{jl} + g_{il}g_{jk} - \lambda g_{ij}g_{kl}).$$

The value λ parametrizes the distance between metrics in $\mathcal{M}(M^3)$, and may take any real value except $\lambda = \frac{2}{3}$, for which the supermetric is *singular*.

- **Lund–Regge supermetric**
  The **Lund–Regge supermetric** (or **simplicial supermetric**) is an analog of the **DeWitt supermetric**, used to measure the distances between *simplicial 3-geometries* in a *simplicial configuration space*.
  More exactly, given a closed *simplicial* 3D manifold M^3 consisting of several *tetrahedra* (i.e., *3-simplices*), a *simplicial geometry* on M^3 is fixed by an assignment of values to the squared edge lengths of M^3, and a flat Riemannian Geometry to the interior of each tetrahedron consistent with those values.
  The squared edge lengths should be positive and constrained by the triangle inequalities and their analogs for the tetrahedra, i.e., all squared measures (lengths, areas, volumes) must be nonnegative (cf. **tetrahedron inequality** in Chap. 3).
  The set $\mathcal{T}(M^3)$ of all simplicial geometries on M^3 is called a *simplicial configuration space*. The Lund–Regge supermetric $((G_{mn}))$ on $\mathcal{T}(M^3)$ is induced from the DeWitt supermetric on $\mathcal{M}(M^3)$, using for representations of points in $\mathcal{T}(M^3)$ such metrics in $\mathcal{M}(M^3)$ which are piecewise flat in the tetrahedra.
- **Space of Lorentz metrics**
  Let M^n be an n-dimensional compact manifold, and $\mathcal{L}(M^n)$ the set of all **Lorentz metrics** (i.e., the pseudo-Riemannian metrics of signature $(n-1, 1)$) on M^n.
  Given a Riemannian metric g on M^n, one can identify the vector space $S^2(M^n)$ of all symmetric 2-tensors with the vector space of endomorphisms of the tangent to M^n which are symmetric with respect to g. In fact, if $\tilde{h}$ is the endomorphism associated to a tensor h, then the distance on $S^2(M^n)$ is given by

$$d_g(h,t) = \sup_{x \in M^n} \sqrt{tr(\tilde{h}_x - \tilde{t}_x)^2}.$$

  The set $\mathcal{L}(M^n)$ taken with the distance d_g is an open subset of $S^2(M^n)$ called the **space of Lorentz metrics**. Cf. **manifold triangulation metric** in Chap. 9.
- **Perelman supermetric proof**
  The *Thurston's Geometrization Conjecture* is that, after two well-known splittings, any 3D manifold admits, as remaining components, only one of eight *Thurston model geometries*. If true, this conjecture implies the validity of the famous *Poincaré Conjecture* of 1904, that any 3-manifold, in which every simple closed curve can be deformed continuously to a point, is homeomorphic to the 3-sphere.
  In 2002, Perelman gave a gapless "sketch of an eclectic proof" of Thurston's conjecture using a kind of supermetric approach to the space of all Riemannian metrics on a given smooth 3-manifold. In a *Ricci flow* the distances decrease in directions of positive curvature since the metric is time-dependent. Perelman's modification of the standard Ricci flow permitted systematic elimination of arising singularities.

7.2 Riemannian Metrics in Information Theory

Some special Riemannian metrics are commonly used in Information Theory. A list of such metrics is given below.

- **Thermodynamic metrics**
  Given the space of all *extensive* (additive in magnitude, mechanically conserved) thermodynamic variables of a system (energy, entropy, amounts of materials), a **thermodynamic metric** is a Riemannian metric on the manifold of equilibrium states defined as the 2nd derivative of one extensive quantity, usually entropy or energy, with respect to the other extensive quantities. This information geometric approach provides a geometric description of thermodynamic systems in equilibrium.
  The **Ruppeiner metric** (Ruppeiner, 1979) is defined by the *line element* $ds_R^2 = g_{ij}^R dx^i dx^j$, where the matrix $((g_{ij}^R))$ of the symmetric metric tensor is a negative *Hessian* (the matrix of 2nd order partial derivatives) of the entropy function S:

$$g_{ij}^R = -\partial_i \partial_j S(M, N^a).$$

  Here M is the internal energy (which is the mass in black hole applications) of the system and N^a refer to other extensive parameters such as charge, angular momentum, volume, etc. This metric is flat if and only if the statistical mechanical system is noninteracting, while curvature singularities are a signal of critical behavior, or, more precisely, of divergent **correlation lengths** (cf. Chap. 24).
  The **Weinhold metric** (Weinhold, 1975) is defined by $g_{ij}^W = \partial_i \partial_j M(S, N^a)$.
  The Ruppeiner and Weinhold metrics are *conformally equivalent* (cf. **conformal metric**) via $ds^2 = g_{ij}^R dM^i dM^j = \frac{1}{T} g_{ij}^W dS^i dS^j$, where T is the temperature.
  The **thermodynamic length** in Chap. 24 is a path function that measures the distance along a path in the state space.
- **Fisher information metric**
  In Statistics, Probability, and Information Geometry, the **Fisher information metric** is a Riemannian metric for a statistical differential manifold (see, for example, [Amar85, Frie98]). Formally, let $p_\theta = p(x, \theta)$ be a family of densities, indexed by n parameters $\theta = (\theta_1, \dots, \theta_n)$ which form the *parameter manifold* P.
  The **Fisher information metric** $g = g_\theta$ on P is a Riemannian metric, defined by the *Fisher information matrix* $((I(\theta)_{ij}))$, where

$$I(\theta)_{ij} = \mathbb{E}_\theta \left[\frac{\partial \ln p_\theta}{\partial \theta_i} \cdot \frac{\partial \ln p_\theta}{\partial \theta_j} \right] = \int \frac{\partial \ln p(x,\theta)}{\partial \theta_i} \frac{\partial \ln p(x,\theta)}{\partial \theta_j} p(x,\theta) dx.$$

  It is a symmetric bilinear form which gives a classical measure (*Rao measure*) for the statistical distinguishability of distribution parameters.

Putting $i(x,\theta) = -\ln p(x,\theta)$, one obtains an equivalent formula

$$I(\theta)_{ij} = \mathbb{E}_\theta\left[\frac{\partial^2 i(x,\theta)}{\partial\theta_i\partial\theta_j}\right] = \int \frac{\partial^2 i(x,\theta)}{\partial\theta_i\partial\theta_j} p(x,\theta)dx.$$

In a coordinate-free language, we get

$$I(\theta)(u,v) = \mathbb{E}_\theta\left[u(\ln p_\theta)\cdot v(\ln p_\theta)\right],$$

where u and v are vectors tangent to the parameter manifold P, and $u(\ln p_\theta) = \frac{d}{dt}\ln p_{\theta+tu|t=0}$ is the derivative of $\ln p_\theta$ along the direction u.
A *manifold of densities* M is the image of the parameter manifold P under the mapping $\theta \to p_\theta$ with certain regularity conditions. A vector u tangent to this manifold is of the form $u = \frac{d}{dt}p_{\theta+tu|t=0}$, and the Fisher information metric $g = g_p$ on M, obtained from the metric g_θ on P, can be written as

$$g_p(u,v) = \mathbb{E}_p\left[\frac{u}{p}\cdot\frac{v}{p}\right].$$

- **Fisher–Rao metric**
Let $\mathcal{P}_n = \{p \in \mathbb{R}^n : \sum_{i=1}^n p_i = 1, p_i > 0\}$ be the simplex of strictly positive probability vectors. An element $p \in \mathcal{P}_n$ is a density of the n-point set $\{1,\dots,n\}$ with $p(i) = p_i$. An element u of the tangent space $T_p(\mathcal{P}_n) = \{u \in \mathbb{R}^n : \sum_{i=1}^n u_i = 0\}$ at a point $p \in \mathcal{P}_n$ is a function on $\{1,\dots,n\}$ with $u(i) = u_i$.
The **Fisher–Rao metric** g_p on $\mathcal{P}_n$ is a Riemannian metric defined by

$$g_p(u,v) = \sum_{i=1}^n \frac{u_i v_i}{p_i}$$

for any $u, v \in T_p(\mathcal{P}_n)$, i.e., it is the **Fisher information metric** on $\mathcal{P}_n$.
The Fisher–Rao metric is the unique (up to a constant factor) Riemannian metric on $\mathcal{P}_n$, contracting under stochastic maps [Chen72].
This metric is isometric, by $p \to 2(\sqrt{p_1},\dots,\sqrt{p_n})$, with the standard metric on an open subset of the sphere of radius two in $\mathbb{R}^n$. This identification allows one to obtain on $\mathcal{P}_n$ the **geodesic distance**, called the **Rao distance**, by

$$2\arccos\left(\sum_i p_i^{1/2} q_i^{1/2}\right).$$

The Fisher–Rao metric can be extended to the set $\mathcal{M}_n = \{p \in \mathbb{R}^n, p_i > 0\}$ of all finite strictly positive measures on the set $\{1,\dots,n\}$. In this case, the geodesic distance on $\mathcal{M}_n$ can be written as

$$2\left(\sum_i (\sqrt{p_i} - \sqrt{q_i})^2\right)^{1/2}$$

for any $p, q \in \mathcal{M}_n$ (cf. **Hellinger metric** in Chap. 14).

- **Monotone metrics**
  Let M_n be the set of all complex $n \times n$ matrices. Let $\mathcal{M} \subset M_n$ be the manifold of all such positive-definite matrices. Let $\mathcal{D} \subset \mathcal{M}$, $\mathcal{D} = \{\rho \in \mathcal{M} : \mathrm{Tr}\rho = 1\}$, be the submanifold of all *density matrices*. It is the space of faithful states of an n-level quantum system; cf. **distances between quantum states** in Chap. 24.
  The tangent space of $\mathcal{M}$ at $\rho \in \mathcal{M}$ is $T_\rho(\mathcal{M}) = \{x \in M_n : x = x^*\}$, i.e., the set of all $n \times n$ *Hermitian matrices*. The tangent space $T_\rho(\mathcal{D})$ at $\rho \in \mathcal{D}$ is the subspace of *traceless* (i.e., with trace 0) matrices in $T_\rho(\mathcal{M})$.
  A Riemannian metric λ on $\mathcal{M}$ is called **monotone metric** if the inequality

$$\lambda_{h(\rho)}(h(u), h(u)) \leq \lambda_\rho(u, u)$$

  holds for any $\rho \in \mathcal{M}$, any $u \in T_\rho(\mathcal{M})$, and any *stochastic*, i.e., completely positive trace preserving mapping h.
  It was proved in [Petz96] that λ is monotone if and only if it can be written as

$$\lambda_\rho(u, v) = \mathrm{Tr}\ uJ_\rho(v),$$

  where J_ρ is an operator of the form $J_\rho = \frac{1}{f(L_\rho/R_\rho)R_\rho}$. Here L_ρ and R_ρ are the left and the right multiplication operators, and $f : (0, \infty) \to \mathbb{R}$ is an operator monotone function which is *symmetric*, i.e., $f(t) = tf(t^{-1})$, and *normalized*, i.e., $f(1) = 1$. Then $J_\rho(v) = \rho^{-1}v$ if v and ρ are commute, i.e., any monotone metric is equal to the **Fisher information metric** on commutative submanifolds.
  The **Bures metric** (or *statistical metric*) is the smallest monotone metric, obtained for $f(t) = \frac{1+t}{2}$. In this case $J_\rho(v) = g$, $\rho g + g\rho = 2v$, is the *symmetric logarithmic derivative*. For any $\rho_1, \rho_2 \in \mathcal{M}$ the **geodesic distance** defined by the Bures metric, (cf. **Bures length** in Chap. 24) can be written as

$$2\sqrt{\mathrm{Tr}(\rho_1) + \mathrm{Tr}(\rho_2) - 2\mathrm{Tr}(\sqrt{\sqrt{\rho_1}\rho_2\sqrt{\rho_1}})}.$$

  On the submanifold $\mathcal{D} = \{\rho \in \mathcal{M} : \mathrm{Tr}\rho = 1\}$ of density matrices it has the form

$$2 \arccos \mathrm{Tr}(\sqrt{\sqrt{\rho_1}\rho_2\sqrt{\rho_1}}).$$

  The **right logarithmic derivative metric** (or *RLD-metric*) is the greatest monotone metric, corresponding to the function $f(t) = \frac{2t}{1+t}$. In this case $J_\rho(v) = \frac{1}{2}(\rho^{-1}v + v\rho^{-1})$ is the *right logarithmic derivative*.
  The **Bogolubov–Kubo–Mori metric** (or *BKM-metric*) is obtained for $f(x) = \frac{x-1}{\ln x}$. It can be written as $\lambda_\rho(u, v) = \frac{\partial^2}{\partial s \partial t}\mathrm{Tr}(\rho + su)\ln(\rho + tv)|_{s,t=0}$.
- **Wigner–Yanase–Dyson metrics**
  The **Wigner–Yanase–Dyson metrics** (or *WYD-metrics*) form a family of metrics on the manifold $\mathcal{M}$ of all complex positive-definite $n \times n$ matrices defined by

$$\lambda_\rho^\alpha(u,v) = \frac{\partial^2}{\partial t\,\partial s}\mathrm{Tr} f_\alpha(\rho + tu) f_{-\alpha}(\rho + sv)|_{s,t=0},$$

where $f_\alpha(x) = \frac{2}{1-\alpha}x^{\frac{1-\alpha}{2}}$, if $\alpha \neq 1$, and is $\ln x$, if $\alpha = 1$. These metrics are monotone for $\alpha \in [-3, 3]$. For $\alpha = \pm 1$ one obtains the **Bogolubov–Kubo–Mori metric**; for $\alpha = \pm 3$ one obtains the **right logarithmic derivative metric**.
The **Wigner–Yanase metric** (or *WY-metric*) is λ_ρ^0, the smallest metric in the family. It can be written as $\lambda_\rho(u,v) = 4\mathrm{Tr}\ u(\sqrt{L_\rho} + \sqrt{R_\rho})^2(v)$.

- **Connes metric**
Roughly, the **Connes metric** is a generalization (from the space of all probability measures of a set X, to the *state space* of any *unital C^*-algebra*) of the **transportation distance** (Chap. 14) defined via *Lipschitz seminorm*.
Let M^n be a smooth n-dimensional manifold. Let $A = C^\infty(M^n)$ be the (commutative) algebra of smooth complex-valued functions on M^n, represented as multiplication operators on the Hilbert space $H = L^2(M^n, S)$ of square integrable sections of the spinor bundle on M^n by $(f\xi)(p) = f(p)\xi(p)$ for all $f \in A$ and for all $\xi \in H$.
Let D be the *Dirac operator*. Let the commutator $[D, f]$ for $f \in A$ be the *Clifford multiplication* by the gradient ∇f, so that its operator norm $||.||$ in H is given by $||[D, f]|| = \sup_{p\in M^n} ||\nabla f||$.
The **Connes metric** is the **intrinsic metric** on M^n, defined by
$$\sup_{f\in A, ||[D,f]||\leq 1} |f(p) - f(q)|.$$
This definition can also be applied to discrete spaces, and even generalized to *C^*-algebras*; cf. **Rieffel metric space**. In particular, for a labeled connected *locally finite* graph $G = (V, E)$ with the vertex-set $V = \{v_1, \dots, v_n, \dots\}$, the Connes metric on V is defined, for any $v_i, v_j \in V$, by $\sup_{||[D,f]||=||df||\leq 1} |f_{v_i} - f_{v_j}|$, where $\{f = \sum f_{v_i} v_i : \sum |f_{v_i}|^2 < \infty\}$ is the set of formal sums f, forming a Hilbert space, and $||\,[D, f]\,||$ is $\sup_i (\sum_{k=1}^{deg(v_i)} (f_{v_k} - f_{v_i})^2)^{\frac{1}{2}}$.
- **Rieffel metric space**
Let V be a *normed space* (or, more generally, a **locally convex topological vector space**, cf. Chap. 2), and let V' be its **continuous dual space**, i.e., the set of all continuous linear functionals f on V. The *weak-* topology* on V' is defined as the weakest (i.e., with the fewest open sets) topology on V' such that, for every $x \in V$, the map $F_x : V' \to \mathbb{R}$ defined by $F_x(f) = f(x)$ for all $f \in V'$, remains continuous.
An *order-unit space* is a *partially ordered* real (complex) vector space $(A, \preceq)$ with a special distinguished element e (*order unit*) satisfying the following properties:

1. For any $a \in A$, there exists $r \in \mathbb{R}$ with $a \preceq re$;
2. If $a \in A$ and $a \preceq re$ for all positive $r \in \mathbb{R}$, then $a \preceq 0$ (*Archimedean property*).

The main example of an order-unit space is the vector space of all self-adjoint elements in a *unital C^*-algebra* with the identity element being the order unit. Here a *C^*-algebra* is a *Banach algebra* over $\mathbb{C}$ equipped with a special *involution*. It is called *unital* if it has a *unit* (multiplicative identity element); such C^*-algebras are also called, roughly, *compact noncommutative topological spaces*.
Main example of a unital C^*-algebra is the complex algebra of linear operators on a complex **Hilbert space** which is topologically closed in the norm topology of operators, and is closed under the operation of taking adjoints of operators.
The *state space* of an order-unit space $(A, \preceq, e)$ is the set $S(A) = \{f \in A'_+ : ||f|| = 1\}$ of *states*, i.e., continuous linear functionals f with $||f|| = f(e) = 1$.
A **Rieffel** (or *compact quantum* as in Rieffel, 1999) **metric space** is a pair $(A, ||.||_{Lip})$, where $(A, \preceq, e)$ is an order-unit space, and $||.||_{Lip}$ is a $[0, +\infty]$-valued seminorm on A (generalizing the *Lipschitz seminorm*) for which it hold:

1. For $a \in A$, $||a||_{Lip} = 0$ holds if and only if $a \in \mathbb{R}e$;
2. the metric $d_{Lip}(f, g) = \sup_{a \in A: ||a||_{Lip} \leq 1} |f(a) - g(a)|$ generates on the state space $S(A)$ its weak-* topology.

So, $(S(A), d_{Lip})$ is a usual metric space. If the order-unit space $(A, \preceq, e)$ is a C^*-algebra, then d_{Lip} is the **Connes metric**, and if, moreover, the C^*-algebra is noncommutative, $(S(A), d_{Lip})$ is called a **noncommutative metric space**.
The term *quantum* is due to the belief that the Planck-scale geometry of *space-time* comes from such C^*-algebras; cf. **quantum space-time** in Chap. 24.
Kuperberg and Weaver, 2010, proposed a new definition of *quantum metric space*, in the setting of *von Neumann algebras*.

7.3 Hermitian Metrics and Generalizations

A *vector bundle* is a geometrical construct where to every point of a *topological space* M we attach a vector space so that all those vector spaces "glued together" form another topological space E. A continuous mapping $\pi : E \to M$ is called a *projection* E on M. For every $p \in M$, the vector space $\pi^{-1}(p)$ is called a *fiber* of the vector bundle.

A *real (complex) vector bundle* is a vector bundle $\pi : E \to M$ whose fibers $\pi^{-1}(p)$, $p \in M$, are real (complex) vector spaces.

In a real vector bundle, for every $p \in M$, the fiber $\pi^{-1}(p)$ locally looks like the vector space $\mathbb{R}^n$, i.e., there is an *open neighborhood* U of p, a natural number n, and a homeomorphism $\varphi : U \times \mathbb{R}^n \to \pi^{-1}(U)$ such that, for all $x \in U$ and $v \in \mathbb{R}^n$, one has $\pi(\varphi(x, v)) = v$, and the mapping $v \to \varphi(x, v)$ yields an isomorphism between $\mathbb{R}^n$ and $\pi^{-1}(x)$. The set U, together with φ, is called a *local trivialization* of the bundle.

If there exists a "global trivialization", then a real vector bundle $\pi : M \times \mathbb{R}^n \to M$ is called *trivial*. Similarly, in a complex vector bundle, for every $p \in M$, the fiber $\pi^{-1}(p)$ locally looks like the vector space $\mathbb{C}^n$. The basic example of such bundle is the trivial bundle $\pi : U \times \mathbb{C}^n \to U$, where U is an open subset of $\mathbb{R}^k$.

Important special cases of a real vector bundle are the *tangent bundle* $T(M^n)$ and the *cotangent bundle* $T^*(M^n)$ of a *real n-dimensional manifold* $M^n_{\mathbb{R}} = M^n$. Important special cases of a complex vector bundle are the tangent bundle and the cotangent bundle of a *complex n-dimensional manifold*.

Namely, a *complex n-dimensional manifold* $M^n_{\mathbb{C}}$ is a *topological space* in which every point has an open neighborhood homeomorphic to an open set of the n-dimensional complex vector space $\mathbb{C}^n$, and there is an atlas of charts such that the change of coordinates between charts is analytic. The (complex) tangent bundle $T_{\mathbb{C}}(M^n_{\mathbb{C}})$ of a complex manifold $M^n_{\mathbb{C}}$ is a vector bundle of all (complex) *tangent spaces* of $M^n_{\mathbb{C}}$ at every point $p \in M^n_{\mathbb{C}}$. It can be obtained as a *complexification* $T_{\mathbb{R}}(M^n_{\mathbb{R}}) \otimes \mathbb{C} = T(M^n) \otimes \mathbb{C}$ of the corresponding real tangent bundle, and is called the *complexified tangent bundle* of $M^n_{\mathbb{C}}$.

The *complexified cotangent bundle* of $M^n_{\mathbb{C}}$ is obtained similarly as $T^*(M^n) \otimes \mathbb{C}$. Any complex n-dimensional manifold $M^n_{\mathbb{C}} = M^n$ can be regarded as a real $2n$-dimensional manifold equipped with a *complex structure* on each tangent space.

A *complex structure* on a real vector space V is the structure of a complex vector space on V that is compatible with the original real structure. It is completely determined by the operator of multiplication by the number i, the role of which can be taken by an arbitrary linear transformation $J : V \to V$, $J^2 = -id$, where id is the *identity mapping*.

A *connection* (or *covariant derivative*) is a way of specifying a derivative of a *vector field* along another vector field in a vector bundle. A **metric connection** is a linear connection in a vector bundle $\pi : E \to M$, equipped with a bilinear form in the fibers, for which parallel displacement along an arbitrary piecewise-smooth curve in M preserves the form, that is, the *scalar product* of two vectors remains constant under parallel displacement.

In the case of a nondegenerate symmetric bilinear form, the metric connection is called the *Euclidean connection*. In the case of nondegenerate antisymmetric bilinear form, the metric connection is called the *symplectic connection*.

- **Bundle metric**
  A **bundle metric** is a metric on a vector bundle.
- **Hermitian metric**
  A **Hermitian metric** on a complex vector bundle $\pi : E \to M$ is a collection of *Hermitian inner products* (i.e., positive-definite symmetric sesquilinear forms) on every fiber $E_p = \pi^{-1}(p)$, $p \in M$, that varies smoothly with the point p in M. Any complex vector bundle has a Hermitian metric.
  The basic example of a vector bundle is the trivial bundle $\pi : U \times \mathbb{C}^n \to U$, where U is an open set in $\mathbb{R}^k$. In this case a Hermitian inner product on $\mathbb{C}^n$, and hence, a Hermitian metric on the bundle $\pi : U \times \mathbb{C}^n \to U$, is defined by

$$\langle u, v \rangle = u^T H \overline{v},$$

where H is a *positive-definite Hermitian matrix*, i.e., a complex $n \times n$ matrix such that $H^* = \overline{H}^T = H$, and $\overline{v}^T H v > 0$ for all $v \in \mathbb{C}^n \backslash \{0\}$. In the simplest case, one has $\langle u, v \rangle = \sum_{i=1}^n u_i \overline{v}_i$.

An important special case is a Hermitian metric h on a complex manifold M^n, i.e., on the complexified tangent bundle $T(M^n) \otimes \mathbb{C}$ of M^n. This is the Hermitian analog of a Riemannian metric. In this case $h = g + iw$, and its real part g is a Riemannian metric, while its imaginary part w is a nondegenerate antisymmetric bilinear form, called a *fundamental form*. Here $g(J(x), J(y)) = g(x, y)$, $w(J(x), J(y)) = w(x, y)$, and $w(x, y) = g(x, J(y))$, where the operator J is an operator of complex structure on M^n; as a rule, $J(x) = ix$. Any of the forms g, w determines h uniquely.

The term *Hermitian metric* can also refer to the corresponding Riemannian metric g, which gives M^n a Hermitian structure.

On a complex manifold, a Hermitian metric h can be expressed in local coordinates by a *Hermitian symmetric tensor* $((h_{ij}))$:

$$h = \sum_{i,j} h_{ij} dz_i \otimes d\overline{z}_j,$$

where $((h_{ij}))$ is a positive-definite Hermitian matrix. The associated fundamental form w is then written as $w = \frac{i}{2} \sum_{i,j} h_{ij} dz_i \wedge d\overline{z}_j$. A *Hermitian manifold* (or *Hermitian space*) is a complex manifold equipped with a Hermitian metric.

- **Kähler metric**

  A **Kähler metric** (or *Kählerian metric*) is a Hermitian metric $h = g + iw$ on a complex manifold M^n whose fundamental form w is *closed*, i.e., $dw = 0$ holds. A *Kähler manifold* is a complex manifold equipped with a Kähler metric.

  If h is expressed in local coordinates, i.e., $h = \sum_{i,j} h_{ij} dz_i \otimes d\overline{z}_j$, then the associated fundamental form w can be written as $w = \frac{i}{2} \sum_{i,j} h_{ij} dz_i \wedge d\overline{z}_j$, where $\wedge$ is the *wedge product* which is antisymmetric, i.e., $dx \wedge dy = -dy \wedge dx$ (hence, $dx \wedge dx = 0$).

  In fact, w is a *differential 2-form* on M^n, i.e., a tensor of rank 2 that is antisymmetric under exchange of any pair of indices: $w = \sum_{i,j} f_{ij} dx^i \wedge dx^j$, where f_{ij} is a function on M^n. The *exterior derivative* dw of w is defined by $dw = \sum_{i,j} \sum_k \frac{\partial f_{ij}}{\partial x_k} dx_k \wedge dx_i \wedge dx_k$. If $dw = 0$, then w is a *symplectic* (i.e., closed nondegenerate) differential 2-form. Such differential 2-forms are called *Kähler forms*.

  The metric on a Kähler manifold locally satisfies $h_{ij} = \frac{\partial^2 K}{\partial z_i \partial \overline{z}_j}$. for some function K, called the *Kähler potential*. The term *Kähler metric* can also refer to the corresponding Riemannian metric g, which gives M^n a Kähler structure. Then a Kähler manifold is defined as a complex manifold which carries a Riemannian metric and a Kähler form on the underlying real manifold.

- **Hessian metric**
Given a smooth f on an open subset of a real vector space, the associated **Hessian metric** is defined by

$$g_{ij} = \frac{\partial^2 f}{\partial x_i \partial x_j}.$$

A Hessian metric is also called an **affine Kähler metric** since a Kähler metric on a complex manifold has an analogous description as $\frac{\partial^2 f}{\partial z_i \partial \overline{z}_j}$.
- **Calabi–Yau metric**
The **Calabi–Yau metric** is a **Kähler metric** which is **Ricci-flat**.
A *Calabi–Yau manifold* (or *Calabi–Yau space*) is a simply connected complex manifold equipped with a Calabi–Yau metric. It can be considered as a $2n$-dimensional (6D being particularly interesting) smooth manifold with holonomy group (i.e., the set of linear transformations of tangent vectors arising from parallel transport along closed loops) in the special unitary group.
- **Kähler–Einstein metric**
A **Kähler–Einstein metric** is a **Kähler metric** on a complex manifold M^n whose *Ricci curvature tensor* is proportional to the metric tensor. This proportionality is an analog of the *Einstein field equation* in the General Theory of Relativity.
A *Kähler–Einstein manifold* (or *Einstein manifold*) is a complex manifold equipped with a Kähler–Einstein metric. In this case the Ricci curvature tensor, seen as an operator on the tangent space, is just multiplication by a constant.
Such a metric exists on any *domain* $D \subset \mathbb{C}^n$ that is bounded and *pseudo-convex*. It can be given by the *line element*

$$ds^2 = \sum_{i,j} \frac{\partial^2 u(z)}{\partial z_i \partial \overline{z}_j} dz_i d\overline{z}_j,$$

where u is a solution to the *boundary value problem*: $det(\frac{\partial^2 u}{\partial z_i \partial \overline{z}_j}) = e^{2u}$ on D, and $u = \infty$ on ∂D. The Kähler–Einstein metric is a complete metric. On the *unit disk* $\Delta = \{z \in \mathbb{C} : |z| < 1\}$ it is coincides with the **Poincaré metric**.
Let h be the Einstein metric on a smooth compact manifold M^{n-1} without boundary, having scalar curvature $(n-1)(n-2)$. A **generalized Delaunay metric** on $\mathbb{R} \times M^{n-1}$ is (Delay, 2010) of the form $g = u^{\frac{4}{n-2}}(dy^2 + h)$, where $u = u(y) > 0$ is a periodic solution of $u'' - \frac{(n-2)^2}{4}u + \frac{n(n-2)}{4}u^{\frac{n+2}{n-2}} = 0$.
There is one parameter family of constant positive curvature conformal metrics on $\mathbb{R} \times \mathbb{S}^{n-1}$, referred to as **Delaunay metric**; cf. **Kottler metric** in Chap. 26.
- **Hodge metric**
The **Hodge metric** is a **Kähler metric** whose *fundamental form* w defines an integral cohomology class or, equivalently, has integral periods.

A *Hodge manifold* (or *Hodge variety*) is a complex manifold equipped with a Hodge metric. A compact complex manifold is a Hodge manifold if and only if it is isomorphic to a smooth algebraic subvariety of some complex projective space.

- **Fubini–Study metric**

The **Fubini–Study metric** (or *Cayley–Fubini–Study metric*) is a **Kähler metric** on a *complex projective space* $\mathbb{C}P^n$ defined by a *Hermitian inner product* $\langle , \rangle$ in $\mathbb{C}^{n+1}$. It is given by the *line element*

$$ds^2 = \frac{\langle x, x\rangle\langle dx, dx\rangle - \langle x, d\overline{x}\rangle\langle\overline{x}, dx\rangle}{\langle x, x\rangle^2}.$$

The **Fubini–Study distance** between points $(x_1 : \ldots : x_{n+1})$ and $(y_1 : \ldots : y_{n+1}) \in \mathbb{C}P^n$, where $x = (x_1, \ldots, x_{n+1})$ and $y = (y_1, \ldots, y_{n+1}) \in \mathbb{C}^{n+1}\backslash\{0\}$, is equal to

$$\arccos\frac{|\langle x, y\rangle|}{\sqrt{\langle x, x\rangle\langle y, y\rangle}}.$$

The Fubini–Study metric is a **Hodge metric**. The space $\mathbb{C}P^n$ endowed with this metric is called a *Hermitian elliptic space* (cf. **Hermitian elliptic metric**).

- **Bergman metric**

The **Bergman metric** is a **Kähler metric** on a bounded *domain* $D \subset \mathbb{C}^n$ defined, for the *Bergman kernel* $K(z, u)$, by the *line element*

$$ds^2 = \sum_{i,j}\frac{\partial^2 \ln K(z, z)}{\partial z_i \partial\overline{z}_j}dz_i d\overline{z}_j.$$

It is a **biholomorhically invariant metric** on D, and it is complete if D is homogeneous. For the *unit disk* $\Delta = \{z \in \mathbb{C} : |z| < 1\}$ the Bergman metric coincides with the **Poincaré metric**; cf. also **Bergman p-metric** in Chap. 13.

The set of all analytic functions $f \neq 0$ of class $L_2(D)$ with respect to the Lebesgue measure, forms the **Hilbert space** $L_{2,a}(D) \subset L_2(D)$ with an orthonormal basis $(\phi_i)_i$. The *Bergman kernel* is a function in the domain $D \times D \subset \mathbb{C}^{2n}$, defined by $K_D(z, u) = K(z, u) = \sum_{i=1}^{\infty}\phi_i(z)\overline{\phi_i(u)}$.

The **Skwarczynski distance** is defined by

$$(1 \frac{|K(z, u)|}{\sqrt{K(z, z)}\sqrt{K(u, u)}})^{\frac{1}{2}}.$$

- **Hyper-Kähler metric**

A **hyper-Kähler metric** is a Riemannian metric g on a $4n$-dimensional *Riemannian manifold* which is compatible with a quaternionic structure on the tangent bundle of the manifold.

Thus, the metric g is Kählerian with respect to 3 Kähler structures (I, w_I, g), (J, w_J, g), and (K, w_K, g), corresponding to the complex structures, as endomorphisms of the tangent bundle, which satisfy the quaternionic relationship

$$I^2 = J^2 = K^2 = IJK = -JIK = -1.$$

A *hyper-Kähler manifold* is a Riemannian manifold equipped with a hyper-Kähler metric. manifolds are Ricci-flat. Compact 4D hyper-Kähler manifolds are called K_3-*surfaces*; they are studied in Algebraic Geometry.

- **Calabi metric**
  The **Calabi metric** is a **hyper-Kähler metric** on the cotangent bundle $T^*(\mathbb{C}P^{n+1})$ of a *complex projective space* $\mathbb{C}P^{n+1}$.
  For $n = 4k + 4$, this metric can be given by the *line element*

$$ds^2 = \frac{dr^2}{1 - r^{-4}} + \frac{1}{4} r^2(1 - r^{-4})\lambda^2 + r^2(\nu_1^2 + \nu_2^2) + \frac{1}{2}(r^2 - 1)(\sigma_{1\alpha}^2 + \sigma_{2\alpha}^2) + \frac{1}{2}(r^2 + 1)(\Sigma_{1\alpha}^2 + \Sigma_{2\alpha}^2),$$

  where $(\lambda, \nu_1, \nu_2, \sigma_{1\alpha}, \sigma_{2\alpha}, \Sigma_{1\alpha}, \Sigma_{2\alpha})$, with α running over k values, are left-invariant *one-forms* (i.e., linear real-valued functions) on the coset $SU(k + 2)/U(k)$. Here $U(k)$ is the *unitary group* consisting of complex $k \times k$ *unitary matrices*, and $SU(k)$ is the *special unitary group* of complex $k \times k$ unitary matrices with determinant 1.
  For $k = 0$, the Calabi metric coincides with the **Eguchi–Hanson metric**.
- **Stenzel metric**
  The **Stenzel metric** is a **hyper-Kähler metric** on the cotangent bundle $T^*(S^{n+1})$ of a sphere S^{n+1}.
- **$SO(3)$-invariant metric**
  An **$SO(3)$-invariant metric** is a 4D 4-dimensional hyper-Kähler metric with the *line element* given, in the Bianchi type *IX* formalism (cf. **Bianchi metrics** in Chap. 26) by

$$ds^2 = f^2(t)dt^2 + a^2(t)\sigma_1^2 + b^2(t)\sigma_2^2 + c^2(t)\sigma_3^2,$$

  where the invariant *one-forms* $\sigma_1, \sigma_2, \sigma_3$ of $SO(3)$ are expressed in terms of *Euler angles* θ, ψ, ϕ as $\sigma_1 = \frac{1}{2}(\sin\psi d\theta - \sin\theta\cos\psi d\phi)$, $\sigma_2 = -\frac{1}{2}(\cos\psi d\theta + \sin\theta\sin\psi d\phi)$, $\sigma_3 = \frac{1}{2}(d\psi + \cos\theta d\phi)$, and the normalization has been chosen so that $\sigma_i \wedge \sigma_j = \frac{1}{2}\epsilon_{ijk} d\sigma_k$. The coordinate t of the metric can always be chosen so that $f(t) = \frac{1}{2}abc$, using a suitable reparametrization.
- **Atiyah–Hitchin metric**
  The **Atiyah–Hitchin metric** is a **complete regular $SO(3)$-invariant metric** with the *line element*

$$ds^2 = \frac{1}{4}a^2b^2c^2\left(\frac{dk}{k(1-k^2)K^2}\right)^2 + a^2(k)\sigma_1^2 + b^2(k)\sigma_2^2 + c^2(k)\sigma_3^2,$$

where a, b, c are functions of k, $ab = -K(k)(E(k) - K(k))$, $bc = -K(k)(E(k) - (1-k^2)K(k))$, $ac = -K(k)E(k)$, and $K(k)$, $E(k)$ are the complete elliptic integrals, respectively, of the first and second kind, with $0 < k < 1$. The coordinate t is given by the change of variables $t = -\frac{2K(1-k^2)}{\pi K(k)}$ up to an additive constant.

- **Taub–NUT metric**
  The **Taub–NUT metric** (cf. also Chap. 26) is a **complete regular** $SO(3)$-**invariant metric** with the *line element*

$$ds^2 = \frac{1}{4}\frac{r+m}{r-m}dr^2 + (r^2 - m^2)(\sigma_1^2 + \sigma_2^2) + 4m^2\frac{r-m}{r+m}\sigma_3^2,$$

  where m is the relevant moduli parameter, and the coordinate r is related to t by $r = m + \frac{1}{2mt}$. NUT manifold was discovered in Ehlers, 1957, and rediscovered in Newman–Tamburino–Unti, 1963; it is closely related to the metric in Taub, 1951.
- **Eguchi–Hanson metric**
  The **Eguchi–Hanson metric** is a **complete regular** $SO(3)$**-invariant metric** with the *line element*

$$ds^2 = \frac{dr^2}{1-\left(\frac{a}{r}\right)^4} + r^2\left(\sigma_1^2 + \sigma_2^2 + \left(1 - \left(\frac{a}{r}\right)^4\right)\sigma_3^2\right),$$

  where a is the moduli parameter, and the coordinate r is $a\sqrt{\coth(a^2t)}$.
  The Eguchi–Hanson metric coincides with the 4D **Calabi metric**.
- **Complex Finsler metric**
  A **complex Finsler metric** is an upper semicontinuous function $F : T(M^n) \to \mathbb{R}_+$ on a complex manifold M^n with the analytic tangent bundle $T(M^n)$ satisfying the following conditions:

  1. F^2 is smooth on $\check{M}^n$, where $\check{M}^n$ is the complement in $T(M^n)$ of the zero section;
  2. $F(p,x) > 0$ for all $p \in M^n$ and $x \in \check{M}_p^n$;
  3. $F(p,\lambda x) = |\lambda|F(p,x)$ for all $p \in M^n$, $x \in T_p(M^n)$, and $\lambda \in \mathbb{C}$.

  The function $G = F^2$ can be locally expressed in terms of the coordinates $(p_1, \dots, p_n, x_1, \dots, x_n)$; the *Finsler metric tensor* of the complex Finsler metric is given by the matrix$((G_{ij})) = ((\frac{1}{2}\frac{\partial^2 F^2}{\partial x_i \partial \overline{x}_j}))$, called the *Levi matrix*. If the matrix $((G_{ij}))$ is positive-definite, the complex Finsler metric F is called *strongly pseudo-convex*.

- **Distance-decreasing semimetric**
  Let d be a semimetric which can be defined on some class $\mathcal{M}$ of complex manifolds containing the *unit disk* $\Delta = \{z \in \mathbb{C} : |z| < 1\}$. It is called **distance-decreasing** if, for any analytic mapping $f : M_1 \to M_2$ with $M_1, M_2 \in \mathcal{M}$, the inequality $d(f(p), f(q)) \leq d(p, q)$ holds for all $p, q \in M_1$.
  The **Carathéodory semimetric** F_C, **Sibony semimetric** F_S, **Azukawa semimetric** F_A and **Kobayashi semimetric** F_K are distance-decreasing with F_C and F_K being the smallest and the greatest distance-decreasing semimetrics. They are generalizations of the **Poincaré metric** to higher-dimensional domains, since $F_C = F_K$ is the **Poincaré metric** on the unit disk Δ, and $F_C = F_K \equiv 0$ on $\mathbb{C}^n$. It holds $F_C(z, u) \leq F_S(z, u) \leq F_A(z, u) \leq F_K(z, u)$ for all $z \in D$ and $u \in \mathbb{C}^n$. If D is convex, then all these metrics coincide.
- **Biholomorphically invariant semimetric**
  A *biholomorphism* is a bijective *holomorphic* (complex differentiable in a neighborhood of every point in its domain) function whose inverse is also holomorphic.
  A semimetric $F(z, u) : D \times \mathbb{C}^n \to [0, \infty]$ on a domain D in $\mathbb{C}^n$ is called **biholomorphically invariant** if $F(z, u) = |\lambda| F(z, u)$ for all $\lambda \in \mathbb{C}$, and $F(z, u) = F(f(z), f'(z)u)$ for any biholomorphism $f : D \to D'$.
  Invariant metrics, including the **Carathéodory**, **Kobayashi**, **Sibony**, **Azukawa**, **Bergman**, and **Kähler–Einstein** metrics, play an important role in Complex Function Theory, Complex Dynamics and Convex Geometry. The first four metrics are used mostly because they are **distance-decreasing**. But they are almost never Hermitian. On the other hand, the Bergman metric and the Kähler–Einstein metric are Hermitian (in fact, Kählerian), but, in general, not distance-decreasing.
  The **Wu metric** (Cheung and Kim, 1996) is an invariant non-Kähler Hermitian metric on a complex manifold M^n which is distance-decreasing, up to a fixed constant factor, for any holomorphic mapping between two complex manifolds.
- **Kobayashi metric**
  Let D be a *domain* in $\mathbb{C}^n$. Let $\mathcal{O}(\Delta, D)$ be the set of all analytic mappings $f : \Delta \to D$, where $\Delta = \{z \in \mathbb{C} : |z| < 1\}$ is the *unit disk*.
  The **Kobayashi metric** (or **Kobayashi–Royden metric**) F_K is a **complex Finsler metric** defined, for all $z \in D$ and $u \in \mathbb{C}^n$, by

  $$F_K(z, u) = \inf\{\alpha > 0 : \exists f \in \mathcal{O}(\Delta, D), f(0) = z, \alpha f'(0) = u\}.$$

  Given a complex manifold M^n, the **Kobayashi semimetric** F_K is defined by

  $$F_K(p, u) = \inf\{\alpha > 0 : \exists f \in \mathcal{O}(\Delta, M^n), f(0) = p, \alpha f'(0) = u\}$$

  for all $p \in M^n$ and $u \in T_p(M^n)$.
  $F_K(p, u)$ is a seminorm of the tangent vector u, called the *Kobayashi seminorm*. F_K is a metric if M^n is *taut*, i.e., $\mathcal{O}(\Delta, M^n)$ is a *normal family* (every sequence has a subsequence which either converge or diverge compactly).

The Kobayashi semimetric is an infinitesimal form of the **Kobayashi semidistance** (or *Kobayashi pseudo-distance*, 1967) K_{M^n} on M^n, defined as follows. Given $p, q \in M^n$, a *chain of disks* α from p to q is a collection of points $p = p^0, p^1, \dots, p^k = q$ of M^n, pairs of points $a^1, b^1; \dots; a^k, b^k$ of the unit disk Δ, and analytic mappings $f_1, \dots f_k$ from Δ into M^n, such that $f_j(a^j) = p^{j-1}$ and $f_j(b^j) = p^j$ for all j.
The length $l(\alpha)$ of a chain α is the sum $d_P(a^1, b^1) + \dots + d_P(a^k, b^k)$, where d_P is the Poincaré metric. The Kobayashi semimetric K_{M^n} on M^n is defined by

$$K_{M^n}(p, q) = \inf_{\alpha} l(\alpha),$$

where the infimum is taken over all lengths $l(\alpha)$ of chains of disks α from p to q.
Given a complex manifold M^n, the **Kobayashi–Busemann semimetric** on M^n is the double dual of the **Kobayashi semimetric**. It is a metric if M^n is taut.

- **Carathéodory metric**
Let D be a *domain* in $\mathbb{C}^n$. Let $\mathcal{O}(D, \Delta)$ be the set of all analytic mappings $f : D \to \Delta$, where $\Delta = \{z \in \mathbb{C} : |z| < 1\}$ is the *unit disk*.
The **Carathéodory metric** F_C is a **complex Finsler metric** defined by

$$F_C(z, u) = \sup\{|f'(z)u| : f \in \mathcal{O}(D, \Delta)\}$$

for any $z \in D$ and $u \in \mathbb{C}^n$.
Given a complex manifold M^n, the **Carathéodory semimetric** F_C is defined by

$$F_C(p, u) = \sup\{|f'(p)u| : f \in \mathcal{O}(M^n, \Delta)\}$$

for all $p \in M^n$ and $u \in T_p(M^n)$. F_C is a metric if M^n is *taut*, i.e., every sequence in $\mathcal{O}(\Delta, M^n)$ has a subsequence which either converge or diverge compactly.
The **Carathéodory semidistance** (or *Carathéodory pseudo-distance*, 1926) C_{M^n} is a semimetric on a complex manifold M^n, defined by

$$C_{M^n}(p, q) = \sup\{d_P(f(p), f(q)) : f \in \mathcal{O}(M^n, \Delta)\},$$

where d_P is the Poincaré metric.
In general, the integrated semimetric of the infinitesimal Carathéodory semimetric is **internal** for the Carathéodory semidistance, but does not equal to it.

- **Azukawa semimetric**
Let D be a *domain* in $\mathbb{C}^n$. Let $K_D(z)$ be the set of all *logarithmically plurisubharmonic* functions $f : D \to [0, 1]$ such that there exist $M, r > 0$ with $f(u) \le M||u - z||_2$ for all $u \in B(z, r) \subset D$; here $||.||_2$ is the l_2-*norm* on $\mathbb{C}^n$, and $B(z, r) = \{x \in \mathbb{C}^n : ||z - x||_2 < r\}$. Let $g_D(z, u)$ be $\sup\{f(u) : f \in K_D(z)\}$.
The **Azukawa semimetric** F_A is a **complex Finsler metric** defined by

$$F_A(z,u) = \overline{\lim}_{\lambda \to 0} \frac{1}{|\lambda|} g_D(z, z + \lambda u)$$

for all $z \in D$ and $u \in \mathbb{C}^n$.

The Azukawa metric is an infinitesimal form of the **Azukawa semidistance**.

- **Sibony semimetric**

  Let D be a *domain* in $\mathbb{C}^n$. Let $K_D(z)$ be the set of all *logarithmically plurisubharmonic* functions $f : D \to [0, 1)$ such that there exist $M, r > 0$ with $f(u) \leq M||u - z||_2$ for all $u \in B(z, r) = \{x \in \mathbb{C}^n : ||z - x||_2 < r\} \subset D$. Let $C^2_{loc}(z)$ be the set of all functions of class C^2 on some open neighborhood of z.

  The **Sibony semimetric** F_S is a **complex Finsler semimetric** defined by

$$F_S(z,u) = \sup_{f \in K_D(z) \cap C^2_{loc}(z)} \sqrt{\sum_{i,j} \frac{\partial^2 f}{\partial z_i \partial \overline{z}_j}(z) u_i \overline{u}_j}$$

  for all $z \in D$ and $u \in \mathbb{C}^n$.

  The Sibony semimetric is an infinitesimal form of the **Sibony semidistance**.

- **Teichmüller metric**

  A *Riemann surface* R is a one-dimensional complex manifold. Two Riemann surfaces R_1 and R_2 are called *conformally equivalent* if there exists a bijective analytic function (i.e., a conformal homeomorphism) from R_1 into R_2. More precisely, consider a fixed closed Riemann surface R_0 of a given *genus* $g \geq 2$. For a closed Riemann surface R of genus g, one can construct a pair (R, f), where $f : R_0 \to R$ is a homeomorphism. Two pairs (R, f) and (R_1, f_1) are called *conformally equivalent* if there exists a conformal homeomorphism $h : R \to R_1$ such that the mapping $(f_1)^{-1} \cdot h \cdot f : R_0 \to R_0$ is homotopic to the identity.

  An *abstract Riemann surface* $R^* = (R, f)^*$ is the equivalence class of all Riemann surfaces, conformally equivalent to R. The set of all equivalence classes is called the *Teichmüller space* $T(R_0)$ of the surface R_0.

  For closed surfaces R_0 of given genus g, the spaces $T(R_0)$ are isometrically isomorphic, and one can speak of the *Teichmüller space* T_g *of surfaces of genus* g. T_g is a complex manifold. If R_0 is obtained from a compact surface of genus $g \geq 2$ by removing n points, then the complex dimension of T_g is $3g - 3 + n$.

  The **Teichmüller metric** is a metric on T_g defined by

$$\frac{1}{2} \inf_h \ln K(h)$$

  for any $R_1^*, R_2^* \in T_g$, where $h : R_1 \to R_2$ is a quasi-conformal homeomorphism, homotopic to the identity, and $K(h)$ is the *maximal dilation* of h. In fact, there exists a unique extremal mapping, called the *Teichmüller mapping* which minimizes the maximal dilation of all such h, and the distance between R_1^* and R_2^* is equal to $\frac{1}{2}\ln K$, where the constant K is the dilation of the Teichmüller mapping.

In terms of the *extremal length* $ext_{R^*}(\gamma)$, the distance between R_1^* and R_2^* is

$$\frac{1}{2}\ln\sup_{\gamma}\frac{ext_{R_1^*}(\gamma)}{ext_{R_2^*}(\gamma)},$$

where the supremum is taken over all simple closed curves on R_0.

The Teichmüller space T_g, with the Teichmüller metric on it, is a **geodesic** metric space (moreover, a **straight** G**-space**) but it is neither **Gromov hyperbolic**, nor a **Busemann convex** metric space.

The **Thurston quasi-metric** on the *Teichmüller space* T_g is defined by

$$\frac{1}{2}\inf_{h}\ln||h||_{Lip}$$

for any $R_1^*, R_2^* \in T_g$, where $h: R_1 \to R_2$ is a quasi-conformal homeomorphism, homotopic to the identity, and $||.||_{Lip}$ is the *Lipschitz norm* on the set of all injective functions $f: X \to Y$ defined by $||f||_{Lip} = \sup_{x,y\in X, x\neq y}\frac{d_Y(f(x),f(y))}{d_X(x,y)}$.

The *moduli space* R_g of conformal classes of Riemann surfaces of genus g is obtained by factorization of T_g by some countable group of automorphisms of it, called the *modular group*. The **Zamolodchikov metric**, defined (1986) in terms of *exactly marginal operators*, is a natural metric on the conformal moduli spaces.

Liu, Sun and Yau, 2005, showed that all known complete metrics on the Teichmüller space and moduli space (including **Teichmüller metric**, **Bergman metric**, *Cheng–Yau–Mok* **Kähler–Einstein metric**, **Carathéodory metric**, *McMullen metric*) are equivalent since they are **quasi-isometric** (cf. Chap. 1) to the *Ricci metric* and the *perturbed Ricci metric* introduced by them.

- **Weil–Petersson metric**

  The **Weil–Petersson metric** is a **Kähler metric** on the Teichmüller space $T_{g,n}$ of abstract Riemann surfaces of genus g with n punctures and negative Euler characteristic. This metric has negative Ricci curvature; it is **geodesically convex** (cf. Chap. 1) and not complete.

  The **Weil–Peterson metric** is **Gromov hyperbolic** if and only if (Brock and Farb, 2006) the complex dimension $3g-3+n$ of $T_{g,n}$ is at most two.

- **Gibbons–Manton metric**

  The **Gibbons–Manton metric** is a $4n$-dimensional **hyper-Kähler metric** on the moduli space of n*-monopoles* which admits an isometric action of the n-dimensional torus T^n. It is a hyper-Kähler quotient of a flat quaternionic vector space.

- **Metrics on determinant lines**
  Let M^n be an n-dimensional compact smooth manifold, and let F be a flat vector bundle over M^n. Let $H^{\bullet}(M^n, F) = \oplus_{i=0}^{n} H^i(M^n, F)$ be the *de Rham cohomology* of M^n with coefficients in F. Given an n-dimensional vector space V, the *determinant line det* V of V is defined as the top exterior power of V, i.e., $det\ V = \wedge^n V$. Given a finite-dimensional graded vector space $V = \oplus_{i=0}^{n} V_i$, the determinant line of V is defined as the tensor product $det\ V = \otimes_{i=0}^{n}(det V_i)^{(-1)^i}$. Thus, the determinant line $det H^{\bullet}(M^n, F)$ of the cogomology $H^{\bullet}(M^n, F)$ can be written as $det H^{\bullet}(M^n, F) = \otimes_{i=0}^{n}(det H^i(M^n, F))^{(-1)^i}$.
  The **Reidemeister metric** is a metric on $det H^{\bullet}(M^n, F)$, defined by a given smooth triangulation of M^n, and the classical *Reidemeister–Franz torsion*.
  Let g^F and $g^{T(M^n)}$ be smooth metrics on the vector bundle F and tangent bundle $T(M^n)$, respectively. These metrics induce a canonical **L_2-metric** $h^{H^{\bullet}(M^n,F)}$ on $H^{\bullet}(M^n, F)$. The **Ray–Singler metric** on $det H^{\bullet}(M^n, F)$ is defined as the product of the metric induced on $det H^{\bullet}(M^n, F)$ by $h^{H^{\bullet}(M^n,F)}$ with the *Ray–Singler analytic torsion*. The **Milnor metric** on $det H^{\bullet}(M^n, F)$ can be defined in a similar manner using the *Milnor analytic torsion*. If g^F is flat, the above two metrics coincide with the Reidemeister metric. Using a co-Euler structure, one can define a *modified Ray–Singler metric* on $det H^{\bullet}(M^n, F)$.
  The **Poincaré–Reidemeister metric** is a metric on the cohomological determinant line $det H^{\bullet}(M^n, F)$ of a closed connected oriented odd-dimensional manifold M^n. It can be constructed using a combination of the Reidemeister torsion with the Poincaré duality. Equivalently, one can define the *Poincaré–Reidemeister scalar product* on $det H^{\bullet}(M^n, F)$ which completely determines the Poincaré–Reidemeister metric but contains an additional sign or phase information.
  The **Quillen metric** is a metric on the inverse of the cohomological determinant line of a compact Hermitian one-dimensional complex manifold. It can be defined as the product of the L_2-metric with the Ray-Singler analytic torsion.
- **Kähler supermetric**
  The **Kähler supermetric** is a generalization of the **Kähler metric** for the case of a *supermanifold*. A *supermanifold* is a generalization of the usual manifold with *fermonic* as well as *bosonic* coordinates. The bosonic coordinates are ordinary numbers, whereas the fermonic coordinates are *Grassmann numbers*.
  Here the term *supermetric* differs from the one used in this chapter.
- **Hofer metric**
  A *symplectic manifold* (M^n, w), $n = 2k$, is a smooth even-dimensional manifold M^n equipped with a *symplectic form*, i.e., a closed nondegenerate 2-form, w.
  A *Lagrangian manifold* is a k-dimensional smooth submanifold L^k of a symplectic manifold (M^n, w), $n = 2k$, such that the form w vanishes identically on L^k, i.e., for any $p \in L^k$ and any $x, y \in T_p(L^k)$, one has $w(x, y) = 0$.

Let $L(M^n, \Delta)$ be the set of all Lagrangian submanifolds of a closed symplectic manifold (M^n, w), diffeomorphic to a given Lagrangian submanifold Δ. A smooth family $\alpha = \{L_t\}_t$, $t \in [0, 1]$, of Lagrangian submanifolds $L_t \in L(M^n, \Delta)$ is called an *exact path* connecting L_0 and L_1, if there exists a smooth mapping $\Psi : \Delta \times [0, 1] \to M^n$ such that, for every $t \in [0, 1]$, one has $\Psi(\Delta \times \{t\}) = L_t$, and $\Psi * w = dH_t \wedge dt$ for some smooth function $H : \Delta \times [0, 1] \to \mathbb{R}$. The *Hofer length* $l(\alpha)$ of an exact path α is defined by $l(\alpha) = \int_0^1 \{\max_{p\in\Delta} H(p, t) - \min_{p\in\Delta} H(p, t)\}dt$.
The **Hofer metric** on the set $L(M^n, \Delta)$ is defined by

$$\inf_{\alpha} l(\alpha)$$

for any $L_0, L_1 \in L(M^n, \Delta)$, where the infimum is taken over all exact paths on $L(M^n, \Delta)$, that connect L_0 and L_1.
The Hofer metric can be defined similarly on the group $Ham(M^n, w)$ of *Hamiltonian diffeomorphisms* of a closed symplectic manifold (M^n, w), whose elements are *time-one mappings* of *Hamiltonian flows* ϕ_t^H: it is $\inf_\alpha l(\alpha)$, where the infimum is taken over all smooth paths $\alpha = \{\phi_t^H\}$, $t \in [0, 1]$, connecting ϕ and ψ.

- **Sasakian metric**
  A **Sasakian metric** is a metric on a *contact manifold*, naturally adapted to the *contact structure*.
  A contact manifold equipped with a Sasakian metric is called a *Sasakian space*, and it is an odd-dimensional analog of a *Kähler manifold*. The scalar curvature of a Sasakian metric which is also **Einstein metric**, is positive.
- **Cartan metric**
  A *Killing form* (or *Cartan–Killing form*) on a finite-dimensional *Lie algebra* Ω over a field $\mathbb{F}$ is a symmetric bilinear form

$$B(x, y) = \mathrm{Tr}(ad_x \cdot ad_y),$$

where Tr denotes the trace of a linear operator, and ad_x is the image of x under *the adjoint representation* of Ω, i.e., the linear operator on the vector space Ω defined by the rule $z \to [x, z]$, where $[,]$ is the Lie bracket.
Let $e_1, \dots e_n$ be a basis for the Lie algebra Ω, and $[e_i, e_j] = \sum_{k=1}^n \gamma_{ij}^k e_k$, where γ_{ij}^k are corresponding *structure constants*. Then the Killing form is given by

$$B(x_i, x_j) = g_{ij} = \sum_{k,l=1}^{n} \gamma_{il}^k \gamma_{ik}^l.$$

In Theoretical Physics, the **metric tensor** $((g_{ij}))$ is called a **Cartan metric**.

Chapter 8
Distances on Surfaces and Knots

8.1 General Surface Metrics

A *surface* is a real 2D (two-dimensional) *manifold* M^2, i.e., a **Hausdorff space**, each point of which has a neighborhood which is homeomorphic to a plane $\mathbb{E}^2$, or a closed half-plane (cf. Chap. 7).

A compact orientable surface is called *closed* if it has no boundary, and it is called a *surface with boundary*, otherwise. There are compact nonorientable surfaces (closed or with boundary); the simplest such surface is the *Möbius strip*. Noncompact surfaces without boundary are called *open*.

Any closed connected surface is homeomorphic to either a sphere with, say, g (cylindric) handles, or a sphere with, say, g *cross-caps* (i.e., caps with a twist like Möbius strip in them). In both cases the number g is called the *genus* of the surface. In the case of handles, the surface is orientable; it is called a *torus* (doughnut), *double torus*, and *triple torus* for $g = 1, 2$ and 3, respectively. In the case of cross-caps, the surface is nonorientable; it is called the *real projective plane*, *Klein bottle*, and *Dyck's surface* for $g = 1, 2$ and 3, respectively. The genus is the maximal number of disjoint simple closed curves which can be cut from a surface without disconnecting it (the *Jordan curve theorem* for surfaces).

The *Euler–Poincaré characteristic* of a surface is (the same for all polyhedral decompositions of a given surface) the number $\chi = v - e + f$, where v, e and f are, respectively, the number of vertices, edges and faces of the decomposition. Then $\chi = 2 - 2g$ if the surface is orientable, and $\chi = 2 - g$ if not. Every surface with boundary is homeomorphic to a sphere with an appropriate number of (disjoint) *holes* (i.e., what remains if an open disk is removed) and handles or cross-caps. If h is the number of holes, then $\chi = 2 - 2g - h$ holds if the surface is orientable, and $\chi = 2 - g - h$ if not.

The *connectivity number* of a surface is the largest number of closed cuts that can be made on the surface without separating it into two or more parts. This number is equal to $3 - \chi$ for closed surfaces, and $2 - \chi$ for surfaces with boundaries. A surface

M.M. Deza, E. Deza, *Encyclopedia of Distances*,
DOI 10.1007/978-3-662-44342-2_8

with connectivity number 1, 2 and 3 is called, respectively, *simply, doubly* and *triply connected*. A sphere is simply connected, while a torus is triply connected.

A surface can be considered as a metric space with its own **intrinsic metric**, or as a figure in space. A surface in $\mathbb{E}^3$ is called *complete* if it is a **complete** metric space with respect to its intrinsic metric.

Useful *shape-aware* (preserved by isomorphic deformations of the surface) distances on the interior of a surface mesh can be defined by isometric embedding of the surface into a suitable high-dimensional Euclidean space; for example, **diffusion metric** (cf. Chap. 15 and **histogram diffusion distance** from Chap. 21) and Rustamov et al., 2009.

A surface is called *differentiable*, *regular*, or *analytic*, respectively, if in a neighborhood of each of its points it can be given by an expression

$$r = r(u, v) = r(x_1(u, v), x_2(u, v), x_3(u, v)),$$

where the *position vector* $r = r(u, v)$ is a differentiable, *regular* (i.e., a sufficient number of times differentiable), or *real analytic*, respectively, vector function satisfying the condition $r_u \times r_v \neq 0$.

Any regular surface has the intrinsic metric with the *line element* (or *first fundamental form*)

$$ds^2 = dr^2 = E(u, v)du^2 + 2F(u, v)du\,dv + G(u, v)dv^2,$$

where $E(u, v) = \langle r_u, r_u \rangle$, $F(u, v) = \langle r_u, r_v \rangle$, $G(u, v) = \langle r_v, r_v \rangle$. The length of a curve defined on the surface by the equations $u = u(t)$, $v = v(t)$, $t \in [0, 1]$, is computed by

$$\int_0^1 \sqrt{Eu'^2 + 2Fu'v' + Gv'^2}dt,$$

and the distance between any points $p, q \in M^2$ is defined as the infimum of the lengths of all curves on M^2, connecting p and q. A **Riemannian metric** is a generalization of the first fundamental form of a surface.

For surfaces, two kinds of *curvature* are considered: *Gaussian curvature*, and *mean curvature*. To compute these curvatures at a given point of the surface, consider the intersection of the surface with a plane, containing a fixed *normal vector*, i.e., a vector which is perpendicular to the surface at this point. This intersection is a plane curve. The *curvature* k of this plane curve is called the *normal curvature* of the surface at the given point. If we vary the plane, the normal curvature k will change, and there are two extremal values, the *maximal curvature* k_1, and the *minimal curvature* k_2, called the *principal curvatures* of the surface. A curvature is taken to be *positive* if the curve turns in the same direction as the surface's chosen normal, otherwise it is taken to be *negative*.

The *Gaussian curvature* is $K = k_1 k_2$ (it can be given entirely in terms of the first fundamental form). The *mean curvature* is $H = \frac{1}{2}(k_1 + k_2)$.

A *minimal surface* is a surface with mean curvature zero or, equivalently, a surface of minimum area subject to constraints on the location of its boundary.

A *Riemann surface* is a one-dimensional *complex manifold*, or a 2D real manifold with a complex structure, i.e., in which the local coordinates in neighborhoods of points are related by complex analytic functions. It can be thought of as a deformed version of the complex plane. All Riemann surfaces are orientable. Closed Riemann surfaces are geometrical models of *complex algebraic curves*. Every connected Riemann surface can be turned into a *complete* 2D *Riemannian manifold* with constant curvature $-1, 0$, or 1. The Riemann surfaces with curvature -1 are called *hyperbolic*, and the *unit disk* $\Delta = \{z \in \mathbb{C} : |z| < 1\}$ is the canonical example. The Riemann surfaces with curvature 0 are called *parabolic*, and $\mathbb{C}$ is a typical example. The Riemann surfaces with curvature 1 are called *elliptic*, and the *Riemann sphere* $\mathbb{C} \cup \{\infty\}$ is a typical example.

- **Regular metric**

  The intrinsic metric of a surface is **regular** if it can be specified by the *line element*

  $$ds^2 = E du^2 + 2F du dv + G dv^2,$$

  where the coefficients of the form ds^2 are regular functions.

  Any regular surface, given by an expression $r = r(u, v)$, has a regular metric with the *line element* ds^2, where $E(u, v) = \langle r_u, r_u \rangle$, $F(u, v) = \langle r_u, r_v \rangle$, $G(u, v) = \langle r_v, r_v \rangle$.
- **Analytic metric**

  The intrinsic metric on a surface is **analytic** if it can be specified by the *line element*

  $$ds^2 = E du^2 + 2F du dv + G dv^2,$$

  where the coefficients of the form ds^2 are real analytic functions.

  Any analytic surface, given by an expression $r = r(u, v)$, has an analytic metric with the *line element* ds^2, where $E(u, v) = \langle r_u, r_u \rangle$, $F(u, v) = \langle r_u, r_v \rangle$, $G(u, v) = \langle r_v, r_v \rangle$.
- **Metric of nonpositive curvature**

  A **metric of nonpositive curvature** is the intrinsic metric on a *saddle-like surface*. A *saddle-like surface* is a generalization of a surface of negative curvature: a twice continuously-differentiable surface is a saddle-like surface if and only if at each point of the surface its Gaussian curvature is nonpositive.

  These surfaces can be seen as antipodes of *convex surfaces*, but they do not form such a natural class of surfaces as do convex surfaces.

  A **metric of negative curvature** is the intrinsic metric on a *surface of negative curvature*, i.e., a surface in $\mathbb{E}^3$ that has negative Gaussian curvature at every point.

A surface of negative curvature locally has a saddle-like structure. The intrinsic geometry of a surface of constant negative curvature (in particular, of a *pseudo-sphere*) locally coincides with the geometry of the *Lobachevsky plane*. There exists no surface in $\mathbb{E}^3$ whose intrinsic geometry coincides completely with the geometry of the Lobachevsky plane (i.e., a complete regular surface of constant negative curvature).

- **Metric of nonnegative curvature**
  A **metric of nonnegative curvature** is the intrinsic metric on a *convex surface*.
  A *convex surface* is a *domain* (i.e., a connected open set) on the boundary of a *convex body* in $\mathbb{E}^3$ (in some sense, it is an antipode of a saddle-like surface).
  The entire boundary of a convex body is called a *complete convex surface*. If the body is finite (bounded), the complete convex surface is called *closed*. Otherwise, it is called *infinite* (an infinite convex surface is homeomorphic to a plane or to a circular cylinder).
  Any convex surface M^2 in $\mathbb{E}^3$ is a *surface of bounded curvature*. The *total Gaussian curvature* $w(A) = \int\int_A K(x)d\sigma(x)$ of a set $A \subset M^2$ is always nonnegative (here $\sigma(.)$ is the *area*, and $K(x)$ is the *Gaussian curvature* of M^2 at a point x), i.e., a convex surface can be seen as a *surface of nonnegative curvature*.
  The intrinsic metric of a convex surface is a **convex metric** (not to be confused with **metric convexity** from Chap. 1) in the sense of Surface Theory, i.e., it displays the *convexity condition*: the sum of the angles of any triangle whose sides are shortest curves is not less that π.
  A **metric of positive curvature** is the intrinsic metric on a *surface of positive curvature*, i.e., a surface in $\mathbb{E}^3$ that has positive Gaussian curvature at every point.
- **Metric with alternating curvature**
  A **metric with alternating curvature** is the intrinsic metric on a surface with alternating (positive or negative) Gaussian curvature.
- **Flat metric**
  A **flat metric** is the intrinsic metric on a *developable surface*, i.e., a surface, on which the Gaussian curvature is everywhere zero. Cf. **flat space** in Chap. 1.
  In general, a Riemannian metric on a surface is locally Euclidean up to a third order error (distortion of metric) measured by the Gaussian curvature.
- **Metric of bounded curvature**
  A **metric of bounded curvature** is the intrinsic metric ρ on a *surface of bounded curvature*.
  A surface M^2 with an intrinsic metric ρ is called a *surface of bounded curvature* if there exists a sequence of **Riemannian metrics** ρ_n defined on M^2, such that $\rho_n \to \rho$ uniformly for any compact set $A \subset M^2$, and the sequence $|w_n|(A)$ is bounded, where $|w|_n(A) = \int\int_A |K(x)|d\sigma(x)$ is the *total absolute curvature* of the metric ρ_n (here $K(x)$ is the Gaussian curvature of M^2 at a point x, and $\sigma(.)$ is the *area*).
- **Λ-Metric**
  A **Λ-metric** (or *metric of type* Λ) is a **complete** metric on a surface with curvature bounded from above by a negative constant.

A Λ-metric does not have embeddings into $\mathbb{E}^3$. It is a generalization of the result in Hilbert, 1901: no complete regular surface of constant negative curvature (i.e., a surface whose intrinsic geometry is the geometry of the Lobachevsky plane) exists in $\mathbb{E}^3$.

- **(h, Δ)-metric**
  A **(h, Δ)-metric** is a metric on a surface with a slowly-changing negative curvature.
  A **complete** (h, Δ)-metric does not permit a regular *isometric embedding* in three-dimensional Euclidean space (cf. **Λ-metric**).
- **G-distance**
  A connected set G of points on a surface M^2 is called a *geodesic region* if, for each point $x \in G$, there exists a *disk* $B(x, r)$ with center at x, such that $B_G = G \cap B(x, r)$ has one of the following forms: $B_G = B(x, r)$ (x is a *regular interior point* of G); B_G is a semidisk of $B(x, r)$ (x is a *regular boundary point* of G); B_G is a sector of $B(x, r)$ other than a semidisk (x is an *angular point* of G); B_G consists of a finite number of sectors of $B(x, r)$ with no common points except x (a *nodal point* of G).
  The **G-distance** between any x and $y \in G$ is the greatest lower bound of the lengths of all rectifiable curves connecting x and $y \in G$ and completely contained in G.
- **Conformally invariant metric**
  Let R be a Riemann surface. A *local parameter* (or *local uniformizing parameter*, *local uniformizer*) is a complex variable z considered as a continuous function $z_{p_0} = \phi_{p_0}(p)$ of a point $p \in R$ which is defined everywhere in some neighborhood (*parametric neighborhood*) $V(p_0)$ of a point $p_0 \in R$ and which realizes a homeomorphic mapping (*parametric mapping*) of $V(p_0)$ onto the disk (*parametric disk*) $\Delta(p_0) = \{z \in \mathbb{C} : |z| < r(p_0)\}$, where $\phi_{p_0}(p_0) = 0$. Under a parametric mapping, any point function $g(p)$ defined in the parametric neighborhood $V(p_0)$, goes into a function of the local parameter z: $g(p) = g(\phi_{p_0}^{-1}(z)) = G(z)$.
  A **conformally invariant metric** is a differential $\rho(z)|dz|$ on the Riemann surface R which is invariant with respect to the choice of the local parameter z. Thus, to each local parameter z ($z : U \to \overline{\mathbb{C}}$) a function $\rho_z : z(U) \to [0, \infty]$ is associated such that, for any local parameters z_1 and z_2, we have
  $$\frac{\rho_{z_2}(z_2(p))}{\rho_{z_1}(z_1(p))} = \left|\frac{dz_1(p)}{dz_2(p)}\right| \text{ for any } p \in U_1 \cap U_2.$$
  Every linear differential $\lambda(z)dz$ and every *quadratic differential* $Q(z)dz^2$ induce conformally invariant metrics $|\lambda(z)||dz|$ and $|Q(z)|^{1/2}|dz|$, respectively (cf. **Q-metric**).
- **Q-metric**
  An **Q-metric** is a **conformally invariant metric** $\rho(z)|dz| = |Q(z)|^{1/2}|dz|$ on a Riemann surface R defined by a *quadratic differential* $Q(z)dz^2$.

A *quadratic differential* $Q(z)dz^2$ is a nonlinear differential on a Riemann surface R which is invariant with respect to the choice of the local parameter z. Thus, to each local parameter z ($z : U \to \overline{\mathbb{C}}$) a function $Q_z : z(U) \to \overline{\mathbb{C}}$ is associated such that, for any local parameters z_1 and z_2, we have

$$\frac{Q_{z_2}(z_2(p))}{Q_{z_1}(z_1(p))} = \left(\frac{dz_1(p)}{dz_2(p)}\right)^2 \quad \text{for any } p \in U_1 \cap U_2.$$

- **Extremal metric**
  Let Γ be a family of locally rectifiable curves on a Riemann surface R and let P be a class of **conformally invariant metrics** $\rho(z)|dz|$ on R such that $\rho(z)$ is square-integrable in the z-plane for every local parameter z, and the following Lebesgue integrals are not simultaneously equal to 0 or ∞:

  $$A_\rho(R) = \int\int_R \rho^2(z)dx\,dy \text{ and } L_\rho(\Gamma) = \inf_{\gamma \in \Gamma} \int_y \rho(z)|dz|.$$

  The *modulus of the family of curves* Γ is defined by

  $$M(\Gamma) = \inf_{\rho \in P} \frac{A_\rho(R)}{(L_\rho(\Gamma))^2}.$$

  The *extremal length of the family of curves* Γ is the reciprocal of $M(\Gamma)$.
  Let P_L be the subclass of P such that, for any $\rho(z)|dz| \in P_L$ and any $\gamma \in \Gamma$, one has $\int_\gamma \rho(z)|dz| \geq 1$. If $P_L \neq \emptyset$, then $M(\Gamma) = \inf_{\rho \in P_L} A_\rho(R)$. Every metric from P_L is called an *admissible metric* for the modulus on Γ. If there exists ρ^* for which

  $$M(\Gamma) = \inf_{\rho \in P_L} A_\rho(R) = A_{\rho^*}(R),$$

  the metric $\rho^*|dz|$ is called an **extremal metric** for the modulus on Γ. It is a **conformally invariant metric**.
- **Fréchet surface metric**
  Let (X, d) be a metric space, M^2 a compact 2D manifold, f a continuous mapping $f : M^2 \to X$, called a *parametrized surface*, and $\sigma : M^2 \to M^2$ a homeomorphism of M^2 onto itself. Two parametrized surfaces f_1 and f_2 are called *equivalent* if $\inf_\sigma \max_{p \in M^2} d(f_1(p), f_2(\sigma(p))) = 0$, where the infimum is taken over all possible homeomorphisms σ. A class f^* of parametrized surfaces, equivalent to f, is called a *Fréchet surface*. It is a generalization of the notion of a surface in Euclidean space to the case of an arbitrary metric space (X, d).
  The **Fréchet surface metric** on the set of all Fréchet surfaces is defined by

  $$\inf_\sigma \max_{p \in M^2} d(f_1(p), f_2(\sigma(p)))$$

for any Fréchet surfaces f_1^* and f_2^*, where the infimum is taken over all possible homeomorphisms σ. Cf. the **Fréchet metric** in Chap. 1.

- **Hempel metric**
  A *handlebody* of genus g is the boundary sum of g copies of a solid torus; it is homeomorphic to the closure of a regular neighborhood of some finite graph in $\mathbb{R}^3$. Given a closed orientable 3-manifold M, its *Heegaard splitting* (of genus g) is $M = A \cup_P B$ where A, B are genus g handlebodies in M such that $M = A \cup B$ and $A \cap B = \partial A = \partial B = P$. Then P is called a (genus g) *Heegaard surface* of M. In knot applications, Heegaard splitting of the *exterior of* a knot K (the complement of an open solid torus knotted like K) are considered.
  Two embedded curves are *isotopic* if there exists a continuous deformation of one embedding to another through a path of embeddings. Given a closed connected orientable surface S of genus at least two, let $C(S) = (V, E)$ denotes the graph whose vertices are isotopy classes of *essential* (not bounding disk on the surface) simple closed curves and whose edges are drawn between vertices with disjoint representative curves. This graph is connected. For any subsets of vertices $X, Y \subset V$, denote by $d_S(X, Y)$ their **set-to-set distance** $\min d_S(x, y) : x \in X, y \in Y$, where $d_S(x, y)$ is the **path metric** of $C(S)$.
  If S is the boundary of a handlebody H, let $M(H)$ denotes the set of vertices with representatives bounding *meridian disks* D of H, i.e., such that ∂D are essential simple closed curves in ∂H. The **Hempel distance** of a (genus $g \geq 2$) Heegaard splitting $M = A \cup_P B$ is defined (Hempel, 2001) to be $d_P(M(A), M(B))$.
  A Heegaard splitting $M = A \cup_P B$ is *stabilized*, if there are meridian disks D_A, D_B of A, B respectively such that ∂D_A and ∂D_B intersects transversely in a single point. The **Reidemeister–Singer distance** between two Heegaard surfaces/splittings is the minimal number of *stabilizations* (roughly, additions of a "trivial" handle) and *destabilizations* (inverse operation) relating them.

8.2 Intrinsic Metrics on Surfaces

In this section we list intrinsic metrics, given by their *line elements* (which, in fact, are 2D **Riemannian metrics**), for some selected surfaces.

- **Quadric metric**
  A *quadric* (or *quadratic surface, surface of second-order*) is a set of points in $\mathbb{E}^3$, whose coordinates in a Cartesian coordinate system satisfy an algebraic equation of degree two. There are 17 classes of such surfaces. Among them are: *ellipsoids, one-sheet* and *two-sheet hyperboloids, elliptic paraboloids, hyperbolic paraboloids, elliptic, hyperbolic* and *parabolic cylinders*, and *conical surfaces*.

For example, a *cylinder* can be given by the following parametric equations:

$$x_1(u,v) = a\cos v,\ x_2(u,v) = a\sin v,\ x_3(u,v) = u.$$

The intrinsic metric on it is given by the *line element*

$$ds^2 = du^2 + a^2dv^2.$$

An *elliptic cone* (i.e., a cone with elliptical cross-section) has the following equations:

$$x_1(u,v) = a\frac{h-u}{h}\cos v,\ x_2(u,v) = b\frac{h-u}{h}\sin v,\ x_3(u,v) = u,$$

where h is the *height*, a is the *semi-major axis*, and b is the *semi-minor axis* of the cone. The intrinsic metric on it is given by the *line element*

$$ds^2 = \frac{h^2 + a^2\cos^2 v + b^2\sin^2 v}{h^2}du^2 + 2\frac{(a^2-b^2)(h-u)\cos v\sin v}{h^2}du\,dv + \\ + \frac{(h-u)^2(a^2\sin^2 v + b^2\cos^2 v)}{h^2}dv^2.$$

- **Sphere metric**
  A *sphere* is a *quadric*, given by the Cartesian equation $(x_1-a)^2 + (x_2-b)^2 + (x_3-c)^2 = r^2$, where the point (a,b,c) is the *center* of the sphere, and $r > 0$ is the *radius* of the sphere. The sphere of radius r, centered at the origin, can be given by the following parametric equations:

  $$x_1(\theta,\phi) = r\sin\theta\cos\phi,\ x_2(\theta,\phi) = r\sin\theta\sin\phi,\ x_3(\theta,\phi) = r\cos\theta,$$

  where the *azimuthal angle* $\phi \in [0, 2\pi)$, and the *polar angle* $\theta \in [0,\pi]$.
  The intrinsic metric on it (in fact, the 2D **spherical metric**) is given by the *line element*

  $$ds^2 = r^2d\theta^2 + r^2\sin^2\theta d\phi^2.$$

  A sphere of radius r has constant positive Gaussian curvature equal to r.
- **Ellipsoid metric**
  An *ellipsoid* is a *quadric* given by the Cartesian equation $\frac{x_1^2}{a^2} + \frac{x_2^2}{b^2} + \frac{x_3^2}{c^2} = 1$, or by the following parametric equations:

  $$x_1(\theta,\phi) = a\cos\phi\sin\theta,\ x_2(\theta,\phi) = b\sin\phi\sin\theta,\ x_3(\theta,\phi) = c\cos\theta,$$

  where the *azimuthal angle* $\phi \in [0, 2\pi)$, and the *polar angle* $\theta \in [0,\pi]$.

The intrinsic metric on it is given by the *line element*

$$ds^2 = (b^2\cos^2\phi + a^2\sin^2\phi)\sin^2\theta d\phi^2 + (b^2 - a^2)\cos\phi\sin\phi\cos\theta\sin\theta d\theta d\phi +$$
$$+((a^2\cos^2\phi + b^2\sin^2\phi)\cos^2\theta + c^2\sin^2\theta)d\theta^2.$$

- **Spheroid metric**
A *spheroid* is an *ellipsoid* having two axes of equal length. It is also a *rotation surface*, given by the following parametric equations:

$$x_1(u,v) = a\sin v\cos u,\ x_2(u,v) = a\sin v\sin u,\ x_3(u,v) = c\cos v,$$

where $0 \le u < 2\pi$, and $0 \le v \le \pi$.
The intrinsic metric on it is given by the *line element*

$$ds^2 = a^2\sin^2 v\,du^2 + \frac{1}{2}(a^2 + c^2 + (a^2 - c^2)\cos(2v))dv^2.$$

- **Hyperboloid metric**
A *hyperboloid* is a *quadric* which may be one- or two-sheeted.
The one-sheeted hyperboloid is a *surface of revolution* obtained by rotating a hyperbola about the perpendicular bisector to the line between the foci, while the two-sheeted hyperboloid is a surface of revolution obtained by rotating a hyperbola about the line joining the foci.
The one-sheeted circular hyperboloid, oriented along the x_3 axis, is given by the Cartesian equation $\frac{x_1^2}{a^2} + \frac{x_2^2}{a^2} - \frac{x_3^2}{c^2} = 1$, or by the following parametric equations:

$$x_1(u,v) = a\sqrt{1+u^2}\cos v,\ x_2(u,v) = a\sqrt{1+u^2}\sin v,\ x_3(u,v) = cu,$$

where $v \in [0, 2\pi)$. The intrinsic metric on it is given by the *line element*

$$ds^2 = \left(c^2 + \frac{a^2u^2}{u^2+1}\right)du^2 + a^2(u^2+1)\,dv^2.$$

- **Rotation surface metric**
A *rotation surface* (or *surface of revolution*) is a surface generated by rotating a 2D curve about an axis. It is given by the following parametric equations:

$$x_1(u,v) = \phi(v)\cos u,\ x_2(u,v) = \phi(v)\sin u,\ x_3(u,v) = \psi(v).$$

The intrinsic metric on it is given by the *line element*

$$ds^2 = \phi^2 du^2 + (\phi'^2 + \psi'^2)\,dv^2.$$

- **Pseudo-sphere metric**
A *pseudo-sphere* is a half of the *rotation surface* generated by rotating a *tractrix* about its asymptote. It is given by the following parametric equations:

$$x_1(u,v) = \text{sech}u \cos v,\ \ x_2(u,v) = \text{sech}u \sin v,\ \ x_3(u,v) = u - \tanh u,$$

where $u \geq 0$, and $0 \leq v < 2\pi$. The intrinsic metric on it is given by the *line element*

$$ds^2 = \tanh^2 u\, du^2 + \text{sech}^2 u\, dv^2.$$

The pseudo-sphere has constant negative Gaussian curvature equal to -1, and in this sense is an analog of the sphere which has constant positive Gaussian curvature.

- **Torus metric**
A *torus* is a surface having genus one. A torus azimuthally symmetric about the x_3 axis is given by the Cartesian equation $(c - \sqrt{x_1^2 + x_2^2})^2 + x_3^2 = a^2$, or by the following parametric equations:

$$x_1(u,v) = (c + a\cos v)\cos u,\ \ x_2(u,v) = (c + a\cos v)\sin u,\ \ x_3(u,v) = a\sin v,$$

where $c > a$, and $u, v \in [0, 2\pi)$.
The intrinsic metric on it is given by the *line element*

$$ds^2 = (c + a\cos v)^2 du^2 + a^2 dv^2.$$

For toroidally confined plasma, such as in magnetic confinement fusion, the coordinates u, v and a correspond to the directions called , respectively, *toroidal* (long, as lines of latitude, way around the torus), *poloidal* (short way around the torus) and *radial*. The **poloidal distance**, used in plasma context, is the distance in the poloidal direction.

- **Helical surface metric**
A *helical surface* (or *surface of screw motion*) is a surface described by a plane curve γ which, while rotating around an axis at a uniform rate, also advances along that axis at a uniform rate. If γ is located in the plane of the axis of rotation x_3 and is defined by the equation $x_3 = f(u)$, the position vector of the helical surface is

$$r = (u\cos v,\ u\sin v,\ f(u) = hv),\ \ h = \text{const},$$

and the intrinsic metric on it is given by the *line element*

$$ds^2 = (1 + f'^2)du^2 + 2hf'du\,dv + (u^2 + h^2)dv^2.$$

If $f = const$, one has a *helicoid*; if $h = 0$, one has a *rotation surface*.

- **Catalan surface metric**
The *Catalan surface* is a *minimal surface*, given by the following equations:

$$x_1(u,v) = u - \sin u \cosh v,\ x_2(u,v) = 1 - \cos u \cosh v,\ x_3(u,v) = 4 \sin\left(\frac{u}{2}\right) \sinh\left(\frac{v}{2}\right).$$

The intrinsic metric on it is given by the *line element*

$$ds^2 = 2\cosh^2\left(\frac{v}{2}\right)(\cosh v - \cos u)du^2 + 2\cosh^2(\frac{v}{2})(\cosh v - \cos u)\, dv^2.$$

- **Monkey saddle metric**
The *monkey saddle* is a surface, given by the Cartesian equation $x_3 = x_1(x_1^2 - 3x_2^2)$, or by the following parametric equations:

$$x_1(u,v) = u,\ x_2(u,v) = v,\ x_3(u,v) = u^3 - 3uv^2.$$

This is a surface which a monkey can straddle with both legs and his tail. The intrinsic metric on it is given by the *line element*

$$ds^2 = (1 + (su^2 - 3v^2)^2)du^2 - 2(18uv(u^2 - v^2))du\, dv + (1 + 36u^2v^2)dv^2).$$

- **Distance-defined surfaces and curves**
We give below examples of plane curves and surfaces which are the loci of points with given value of some function of their Euclidean distances to the given objects.
A *parabola* is the locus of all points in $\mathbb{R}^2$ that are equidistant from the given point (*focus*) and given line (*directrix*) on the plane.
A *hyperbola* is the locus of all points in $\mathbb{R}^2$ such that the ratio of their distances to the given point and line is a constant (*eccentricity*) greater than 1. It is also the locus of all points in $\mathbb{R}^2$ such that the absolute value of the difference of their distances to the two given foci is constant.
An *ellipse* is the locus of all points in $\mathbb{R}^2$ such that the sum of their distances to the two given points (*foci*) is constant; cf. **elliptic orbit distance** in Chap. 25. A *circle* is an ellipse in which the two foci are coincident.
A *Cassini oval* is the locus of all points in $\mathbb{R}^2$ such that the product of their distances to two given points is a constant k. If the distance between two points is $2\sqrt{k}$, then such oval is called a *lemniscate of Bernoulli*.
A *circle of Appolonius* is the locus of points in $\mathbb{R}^2$ such that the ratio of their distances to the first and second given points is constant.
A *Cartesian oval* is the locus of points in $\mathbb{R}^2$ such that their distances r_1, r_2 to the foci $(-1, 0), (1, 0)$ are related linearly by $ar_1 + br_2 = 1$. The cases $a = b, a = -b$ and $a = \frac{1}{2}$ (or $b = \frac{1}{2}$) correspond to the ellipse, hyperbola and *limaçon of Pascal*, respectively.

A *Cassinian curve* is the locus of all points in $\mathbb{R}^2$ such that the product of their distances to n given points (*poles*) is constant. If the poles form a regular n-gon, then this (algebraic of degree $2n$) curve is a *sinusoidal spiral* given also by polar equation $r^n = 2\cos(n\theta)$, and the case $n = 3$ corresponds to the *Kiepert curve*.
Farouki and Moon, 2000, considered many other multipolar generalizations of above curves. For example, their *trifocal ellipse* is the locus of all points in $\mathbb{R}^2$ (seen as the complex plane) such that the sum of their distances to the three cube roots of unity is a constant k. If $k = 2\sqrt{3}$, the curve pass through (and is singular at) the three poles.
In $\mathbb{R}^3$, a surface, rotationally symmetric about an axis, is a locus defined via Euclidean distances of its points to the two given poles belonging to this axis. For example, a *spheroid* (or *ellipsoid of revolution*) is a quadric obtained by rotating an ellipse about one of its principal axes.
It is a sphere, if this ellipse is a circle. If the ellipse is rotated about its major axis, the result is an elongated (as the rugby ball) spheroid which is the locus of all points in $\mathbb{R}^3$ such that the sum of their distances to the two given points is constant. The rotation about its minor axis results in a flattened spheroid (as the Earth) which is the locus of all points in $\mathbb{R}^3$ such that the sum of the distances to the closest and the farthest points of given circle is constant.
A *hyperboloid of revolution of two sheets* is a quadric obtained by revolving a hyperbola about its semi-major (real) axis. Such hyperboloid with axis AB is the locus of all points in $\mathbb{R}^3$ such that the absolute value of the difference of their distances to the points A and B is constant.
Any point in $\mathbb{R}^n$ is uniquely defined by its Euclidean distances to the vertices of a nondegenerated n-simplex. If a surface which is not rotationally symmetric about an axis, is a locus in $\mathbb{R}^3$ defined via distances of its points to the given poles, then three noncollinear poles is needed, and the surface is symmetric with respect to reflexion in the plane defined by the three poles.

8.3 Distances on Knots

A *knot* is a closed, self-nonintersecting curve that is embedded in S^3. The *trivial knot* (or *unknot*) O is a closed loop that is not knotted. A knot can be generalized to a link which is a set of disjoint knots. Every link has its *Seifert surface*, i.e., a compact oriented surface with the given link as boundary.

Two knots (links) are called *equivalent* if one can be smoothly deformed into another. Formally, a link is defined as a smooth one-dimensional *submanifold* of the 3-sphere S^3; a knot is a link consisting of one component; two links L_1 and L_2 are called *equivalent* if there exists an orientation-preserving homeomorphism $f : S^3 \to S^3$ such that $f(L_1) = L_2$.

All the information about a knot can be described using a *knot diagram*. It is a projection of a knot onto a plane such that no more than two points of the knot are projected to the same point on the plane, and at each such point it is indicated which strand is closest to the plane, usually by erasing part of the lower strand. Two different knot diagrams may both represent the same knot. Much of Knot Theory is devoted to telling when two knot diagrams represent the same knot.

An *unknotting operation* is an operation which changes the overcrossing and the undercrossing at a double point of a given knot diagram. The *unknotting number* of a knot K is the minimum number of unknotting operations needed to deform a diagram of K into that of the trivial knot, where the minimum is taken over all diagrams of K. Roughly, the unknotting number is the smallest number of times a knot K must be passed through itself to untie it. An ♯*-unknotting operation* in a diagram of a knot K is an analog of the unknotting operation for a ♯*-part* of the diagram consisting of two pairs of parallel strands with one of the pair overcrossing another. Thus, an ♯-unknotting operation changes the overcrossing and the undercrossing at each vertex of obtained quadrangle.

- **Gordian distance**
  The **Gordian distance** is a metric on the set of all knots defined, for given knots K and K', as the minimum number of unknotting operations needed to deform a diagram of K into that of K', where the minimum is taken over all diagrams of K from which one can obtain diagrams of K'. The unknotting number of K is equal to the Gordian distance between K and the trivial knot O.
  Let rK be the knot obtained from K by taking its mirror image, and let $-K$ be the knot with the reversed orientation. The **positive reflection distance** $Ref_+(K)$ is the Gordian distance between K and rK. The **negative reflection distance** $Ref_-(K)$ is the Gordian distance between K and $-rK$. The **inversion distance** $Inv(K)$ is the Gordian distance between K and $-K$.
  The Gordian distance is the case $k = 1$ of the C_k**-distance** which is the minimum number of C_k*-moves* needed to transform K into K'; Habiro, 1994 and Goussarov, 1995, independently proved that, for $k > 1$, it is finite if and only if both knots have the same *Vassiliev invariants of order less than* k. A C_1-move is a single crossing change, a C_2-move (or *delta-move*) is a simultaneous crossing change for 3 arcs forming a triangle. C_2- and C_3-distances are called **delta distance** and **clasp-pass distance**, respectively.
- **♯-Gordian distance**
  The **♯-Gordian distance** (see, for example, [Mura85]) is a metric on the set of all knots defined, for given knots K and K', as the minimum number of ♯-unknotting operations needed to deform a diagram of K into that of K', where the minimum is taken over all diagrams of K from which one can obtain diagrams of K'.
  Let rK be the knot obtained from K by taking its mirror image, and let $-K$ be the knot with the reversed orientation. The **positive ♯-reflection distance** $Ref^\sharp_+(K)$ is the ♯-Gordian distance between K and rK. The **negative ♯-reflection distance** $Ref^\sharp_-(K)$ is the ♯-Gordian distance between K and $-rK$. The **♯-inversion distance** $Inv^\sharp(K)$ is the ♯-Gordian distance between K and $-K$.

- **Knot complement hyperbolic metric**
  The *complement* of a knot K (or a link L) is $S^3\backslash K$ (or $S^3\backslash L$, respectively).
  A knot (or, in general, a link) is called *hyperbolic* if its complement supports a complete Riemannian metric of constant curvature -1. In this case, the metric is called a **knot** (or **link**) **complement hyperbolic metric**, and it is unique.
  A knot is hyperbolic if and only if (Thurston, 1978) it is not a *satellite knot* (then it supports a complete locally homogeneous Riemannian metric) and not a *torus knot* (does not lie on a trivially embedded torus in S^3). The complement of any nontrivial knot supports a complete nonpositively curved Riemannian metric.

Chapter 9
Distances on Convex Bodies, Cones, and Simplicial Complexes

9.1 Distances on Convex Bodies

A *convex body* in the n-dimensional Euclidean space $\mathbb{E}^n$ is a convex *compact connected* subset of $\mathbb{E}^n$. It is called *solid* (or *proper*) if it has nonempty interior. Let K denote the space of all convex bodies in $\mathbb{E}^n$, and let K_p be the subspace of all proper convex bodies. Given a set $X \subset \mathbb{E}^n$, its *convex hull* $conv(X)$ is the minimal *convex* set containing X.

Any metric space (K, d) on K is called a *metric space of convex bodies*. Such spaces, in particular the metrization by the **Hausdorff metric**, or by the **symmetric difference metric**, play a basic role in Convex Geometry (see, for example, [Grub93]).

For $C, D \in K \backslash \{\emptyset\}$, the *Minkowski addition* and the *Minkowski nonnegative scalar multiplication* are defined by $C + D = \{x + y : x \in C, y \in D\}$, and $\alpha C = \{\alpha x : x \in C\}, \alpha \geq 0$, respectively. The Abelian semigroup $(K, +)$ equipped with nonnegative scalar multiplication operators can be considered as a *convex cone*.

The *support function* $h_C : S^{n-1} \rightarrow \mathbb{R}$ of $C \in K$ is defined by $h_C(u) = \sup\{\langle u, x\rangle : x \in C\}$ for any $u \in S^{n-1}$, where S^{n-1} is the $(n-1)$-dimensional *unit sphere* in $\mathbb{E}^n$, and $\langle , \rangle$ is the *inner product* in $\mathbb{E}^n$. The *width* $w_C(u)$ is $h_C(u) + h_C(-u) = h_{C-C}(u)$. It is the perpendicular distance between the parallel supporting hyperplanes perpendicular to given direction. The *mean width* is the average of width over all directions in S^{n-1}.

- **Area deviation**
  The **area deviation** (or **template metric**) is a metric on the set K_p in $\mathbb{E}^2$ (i.e., on the set of plane convex disks) defined by
  $$A(C \triangle D),$$
  where $A(.)$ is the *area*, and $\triangle$ is the *symmetric difference*. If $C \subset D$, then it is equal to $A(D) - A(C)$.

M.M. Deza, E. Deza, *Encyclopedia of Distances*,
DOI 10.1007/978-3-662-44342-2_9

- **Perimeter deviation**
The **perimeter deviation** is a metric on K_p in $\mathbb{E}^2$ defined by

$$2p(conv(C \cup D)) - p(C) - p(D),$$

where $p(.)$ is the *perimeter*. In the case $C \subset D$, it is equal to $p(D) - p(C)$.
- **Mean width metric**
The **mean width metric** is a metric on K_p in $\mathbb{E}^2$ defined by

$$v2W(conv(C \cup D)) - W(C) - W(D),$$

where $W(.)$ is the *mean width*: $W(C) = p(C)/\pi$, and $p(.)$ is the *perimeter*.
- **Florian metric**
The **Florian metric** is a metric on K defined by

$$\int_{S^{n-1}} |h_C(u) - h_D(u)| d\sigma(u) = ||h_C - h_D||_1.$$

It can be expressed in the form $2S(conv(C \cup D)) - S(C) - S(D)$ for $n = 2$ (cf. **perimeter deviation**); it can be expressed also in the form $nk_n(2W(conv(C \cup D)) - W(C) - W(D))$ for $n \geq 2$ (cf. **mean width metric**).
Here $S(.)$ is the *surface area*, k_n is the *volume* of the *unit ball* $\overline{B}^n$ of $\mathbb{E}^n$, and $W(.)$ is the *mean width*: $W(C) = \frac{1}{nk_n} \int_{S^{n-1}} (h_C(u) + h_C(-u)) d\sigma(u)$.
- **McClure–Vitale metric**
Given $1 \leq p \leq \infty$, the **McClure–Vitale metric** is a metric on K, defined by

$$\left(\int_{S^{n-1}} |h_C(u) - h_D(u)|^p d\sigma(u) \right)^{\frac{1}{p}} = ||h_C - h_D||_p.$$

- **Pompeiu–Hausdorff–Blaschke metric**
The **Pompeiu–Hausdorff–Blaschke metric** is a metric on K defined by

$$\max\{\sup_{x \in C} \inf_{y \in D} ||x - y||_2, \sup_{y \in D} \inf_{x \in C} ||x - y||_2\},$$

where $||.||_2$ is the Euclidean norm on $\mathbb{E}^n$.
In terms of support functions and using Minkowski addition, this metric is

$$\sup_{u \in S^{n-1}} |h_C(u) - h_D(u)| = ||h_C - h_D||_\infty = \inf\{\lambda \geq 0 : C \subset D + \lambda \overline{B}^n, D \subset C + \lambda \overline{B}^n\},$$

where $\overline{B}^n$ is the *unit ball* of $\mathbb{E}^n$. This metric can be defined using any norm on $\mathbb{R}^n$ and for the space of bounded closed subsets of any metric space.

- **Pompeiu–Eggleston metric**
The **Pompeiu–Eggleston metric** is a metric on K defined by

$$\sup_{x \in C} \inf_{y \in D} ||x-y||_2 + \sup_{y \in D} \inf_{x \in C} ||x-y||_2,$$

where $||.||_2$ is the Euclidean norm on $\mathbb{E}^n$.
In terms of support functions and using Minkowski addition, this metric is

$$\max\{0, \sup_{u \in S^{n-1}} (h_C(u) - h_D(u))\} + \max\{0, \sup_{u \in S^{n-1}} (h_D(u) - h_C(u))\} =$$
$$= \inf\{\lambda \geq 0 : C \subset D + \lambda \overline{B}^n\} + \inf\{\lambda \geq 0 : D \subset C + \lambda \overline{B}^n\},$$

where $\overline{B}^n$ is the *unit ball* of $\mathbb{E}^n$. This metric can be defined using any norm on $\mathbb{R}^n$ and for the space of bounded closed subsets of any metric space.
- **Sobolev distance**
The **Sobolev distance** is a metric on K defined by

$$||h_C - h_D||_w,$$

where $||.||_w$ is the *Sobolev* 1*-norm* on the set $G_{S^{n-1}}$ of all real continuous functions on the *unit sphere* S^{n-1} of $\mathbb{E}^n$.
The *Sobolev* 1*-norm* is defined by $||f||_w = \langle f, f \rangle_w^{1/2}$, where $\langle , \rangle_w$ is an *inner product* on $G_{S^{n-1}}$, given by

$$\langle f, g \rangle_w = \int_{S^{n-1}} (fg + \nabla_s(f, g)) dw_0, \quad w_0 = \frac{1}{n \cdot k_n} w,$$

where $\nabla_s(f, g) = \langle grad_s f, grad_s g \rangle$, $\langle , \rangle$ is the *inner product* in $\mathbb{E}^n$, and $grad_s$ is the *gradient* on S^{n-1} (see [ArWe92]).
- **Shephard metric**
The **Shephard metric** is a metric on K_p defined by

$$\ln(1 + 2 \inf\{\lambda \geq 0 : C \subset D + \lambda(D - D), D \subset C + \lambda(C - C)\}).$$

- **Nikodym metric**
The **Nikodym metric** (or **volume of symmetric difference**, **Dinghas distance**) is a metric on K_p defined by

$$V(C \triangle D) = \int (1_{x \in C} - 1_{x \in D})^2 dx,$$

where $V(.)$ is the *volume* (i.e., the Lebesgue n-dimensional measure), and $\triangle$ is the *symmetric difference*. For $n = 2$, one obtains the **area deviation**.

Normalized volume of symmetric difference is a variant of **Steinhaus distance** defined by

$$\frac{V(C \triangle D)}{V(C \cup D)}.$$

- **Eggleston distance**
The **Eggleston distance** (or **symmetric surface area deviation**) is a distance on K_p defined by

$$S(C \cup D) - S(C \cap D),$$

where $S(.)$ is the *surface area*. It is not a metric.
- **Asplund metric**
The **Asplund metric** is a metric on the space $K_p/\approx$ of affine-equivalence classes in K_p defined by

$$\ln \inf\{\lambda \geq 1 : \exists T : \mathbb{E}^n \to \mathbb{E}^n \text{ affine}, x \in \mathbb{E}^n, C \subset T(D) \subset \lambda C + x\}$$

for any equivalence classes C^* and D^* with the representatives C and D, respectively.
- **Macbeath metric**
The **Macbeath metric** is a metric on the space $K_p/\approx$ of affine-equivalence classes in K_p defined by

$$\ln \inf\{|\det T \cdot P| : \exists T, P : \mathbb{E}^n \to \mathbb{E}^n \text{ regular affine}, C \subset T(D), D \subset P(C)\}$$

for any equivalence classes C^* and D^* with the representatives C and D, respectively.
Equivalently, it can be written as $\ln \delta(C, D) + \ ln\delta(D, C)$, where $\delta(C, D) = \inf_T\{\frac{V(T(D))}{V(C)}; C \subset T(D)\}$, and T is a regular affine mapping of $\mathbb{E}^n$ onto itself.
- **Banach–Mazur metric**
The **Banach–Mazur metric** is a metric on the space $K_{po}/\sim$ of the equivalence classes of proper 0-symmetric convex bodies with respect to linear transformations defined by

$$\ln \inf\{\lambda \geq 1 : \exists T : \mathbb{E}^n \to \mathbb{E}^n \text{ linear}, C \subset T(D) \subset \lambda C\}$$

for any equivalence classes C^* and D^* with the representatives C and D, respectively.
It is a special case of the **Banach–Mazur distance** (cf. Chap. 1).
- **Separation distance**
The **separation distance** between two disjoint convex bodies C and D in $\mathbb{E}^n$ (in general, between any two disjoint subsets) $\mathbb{E}^n$) is (Buckley, 1985) their Euclidean

set-set distance $\inf\{||x-y||_2 : x \in C, y \in D\}$, while $\sup\{||x-y||_2 : x \in C, y \in D\}$ is their **spanning distance**.

- **Penetration depth distance**
  The **penetration depth distance** between two interpenetrating convex bodies C and D in $\mathbb{E}^n$ (in general, between any two interpenetrating subsets of $\mathbb{E}^n$) is (Cameron–Culley, 1986) defined as the minimum *translation distance* that one body undergoes to make the interiors of C and D disjoint:

$$\min\{||t||_2 : interior(C + t) \cap D = \emptyset\}.$$

  Keerthi–Sridharan, 1991, considered $||t||_1$- and $||t||_\infty$-analogs of this distance. Cf. **penetration distance** in Chap. 23 and **penetration depth** in Chap. 24.
- **Growth distances**
  Let $C, D \in K_p$ be two compact convex proper bodies. Fix their *seed points* $p_C \in int C$ and $p_D \in int D$; usually, they are the centroids of C and D. The *growth function* $g(C, D)$ is the minimal number $\lambda > 0$, such that

$$(\{p_C\} + \lambda(C\backslash\{p_C\})) \cap (\{p_D\} + \lambda(D\backslash\{p_D\})) \neq \emptyset.$$

  It is the amount objects must be grown if $g(C, D) > 1$ (i.e., $C \cap D = \emptyset$), or contracted if $g(C, D) > 1$ (i.e., $int C \cap int D \neq \emptyset$) from their internal seed points until their surfaces just touch. The **growth separation distance** $d_S(C, D)$ and the **growth penetration distance** $d_P(C, D)$ [OnGi96] are defined as

$$d_S(C, D) = \max\{0, r_{CD}(g(C, D) - 1)\} \text{ and } d_P(C, D) = \max\{0, r_{CD}(1 - g(C, D))\},$$

  where r_{CD} is the scaling coefficient (usually, the sum of radii of circumscribing spheres for the sets $C\backslash\{p_C\}$ and $D\backslash\{p_D\}$).
  The *one-sided growth distance* between disjoint C and D (Leven–Sharir, 1987) is

$$-1 + \min \lambda > 0 : (\{p_C\} + \lambda\{(C\backslash\{p_C\})) \cap D \neq \emptyset\}.$$

- **Minkowski difference**
  The **Minkowski difference** on the set of all compact subsets, in particular, on the set of all *sculptured objects* (or *free form objects*), of $\mathbb{R}^3$ is defined by

$$A - B = \{x - y : x \in A, y \in B\}.$$

  If we consider object B to be free to move with fixed orientation, the Minkowski difference is a set containing all the translations that bring B to intersect with A. The closest point from the Minkowski difference boundary, $\partial(A - B)$, to the origin gives the **separation distance** between A and B.

If both objects intersect, the origin is inside of their Minkowski difference, and the obtained distance can be interpreted as a **penetration depth distance**.

- **Demyanov distance**
Given $C \in K_p$ and $u \in S^{n-1}$, denote, if $|\{c \in C : \langle u, c\rangle = h_C(u)\}| = 1$, this unique point by $y(u, C)$ (*exposed point of* C *in direction* u).
The *Demyanov difference* $A \ominus B$ of two subsets $A, B \in K_p$ is the closure of

$$conv(\cup_{T(A)\cap T(B)}\{y(u, A) - y(u, B)\}),$$

where $T(C) = \{u \in S^{n-1} : |\{c \in C : \langle u, c\rangle = h_C(u)\}| = 1\}$.
The **Demyanov distance** between two subsets $A, B \in K_p$ is defined by

$$||A \ominus B|| = \max_{c \in A \ominus B} ||c||_2.$$

It is shown in [BaFa07] that $||A \ominus B|| = \sup_\alpha ||St_\alpha(A) - St_\alpha(M)||_2$, where $St_\alpha(C)$ is a *generalized Steiner point* and the supremum is over all "sufficiently smooth" probabilistic measures α.

- **Maximum polygon distance**
The **maximum polygon distance** is a distance between two convex polygons $P = (p_1, \dots, p_n)$ and $Q = (q_1, \dots, q_m)$ defined by

$$\max_{i,j} ||p_i - q_j||_2, \; i \in \{1, \dots, n\}, \; j \in \{1, \dots, m\}.$$

- **Grenander distance**
Let $P = (p_1, \dots, p_n)$ and $Q = (q_1, \dots, q_m)$ be two disjoint convex polygons, and let $L(p_i, q_j), L(p_l, q_m)$ be two intersecting *critical support lines* for P and Q. Then the **Grenander distance** between P and Q is defined by

$$||p_i - q_j||_2 + ||p_l - q_m||_2 - \Sigma(p_i, p_l) - \Sigma(g_j, q_m),$$

where $||.||_2$ is the Euclidean norm, and $\Sigma(p_i, p_l)$ is the sum of the edges lengths of the polynomial chain $p_i, \dots, p_l$.
Here $P = (p_1, \dots, p_n)$ is a convex polygon with the vertices in standard form, i.e., the vertices are specified according to Cartesian coordinates in a clockwise order, and no three consecutive vertices are collinear. A line L is a *line of support* of P if the interior of P lies completely to one side of L.
Given two disjoint polygons P and Q, the line $L(p_i, q_j)$ is a *critical support line* if it is a line of support for P at p_i, a line of support for Q at q_j, and P and Q lie on opposite sides of $L(p_i, q_j)$. In general, a chord $[a, b]$ of a convex body C is called its **affine diameter** if there is a pair of different hyperplanes each containing one of the endpoints a, b and supporting C.

9.2 Distances on Cones

A *convex cone* C in a real vector space V is a subset C of V such that $C + C \subset C$, $\lambda C \subset C$ for any $\lambda \geq 0$. A cone C induces a *partial order* on V by

$$x \preceq y \text{ if and only if } y - x \in C.$$

The order $\preceq$ respects the vector structure of V, i.e., if $x \preceq y$ and $z \preceq u$, then $x + z \preceq y + u$, and if $x \preceq y$, then $\lambda x \preceq \lambda y$, $\lambda \in \mathbb{R}$, $\lambda \geq 0$. Elements $x, y \in V$ are called *comparable* and denoted by $x \sim y$ if there exist positive real numbers α and β such that $\alpha y \preceq x \preceq \beta y$. Comparability is an equivalence relation; its equivalence classes (which belong to C or to $-C$) are called *parts* (or *components, constituents*).

Given a convex cone C, a subset $S = \{x \in C : T(x) = 1\}$, where $T : V \to \mathbb{R}$ is a positive linear functional, is called a *cross-section* of C. A convex cone C is called *almost Archimedean* if the closure of its restriction to any 2D subspace is also a cone.

A convex cone C is called *pointed* if $C \cup (-C) = \{0\}$ and *solid* if $int\ C \neq \emptyset$.

- **Koszul–Vinberg metric**
  Given an open pointed convex cone C, let C^* be its dual cone.
  The **Koszul–Vinberg metric** on C (Koszul, 1965, and Vinberg, 1963) is an affine invariant Riemannian metric defined as the Hessian $g = d^2\psi_C$, where $\psi_C(x) = -\log \int_{C^*} e^{-(\epsilon,x)} d\epsilon$ for any $x \in C$.
  The Hessian of the *entropy* (Legendre transform of $\psi_C(x)$) defines a metric on C^*, which (Barbaresco, 2014) is equivalent to the **Fisher–Rao metric** (Sect. 7.2).
- **Invariant distances on symmetric cones**
  An open convex cone C in an Euclidean space V is said to be *homogeneous* if its group of linear automorhisms $G = \{g \in GL(V) : g(C) = C\}$ act transitively on C. If, moreover, $\overline{C}$ is pointed and C is self-dual with respect to the given inner product $\langle , \rangle$, then it is called a *symmetric cone*. Any symmetric cone is a Cartesian product of such cones of only five types: the cones $Sym(n, \mathbb{R})^+$, $Her(n, \mathbb{C})^+$ (cf. Chap. 12), $Her(n, \mathbb{H})^+$ of positive-definite Hermitian matrices with real, complex or quaternion entries, the *Lorentz cone* (or *forward light cone*) $\{(t, x_1, \dots, x_n) \in \mathbb{R}^{n+1} : t^2 > x_1^2 + \cdots + x_n^2\}$ and 27-dimensional exceptional cone of 3×3 positive-definite matrices over the octonions $\mathbb{O}$. An $n \times n$ quaternion matrix A can be seen as a $2n \times 2n$ complex matrix A'; so, $A \in Her(n, \mathbb{H})^+$ means $A' \in Her(2n, \mathbb{C})^+$.
  Let V be an *Euclidean Jordan algebra*, i.e., a finite-dimensional *Jordan algebra* (commutative algebra satisfying $x^2(xy) = x(x^2y)$ and having a multiplicative identity e) equipped with an *associative* ($\langle xy, z\rangle = \langle y, xz\rangle$) inner product $\langle , \rangle$. Then the set of square elements of V is a symmetric cone, and every symmetric cone arises in this way. Denote $P(x)y = 2x(xy) - x^2y$ for any $x, y \in C$.

For example, for $C = PD_n(\mathbb{R})$, the group G is $GL(n, \mathbb{R})$, the inner product is $\langle X, Y\rangle = \text{Tr}(XY)$, the Jordan product is $\frac{1}{2}(XY+YX)$, and $P(X)Y = XYX$, where the multiplication on the right hand side is the usual matrix multiplication.

If r is the rank of V, then for any $x \in V$ there is a complete set of orthogonal primitive idempotents $c_1, \dots, c_r \neq 0$ (i.e., $c_i^2 = c_i$, c_i indecomposable, $c_i c_j = 0$ if $i \neq j$, $\sum_{i=1}^{r} c_i = e$) and real numbers $\lambda_1, \dots, \lambda_r$, called *eigenvalues* of x, such that $x = \sum_{i=1}^{r} \lambda_i c_i$. Let $x, y \in C$ and $\lambda_1, \dots, \lambda_r$ be the eigenvalues of $P(x^{-\frac{1}{2}})y$. Lim, 2001, defined following three G-invariant distances on any symmetric cone C:

$$d_R = \left(\sum_{1\leq i\leq r} \ln^2 \lambda_i\right)^{\frac{1}{2}}, \quad d_F = \max_{1\leq i\leq r} \ln|\lambda_i|, \quad d_H = \ln(\max_{1\leq i\leq r} \lambda_i (\min_{1\leq i\leq r} \lambda_i)^{-1}).$$

For above distances, the geometric mean $P(x^{\frac{1}{2}})(P(x^{-\frac{1}{2}}y)^{\frac{1}{2}}$ is the midpoint of x and y. The distances $d_R(x, y)$, $d_F(x, y)$ are the intrinsic metrics of G-invariant Riemannian and Finsler metrics on C. The Riemannian geodesic curve $\alpha(t) = P(x^{\frac{1}{2}})(P(x^{-\frac{1}{2}}y)^t$ is one of infinitely many shortest Finsler curves passing through x and y. The space $(C, d_R(x, y))$ is an **Hadamard space** (cf. Chap. 6), while $(C, d_F(x, y))$ is not. The distance $d_F(x, y)$ is the **Thompson's part metric** on C, and the distance $d_H(x, y)$ is the **Hilbert projective semimetric** on C which is a complete metric on the unit sphere on C.

- **Thompson's part metric**
  Given a convex cone C in a real Banach space V, the **Thompson's part metric** on a *part* $K \subset C\backslash\{0\}$ is defined (Thompson, 1963) by

$$\log \max\{m(x, y), m(y, x)\}$$

  for any $x, y \in K$, where $m(x, y) = \inf\{\lambda \in \mathbb{R} : y \preceq \lambda x\}$.
  If C is *almost Archimedean*, then K equipped with this metric is a **complete** metric space. If C is finite-dimensional, then one obtains a **chord space** (cf. Chap. 6). The *positive cone* $\mathbb{R}^n_+ = \{(x_1, \dots, x_n) : x_i \geq 0 \text{ for } 1 \leq i \leq n\}$ equipped with this metric is isometric to a *normed space* which can be seen as being flat. The same holds for the **Hilbert projective semimetric** on $\mathbb{R}^n_+$.
  If C is a closed solid cone in $\mathbb{R}^n$, then *int* C can be seen as an n-dimensional manifold M^n. If for any tangent vector $v \in T_p(M^n)$, $p \in M^n$, we define a norm $||v||_p^T = \inf\{\alpha > 0 : -\alpha p \preceq v \preceq \alpha p\}$, then the length of any piecewise differentiable curve $\gamma : [0, 1] \to M^n$ is $l(\gamma) = \int_0^1 ||\gamma'(t)||^T_{\gamma(t)} dt$, and the distance between x and y is $\inf_\gamma l(\gamma)$, where the infimum is taken over all such curves γ with $\gamma(0) = x$, $\gamma(1) = y$.
- **Hilbert projective semimetric**
  Given a pointed closed convex cone C in a real Banach space V, the **Hilbert projective semimetric** on $C\backslash\{0\}$ is defined (Bushell, 1973), for $x, y \in C\backslash\{0\}$, by

$$h(x, y) = \log(m(x, y)m(y, x)),$$

where $m(x, y) = \inf\{\lambda \in \mathbb{R} : y \preceq \lambda x\}$; it holds $\frac{1}{m(y,x)} = \sup\{\lambda \in \mathbb{R} : \lambda y \preceq x\}$. This semimetric is finite on the interior of C and $h(\lambda x, \lambda' y) = h(x, y)$ for $\lambda, \lambda' > 0$. So, $h(x, y)$ is a metric on the *projectivization* of C, i.e., the space of rays of this cone.

If C is finite-dimensional, and S is a *cross-section* of C (in particular, $S = \{x \in C : ||x|| = 1\}$, where $||.||$ is a norm on V), then, for any distinct points $x, y \in S$, it holds $h(x, y) = |\ln(x, y, z, t)|$, where z, t are the points of the intersection of the line $l_{x,y}$ with the boundary of S, and (x, y, z, t) is the **cross-ratio** of x, y, z, t. Cf. the **Hilbert projective metric** in Chap. 6.

If C is finite-dimensional and *almost Archimedean*, then each part of C is a **chord space** (cf. Chap. 6) under the Hilbert projective semimetric. On the *Lorentz cone* $L = \{x = (t, x_1, \dots, x_n) \in \mathbb{R}^{n+1} : t^2 > x_1^2 + \dots + x_n^2\}$, this semimetric is isometric to the n-dimensional *hyperbolic space*. On the hyperbolic subspace $H = \{x \in L : \det(x) = 1\}$, it holds $h(x, y) = 2d(x, y)$, where $d(x, y)$ is the **Thompson's part metric** which is (on H) the usual hyperbolic distance $\operatorname{arccosh}\langle x, y\rangle$.

If C is a closed solid cone in $\mathbb{R}^n$, then *int* C can be seen as an n**-manifold** M^n (Chap. 2). If for any tangent vector $v \in T_p(M^n)$, $p \in M^n$, we define a seminorm $||v||_p^H = m(p, v) - m(v, p)$, then the length of any piecewise differentiable curve $\gamma : [0, 1] \to M^n$ is $l(\gamma) = \int_0^1 ||\gamma'(t)||_{\gamma(t)}^H dt$, and $h(x, y) = \inf_\gamma l(\gamma)$, where the infimum is taken over all such curves γ with $\gamma(0) = x$ and $\gamma(1) = y$.

- **Bushell metric**

  Given a convex cone C in a real Banach space V, the **Bushell metric** on the set $S = \{x \in C : \sum_{i=1}^n |x_i| = 1\}$ (in general, on any *cross-section* of C) is defined by

$$\frac{1 - m(x, y) \cdot m(y, x)}{1 + m(x, y) \cdot m(y, x)}$$

  for any $x, y \in S$, where $m(x, y) = \inf\{\lambda \in \mathbb{R} : y \preceq \lambda x\}$. In fact, it is equal to $\tanh(\frac{1}{2} h(x, y))$, where h is the **Hilbert projective semimetric**.

- k**-oriented distance**

  A *simplicial cone* C in $\mathbb{R}^n$ is defined as the intersection of n (open or closed) half-spaces, each of whose supporting planes contain the origin 0. For any set M of n points on the *unit sphere*, there is a unique simplicial cone C that contains these points. The *axes* of the cone C can be constructed as the set of the n rays, where each ray originates at the origin, and contains one of the points from M.

  Given a *partition* $\{C_1, \dots, C_k\}$ of $\mathbb{R}^n$ into a set of simplicial cones $C_1, \dots, C_k$, the k**-oriented distance** is a metric on $\mathbb{R}^n$ defined by

$$d_k(x - y)$$

for all $x, y \in \mathbb{R}^n$, where, for any $x \in C_i$, the value $d_k(x)$ is the length of the shortest path from the origin 0 to x traveling only in directions parallel to the axes of C_i.

- **Cones over metric space**
A **cone over a metric space** (X, d) is the quotient space $Con(X, d) = (X \times [0, 1])/(X \times \{0\})$ obtained from the product $X \times \mathbb{R}_{\geq 0}$ by collapsing the *fiber* (subspace $X \times \{0\}$) to a point (the apex of the cone). Cf. **metric cone** in Chap. 1.
The *Euclidean cone over the metric space* (X, d) is the cone $Con(X, d)$ with a metric d defined, for any $(x, t), (y, s) \in Con(X, d)$, by

$$\sqrt{t^2 + s^2 - 2ts\cos(\min\{d(x, y), \pi\})}.$$

If (X, d) is a compact metric space with diameter < 2, the **Krakus metric** is a metric on $Con(X, d)$ defined, for any $(x, t), (y, s) \in Con(X, d)$, by

$$\min\{s, t\}d(x, y) + |t - s|.$$

The cone $Con(X, d)$ with the Krakus metric admits a unique *midpoint* for each pair of its points if (X, d) has this property.
If M^n is a manifold with a pseudo-Riemannian metric g, one can consider a metric $dr^2 + r^2 g$ (in general, a metric $\frac{1}{k}dr^2 + r^2 g$, $k \neq 0$) on $Con(M^n) = M^n \times \mathbb{R}_{>0}$. For example, $Con(M^n) = \mathbb{R}^n \setminus \{0\}$ if (M^n, g) is the unit sphere in $\mathbb{R}^n$.
A *spherical cone* (or *suspension*) $\Sigma(X)$ over a metric space (X, d) is the quotient of the product $X \times [0, a]$ obtained by identifying all points in the fibers $X \times \{0\}$ and $X \times \{a\}$. If (X, d) is a **length space** (cf. Chap. 6) with $diam(X) \leq \pi$, and $a = \pi$, the **suspension metric** on $\Sigma(X)$ is defined, for any $(x, t), (y, s) \in \Sigma(X)$, by

$$\arccos(\cos t \cos s + \sin t \sin s \cos d(x, y)).$$

9.3 Distances on Simplicial Complexes

An r-dimensional *simplex* (or *geometrical simplex, hypertetrahedron*) is the *convex hull* of $r + 1$ points of $\mathbb{E}^n$ which do not lie in any $(r - 1)$-plane. The boundary of an r-simplex has $r + 1$ 0-*faces* (polytope vertices), $\frac{r(r+1)}{2}$ 1-*faces* (polytope edges), and $\binom{r+1}{i+1}$ i-*faces*, where $\binom{r}{i}$ is the binomial coefficient. The *content* (i.e., the *hypervolume*) of a simplex can be computed using the *Cayley–Menger determinant*. The regular simplex of dimension r is denoted by α_r. *Simplicial depth of a point* $p \in \mathbb{E}^n$ *relative to a set* $P \subset \mathbb{E}^n$ is the number of simplices S, generated by $(n + 1)$-subsets of P and containing p.

Roughly, a *geometrical simplicial complex* is a space with a *triangulation*, i.e., a decomposition of it into closed simplices such that any two simplices either do not intersect or intersect only along a common face.

An *abstract simplicial complex* S is a set, whose elements are called *vertices*, in which a family of finite nonempty subsets, called *simplices*, is distinguished, such that every nonempty subset of a simplex s is a simplex, called a *face* of s, and every one-element subset is a simplex. A simplex is called i-dimensional if it consists of $i+1$ vertices. The *dimension* of S is the maximal dimension of its simplices. For every simplicial complex S there exists a triangulation of a polyhedron whose simplicial complex is S. This geometric simplicial complex, denoted by *GS*, is called the *geometric realization* of S.

- **Vietoris–Rips complex**
  Given a metric space (X, d) and distance δ, their **Vietoris–Rips complex** is an abstract simplicial complex, the simplexes of which are the finite subsets M of (X, d) having diameter at most δ; the dimension of a simplex defined by M is $|M| - 1$.
- **Simplicial metric**
  Given an abstract simplicial complex S, the points of geometric simplicial complex *GS*, realizing S, can be identified with the functions $\alpha : S \to [0, 1]$ for which the set $\{x \in S : \alpha(x) \neq 0\}$ is a simplex in S, and $\sum_{x \in S} \alpha(x) = 1$. The number $\alpha(x)$ is called the x-*th barycentric coordinate* of α.
  The **simplicial metric** on *GS* (Lefschetz, 1939) is the Euclidean metric on it:
  $$\sqrt{\sum_{x \in S}(\alpha(x) - \beta(x))^2}.$$
  Tukey, 1939, found another metric on *GS*, topologically equivalent to a simplicial one. His **polyhedral metric** is the **intrinsic metric**, defined as the infimum of the lengths of the polygonal lines joining the points α and β such that each link is within one of the simplices. An example of a polyhedral metric is the intrinsic metric on the surface of a convex polyhedron in $\mathbb{E}^3$.
- **Polyhedral space**
  A Euclidean **polyhedral space** is a simplicial complex with a **polyhedral metric**. Every simplex is a **flat space** (a metric space locally isometric to some $\mathbb{E}^n$; cf. Chap. 1), and the metrics of any two simplices coincide on their intersection. The metric is the maximal metric not exceeding the metrics of simplices.
  If such a space is an n-**manifold** (Chap. 2), a point in it is a **metric singularity** if it has no neighborhood isometric to an open subset of $\mathbb{E}^n$.
  A polyhedral metric on a simplicial complex in a space of constant (positive or negative) curvature results in *spherical* and *hyperbolic polyhedral spaces*.
  The *dimension* of a polyhedral space is the maximal dimension of simplices used to glue it. **Metric graphs** (Chap. 15) are just one-dimensional polyhedral spaces.

The surface of a convex polyhedron is a 2D polyhedral space. A polyhedral metric d on a triangulated surface is a **circle-packing metric** (Thurston, 1985) if there exists a vertex-weighting $w(x) > 0$ with $d(x, y) = w(x) + w(y)$ for any edge xy.

- **Manifold edge-distance**
  A (boundaryless) *combinatorial n-manifold* is an abstract n-dimensional simplicial complex M^n in which the *link* of each r-simplex is an $(n - r - 1)$-sphere. The category of such spaces is equivalent to the category of piecewise-linear (PL) manifolds.
  The *link* of a simplex S is $Cl(Star_S) - Star_S$, where $Star_S$ is the set of all simplices in M^n having a face S, and $Cl(Star_S)$ is the smallest simplicial subcomplex of M^n containing $Star_S$.
  The **edge-distance** between vertices $u, v \in M^n$ is the minimum number of edges needed to connect them.
- **Manifold triangulation metric**
  Let M^n be a compact PL (piecewise-linear) n-dimensional manifold. A *triangulation* of M^n is a simplicial complex such that its corresponding polyhedron is PL-homeomorphic to M^n. Let T_{M^n} be the set of all *combinatorial types* of triangulations, where two triangulations are equivalent if they are simplicially isomorphic.
  Every such triangulation can be seen as a metric on the smooth manifold M if one assigns the unit length for any of its 1-dimensional simplices; so, T_{M^n} can be seen as a discrete analog of the space of Riemannian structures, i.e., isometry classes of Riemannian metrics on M^n.
  A **manifold triangulation metric** between two triangulations x and y is (Nabutovsky and Ben-Av, 1993) an **editing metric** on T_{M^n}, i.e., the minimal number of elementary moves, from a given finite list of operations, needed to obtain y from x.
  For example, the *bistellar move* consists of replacing a subcomplex of a given triangulation, which is simplicially isomorphic to a subcomplex of the boundary of the standard $(n + 1)$-simplex, by the complementary subcomplex of the boundary of an $(n + 1)$-simplex, containing all remaining n-simplices and their faces. Every triangulation can be obtained from any other one by a finite sequence of bistellar moves.
- **Polyhedral chain metric**
  An r-dimensional *polyhedral chain* A in $\mathbb{E}^n$ is a linear expression $\sum_{i=1}^{m} d_i t_i^r$, where, for any i, the value t_i^r is an r-dimensional simplex of $\mathbb{E}^n$. The *boundary* ∂A of a chain AD is the linear combination of boundaries of the simplices in the chain. The boundary of an r-dimensional chain is an $(r - 1)$-dimensional chain.
  A **polyhedral chain metric** is a **norm metric** $||A - B||$ on the set $C_r(\mathbb{E}^n)$ of all r-dimensional polyhedral chains. As a norm $||.||$ on $C_r(\mathbb{E}^n)$ one can take:

  1. The *mass* of a polyhedral chain, i.e., $|A| = \sum_{i=1}^{m} |d_i||t_i^r|$, where $|t^r|$ is the volume of the cell t_i^r;

2. The *flat norm* of a polyhedral chain, i.e., $|A|^\flat = \inf_D\{|A - \partial D| + |D|\}$, where the infimum is taken over all $(r+1)$-dimensional polyhedral chains;
3. The *sharp norm* of a polyhedral chain, i.e.,

$$|A|^\sharp = \inf\left(\frac{\sum_{i=1}^m |d_i||t_i^r||v_i|}{r+1} + |\sum_{i=1}^m d_i T_{v_i} t_i^r|^\flat\right),$$

where the infimum is taken over all *shifts* v (here $T_v t^r$ is the cell obtained by shifting t^r by a vector v of length $|v|$). A flat chain of finite mass is a sharp chain. If $r = 0$, than $|A|^\flat = |A|^\sharp$.

The metric space of *polyhedral co-chains* (i.e., linear functions of polyhedral chains) can be defined similarly. As a norm of a polyhedral co-chain X one can take:

1. The *co-mass* of a polyhedral co-chain, i.e., $|X| = \sup_{|A|=1}|X(A)|$, where $X(A)$ is the value of the co-chain X on a chain A;
2. The *flat co-norm* of a polyhedral co-chain, i.e., $|X|^\flat = \sup_{|A|^\flat=1}|X(A)|$;
3. The *sharp co-norm* of a polyhedral co-chain, i.e., $|X|^\sharp = \sup_{|A|^\sharp=1}|X(A)|$.

Part III
Distances in Classical Mathematics

Chapter 10
Distances in Algebra

10.1 Group Metrics

A *group* $(G, \cdot, e)$ is a set G of elements with a binary operation $\cdot$, called the *group operation*, that together satisfy the four fundamental properties of *closure* ($x \cdot y \in G$ for any $x, y \in G$), *associativity* ($x \cdot (y \cdot z) = (x \cdot y) \cdot z$ for any $x, y, z \in G$), the *identity property* ($x \cdot e = e \cdot x = x$ for any $x \in G$), and the *inverse property* (for any $x \in G$, there exists an element $x^{-1} \in G$ such that $x \cdot x^{-1} = x^{-1} \cdot x = e$).

In additive notation, a group $(G, +, 0)$ is a set G with a binary operation $+$ such that the following properties hold: $x + y \in G$ for any $x, y \in G$, $x + (y + z) = (x + y) + z$ for any $x, y, z \in G$, $x + 0 = 0 + x = x$ for any $x \in G$, and, for any $x \in G$, there exists an element $-x \in G$ such that $x + (-x) = (-x) + x = 0$.

A group $(G, \cdot, e)$ is called *finite* if the set G is finite. A group $(G, \cdot, e)$ is called *Abelian* if it is *commutative*, i.e., $x \cdot y = y \cdot x$ for any $x, y \in G$.

Most metrics considered in this section are **group norm metrics** on a group $(G, \cdot, e)$, defined by

$$||x \cdot y^{-1}||$$

(or, sometimes, by $||y^{-1} \cdot x||$), where $||.||$ is a *group norm*, i.e., a function $||.|| : G \to \mathbb{R}$ such that, for any $x, y \in G$, we have the following properties:

1. $||x|| \geq 0$, with $||x|| = 0$ if and only if $x = e$;
2. $||x|| = ||x^{-1}||$;
3. $||x \cdot y|| \leq ||x|| + ||y||$ (*triangle inequality*).

In additive notation, a group norm metric on a group $(G, +, 0)$ is defined by $||x + (-y)|| = ||x - y||$, or, sometimes, by $||(-y) + x||$.

The simplest example of a group norm metric is the **bi-invariant ultrametric** (sometimes called the *Hamming metric*) $||x \cdot y^{-1}||_H$, where $||x||_H = 1$ for $x \neq e$, and $||e||_H = 0$.

M.M. Deza, E. Deza, *Encyclopedia of Distances*,
DOI 10.1007/978-3-662-44342-2_10

- **Bi-invariant metric**
A metric (in general, a semimetric) d on a group $(G, \cdot, e)$ is called **bi-invariant** if

$$d(x, y) = d(x \cdot z, y \cdot z) = d(z \cdot x, z \cdot y)$$

for any $x, y, z \in G$ (cf. **translation invariant metric** in Chap. 5). Any **group norm metric** on an Abelian group is bi-invariant.
A metric (in general, a semimetric) d on a group $(G, \cdot, e)$ is called a **right-invariant metric** if $d(x, y) = d(x \cdot z, y \cdot z)$ for any $x, y, z \in G$, i.e., the operation of right multiplication by an element z is a **motion** of the metric space (G, d). Any group norm metric defined by $||x \cdot y^{-1}||$, is right-invariant.
A metric (in general, a semimetric) d on a group $(G, \cdot, e)$ is called a **left-invariant metric** if $d(x, y) = d(z \cdot x, z \cdot y)$ holds for any $x, y, z \in G$, i.e., the operation of left multiplication by an element z is a motion of the metric space (G, d). Any group norm metric defined by $||y^{-1} \cdot x||$, is left-invariant.
Any right-invariant or left-invariant (in particular, bi-invariant) metric d on G is a group norm metric, since one can define a group norm on G by $||x|| = d(x, 0)$.
- **G-invariant metric**
Given a metric space (X, d) and an action $g(x)$ of a group G on it, the metric d is called **G-invariant** (under this action) if for all $x, y \in X, g \in G$ it holds

$$d(g(x), g(y)) = d(x, y).$$

For every G-invariant metric d_X on X and every point $x \in X$, the function

$$d_G(g_1, g_2) = d_X(g_1(x), g_2(x))$$

is a **left-invariant metric** on G. This metric is called **orbit metric** in [BBI01], since it is the restriction of d on the orbit Gx, which can be identified with G.
- **Positively homogeneous distance**
A distance d on an Abelian group $(G, +, 0)$ is called **positively homogeneous** if

$$d(mx, my) = md(x, y)$$

for all $x, y \in G$ and all $m \in \mathbb{N}$, where mx is the sum of m terms all equal to x.
- **Translation discrete metric**
A **group norm metric** (in general, a group norm semimetric) on a group $(G, \cdot, e)$ is called **translation discrete** if the *translation distances* (or *translation numbers*)

$$\tau_G(x) = \lim_{n \to \infty} \frac{||x^n||}{n}$$

of the *nontorsion elements* x (i.e., such that $x^n \neq e$ for any $n \in \mathbb{N}$) of the group with respect to that metric are bounded away from zero.
If the numbers $\tau_G(x)$ are just nonzero, such a group norm metric is called a **translation proper metric**.

- **Word metric**
Let $(G, \cdot, e)$ be a finitely-generated group with a set A of generators (i.e., A is finite, and every element of G can be expressed as a product of finitely many elements A and their inverses). The *word length* $w_W^A(x)$ of an element $x \in G \backslash \{e\}$ is defined by

$$w_W^A(x) = \inf\{r : x = a_1^{\epsilon_1} \dots a_r^{\epsilon_r}, a_i \in A, \epsilon_i \in \{\pm 1\}\} \text{ and } w_W^A(e) = 0.$$

The **word metric** d_W^A *associated with* A is a **group norm metric** on G defined by

$$w_W^A(x \cdot y^{-1}).$$

As the word length w_W^A is a *group norm* on G, d_W^A is **right-invariant**. Sometimes it is defined as $w_W^A(y^{-1} \cdot x)$, and then it is **left-invariant**. In fact, d_W^A is the maximal metric on G that is right-invariant, and such that the distance from any element of A or A^{-1} to the identity element e is equal to one.
If A and B are two finite sets of generators of the group $(G, \cdot, e)$, then the identity mapping between the metric spaces (G, d_W^A) and (G, d_W^B) is a **quasi-isometry**, i.e., the word metric is unique up to quasi-isometry.
The word metric is the **path metric** of the *Cayley graph* Γ of $(G, \cdot, e)$, constructed with respect to A. Namely, Γ is a graph with the vertex-set G in which two vertices x and $y \in G$ are connected by an edge if and only if $y = a^{\epsilon}x$, $\epsilon = \pm 1$, $a \in A$.

- **Weighted word metric**
Let $(G, \cdot, e)$ be a finitely-generated group with a set A of generators. Given a bounded *weight function* $w : A \to (0, \infty)$, the *weighted word length* $w_{WW}^A(x)$ of an element $x \in G \backslash \{e\}$ is defined by $w_{WW}^A(e) = 0$ and

$$w_{WW}^A(x) = \inf\left\{\sum_{i=1}^{t} w(a_i), t \in \mathbb{N} : x = a_1^{\epsilon_1} \dots a_t^{\epsilon_t}, a_i \in A, \epsilon_i \in \{\pm 1\}\right\}.$$

The **weighted word metric** d_{WW}^A *associated with* A is a **group norm metric** on G defined by

$$w_{WW}^A(x \cdot y^{-1}).$$

As the weighted word length w_{WW}^A is a *group norm* on G, d_{WW}^A is **right-invariant**. Sometimes it is defined as $w_{WW}^A(y^{-1} \cdot x)$, and then it is **left-invariant**.
The metric d_{WW}^A is the supremum of semimetrics d on G with the property that $d(e, a) \leq w(a)$ for any $a \in A$.

The metric d^A_{WW} is a **coarse-path metric**, and every right-invariant coarse path metric is a weighted word metric up to **coarse isometry**.
The metric d^A_{WW} is the **path metric** of the *weighted Cayley graph* Γ_W of $(G, \cdot, e)$ constructed with respect to A. Namely, Γ_W is a weighted graph with the vertex-set G in which two vertices x and $y \in G$ are connected by an edge with the weight $w(a)$ if and only if $y = a^{\epsilon}x$, $\epsilon = \pm 1$, $a \in A$.

- **Interval norm metric**
An **interval norm metric** is a **group norm metric** on a finite group $(G, \cdot, e)$ defined by
$$||x \cdot y^{-1}||_{int},$$
where $||.||_{int}$ is an *interval norm* on G, i.e., a *group norm* such that the values of $||.||_{int}$ form a set of consecutive integers starting with 0.
To each interval norm $||.||_{int}$ corresponds an ordered *partition* $\{B_0, \dots, B_m\}$ of G with $B_i = \{x \in G : ||x||_{int} = i\}$; cf. **Sharma–Kaushik distance** in Chap. 16.
The *Hamming* and *Lee* norms are special cases of interval norm. A *generalized Lee norm* is an interval norm for which each class has a form $B_i = \{a, a^{-1}\}$.
- **C-metric**
A **C-metric** d is a metric on a group $(G, \cdot, e)$ satisfying the following conditions:

  1. The values of d form a set of consecutive integers starting with 0;
  2. The cardinality of the sphere $B(x, r) = \{y \in G : d(x, y) = r\}$ is independent of the particular choice of $x \in G$.

  The **word metric**, the **Hamming metric**, and the **Lee metric** are C-metrics. Any **interval norm metric** is a C-metric.
- **Order norm metric**
Let $(G, \cdot, e)$ be a finite Abelian group. Let $ord(x)$ be the *order* of an element $x \in G$, i.e., the smallest positive integer n such that $x^n = e$. Then the function $||.||_{ord} : G \to \mathbb{R}$ defined by $||x||_{ord} = \ln ord(x)$, is a *group norm* on G, called the *order norm*.
The **order norm metric** is a **group norm metric** on G, defined by
$$||x \cdot y^{-1}||_{ord}.$$
- **Monomorphism norm metric**
Let $(G, +, 0)$ be a group. Let $(H, \cdot, e)$ be a group with a *group norm* $||.||_H$. Let $f : G \to H$ be a *monomorphism* of groups G and H, i.e., an injective function such that $f(x + y) = f(x) \cdot f(y)$ for any $x, y \in G$. Then the function $||.||^f_G : G \to \mathbb{R}$ defined by $||x||^f_G = ||f(x)||_H$, is a *group norm* on G, called the *monomorphism norm*.

The **monomorphism norm metric** is a **group norm metric** on G defined by

$$||x - y||_G^f.$$

- **Product norm metric**

Let $(G, +, 0)$ be a group with a *group norm* $||.||_G$. Let $(H, \cdot, e)$ be a group with a group norm $||.||_H$. Let $G \times H = \{\alpha = (x, y) : x \in G, y \in H\}$ be the Cartesian product of G and H, and $(x, y) \cdot (z, t) = (x + z, y \cdot t)$.
Then the function $||.||_{G\times H} : G\times H \to \mathbb{R}$ defined by $||\alpha||_{G\times H} = ||(x, y)||_{G\times H} = ||x||_G + ||y||_H$, is a group norm on $G \times H$, called the *product norm*.
The **product norm metric** is a **group norm metric** on $G \times H$ defined by

$$||\alpha \cdot \beta^{-1}||_{G\times F}.$$

On the Cartesian product $G \times H$ of two finite groups with the *interval norms* $||.||_G^{int}$ and $||.||_H^{int}$, an interval norm $||.||_{G\times H}^{int}$ can be defined. In fact, $||\alpha||_{G\times H}^{int} = ||(x, y)||_{G\times H}^{int} = ||x||_G + (m + 1)||y||_H$, where $m = \max_{a\in G} ||a||_G^{int}$.

- **Quotient norm metric**

Let $(G, \cdot, e)$ be a group with a *group norm* $||.||_G$. Let $(N, \cdot, e)$ be a *normal subgroup* of $(G, \cdot, e)$, i.e., $xN = Nx$ for any $x \in G$. Let $(G/N, \cdot, eN)$ be the *quotient group* of G, i.e., $G/N = \{xN : x \in G\}$ with $xN = \{x \cdot a : a \in N\}$, and $xN \cdot yN = xyN$. Then the function $||.||_{G/N} : G/N \to \mathbb{R}$ defined by $||xN||_{G/N} = \min_{a\in N} ||xa||_X$, is a group norm on G/N, called the *quotient norm*.
A **quotient norm metric** is a **group norm metric** on G/N defined by

$$||xN \cdot (yN)^{-1}||_{G/N} = ||xy^{-1}N||_{G/N}.$$

If $G = \mathbb{Z}$ with the norm being the absolute value, and $N = m\mathbb{Z}$, $m \in \mathbb{N}$, then the quotient norm on $\mathbb{Z}/m\mathbb{Z} = \mathbb{Z}_m$ coincides with the *Lee norm*.
If a metric d on a group $(G, \cdot, e)$ is **right-invariant**, then for any normal subgroup $(N, \cdot, e)$ of $(G, \cdot, e)$ the metric d induces a right-invariant metric (in fact, the **Hausdorff metric**) d^* on G/N by

$$d^*(xN, yN) = \max\{\max_{b\in yN} \min_{a\in xN} d(a, b), \max_{a\in xN} \min_{b\in yN} d(a, b)\}.$$

- **Commutation distance**

Let $(G, \cdot, e)$ be a finite non-Abelian group. Let $Z(G) = \{c \in G : x \cdot c = c \cdot x \text{ for any } x \in G\}$ be the *center* of G.
The *commutation graph* of G is defined as a graph with the vertex-set G in which distinct elements $x, y \in G$ are connected by an edge whenever they *commute*, i.e., $x \cdot y = y \cdot x$. (Darafsheh, 2009, consider noncommuting graph on $G \setminus Z(G)$.)

Any two noncommuting elements $x, y \in G$ are connected in this graph by the path x, c, y, where c is any element of $Z(G)$ (for example, e). A path $x = x^1, x^2, \dots, x^k = y$ in the commutation graph is called an $(x - y)$ *N-path* if $x^i \notin Z(G)$ for any $i \in \{1, \dots, k\}$. In this case the elements $x, y \in G \backslash Z(G)$ are called *N-connected*.

The **commutation distance** (see [DeHu98]) d is an extended distance on G defined by the following conditions:

1. $d(x, x) = 0$;
2. $d(x, y) = 1$ if $x \neq y$, and $x \cdot y = y \cdot x$;
3. $d(x, y)$ is the minimum length of an $(x - y)$ N-path for any N-connected elements x and $y \in G \backslash Z(G)$;
4. $d(x, y) = \infty$ if $x, y \in G \backslash Z(G)$ are not connected by any N-path.

Given a group G and a G-conjugacy class X in it, Bates–Bundy–Perkins–Rowley in 2003, 2004, 2007, 2008 considered *commuting graph* (X, E) whose vertex set is X and distinct vertices $x, y \in X$ are joined by an edge $e \in E$ whenever they commute.

- **Modular distance**
Let $(\mathbb{Z}_m, +, 0)$, $m \geq 2$, be a finite *cyclic group*. Let $r \in \mathbb{N}$, $r \geq 2$. The *modular r-weight* $w_r(x)$ of an element $x \in \mathbb{Z}_m = \{0, 1, \dots, m\}$ is defined as $w_r(x) = \min\{w_r(x), w_r(m - x)\}$, where $w_r(x)$ is the *arithmetic r-weight* of the integer x. The value $w_r(x)$ can be obtained as the number of nonzero coefficients in the *generalized nonadjacent form* $x = e_n r^n + \dots e_1 r + e_0$ with $e_i \in \mathbb{Z}$, $|e_i| < r$, $|e_i + e_{i+1}| < r$, and $|e_i| < |e_{i+1}|$ if $e_i e_{i+1} < 0$. Cf. **arithmetic r-norm metric** in Chap. 12.
The **modular distance** is a distance on $\mathbb{Z}_m$, defined by
$$w_r(x - y).$$
The modular distance is a metric for $w_r(m) = 1$, $w_r(m) = 2$, and for several special cases with $w_r(m) = 3$ or 4. In particular, it is a metric for $m = r^n$ or $m = r^n - 1$; if $r = 2$, it is a metric also for $m = 2^n + 1$ (see, for example, [Ernv85]).
The most popular metric on $\mathbb{Z}_m$ is the **Lee metric** defined by $||x - y||_{Lee}$, where $||x||_{Lee} = \min\{x, m - x\}$ is the *Lee norm* of an element $x \in \mathbb{Z}_m$.
- **G-norm metric**
Consider a finite field $\mathbb{F}_{p^n}$ for a prime p and a natural number n. Given a compact convex centrally-symmetric body G in $\mathbb{R}^n$, define the *G-norm* of an element $x \in \mathbb{F}_{p^n}$ by $||x||_G = \inf\{\mu \geq 0 : x \in p\mathbb{Z}^n + \mu G\}$.
The **G-norm metric** is a **group norm metric** on $\mathbb{F}_{p^n}$ defined by
$$||x \cdot y^{-1}||_G.$$

- **Permutation norm metric**

Given a finite metric space (X, d), the **permutation norm metric** is a **group norm metric** on the group $(Sym_X, \cdot, id)$ of all permutations of X (*id* is the *identity mapping*) defined by

$$||f \cdot g^{-1}||_{Sym},$$

where the *group norm* $||.||_{Sym}$ on Sym_X is given by $||f||_{Sym} = \max_{x \in X} d(x, f(x))$.

- **Metric of motions**

Let (X, d) be a metric space, and let $p \in X$ be a fixed element of X.

The **metric of motions** (see [Buse55]) is a metric on the group $(\Omega, \cdot, id)$ of all **motions** of (X, d) (*id* is the *identity mapping*) defined by

$$\sup_{x \in X} d(f(x), g(x)) \cdot e^{-d(p,x)}$$

for any $f, g \in \Omega$ (cf. **Busemann metric of sets** in Chap. 3). If the space (X, d) is bounded, a similar metric on Ω can be defined as

$$\sup_{x \in X} d(f(x), g(x)).$$

Given a semimetric space (X, d), the **semimetric of motions** on $(\Omega, \cdot, id)$ is

$$d(f(p), g(p)).$$

- **General linear group semimetric**

Let $\mathbb{F}$ be a locally compact nondiscrete *topological field*. Let $(\mathbb{F}^n, ||.||_{\mathbb{F}^n})$, $n \geq 2$, be a *normed vector space* over $\mathbb{F}$. Let $||.||$ be the *operator norm* associated with the normed vector space $(\mathbb{F}^n, ||.||_{\mathbb{F}^n})$. Let $GL(n, \mathbb{F})$ be the *general linear group* over $\mathbb{F}$. Then the function $|.|_{op} : GL(n, \mathbb{F}) \to \mathbb{R}$ defined by $|g|_{op} = \sup\{|\ln ||g|| |, |\ln ||g^{-1}|| |\}$, is a seminorm on $GL(n, \mathbb{F})$.

The **general linear group semimetric** on the group $GL(n, \mathbb{F})$ is defined by

$$|g \cdot h^{-1}|_{op}.$$

It is a **right-invariant** semimetric which is unique, up to **coarse isometry**, since any two norms on $\mathbb{F}^n$ are **bi-Lipschitz equivalent**.

- **Generalized torus semimetric**

Let $(T, \cdot, e)$ be a *generalized torus*, i.e., a *topological group* which is isomorphic to a direct product of n multiplicative groups $\mathbb{F}_i^*$ of locally compact nondiscrete *topological fields* $\mathbb{F}_i$. Then there is a proper continuous homomorphism $v : T \to \mathbb{R}^n$, namely, $v(x_1, \dots, x_n) = (v_1(x_1), \dots, v_n(x_n))$, where $v_i : \mathbb{F}_i^* \to \mathbb{R}$ are proper continuous homomorphisms from the $\mathbb{F}_i^*$ to the additive group $\mathbb{R}$, given by the

logarithm of the *valuation*. Every other proper continuous homomorphism $v' : T \to \mathbb{R}^n$ is of the form $v' = \alpha \cdot v$ with $\alpha \in GL(n, \mathbb{R})$. If $||.||$ is a norm on $\mathbb{R}^n$, one obtains the corresponding seminorm $||x||_T = ||v(x)||$ on T.
The **generalized torus semimetric** is defined on the group $(T, \cdot, e)$ by

$$||xy^{-1}||_T = ||v(xy^{-1})|| = ||v(x) - v(y)||.$$

- **Stable norm metric**
Given a Riemannian manifold (M, g), the **stable norm metric** is a **group norm metric** on its *real homology group* $H_k(M, \mathbb{R})$ defined by the following *stable norm* $||h||_s$: the infimum of the Riemannian k-volumes of real cycles representing h.
The Riemannian manifold $(\mathbb{R}^n, g)$ is within finite **Gromov–Hausdorff distance** (cf. Chap. 1) from an n-dimensional normed vector space $(\mathbb{R}^n, ||.||_s)$.
If (M, g) is a compact connected oriented Riemannian manifold, then the manifold $H_1(M, \mathbb{R})/H_1(M, \mathbb{R})$ with metric induced by $||.||_s$ is called the *Albanese torus* (or *Jacobi torus*) of (M, g). This **Albanese metric** is a **flat metric** (cf. Chap. 8).
- **Heisenberg metric**
Let $(H, \cdot, e)$ be the (real) *Heisenberg group* $\mathcal{H}^n$, i.e., a group on the set $H = \mathbb{R}^n \times \mathbb{R}^n \times \mathbb{R}$ with the group law $h \cdot h' = (x, y, t) \cdot (x', y', t') = (x + x', y + y', t + t' + 2\sum_{i=1}^{n}(x'_i y_i - x_i y'_i)$, and the identity $e = (0, 0, 0)$. Let $|.|_{Heis}$ be the *Heisenberg gauge* (Cygan, 1978) on $\mathcal{H}^n$ defined by $|h|_{Heis} = |(x, y, t)|_{Heis} = ((\sum_{i=1}^{n}(x_i^2 + y_i^2))^2 + t^2)^{1/4}$.
The **Heisenberg metric** (or **Korányi metric**, **Cygan metric**, **gauge metric**) d_{Heis} is a **group norm metric** on $\mathcal{H}^n$ defined by

$$|x^{-1} \cdot y|_{Heis}.$$

One can identify the Heisenberg group $\mathcal{H}^{n-1} = \mathbb{C}^{n-1} \times \mathbb{R}$ with $\partial\mathbb{H}^n_{\mathbb{C}} \setminus \{\infty\}$, where $\mathbb{H}^n_{\mathbb{C}}$ is the Hermitian (i.e., complex) hyperbolic n-space, and ∞ is any point of its boundary $\partial\mathbb{H}^n_{\mathbb{C}}$. So, the usual hyperbolic metric of $\mathbb{H}^{n+1}_{\mathbb{C}}$ induces a metric on $\mathcal{H}^n$.
The **Hamenstädt distance** on $\partial\mathbb{H}^n_{\mathbb{C}} \setminus \{\infty\}$ (Hersonsky–Paulin, 2004) is $\frac{1}{\sqrt{2}} d_{Heis}$.
Sometimes, the term *Cygan metric* is reserved for the extension of the metric d_{Heis} on whole $\mathbb{H}^n_{\mathbb{C}}$ and (Apanasov, 2004) for its generalization (via the *Carnot group* $\mathbb{F}^{n-1} \times Im\mathbb{F}$) on $\mathbb{F}$-hyperbolic spaces $\mathbb{H}^n_{\mathbb{F}}$ over numbers $\mathbb{F}$ that can be complex numbers, or quaternions or, for $n = 2$, octonions. Also, the generalization of d_{Heis} on Carnot groups *of Heisenberg type* is called the *Cygan metric*.
The second natural metric on $\mathcal{H}^n$ is the **Carnot–Carathéodory metric** (or **CC metric**, **sub-Riemannian metric**; cf. Chap. 7) d_C defined as the **length metric** (cf. Chap. 6) using *horizontal vector fields* on $\mathcal{H}^n$. This metric is the **internal metric** (cf. Chap. 4) corresponding to d_{Heis}.

The metric d_{Heis} is **bi-Lipschitz equivalent** with d_C but not with any Riemannian distance and, in particular, not with any Euclidean metric. For both metrics, the Heisenberg group $\mathcal{H}^n$ is a **fractal** since its **Hausdorff dimension**, $2n + 2$, is strictly greater than its **topological dimension**, $2n + 1$.

- **Metric between intervals**

  Let G be the set of all intervals $[a, b]$ of $\mathbb{R}$. The set G forms semigroups $(G, +)$ and $(G, \cdot)$ under addition $I + J = \{x + y : x \in I, y \in J\}$ and under multiplication $I \cdot J = \{x \cdot y : x \in I, y \in J\}$, respectively.

  The **metric between intervals** is a metric on G, defined by

  $$\max\{|I|, |J|\}$$

  for all $I, J \in G$, where, for $K = [a, b]$, one has $|K| = |a - b|$.

- **Metric between games**

  Consider *positional games*, i.e., two-player nonrandom games of perfect information with real-valued outcomes. Play is alternating with a nonterminated game having move options for both players. Real-world examples include Chess, Go and Tic-Tac-Toe. Formally, let $F_{\mathbb{R}}$ be the universe of games defined inductively as follows:

  1. Every real number $r \in \mathbb{R}$ belongs to $F_{\mathbb{R}}$ and is called an *atomic game*.
  2. If $A, B \subset F_{\mathbb{R}}$ with $1 \le |A|, |B| < \infty$, then $\{A|B\} \in F_{\mathbb{R}}$ (*nonatomic game*).

  Write any game $G = \{A|B\}$ as $\{G^L|G^R\}$, where $G^L = A$ and $G^R = B$ are the set of left and right moves of G, respectively.

  $F_{\mathbb{R}}$ becomes a commutative semigroup under the following addition operation:

  1. If p and q are atomic games, then $p + q$ is the usual addition in $\mathbb{R}$.
  2. $p + \{g_{l_1}, \dots | g_{r_1}, \dots\} = \{g_{l_1} + p, \dots | g_{r_1} + p, \dots\}$.
  3. If G and H are both nonatomic, then $\{G^L|G^R\} + \{H^L|H^R\} = \{I^L|I^R\}$, where $I^L = \{g_l + H, G + h_l : g_l \in G^L, h_l \in H^L\}$ and $I^R = \{g_r + H, G + h_r : g_r \in G^R, h_r \in H^R\}$.

  For any game $G \in F_{\mathbb{R}}$, define the optimal outcomes $\overline{L}(G)$ and $\overline{R}(G)$ (if both players play optimally with Left and Right starting, respectively) as follows: $\overline{L}(p) = \overline{R}(p) = p$ and $\overline{L}(G) = \max\{\overline{R}(g_l) : g_l \in G^L\}$, $\overline{R}(G) = \max\{\overline{L}(g_r) : g_r \in G^R\}$.

  The **metric between games** G and H defined by Ettinger, 2000, is the following **extended metric** on $F_{\mathbb{R}}$:

  $$\sup_X |\overline{L}(G + X) - \overline{L}(H + X)| = \sup_X |\overline{R}(G + X) - \overline{R}(H + X)|.$$

- **Helly semimetric**

  Consider a game $(\mathcal{A}, \mathcal{B}, H)$ between players A and B with *strategy sets* $\mathcal{A}$ and $\mathcal{B}$, respectively. Here $H = H(\cdot, \cdot)$ is the *payoff function*, i.e., if player A plays $a \in \mathcal{A}$ and player B plays $b \in \mathcal{B}$, then A pays H(a,b) to B. A player's *strategy set*

is the set of available to him *pure strategies*, i.e., complete algorithms for playing the game, indicating the move for every possible situation throughout it.
The **Helly semimetric** between strategies $a_1 \in \mathcal{A}$ and $a_2 \in \mathcal{A}$ of A is defined by

$$\sup_{b \in \mathcal{B}} |H(a_1, b) - H(a_2, b)|.$$

- **Factorial ring semimetric**
  Let $(A, +, \cdot)$ be a *factorial ring*, i.e., an *integral domain* (nonzero commutative ring with no nonzero zero divisors), in which every nonzero nonunit element can be written as a product of (nonunit) irreducible elements, and such factorization is unique up to permutation.
  The **factorial ring semimetric** is a semimetric on the set $A \backslash \{0\}$, defined by

$$\ln \frac{lcm(x, y)}{gcd(x, y)},$$

  where $lcm(x, y)$ is the *least common multiple*, and $gcd(x, y)$ is the *greatest common divisor* of elements $x, y \in A \backslash \{0\}$.
- **Frankild–Sather-Wagstaff metric**
  Let $\mathcal{G}(R)$ be the set of isomorphism classes, up to a shift, of semidualizing complexes over a local Noetherian commutative ring R. An R-*complex* is a particular sequence of R-module homomorphisms; see [FrSa07]) for exact definitions.
  The **Frankild–Sather-Wagstaff metric** [FrSa07] is a metric on $\mathcal{G}(R)$ defined, for any classes $[K], [L] \in \mathcal{G}(R)$, as the infimum of the *lengths* of chains of pairwise comparable elements starting with $[K]$ and ending with $[L]$.

10.2 Metrics on Binary Relations

A *binary relation* R on a set X is a subset of $X \times X$; it is the arc-set of the directed graph (X, R) with the vertex-set X.

A binary relation R which is *symmetric* ($(x, y) \in R$ implies $(y, x) \in R$), *reflexive* (all $(x, x) \in R$), and *transitive* ($(x, y), (y, z) \in R$ imply $(x, z) \in R$) is called an *equivalence relation* or a *partition* (of X into equivalence classes). Any q-*ary sequence* $x = (x_1, \dots, x_n)$, $q \geq 2$ (i.e., with $0 \leq x_i \leq q - 1$ for $1 \leq i \leq n$), corresponds to the partition $\{B_0, \dots, B_{q-1}\}$ of $V_n = \{1, \dots, n\}$, where $B_j = \{1 \leq i \leq n : x_i = j\}$ are the equivalence classes.

A binary relation R which is *antisymmetric* ($(x, y), (y, x) \in R$ imply $x = y$), reflexive, and transitive is called a *partial order*, and the pair (X, R) is called a *poset* (partially ordered set). A partial order R on X is denoted also by $\preceq$ with $x \preceq y$ if and only if $(x, y) \in R$. The order $\preceq$ is called *linear* if any elements $x, y \in X$ are *compatible*, i.e., $x \preceq y$ or $y \preceq x$.

A poset $(L, \preceq)$ is called a *lattice* if every two elements $x, y \in L$ have the *join* $x \vee y$ and the *meet* $x \wedge y$. All partitions of X form a lattice by refinement; it is a sublattice of the lattice (by set-inclusion) of all binary relations.

- **Kemeny distance**
  The **Kemeny distance** between binary relations R_1 and R_2 on a set X is the **Hamming metric** $|R_1 \triangle R_2|$. It is twice the minimal number of inversions of pairs of adjacent elements of X which is necessary to obtain R_2 from R_1.
  If R_1, R_2 are *partitions*, then the Kemeny distance coincides with the **Mirkin–Tcherny distance**, and $1 - \frac{|R_1 \triangle R_2|}{n(n-1)}$ is the *Rand index*.
  If binary relations R_1, R_2 are *linear orders* (or *permutations*) on the set X, then the Kemeny distance coincides with the **Kendall τ distance** (Chap. 11).
- **Drápal–Kepka distance**
  The **Drápal–Kepka distance** between distinct *quasigroups* (differing from groups in that they need not be associative) $(X, +)$ and $(X, \cdot)$ is the **Hamming metric** $|\{(x, y) : x + y \neq x \cdot y\}|$ between their *Cayley tables.*
  For finite nonisomorphic groups, this distance is (Ivanyos, Le Gall and Yoshida, 2012) at least $2(\frac{|X|}{3})^2$ with equality (Drápal, 2003) for some 3-groups.
- **Metrics between partitions**
  Let X be a finite set of cardinality $n = |X|$, and let A, B be nonempty subsets of X. Let P_X be the set of partitions of X, and $P, Q \in P_X$. Let $P_1, \dots, P_q$ be *blocks* in the partition P, i.e., the pairwise disjoint sets such that $X = P_1 \cup \dots \cup P_q$, $q \geq 2$. Let $P \vee Q$ be the *join* of P and Q, and $P \wedge Q$ the *meet* of P and Q in the *lattice* $\mathbb{P}_X$ *of partitions* of X.
  Consider the following *editing operations* on partitions:

  - An *augmentation* transforms a partition P of $A \backslash \{B\}$ into a partition of A by either including the objects of B in a block, or including B as a new block;
  - An *removal* transforms a partition P of A into a partition of $A \backslash \{B\}$ by deleting the objects in B from each block that contains them;
  - A *division* transforms one partition P into another by the simultaneous removal of B from P_i (where $B \subset P_i$, $B \neq P_i$), and augmentation of B as a new block;
  - A *merging* transforms one partition P into another by the simultaneous removal of B from P_i (where $B = P_i$), and augmentation of B to P_j (where $j \neq i$);
  - A *transfer* transforms one partition P into another by the simultaneous removal of B from P_i (where $B \subset P_i$), and augmentation of B to P_j (where $j \neq i$).

Define (see, say, [Day81]), using above operations, the following metrics on P_X:

1. The minimum number of augmentations and removals of single objects needed to transform P into Q;
2. The minimum number of divisions, mergings, and transfers of single objects needed to transform P into Q;
3. The minimum number of divisions, mergings, and transfers needed to transform P into Q;
4. The minimum number of divisions and mergings needed to transform P into Q; in fact, it is equal to $|P| + |Q| - 2|P \vee Q|$;
5. $\sigma(P) + \sigma(Q) - 2\sigma(P \wedge Q)$, where $\sigma(P) = \sum_{P_i \in P} |P_i|(|P_i| - 1)$;
6. $e(P) + e(Q) - 2e(P \wedge Q)$, where $e(P) = \log_2 n + \sum_{P_i \in P} \frac{|P_i|}{n} \log_2 \frac{|P_i|}{n}$;
7. $2n - \sum_{P_i \in P} \max_{Q_j \in Q} |P_i \cap Q_j| - \sum_{Q_j \in Q} \max_{P_i \in P} |P_i \cap Q_j|$ (van Dongen, 2000).

The **Reignier distance** is the minimum number of elements that must be moved between the blocks of partition P in order to transform it into Q. Cf. **Earth Mover's distance** in Chap. 21 and the above metric 2. Cf. also Wagner–Wagner, 2007, for an overview of other distances between partitions (clusterings).

10.3 Metrics on Semilattices

Consider a poset $(L, \preceq)$. The *meet* (or *infimum*) $x \wedge y$ (if it exists) of two elements x and y is the unique element satisfying $x \wedge y \preceq x, y$, and $z \preceq x \wedge y$ if $z \preceq x, y$. The *join* (or *supremum*) $x \vee y$ (if it exists) is the unique element such that $x, y \preceq x \vee y$, and $x \vee y \preceq z$ if $x, y \preceq z$. A poset $(L, \preceq)$ is called a *lattice* if every its elements x, y have the join $x \vee y$ and the meet $x \wedge y$. A poset is a *meet* (or *lower*) *semilattice* if only the meet-operation is defined. A poset is a *join* (or *upper*) *semilattice* if only the join-operation is defined.

A lattice $\mathbb{L} = (L, \preceq, \vee, \wedge)$ is called a *semimodular lattice* if the *modularity relation xMy* is symmetric: *xMy* implies *yMx* for any $x, y \in L$. Here two elements x and y are said to constitute a *modular pair*, in symbols *xMy*, if $x \wedge (y \vee z) = (x \wedge y) \vee z$ for any $z \preceq x$. A lattice $\mathbb{L}$ in which every pair of elements is modular, is called a *modular lattice*.

Given a lattice $\mathbb{L}$, a function $v : L \to \mathbb{R}_{\geq 0}$, satisfying $v(x \vee y) + v(x \wedge y) \leq v(x) + v(y)$ for all $x, y \in L$, is called a *subvaluation* on $\mathbb{L}$. A subvaluation v is *isotone* if $v(x) \leq v(y)$ whenever $x \preceq y$, and it is *positive* if $v(x) < v(y)$ whenever $x \preceq y, x \neq y$. A subvaluation v is called a *valuation* if it is isotone and $v(x \vee y) + v(x \wedge y) = v(x) + v(y)$ for all $x, y \in L$.

- **Lattice valuation metric**
  Let $\mathbb{L} = (L, \preceq, \vee, \wedge)$ be a lattice, and let v be an isotone subvaluation on $\mathbb{L}$. The *lattice subvaluation semimetric* d_v on L is defined by

$$2v(x \vee y) - v(x) - v(y).$$

  (It can be defined also on some semilattices.) If v is a positive subvaluation on $\mathbb{L}$, one obtains a metric, called the **lattice subvaluation metric**. If v is a valuation, d_v is called the *valuation semimetric* and can be written as

$$v(x \vee y) - v(x \wedge y) = v(x) + v(y) - 2v(x \wedge y).$$

  If v is a positive valuation on $\mathbb{L}$, one obtains a metric, called the **lattice valuation metric**, and the lattice is called a **metric lattice**.
  If $L = \mathbb{N}$ (the set of positive integers), $x \vee y = lcm(x, y)$ (least common multiple), $x \wedge y = gcd(x, y)$ (greatest common divisor), and the positive valuation $v(x) = \ln x$, then $d_v(x, y) = \ln \frac{lcm(x,y)}{gcd(x,y)}$.
  This metric can be generalized on any *factorial ring* equipped with a positive valuation v such that $v(x) \geq 0$ with equality only for the multiplicative unit of the ring, and $v(xy) = v(x) + v(y)$. Cf. **factorial ring semimetric**.
- **Finite subgroup metric**
  Let $(G, \cdot, e)$ be a group. Let $\mathbb{L} = (L, \subset, \cap)$ be the meet semilattice of all finite subgroups of the group $(G, \cdot, e)$ with the meet $X \cap Y$ and the valuation $v(X) = \ln |X|$.
  The **finite subgroup metric** is a **valuation metric** on L defined by

$$v(X) + v(Y) - 2v(X \wedge Y) = \ln \frac{|X||Y|}{(|X \cap Y|)^2}.$$

- **Join semilattice distances**
  Let $\mathbb{L} = (L, \preceq, \vee)$ be a join semilattice, finite or infinite, such that every maximal chain in every interval $[x, y]$ is finite. For $x \preceq y$, the *height* $h(x, y)$ *of* y *above* x is the least cardinality of a finite maximal (by inclusion) chain of $[x, y]$ minus 1. Call the join semilattice $\mathbb{L}$ *semimodular* if for all $x, y \in L$, whenever there exists an element z covered by both x and y, the join $x \vee y$ covers both x and y, or, in other words, whenever elements x, y have a common lower bound z, it holds $h(x, x \vee y) \leq h(z, y)$. Any *tree* (i.e., all intervals $[x, z]$ are finite, each pair x, y of uncomparable elements have a least common upper bound $x \vee y$ but they never have a common lower bound) is semimodular. Consider the following distances on L:
  $d_{\text{path}}(x, y)$ is the path metric of the *Hasse diagram* of $(L, \preceq)$, i.e., a graph with vertex-set L and an edge between two elements if they are comparable.
  $d_{\text{a.path}}(x, y)$ is the smallest number of the form $h(x, z) + h(y, z)$, where z is a common upper bound of x and y, i.e., it is the **ancestral path distance**; cf. **pedigree-based distances** in Chap. 23. This and next distance reflect the way

how Roman civil law and medieval canon law, respectively, measured degree of kinship.
$d_{max}(x, y)$ is defined by $\max(h(x, x \vee y), h(y, x \vee y))$.
It holds $d_{a.path}(x, y) \geq d_{path}(x, y) \geq d_{max}(x, y)$. Foldes, 2013, proved that $d_{max}(x, y)$ is a metric if $\mathbb{L}$ is semimodular and that $d_{a.path}(x, y)$ is a metric if and only if $\mathbb{L}$ is semimodular, in which case $d_{a.path}(x, y) = d_{path}(x, y)$.

- **Gallery distance of flags**
Let $\mathbb{L}$ be a lattice. A *chain* C in $\mathbb{L}$ is a subset of L which is *linearly ordered*, i.e., any two elements of C are compatible. A *flag* is a chain in $\mathbb{L}$ which is maximal with respect to inclusion. If $\mathbb{L}$ is a semimodular lattice, containing a finite flag, then $\mathbb{L}$ has a unique minimal and a unique maximal element, and any two flags C, D in $\mathbb{L}$ have the same cardinality, $n + 1$. Then n is the *height* of the lattice $\mathbb{L}$. Two flags C, D are called *adjacent* if either they are equal or D contains exactly one element not in C. A *gallery* from C to D of length m is a sequence of flags $C = C_0, C_1, \dots, C_m = D$ such that C_{i-1} and C_i are adjacent for $i = 1, \dots, m$.
A **gallery distance of flags** (see [Abel91]) is a distance on the set of all flags of a semimodular lattice $\mathbb{L}$ with finite height defined as the minimum of lengths of galleries from C to D. It can be written as
$$|C \vee D| - |C| = |C \vee D| - |D|,$$
where $C \vee D = \{c \vee d : c \in C, d \in D\}$ is the subsemilattice generated by C and D. This distance is the **gallery metric** of the *chamber system* consisting of flags.

- **Scalar and vectorial metrics**
Let $\mathbb{L} = (L, \leq, \max, \min)$ be a lattice with the join $\max\{x, y\}$, and the meet $\min\{x, y\}$ on a set $L \subset [0, \infty)$ which has a fixed number a as the greatest element and is closed under *negation*, i.e., for any $x \in L$, one has $\overline{x} = a - x \in L$.
The **scalar metric** d on L is defined, for $x \neq y$, by
$$d(x, y) = \max\{\min\{x, \overline{y}\}, \min\{\overline{x}, y\}\}.$$
The **scalar metric** d^* on $L^* = L \cup \{*\}, * \notin L$, is defined, for $x \neq y$, by
$$d^*(x, y) = \begin{cases} d(x, y), & \text{if} \quad x, y \in L, \\ \max\{x, \overline{x}\}, & \text{if} \;\; y = *, x \neq *, \\ \max\{y, \overline{y}\}, & \text{if} \;\; x = *, y \neq *. \end{cases}$$
Given a norm $||.||$ on $\mathbb{R}^n$, $n \geq 2$, the **vectorial metric** on L^n is defined by
$$||(d(x_1, y_1), \dots, d(x_n, y_n))||,$$

and the **vectorial metric** on $(L^*)^n$ is defined by

$$||(d^*(x_1, y_1), \dots, d^*(x_n, y_n))||.$$

The vectorial metric on $L_2^n = \{0, 1\}^n$ with l_1-*norm* on $\mathbb{R}^n$ is the **Fréchet–Nikodym–Aronszyan distance**. The vectorial metric on $L_m^n = \{0, \frac{1}{m-1}, \dots, \frac{m-2}{m-1}, 1\}^n$ with l_1-*norm* on $\mathbb{R}^n$ is the **Sgarro m-valued metric**. The vectorial metric on $[0, 1]^n$ with l_1-*norm* on $\mathbb{R}^n$ is the **Sgarro fuzzy metric**. If L is L_m or $[0, 1]$, and $x = (x_1, \dots, x_n, x_{n+1}, \dots, x_{n+r})$, $y = (y_1, \dots, y_n, *, \dots, *)$, where $*$ stands in r places, then the vectorial metric between x and y is the **Sgarro metric** (see, for example, [CSY01]).

- **Metrics on Riesz space**

A *Riesz space* (or *vector lattice*) is a partially ordered vector space $(V_{Ri}, \preceq)$ in which the following conditions hold:

1. The vector space structure and the partial order structure are compatible: $x \preceq y$ implies $x + z \preceq y + z$, and $x \succ 0$, $\lambda \in \mathbb{R}$, $\lambda > 0$ implies $\lambda x \succ 0$;
2. For any two elements $x, y \in V_{Ri}$ there exists the join $x \vee y \in V_{Ri}$ (in particular, the join and the meet of any finite set of elements from V_{Ri} exist).

The **Riesz norm metric** is a **norm metric** on V_{Ri} defined by

$$||x - y||_{Ri},$$

where $||.||_{Ri}$ is a *Riesz norm*, i.e., a *norm* on V_{Ri} such that, for any $x, y \in V_{Ri}$, the inequality $|x| \leq |y|$, where $|x| = (-x) \vee (x)$, implies $||x||_{Ri} \leq ||y||_{Ri}$.
The space $(V_{Ri}, ||.||_{Ri})$ is called a *normed Riesz space*. In the case of completeness it is called a *Banach lattice*. All Riesz norms on a Banach lattice are equivalent.
An element $e \in V_{Ri}^+ = \{x \in V_{Ri} : x \succ 0\}$ is called a *strong unit* of V_{Ri} if for each $x \in V_{Ri}$ there exists $\lambda \in \mathbb{R}$ such that $|x| \preceq \lambda e$. If a Riesz space V_{Ri} has a strong unit e, then $||x|| = \inf\{\lambda \in \mathbb{R} : |x| \preceq \lambda e\}$ is a Riesz norm, and one obtains on V_{Ri} a Riesz norm metric

$$\inf\{\lambda \in \mathbb{R} : |x - y| \preceq \lambda e\}.$$

A *weak unit* of V_{Ri} is an element e of V_{Ri}^+ such that $e \wedge |x| = 0$ implies $x = 0$. A Riesz space V_{Ri} is called *Archimedean* if, for any two $x, y \in V_{Ri}^+$, there exists a natural number n, such that $nx \preceq y$. The **uniform metric** on an Archimedean Riesz space with a weak unit e is defined by

$$\inf\{\lambda \in \mathbb{R} : |x - y| \wedge e \preceq \lambda e\}.$$

- **Machida metric**

For a fixed integer $k \geq 2$ and the set $V_k = \{0, 1, \dots, k-1\}$, let $O_k^{(n)}$ be the set of all n-ary functions from $(V_k)^n$ into V_k and $O_k = \cup_{n=1}^{\infty} O_k^{(n)}$. Let Pr_k be the set of all *projections* pr_i^n over V_k, where $pr_i^n(x_1, \dots, x_i, \dots, x_n) = x_i$ for any $x_1, \dots, x_n \in V_k$.

A *clone over* V_k is a subset C of O_k containing Pr_k and closed under (functional) composition. The set L_k of all clones over V_k is a lattice. The *Post lattice* L_2 defined over Boolean functions, is countable but any L_k with $k \geq 3$ is not. For $n \geq 1$ and a clone $C \in L_k$, let $C^{(n)}$ denote *n-slice* $C \cap O_k^{(n)}$.

For any two clones $C_1, C_2 \in L_k$, Machida, 1998, defined the distance to be 0 if $C_1 = C_2$ and $(\min\{n : C_1^{(n)} \neq C_2^{(n)}\})^{-1}$, otherwise. The lattice L_k of clones with this distance is a compact ultrametric space. Cf. **Baire metric** in Chap. 11.

Chapter 11
Distances on Strings and Permutations

An *alphabet* is a finite set $\mathcal{A}$, $|\mathcal{A}| \geq 2$, elements of which are called *characters* (or *symbols*). A *string* (or *word*) is a sequence of characters over a given finite alphabet $\mathcal{A}$. The set of all finite strings over the alphabet $\mathcal{A}$ is denoted by $W(\mathcal{A})$. Examples of real world applications, using distances and similarities of string pairs, are Speech Recognition, Bioinformatics, Information Retrieval, Machine Translation, Lexicography, Dialectology.

A *substring* (or *factor*, *chain*, *block*) of the string $x = x_1 \dots x_n$ is any contiguous subsequence $x_i x_{i+1} \dots x_k$ with $1 \leq i \leq k \leq n$. A *prefix* of a string x is any its substring starting with x_1; a *suffix* is any its substring finishing with x_n. If a string is a part of a text, then the *delimiters* (a space, a dot, a comma, etc.) are added to $\mathcal{A}$.

A *vector* is any finite sequence consisting of real numbers, i.e., a finite string over the *infinite alphabet* $\mathbb{R}$. A *frequency vector* (or *discrete probability distribution*) is any string $x_1 \dots x_n$ with all $x_i \geq 0$ and $\sum_{i=1}^{n} x_i = 1$. A *permutation* (or *ranking*) is any string $x_1 \dots x_n$ with all x_i being different numbers from $\{1, \dots, n\}$.

An *editing operation* is an operation on strings, i.e., a *symmetric binary relation* on the set of all considered strings. Given a set of editing operations $\mathcal{O} = \{O_1, \dots, O_m\}$, the corresponding **editing metric** (or *unit cost edit distance*) between strings x and y is the minimum number of editing operations from $\mathcal{O}$ needed to obtain y from x. It is the **path metric** of a graph with the vertex-set $W(\mathcal{A})$ and xy being an edge if y can be obtained from x by one of the operations from $\mathcal{O}$.

In some applications, a *cost function* is assigned to each type of editing operation; then the editing distance is the minimal total cost of transforming x into y. Given a set of editing operations $\mathcal{O}$ on strings, the corresponding **necklace editing metric** between cyclic strings x and y is the minimum number of editing operations from $\mathcal{O}$ needed to obtain y from x, minimized over all rotations of x.

The main editing operations on strings are:

- *Character indel*, i.e., insertion or deletion of a character;
- *Character replacement*;
- *Character swap*, i.e., an interchange of adjacent characters;

M.M. Deza, E. Deza, *Encyclopedia of Distances*,
DOI 10.1007/978-3-662-44342-2_11

- *Substring move*, i.e., transforming, say, the string $x = x_1 \dots x_n$ into the string $x_1 \dots x_{i-1}\mathbf{x_j} \dots \mathbf{x_{k-1}} x_i \dots x_{j-1} x_k \dots x_n$;
- *Substring copy*, i.e., transforming, say, $x = x_1 \dots x_n$ into $x_1 \dots x_{i-1}\mathbf{x_j} \dots \mathbf{x_{k-1}}$ $x_i \dots x_n$;
- *Substring uncopy*, i.e., the removal of a substring provided that a copy of it remains in the string.

We list below the main distances on strings. However, some string distances will appear in Chaps. 15, 21 and 23, where they fit better, with respect to the needed level of generalization or specification.

11.1 Distances on General Strings

- **Levenstein metric**
  The **Levenstein metric** (or **edit distance**, *Hamming+Gap metric*, *shuffle-Hamming distance*) is (Levenstein, 1965) an editing metric on $W(\mathcal{A})$, obtained for $\mathcal{O}$ consisting of only character replacements and indels.
  The Levenstein metric between strings $x = x_1 \dots x_m$ and $y = y_1 \dots y_n$ is

  $$d_L(x, y) = \min\{d_H(x^*, y^*)\},$$

  where x^*, y^* are strings of length k, $k \geq \max\{m, n\}$, over the alphabet $\mathcal{A}^* = \mathcal{A} \cup \{*\}$ so that, after deleting all new characters $*$, strings x^* and y^* shrink to x and y, respectively. Here, the *gap* is the new symbol $*$, and x^*, y^* are *shuffles* of strings x and y with strings consisting of only $*$.
  The *Levenstein similarity* is $1 - \frac{d_L(x,y)}{\max\{m,n\}}$.
  The **Damerau–Levenstein metric** (Damerau, 1964) is an editing metric on $W(\mathcal{A})$, obtained for $\mathcal{O}$ consisting only of character replacements, indels and transpositions. In the Levenstein metric, a transposition corresponds to two editing operations: one insertion and one deletion.
  The **constrained edit distance** (Oomen, 1986) is the Levenstein metric, but the ranges for the number of replacements, insertions and deletions are specified.
- **Editing metric with moves**
  The **editing metric with moves** is an editing metric on $W(\mathcal{A})$ [Corm03], obtained for $\mathcal{O}$ consisting of only substring moves and indels.
- **Editing compression metric**
  The **editing compression metric** is an editing metric on $W(\mathcal{A})$ [Corm03], obtained for $\mathcal{O}$ consisting of only indels, copy and uncopy operations.
- **Indel metric**
  The **indel metric** is an editing metric on $W(\mathcal{A})$, obtained for $\mathcal{O}$ consisting of only indels. It is an analog of the **Hamming metric** $|X \Delta Y|$ between sets X and Y. For strings $x = x_1 \dots x_m$ and $y = y_1 \dots y_n$ it is $m + n - 2LCS(x, y)$, where the similarity $LCS(x, y)$ is the length of the longest common subsequence of x and y.

The **factor distance** on $W(\mathcal{A})$ is $m + n - 2LCF(x, y)$, where the similarity $LCF(x, y)$ is the length of the longest common substring (factor) of x and y. The *LCS ratio* and the *LCF ratio* are the similarities on $W(\mathcal{A})$ defined by $\frac{LCS(x,y)}{\min\{m,n\}}$ and $\frac{LCF(x,y)}{\min\{m,n\}}$, respectively; sometimes, the denominator is $\max\{m, n\}$ or $\frac{m+n}{2}$.

- **Swap metric**
  The **swap metric** (or *interchange distance*, *Dodson distance*) is an editing metric on $W(\mathcal{A})$, obtained for $\mathcal{O}$ consisting only of character swaps, i.e., it is the minimum number of interchanges of adjacent pairs of symbols, converting x into y.
- **Antidistance**
  There are $(n-1)!$ *circular permutations*, i.e., cyclic orders, of a set X of size n. The **antidistance** between circular permutations x and y is the **swap metric** between x and the reversal of y.
  Also, given complex $n \times n$ matrices A and B, the *unitary similarity orbit* through B is $\sup_{U \in \mathbb{U}_n} ||U^*BU||_\infty$, where $U \in \mathbb{U}_n$ is the group of unitary matrices. Ando, 1996, define *anti-distance* between A and this orbit as $\sup_{U \in \mathbb{U}_n} ||A - U^*BU||_\infty$.
  Also, given a simple connected graph (V, E), we assign directions to edges and the weight of each edge (either 1 or -1) depending on the direction of the traverse. Iravanian, 2012, define *anti-distance* $d(u, v) = -d(v, u)$ between vertices as the weighted average length of all simple paths from u to v.
- **Edit distance with costs**
  Given a set of editing operations $\mathcal{O} = \{O_1, \dots, O_m\}$ and a *weight* (or *cost function*) $w_i \geq 0$, assigned to each type O_i of operation, the **edit distance with costs** between strings x and y is the minimal total cost of an *editing path* between them, i.e., the minimal sum of weights for a sequence of operations transforming x into y.
  The **normalized edit distance** between strings x and y (Marzal–Vidal, 1993) is the minimum, over all editing paths P between them, of $\frac{W(P)}{L(P)}$, where $W(P)$ and $L(P)$ are the total cost and the length of the editing path P.
- **Transduction edit distances**
  The **Levenstein metric** with costs between strings x and y is modeled in [RiYi98] as a memoryless stochastic transduction between x and y.
  Each step of transduction generates either a character replacement pair (a, b), a deletion pair $(a, \emptyset)$, an insertion pair $(\emptyset, b)$, or the specific termination symbol t according to a probability function $\delta : E \cup \{t\} \to [0, 1]$, where E is the set of all possible above pairs. Such a transducer induces a probability function on the set of all sequences of operations.
  The **transduction edit distances** between strings x and y are ([RiYi98]) $\ln p$ of the following probabilities p:
  - for the **Viterbi edit distance**, the probability of the most likely sequence of editing operations transforming x into y;
  - for the **stochastic edit distance**, the probability of the string pair (x, y).

This model allows one to learn, in order to reduce error rate, the edit costs for the Levenstein metric from a corpus of examples (training set of string pairs). This learning is automatic; it reduces to estimating the parameters of above transducer.

- **Bag distance**
The **bag distance** (or *multiset metric*, *counting filter*) is a metric on $W(\mathcal{A})$ defined (Navarro, 1997) by

$$\max\{|X \backslash Y|, |Y \backslash X|\}$$

for any strings x and y, where X and Y are the *bags of symbols* (multisets of characters) in strings x and y, respectively, and, say, $|X \backslash Y|$ counts the number of elements in the multiset $X \backslash Y$. It is a (computationally) cheap approximation of the **Levenstein metric**. Cf. **metrics between multisets** in Chap. 1.

- **Marking metric**
The **marking metric** is a metric on $W(\mathcal{A})$ [EhHa88] defined by

$$\ln_2 ((diff(x, y) + 1)(diff(y, x) + 1))$$

for any strings $x = x_1 \dots x_m$ and $y = y_1 \dots y_n$, where $diff(x, y)$ is the minimal cardinality $|M|$ of a subset $M \subset \{1, \dots, m\}$ such that any substring of x, not containing any x_i with $i \in M$, is a substring of y.
Another metric defined in [EhHa88], is $\ln_2(diff(x, y) + diff(y, x) + 1)$.

- **Transformation distance**
The **transformation distance** is an **editing distance with costs** on $W(\mathcal{A})$ (Varre–Delahaye–Rivals, 1999) obtained for $\mathcal{O}$ consisting only of substring copy, uncopy and substring indels. The distance between strings x and y is the minimal cost of transformation x into y using these operations, where the cost of each operation is the length of its description.
For example, the description of the copy requires a binary code specifying the type of operation, an offset between the substring locations in x and in y, and the length of the substring. A code for insertion specifies the type of operation, the length of the substring and the sequence of the substring.

- **L_1-rearrangement distance**
The **L_1-rearrangement distance** (Amir et al., 2007) between strings $x = x_1 \dots x_m$ and $y = y_1 \dots y_m$ is defined by

$$\min_{\pi} \sum_{i=1}^{m} |i - \pi(i)|,$$

where $\pi : \{1, \dots, m\} \to \{1, \dots, m\}$ is a permutation transforming x into y; if there are no such permutations, the distance is equal to ∞.
The **L_∞-rearrangement distance** (Amir et al., 2007) between x and y is $\min_\pi \max_{1 \leq i \leq m} |i - \pi(i)|$ and it is ∞ if such a permutation does not exist.
Cf. **genome rearrangement distances** in Chap. 23.

- **Normalized information distance**
The **normalized information distance** d between two binary strings x and y is a symmetric function on $W(\{0,1\})$ [LCLMV04] defined by

$$\frac{\max\{K(x|y^*), K(y|x^*)\}}{\max\{K(x), K(y)\}}.$$

Here, for binary strings u and v, u^* is a shortest binary program to compute u on an appropriate (i.e., using a *Turing-complete* language) universal computer, the *Kolmogorov complexity* (or *algorithmic entropy*) $K(u)$ is the length of u^* (the ultimate compressed version of u), and $K(u|v)$ is the length of the shortest program to compute u if v is provided as an auxiliary input.
The function $d(x,y)$ is a metric up to small error term: $d(x,x) = O((K(x))^{-1})$, and $d(x,z) - d(x,y) - d(y,z) = O((\max\{K(x), K(y), K(z)\})^{-1})$. (Cf. $d(x,y)$ the **information metric** (or *entropy metric*) $H(X|Y)+H(Y|X)$ between stochastic sources X and Y.)
The Kolmogorov complexity is uncomputable and depends on the chosen computer language; so, instead of $K(u)$, were proposed the *minimum message length* (shortest overall message) by Wallace, 1968, and the *minimum description length* (largest compression of data) by Rissanen, 1978.
The **normalized compression distance** is a distance on $W(\{0,1\})$ [LCLMV04, BGLVZ98] defined by

$$\frac{C(xy) - \min\{C(x), C(y)\}}{\max\{C(x), C(y)\}}$$

for any binary strings x and y, where $C(x), C(y)$, and $C(xy)$ denote the size of the compression (by fixed compressor C, such as gzip, bzip2, or PPMZ) of strings x, y, and their *concatenation* xy. This distance is not a metric. It is an approximation of the normalized information distance. A similar distance is defined by $\frac{C(xy)}{C(x)+C(y)} - \frac{1}{2}$.
- **Lempel–Ziv distance**
The **Lempel–Ziv distance** between two binary strings x and y of length n is

$$\max\{\frac{LZ(x|y)}{LZ(x)}, \frac{LZ(y|x)}{LZ(y)}\},$$

where $LZ(x) = \frac{|P(x)|\log|P(x)|}{n}$ is the *Lempel–Ziv complexity* of x, approximating its *Kolmogorov complexity* $K(x)$. Here $P(x)$ is the set of nonoverlapping substrings into which x is parsed sequentially, so that the new substring is not yet contained in the set of substrings generated so far. For example, such a *Lempel–Ziv parsing* for $x = 001100101010011$ is $0|01|1|00|10|101|001|11$. Now, $LZ(x|y) = \frac{|P(x)\setminus P(y)|\log|P(x)\setminus P(y)|}{n}$.

- **Anthony–Hammer similarity**
The **Anthony–Hammer similarity** between a binary string $x = x_1 \dots x_n$ and the set Y of binary strings $y = y_1 \dots y_n$ is the maximal number m such that, for every m-subset $M \subset \{1, \dots, n\}$, the substring of x, containing only x_i with $i \in M$, is a substring of some $y \in Y$ containing only y_i with $i \in M$.
- **Jaro similarity**
Given strings $x = x_1 \dots x_m$ and $y = y_1 \dots y_n$, call a character x_i *common with* y if $x_i = y_j$, where $|i - j| \leq \frac{\min\{m,n\}}{2}$. Let $x' = x'_1 \dots x'_{m'}$ be all the characters of x which are common with y (in the same order as they appear in x), and let $y' = y'_1 \dots y'_{n'}$ be the analogic string for y.
The **Jaro similarity** $Jaro(x, y)$ between strings x and y is defined by
$$\frac{1}{3}\left(\frac{m'}{m} + \frac{n'}{n} + \frac{|\{1 \leq i \leq \min\{m', n'\} : x'_i = y'_i\}|}{\min\{m', n'\}}\right).$$
This and following two similarities are used in Record Linkage.
- **Jaro–Winkler similarity**
The **Jaro–Winkler similarity** between strings x and y is defined by
$$Jaro(x, y) + \frac{\max\{4, LCP(x, y)\}}{10}(1 - Jaro(x, y)),$$
where $Jaro(x, y)$ is the **Jaro similarity**, and $LCP(x, y)$ is the length of the longest common prefix of x and y.
- **q-gram similarity**
Given an integer $q \geq 1$ (usually, q is 2 or 3), the **q-gram similarity** between strings x and y is defined by
$$\frac{2q(x, y)}{q(x) + q(y)},$$
where $q(x)$, $q(y)$ and $q(x, y)$ are the sizes of multisets of all *q-grams* (substrings of length q) occurring in x, y and both of them, respectively.
Sometimes, $q(x, y)$ is divided not by the average of $q(x)$ and $q(y)$, as above, but by their minimum, maximum or *harmonic mean* $\frac{2q(x)q(y)}{q(x)+q(y)}$. Cf. **metrics between multisets** in Chap. 1 and, in Chap. 17, **Dice similarity**, **Simpson similarity**, **Braun–Blanquet similarity** and **Anderberg similarity**.
The q-gram similarity is an example of **token-based similarities**, i.e., ones defined in terms of *tokens* (selected substrings or words). Here tokens are q-grams. A generic **dictionary-based metric** between strings x and y is $|D(x) \Delta D(y)|$, where $D(z)$ denotes the full *dictionary* of z, i.e., the set of all of its substrings.
- **Prefix-Hamming metric**
The **prefix-Hamming metric** between strings $x = x_1 \dots x_m$ and $y = y_1 \dots y_n$ is

$$(\max\{m,n\} - \min\{m,n\}) + |\{1 \le i \le \min\{m,n\} : x_i \ne y_i\}|.$$

- **Weighted Hamming metric**
If $(\mathcal{A}, d)$ is a metric space, then the **weighted Hamming metric** between strings $x = x_1 \dots x_m$ and $y = y_1 \dots y_m$ is defined by

$$\sum_{i=1}^{m} d(x_i, y_i).$$

The term *weighted Hamming metric* (or *weighted Hamming distance*) is also used for $\sum_{1 \le i \le m, x_i \ne y_i} w_i$, where, for any $1 \le i \le m$, $w(i) > 0$ is its *weight*.
- **Fuzzy Hamming distance**
If $(\mathcal{A}, d)$ is a metric space, the **fuzzy Hamming distance** between strings $x = x_1 \dots x_m$ and $y = y_1 \dots y_m$ is an **editing distance with costs** on $W(\mathcal{A})$ obtained for $\mathcal{O}$ consisting of only indels, each of fixed cost $q > 0$, and *character shifts* (i.e., moves of 1-character substrings), where the cost of replacement of i by j is a function $f(|i - j|)$. This distance is the minimal total cost of transforming x into y by these operations. Bookstein–Klein–Raita, 2001, introduced this distance for Information Retrieval and proved that it is a metric if f is a monotonically increasing concave function on integers vanishing only at 0.
The case $f(|i - j|) = C|i - j|$, where $C > 0$ is a constant and $|i - j|$ is a time shift, corresponds to the Victor–Purpura **spike train distance** in Chap. 23.
Ralescu, 2003, introduced, for Image Retrieval, another **fuzzy Hamming distance** on $\mathcal{R}^m$. The **Ralescu distance** between two strings $x = x_1 \dots x_m$ and $y = y_1 \dots y_m$ is the fuzzy cardinality of the difference fuzzy set $D_\alpha(x, y)$ (where α is a parameter) with membership function

$$\mu_i = 1 - e^{-\alpha(x_i - y_i)^2}, 1 \le i \le m.$$

The *nonfuzzy cardinality of the fuzzy set* $D_\alpha(x, y)$ approximating its fuzzy cardinality is $|\{1 \le i \le m : \mu_i > \frac{1}{2}\}|$.
- **Needleman–Wunsch–Sellers metric**
If $(\mathcal{A}, d)$ is a metric space, the **Needleman–Wunsch–Sellers metric** (or *global alignment metric*) is an **editing distance with costs** on $W(\mathcal{A})$ [NeWu70], obtained for $\mathcal{O}$ consisting of only indels, each of fixed cost $q > 0$, and character replacements, where the cost of replacement of i by j is $d(i, j)$. This metric is the minimal total cost of transforming x into y by these operations. It is

$$\min\{d_{wH}(x^*, y^*)\},$$

where x^*, y^* are strings of length k, $k \ge \max\{m, n\}$, over the alphabet $\mathcal{A}^* = \mathcal{A} \cup \{*\}$, so that, after deleting all new characters $*$, strings x^* and y^* shrink to x and y, respectively. Here $d_{wH}(x^*, y^*)$ is the **weighted Hamming metric** between x^* and y^* with weight $d(x_i^*, y_i^*) = q$ (i.e., the editing operation is an indel) if one of x_i^*, y_i^* is $*$, and $d(x_i^*, y_i^*) = d(i, j)$, otherwise.

The **Gotoh–Smith–Waterman distance** (or *string distance with affine gaps*) is a more specialized editing metric with costs (see [Goto82]). It discounts mismatching parts at the beginning and end of the strings x, y, and introduces two indel costs: one for starting an *affine gap* (contiguous block of indels), and another one (lower) for extending a gap.

- **Duncan metric**
Consider the set X of all strictly increasing infinite sequences $x = \{x_n\}_n$ of positive integers. Define $N(n, x)$ as the number of elements in $x = \{x_n\}_n$ which are less than n, and $\delta(x)$ as the *density* of x, i.e., $\delta(x) = \lim_{n\to\infty} \frac{N(n,x)}{n}$. Let Y be the subset of X consisting of all sequences $x = \{x_n\}_n$ for which $\delta(x) < \infty$. The **Duncan metric** is a metric on Y defined, for $x \neq y$, by

$$\frac{1}{1 + LCP(x, y)} + |\delta(x) - \delta(y)|,$$

where $LCP(x, y)$ is the length of the longest common prefix of x and y.
- **Martin metric**
The **Martin metric** d^a between strings $x = x_1 \dots x_m$ and $y = y_1 \dots y_n$ is

$$|2^{-m} - 2^{-n}| + \sum_{t=1}^{\max\{m,n\}} \frac{a_t}{|\mathcal{A}|^t} \sup_z |k(z, x) - k(z, y)|,$$

where z is any string of length t, $k(z, x)$ is the *Martin kernel* of a *Markov chain* $M = \{M_t\}_{t=0}^{\infty}$, and the sequence $a \in \{a = \{a_t\}_{t=0}^{\infty} : a_t > 0, \sum_{t=1}^{\infty} a_t < \infty\}$ is a parameter.
- **Baire metric**
The **Baire metric** is an ultrametric between strings x and y defined, for $x \neq y$, by

$$\frac{1}{1 + LCP(x, y)},$$

where $LCP(x, y)$ is the length of the longest common prefix of strings (finite or infinite) x and y. Cf. **Baire space** in Chap. 2.
Given an infinite *cardinal number* κ and a set A of cardinality κ, the Cartesian product of countably many copies of A endowed with above ultrametric $\frac{1}{1+LCP(x,y)}$ is called the **Baire space of weight** κ and denoted by $B(\kappa)$. In particular, $B(\aleph_0)$ (called the *Baire 0-dimensional space*) is homeomorphic to the space *Irr* of irrationals with **continued fraction metric** (cf. Chap. 12).
- **Generalized Cantor metric**
The **generalized Cantor metric** (or, sometimes, *Baire distance*) is an ultrametric between infinite strings x and y defined, for $x \neq y$, by

$$a^{1+LCP(x,y)},$$

where a is a fixed number from the interval $(0, 1)$, and $LCP(x, y)$ is the length of the longest common prefix of x and y.
This ultrametric space is **compact**. In the case $a = \frac{1}{2}$, this metric was considered on a remarkable **fractal**, the *Cantor set*; cf. **Cantor metric** in Chap. 18. Another important case is $a = \frac{1}{e} \approx 0.367879441$.
Comyn–Dauchet, 1985, and Kwiatkowska, 1990, introduced some analogs of generalized Cantor metric for *traces*, i.e., equivalence classes of strings with respect to a congruence relation identifying strings x, y that are identical up to permutation of concurrent actions ($xy = yx$).

- **Parikh distance**
Given an ordered alphabet $\mathcal{A} = \{a_1, \dots, a_k\}$, the **Parikh distance** between words x and y over it is the **Manhattan metric** $\sum_{i=1}^{k} |x_i - y_i|$ between their *Parikh maps* (or *commutative images*) $P(x)$ and $P(y)$, where, for a word w, w_i denotes the number of occurrences of a_i in w and $P(w)$ is $(w_1, \dots, w_k)$.
- **Parentheses string metrics**
Let P_n be the set of all strings on the alphabet $\{(,)\}$ generated by a grammar and having n open and n closed parentheses. A **parentheses string metric** is an editing metric on P_n corresponding to a given set of editing operations.
For example, the **Monjardet metric** (Monjardet, 1981) between two strings $x, y \in P_n$ is the minimum number of adjacent parentheses interchanges ("()" to ")(" or ")(" to "()") needed to obtain y from x. It is the **Manhattan** metric between their *representations* p_x and p_x, where $p_z = (p_z(1), \dots, p_z(n))$ and $p_z(i)$ is the number of open parentheses written before the i-th closed parentheses of $z \in P_n$.
There is a bijection between parentheses strings and binary trees; cf. the **tree rotation distance** in Chap. 15.
- **Dehornoy–Autord distance**
The **Dehornoy–Autord distance** (2010) between two shortest expressions x and y of a permutation as a product of transpositions t_i, is the minimal, needed to get x from y, number of *braid relations*: $t_i t_j t_i = t_j t_i t_j$ with $|i - j| = 1$ and $t_i t_j = t_j t_i$ with $|i - j| \geq 2$.
This distance can be extended to the decompositions of any given *positive braid* in terms of *Artin's generators*. The permutations corresponds to the *simple braids* which are the divisors of *Garside's fundamental braid* in the *braid monoid*.
- **Schellenkens complexity quasi-metric**
The **Schellenkens complexity quasi-metric** between infinite strings $x = (x_i)$ and $y = (y_i)$ $(i = 0, 1, \dots,)$ over $\mathbb{R}_{\geq 0}$ with $\sum_{i=0}^{\infty} 2^{-i} \frac{1}{x_i} < \infty$ (seen as complexity functions) is defined (Schellenkens, 1995) by

$$\sum_{i=0}^{\infty} 2^{-i} \max\{0, \frac{1}{x_i} - \frac{1}{y_i}\}.$$

- **Graev metrics**
Let (X, d) be a metric space. Let $\overline{X} = X \cup X' \cup \{e\}$, where $X' = \{x' : x \in X\}$ is a disjoint copy of X, and $e \notin X \cup X'$. We use the notation $(e')' = e$ and $(x')' = x$ for any $x \in X$; also, the letters x, y, x_i, y_i will denote elements of $\overline{X}$. Let $(\overline{X}, D)$ be a metric space such that $D(x, y) = D(x', y') = d(x, y)$, $D(x, e) = D(x', e)$ and $D(x, y') = D(x', y)$ for all $x, y \in X$.
Denote by $W(X)$ the set of all words over $\overline{X}$ and, for each word $w \in W(X)$, denote by $l(w)$ its length. A word $w \in W(X)$ is called *irreducible* if $w = e$ or $w = x_0 \dots x_n$, where $x_i \neq e$ and $x_{i+1} \neq x_i'$ for $0 \leq i < n$.
For each word w over $\overline{X}$, denote by $\hat{w}$ the unique irreducible word obtained from w by successively replacing any occurrence of xx' in w by e and eliminating e from any occurrence of the form $w_1 e w_2$, where $w_1 = w_2 - \emptyset$ is excluded.
Denote by $F(X)$ the set of all irreducible words over $\overline{X}$ and, for $u, v \in F(X)$, define $u \cdot v = w'$, where w is the concatenation of words u and v. Then $F(X)$ becomes a group; its identity element is the (nonempty) word e.
For any two words $v = x_0 \dots x_n$ and $u = y_0 \dots y_n$ over $\overline{X}$ of the same length, let $\rho(v, u) = \sum_{i=0}^{n} D(x_i, y_i)$. The **Graev metric** between two irreducible words $u = u, v \in F(X)$ is defined [DiGa07] by

$$\inf\{\rho(u^*, v^*) : u^*, v^* \in W(X),\ l(u^*) = l(v^*),\ \widehat{u^*} = u, \widehat{v^*} = v\}.$$

Graev proved that this metric is **bi-invariant metric** on $F(X)$ and that $F(X)$ is a topological group in the topology induced by it.
- **String-induced alphabet distance**
Let $a = (a_1, \dots, a_m)$ be a finite string over alphabet X, $|X| = n \geq 2$. Let $A(x) = \{1 \leq i \leq m : a_i = x\} \neq \emptyset$ for any $x \in X$.
The **string-induced distance** between symbols $x, y \in X$ is the **set-set distance** (cf. Chap. 1) defined by

$$d_a(x, y) = \min\{|i - j| : i \in A(x), j \in A(y)\}.$$

A k**-radius sequence** (Jaromczyk and Lonc, 2004) is a string a over X with $\max_{x,y \in X} d_a(x, y) \leq k$, i.e., any two symbols (say, large digital images) occur in some window (say, memory cache) of length $k + 1$. Minimal length m corresponds to most efficient pipelining of images when no more than $k + 1$ of them can be placed in main memory in any given time.

11.2 Distances on Permutations

A *permutation* (or *ranking*) is any string $x_1 \dots x_n$ with all x_i being different numbers from $\{1, \dots, n\}$; a *signed permutation* is any string $x_1 \dots x_n$ with all $|x_i|$ being different numbers from $\{1, \dots, n\}$. Denote by $(Sym_n, \cdot, id)$ the group of all permutations of the set $\{1, \dots, n\}$, where *id* is the *identity mapping*.

The restriction, on the set Sym_n of all n-permutation vectors, of any metric on $\mathbb{R}^n$ is a metric on Sym_n; the main example is the l_p**-metric** $(\sum_{i=1}^{n} |x_i - y_i|^p)^{\frac{1}{p}}$, $p \geq 1$.

The main editing operations on permutations are:

- *Block transposition*, i.e., a substring move;
- *Character move*, i.e., a transposition of a block consisting of only one character;
- *Character swap*, i.e., interchanging of any two adjacent characters;
- *Character exchange*, i.e., interchanging of any two characters (in Group Theory, it is called *transposition*);
- *One-level character exchange*, i.e., exchange of characters x_i and x_j, $i < j$, such that, for any k with $i < k < j$, either $\min\{x_i, x_j\} > x_k$, or $x_k > \max\{x_i, x_j\}$;
- *Block reversal*, i.e., transforming, say, the permutation $x = x_1 \dots x_n$ into the permutation $x_1 \dots x_{i-1}\mathbf{x_j x_{j-1} \dots x_{i+1} x_i} x_{j+1} \dots x_n$ (so, a swap is a reversal of a block consisting only of two characters);
- *Signed reversal*, i.e., a reversal in signed permutation, followed by multiplication on -1 of all characters of the reversed block.

Below we list the most used editing and other metrics on Sym_n.

- **Hamming metric on permutations**
  The **Hamming metric on permutations** d_H is an editing metric on Sym_n, obtained for $\mathcal{O}$ consisting of only character replacements. It is a **bi-invariant** metric. Also, $n - d_H(x, y)$ is the number of fixed points of xy^{-1}.
- **Spearman ρ distance**
  The **Spearman ρ distance** is the Euclidean metric on Sym_n:

$$\sqrt{\sum_{i=1}^{n}(x_i - y_i)^2}.$$

  Its square is a 2-**near-metric**. Cf. **Spearman ρ rank correlation** in Chap. 17.
- **Spearman footrule distance**
  The **Spearman footrule distance** is the l_1-*metric* on Sym_n:

$$\sum_{i=1}^{n}|x_i - y_i|.$$

  Cf. **Spearman footrule similarity** in Chap. 17.
  Both above Spearman distances are **bi-invariant**.
- **Kendall τ distance**
  The **Kendall τ distance** (or *inversion metric*, permutation **swap metric**, *bubble-sort distance*) I is an editing metric on Sym_n, obtained for $\mathcal{O}$ consisting only of character swaps.
  In terms of Group Theory, $I(x, y)$ is the number of adjacent transpositions needed to obtain x from y. Also, $I(x, y)$ is the number of *relative inversions* of x and y, i.e., pairs (i, j), $1 \leq i < j \leq n$, with $(x_i - x_j)(y_i - y_j) < 0$. Cf. **Kendall τ rank correlation** in Chap. 17.

In [BCFS97] the following metrics, associated with $I(x, y)$, were given:

1. $\min_{z \in Sym_n}(I(x, z) + I(z^{-1}, y^{-1}))$;
2. $\max_{z \in Sym_n} I(zx, zy)$;
3. $\min_{z \in Sym_n} I(zx, zy) = T(x, y)$, where T is the **Cayley metric**;
4. editing metric with $\mathcal{O}$ consisting only of one-level character exchanges.

- **Daniels–Guilbaud semimetric**
  The **Daniels–Guilbaud semimetric** (see [Monj98]) is defined, for any $x, y \in Sym_n$, as the number of triples (i, j, k), $1 \leq i < j < k \leq n$, such that (x_i, x_j, x_k) is not a cyclic shift of (y_i, y_j, y_k). So, it is 0 if and only if x is a cyclic shift of y.
- **Cayley metric**
  The **Cayley metric** (or **transposition distance**) T is an editing metric on Sym_n, obtained for $\mathcal{O}$ consisting only of character exchanges. In terms of Group Theory, $T(x, y)$ is the minimum number of transpositions needed to obtain x from y.
  The metric T is **bi-invariant**. Also, $n - T(x, y)$ is the number of cycles in xy^{-1}, and, for the **Hamming metric on permutations**, $d_H(x, y) - T(x, y)$ is the number of cycles with length at least 2 in xy^{-1}.
- **Ulam metric**
  The **Ulam metric** (or **permutation editing metric**) U is an editing metric on Sym_n, obtained for $\mathcal{O}$ consisting only of character moves. It is the half of the **indel metric** on Sym_n.
  Also, $n - U(x, y) = LCS(x, y) = LIS(xy^{-1})$, where $LCS(x, y)$ is the length of the longest common subsequence (not necessarily a substring) of x and y, while $LIS(z)$ is the length of the longest increasing subsequence of $z \in Sym_n$.
  This and the preceding six metrics are **right-invariant**.
- **Reversal metric**
  The **reversal metric** is an editing metric on Sym_n, obtained for $\mathcal{O}$ consisting only of block reversals.
- **Signed reversal metric**
  The **signed reversal metric** (Sankoff, 1989) is an editing metric on the set of all $2^n n!$ signed permutations of the set $\{1, \ldots, n\}$, obtained for $\mathcal{O}$ consisting only of signed reversals.
  This metric is used in Biology, where a signed permutation represents a single-chromosome genome, seen as a permutation of genes (along the chromosome) each having a direction (so, a sign + or −).
- **Chain metric**
  The **chain metric** (or *rearrangement metric*) is a metric on Sym_n [Page65] defined, for any $x, y \in Sym_n$, as the minimum number, minus 1, of chains (substrings) $y'_1, \ldots, y'_t$ of y, such that x can be *parsed* (concatenated) into, i.e., $x = y'_1 \ldots y'_t$.
- **Lexicographic metric**
  The **lexicographic metric** (Golenko–Ginzburg, 1973) is a metric on Sym_n:

$$|N(x) - N(y)|,$$

where $N(x)$ is the ordinal number of the position (among $1, \ldots, n!$) occupied by the permutation x in the *lexicographic ordering* of the set Sym_n.
In the *lexicographic ordering* of Sym_n, $x = x_1 \ldots x_n \prec y = y_1 \ldots y_n$ if there exists $1 \leq i \leq n$ such that $x_1 = x_1, \ldots, x_{i-1} = y_{i-1}$, but $x_i < y_i$.

- **Fréchet permutation metric**
The **Fréchet permutation metric** is the **Fréchet product metric** (cf. Chap. 4) on the set Sym_∞ of permutations of positive integers defined by

$$\sum_{i=1}^{\infty} \frac{1}{2^i} \frac{|x_i - y_i|}{1 + |x_i - y_i|}.$$

- **Distance-rationalizable voting rule**
Let $e = (\pi_1, \ldots, \pi_m)$ be a finite string over alphabet Sym_n; it can be seen as an election in which, for each $i, 1 \leq i \leq m$, the *voter* v_i give the ranking $\pi_i = (\pi_i(c_1), \ldots, \pi_i(c_n))$ on the set $C = \{c_1, \ldots, c_n)$ of *candidates*. Let $X = Sym_n^m$ be the set of all possible elections with m voters in each.
A *voting rule* is any map $R : X \to P(C)$ assigning to each election e a set $R(e) \subset C$ of its *R-winners*. For example, the winners of *plurality rule* are candidates with the largest number of first-place votes. A candidate is a *unanimity winner* if all voters rank him first. A candidate c_i is a *Condorcet winner* if for each $c_j \in C \setminus \{c_i\}$, a strict majority of voters prefer c_i to c_j. A candidate is a *Dodson winner* if the number of swaps of adjacent candidates in the rankings by voters after which he became a Condorcet winner, is minimal. So, $|R(e)| \leq 1$ for elections with unanimity or Condorcet rule, and $|R(e)| \geq 1$ for plurality or Dodson rule.
A *consensus class* is a pair (Y, W), where $Y \subset X$ is a set of elections and W is a voting rule with unique (Y, W)*-winner* (i.e., $|W(e)| = 1$) for all $e \in Y$. Let $\mathcal{U}$ and $\mathcal{C}$ denote the consensus classes of all elections having the Condorcet winner and the unanimity winner, respectively.
Given a distance d on X and consensus class (Y, W), the voting rule R is called (Meskanen–Nurmi, 2008, and Elkind–Faliszewski–Slinko, 2009) **$(d; (Y, W))$-distance-rationalizable** if, for each election e, a candidate c_i is its R-winner if and only if he is the (Y, W)-winner in a d-closest election in Y.
The plurality rule is $(d_H; \mathcal{U})$-rationalizable, where $d_H(e, e')$ is the **Hamming distance** $|\{i \leq i \leq m : \pi_i \neq \pi'_i\}|$. The Dodson rule is $(d_{sw}; \mathcal{C})$-rationalizable, where $d_{sw}(e, e') = \sum_{1 \leq i \leq m} d_{sw}(\pi_i, \pi'_i)$ and d_{sw} on rankings is the **swap metric**. Similar framework (minimization of an *aggregation function* of distances between a collective opinion and the individual judgements) is used in *distance-based judgement aggregation* and in general *distance-based semantics* for decision or choice.

Chapter 12
Distances on Numbers, Polynomials, and Matrices

12.1 Metrics on Numbers

Here we consider the most important metrics on the classical number systems: the semiring $\mathbb{N}$ of natural numbers, the ring $\mathbb{Z}$ of integers, and the fields $\mathbb{Q}$, $\mathbb{R}$, $\mathbb{C}$ of rational, real, complex numbers, respectively. We consider also the algebra $\mathcal{Q}$ of quaternions.

- **Metrics on natural numbers**
  There are several well-known metrics on the set $\mathbb{N}$ of natural numbers:

  1. $|n - m|$; the restriction of the **natural metric** (from $\mathbb{R}$) on $\mathbb{N}$;
  2. $p^{-\alpha}$, where α is the highest power of a given prime number p dividing $m - n$, for $m \neq n$ (and equal to 0 for $m = n$); the restriction of the **p-adic metric** (from $\mathbb{Q}$) on $\mathbb{N}$;
  3. $\ln \frac{lcm(m,n)}{gcd(m,n)}$; an example of the **lattice valuation metric**;
  4. $w_r(n - m)$, where $w_r(n)$ is the *arithmetic r-weight* of n; the restriction of the **arithmetic r-norm metric** (from $\mathbb{Z}$) on $\mathbb{N}$;
  5. $\frac{|n-m|}{mn}$ (cf. **M-relative metric** in Chap. 5);
  6. $1 + \frac{1}{m+n}$ for $m \neq n$ (and equal to 0 for $m = n$); the **Sierpinski metric**.

  Most of these metrics on $\mathbb{N}$ can be extended on $\mathbb{Z}$. Moreover, any one of the above metrics can be used in the case of an arbitrary countable set X. For example, the **Sierpinski metric** is defined, in general, on a countable set $X = \{x_n : n \in \mathbb{N}\}$ by $1 + \frac{1}{m+n}$ for all $x_m, x_n \in X$ with $m \neq n$ (and is equal to 0, otherwise).
- **Arithmetic r-norm metric**
  Let $r \in \mathbb{N}, r \geq 2$. The *modified r-ary form* of an integer x is a representation

$$x = e_n r^n + \cdots + e_1 r + e_0,$$

  where $e_i \in \mathbb{Z}$, and $|e_i| < r$ for all $i = 0, \ldots, n$.

M.M. Deza, E. Deza, *Encyclopedia of Distances*,
DOI 10.1007/978-3-662-44342-2_12

An r-ary form is called *minimal* if the number of nonzero coefficients is minimal. The minimal form is not unique, in general. But if the coefficients e_i, $0 \le i \le n-1$, satisfy the conditions $|e_i + e_{i+1}| < r$, and $|e_i| < |e_{i+1}|$ if $e_i e_{i+1} < 0$, then the above form is unique and minimal; it is called the *generalized nonadjacent form*.

The *arithmetic r-weight* $w_r(x)$ of an integer x is the number of nonzero coefficients in a *minimal r-ary form* of x, in particular, in the generalized nonadjacent form. The **arithmetic r-norm metric** on $\mathbb{Z}$ (see, for example, [Ernv85]) is defined by

$$w_r(x-y).$$

- **Distance between consecutive primes**

  The **distance between consecutive primes** (or *prime gap, prime difference function*) is the difference $g_n = p_{n+1} - p_n$ between two successive prime numbers.

  It holds $g_n \le p_n$, $\overline{\lim}_{n\to\infty} g_n = \infty$ and (Zhang, 2013) $\underline{\lim}_{n\to\infty} g_n < 7 \times 10^7$, improved to ≤ 246 (conjecturally, to ≤ 6) by Polymath8, 2014. There is no $\lim_{n\to\infty} g_n$ but $g_n \approx \ln p_n$ for the average g_n.

  Open *Polignac's conjecture*: for any $k \ge 1$, there are infinitely many n with $g_n = 2k$; the case $k = 1$ (i.e., that $\underline{\lim}_{n\to\infty} g_n = 2$ holds) is the *twin prime conjecture*.
- **Distance Fibonacci numbers**

  Fibonacci numbers are defined by the recurrence $F_n = F_{n-1} + F_{n-2}$ for $n \ge 2$ with initial terms $F_0 = 0$ and $F_1 = 1$. **Distance Fibonacci numbers** are three following generalizations of them in the distance sense, considered by Wloch et al..

  Kwaśnik–Wloch, 2000: $F(k,n) = F(k,n-1) + F(k,n-k)$ for $n > k$ and $F(k,n) = n+1$ for $n \le k$.

  Bednarz et al., 2012: $Fd(k,n) = Fd(k,n-k+1) + Fd(k,n-k)$ for $n \ge k > 1$ and $Fd(k,n) = 1$ for $0 \le n < k$.

  Wloch et al., 2013: $F_2(k,n) = F_2(k,n-2) + F_2(k,n-k)$ for $n \ge k \ge 1$ and $F_2(k,n) = 1$ for $0 \le n < k$.
- **p-adic metric**

  Let p be a prime number. Any nonzero rational number x can be represented as $x = p^{\alpha}\frac{c}{d}$, where c and d are integers not divisible by p, and α is a unique integer. The *p-adic norm* of x is defined by $|x|_p = p^{-\alpha}$. Moreover, $|0|_p = 0$ is defined.

  The p-**adic metric** is a **norm metric** on the set $\mathbb{Q}$ of rational numbers defined by

  $$|x-y|_p.$$

  This metric forms the basis for the algebra of p-adic numbers. The **Cauchy completions** of the metric spaces $(\mathbb{Q}, |x-y|_p)$ and $(\mathbb{Q}, |x-y|)$ with the **natural**

metric $|x - y|$ give the fields $\mathbb{Q}_p$ of *p-adic numbers* and $\mathbb{R}$ of real numbers, respectively.

The **Gajić metric** is an **ultrametric** on the set $\mathbb{Q}$ of rational numbers defined, for $x \neq y$ (via the integer part $\lfloor z \rfloor$ of a real number z), by

$$\inf\{2^{-n} : n \in \mathbb{Z},\ \lfloor 2^n(x-e) \rfloor = \lfloor 2^n(y-e) \rfloor\},$$

where e is any fixed irrational number. This metric is **equivalent** to the **natural metric** $|x - y|$ on $\mathbb{Q}$.

- **Continued fraction metric on irrationals**

  The **continued fraction metric on irrationals** is a complete metric on the set *Irr* of irrational numbers defined, for $x \neq y$, by

$$\frac{1}{n},$$

  where n is the first index for which the continued fraction expansions of x and y differ. This metric is **equivalent** to the **natural metric** $|x - y|$ on *Irr* which is noncomplete and disconnected. Also, the *Baire 0-dimensional space* $B(\aleph_0)$ (cf. **Baire metric** in Chap. 11) is homeomorphic to *Irr* endowed with this metric.

- **Natural metric**

  The **natural metric** (or **absolute value metric, line metric,** *the distance between numbers*) is a metric on $\mathbb{R}$ defined by

$$|x - y| = \begin{cases} y - x, \text{ if } x - y < 0, \\ x - y, \text{ if } x - y \geq 0. \end{cases}$$

  On $\mathbb{R}$ all l_p**-metrics** coincide with the natural metric. The metric space $(\mathbb{R}, |x - y|)$ is called the *real line* (or *Euclidean line*).

  There exist many other metrics on $\mathbb{R}$ coming from $|x - y|$ by some **metric transform** (cf. Chap. 4). For example: $\min\{1, |x-y|\}$, $\frac{|x-y|}{1+|x-y|}$, $|x|+|x-y|+|y|$ (for $x \neq y$) and, for a given $0 < \alpha < 1$, the **generalized absolute value metric** $|x - y|^\alpha$.

  Some authors use $|x - y|$ as the *Polish notation* (parentheses-free and computer-friendly) of the distance function in any metric space.

- **Zero bias metric**

  The **zero bias metric** is a metric on $\mathbb{R}$ defined by

$$1 + |x - y|$$

  if one and only one of x and y is strictly positive, and by

$$|x - y|,$$

  otherwise, where $|x - y|$ is the **natural metric** (see, for example, [Gile87]).

- **Sorgenfrey quasi-metric**
  The **Sorgenfrey quasi-metric** is a quasi-metric d on $\mathbb{R}$ defined by

$$y - x$$

  if $y \geq x$, and equal to 1, otherwise. Some similar quasi-metrics on $\mathbb{R}$ are:

  1. $d_1(x, y) = \max\{y - x, 0\}$ (in general, $\max\{f(y) - f(x), 0\}$ is a quasi-metric on a set X if $f : X \to \mathbb{R}_{\geq 0}$ is an injective function);
  2. $d_2(x, y) = \min\{y - x, 1\}$ if $y \geq x$, and equal to 1, otherwise;
  3. $d_3(x, y) = y - x$ if $y \geq x$, and equal to $a(x - y)$ (for fixed $a > 0$), otherwise;
  4. $d_4(x, y) = e^y - e^x$ if $y \geq x$, and equal to $e^{-y} - e^{-x}$ otherwise.

- **Real half-line quasi-semimetric**
  The **real half-line quasi-semimetric** is defined on the half-line $\mathbb{R}_{>0}$ by

$$\max\left\{0, \ln \frac{y}{x}\right\}.$$

- **Janous–Hametner metric**
  The **Janous–Hametner metric** is defined on the half-line $\mathbb{R}_{>0}$ by

$$\frac{|x - y|}{(x + y)^t},$$

  where $t = -1$ or $0 \leq t \leq 1$, and $|x - y|$ is the **natural metric**.
- **Extended real line metric**
  An **extended real line metric** is a metric on $\mathbb{R} \cup \{+\infty\} \cup \{-\infty\}$. The main example (see, for example, [Cops68]) of such metric is given by

$$|f(x) - f(y)|,$$

  where $f(x) = \frac{x}{1+|x|}$ for $x \in \mathbb{R}$, $f(+\infty) = 1$, and $f(-\infty) = -1$.
  Another metric, commonly used on $\mathbb{R} \cup \{+\infty\} \cup \{-\infty\}$, is defined by

$$|\arctan x - \arctan y|,$$

  where $-\frac{1}{2}\pi < \arctan x < \frac{1}{2}\pi$ for $-\infty < x < \infty$, and $\arctan(\pm\infty) = \pm\frac{1}{2}\pi$.
- **Complex modulus metric**
  The **complex modulus metric** on the set $\mathbb{C}$ of complex numbers is defined by

$$|z - u|,$$

  where, for any $z = z_1 + z_2 i \in \mathbb{C}$, the number $|z| = \sqrt{z\bar{z}} = \sqrt{z_1^2 + z_2^2}$ is the *complex modulus*. The *complex argument* θ is defined by $z = |z|(\cos(\theta) + i \sin(\theta))$.

The metric space $(\mathbb{C}, |z-u|)$ is called the *complex* (or *Wessel–Argand*) *plane*. It is isometric to the Euclidean plane $(\mathbb{R}^2, ||x-y||_2)$. So, the metrics on $\mathbb{R}^2$, given in Chaps. 19 and 5, can be seen as metrics on $\mathbb{C}$. For example, the **British Rail metric** on $\mathbb{C}$ is $|z|+|u|$ for $z \neq u$. The p**-relative** (if $1 \leq p < \infty$) and **relative metric** (if $p=\infty$) on $\mathbb{C}$ are defined for $|z|+|u| \neq 0$ respectively, by

$$\frac{|z-u|}{\sqrt[p]{|z|^p+|u|^p}} \text{ and } \frac{|z-u|}{\max\{|z|,|u|\}}.$$

- **$\mathbb{Z}(\eta_m)$-related norm metrics**
  A *Kummer* (or *cyclotomic*) *ring* $\mathbb{Z}(\eta_m)$ is a subring of the ring $\mathbb{C}$ (and an extension of the ring $\mathbb{Z}$), such that each of its elements has the form $\sum_{j=0}^{m-1} a_j \eta_m^j$, where η_m is a primitive m-th root $\exp(\frac{2\pi i}{m})$ of unity, and all a_j are integers.
  The *complex modulus* $|z|$ of $z=a+b\eta_m \in \mathbb{C}$ is defined by

$$|z|^2 = z\overline{z} = a^2 + (\eta_m + \overline{\eta_m})ab + b^2 = a^2 + 2ab\cos\left(\frac{2\pi i}{m}\right) + b^2.$$

  Then $(a+b)^2 = q^2$ for $m=2$ (or 1), a^2+b^2 for $m=4$, and a^2+ab+b^2 for $m=6$ (or 3), i.e., for the ring $\mathbb{Z}$ of usual integers, $\mathbb{Z}(i)$ of *Gaussian integers* and $\mathbb{Z}(\rho)$ of *Eisenstein–Jacobi* (or *EJ*) *integers*.
  The set of units of $\mathbb{Z}(\eta_m)$ contain $\eta_m^j, 0 \leq j \leq m-1$; for $m=5$ and $m \geq 6$, units of infinite order appear also, since $\cos(\frac{2\pi i}{m})$ is irrational. For $m=2,4,6$, the set of units is $\{\pm 1\}, \{\pm 1, \pm i\}, \{\pm 1, \pm\rho, \pm\rho^2\}$, where $i=\eta_4$ and $\rho=\eta_6=\frac{1+i\sqrt{3}}{2}$.
  The norms $|z| = \sqrt{a^2+b^2}$ and $||z||_i = |a|+|b|$ for $z=a+bi \in \mathbb{C}$ give rise to the **complex modulus** and i**-Manhattan** metrics on $\mathbb{C}$. They coincide with the Euclidean (l_2-) and Manhattan (l_1-) metrics, respectively, on $\mathbb{R}^2$ seen as the complex plane. The restriction of the i-Manhattan metric on $\mathbb{Z}(i)$ is the path metric of the square grid $\mathbb{Z}^2$ of $\mathbb{R}^2$; cf. **grid metric** in Chap. 19.
  The ρ**-Manhattan metric** on $\mathbb{C}$ is defined by the norm $||z||_\rho$, i.e.,

$$\min\{|a|+|b|+|c| : z = a+b\rho+c\rho^2\}$$
$$= \min\{|a|+|b|, |a+b|+|b|, |a+b|+|a| : z = a+b\rho\}.$$

  The restriction of the ρ-Manhattan metric on $\mathbb{Z}(\rho)$ is the path metric of the triangular grid of $\mathbb{R}^2$ (seen as the *hexagonal lattice* $A_2 = \{(a,b,c) \in \mathbb{Z}^3 : a+b+c=0\}$), i.e., the **hexagonal metric** (Chap. 19).
  Let f denote either i or $\rho = \frac{1+i\sqrt{3}}{2}$. Given a $\pi \in \mathbb{Z}(f) \setminus \{0\}$ and $z, z' \in \mathbb{Z}(f)$, we write $z \equiv z' \pmod{\pi}$ if $z - z' = \delta\pi$ for some $\delta \in \mathbb{Z}(f)$. For the quotient ring $\mathbb{Z}_\pi(f) = \{z \pmod{\pi} : z \in \mathbb{Z}(f)\}$, it holds $|\mathbb{Z}_\pi(f)| = ||\pi||_f^2$.
  Call two congruence classes $z \pmod{\pi}$ and $z' \pmod{\pi}$ *adjacent* if $z - z' \equiv f^j \pmod{\pi}$ for some j. The resulting graph on $\mathbb{Z}_\pi(f)$ called a *Gaussian network* or *EJ network* if, respectively, $f=i$ or $f=\rho$. The path metrics

of these networks coincide with their norm metrics, defined (Fan–Gao, 2004) for $z(\mod \pi)$ and $z'(\mod \pi)$, by

$$\min ||u||_f : u \in z - z' (\mod \pi).$$

These metrics are different from the previously defined [Hube94a, Hube94b] distance on $\mathbb{Z}_\pi(f)$: $||v||_f$, where $v \in z - z' (\mod \pi)$ is selected by minimizing the complex modulus. For $f = i$, this is the **Mannheim distance** (Chap. 16), which is not a metric.

- **Chordal metric**
The **chordal metric** d_χ is a metric on the set $\overline{\mathbb{C}} = \mathbb{C} \cup \{\infty\}$ defined by

$$d_\chi(z,u) = \frac{2|z-u|}{\sqrt{1+|z|^2}\sqrt{1+|u|^2}} \text{ and } d_\chi(z,\infty) = \frac{2}{\sqrt{1+|z|^2}}$$

for all $u, z \in \mathbb{C}$ (cf. M**-relative metric** in Chap. 5).
The metric space $(\overline{\mathbb{C}}, d_\chi)$ is called the *extended complex plane*. It is homeomorphic and conformally equivalent to the *Riemann sphere*, i.e., the *unit sphere* $S^2 = \{(x_1,x_2,x_3) \in \mathbb{E}^3 : x_1^2 + x_2^2 + x_3^2 = 1\}$ (considered as a metric subspace of $\mathbb{E}^3$), onto which $(\overline{\mathbb{C}}, d_\chi)$ is one-to-one mapped under stereographic projection. The plane $\overline{\mathbb{C}}$ can be identified with the plane $x_3 = 0$ such that the and imaginary axes coincide with the x_1 and x_2 axes. Under stereographic projection, each point $z \in \mathbb{C}$ corresponds to the point $(x_1,x_2,x_3) \in S^2$, where the ray drawn from the "north pole" $(0,0,1)$ to the point z meets the sphere S^2; the "north pole" corresponds to the point at ∞. The chordal (spherical) metric between two points $p, q \in S^2$ is taken to be the distance between their preimages $z, u \in \overline{\mathbb{C}}$.
The chordal metric can be defined equivalently on $\overline{\mathbb{R}}^n = \mathbb{R}^n \cup \{\infty\}$:

$$d_\chi(x,y) = \frac{2||x-y||_2}{\sqrt{1+||x||_2^2}\sqrt{1+||y||_2^2}} \text{ and } d_\chi(x,\infty) = \frac{2}{\sqrt{1+||x||_2^2}}.$$

The restriction of the metric d_χ on $\mathbb{R}^n$ is a **Ptolemaic metric**; cf. Chap. 1.
Given $\alpha > 0$, $\beta \geq 0$, $p \geq 1$, the **generalized chordal metric** is a metric on $\mathbb{C}$ (in general, on $(\mathbb{R}^n, ||.||_2)$ and even on any *Ptolemaic space* $(V, ||.||)$), defined by

$$\frac{|z-u|}{\sqrt[p]{\alpha+\beta|z|^p} \cdot \sqrt[p]{\alpha+\beta|u|^p}}.$$

- **Metrics on quaternions**
Quaternions are members of a noncommutative division algebra $\mathcal{Q}$ over the field $\mathbb{R}$, geometrically realizable in $\mathbb{R}^4$ [Hami66]. Formally,

$$\mathcal{Q} = \{q = q_1 + q_2 i + q_3 j + q_4 k : q_i \in \mathbb{R}\},$$

where the *basic units* $1, i, j, k \in \mathcal{Q}$ satisfy $i^2 = j^2 = k^2 = -1$ and $ij = -ji = k$.

The *quaternion norm* is defined by $||q|| = \sqrt{q\overline{q}} = \sqrt{q_1^2 + q_2^2 + q_3^2 + q_4^2}$, where $\overline{q} = q_1 - q_2 i - q_3 j - q_4 k$. The **quaternion metric** is the norm metric $||q - q'||$ on $\mathcal{Q}$.

The set of all *Lipschitz integers* and *Hurwitz integers* are defined, respectively, by

$$L = \{q_1 + q_2 i + q_3 j + q_4 k : q_i \in \mathbb{Z}\} \text{ and}$$

$$H = \{q_1 + q_2 i + q_3 j + q_4 k : \text{all } q_i \in \mathbb{Z} \text{ or all } q_i + \frac{1}{2} \in \mathbb{Z}\}.$$

A quaternion $q \in L$ is *irreducible* (i.e., $q = q'q''$ implies $\{q', q''\} \cap \{\pm 1, \pm i, \pm j, \pm k\} \neq \emptyset$) if and only if $||q||$ is a prime. Given an irreducible $\pi \in L$ and $q, q' \in H$, we write $q \equiv q' (\mod \pi)$ if $q - q' = \delta\pi$ for some $\delta \in L$.

For the rings $L_\pi = \{q (\mod \pi) : q \in L\}$ and $H_\pi = \{q (\mod \pi) : q \in H\}$ it holds $|L_\pi| = ||\pi||^2$ and $|H_\pi| = 2||\pi||^2 - 1$.

The **quaternion Lipschitz metric** on L_π is defined (Martinez et al., 2009) by

$$d_L(\alpha, \beta) = \min \sum_{1 \le s \le 4} |q_s| : \alpha - \beta \equiv q_1 + q_2 i + q_3 j + q_4 k (\mod \pi).$$

The ring H is additively generated by its subring L and $w = \frac{1}{2}(1 + i + j + k)$. The **Hurwitz metric** on the ring H_π is defined (Guzëltepe, 2013) by

$$d_H(\alpha, \beta) = \min \sum_{1 \le s \le 5} |q_s| : \alpha - \beta \equiv q_1 + q_2 i + q_3 j + q_4 k + q_5 w (\mod \pi).$$

Cf. the **hyper-Kähler** and **Gibbons–Manton** metrics in Sect. 7.3 and the **unit quaternions** and **joint angle** metrics in Sect. 18.3.

12.2 Metrics on Polynomials

A *polynomial* is a sum of powers in one or more variables multiplied by coefficients. A *polynomial in one variable* (or *monic polynomial*) with constant real (complex) coefficients is given by $P = P(z) = \sum_{k=0}^{n} a_k z^k$, $a_k \in \mathbb{R}$ ($a_k \in \mathbb{C}$). The set $\mathcal{P}$ of all real (complex) polynomials forms a ring $(\mathcal{P}, +, \cdot, 0)$. It is also a vector space over $\mathbb{R}$ (over $\mathbb{C}$).

- **Polynomial norm metric**
  A **polynomial norm metric** is a **norm metric** on the vector space $\mathcal{P}$ of all real (complex) polynomials defined by

$$||P - Q||,$$

  where $||.||$ is a *polynomial norm*, i.e., a function $||.|| : \mathcal{P} \to \mathbb{R}$ such that, for all $P, Q \in \mathcal{P}$ and for any scalar k, we have the following properties:

  1. $||P|| \geq 0$, with $||P|| = 0$ if and only if $P \equiv 0$;
  2. $||kP|| = |k|||P||$;
  3. $||P + Q|| \leq ||P|| + ||Q||$ (triangle inequality).

  The l_p*-norm* and L_p*-norm* of a polynomial $P(z) = \sum_{k=0}^{n} a_k z^k$ are defined by

$$||P||_p = (\sum_{k=0}^{n} |a_k|^p)^{1/p} \text{ and } ||P||_{L_p} = (\int_0^{2\pi} |P(e^{i\theta})|^p \frac{d\theta}{2\pi})^{\frac{1}{p}} \text{ for } 1 \leq p < \infty,$$

$$||P||_\infty = \max_{0 \leq k \leq n} |a_k| \text{ and } ||P||_{L_\infty} = \sup_{|z|=1} |P(z)| \text{ for } p = \infty.$$

  The values $||P||_1$ and $||P||_\infty$ are called the *length* and *height* of polynomial P.
- **Distance from irreducible polynomials**
  For any field $\mathbb{F}$, a polynomial with coefficients in $\mathbb{F}$ is said to be *irreducible over* $\mathbb{F}$ if it cannot be factored into the product of two nonconstant polynomials with coefficients in $\mathbb{F}$. Given a metric d on the polynomials over $\mathbb{F}$, the **distance** (of a given polynomial $P(z)$) **from irreducible polynomials** is $d_{ir}(P) = \inf d(P, Q)$, where $Q(z)$ is any irreducible polynomial of the same degree over $\mathbb{F}$.
  Polynomial conjecture of Turán, 1967, is that there exists a constant C with $d_{ir}(P) \leq C$ for every polynomial P over $\mathbb{Z}$, where $d(P, Q)$ is the *length* $||P - Q||_1$ of $P - Q$.
  Lee–Ruskey–Williams, 2007, conjectured that there exists a constant C with $d_{ir}(P) \leq C$ for every polynomial P over the Galois field $\mathbb{F}_2$, where $d(P, Q)$ is the **Hamming distance** between the $(0, 1)$-sequences of coefficients of P and Q.
- **Bombieri metric**
  The **Bombieri metric** (or **polynomial bracket metric**) is a **polynomial norm metric** on the set $\mathcal{P}$ of all real (complex) polynomials defined by

$$[P - Q]_p,$$

  where $[.]_p$, $0 \leq p \leq \infty$, is the *Bombieri p-norm*.
  For a polynomial $P(z) = \sum_{k=0}^{n} a_k z^k$ it is defined by

$$[P]_p = \left(\sum_{k=0}^{n} \binom{n}{k}^{1-p} |a_k|^p \right)^{\frac{1}{p}}.$$

- **Metric space of roots**
The **metric space of roots** is (Ćurgus–Mascioni, 2006) the space (X, d) where X is the family of all multisets of complex numbers with n elements and the distance between multisets $U = \{u_1, \ldots, u_n\}$ and $V = \{v_1, \ldots, v_n\}$ is defined by the following analog of the **Fréchet metric**:

$$\min_{\tau \in Sym_n} \max_{1 \le j \le n} |u_j - v_{\tau(j)}|,$$

where τ is any permutation of $\{1, \ldots, n\}$. Here the set of roots of some monic complex polynomial of degree n is considered as a multiset with n elements. Cf. **metrics between multisets** in Chap. 1.
The function assigning to each polynomial the multiset of its roots is a **homeomorphism** between the metric space of all monic complex polynomials of degree n with the **polynomial norm metric** l_∞ and the metric space of roots.

12.3 Metrics on Matrices

An $m \times n$ matrix $A = ((a_{ij}))$ over a field $\mathbb{F}$ is a table consisting of m rows and n columns with the entries a_{ij} from $\mathbb{F}$. The set of all $m \times n$ matrices with real (complex) entries is denoted by $M_{m,n}$ or $\mathbb{R}^{m \times n}$ ($\mathbb{C}^{m \times n}$). It forms a *group* $(M_{m,n}, +, 0_{m,n})$, where $((a_{ij})) + ((b_{ij})) = ((a_{ij} + b_{ij}))$, and the matrix $0_{m,n} \equiv 0$. It is also an mn-dimensional vector space over $\mathbb{R}$ ($\mathbb{C}$).

The *transpose* of a matrix $A = ((a_{ij})) \in M_{m,n}$ is the matrix $A^T = ((a_{ji})) \in M_{n,m}$. A $m \times n$ matrix A is called a *square matrix* if $m = n$, and a *symmetric matrix* if $A = A^T$. The *conjugate transpose* (or *adjoint*) of a matrix $A = ((a_{ij})) \in M_{m,n}$ is the matrix $A^* = ((\overline{a}_{ji})) \in M_{n,m}$. An *Hermitian matrix* is a complex square matrix A with $A = A^*$.

The set of all square $n \times n$ matrices with real (complex) entries is denoted by M_n. It forms a *ring* $(M_n, +, \cdot, 0_n)$, where $+$ and 0_n are defined as above, and $((a_{ij})) \cdot ((b_{ij})) = ((\sum_{k=1}^{n} a_{ik} b_{kj}))$. It is also an n^2-dimensional vector space over $\mathbb{R}$ (over $\mathbb{C}$). The *trace* of a square $n \times n$ matrix $A = ((a_{ij}))$ is defined by $\mathrm{Tr}(A) = \sum_{i=1}^{n} a_{ii}$.

The *identity matrix* is $1_n = ((c_{ij}))$ with $c_{ii} = 1$, and $c_{ij} = 0$, $i \neq j$. An *unitary matrix* $U = ((u_{ij}))$ is a square matrix defined by $U^{-1} = U^*$, where U^{-1} is the *inverse matrix* of U, i.e., $UU^{-1} = 1_n$. A matrix $A \in M_{m,n}$ is *orthonormal* if $A^* A = 1_n$. A matrix $A \in \mathbb{R}^{n \times n}$ is *orthogonal* if $A^T = A^{-1}$, *normal* if $A^T A = AA^T$ and *singular* if its determinant is 0.

If for a matrix $A \in M_n$ there is a vector x such that $Ax = \lambda x$ for some scalar λ, then λ is called an *eigenvalue* of A with corresponding *eigenvector* x. Given a matrix $A \in \mathbb{C}^{m \times n}$, its *singular values* $s_i(A)$ are defined as $\sqrt{\lambda(A^* A)}$. A real matrix A is *positive-definite* if $v^T A v > 0$ for all nonzero real vectors v; it holds if and only if all eigenvalues of $A_H = \frac{1}{2}(A + A^T)$ are positive. An Hermitian matrix A is *positive-definite* if $v^* A v > 0$ for all nonzero complex vectors v; it holds if and only if all $\lambda(A)$ are positive.

The *mixed states* of a n-dimensional *quantum system* are described by their *density matrices*, i.e., positive-semidefinite Hermitian $n \times n$ matrices of trace 1. The set of such matrices is convex, and its extremal points describe the *pure states*. Cf. **monotone metrics** in Chap. 7 and **distances between quantum states** in Chap. 24.

- **Matrix norm metric**
  A **matrix norm metric** is a **norm metric** on the set $M_{m,n}$ of all real (complex) $m \times n$ matrices defined by

$$||A - B||,$$

  where $||.||$ is a *matrix norm*, i.e., a function $||.|| : M_{m,n} \to \mathbb{R}$ such that, for all $A, B \in M_{m,n}$, and for any scalar k, we have the following properties:

  1. $||A|| \geq 0$, with $||A|| = 0$ if and only if $A = 0_{m,n}$;
  2. $||kA|| = |k|||A||$;
  3. $||A + B|| \leq ||A|| + ||B||$ (triangle inequality).
  4. $||AB|| \leq ||A|| \cdot ||B||$ (*submultiplicativity*).

  All matrix norm metrics on $M_{m,n}$ are equivalent. The simplest example of such metric is the **Hamming metric** on $M_{m,n}$ (in general, on the set $M_{m,n}(\mathbb{F})$ of all $m \times n$ matrices with entries from a field $\mathbb{F}$) defined by $||A - B||_H$, where $||A||_H$ is the *Hamming norm* of $A \in M_{m,n}$, i.e., the number of nonzero entries in A. Example of a *generalized* (i.e., not submultiplicative one) *matrix norm* is the *max element norm* $||A = ((a_{ij}))||\max = \max_{i,j} |a_{ij}|$; but $\sqrt{mn}||A||_{\max}$ is a matrix norm.
- **Natural norm metric**
  A **natural** (or *operator, induced*) **norm metric** is a **matrix norm metric** on the set M_n defined by

$$||A - B||_{\text{nat}},$$

  where $||.||_{\text{nat}}$ is a *natural* (or *operator, induced*) *norm* on M_n, induced by the vector norm $||x||$, $x \in \mathbb{R}^n$ ($x \in \mathbb{C}^n$), is a matrix norm defined by

$$||A||_{\text{nat}} = \sup_{||x|| \neq 0} \frac{||Ax||}{||x||} = \sup_{||x||=1} ||Ax|| = \sup_{||x|| \leq 1} ||Ax||.$$

  The natural norm metric can be defined in similar way on the set $M_{m,n}$ of all $m \times n$ real (complex) matrices: given vector norms $||.||_{\mathbb{R}^m}$ on $\mathbb{R}^m$ and $||.||_{\mathbb{R}^n}$ on $\mathbb{R}^n$, the *natural norm* $||A||_{\text{nat}}$ of a matrix $A \in M_{m,n}$, induced by $||.||_{\mathbb{R}^n}$ and $||.||_{\mathbb{R}^m}$, is a matrix norm defined by $||A||_{\text{nat}} = \sup_{||x||_{\mathbb{R}^n}=1} ||Ax||_{\mathbb{R}^m}$.
- **Matrix p-norm metric**
  A **matrix p-norm metric** is a **natural norm metric** on M_n defined by

$$||A - B||^p_{\text{nat}},$$

where $||.||^p_{\text{nat}}$ is the *matrix* (or *operator*) *p-norm*, i.e., a *natural norm*, induced by the vector l_p*-norm*, $1 \leq p \leq \infty$:

$$||A||^p_{\text{nat}} = \max_{||x||_p=1} ||Ax||_p, \text{ where } ||x||_p = (\sum_{i=1}^{n} |x_i|^p)^{1/p}.$$

The **maximum absolute column** and **maximum absolute row metric** are the **matrix 1-norm** and **matrix ∞-norm metric** on M_n. For a matrix $A = ((a_{ij})) \in M_n$, the *maximum absolute column* and *maximum absolute row sum norm* are

$$||A||^1_{\text{nat}} = \max_{1 \leq j \leq n} \sum_{i=1}^{n} |a_{ij}| \text{ and } |A||^{\infty}_{\text{nat}} = \max_{1 \leq i \leq n} \sum_{j=1}^{n} |a_{ij}|.$$

The **spectral norm metric** is the **matrix 2-norm metric** $||A-B||^2_{\text{nat}}$ on M_n. The matrix 2-norm $||.||^2_{\text{nat}}$, induced by the vector l_2*-norm*, is also called the *spectral norm* and denoted by $||.||_{sp}$. For a symmetric matrix $A = ((a_{ij})) \in M_n$, it is

$$||A||_{sp} = s_{\max}(A) = \sqrt{\lambda_{\max}(A^* A)},$$

where $A^* = ((\overline{a}_{ji}))$, while $s_{\max}$ and $\lambda_{\max}$ are largest singular value and eigenvalue.

- **Frobenius norm metric**
  The **Frobenius norm metric** is a **matrix norm metric** on $M_{m,n}$ defined by

$$||A - B||_{Fr},$$

where $||.||_{Fr}$ is the *Frobenius* (or *Hilbert–Schmidt*) *norm*. For $A = ((a_{ij}))$, it is

$$||A||_{Fr} = \sqrt{\sum_{i,j} |a_{ij}|^2} = \sqrt{\text{Tr}(A^* A)} = \sqrt{\sum_{1 \leq i \leq \text{rank}(A)} \lambda_i} = \sqrt{\sum_{1 \leq i \leq \text{rank}(A)} s_i^2},$$

where λ_i, s_i are the eigenvalues and singular values of A.
This norm is strictly convex, is a differentiable function of its elements a_{ij} and is the only unitarily invariant norm among $||A||_p = (\sum_{i=1}^{m} \sum_{j=1}^{n} |a_{ij}|^p)^{\frac{1}{p}}$, $p \geq 1$.
The **trace norm metric** is a matrix norm metric on $M_{m,n}$ defined by

$$||A - B||_{tr},$$

where $||.||_{tr}$ is the *trace norm* (or *nuclear norm*) on $M_{m,n}$ defined by

$$||A||_{tr} = \sum_{i=1}^{\min\{m,n\}} s_i(A) = \text{Tr}(\sqrt{A^*A}).$$

- **Schatten norm metric**
  Given $1 \leq p < \infty$, the **Schatten norm metric** is a **matrix norm metric** on $M_{m,n}$ defined by

$$||A - B||^p_{Sch},$$

  where $||.||^p_{Sch}$ is the *Schatten p-norm* on $M_{m,n}$. For a matrix $A \in M_{m,n}$, it is defined as the p-th root of the sum of the p-th powers of all its *singular values*:

$$||A||^p_{Sch} = (\sum_{i=1}^{\min\{m,n\}} s_i^p(A))^{\frac{1}{p}}.$$

  For $p = \infty$, 2 and 1, one obtains the **spectral norm metric**, **Frobenius norm metric** and **trace norm metric**, respectively.
- **(c, p)-norm metric**
  Let $k \in \mathbb{N}$, $k \leq \min\{m, n\}$, $c \in \mathbb{R}^k$, $c_1 \geq c_2 \geq \cdots \geq c_k > 0$, and $1 \leq p < \infty$. The (c, p)**-norm metric** is a **matrix norm metric** on $M_{m,n}$ defined by

$$||A - B||^k_{(c,p)},$$

  where $||.||^k_{(c,p)}$ is the (c, p)*-norm* on $M_{m,n}$. For a matrix $A \in M_{m,n}$, it is defined by

$$||A||^k_{(c,p)} = (\sum_{i=1}^{k} c_i s_i^p(A))^{\frac{1}{p}},$$

  where $s_1(A) \geq s_2(A) \geq \cdots \geq s_k(A)$ are the first k *singular values* of A.
  If $p = 1$, it is the *c-norm*. If, moreover, $c_1 = \cdots = c_k = 1$, it is the *Ky Fan k-norm*.
- **Ky Fan k-norm metric**
  Given $k \in \mathbb{N}$, $k \leq \min\{m, n\}$, the **Ky Fan k-norm metric** is a **matrix norm metric** on $M_{m,n}$ defined by

$$||A - B||^k_{KF},$$

  where $||.||^k_{KF}$ is the *Ky Fan k-norm* on $M_{m,n}$. For a matrix $A \in M_{m,n}$, it is defined as the sum of its first k *singular values*:

$$||A||_{KF}^{k} = \sum_{i=1}^{k} s_i(A).$$

For $k = 1$ and $k = \min\{m, n\}$, one obtains the **spectral** and **trace** norm metrics.

- **Cut norm metric**
The **cut norm metric** is a **matrix norm metric** on $M_{m,n}$ defined by

$$||A - B||_{cut},$$

where $||.||_{cut}$ is the *cut norm* on $M_{m,n}$ defined, for a matrix $A = ((a_{ij})) \in M_{m,n}$, as:

$$||A||_{cut} = \max_{I \subset \{1,\dots,m\}, J \subset \{1,\dots,n\}} |\sum_{i \in I, j \in J} a_{ij}|.$$

Cf. in Chap. 15 the **rectangle distance on weighted graphs** and the **cut semimetric**, but the **weighted cut metric** in Chap. 19 is not related.

- **Matrix nearness problems**
A norm $||.||$ is *unitarily invariant* on $M_{m,n}$ if $||B|| = ||UBV||$ for all $B \in M_{m,n}$ and all unitary matrices U, V. All *Schatten p-norms* are
Given a unitarily invariant norm $||.||$ on $M_{m,n}$, a matrix property $\mathcal{P}$ defining a subspace or compact subset of $M_{m,n}$ (so that $d_{||.||}(A, \mathcal{P})$ below is well defined) and a matrix $A \in M_{m,n}$, then the *distance to* $\mathcal{P}$ is the **point-set distance** on $M_{m,n}$

$$d(A) = d_{||.||}(A, \mathcal{P}) = \min\{||E|| : A + E \text{ has property } \mathcal{P}\}.$$

A **matrix nearness problem** is [High89] to find an explicit formula for $d(A)$, the $\mathcal{P}$*-closest matrix* (or matrices) $X_{||.||}(A) = A + E$, satisfying the above minimum, and efficient algorithms for computing $d(A)$ and $X_{||.||}(A)$. The *componentwise nearness problem* is to find $d'(A) = \min\{\epsilon : |E| \leq \epsilon|A|, A+E \text{ has property } \mathcal{P}\}$, where $|B| = ((|b_{ij}|))$ and the matrix inequality is interpreted componentwise.
The most used norms for $B = ((b_{ij}))$ are the *Schatten* 2- and ∞-*norms* (cf. **Schatten norm metric**): the *Frobenius norm* $||B||_{Fr} = \sqrt{\mathrm{Tr}(B^*B)} = \sqrt{\sum_{1 \leq i \leq \mathrm{rank}(B)} s_i^2}$ and the *spectral norm* $||B||_{sp} = \sqrt{\lambda_{\max}(B^*B)} = s_1(B)$.
Examples of closest matrices $X = X_{||.||}(A, \mathcal{P})$ follow.
Let $A \in \mathbb{C}^{n \times n}$. Then $A = A_H + A_S$, where $A_H = \frac{1}{2}(A + A^*)$ is Hermitian and $A_H = \frac{1}{2}(A - A^*)$ is *skew-Hermitian* (i.e., $A_H^* = -A_H$). Let $A = U\Sigma V^*$ be a *singular value decomposition* (SVD) of A, i.e., $U \in M_m$ and $V^* \in M_n$ are unitary, while $\Sigma = \mathrm{diag}(s_1, s_2, \dots, s_{\min\{m,n\}})$ is an $m \times n$ diagonal matrix with $s_1 \geq s_2 \geq \dots \geq s_{\mathrm{rank}(A)} > 0 = \dots = 0$. Fan and Hoffman, 1955, showed that, for any unitarily invariant norm, A_H, A_S, UV^* are closest Hermitian (symmetric), skew-Hermitian (skew-symmetric) and unitary

(orthogonal) matrices, respectively. Such matrix $X_{Fr}(A)$ is a unique minimizer in all three cases.
Let $A \in \mathbb{R}^{n \times n}$. Gabriel, 1979, found the closest normal matrix $X_{Fr}(A)$. Higham found in 1988 a unique closest symmetric positive-semidefinite matrix $X_{Fr}(A)$ and, in 2001, the closest matrix of this type with unit diagonal (i.e., ab correlation matrix).
Given a SVD $A = U\Sigma V^*$ of A, let A_k denote $U\Sigma_k V^*$, where Σ_k is a diagonal matrix diag$(s_1, s_2, \dots, s_k, 0, \dots, 0)$ containing the largest k singular values of A. Then (Mirsky, 1960) A_k achieves $\min_{\text{rank}(A+E) \leq k} ||E||$ for any unitarily invariant norm. So, $||A - A_k||_{Fr} = \sqrt{\sum_{i=k+1}^{\text{rank}(A)} s_i^2}$ (Eckart–Young, 1936) and $||A - A_k||_{sp} = s_{\max}(A - A_k) = s_{k+1}(A)$. A_k is a unique minimizer $X_{Fr}(A)$ if $s_k > s_{k+1}$.
Let $A \in \mathbb{R}^{n \times n}$ be nonsingular. Then its **distance to singularity** $d(A, Sing) = \min\{||E|| : A + E \text{ is singular}\}$ is, for both above norms, $s_n(A) = \frac{1}{s_1(A^{-1})} = \frac{1}{||A^{-1}||_{sp}} = \sup\{\delta : \delta\mathbb{B}_{\mathbb{R}^n} \subseteq A\mathbb{B}_{\mathbb{R}^n}\}$; here $\mathbb{B}_{\mathbb{R}^n} = \{x \in \mathbb{R}^n : ||x|| \leq 1\}$.
Given a closed convex cone $C \subseteq \mathbb{R}^n$, call a matrix $A \in \mathbb{R}^{m \times n}$ *feasible* if $\{Ax : x \in C\} = \mathbb{R}^m$; so, for $m = n$ and $C = \mathbb{R}^n$, feasibly means nonsingularity. Renegar, 1995, showed that, for feasible matrix A, its **distance to infeasibility** $\min\{||E||_{\text{nat}} : A + E \text{ is not feasible}\}$ is $\sup\{\delta : \delta\mathbb{B}_{\mathbb{R}^m} \subseteq A(\mathbb{B}_{\mathbb{R}^n} \cap C)\}$.
Lewis, 2003, generalized this by showing that, given two real normed spaces X, Y and a surjective *convex process* (or *set valued sublinear mapping*) F from X to Y, i.e., a multifunction for which $\{(x, y) : y \in F(x)\}$ is a closed convex cone, it holds

$$\min\{||E||_{\text{nat}} : E \text{ is any linear map } X \to Y, F+E \text{ is not surjective}\} = \frac{1}{||F^{-1}||_{\text{nat}}}.$$

Donchev et al. 2002, extended this, computing **distance to irregularity**; cf. **metric regularity** (Chap. 1). Cf. the above four *distances to ill-posedness* with **distance to uncontrollability** (Chap. 18) and **distances from symmetry** (Chap. 21).

- ***Sym*$(n, \mathbb{R})^+$ and *Her*$(n, \mathbb{C})^+$ metrics**
  Let $Sym(n, \mathbb{R})^+$ and $Her(n, \mathbb{C})^+$ be the cones of $n \times n$ symmetric real and Hermitian complex positive-definite $n \times n$ matrices. The $Sym(n, \mathbb{R})^+$ **metric** is defined, for any $A, B \in Sym(n, \mathbb{R})^+$, as

$$\left(\sum_{i=1}^{n} \log^2 \lambda_i\right)^{\frac{1}{2}},$$

  where $\lambda_1, \dots, \lambda_n$ are the *eigenvalues* of the matrix $A^{-1}B$ (the same as those of $A^{-\frac{1}{2}}BA^{-\frac{1}{2}}$). It is the **Riemannian distance**, arising from the Riemannian metric $ds^2 = \text{Tr}((A^{-1}(dA))^2)$. This metric was rediscovered in Förstner–Moonen, 1999, and Pennec et al., 2004, via *generalized eigenvalue problem*: $det(\lambda A - B) = 0$.

The $Her(n, \mathbb{C})^+$ **metric** is defined, for any $A, B \in Her(n, \mathbb{C})^+$, by

$$d_R(A, B) = || \log(A^{-\frac{1}{2}} B A^{-\frac{1}{2}}) ||_{Fr},$$

where $||H||_{Fr} = (\sum_{i,j} |h_{ij}|^2)^{\frac{1}{2}}$ is the *Frobenius norm* of the matrix $H = ((h_{ij}))$. It is the **Riemannian distance** arising from the Riemannian metric of nonpositive curvature, defined locally (at H) by $ds = ||H^{-\frac{1}{2}} dH \, H^{-\frac{1}{2}}||_{Fr}$. In other words, this distance is the **geodesic distance**

$$\inf\{L(\gamma) : \gamma \text{ is a (differentiable) path from A to B}\},$$

where $L(\gamma) = \int_A^B ||\gamma^{-\frac{1}{2}}(t)\gamma'(t)\gamma^{-\frac{1}{2}}(t)||_{Fr} dt$ and the geodesic $[A, B]$ is parametrized by $\gamma(t) = A^{\frac{1}{2}}(A^{-\frac{1}{2}} B A^{-\frac{1}{2}})^t A^{\frac{1}{2}}$ in the sense that $d_R(A, \gamma(t)) = t d_R(A, B)$ for each $t \in [0, 1]$. In particular, the geodesic midpoint $\gamma(\frac{1}{2})$ of $[A, B]$ can be seen as the *geometric mean* of two positive-definite matrices A and B.
The space $(Her(n, \mathbb{C})^+, d_R))$ is an **Hadamard** (i.e., complete and CAT(0)) **space**, cf. Chap. 6. But $Her(n, \mathbb{C})^+$ is not complete with respect to matrix norms; it has a boundary consisting of the singular positive-semidefinite matrices.
Above $Sym(n, \mathbb{R})^+$ and $Her(n, \mathbb{C})^+$ metrics are the special cases of the distance $d_R(x, y)$ among **invariant distances on symmetric cones** in Chap. 9.
Cf. also, in Chap. 24, the **trace distance** on all Hermitian of trace 1 positive-definite $n \times n$ matrices and in Chap. 7, the **Wigner–Yanase–Dyson metrics** on all complex positive-definite $n \times n$ matrices.
The **Bartlett distance** between two matrices $A, B \in Her(n, \mathbb{C})^+$, is defined (Conradsen et al., 2003, for radar applications) by

$$\ln\left(\frac{(det(A + B))^2}{4det(A)det(B)}\right).$$

- **Siegel distance**
  The *Siegel half-plane* is the set SH_n of $n \times n$ matrices $Z = X + iY$, where X, Y are symmetric or Hermitian and Y is positive-definite. The **Siegel–Hua metric** (Siegel, 1943, and independently, Hua, 1944) on SH_n is defined by

  $$ds^2 = \text{Tr}(Y^{-1}(dZ)Y^{-1}(d\overline{Z})).$$

  It is unique metric preserved by any automorphism of SH_n. The Siegel–Hua metric on the *Siegel disk* $SD_n = \{W = (Z - iI)(Z + iI)^{-1} : Z \in SH_n\}$ is defined by

  $$ds^2 = \text{Tr}((I - WW^*)^{-1} dW (I - W^*W)^{-1} dW^*).$$

  For n=1, the Siegel–Hua metric is the **Poincaré metric** (cf. Chap. 6) on the *Poincaré half-plane* SH_1 and the *Poincaré disk* SD_1, respectively.

Let $A_n = \{Z = iY : Y > 0\}$ be the imaginary axe on the Siegel half-plane. The Siegel–Hua metric on A_n is the Riemannian **trace metric** $ds^2 = \mathrm{Tr}((P^{1}dP)^2)$. The corresponding distances are $Sym(n, \mathbb{R})^+$ **metric** or $Her(n, \mathbb{C})^+$ **metric**.
The **Siegel distance** $d_{Siegel}(Z_1, Z_2)$ on $SH_n \setminus A_n$ is defined by

$$d^2_{Siegel}(Z_1, Z_2) = \sum_{i=1}^{n} \log^2 \left(\frac{1 + \sqrt{\lambda_i}}{1 - \sqrt{\lambda_i}} \right);$$

$\lambda_1, \dots, \lambda_n$ are the *eigenvalues* of the matrix $(Z_1 - Z_2)(Z_1 - \overline{Z_2}) - 1(\overline{Z_1} - \overline{Z_2})(\overline{Z_1} - Z_2)^{-1}$.

- **Barbaresco metrics**
Let $z(k)$ be a complex temporal (discrete time) *stationary* signal, i.e., its mean value is constant and its *covariance function* $\mathbb{E}[z(k_1)z^*(k_2)]$ is only a function of $k_1 - k_2$. Such signal can be represented by its covariance $n \times n$ matrix $R = ((r_{ij}))$, where $r_{ij} = \mathbb{E}[z(i), z * (j)] = \mathbb{E}[z(n)z * (n - i + j)]$. It is a positive-definite *Toeplitz* (i.e. diagonal-constant) Hermitian matrix. In radar applications, such matrices represent the Doppler spectra of the signal. Matrices R admit a parametrization (complex ARM, i.e., m-th order autoregressive model) by *partial autocorrelation coefficients* defined recursively as the complex correlation between the forward and backward prediction errors of the $(m-1)$-th order complex ARM.
Barbaresco [Barb12] defined, via this parametrization, a **Bergman metric** (cf. Chap. 7) on the bounded domain $\mathbb{R} \times D_n \subset \mathbb{C}^n$ of above matrices R; here D is a *Poincaré disk*. He also defined a related **Kähler metric** on $M \times S_n$, where M is the set of positive-definite Hermitian matrices and SD_n is the *Siegel disk* (cf. **Siegel distance**). Such matrices represent spatiotemporal stationary signals, i.e., in radar applications, the Doppler spectra and spatial directions of the signal.
Cf. **Ruppeiner metric** (Chap. 7) and **Martin cepstrum distance** (Chap. 21).
- **Distances between graphs of matrices**
The *graph* $G(A)$ *of a complex* $m \times n$ *matrix* A is the *range* (i.e., the span of columns) of the matrix $R(A) = ([IA^T])^T$. So, $G(A)$ is a subspace of $\mathbb{C}^{m+n}$ of all vectors v, for which the equation $R(A)x = v$ has a solution.
A **distance between graphs of matrices** A and B is a distance between the subspaces $G(A)$ and $G(B)$. It can be an **angle distance between subspaces** or, for example, the following distance (cf. also the **Kadets distance** in Chap. 1 and the **gap metric** in Chap. 18).
The **spherical gap distance** between subspaces A and B is defined by

$$\max\{\max_{x \in S(A)} d_E(x, S(B)), \max_{y \in S(B)} d_E(y, S(A))\},$$

where $S(A), S(B)$ are the unit spheres of the subspaces A, B, $d(z, C)$ is the **point-set distance** $\inf_{y \in C} d(z, y)$ and $d_E(z, y)$ is the Euclidean distance.

- **Angle distances between subspaces**
  Consider the *Grassmannian space* $G(m,n)$ of all n-dimensional subspaces of Euclidean space $\mathbb{E}^m$; it is a compact *Riemannian manifold* of dimension $n(m-n)$. Given two subspaces $A, B \in G(m,n)$, the *principal angles* $\frac{\pi}{2} \geq \theta_1 \geq \cdots \geq \theta_n \geq 0$ between them are defined, for $k = 1, \dots, n$, inductively by

$$\cos\theta_k = \max_{x \in A} \max_{y \in B} x^T y = (x^k)^T y^k$$

  subject to the conditions $||x||_2 = ||y||_2 = 1$, $x^T x^i = 0$, $y^T y^i = 0$, for $1 \leq i \leq k-1$, where $||.||_2$ is the Euclidean norm.
  The principal angles can also be defined in terms of orthonormal matrices Q_A and Q_B spanning subspaces A and B, respectively: in fact, n ordered singular values of the matrix $Q_A Q_B \in M_n$ can be expressed as cosines $\cos\theta_1, \dots, \cos\theta_n$.
  The **geodesic distance** between subspaces A and B is (Wong, 1967) defined by

$$\sqrt{2\sum_{i=1}^{n} \theta_i^2}.$$

  The **Martin distance** between subspaces A and B is defined by

$$\sqrt{\ln \prod_{i=1}^{n} \frac{1}{\cos^2\theta_i}}.$$

  In the case when the subspaces represent ARMs (*autoregressive models*), the Martin distance can be expressed in terms of the *cepstrum* of the autocorrelation functions of the models. Cf. the **Martin cepstrum distance** in Chap. 21.
  The **Asimov distance** between subspaces A and B is defined by θ_1. It can be expressed also in terms of the **Finsler metric** on the manifold $G(m,n)$.
  The **gap distance** between subspaces A and B is defined by $\sin\theta_1$. It is the l_2-*norm* of the difference of the *orthogonal projectors* onto A and B. Many versions of this distance are used in Control Theory (cf. **gap metric** in Chap. 18).
  The **Frobenius distance** between subspaces A and B is defined by

$$\sqrt{2\sum_{i=1}^{n} \sin^2\theta_i}.$$

  It is the *Frobenius norm* of the difference of above projectors onto A and B.
  Similar distances $\sqrt{\sum_{i=1}^{n} \sin^2\theta_i}$, $2\sin(\frac{\theta_1}{2})$, $\sqrt{1 - \prod_{i=1}^{n} \cos^2\theta_i}$ and arccos $(\prod_{i=1}^{n} \cos\theta_i)$ are called the **chordal distance**, **chordal 2-norm distance**, **Binet–Cauchy distance** and *Fubini–Study distance* (cf. Chap. 7), respectively.

- **Larsson–Villani metric**
  Let A and B be two arbitrary orthonormal $m \times n$ matrices of full rank, and let θ_{ij} be the angle between the i-th column of A and the j-th column of B.
  We call **Larsson–Villani metric** the distance between A and B (used by Larsson and Villani, 2000, for multivariate models) the square of which is defined by

$$n - \sum_{i=1}^{n}\sum_{j=1}^{n} \cos^2 \theta_{ij}.$$

  The square of usual Euclidean distance between A and B is $2(1 - \sum_{i=1}^{n} \cos \theta_{ii})$. For $n = 1$, above two distances are $\sin \theta$ and $\sqrt{2(1 - \cos \theta)}$, respectively.
- **Lerman metric**
  Given a finite set X and real symmetric $|X| \times |X|$ matrices $((d_1(x, y)))$, $((d_2(x, y)))$ with $x, y \in X$, their **Lerman semimetric** (cf. **Kendall τ distance** on permutations in Chap. 11) is defined by

$$|\{(\{x,y\},\{u,v\}) : (d_1(x,y) - d_1(u,v))(d_2(x,y) - d_2(u,v)) < 0\}| \binom{|X|+1}{2}^{-2},$$

  where $(\{x, y\}, \{u, v\})$ is any pair of unordered pairs of elements x, y, u, v from X. Similar **Kaufman semimetric** between $((d_1(x, y)))$ and $((d_2(x, y)))$ is

$$\frac{|\{(\{x,y\},\{u,v\}) : (d_1(x,y) - d_1(u,v))(d_2(x,y) - d_2(u,v)) < 0\}|}{|\{(\{x,y\},\{u,v\}) : (d_1(x,y) - d_1(u,v))(d_2(x,y) - d_2(u,v)) \neq 0\}|}.$$

Chapter 13
Distances in Functional Analysis

Functional Analysis is the branch of Mathematics concerned with the study of spaces of functions. This usage of the word *functional* goes back to the calculus of variations which studies functions whose argument is a function. In the modern view, Functional Analysis is seen as the study of complete *normed vector spaces*, i.e., **Banach spaces**.

For any real number $p \geq 1$, an example of a Banach space is given by **L_p-space** of all Lebesgue-measurable functions whose absolute value's p-th power has finite integral.

A **Hilbert space** is a Banach space in which the norm arises from an *inner product*. Also, in Functional Analysis are considered *continuous linear operators* defined on Banach and Hilbert spaces.

13.1 Metrics on Function Spaces

Let $I \subset \mathbb{R}$ be an *open interval* (i.e., a nonempty connected open set) in $\mathbb{R}$. A real function $f : I \to \mathbb{R}$ is called *real analytic* on I if it agrees with its *Taylor series* in an *open neighborhood* U_{x_0} of every point $x_0 \in I$: $f(x) = \sum_{n=0}^{\infty} \frac{f^{(n)}(x_0)}{n!}(x - x_0)^n$ for any $x \in U_{x_0}$. Let $D \subset \mathbb{C}$ be a *domain* (i.e., a *convex* open set) in $\mathbb{C}$.

A complex function $f : D \to \mathbb{C}$ is called *complex analytic* (or, simply, *analytic*) on D if it agrees with its Taylor series in an open neighborhood of every point $z_0 \in D$. A complex function f is analytic on D if and only if it is *holomorphic* on D, i.e., if it has a complex derivative $f'(z_0) = \lim_{z \to z_0} \frac{f(z)-f(z_0)}{z-z_0}$ at every point $z_0 \in D$.

- **Integral metric**
  The **integral metric** is the L_1-*metric* on the set $C_{[a,b]}$ of all continuous real (complex) functions on a given segment $[a, b]$ defined by

M.M. Deza, E. Deza, *Encyclopedia of Distances*,
DOI 10.1007/978-3-662-44342-2_13

$$\int_a^b |f(x) - g(x)|dx.$$

The corresponding metric space is abbreviated by $C^1_{[a,b]}$. It is a Banach space.
In general, for any **compact** topological space X, the integral metric is defined on the set of all continuous functions $f : X \to \mathbb{R}$ ($\mathbb{C}$) by $\int_X |f(x) - g(x)|dx$.

- **Uniform metric**
  The **uniform metric** (or **sup metric**) is the L_∞**-metric** on the set $C_{[a,b]}$ of all real (complex) continuous functions on a given segment $[a, b]$ defined by

$$\sup_{x \in [a,b]} |f(x) - g(x)|.$$

  The corresponding metric space is abbreviated by $C^\infty_{[a,b]}$. It is a Banach space.
  A generalization of $C^\infty_{[a,b]}$ is the *space of continuous functions* $C(X)$, i.e., a metric space on the set of all continuous (more generally, bounded) functions $f : X \to \mathbb{C}$ of a topological space X with the L_∞-metric $\sup_{x \in X} |f(x) - g(x)|$.
  In the case of the metric space $C(X, Y)$ of continuous (more generally, bounded) functions $f : X \to Y$ from one **metric compactum** (X, d_X) to another (Y, d_Y), the sup metric between two functions $f, g \in C(X, Y)$ is defined by $\sup_{x \in X} d_Y(f(x), g(x))$.
  The metric space $C^\infty_{[a,b]}$, as well as the metric space $C^1_{[a,b]}$, are two of the most important cases of the metric space $C^p_{[a,b]}$, $1 \le p \le \infty$, on the set $C_{[a,b]}$ with the L_p-metric $(\int_a^b |f(x) - g(x)|^p dx)^{\frac{1}{p}}$. The space $C^p_{[a,b]}$ is an example of an L_p*-space*.
- **Dogkeeper distance**
  Given a metric space (X, d), the **dogkeeper distance** is a metric on the set of all functions $f : [0, 1] \to X$, defined by

$$\inf_\sigma \sup_{t \in [0,1]} d(f(t), g(\sigma(t))),$$

  where $\sigma : [0, 1] \to [0, 1]$ is a continuous, monotone increasing function such that $\sigma(0) = 0$, $\sigma(1) = 1$. This metric is a special case of the **Fréchet metric**.
  For the case, when (X, d) is Euclidean space $\mathbb{R}^n$, this metric is the original (1906) **Fréchet distance** between parametric curves $f, g : [0, 1] \to \mathbb{R}^n$. This distance can be seen as the length of the shortest leash that is sufficient for the man and the dog to walk their paths f and g from start to end. For example, the Fréchet distance between two concentric circles of radius r_1 and r_2 respectively is $|r_1 - r_2|$.
  The **discrete Fréchet distance** (or *coupling distance*, Eiter and Mannila, 1994) is an approximation of the Fréchet metric for polygonal curves f and g. It considers only positions of the leash where its endpoints are located at vertices of f and g. So, this distance is the minimum, over all order-preserving pairings of vertices in f and g, of the maximal Euclidean distance between paired vertices.

- **Bohr metric**

Let $\mathbb{R}$ be a metric space with a metric ρ. A continuous function $f : \mathbb{R} \to \mathbb{R}$ is called *almost periodic* if, for every $\epsilon > 0$, there exists $l = l(\epsilon) > 0$ such that every interval $[t_0, t_0 + l(\epsilon)]$ contains at least one number τ for which $\rho(f(t), f(t+\tau)) < \epsilon$ for $-\infty < t < +\infty$.

The **Bohr metric** is the **norm metric** $||f-g||$ on the set AP of all almost periodic functions defined by the norm

$$||f|| = \sup_{-\infty<t<+\infty} |f(t)|.$$

It makes AP a Banach space. Some generalizations of almost periodic functions were obtained using other norms; cf. **Stepanov distance**, **Weyl distance**, **Besicovitch distance** and **Bochner metric**.

- **Stepanov distance**

The **Stepanov distance** is a distance on the set of all measurable functions $f : \mathbb{R} \to \mathbb{C}$ with summable p-th power on each bounded integral, defined by

$$\sup_{x\in\mathbb{R}} \left(\frac{1}{l} \int_x^{x+l} |f(x) - g(x)|^p dx \right)^{1/p}.$$

The **Weyl distance** is a distance on the same set defined by

$$\lim_{l\to\infty} \sup_{x\in\mathbb{R}} \left(\frac{1}{l} \int_x^{x+l} |f(x) - g(x)|^p dx \right)^{1/p}.$$

- **Besicovitch distance**

The **Besicovitch distance** is a distance on the set of all measurable functions $f : \mathbb{R} \to \mathbb{C}$ with summable p-th power on each bounded integral defined by

$$\left(\overline{\lim}_{T\to\infty} \frac{1}{2T} \int_{-T}^{T} |f(x) - g(x)|^p dx \right)^{1/p}.$$

The *generalized Besicovitch almost periodic functions* correspond to this distance.

- **Bochner metric**

Given a measure space $(\Omega, \mathcal{A}, \mu)$, a Banach space $(V, ||.||_V)$, and $1 \leq p \leq \infty$, the *Bochner space* (or *Lebesgue–Bochner space*) $L^p(\Omega, V)$ is the set of all measurable functions $f : \Omega \to V$ such that $||f||_{L^p(\Omega,V)} \leq \infty$.

Here the *Bochner norm* $||f||_{L^p(\Omega,V)}$ is defined by $(\int_\Omega ||f(\omega)||_V^p d\mu(\omega))^{\frac{1}{p}}$ for $1 \leq p < \infty$, and, for $p = \infty$, by $\text{ess sup}_{\omega\in\Omega} ||f(\omega)||_V$.

- **Bergman p-metric**
Given $1 \le p < \infty$, let $L_p(\Delta)$ be the L_p-space of Lebesgue measurable functions f on the *unit disk* $\Delta = \{z \in \mathbb{C} : |z| < 1\}$ with $||f||_p = \left(\int_\Delta |f(z)|^p \mu(dz)\right)^{\frac{1}{p}} < \infty$.
The *Bergman space* $L_p^a(\Delta)$ is the subspace of $L_p(\Delta)$ consisting of analytic functions, and the **Bergman p-metric** is the **L_p-metric** on $L_p^a(\Delta)$ (cf. **Bergman metric** in Chap. 7). Any Bergman space is a Banach space.
- **Bloch metric**
The *Bloch space* B on the *unit disk* $\Delta = \{z \in \mathbb{C} : |z| < 1\}$ is the set of all analytic functions f on Δ such that $||f||_B = \sup_{z\in\Delta}(1 - |z|^2)|f'(z)| < \infty$. Using the complete *seminorm* $||.||_B$, a norm on B is defined by
$$||f|| = |f(0)| + ||f||_B.$$
The **Bloch metric** is the **norm metric** $||f - g||$ on B. It makes B a Banach space.
- **Besov metric**
Given $1 < p < \infty$, the *Besov space* B_p on the *unit disk* $\Delta = \{z \in \mathbb{C} : |z| < 1\}$ is the set of all analytic functions f in Δ such that $||f||_{B_p} = \left(\int_\Delta (1 - |z|^2)^p |f'(z)|^p d\lambda(z)\right)^{\frac{1}{p}} < \infty$, where $d\lambda(z) = \frac{\mu(dz)}{(1-|z|^2)^2}$ is the Möbius invariant measure on Δ. Using the complete *seminorm* $||.||_{B_p}$, the *Besov norm* on B_p is defined by
$$||f|| = |f(0)| + ||f||_{B_p}.$$
The **Besov metric** is the **norm metric** $||f - g||$ on B_p.
It makes B_p a Banach space. The set B_2 is the classical *Dirichlet space* of functions analytic on Δ with square integrable derivative, equipped with the **Dirichlet metric**. The *Bloch space* B can be considered as B_∞.
- **Hardy metric**
Given $1 \le p < \infty$, the *Hardy space* $H^p(\Delta)$ is the class of functions, analytic on the *unit disk* $\Delta = \{z \in \mathbb{C} : |z| < 1\}$, and satisfying the following growth condition for the *Hardy norm* $||.||_{H^p}$:
$$||f||_{H^p(\Delta)} = \sup_{0<r<1} \left(\frac{1}{2\pi}\int_0^{2\pi} |f(re^{i\theta})|^p d\theta\right)^{\frac{1}{p}} < \infty.$$
The **Hardy metric** is the **norm metric** $||f - g||_{H^p(\Delta)}$ on $H^p(\Delta)$. It makes $H^p(\Delta)$ a Banach space.
In Complex Analysis, the Hardy spaces are analogs of the L_p-spaces of Functional Analysis. Such spaces are applied in Mathematical Analysis itself, and also in Scattering Theory and Control Theory (cf. Chap. 18).

- **Part metric**
  The **part metric** is a metric on a *domain* D of $\mathbb{R}^2$ defined for any $x, y \in \mathbb{R}^2$ by

$$\sup_{f \in H^+} \left| \ln\left(\frac{f(x)}{f(y)}\right) \right|,$$

  where H^+ is the set of all positive *harmonic functions* on the domain D.
  A twice-differentiable real function $f : D \to \mathbb{R}$ is called *harmonic* on D if its *Laplacian* $\Delta f = \frac{\partial^2 f}{\partial x_1^2} + \frac{\partial^2 f}{\partial x_2^2}$ vanishes on D.
- **Orlicz metric**
  Let $M(u)$ be an even convex function of a real variable which is increasing for u positive, and $\lim_{u \to 0} u^{-1}M(u) = \lim_{u \to \infty} u(M(u))^{-1} = 0$. In this case the function $p(v) = M'(v)$ does not decrease on $[0, \infty)$, $p(0) = \lim_{v \to 0} p(v) = 0$, and $p(v) > 0$ when $v > 0$. Writing $M(u) = \int_0^{|u|} p(v)dv$, and defining $N(u) = \int_0^{|u|} p^{-1}(v)dv$, one obtains a pair $(M(u), N(u))$ of *complementary functions*.
  Let $(M(u), N(u))$ be a pair of complementary functions, and let G be a bounded closed set in $\mathbb{R}^n$. The *Orlicz space* $L_M^*(G)$ is the set of Lebesgue-measurable functions f on G satisfying the following growth condition for the *Orlicz norm* $||f||_M$:

$$||f||_M = \sup\left\{\int_G f(t)g(t)dt : \int_G N(g(t))dt \le 1\right\} < \infty.$$

  The **Orlicz metric** is the norm metric $||f - g||_M$ on $L_M^*(G)$. It makes $L_M^*(G)$ a Banach space [Orli32].
  When $M(u) = u^p, 1 < p < \infty$, $L_M^*(G)$ coincides with the space $L_p(G)$, and, up to scalar factor, the L_p*-norm* $||f||_p$ coincides with $||f||_M$.
  The Orlicz norm is equivalent to the *Luxemburg norm* $||f||_{(M)} = \inf\{\lambda > 0 : \int_G M(\lambda^{-1}f(t))dt \le 1\}$; in fact, $||f||_{(M)} \le ||f||_M \le 2||f||_{(M)}$.
- **Orlicz–Lorentz metric**
  Let $w : (0, \infty) \to (0, \infty)$ be a nonincreasing function. Let $M : [0, \infty) \to [0, \infty)$ be a nondecreasing and convex function with $M(0) = 0$. Let G be a bounded closed set in $\mathbb{R}^n$.
  The *Orlicz–Lorentz space* $L_{w,M}(G)$ is the set of all Lebesgue-measurable functions f on G satisfying the following growth condition for the *Orlicz–Lorentz norm* $||f||_{w,M}$:

$$||f||_{w,M} = \inf\left\{\lambda > 0 : \int_0^\infty w(x)M\left(\frac{f^*(x)}{\lambda}\right)dx \le 1\right\} < \infty,$$

  where $f^*(x) = \sup\{t : \mu(|f| \ge t) \ge x\}$ is the *nonincreasing rearrangement* of f.
  The **Orlicz–Lorentz metric** is the **norm metric** $||f - g||_{w,M}$ on $L_{w,M}(G)$. It makes $L_{w,M}(G)$ a Banach space.

The Orlicz–Lorentz space is a generalization of the *Orlicz space* $L^*_M(G)$ (cf. **Orlicz metric**), and the *Lorentz space* $L_{w,q}(G)$, $1 \leq q < \infty$, of all Lebesgue-measurable functions f on G satisfying the following growth condition for the *Lorentz norm*:

$$||f||_{w,q} = \left(\int_0^\infty w(x)(f^*(x))^q\right)^{\frac{1}{q}} < \infty.$$

- **Hölder metric**
Let $L^\alpha(G)$ be the set of all bounded continuous functions f defined on a subset G of $\mathbb{R}^n$, and satisfying the *Hölder condition* on G. Here, a function f satisfies the *Hölder condition* at a point $y \in G$ with *index* (or *order*) α, $0 < \alpha \leq 1$, and with coefficient $A(y)$, if $|f(x)-f(y)| \leq A(y)|x-y|^\alpha$ for all $x \in G$ sufficiently close to y.
If $A = \sup_{y\in G}(A(y)) < \infty$, the Hölder condition is called *uniform* on G, and A is called the *Hölder coefficient* of G. The quantity $|f|_\alpha = \sup_{x,y\in G} \frac{|f(x)-f(y)|}{|x-y|^\alpha}$, $0 \leq \alpha \leq 1$, is called the *Hölder α-seminorm* of f, and the *Hölder norm* of f is defined by

$$||f||_{L^\alpha(G)} = \sup_{x\in G}|f(x)| + |f|_\alpha.$$

The **Hölder metric** is the **norm metric** $||f-g||_{L^\alpha(G)}$ on $L^\alpha(G)$. It makes $L^\alpha(G)$ a Banach space.
- **Sobolev metric**
The *Sobolev space* $W^{k,p}$ is a subset of an L_p-space such that f and its derivatives up to order k have a finite *L_p-norm*. Formally, given a subset G of $\mathbb{R}^n$, define

$$W^{k,p} = W^{k,p}(G) = \{f \in L_p(G) : f^{(i)} \in L_p(G), 1 \leq i \leq k\},$$

where $f^{(i)} = \partial^{\alpha_1}_{x_1} \dots \partial^{\alpha_n}_{x_n} f$, $\alpha_1 + \dots + \alpha_n = i$, and the derivatives are taken in a weak sense. The *Sobolev norm* on $W^{k,p}$ is defined by

$$||f||_{k,p} = \sum_{i=0}^k ||f^{(i)}||_p.$$

In fact, it is enough to take only the first and last in the sequence, i.e., the norm defined by $||f||_{k,p} = ||f||_p + ||f^{(k)}||_p$ is equivalent to the norm above.
For $p = \infty$, the Sobolev norm is equal to the *essential supremum* of $|f|$: $||f||_{k,\infty} = ess\sup_{x\in G}|f(x)|$, i.e., it is the infimum of all numbers $a \in \mathbb{R}$ for which $|f(x)| > a$ on a set of measure zero.
The **Sobolev metric** is the **norm metric** $||f-g||_{k,p}$ on $W^{k,p}$. It makes $W^{k,p}$ a Banach space.

The Sobolev space $W^{k,2}$ is denoted by H^k. It is a Hilbert space for the *inner product* $\langle f, g\rangle_k = \sum_{i=1}^{k}\langle f^{(i)}, g^{(i)}\rangle_{L_2} = \sum_{i=1}^{k}\int_G f^{(i)}\overline{g}^{(i)}\mu(d\omega)$.

- **Variable exponent space metrics**

  Let G be a nonempty open subset of $\mathbb{R}^n$, and let $p : G \to [1,\infty)$ be a measurable bounded function, called a *variable exponent*. The *variable exponent Lebesgue space* $L_{p(.)}(G)$ is the set of all measurable functions $f : G \to \mathbb{R}$ for which the *modular* $\varrho_{p(.)}(f) = \int_G |f(x)|^{p(x)}dx$ is finite. The *Luxemburg norm* on this space is defined by

$$||f||_{p(.)} = \inf\{\lambda > 0 : \varrho_{p(.)}(f/\lambda) \le 1\}.$$

  The **variable exponent Lebesgue space metric** is the **norm metric** $||f-g||_{p(.)}$ on $L_{p(.)}(G)$.

  A *variable exponent Sobolev space* $W^{1,p(.)}(G)$ is a subspace of $L_{p(.)}(G)$ consisting of functions f whose distributional gradient exists almost everywhere and satisfies the condition $|\nabla f| \in L_{p(.)}(G)$. The norm

$$||f||_{1,p(.)} = ||f||_{p(.)} + ||\nabla f||_{p(.)}$$

  makes $W^{1,p(.)}(G)$ a Banach space. The **variable exponent Sobolev space metric** is the norm metric $||f-g||_{1,p(.)}$ on $W^{1,p(.)}$.

- **Schwartz metric**

  The *Schwartz space* (or *space of rapidly decreasing functions*) $S(\mathbb{R}^n)$ is the class of all *Schwartz functions*, i.e., infinitely-differentiable functions $f : \mathbb{R}^n \to \mathbb{C}$ that decrease at infinity, as do all their derivatives, faster than any inverse power of x. More precisely, f is a Schwartz function if we have the following growth condition:

$$||f||_{\alpha\beta} = \sup_{x\in\mathbb{R}^n}\left|x_1^{\beta_1}\dots x_n^{\beta_n}\frac{\partial^{\alpha_1+\dots+\alpha_n}f(x_1,\dots,x_n)}{\partial x_1^{\alpha_1}\dots\partial x_n^{\alpha_n}}\right| < \infty$$

  for any nonnegative integer vectors α and β. The family of *seminorms* $||.||_{\alpha\beta}$ defines a **locally convex** topology of $S(\mathbb{R}^n)$ which is **metrizable** and complete. The **Schwartz metric** is a metric on $S(\mathbb{R}^n)$ which can be obtained using this topology (cf. **countably normed space** in Chap. 2).

  The corresponding metric space on $S(\mathbb{R}^n)$ is a *Fréchet space* in the sense of Functional Analysis, i.e., a locally convex *F-space*.

- **Bregman quasi-distance**

  Let $G \subset \mathbb{R}^n$ be a closed set with the nonempty interior G^0. Let f be a *Bregman function with zone* G.

  The **Bregman quasi-distance** $D_f : G \times G^0 \to \mathbb{R}_{\ge 0}$ is defined by

$$D_f(x,y) = f(x) - f(y) - \langle\nabla f(y), x-y\rangle,$$

where $\nabla f = (\frac{\partial f}{\partial x_1}, \dots \frac{\partial f}{\partial x_n})$. $D_f(x, y) = 0$ if and only if $x = y$. Also $D_f(x, y) + D_f(y, z) - D_f(x, z) = \langle \nabla f(z) - \nabla f(y), x - y \rangle$ but, in general, D_f does not satisfy the triangle inequality, and is not symmetric.
A real-valued function f whose effective domain contains G is called a *Bregman function with zone* G if the following conditions hold:

1. f is continuously differentiable on G^0;
2. f is strictly convex and continuous on G;
3. For all $\delta \in \mathbb{R}$ the *partial level sets* $\Gamma(x, \delta) = \{y \in G^0 : D_f(x, y) \leq \delta\}$ are bounded for all $x \in G$;
4. If $\{y_n\}_n \subset G^0$ converges to y^*, then $D_f(y^*, y_n)$ converges to 0;
5. If $\{x_n\}_n \subset G$ and $\{y_n\}_n \subset G^0$ are sequences such that $\{x_n\}_n$ is bounded, $\lim_{n\to\infty} y_n = y^*$, and $\lim_{n\to\infty} D_f(x_n, y_n) = 0$, then $\lim_{n\to\infty} x_n = y^*$.

When $G = \mathbb{R}^n$, a sufficient condition for a strictly convex function to be a Bregman function has the form: $\lim_{||x||\to\infty} \frac{f(x)}{||x||} = \infty$.

13.2 Metrics on Linear Operators

A *linear operator* is a function $T : V \to W$ between two vector spaces V, W over a field $\mathbb{F}$, that is compatible with their linear structures, i.e., for any $x, y \in V$ and any scalar $k \in \mathbb{F}$, we have the following properties: $T(x + y) = T(x) + T(y)$, and $T(kx) = kT(x)$.

- **Operator norm metric**
  Consider the set of all linear operators from a *normed space* $(V, ||.||_V)$ into a normed space $(W, ||.||_W)$. The *operator norm* $||T||$ of a *linear operator* $T : V \to W$ is defined as the largest value by which T stretches an element of V, i.e.,

$$||T|| = \sup_{||v||_V \neq 0} \frac{||T(v)||_W}{||v||_V} = \sup_{||v||_V = 1} ||T(v)||_W = \sup_{||v||_V \leq 1} ||T(v)||_W.$$

  A linear operator $T : V \to W$ from a normed space V into a normed space W is called *bounded* if its operator norm is finite. For normed spaces, a linear operator is bounded if and only if it is *continuous*.
  The **operator norm metric** is a **norm metric** on the set $B(V, W)$ of all bounded linear operators from V into W, defined by

$$||T - P||.$$

  The space $(B(V, W), ||.||)$ is called the *space of bounded linear operators*. This metric space is **complete** if W is. If $V = W$ is complete, the space $B(V, V)$ is a *Banach algebra*, as the operator norm is a *submultiplicative norm*.

A linear operator $T : V \to W$ from a Banach space V into another Banach space W is called *compact* if the image of any bounded subset of V is a relatively compact subset of W. Any compact operator is bounded (and, hence, continuous). The space $(K(V, W), ||.||)$ on the set $K(V, W)$ of all compact operators from V into W with the operator norm $||.||$ is called the *space of compact operators*.

- **Nuclear norm metric**

  Let $B(V, W)$ be the space of all bounded linear operators mapping a Banach space $(V, ||.||_V)$ into another Banach space $(W, ||.||_W)$. Let the *Banach dual* of V be denoted by V', and the value of a functional $x' \in V'$ at a vector $x \in V$ by $\langle x, x' \rangle$.

  A linear operator $T \in B(V, W)$ is called a *nuclear operator* if it can be represented in the form $x \mapsto T(x) = \sum_{i=1}^{\infty} \langle x, x_i' \rangle y_i$, where $\{x_i'\}_i$ and $\{y_i\}_i$ are sequences in V' and W, respectively, such that $\sum_{i=1}^{\infty} ||x_i'||_{V'} ||y_i||_W < \infty$. This representation is called *nuclear*, and can be regarded as an expansion of T as a sum of operators of rank 1 (i.e., with one-dimensional range). The *nuclear norm* of T is defined as

  $$||T||_{nuc} = \inf \sum_{i=1}^{\infty} ||x_i'||_{V'} ||y_i||_W,$$

  where the infimum is taken over all possible nuclear representations of T.

  The **nuclear norm metric** is the **norm metric** $||T - P||_{nuc}$ on the set $N(V, W)$ of all nuclear operators mapping V into W. The space $(N(V, W), ||.||_{nuc})$, called the *space of nuclear operators*, is a Banach space.

  A *nuclear space* is defined as a **locally convex** space for which all continuous linear functions into an arbitrary Banach space are nuclear operators. A nuclear space is constructed as a projective limit of Hilbert spaces H_α with the property that, for each $\alpha \in I$, one can find $\beta \in I$ such that $H_\beta \subset H_\alpha$, and the embedding operator $H_\beta \ni x \to x \in H_\alpha$ is a *Hilbert–Schmidt operator*. A *normed space* is nuclear if and only if it is finite-dimensional.

- **Finite nuclear norm metric**

  Let $F(V, W)$ be the space of all *linear operators of finite rank* (i.e., with finite-dimensional range) mapping a Banach space $(V, ||.||_V)$ into another Banach space $(W, ||.||_W)$. A linear operator $T \in F(V, W)$ can be represented in the form $x \mapsto T(x) = \sum_{i=1}^{n} \langle x, x_i' \rangle y_i$, where $\{x_i'\}_i$ and $\{y_i\}_i$ are sequences in V' (*Banach dual* of V) and W, respectively, and $\langle x, x' \rangle$ is the value of a functional $x' \in V'$ at a vector $x \in V$. The *finite nuclear norm* of T is defined as

  $$||T||_{fnuc} = \inf \sum_{i=1}^{n} ||x_i'||_{V'} ||y_i||_W,$$

  where the infimum is taken over all possible finite representations of T.

The **finite nuclear norm metric** is the **norm metric** $||T - P||_{fnuc}$ on $F(V, W)$. The space $(F(V, W), ||.||_{fnuc})$ is called the *space of operators of finite rank*. It is a dense linear subspace of the *space of nuclear operators* $N(V, W)$.

- **Hilbert–Schmidt norm metric**
Consider the set of all linear operators from a Hilbert space $(H_1, ||.||_{H_1})$ into a Hilbert space $(H_2, ||.||_{H_2})$. The *Hilbert–Schmidt norm* $||T||_{HS}$ of a linear operator $T : H_1 \to H_2$ is defined by

$$||T||_{HS} = \left(\sum_{\alpha \in I} ||T(e_\alpha)||_{H_2}^2\right)^{1/2},$$

where $(e_\alpha)_{\alpha \in I}$ is an orthonormal basis in H_1. A linear operator $T : H_1 \to H_2$ is called a *Hilbert–Schmidt operator* if $||T||_{HS}^2 < \infty$.
The **Hilbert–Schmidt norm metric** is the **norm metric** $||T - P||_{HS}$ on the set $S(H_1, H_2)$ of all Hilbert–Schmidt operators from H_1 into H_2. In Euclidean space $||.||_{HS}$ is also called *Frobenius norm*; cf. **Frobenius norm metric** in Chap. 12.
For $H_1 = H_2 = H$, the algebra $S(H, H) = S(H)$ with the Hilbert–Schmidt norm is a *Banach algebra*. It contains operators of finite rank as a dense subset, and is contained in the space $K(H)$ of compact operators. An *inner product* $\langle , \rangle_{HS}$ on $S(H)$ is defined by $\langle T, P\rangle_{HS} = \sum_{\alpha \in I} \langle T(e_\alpha), P(e_\alpha)\rangle$, and $||T||_{HS} = \langle T, T\rangle_{HS}^{1/2}$. So, $S(H)$ is a Hilbert space (independent of the chosen basis $(e_\alpha)_{\alpha \in I}$).

- **Trace-class norm metric**
Given a Hilbert space H, the *trace-class norm* of a linear operator $T : H \to H$ is

$$||T||_{tc} = \sum_{\alpha \in I} \langle |T|(e_\alpha), e_\alpha\rangle,$$

where $|T|$ is the *absolute value* of T in the *Banach algebra* $B(H)$ of all bounded operators from H into itself, and $(e_\alpha)_{\alpha \in I}$ is an orthonormal basis of H.
An operator $T : H \to H$ is called a *trace-class operator* if $||T||_{tc} < \infty$. Any such operator is the product of two *Hilbert–Schmidt operators*.
The **trace-class norm metric** is the **norm metric** $||T - P||_{tc}$ on the set $L(H)$ of all trace-class operators from H into itself.
The set $L(H)$ with the norm $||.||_{tc}$ forms a Banach algebra which is contained in the algebra $K(H)$ (of all compact operators from H into itself), and contains the algebra $S(H)$ of all Hilbert–Schmidt operators from H into itself.

- **Schatten p-class norm metric**
Let $1 \leq p < \infty$. Given a separable Hilbert space H, the *Schatten p-class norm* of a compact linear operator $T : H \to H$ is defined by

$$||T||_{Sch}^p = \left(\sum_n |s_n|^p\right)^{\frac{1}{p}},$$

where $\{s_n\}_n$ is the sequence of *singular values* of T. A compact operator $T : H \to H$ is called a *Schatten p-class operator* if $||T||_{Sch}^p < \infty$.
The **Schatten p-class norm metric** is the **norm metric** $||T - P||_{Sch}^p$ on the set $S_p(H)$ of all Schatten p-class operators from H onto itself. The set $S_p(H)$ with the norm $||.||_{Sch}^p$ forms a Banach space. $S_1(H)$ is the *trace-class* of H, and $S_2(H)$ is the *Hilbert–Schmidt class* of H. Cf. **Schatten norm metric** (in Chap. 12) for which **trace** and **Frobenius** norm metrics are cases $p = 1$ and $p = 2$, respectively.

- **Continuous dual space**
For any vector space V over some field, its *algebraic dual space* is the set of all linear functionals on V.
Let $(V, ||.||)$ be a *normed vector space*. Let V' be the set of all *continuous* linear functionals T from V into the base field ($\mathbb{R}$ or $\mathbb{C}$). Let $||.||'$ be the *operator norm* on V' defined by

$$||T||' = \sup_{||x|| \le 1} |T(x)|.$$

The space $(V', ||.||')$ is a Banach space which is called the **continuous dual** (or *Banach dual*) of $(V, ||.||)$.
The continuous dual of the metric space l_p^n (l_p^∞) is l_q^n (l_q^∞, respectively), where q is defined by $\frac{1}{p} + \frac{1}{q} = 1$. The continuous dual of l_1^n (l_1^∞) is l_∞^n (l_∞^∞, respectively).

- **Distance constant of operator algebra**
Let $\mathcal{A}$ be an subalgebra of $B(H)$, the algebra of all bounded operators on a Hilbert space H. For any operator $T \in B(H)$, let P be a projection, $P^\perp$ be its orthogonal complement and $\beta(T, \mathcal{A}) = \sup\{||P^\perp TP|| : P^\perp \mathcal{A} P = (0)\}$.
Let $dist(T, \mathcal{A}) = \inf_{A \in \mathcal{A}} ||T - A||$ be the *distance of T to algebra $\mathcal{A}$*; cf. **matrix nearness problems** in Chap. 12. It holds $dist(T, \mathcal{A}) \ge \beta(T, \mathcal{A})$.
The algebra $\mathcal{A}$ is *reflexive* if $\beta(T, \mathcal{A}) = 0$ implies $T \in \mathcal{A}$; it is *hyperreflexive* if there exists a constant $C \ge 1$ such that, for any operator $T \in B(H)$, it holds

$$dist(T, \mathcal{A}) \le C\beta(T, \mathcal{A}).$$

The smallest such C is called the **distance constant** of the algebra $\mathcal{A}$.
In the case of a reflexive algebra of matrices with nonzero entries specified by a given pattern, the problem of finding the distance constant can be formulated as a matrix-filling problem: given a partially completed matrix, fill in the remaining entries so that the operator norm of the resulting complete matrix is minimized.

Chapter 14
Distances in Probability Theory

A *probability space* is a *measurable space* $(\Omega, \mathcal{A}, P)$, where $\mathcal{A}$ is the set of all measurable subsets of Ω, and P is a measure on $\mathcal{A}$ with $P(\Omega) = 1$. The set Ω is called a *sample space*. An element $a \in \mathcal{A}$ is called an *event*. $P(a)$ is called the *probability* of the event a. The measure P on $\mathcal{A}$ is called a *probability measure*, or *(probability) distribution law*, or simply *(probability) distribution*.

A *random variable* X is a measurable function from a probability space $(\Omega, \mathcal{A}, P)$ into a measurable space, called a *state space* of possible values of the variable; it is usually taken to be $\mathbb{R}$ with the *Borel σ-algebra*, so $X : \Omega \to \mathbb{R}$. The range $\mathcal{X}$ of the variable X is called the *support* of the distribution P; an element $x \in \mathcal{X}$ is called a *state*.

A distribution law can be uniquely described via a *cumulative distribution* (or simply, *distribution*) *function* CDF, which describes the probability that a random value X takes on a value at most x: $F(x) = P(X \leq x) = P(\omega \in \Omega : X(\omega) \leq x)$.

So, any random variable X gives rise to a *probability distribution* which assigns to the interval $[a, b]$ the probability $P(a \leq X \leq b) = P(\omega \in \Omega : a \leq X(\omega) \leq b)$, i.e., the probability that the variable X will take a value in the interval $[a, b]$.

A distribution is called *discrete* if $F(x)$ consists of a sequence of finite jumps at x_i; a distribution is called *continuous* if $F(x)$ is continuous. We consider (as in the majority of applications) only discrete or *absolutely continuous* distributions, i.e., the CDF function $F : \mathbb{R} \to \mathbb{R}$ is *absolutely continuous*. It means that, for every number $\epsilon > 0$, there is a number $\delta > 0$ such that, for any sequence of pairwise disjoint intervals $[x_k, y_k]$, $1 \leq k \leq n$, the inequality $\sum_{1 \leq k \leq n}(y_k - x_k) < \delta$ implies the inequality $\sum_{1 \leq k \leq n} |F(y_k) - F(x_k)| < \epsilon$.

A distribution law also can be uniquely defined via a *probability density* (or *density, probability*) *function* PDF of the underlying random variable. For an absolutely continuous distribution, the CDF is almost everywhere differentiable, and the PDF is defined as the derivative $p(x) = F^{'}(x)$ of the CDF; so, $F(x) = P(X \leq x) = \int_{-\infty}^{x} p(t)dt$, and $\int_{a}^{b} p(t)dt = P(a \leq X \leq b)$. In the discrete case,

M.M. Deza, E. Deza, *Encyclopedia of Distances*,
DOI 10.1007/978-3-662-44342-2_14

the PDF is $\sum_{x_i \le x} p(x_i)$, where $p(x) = P(X = x)$ is the *probability mass function*. But $p(x) = 0$ for each fixed x in any continuous case.

The random variable X is used to "push-forward" the measure P on Ω to a measure dF on $\mathbb{R}$. The underlying probability space is a technical device used to guarantee the existence of random variables and sometimes to construct them.

We usually present the discrete version of probability metrics, but many of them are defined on any measurable space; see [Bass89, Bass13, Cha08]. For a probability distance d on random quantities, the conditions $P(X = Y) = 1$ or equality of distributions imply (and characterize) $d(X, Y) = 0$; such distances are called [Rach91] *compound* or *simple* distances, respectively. Often, some *ground* distance d is given on the state space $\mathcal{X}$ and the presented distance is a lifting of it to a distance on distributions. A quasi-distance between distributions is also called **divergence** or *distance statistic*.

Below we denote $p_X = p(x) = P(X = x)$, $F_X = F(x) = P(X \le x)$, $p(x, y) = P(X = x, Y = y)$. We denote by $\mathbb{E}[X]$ the *expected value* (or *mean*) of the random variable X: in the discrete case $\mathbb{E}[X] = \sum_x xp(x)$, in the continuous case $\mathbb{E}[X] = \int xp(x)dx$.

The *covariance* between the random variables X and Y is $Cov(X, Y) = \mathbb{E}[(X - \mathbb{E}[X])(Y - \mathbb{E}[Y])] = \mathbb{E}[XY] - \mathbb{E}[X]\mathbb{E}[Y]$. The *variance* and *standard deviation* of X are $Var(X) = Cov(X, X)$ and $\sigma(X) = \sqrt{Var(X)}$, respectively. The *correlation* between X and Y is $Corr(X, Y) = \frac{Cov(X,Y)}{\sigma(X)\sigma(Y)}$; cf. Chap. 17.

14.1 Distances on Random Variables

All distances in this section are defined on the set $\mathbf{Z}$ of all random variables with the same support $\mathcal{X}$; here $X, Y \in \mathbf{Z}$.

- ***p*-Average compound metric**
  Given $p \ge 1$, the ***p*-average compound metric** (or *L_p-metric between variables*) is a metric on $\mathbf{Z}$ with $\mathcal{X} \subset \mathbb{R}$ and $\mathbb{E}[|Z|^p] < \infty$ for all $Z \in \mathbf{Z}$ defined by

$$(\mathbb{E}[|X - Y|^p])^{1/p} = (\sum_{(x,y) \in \mathcal{X} \times \mathcal{X}} |x - y|^p p(x, y))^{1/p}.$$

  For $p = 2$ and ∞, it is called, respectively, the *mean-square distance* and *essential supremum distance* between variables.
- **Lukaszyk–Karmovski metric**
  The **Lukaszyk–Karmovski metric** (2001) on $\mathbb{Z}$ with $\mathcal{X} \subset \mathbb{R}$ is defined by

$$\sum_{(x,y) \in \mathcal{X} \times \mathcal{X}} |x - y| p(x) p(y).$$

For continuous random variables, it is defined by $\int_{-\infty}^{+\infty}\int_{-\infty}^{+\infty}|x-y|F(x)F(y)dxdy$. This function can be positive for $X = Y$. Such possibility is excluded, and so, it will be a distance metric, if and only if it holds

$$\int_{-\infty}^{+\infty}\int_{-\infty}^{+\infty}|x-y|\delta(x-\mathbb{E}[X])\delta(y-\mathbb{E}[Y])dxdy = |\mathbb{E}[X]-\mathbb{E}[Y]|.$$

- **Absolute moment metric**
  Given $p \geq 1$, the **absolute moment metric** is a metric on $\mathbf{Z}$ with $\mathcal{X} \subset \mathbb{R}$ and $\mathbb{E}[|Z|^p] < \infty$ for all $Z \in \mathbf{Z}$ defined by

$$|(\mathbb{E}[|X|^p])^{1/p} - (\mathbb{E}[|Y|^p])^{1/p}|.$$

  For $p = 1$ it is called the *engineer metric*.
- **Indicator metric**
  The **indicator metric** is a metric on $\mathbf{Z}$ defined by

$$\mathbb{E}[1_{X\neq Y}] = \sum_{(x,y)\in\mathcal{X}\times\mathcal{X}} 1_{x\neq y}p(x,y) = \sum_{(x,y)\in\mathcal{X}\times\mathcal{X},x\neq y} p(x,y).$$

  (Cf. **Hamming metric** in Chap. 1.)
- **Ky Fan metric** K
  The **Ky Fan metric** K is a metric K on $\mathbf{Z}$, defined by

$$\inf\{\epsilon > 0 : P(|X-Y| > \epsilon) < \epsilon\}.$$

  It is the case $d(x,y) = |X-Y|$ of the **probability distance**.
- **Ky Fan metric** K^*
  The **Ky Fan metric** K^* is a metric on $\mathbf{Z}$ defined by

$$\mathbb{E}\left[\frac{|X-Y|}{1+|X-Y|}\right] = \sum_{(x,y)\in\mathcal{X}\times\mathcal{X}} \frac{|x-y|}{1+|x-y|}p(x,y).$$

- **Probability distance**
  Given a metric space $(\mathcal{X}, d)$, the **probability distance** on $\mathbf{Z}$ is defined by

$$\inf\{\epsilon > 0 : P(d(X,Y) > \epsilon) < \epsilon\}.$$

14.2 Distances on Distribution Laws

All distances in this section are defined on the set $\mathcal{P}$ of all distribution laws such that corresponding random variables have the same range $\mathcal{X}$; here $P_1, P_2 \in \mathcal{P}$.

- **L_p-metric between densities**
The **L_p-metric between densities** is a metric on $\mathcal{P}$ (for a countable $\mathcal{X}$) defined, for any $p \geq 1$, by

$$(\sum_x |p_1(x) - p_2(x)|^p)^{\frac{1}{p}}.$$

For $p = 1$, one half of it is called the **variational distance** (or *total variation distance, Kolmogorov distance*). For $p = 2$, it is the **Patrick–Fisher distance**. The *point metric* $\sup_x |p_1(x) - p_2(x)|$ corresponds to $p = \infty$.
The **Lissak–Fu distance** with parameter $\alpha > 0$ is defined as $\sum_x |p_1(x) - p_2(x)|^\alpha$.
- **Bayesian distance**
The *error probability in classification* is the following error probability of the optimal Bayes rule for the classification into two classes with a priori probabilities $\phi, 1 - \phi$ and corresponding densities p_1, p_2 of the observations:

$$P_e = \sum_x \min(\phi p_1(x), (1 - \phi)p_2(x)).$$

The **Bayesian distance** on $\mathcal{P}$ is defined by $1 - P_e$.
For the classification into m classes with *a priori* probabilities ϕ_i, $1 \leq i \leq m$, and corresponding densities p_i of the observations, the error probability becomes

$$P_e = 1 - \sum_x p(x) \max_i P(C_i|x),$$

where $P(C_i|x)$ is the *a posteriori* probability of the class C_i given the observation x and $p(x) = \sum_{i=1}^m \phi_i P(x|C_i)$. The *general mean distance between m classes C_i* (cf. m-hemimetric in Chap. 3) is defined (Van der Lubbe, 1979) for $\alpha > 0, \beta > 1$ by

$$\sum_x p(x) \left(\sum_i P(C_i|x)^\beta \right)^\alpha.$$

The case $\alpha = 1, \beta = 2$ corresponds to the *Bayesian distance* in Devijver, 1974; the case $\beta = \frac{1}{\alpha}$ was considered in Trouborst et al., 1974.
- **Mahalanobis semimetric**
The **Mahalanobis semimetric** is a semimetric on $\mathcal{P}$ (for $\mathcal{X} \subset \mathbb{R}^n$) defined by

$$\sqrt{(\mathbb{E}_{P_1}[X] - \mathbb{E}_{P_2}[X])^T A(\mathbb{E}_{P_1}[X] - \mathbb{E}_{P_2}[X])}$$

for a given positive-semidefinite matrix A; its square is a **Bregman quasi-distance** (cf. Chap. 13). Cf. also the **Mahalanobis distance** in Chap. 17.

- **Engineer semimetric**
  The **engineer semimetric** is a semimetric on $\mathcal{P}$ (for $\mathcal{X} \subset \mathbb{R}$) defined by

$$|\mathbb{E}_{P_1}[X] - \mathbb{E}_{P_2}[X]| = |\sum_x x(p_1(x) - p_2(x))|.$$

- **Stop-loss metric of order** m
  The **stop-loss metric of order** m is a metric on $\mathcal{P}$ (for $\mathcal{X} \subset \mathbb{R}$) defined by

$$\sup_{t \in \mathbb{R}} \sum_{x \geq t} \frac{(x-t)^m}{m!}(p_1(x) - p_2(x)).$$

- **Kolmogorov–Smirnov metric**
  The **Kolmogorov–Smirnov metric** (or *Kolmogorov metric, uniform metric*) is a metric on $\mathcal{P}$ (for $\mathcal{X} \subset \mathbb{R}$) defined (1948) by

$$\sup_{x \in \mathbb{R}} |P_1(X \leq x) - P_2(X \leq x)|.$$

  This metric is used, for example, in Biology as a measure of sexual dimorphism.
  The **Kuiper distance** on $\mathcal{P}$ is defined by

$$\sup_{x \in \mathbb{R}}(P_1(X \leq x) - P_2(X \leq x)) + \sup_{x \in \mathbb{R}}(P_2(X \leq x) - P_1(X \leq x)).$$

  (Cf. **Pompeiu–Eggleston metric** in Chap. 9.)
  The **Crnkovic–Drachma distance** is defined by

$$\sup_{x \in \mathbb{R}}(P_1(X \leq x) - P_2(X \leq x)) \ln \frac{1}{\sqrt{(P_1(X \leq x)(1 - P_1(X \leq x))}} +$$

$$+ \sup_{x \in \mathbb{R}}(P_2(X \leq x) - P_1(X \leq x)) \ln \frac{1}{\sqrt{(P_1(X \leq x)(1 - P_1(X \leq x))}}.$$

- **Cramér–von Mises distance**
  The **Cramér–von Mises distance** (1928) is defined on $\mathcal{P}$ (for $\mathcal{X} \subset \mathbb{R}$) by

$$\omega^2 = \int_{-\infty}^{+\infty} (P_1(X \leq x) - P_2(X \leq x))^2 dP_2(x).$$

  The **Anderson–Darling distance** (1954) on $\mathcal{P}$ is defined by

$$\int_{-\infty}^{+\infty} \frac{(P_1(X \leq x) - P_2)(X \leq x))^2}{(P_2(X \leq x)(1 - P_2(X \leq x))} dP_2(x).$$

In Statistics, above distances of Kolmogorov–Smirnov, Cramér–von Mises, Anderson–Darling and, below, χ^2**-distance** are the main measures of *goodness of fit* between estimated, P_2, and theoretical, P_1, distributions.
They and other distances were generalized (for example by Kiefer, 1955, and Glick, 1969) on *K-sample setting*, i.e., some convenient generalized distances $d(P_1, \dots, P_K)$ were defined. Cf. **m-hemimetric** in Chap. 3.

- **Energy distance**
  The **energy distance** (Székely, 2005) between cumulative density functions $F(X)$, $F(Y)$ of two independent random vectors $X, Y \in \mathbb{R}^n$ is defined by
  $$d(F(X), F(Y)) = 2\mathbb{E}[||(X - Y||] - \mathbb{E}[||X - X'||] - \mathbb{E}[||(Y - Y'||],$$
  where X, X' are *iid* (independent and identically distributed), Y, Y' are iid and $||.||$ is the length of a vector. Cf. **distance covariance** in Chap. 17.
  It holds $d(F(X), F(Y)) = 0$ if and only if X, Y are iid.
- **Prokhorov metric**
  Given a metric space $(\mathcal{X}, d)$, the **Prokhorov metric** on $\mathcal{P}$ is defined (1956) by
  $$\inf\{\epsilon > 0 : P_1(X \in B) \leq P_2(X \in B^\epsilon) + \epsilon \text{ and } P_2(X \in B) \leq P_1(X \in B^\epsilon) + \epsilon\},$$
  where B is any Borel subset of $\mathcal{X}$, and $B^\epsilon = \{x : d(x, y) < \epsilon, y \in B\}$.
  It is the smallest (over all joint distributions of pairs (X, Y) of random variables X, Y such that the marginal distributions of X and Y are P_1 and P_2, respectively) **probability distance** between random variables X and Y.
- **Levy–Sibley metric**
  The **Levy–Sibley metric** is a metric on $\mathcal{P}$ (for $\mathcal{X} \subset \mathbb{R}$ only) defined by
  $$\inf\{\epsilon > 0 : P_1(X \leq x - \epsilon) - \epsilon \leq P_2(X \leq x) \leq P_1(X \leq x + \epsilon) + \epsilon \text{ for any } x \in \mathbb{R}\}.$$
  It is a special case of the **Prokhorov metric** for $(\mathcal{X}, d) = (\mathbb{R}, |x - y|)$.
- **Dudley metric**
  Given a metric space $(\mathcal{X}, d)$, the **Dudley metric** on $\mathcal{P}$ is defined by
  $$\sup_{f \in F} |\mathbb{E}_{P_1}[f(X)] - \mathbb{E}_{P_2}[f(X)]| = \sup_{f \in F} |\sum_{x \in \mathcal{X}} f(x)(p_1(x) - p_2(x))|,$$
  where $F = \{f : \mathcal{X} \to \mathbb{R}, ||f||_\infty + Lip_d(f) \leq 1\}$, and $Lip_d(f) = \sup_{x,y \in \mathcal{X}, x \neq y} \frac{|f(x) - f(y)|}{d(x,y)}$.
- **Szulga metric**
  Given a metric space $(\mathcal{X}, d)$, the **Szulga metric** (1982) on $\mathcal{P}$ is defined by
  $$\sup_{f \in F} |(\sum_{x \in \mathcal{X}} |f(x)|^p p_1(x))^{1/p} - (\sum_{x \in \mathcal{X}} |f(x)|^p p_2(x))^{1/p}|,$$

where $F = \{f : X \to \mathbb{R}, Lip_d(f) \leq 1\}$, and $Lip_d(f) = \sup_{x,y \in \mathcal{X}, x \neq y} \frac{|f(x)-f(y)|}{d(x,y)}$.

- **Zolotarev semimetric**
The **Zolotarev semimetric** is a semimetric on $\mathcal{P}$, defined (1976) by

$$\sup_{f \in F} \left| \sum_{x \in \mathcal{X}} f(x)(p_1(x) - p_2(x)) \right|,$$

where F is any set of functions $f : \mathcal{X} \to \mathbb{R}$ (in the continuous case, F is any set of such bounded continuous functions); cf. **Szulga metric**, **Dudley metric**.

- **Convolution metric**
Let G be a separable locally compact Abelian group, and let $C(G)$ be the set of all real bounded continuous functions on G vanishing at infinity. Fix a function $g \in C(G)$ such that $|g|$ is integrable with respect to the Haar measure on G, and $\{\beta \in G^* : \hat{g}(\beta) = 0\}$ has empty interior; here G^* is the dual group of G, and $\hat{g}$ is the Fourier transform of g.
The **convolution metric** (or *smoothing metric*) is defined (Yukich, 1985), for any two finite signed Baire measures P_1 and P_2 on G, by

$$\sup_{x \in G} \left| \int_{y \in G} g(xy^{-1})(dP_1 - dP_2)(y) \right|.$$

It can also be seen as the difference $T_{P_1}(g) - T_{P_2}(g)$ of *convolution operators* on $C(G)$ where, for any $f \in C(G)$, the operator $T_P f(x)$ is $\int_{y \in G} f(xy^{-1}) dP(y)$.
In particular, this metric can be defined on the space of probability measures on $\mathbb{R}^n$, where g is a PDF satisfying above conditions.

- **Discrepancy metric**
Given a metric space $(\mathcal{X}, d)$, the **discrepancy metric** on $\mathcal{P}$ is defined by

$$\sup\{|P_1(X \in B) - P_2(X \in B)| : B \text{ is any closed ball}\}.$$

- **Bi-discrepancy semimetric**
The **bi-discrepancy semimetric** (evaluating the proximity of distributions P_1, P_2 over different collections $\mathcal{A}_1, \mathcal{A}_2$ of measurable sets) is defined by

$$D(P_1, P_2) + D(P_2, P_1),$$

where $D(P_1, P_2) = \sup\{\inf\{P_2(C) : B \subset C \in \mathcal{A}_2\} - P_1(B) : B \in \mathcal{A}_1\}$ (*discrepancy*).

- **Le Cam distance**
The **Le Cam distance** (1974) is a semimetric, evaluating the proximity of probability distributions P_1, P_2 (on different spaces $\mathcal{X}_1, \mathcal{X}_2$) and defined as follows:

$$\max\{\delta(P_1, P_2), \delta(P_2, P_1)\},$$

where $\delta(P_1, P_2) = \inf_B \sum_{x_2 \in \mathcal{X}_2} |BP_1(X_2 = x_2) - BP_2(X_2 = x_2)|$ is the *Le Cam deficiency*. Here $BP_1(X_2 = x_2) = \sum_{x_1 \in \mathcal{X}_1} p_1(x_1)b(x_2|x_1)$, where B is a probability distribution over $\mathcal{X}_1 \times \mathcal{X}_2$, and

$$b(x_2|x_1) = \frac{B(X_1 = x_1, X_2 = x_2)}{B(X_1 = x_1)} = \frac{B(X_1 = x_1, X_2 = x_2)}{\sum_{x \in \mathcal{X}_2} B(X_1 = x_1, X_2 = x)}.$$

So, $BP_2(X_2 = x_2)$ is a probability distribution over $\mathcal{X}_2$, since $\sum_{x_2 \in \mathcal{X}_2} b(x_2|x_1) = 1$.

Le Cam distance is not a probabilistic distance, since P_1 and P_2 are defined over different spaces; it is a distance between statistical experiments (models).

- **Skorokhod–Billingsley metric**
  The **Skorokhod–Billingsley metric** is a metric on $\mathcal{P}$, defined by

$$\inf_f \max\left\{ \sup_x |P_1(X \leq x) - P_2(X \leq f(x))|, \sup_x |f(x) - x|, \right.$$
$$\left. \sup_{x \neq y} \left| \ln \frac{f(y) - f(x)}{y - x} \right| \right\},$$

  where $f : \mathbb{R} \to \mathbb{R}$ is any strictly increasing continuous function.
- **Skorokhod metric**
  The **Skorokhod metric** is a metric on $\mathcal{P}$ defined (1956) by

$$\inf\{\epsilon > 0 : \max\{\sup_x |P_1(X < x) - P_2(X \leq f(x))|, \sup_x |f(x) - x|\} < \epsilon\},$$

  where $f : \mathbb{R} \to \mathbb{R}$ is a strictly increasing continuous function.
- **Birnbaum–Orlicz distance**
  The **Birnbaum–Orlicz distance** (1931) is a distance on $\mathcal{P}$ defined by

$$\sup_{x \in \mathbb{R}} f(|P_1(X \leq x) - P_2(X \leq x)|),$$

  where $f : \mathbb{R}_{\geq 0} \to \mathbb{R}_{\geq 0}$ is any nondecreasing continuous function with $f(0) = 0$, and $f(2t) \leq Cf(t)$ for any $t > 0$ and some fixed $C \geq 1$. It is a **near-metric**, since the C**-triangle inequality** $d(P_1, P_2) \leq C(d(P_1, P_3) + d(P_3, P_2))$ holds.
  Birnbaum–Orlicz distance is also used, in Functional Analysis, on the set of all integrable functions on the segment $[0, 1]$, where it is defined by $\int_0^1 H(|f(x) - g(x)|)dx$, where H is a nondecreasing continuous function from $[0, \infty)$ onto $[0, \infty)$ which vanishes at the origin and satisfies the *Orlicz condition*: $\sup_{t>0} \frac{H(2t)}{H(t)} < \infty$.
- **Kruglov distance**
  The **Kruglov distance** (1973) is a distance on $\mathcal{P}$, defined by

$$\int f(P_1(X \leq x) - P_2(X \leq x))dx,$$

where $f : \mathbb{R}_{\geq 0} \to \mathbb{R}_{\geq 0}$ is any even strictly increasing function with $f(0) = 0$, and $f(s+t) \leq C(f(s) + f(t))$ for any $s, t \geq 0$ and some fixed $C \geq 1$. It is a **near-metric**, since the C**-triangle inequality** $d(P_1, P_2) \leq C(d(P_1, P_3) + d(P_3, P_2))$ holds.

- **Bregman divergence**
  Given a differentiable strictly convex function $\phi(p) : \mathbb{R}^n \to \mathbb{R}$ and $\beta \in (0, 1)$, the **skew Jensen** (or *skew Burbea–Rao*) **divergence** on $\mathcal{P}$ is (Basseville–Cardoso, 1995)

$$J_\phi^{(\beta)}(P_1, P_2) = \beta\phi(p_1) + (1-\beta)\phi(p_2) - \phi(\beta p_1 + (1-\beta)p_2).$$

The **Burbea–Rao distance** (1982) is the case $\beta = \frac{1}{2}$ of it, i.e., it is

$$\sum_x \left(\frac{\phi(p_1(x)) + \phi(p_2(x))}{2} - \phi(\frac{p_1(x) + (p_2(x)}{2}) \right).$$

The **Bregman divergence** (1967) is a quasi-distance on $\mathcal{P}$ defined by

$$\sum_x (\phi(p_1(x)) - \phi(p_2(x)) - (p_1(x) - p_2(x))\phi'(p_2(x))) = \lim_{\beta \to 1} \frac{1}{\beta} J_\phi^{(\beta)}(P_1, P_2).$$

The **generalised Kullback–Leibler distance** $\sum_x p_1(x) \ln \frac{p_1(x)}{p_2(x)} - \sum_x (p_1(x) - p_2(x))$ and **Itakura–Saito distance** (cf. Chap. 21) $\sum_x \frac{p_1(x)}{p_2(x)} - \ln \frac{p_1(x)}{p_2(x)} - 1$ are the cases $\phi(p) = \sum_x p(x) \ln p(x) - \sum_x p(x)$ and $\phi(p) = -\sum_x \ln p(x)$ of the Bregman divergence. Cf. **Bregman quasi-distance** in Chap. 13.
Csizár, 1991, proved that the **Kullback–Leibler distance** is the only **Bregman divergence** which is an f**-divergence**.

- f**-divergence**
  Given a convex function $f(t) : \mathbb{R}_{\geq 0} \to \mathbb{R}$ with $f(1) = 0$, $f'(1) = 0$, $f''(1) = 1$, the f**-divergence** (independently, Csizár, 1963, Morimoto, 1963, Ali–Silvey, 1966, Ziv–Zakai, 1973, and Akaike, 1974) on $\mathcal{P}$ is defined by

$$\sum_x p_2(x) f\left(\frac{p_1(x)}{p_2(x)}\right).$$

The cases $f(t) = t \ln t$ and $f(t) = (t-1)^2$ correspond to the **Kullback–Leibler distance** and to the χ^2**-distance** below, respectively. The case $f(t) = |t-1|$ corresponds to the **variational distance**, and the case $f(t) = 4(1-\sqrt{t})$ (as well as $f(t) = 2(t+1) - 4\sqrt{t}$) corresponds to the squared **Hellinger metric**.

Semimetrics can also be obtained, as the square root of the f-divergence, in the cases $f(t) = (t-1)^2/(t+1)$ (the **Vajda–Kus semimetric**), $f(t) = |t^a - 1|^{1/a}$ with $0 < a \leq 1$ (the **generalized Matusita distance**), and $f(t) = \frac{(t^a+1)^{1/a} - 2^{(1-a)/a}(t+1)}{1-1/\alpha}$ (the **Osterreicher semimetric**).

- **α-divergence**
  Given $\alpha \in \mathbb{R}$, the **α-divergence** (independently, Csizár, 1967, Havrda–Charvát, 1967, Cressie–Read, 1984, and Amari, 1985) is defined as $KL(P_1, P_2)$, $KL(P_2, P_1)$ for $\alpha = 1, 0$ and for $\alpha \neq 0, 1$, it is

$$\frac{1}{\alpha(1-\alpha)}\left(1 - \sum_x p_2(x)\left(\frac{p_1(x)}{p_2(x)}\right)^{\alpha}\right).$$

  The **Amari divergence** come from the above by the transformation $\alpha = \frac{1+t}{2}$.
- **Harmonic mean similarity**
  The **harmonic mean similarity** is a similarity on $\mathcal{P}$ defined by

$$2\sum_x \frac{p_1(x)p_2(x)}{p_1(x) + p_2(x)}.$$

- **Fidelity similarity**
  The **fidelity similarity** (or *Bhattacharya coefficient*, *Hellinger affinity*) on $\mathcal{P}$ is

$$\rho(P_1, P_2) = \sum_x \sqrt{p_1(x)p_2(x)}.$$

  Cf. more general **quantum fidelity similarity** in Chap. 24.
- **Hellinger metric**
  In terms of the **fidelity similarity** ρ, the **Hellinger metric** (or **Matusita distance**, *Hellinger–Kakutani metric*) on $\mathcal{P}$ is defined by

$$(\sum_x (\sqrt{p_1(x)} - \sqrt{p_2(x)})^2)^{\frac{1}{2}} = 2\sqrt{1 - \rho(P_1, P_2)}.$$

- **Bhattacharya distance** 1
  In terms of the **fidelity similarity** ρ, the **Bhattacharya distance** 1 (1946) is

$$(\arccos \rho(P_1, P_2))^2$$

  for $P_1, P_2 \in \mathcal{P}$. Twice this distance is the **Rao distance** from Chap. 7. It is used also in Statistics and Machine Learning, where it is called the *Fisher distance*.
  The **Bhattacharya distance** 2 (1943) on $\mathcal{P}$ is defined by

$$-\ln \rho(P_1, P_2).$$

- **χ^2-distance**
  The **χ^2-distance** (or **Pearson χ^2-distance**) is a quasi-distance on $\mathcal{P}$, defined by
  $$\sum_x \frac{(p_1(x) - p_2(x))^2}{p_2(x)}.$$
  The **Neyman χ^2-distance** is a quasi-distance on $\mathcal{P}$, defined by
  $$\sum_x \frac{(p_1(x) - p_2(x))^2}{p_1(x)}.$$
  The half of χ^2-distance is also called *Kagan's divergence*.
  The probabilistic **symmetric χ^2-measure** is a distance on $\mathcal{P}$, defined by
  $$2\sum_x \frac{(p_1(x) - p_2(x))^2}{p_1(x) + p_2(x)}.$$
- **Separation quasi-distance**
  The **separation distance** is a quasi-distance on $\mathcal{P}$ (for a countable $\mathcal{X}$) defined by
  $$\max_x \left(1 - \frac{p_1(x)}{p_2(x)}\right).$$
  (Not to be confused with **separation distance** in Chap. 9.)
- **Kullback–Leibler distance**
  The **Kullback–Leibler distance** (or *relative entropy*, *information deviation*, *information gain*, *KL-distance*) is a quasi-distance on $\mathcal{P}$, defined (1951) by
  $$KL(P_1, P_2) = \mathbb{E}_{P_1}[\ln L] = \sum_x p_1(x) \ln \frac{p_1(x)}{p_2(x)},$$
  where $L = \frac{p_1(x)}{p_2(x)}$ is the *likelihood ratio*. Therefore,
  $$KL(P_1, P_2) = -\sum_x p_1(x) \ln\, p_2(x) + \sum_x p_1(x) \ln\, p_1(x) = H(P_1, P_2) - H(P_1),$$
  where $H(P_1)$ is the *entropy* of P_1, and $H(P_1, P_2)$ is the *cross-entropy* of P_1 and P_2.
  If P_2 is the product of marginals of P_1 (say, $p_2(x, y) = p_1(x)p_1(y)$), the KL-distance $KL(P_1, P_2)$ is called the *Shannon information quantity* and (cf. **Shannon distance**) is equal to $\sum_{(x,y)\in\mathcal{X}\times\mathcal{X}} p_1(x, y) \ln \frac{p_1(x,y)}{p_1(x)p_1(y)}$.
  The **exponential divergence** is defined by $\sum_x p_1(x)(\ln \frac{p_1(x)}{p_2(x)})^2$.

- **Distance to normality**
For a continuous distribution P on $\mathbb{R}$, the *differential entropy* is defined by

$$h(P) = -\int_{-\infty}^{\infty} p(x)\ln p(x)dx.$$

It is $\ln(\delta\sqrt{2\pi e})$ for a *normal* (or *Gaussian*) *distribution* $g_{\delta,\mu}(x) = \frac{1}{\sqrt{2\pi\delta^2}}\exp\left(-\frac{(x-\mu)^2}{2\delta^2}\right)$ with variance δ^2 and mean μ.
The **distance to normality** (or *negentropy*) of P is the **Kullback–Leibler distance** $KL(P,g) = \int_{-\infty}^{\infty} p(x)\ln\left(\frac{p(x)}{g(x)}\right)dx = h(g) - h(P)$, where q is a normal distribution with the same variance as P. So, it is nonnegative and equal to 0 if and only if $P = g$ almost everywhere. Cf. **Shannon distance**.
Also, $h(u_{a,b}) = \ln(b-a)$ for an *uniform distribution* with minimum a and maximum $b > a$, i.e., $u_{a,b}(x) = \frac{1}{b-a}$, if $x \in [a,b]$, and it is 0, otherwise. It holds $h(u_{a,b}) \geq h(P)$ for any distribution P with support contained in $[a,b]$; so, $h(u_{a,b}) - h(P)$ can be called the *distance to uniformity*. Tononi, 2008, used it in his model of consciousness.
- **Jeffrey distance**
The **Jeffrey distance** (or *J-divergence*, *KL2-distance*) is a symmetric version of the **Kullback–Leibler distance** defined (1946) on $\mathcal{P}$ by

$$KL(P_1,P_2) + KL(P_2,P_1) = \sum_x ((p_1(x) - p_2(x))\ln\frac{p_1(x)}{p_2(x)}.$$

The **Aitchison distance** (1986) is defined by $\sqrt{\sum_x (\ln\frac{p_1(x)g(p_1)}{p_2(x)g(p_2)})^2}$, where $g(p) = (\prod_x p(x))^{\frac{1}{n}}$ is the geometric mean of components $p(x)$ of p.
- **Resistor-average distance**
The **resistor-average distance** is (Johnson–Simanović, 2000) a symmetric version of the **Kullback–Leibler distance** on $\mathcal{P}$ which is defined by the harmonic sum

$$\left(\frac{1}{KL(P_1,P_2)} + \frac{1}{KL(P_2,P_1)}\right)^{-1}.$$

- **Jensen–Shannon divergence**
Given a number $\beta \in [0,1]$ and $P_1, P_2 \in \mathcal{P}$, let P_3 denote $\beta P_1 + (1-\beta)P_2$. The **skew divergence** and the **Jensen–Shannon divergence** between P_1 and P_2 are defined on $\mathcal{P}$ as $KL(P_1,P_3)$ and $\beta KL(P_1,P_3) + (1-\beta)KL(P_2,P_3)$, respectively. Here KL is the **Kullback–Leibler distance**; cf. **clarity similarity**.
In terms of *entropy* $H(P) = -\sum_x p(x)\ln p(x)$, the Jensen–Shannon divergence is $H(\beta P_1 + (1-\beta)P_2) - \beta H(P_1) - (1-\beta)H(P_2)$, i.e., the **Jensen divergence** (cf. **Bregman divergence**).

Let $P_3 = \frac{1}{2}(P_1 + P_2)$, i.e., $\beta = \frac{1}{2}$. Then the skew divergence and twice the Jensen–Shannon divergence are called **K-divergence** and **Topsøe distance** (or *information statistics*), respectively. The Topsøe distance is a symmetric version of $KL(P_1, P_2)$. It is not a metric, but its square root is a metric.

- **Clarity similarity**
  The **clarity similarity** is a similarity on $\mathcal{P}$, defined by

$$(KL(P_1, P_3) + KL(P_2, P_3)) - (KL(P_1, P_2) + KL(P_2, P_1)) =$$

$$= \sum_x \left(p_1(x) \ln \frac{p_2(x)}{p_3(x)} + p_2(x) \ln \frac{p_1(x)}{p_3(x)} \right),$$

  where KL is the **Kullback–Leibler distance**, and P_3 is a fixed probability law. It was introduced in [CCL01] with P_3 being the probability distribution of English.
- **Ali–Silvey distance**
  The **Ali–Silvey distance** is a quasi-distance on $\mathcal{P}$ defined by the functional

$$f(\mathbb{E}_{P_1}[g(L)]),$$

  where $L = \frac{p_1(x)}{p_2(x)}$ is the *likelihood ratio*, f is a nondecreasing function on $\mathbb{R}$, and g is a continuous convex function on $\mathbb{R}_{\geq 0}$ (cf. **f-divergence**).
  The case $f(x) = x$, $g(x) = x \ln x$ corresponds to the **Kullback–Leibler distance**; the case $f(x) = -\ln x$, $g(x) = x^t$ corresponds to the **Chernoff distance**.
- **Chernoff distance**
  The **Chernoff distance** (or *Rényi cross-entropy*) on $\mathcal{P}$ is defined (1954) by

$$\max_{t \in (0,1)} D_t(P_1, P_2),$$

  where $0 < t < 1$ and $D_t(P_1, P_2) = -\ln \sum_x (p_1(x))^t (p_2(x))^{1-t}$ (called the *Chernoff coefficient*) which is proportional to the **Rényi distance**.
- **Rényi distance**
  Given $t \in \mathbb{R}$, the **Rényi distance** (or *order t Rényi entropy*, 1961) is a quasi-distance on $\mathcal{P}$ defined as the **Kullback–Leibler distance** $KL(P_1, P_2)$ if $t = 1$, and, otherwise, by

$$\frac{1}{1-t} \ln \sum_x p_2(x) \left(\frac{p_1(x)}{p_2(x)} \right)^t.$$

  For $t = \frac{1}{2}$, one half of the Rényi distance is the **Bhattacharya distance** 2. Cf. **f-divergence** and **Chernoff distance**.

- **Shannon distance**
  Given a *measure space* $(\Omega, \mathcal{A}, P)$, where the set Ω is finite and P is a probability measure, the *entropy* (or *Shannon information entropy*) of a function $f : \Omega \to X$, where X is a finite set, is defined by

$$H(f) = -\sum_{x \in X} P(f = x) \log_a(P(f = x)).$$

  Here $a = 2$, e, or 10 and the unit of entropy is called a *bit*, *nat*, or *dit* (digit), respectively. The function f can be seen as a partition of the measure space.
  For any two such partitions $f : \Omega \to X$ and $g : \Omega \to Y$, denote by $H(f, g)$ the entropy of the partition $(f, g) : \Omega \to X \times Y$ (*joint entropy*), and by $H(f|g)$ the *conditional entropy* (or *equivocation*). Then the **Shannon distance** between f and g is a metric defined by

$$H(f|g) + H(g|f) = 2H(f, g) - H(f) - H(g) = H(f, g) - I(f; g),$$

  where $I(f; g) = H(f) + H(g) - H(f, g)$ is the *Shannon mutual information*.
  If P is the uniform probability law, then Goppa showed that the Shannon distance can be obtained as a limiting case of the **finite subgroup metric**.
  In general, the **information metric** (or **entropy metric**) between two random variables (information sources) X and Y is defined by

$$H(X|Y) + H(Y|X) = H(X, Y) - I(X; Y),$$

  where the *conditional entropy* $H(X|Y)$ is defined by $\sum_{x \in X} \sum_{y \in Y} p(x, y) \ln p(x|y)$, and $p(x|y) = P(X = x|Y = y)$ is the conditional probability.
  The **Rajski distance** (or *normalized information metric*) is defined (Rajski, 1961, for discrete probability distributions X, Y) by

$$\frac{H(X|Y) + H(Y|X)}{H(X, Y)} = 1 - \frac{I(X; Y)}{H(X, Y)}.$$

  It is equal to 1 if X and Y are independent. (Cf., a different one, **normalized information distance** in Chap. 11).
- **Transportation distance**
  Given a metric space $(\mathcal{X}, d)$, the **transportation distance** (and/or, according to Villani, 2009, **Monge–Kantorovich–Wasserstein–Rubinstein–Ornstein–Gini–Dall'Aglio–Mallows–Tanaka distance**) is the metric defined by

$$W_1(P_1, P_2) = \inf \mathbb{E}_S[d(X, Y)] = \inf_S \int_{(X,Y) \in \mathcal{X} \times \mathcal{X}} d(X, Y) dS(X, Y),$$

where the infimum is taken over all joint distributions S of pairs (X, Y) of random variables X, Y such that marginal distributions of X and Y are P_1 and P_2.

For any **separable** metric space $(\mathcal{X}, d)$, this is equivalent to the **Lipschitz distance between measures** $\sup_f \int f d(P_1 - P_2)$, where the supremum is taken over all functions f with $|f(x) - f(y)| \leq d(x, y)$ for any $x, y \in \mathcal{X}$. Cf. **Dudley metric**.

In general, for a Borel function $c : \mathcal{X} \times \mathcal{X} \to \mathbb{R}_{\geq 0}$, the c**-transportation distance** $T_c(P_1, P_2)$ is $\inf \mathbb{E}_S[c(X, Y)]$. It is the minimal total transportation cost if $c(X, Y)$ is the cost of transporting a unit of mass from the location X to the location Y. Cf. the **Earth Mover's distance** (Chap. 21), which is a discrete form of it.

The L_p**-Wasserstein distance** is $W_p = (T_{d^p})^{1/p} = (\inf \mathbb{E}_S[d^p(X, Y)])^{1/p}$. For $(\mathcal{X}, d) = (\mathbb{R}, |x - y|)$, it is also called the L_p*-metric between distribution functions* (CDF) F_i with $F_i^{-1}(x) = \sup_u(P_i(X \leq x) < u)$, and can be written as

$$(\inf \mathbb{E}[|X - Y|^p])^{1/p} = \left(\int_{\mathbb{R}} |F_1(x) - F_2(x)|^p dx\right)^{1/p}$$
$$= \left(\int_0^1 |F_1^{-1}(x) - F_2^{-1}(x)|^p dx\right)^{1/p}.$$

For $p = 1$, this metric is called **Monge–Kantorovich metric** (or **Wasserstein metric**, **Fortet–Mourier metric**, **Hutchinson metric**, **Kantorovich–Rubinstein metric**). For $p = 2$, it is the **Levy–Fréchet metric** (Fréchet, 1957).

- **Ornstein $\overline{d}$-metric**
  The **Ornstein $\overline{d}$-metric** is a metric on $\mathcal{P}$ (for $\mathcal{X} = \mathbb{R}^n$) defined (1974) by

$$\frac{1}{n} \inf \int_{x,y} \left(\sum_{i=1}^{n} 1_{x_i \neq y_i}\right) dS,$$

  where the infimum is taken over all joint distributions S of pairs (X, Y) of random variables X, Y such that marginal distributions of X and Y are P_1 and P_2.
- **Distances between belief assignments**
  In *Bayesian* (or *subjective*, *evidential*) interpretation, a probability can be assigned to any statement, even if no random process is involved, as a way to represent its subjective plausibility, or the degree to which it is supported by the available evidence, or, mainly, degree of belief. Within this approach, *imprecise probability* generalizes Probability Theory to deal with scarce, vague, or conflicting information. The main model is *Dempster–Shafer theory*, which allows evidence to be combined.

Given a set X, a (basic) **belief assignment** is a function $m : P(X) \to [0, 1]$ (where $P(X)$ is the set of all subsets of X) with $m(\emptyset) = 0$ and $\sum_{A \subset P(X)} m(A) = 1$. Probability measures are a special case in which $m(A) > 0$ only for singletons.
For the classic probability $P(A)$, it holds then $\text{Bel}(A) \leq P(A) \leq \text{Pl}(A)$, where the *belief function* and *plausibility function* are defined, respectively, by

$$\text{Bel}(A) = \sum_{B:B \subset A} m(B) \text{ and } \text{Pl}(A) = \sum_{B:B \cap A \neq \emptyset} m(B) = 1 - \text{Bel}(\overline{A}).$$

The original (Dempster, 1967) *conflict factor* between two belief assignments m_1 and m_2 was defined as $c(m_1, m_2) = \sum_{A \cap B = \emptyset} m_1(A)m_2(B)$. This is not a distance since $c(m, m) > 0$. The combination of m_1 and m_2, seen as independent sources of evidence, is defined by $m_1 \oplus m_2(A) = \frac{1}{1-c(m_1,m_2)} \sum_{B \cap C = A} m_1(B)m_2(C)$.
Usually, a distance between m_1 and m_2 estimates the difference between these sources in the form $d_U = |U(m_1) - U(m_2)|$, where U is an uncertainty measure; see Sarabi-Jamab et al., 2013, for a comparison of their performance. In particular, this distance is called:

confusion (Hoehle, 1981) if $U(m) - \sum_A m(A) \log_2 \text{Bel}(A)$;
dissonance (Yager, 1983) if $U(m) = E(m) = -\sum_A m(A) \log_2 \text{Pl}(A)$;
Yager's factor (Eager, 1983) if $U(m) = 1 - \sum_{A \neq \emptyset} \frac{m(A)}{|A|}$;
possibility-based (Smets, 1983) if $U(m) = -\sum_A \log_2 \sum_{B:A \subset B} m(B)$;
U-uncertainty (Dubois–Prade, 1985) if $U(m) = I(m) = -\sum_A m(A) \log_2 |A|$;
Lamata–Moral's (1988) if $U(m) = \log_2(\sum_A m(A)|A|)$ and $U(m) = E(m) + I(m)$;
discord (Klir–Ramer, 1990) if $U(m) = D(m) = -\sum_A m(A) \log_2(1 - \sum_B m(B) \frac{|B \setminus A|}{|B|})$ and a variant: $U(m) = D(m) + I(m)$;
strife (Klir–Parviz, 1992) if $U(m) = -\sum_A m(A) \log_2(\sum_B m(B) \frac{|A \cap B|}{|A|})$;
Pal et al.'s (1993) if $U(m) = G(m) = -\sum_A \log_2 m(A)$ and $U(m) = G(m) + I(m)$;
total conflict (George–Pal, 1996) if $U(m) = \sum_A m(A) \sum_B (m(B)(1 - \frac{|A \cap B|}{|A \cup B|}))$.

Among other distances used are the **cosine distance** $1 - \frac{m_1^T m_2}{||m_1|| ||m_2||}$, the **Mahalanobis distance** $\sqrt{(m_1 - m_2)^T A(m_1 - m_2)}$ for some matrices A, and *pignistic-based* one (Tessem, 1993) $\max_A \{|\sum_{B \neg \emptyset}(m_1(B) - m_2(B) \frac{|A \cap B|}{|B|}|\}$.

Part IV
Distances in Applied Mathematics

Chapter 15
Distances in Graph Theory

A *graph* is a pair $G = (V, E)$, where V is a set, called the set of *vertices* of the graph G, and E is a set of unordered pairs of vertices, called the *edges* of the graph G. A *directed graph* (or *digraph*) is a pair $D = (V, E)$, where V is a set, called the set of *vertices* of the digraph D, and E is a set of ordered pairs of vertices, called *arcs* of the digraph D.

A graph in which at most one edge may connect any two vertices, is called a *simple graph*. If multiple edges are allowed between vertices, the graph is called a *multigraph*. A graph, together with a function which assigns a positive weight to each edge, is called a *weighted graph* or *network*.

The graph is called *finite* (*infinite*) if the set V of its vertices is finite (infinite, respectively). The *order* and *size* of a finite graph (V, E) are $|V|$ and $|E|$, respectively.

A *subgraph* of a graph $G = (V, E)$ is a graph $G' = (V', E')$ with $V' \subset V$ and $E' \subset E$. If G' is a subgraph of G, then G is called a *supergraph* of G'. A subgraph (V', E') of (V, E) is its *induced subgraph* if $E' = \{e = uv \in E : u, v \in V'\}$.

A graph $G = (V, E)$ is called *connected* if, for any $u, v \in V$, there exists a $(u-v)$ *walk*, i.e., a sequence of edges $uw_1 = w_0w_1, w_1w_2, \dots, w_{n-1}w_n = w_{n-1}v$ from E. A $(u - v)$ *path* is a $(u - v)$ walk with distinct edges. A graph is called m-*connected* if there is no set of $m - 1$ edges whose removal disconnects the graph; a connected graph is 1-connected. A digraph $D = (V, E)$ is called *strongly connected* if, for any $u, v \in V$, the *directed* $(u - v)$ and $(v - u)$ paths both exist. A maximal connected subgraph of a graph G is called its *connected component*.

Vertices connected by an edge are called *adjacent*. The *degree* $deg(v)$ of a vertex $v \in V$ of a graph $G = (V, E)$ is the number of its vertices adjacent to v.

A *complete graph* is a graph in which each pair of vertices is connected by an edge. A *bipartite graph* is a graph in which the set V of vertices is decomposed into two disjoint subsets so that no two vertices within the same subset are adjacent. A *simple path* is a simple connected graph in which two vertices have degree 1, and

M.M. Deza, E. Deza, *Encyclopedia of Distances*,
DOI 10.1007/978-3-662-44342-2_15

other vertices (if they exist) have degree 2; the *length* of a path is the number of its edges.

A *cycle* is a *closed simple path*, i.e., a simple connected graph in which every vertex has degree 2. The *circumference* of a graph is the length of the longest cycle in it. A *tree* is a simple connected graph without cycles. A tree having a path from which every vertex has distance ≤ 1 or ≤ 2, is called a *caterpillar* or *lobster*, respectively.

Two graphs which contain the same number of vertices connected in the same way are called *isomorphic*. Formally, two graphs $G = (V(G), E(G))$ and $H = (V(H), E(H))$ are called *isomorphic* if there is a bijection $f : V(G) \to V(H)$ such that, for any $u, v \in V(G)$, $uv \in E(G)$ if and only if $f(u)f(v) \in E(H)$.

We will consider mainly simple finite graphs and digraphs; more exactly, the equivalence classes of such isomorphic graphs.

15.1 Distances on the Vertices of a Graph

- **Path metric**
  The **path metric** (or **graphic metric**, *shortest path metric*) d_{path} is a metric on the vertex-set V of a connected graph $G = (V, E)$ defined, for any $u, v \in V$, as the length of a shortest $(u - v)$ path in G, i.e., a *geodesic*. Examples follow.
  Given an integer $n \geq 1$, the **line metric on** $\{1, \dots, n\}$ in Chap. 1 is the path metric of the path $P_n = \{1, \dots, n\}$. The path metric of the *Cayley graph* Γ of a finitely generated group $(G, \cdot, e)$ is called a **word metric**.
  The **hypercube metric** is the path metric of a *hypercube graph* $H(m, 2)$ with the vertex-set $V = \{0, 1\}^m$, and whose edges are the pairs of vectors $x, y \in \{0, 1\}^m$ such that $|\{i \in \{1, \dots, n\} : x_i \neq y_i\}| = 1$; it is equal to $|\{i \in \{1, \dots, n\} : x_i = 1\} \Delta \{i \in \{1, \dots, n\} : y_i = 1\}|$. The graphic metric space associated with a hypercube graph coincides with a **Hamming cube**, i.e., the metric space $(\{0, 1\}^m, d_{l_1})$.
  The **belt distance** (Garber–Dolbilin, 2010) is the path metric of a *belt graph* $B(P)$ of a polytope P with centrally symmetric facets. The vertices of $B(P)$ are the facets of P and two vertices are connected by an edge if the corresponding facets lie in the same *belt* (the set of all facets of P parallel to a given face of codimension 2).
  The reciprocal path metric is called **geodesic similarity**.
- **Weighted path metric**
  The **weighted path metric** d_{wpath} is a metric on the vertex-set V of a connected weighted graph $G = (V, E)$ with positive edge-weights $(w(e))_{e \in E}$ defined by

$$\min_P \sum_{e \in P} w(e),$$

  where the minimum is taken over all $(u - v)$ paths P in G.

Sometimes, $\frac{1}{w(e)}$ is called the *length* of the edge e. In the theory of electrical networks, the edge-length $\frac{1}{w(e)}$ is identified with the *resistance* of the edge e. The **inverse weighted path metric** is $\min_P \sum_{e \in P} \frac{1}{w(e)}$.

- **Metric graph**

  A **metric** (or *metrized*) **graph** is a connected graph $G = (V, E)$, where edges e are identified with line segments $[0, l(e)]$ of length $l(e)$. Let x_e be the coordinate on the segment $[0, l(e)]$ with vertices corresponding to $x_e = 0, l(e)$; the ends of distinct segments are identified if they correspond to the same vertex of G. A *function* f on G is the $|E|$-tuple of functions $f_e(x_e)$ on the segments.

  A metric graph can be seen as an infinite metric space (X, d), where X is the set of all points on above segments, and the distance between two points is the length of the shortest, along the line segments traversed, path connecting them. Also, it can be seen as one-dimensional Riemannian manifold with singularities. There is a bijection between the metric graphs, the equivalence classes of finite connected edge-weighted graphs and the resistive electrical networks: if an edge e of a metric graph has length $l(e)$, then $\frac{1}{l(e)}$ is the weight of e in the corresponding edge-weighted graph and $l(e)$ is the resistance along e in the corresponding resistive electric circuit. Cf. the **resistance metric**.

  A **quantum graph** is a metric graph equipped with a self-adjoint differential operator (such as a *Laplacian*) acting on functions on the graph. The *Hilbert space* of the graph is $\oplus_{e \in E} L^2([0, w(e)])$, where the inner product of functions is $\langle f, g \rangle = \sum_{e \in E} \int_0^{w(e)} f_e^*(x_e) g_e(x_e) dx_e$.

- **Spin network**

  A **spin network** is (Penrose, 1971) a connected graph (V, E) with edge-weights $(w(e))_{e \in E}$ (*spins*), $w(e) \in \mathbb{N}$, such that for any distinct edges e_1, e_2, e_3 with a common vertex, it holds **spin triangle inequality** $|w(e_1) - w(e_2)| \le w(e_3) \le w(e_1)+w(e_2)$ and *fermion conservation*: $w(e_1)+w(e_2)+w(e_3)$ is an even number.

  The **quantum space-time** (Chap. 24) in *Loop Quantum Gravity* is a network of loops at Planck scale. Loops are represented by adapted spin networks: directed graphs whose arcs are labeled by irreducible representations of a compact Lie group and vertices are labeled by *interwinning operators* from the tensor product of labels on incoming arcs to the tensor product of labels on outgoing arcs. Such networks represent "quantum states" of the gravitational field on a 3D hypersurface.

- **Detour distance**

  Given a connected graph $G = (V, E)$, the **detour distance** is (Chartrand and Zhang, 2004) a metric on the vertex-set V defined, for $u \neq v$, as the length of the longest $(u - v)$ path in G. So, this distance is 1 or $|V| - 1$ if and only if uv is a *bridge* of G or, respectively, G contains a Hamiltonian $(u - v)$ path.

  The **monophonic distance** is (Santhakumaran and Titus, 2011) a distance (in general, not a metric) on the V defined, for $u \neq v$, as the length of a longest *monophonic* (or *minimal*), i.e., containing no chords, $(u - v)$ path in G.

The *height* of a DAG (acyclic digraph) is the number of vertices in a longest directed path.

- **Cutpoint additive metric**
Given a graph $G = (V, E)$, Klein–Zhu, 1998, call a metric d on V **graph-geodetic metric** if, for $u, w, v \in V$, the **triangle equality** $d(u, w) + d(w, v) = d(u, v)$ holds if w is a (u, v)-*gatekeeper*, i.e., w lies on any path connecting u and v. Cf. **metric interval** in Chap. 1. Any gatekeeper is a *cutpoint*, i.e., removing it disconnects G and a *pivotal point*, i.e., it lies on any shortest path between u and v.
Chebotarev, 2010, call a metric d on the vertices of a multigraph without loops **cutpoint additive** if $d(u, w)+d(w, v) = d(u, v)$ holds if and only if w lies on any path connecting u and v. The **resistance metric** is cutpoint additive (Gvishiani and Gurvich, 1992), while the **path metric** is graph-geodetic only (in the weaker Klein–Zhu sense). See also **Chebotarev–Shamis metric**.
- **Graph boundary**
Given a connected graph $G = (V, E)$, a vertex $v \in V$ is (Chartrand et al., 2003) a *boundary vertex* if there exists a *witness*, i.e., a vertex $u \in V$ such that $d(u, v) \geq d(u, w)$ for all neighbors w of v. So, the end-vertices of a longest path are boundary vertices. The **boundary** of G is the set of all boundary vertices.
The *boundary of a subset* $M \subset V$ is the set $\partial M \subset E$ of edges having precisely one endpoint in M. The **isoperimetric number** of G is (Buser, 1978) $\inf \frac{\partial M}{|M|}$, where the infimum is taken over all $M \subset V$ with $2|M| \leq |V|$.
- **Graph diameter**
Given a connected graph $G = (V, E)$, its **graph diameter** is the largest value of the **path metric** between vertices of G.
A connected graph is *vertex-critical* (*edge-critical*) if deleting any vertex (edge) increases its diameter. A graph G of diameter k is *goal-minimal* if for every edge uv, the inequality $d_{G-uv}(x, y) > k$ holds if and only if $\{u, v\} = \{x, y\}$.
If G is m-connected and a is an integer, $0 \leq a < m$, then the a-**fault diameter** of G is the maximal diameter of a subgraph of G induced by $|V| - a$ of its vertices. For $0 < a \leq m$, the a-**wide distance** $d_a(u, v)$ between vertices u and v is the minimum integer l, for which there are at least a internally disjoint $(u - v)$ paths of length at most l in G: cf. **Hsu–Lyuu–Flandrin–Li distance**. The a-*wide diameter* of G is $\max_{u,v \in V} d_a(u, v)$; it is at least the $(a - 1)$-fault diameter of G.
Given a *strong orientation* O of a connected graph $G = (V, E)$, i.e., a strongly connected digraph $D = (V, E')$ with arcs $e' \in E'$ obtained from edges $e \in E$ by orientation O, the *diameter* of D is the maximal length of shortest directed $(u-v)$ path in it. The **oriented diameter** of a graph G is the smallest diameter among strong orientations of G. If it is equal to the diameter of G, then any orientation realizing this equality is called *tight*. For example, a *hypercube graph* $H(m, 2)$ admits a tight orientation if $m \geq 4$ (McCanna, 1988).
- **Path quasi-metric in digraphs**
The **path quasi-metric in digraphs** d_{dpath} is a quasi-metric on the vertex-set V of a strongly connected digraph $D = (V, E)$ defined, for any $u, v \in V$, as the length of a shortest directed $(u - v)$ path in D.

The **circular metric in digraphs** is a metric on the vertex-set V of a strongly connected digraph $D = (V, E)$, defined by $d_{dpath}(u, v) + d_{dpath}(v, u)$.

- **Strong distance in digraphs**
The **strong distance in digraphs** is a metric between vertices v and v of a strongly connected digraph $D = (V, E)$ defined (Chartrand–Erwin–Raines–Zhang, 1999) as the minimum *size* (the number of edges) of a strongly connected subdigraph of D containing v and v. Cf. **Steiner distance of a set**.
- **Υ-metric**
Given a class Υ of connected graphs, the metric d of a metric space (X, d) is called a **Υ-metric** if (X, d) is isometric to a subspace of a metric space (V, d_{wpath}), where $G = (V, E) \in \Upsilon$, and d_{wpath} is the **weighted path metric** on V with positive edge-weight function w; cf. **tree-like metric**.
- **Tree-like metric**
A **tree-like metric** (or **weighted tree metric**) d on a set X is a **Υ-metric** for the class Υ of all trees, i.e., the metric space (X, d) is isometric to a subspace of a metric space (V, d_{wpath}), where $T = (V, E)$ is a tree, and d_{wpath} is the **weighted path metric** on the vertex-set V of T with a positive weight function w. A metric is a tree-like metric if and only if it satisfies the **four-point inequality**.
A metric d on a set X is called a **relaxed tree-like metric** if the set X can be embedded in some (not necessary positively) edge-weighted tree such that, for any $x, y \in X$, $d(x, y)$ is equal to the sum of all edge weights along the (unique) path between corresponding vertices x and y in the tree. A metric is a relaxed tree-like metric if and only if it is a **relaxed four-point inequality metric**.
- **Katz similarity**
Given a connected graph $G = (V, E)$ with positive edge-weight function $w = (w(e))_{e \in E}$, let $V = \{v_1, \dots, v_n\}$. Denote by A the $(n \times n)$-matrix $((a_{ij}))$, where $a_{ij} = a_{ji} = w(ij)$ if ij is an edge, and $a_{ij} = 0$, otherwise. Let I be the identity $(n \times n)$-matrix, and let $t, 0 < t < \frac{1}{\lambda}$, be a parameter, where $\lambda = \max_i |\lambda_i|$ is the *spectral radius* of A and λ_i are the eigenvalues of A. Define the $(n \times n)$-matrix

$$K = ((k_{ij})) = \sum_{i=1}^{\infty} t^i A^i = (I - tA)^{-1} - I.$$

The number k_{ij} is called the **Katz similarity** between v_i and v_j. Katz, 1953, proposed it for evaluating social status.
Chebotarev, 2011, defined, for a similar $(n \times n)$-matrix $((c_{ij})) = \sum_{i=0}^{\infty} t^i A^i = (I - tA)^{-1}$ and connected edge-weighted multigraphs allowing loops, the **walk distance** between vertices v_i and v_j as any positive multiple of $d_t(i, j) = -\ln \frac{c_{ij}}{\sqrt{c_{ii} c_{jj}}}$ (cf. the **Nei standard genetic distance** in Chap. 23). He proved that d_t is a **cutpoint additive metric** and the **path metric** in G coincides with the *short walk distance* $\lim_{t \to 0^+} \frac{d_t}{-\ln t}$ in G, while the **resistance metric** in G coincides with the *long walk distance* $\lim_{t \to \frac{1}{\lambda}^-} \frac{2d_t}{n(t^{-1} - \lambda)}$ in the graph G' obtained from G by attaching weighted loops that provide G' with uniform weighted degrees.

If G is a simple unweighted graph, then A is its adjacency matrix. Let J be the $(n \times n)$-matrix of all ones and let $\mu = \min_i \lambda_i$. Let $N = ((n_{ij})) = \mu(I - J) - A$. Neumaier, 1980, remarked that $((\sqrt{n_{ij}}))$ is a semimetric on the vertices of G.

- **Resistance metric**

  Given a connected graph $G = (V, E)$ with positive edge-weight function $w = (w(e))_{e \in E}$, let us interpret the edge-weights as electrical conductances and their inverses as resistances. For any two different vertices u and v, suppose that a battery is connected across them, so that one unit of a current flows in at u and out in v. The voltage (potential) difference, required for this, is, by Ohm's law, the effective resistance between u and v in an electrical network; it is called the **resistance** (or *electric*) **metric** $\Omega(u, v)$ between them (Sharpe, 1967, Gvishiani–Gurvich, 1987, and Klein–Randic, 1993 [KlRa93]). So, if a potential of one volt is applied across vertices u and v, a current of $\frac{1}{\Omega(u,v)}$ will flow. The number $\frac{1}{\Omega(u,v)}$ is a measure of the *connectivity* between u and v.

  Let $r(u, v) = \frac{1}{w(e)}$ if uv is an edge, and $r(u, v) = 0$, otherwise. Formally,

  $$\Omega(u, v) = (\sum_{w \in V} f(w) r(w, v))^{-1},$$

  where $f : V \to [0, 1]$ is the unique function with $f(u) = 1$, $f(v) = 0$ and $\sum_{z \in V} (f(w) - f(z)) r(w, z) = 0$ for any $w \neq u, v$.

  The resistance metric is a weighted average of the lengths of all $(u - v)$ paths. It is applied when the number of $(u - v)$ paths, for any $u, v \in V$, matters.

  A probabilistic interpretation (Gobel–Jagers, 1974) is: $\Omega(u, v) = (deg(u)Pr(u \to v))^{-1}$, where $deg(u)$ is the degree of the vertex u, and $Pr(u \to v)$ is the probability for a random walk leaving u to arrive at v before returning to u. The expected commuting time between u and v is $2 \sum_{e \in E} w(e) \Omega(u, v)$.

  Then $\Omega(u, v) \leq \min_P \sum_{e \in P} \frac{1}{w(e)}$, where P is any $(u - v)$ path (cf. **inverse weighted path metric**), with equality if and only if such a path P is unique. So, if $w(e) = 1$ for all edges, the equality means that G is a **geodetic graph**, and hence the path and resistance metrics coincide. Also, it holds that $\Omega(u, v) = \frac{|\{t : uv \in t \in T\}|}{|T|}$ if uv is an edge, and $\Omega(u, v) = \frac{|T' - T|}{|T|}$, otherwise, where T, T' are the sets of spanning trees for $G = (V, E)$ and $G' = (V, E \cup \{uv\})$.

  If $w(e) = 1$ for all edges, then $\Omega(u, v) = (g_{uu} + g_{vv}) - (g_{uv} + g_{vu})$, where $((g_{ij}))$ is the Moore–Penrose *generalized inverse* of the *Laplacian matrix* $((l_{ij}))$ of the graph G: here l_{ii} is the degree of vertex i, while, for $i \neq j$, $l_{ij} = 1$ if the vertices i and j are adjacent, and $l_{ij} = 0$, otherwise. A symmetric (for an undirected graph) and positive-semidefinite matrix $((g_{ij}))$ admits a representation KK^T. So, $\Omega(u, v)$ is the squared Euclidean distance between the u-th and v-th rows of K.

  The distance $\sqrt{\Omega(u, v)}$ is a **Mahalanobis distance** (cf. Chap. 17) with a weighting matrix $((g_{ij}))$. So, $\Omega_{u,v} = a_{uv} |((g_{ij}))| a_{uv}$, where a_{uv} are the vectors of zeros except for $+1$ and -1 in the u-th and v-th positions. This distance is called a *diffusion metric* in [CLMNWZ05] because it depends on a random walk.

The number $\frac{1}{2}\sum_{u,v\in V}\Omega(u,v)$ is called the *total resistance* (or *Kirchhoff index*) of G.

- **Hitting time quasi-metric**
Let $G = (V, E)$ be a connected graph. Consider random walks on G, where at each step the walk moves to a vertex randomly with uniform probability from the neighbors of the current vertex. The **hitting** (or *first-passage*) **time quasi-metric** $H(u, v)$ from $u \in V$ to $v \in V$ is the expected number of steps (edges) for a random walk on G beginning at u to reach v for the first time; it is 0 for $u = v$. This quasi-metric is a **weightable quasi-semimetric** (cf. Chap. 1).
The **commuting time metric** is $C(u, v) = H(u, v) + H(v, u)$.
Then $C(u, v) = 2|E|\Omega(u, v)$, where $\Omega(u, v)$ is the **resistance metric** (or *effective resistance*), i.e., 0 if $u = v$ and, otherwise, $\frac{1}{\Omega(u,v)}$ is the current flowing into v, when grounding v and applying a 1 volt potential to u (each edge is seen as a resistor of 1 ohm). Also, $\Omega(u, v) = \sup_{f:V\to\mathbb{R},\,D(f)>0}\frac{(f(u)-f(v))^2}{DE(f)}$, where $DE(f)$ is the *Dirichlet energy* of f, i.e., $\sum_{st\in E}(f(s) - f(t))^2$.
The above setting can be generalized to weighted digraphs $D = (V, E)$ with arc-weights c_{ij} for $ij \in E$ and the *cost* of a directed $(u - v)$ path being the sum of the weights of its arcs. Consider the random walk on D, where at each step the walk moves by arc ij with *reference probability* p_{ij} proportional to $\frac{1}{c_{ij}}$; set $p_{ij} = 0$ if $ij \notin E$. Saerens et al., 2008, defined the *randomized et al.* shortest path quasi-distance $d(u, v)$ on vertices of D as the minimum expected cost of a directed $(u - v)$ path in the probability distribution minimizing the expected cost among all distributions having a fixed **Kullback–Leibler distance** (cf. Chap. 14) with reference probability distribution. In fact, their biased random walk model depends on a parameter $\theta \geq 0$. For $\theta = 0$ and large θ, the distance $d(u, v) + d(v, u)$ become a metric; it is proportional to the commuting time and the usual path metric, respectively.
- **Chebotarev–Shamis metric**
Given $\alpha > 0$ and a connected weighted *multigraph* $G = (V, E; w)$ with positive edge-weight function $w = (w(e))_{e\in E}$, denote by $L = ((l_{ij}))$ the *Laplacian* (or *Kirchhoff*) matrix of G, i.e., $l_{ij} = -w(ij)$ for $i \neq j$ and $l_{ii} = \sum_{j\neq i} w(ij)$. The **Chebotarev–Shamis metric** $d_\alpha(u, v)$ (Chebotarev and Shamis, 2000, called $\frac{1}{2}d_\alpha(u, v)$ **α-forest metric**) between vertices u and v is defined by
$$2q_{uv} - q_{uu} - q_{vv}$$
for the **protometric** $((g_{ij})) = -(I + \alpha L)^{-1}$, where I is the identity matrix.
Chebotarev and Shamis showed that their metric of $G = (V, E; w)$ is the **resistance metric** of another weighted multigraph, $G' = (V', E'; w')$, where $V' = V \cup \{0\}$, $E' = E \cup \{u0 : u \in V\}$, while $w'(e) = \alpha w(e)$ for all $e \in E$ and $w'(u0) = 1$ for all $u \in V$. In fact, there is a bijection between the forests of G and trees of G'. This metric becomes the resistance metric of $G = (V, E; w)$ as $\alpha \to \infty$.
Their **forest metric** (1997) is the case $\alpha = 1$ of the α-forest metric.

Chebotarev, 2010, remarked that $2\ln q_{uv} - \ln q_{uu} - \ln q_{vv}$ is a **cutpoint additive metric** $d''_{\alpha}(u, v)$, i.e., $d''_{\alpha}(u, w) + d''_{\alpha}(w, v) = d''_{\alpha}(u, v)$ holds if and only if w lies on any path connecting u and v. The metric d''_{α} is the **path metric** if $\alpha \to 0^{+}$ and the **resistance metric** if $\alpha \to \infty$.

- **Truncated metric**
  The **truncated metric** is a metric on the vertex-set of a graph, which is equal to 1 for any two adjacent vertices, and is equal to 2 for any nonadjacent different vertices. It is the 2-**truncated metric** for the path metric of the graph. It is the $(1, 2) - B$-**metric** if the degree of any vertex is at most B.
- **Hsu–Lyuu–Flandrin–Li distance**
  Given an *m-connected* graph $G = (V, E)$ and two vertices $u, v \in V$, a *container* $C(u, v)$ *of width* m is a set of m $(u - v)$ paths with any two of them intersecting only in u and v. The *length of a container* is the length of the longest path in it.
  The **Hsu–Lyuu–Flandrin–Li distance** between vertices u and v (Hsu–Lyuu, 1991, and Flandrin–Li, 1994) is the minimum of container lengths taken over all containers $C(u, v)$ of width m. This generalization of the path metric is used in parallel architectures for interconnection networks.
- **Multiply-sure distance**
  The **multiply-sure distance** is a distance on the vertex-set V of an *m-connected* weighted graph $G = (V, E)$, defined, for any $u, v \in V$, as the minimum weighted sum of lengths of m disjoint $(u-v)$ paths. This generalization of the path metric helps when several disjoint paths between two points are needed, for example, in communication networks, where $m - 1$ of $(u - v)$ paths are used to code the message sent by the remaining $(u - v)$ path (see [McCa97]).
- **Cut semimetric**
  A *cut* is a *partition* of a set into two parts. Given a subset S of $V_n = \{1, \dots, n\}$, we obtain the partition $\{S, V_n \backslash S\}$ of V_n. The **cut semimetric** (or **split semimetric**) δ_S defined by this partition, is a semimetric on V_n defined by

$$\delta_S(i, j) = \begin{cases} 1, & \text{if } i \neq j, |S \cap \{i, j\}| = 1, \\ 0, & \text{otherwise}. \end{cases}$$

  Usually, it is considered as a vector in $\mathbb{R}^{|E_n|}$, $E(n) = \{\{i, j\} : 1 \le i < j \le n\}$.
  A *circular cut* of V_n is defined by a subset $S_{[k+1,l]} = \{k + 1, \dots, l\} (\mathrm{mod}\, n) \subset V_n$: if we consider the points $\{1, \dots, n\}$ as being ordered along a circle in that circular order, then $S_{[k+1,l]}$ is the set of its consecutive vertices from $k + 1$ to l. For a circular cut, the corresponding cut semimetric is called a **circular cut semimetric**.
  An **even cut semimetric** (**odd cut semimetric**) is δ_S on V_n with even (odd, respectively) $|S|$. A k-**uniform cut semimetric** is δ_S on V_n with $|S| \in \{k, n - k\}$. An **equicut semimetric** (**inequicut semimetric**) is δ_S on V_n with $|S| \in \{\lfloor \frac{n}{2} \rfloor, \lceil \frac{n}{2} \rceil\}$ ($|S| \notin \{\lfloor \frac{n}{2} \rfloor, \lceil \frac{n}{2} \rceil\}$, respectively); see, for example, [DeLa97].

- **Decomposable semimetric**
  A **decomposable semimetric** is a semimetric on $V_n = \{1, \dots, n\}$ which can be represented as a nonnegative linear combination of **cut semimetrics**. The set of all decomposable semimetrics on V_n is a *convex cone*, called the *cut cone* CUT_n. A semimetric on V_n is decomposable if and only if it is a **finite l_1-semimetric**.
  A **circular decomposable semimetric** is a semimetric on $V_n = \{1, \dots, n\}$ which can be represented as a nonnegative linear combination of **circular cut semimetrics**. A semimetric on V_n is circular decomposable if and only if it is a **Kalmanson semimetric** with respect to the same ordering (see [ChFi98]).
- **Finite l_p-semimetric**
  A **finite l_p-semimetric** d is a semimetric on $V_n = \{1, \dots, n\}$ such that (V_n, d) is a semimetric subspace of the l_p^m-*space* $(\mathbb{R}^m, d_{l_p})$ for some $m \in \mathbb{N}$.
  If, instead of V_n, is taken $X = \{0, 1\}^n$, the metric space (X, d) is called the l_p^n-*cube*. The l_1^n-cube is called a **Hamming cube**; cf. Chap. 4. It is the graphic metric space associated with a hypercube graph $H(n, 2)$, and any subspace of it is called a **partial cube**.
- **Kalmanson semimetric**
  A **Kalmanson semimetric** d *with respect to the ordering* $1, \dots, n$ is a semimetric on $V_n = \{1, \dots, n\}$ which satisfies the condition

$$\max\{d(i, j) + d(r, s), d(i, s) + d(j, r)\} \leq d(i, r) + d(j, s)$$

  for all $1 \leq i \leq j \leq r \leq s \leq n$.
  Equivalently, if the points $\{1, \dots, n\}$ are ordered along a circle C_n in that circular order, then the distance d on V_n is a Kalmanson semimetric if the inequality

$$d(i, r) + d(j, s) \leq d(i, j) + d(r, s)$$

  holds for $i, j, r, s \in V_n$ whenever the segments $[i, j]$, $[r, s]$ are crossing chords of C_n.
  A **tree-like metric** is a Kalmanson metric for some ordering of the vertices of the tree. The Euclidean metric, restricted to the points that form a convex polygon in the plane, is a Kalmanson metric.
- **Multicut semimetric**
  Let $\{S_1, \dots, S_q\}$, $q \geq 2$, be a *partition* of the set $V_n = \{1, \dots, n\}$, i.e., a collection $S_1, \dots, S_q$ of pairwise disjoint subsets of V_n such that $S_1 \cup \dots \cup S_q = V_n$.
  The **multicut semimetric** $\delta_{S_1,\dots,S_q}$ is a semimetric on V_n defined by

$$\delta_{S_1,\dots,S_q}(i, j) = \begin{cases} 0, & \text{if } i, j \in S_h \text{ for some } h, 1 \leq h \leq q, \\ 1, & \text{otherwise.} \end{cases}$$

- **Oriented cut quasi-semimetric**
  Given a subset S of $V_n = \{1, \dots, n\}$, the **oriented cut quasi-semimetric** δ'_S is a quasi-semimetric on V_n defined by
  $$\delta'_S(i,j) = \begin{cases} 1, & \text{if } i \in S, j \notin S, \\ 0, & \text{otherwise.} \end{cases}$$
  Usually, it is considered as the vector of $\mathbb{R}^{|I_n|}$, $I(n) = \{(i,j) : 1 \le i \ne j \le n\}$. The **cut semimetric** δ_S is $\delta'_S + \delta'_{V_n \setminus S}$.
- **Oriented multicut quasi-semimetric**
  Given a *partition* $\{S_1, \dots, S_q\}$, $q \ge 2$, of V_n, the **oriented multicut quasi-semimetric** $\delta'_{S_1,\dots,S_q}$ is a quasi-semimetric on V_n defined by
  $$\delta'_{S_1,\dots,S_n}(i,j) = \begin{cases} 1, & \text{if} \quad i \in S_h, j \in S_m, h < m, \\ 0, & \text{otherwise} \end{cases}.$$

15.2 Distance-Defined Graphs

Below we first give some graphs defined in terms of distances between their vertices. Then some graphs associated with metric spaces are presented.

A graph (V, E) is, say, *distance-invariant* or *distance monotone* if its metric space (V, d_{path}) is **distance invariant** or **distance monotone**, respectively (cf. Chap. 1). The definitions of such graphs, being straightforward subcases of corresponding metric spaces, will be not given below.

- **k-Power of a graph**
  The **k-power** of a graph $G = (V, E)$ is the supergraph $G^k = (V, E')$ of G with edges between all pairs of vertices having path distance at most k.
- **Distance-residual subgraph**
  For a connected finite graph $G = (V, E)$ and a set $M \subset V$ of its vertices, a **distance-residual subgraph** is (Luksic and Pisanski, 2010) a subgraph induced on the set of vertices u of G at the maximal **point-set distance** $\min_{v \in M} d_{\text{path}}(u, v)$ from M. Such a subgraph is called *vertex-residual* if M consists of a vertex, and *edge-residual* if M consists of two adjacent vertices.
- **Isometric subgraph**
  A subgraph H of a graph $G = (V, E)$ is called an **isometric subgraph** if the path metric between any two points of H is the same as their path metric in G.
  A subgraph H is called a *convex subgraph* if it is isometric, and for any $u, v \in H$ every vertex on a shortest $(u - v)$ path belonging to H also belongs to H.
  A subset $M \subset V$ is called *gated* if for every $u \in V \setminus M$ there exists a unique vertex $g \in M$ (called a *gate*) lying on a shortest $(u - v)$ path for every $v \in M$. The subgraph induced by a gated set is a convex subgraph.

- **Retract subgraph**
  A subgraph H of G is called a **retract subgraph** if it is induced by an idempotent **metric mapping** of G into itself, i.e., $f^2 = f : V \to V$ with $d_{\text{path}}(f(u), f(v)) \le d_{\text{path}}(u, v)$ for $u, v \in V$. Any retract subgraph is isometric.
- **Partial cube**
  A **partial cube** is an **isometric subgraph** of a **Hamming cube**, i.e., of a hypercube $H(m, 2)$. Similar topological notion was introduced by Acharya, 1983: any graph (V, E) admits a *set-indexing* $f : V \cup E \to 2^X$ with injective $f|_V, f|_R$ and $f(uv) = f(u)\Delta f(v)$ for any $(uv) \in E$. The *set-indexing number* is $\min |X|$.
- **Median graph**
  A connected graph $G = (V, E)$ is called a **median graph** if, for any three vertices $u, v, w \in V$, there exists a unique vertex that lies simultaneously on a shortest $(u - v)$, $(u - w)$ and $(w - v)$ paths, i.e., (V, d_{path}) is a **median metric space**.
  The median graphs are exactly **retract subgraphs** of hypercubes. Also, they are exactly **partial cubes** such that the vertex-set of any *convex subgraph* is *gated* (cf. **isometric subgraph**).
- **Geodetic graph**
  A graph is called **geodetic** if there exists at most one shortest path between any two of its vertices. A graph is called *strongly geodetic* if there exists at most one path of length less than or equal to the diameter between any two of its vertices.
  A *uniformly geodetic graph* is a connected graph such that the number of shortest paths between any two vertices u and v depends only on $d(u, v)$.
  A graph is a *forest* (disjoint union of trees) if and only if there exists at most one path between any two of its vertices.
  The *geodetic number* of a finite connected graph (V, E) [BuHa90] is $\min |M|$ over sets $M \subset V$ such that any $x \in V$ lies on a shortest $(u - v)$ path with $u, v \in M$.
- **k-geodetically connected graph**
  A k-connected graph is called (Entringer–Jackson–Slater, 1977) **k-geodetically connected** ($k - GC$) if the removal of less than k vertices (or, equivalently, edges) does not affect the **path metric** between any pair of the remaining vertices.
  $2 - GC$ graphs are called *self-repairing*. Cf. **Hsu–Lyuu–Flandrin–Li distance**.
- **Interval distance monotone graph**
  A connected graph $G = (V, E)$ is called **interval distance monotone** if any of its intervals $I_G(u, v)$ induces a *distance monotone graph*, i.e., its path metric is **distance monotone**, cf. Chap. 1.
  A graph is interval distance monotone if and only if (Zhang–Wang, 2007) each of its intervals is isomorphic to either a path, a cycle or a hypercube.
- **Distance-regular graph**
  A connected *regular* (i.e., every vertex has the same degree) graph $G = (V, E)$ of diameter T is called **distance-regular** (or *drg*) if, for every two its vertices u, v and any integers $0 \le i, j \le T$, the number $|\{w \in V : d_{\text{path}}(u, w) = i,$

$d_{\text{path}}(v, w) = j\}|$ depends only on i, j and $k = d_{\text{path}}(u, v)$, but not on the choice of u and v.

A special case of it is a **distance-transitive graph**, i.e., such that its group of automorphisms is transitive, for any $0 \le i \le T$, on the pairs of vertices (u, v) with $d_{\text{path}}(u, v) = i$. An analog of drg is an *edge-regular graph* (Fiol–Carriga, 2001).

Any drg is a **distance-balanced graph** (or *dbg*), i.e., $|W_{u,v}| = |W_{v,u}|$, where $W_{u,v} = \{x \in V : d(x, u) < d(x, v)\}$. Such graph is also called *self-median* since it is exactly one, **metric median** (cf. **eccentricity** in Chap. 1) of which is V. A gbg is called *nicely distance-balanced* if $|W_{u,v}|$ is the same for all edges uv.

Any drg is a **distance degree-regular graph** (i.e., $|\{x \in V : d(x, u) = i\}|$ depends only on i; such graph is also called *strongly distance-balanced*), and a **walk-regular graph** (i.e., the number of closed walks of length i starting at u depends only on i). van Dam–Omidi, 2013, call a graph *strongly walk-regular* if there is an $l \ge 2$ such that the number of walks of length l from u to v depends only on whether the $d(u, v)$ is 0, 1, or $\ge$2; for $l = 2$, it is a *strongly regular graph*, i.e., a drg of diameter 2. A *d-Deza graph* (Gu, 2013) is a regular graph (V, E) in which there are exactly d different values of $|\{w \in V : d(u, w) = d(v, w) = 1\}|$ for distinct $u, v \in V$.

A graph G is a **distance-regularized graph** if for each $u \in V$, if admits an *intersection array at vertex u*, i.e., the numbers $a_i(u) = |G_i(u) \cap G_1(v)|$, $b_i(u) = |G_{i+1}(u) \cap G_1(v)|$ and $c_i(u) = |G_{i-1}(v) \cap G_1(v)|$ depend only on the distance $d(u, v) = i$ and are independent of the choice of the vertex $v \in G_i(u)$. Here, for any i, $G_i(w)$ is the set of all vertices at the distance i from w. Godsil–Shawe-Taylor, 1987, defined such graph and proved that it is either drg or *distance-biregular* (a bipartite one with vertices in the same class having the same intersection array).

A drg is also called a **metric association scheme** or *P-polynomial association scheme*. A finite **polynomial metric space** (cf. Chap. 1) is a special case of it, also called a *(P and Q)-polynomial association scheme*.

- **Distance-regular digraph**

  A strongly connected digraph $D = (V, E)$ is called **distance-regular** (Damerell, 1981) if, for any its vertices u, v with $d_{\text{path}}(u, v) = k$ and for any integer $0 \le i \le k + 1$, the number of vertices w, such that $d_{\text{path}}(u, w) = i$ and $d_{\text{path}}(v, w) = 1$, depends only on k and i, but not on the choice of u and v.

  In order to find interesting classes of distance-regular digraphs with unbounded diameter, the above definition was weakened by two teams in different directions. Call $\overline{d(x, y)} = (d(x, y), d(y, x))$ the **two-way distance in digraph** D. A strongly connected digraph $D = (V, E)$ is called **weakly distance-regular** (Wang and Suzuku, 2003) if, for any its vertices u, v with $\overline{d(u, v)} = (k_1, k_2)$, the number of vertices w, such that $\overline{d(w, u)} = (i_1, i_2)$ and $\overline{d(w, v)} = (j_1, j_2)$, depends only on the values $k_1, k_2, i_1, i_2, j_1, j_2$. Comellas et al., 2004, defined a **weakly distance-regular digraph** as one in which, for any vertices u and v, the number of $u \to v$ walks of every given length only depends on the distance $d(u, v)$.

- **Metrically almost transitive graph**
An *automorphism* of a graph $G = (V, E)$ is a map $g : V \to V$ such that u is adjacent to v if and only if $g(u)$ is adjacent to $g(v)$, for any $u, v \in V$. The set $Aut(G)$ of automorphisms of G is a group with respect to the composition of functions.
A graph G is **metrically almost transitive** (Krön–Möller, 2008) if there is an integer r such that, for any vertex $u \in V$ it holds

$$\cup_{g \in Aut(G)}\{g(\overline{B}(u, r) = \{v \in V : d_{\text{path}}(u, v) \leq r\})\} = V.$$

- **Metric end**
Given an infinite graph $G = (V, E)$, a *ray* is a sequence $(x_0, x_1, \dots)$ of distinct vertices such that x_i and x_{i+1} are adjacent for $i \geq 0$.
Two rays R_1 and R_2 are equivalent whenever it is impossible to find a bounded set of vertices F such that any path from R_1 to R_2 contains an element of F.
Metric ends are defined as equivalence classes of *metric rays* which are rays without infinite, bounded subsets.
- **Graph of polynomial growth**
Let $G = (V, E)$ be a transitive locally finite graph. For a vertex $v \in V$, the *growth function* is defined by

$$f(n) = |\{u \in V : d(u, v) \leq n\}|,$$

and it does not depend on v. Cf. **growth rate of metric space** in Chap. 1.
The graph G is a **graph of polynomial growth** if there are some positive constants k, C such that $f(n) \leq Cn^k$ for all $n \geq 0$. It is a **graph of exponential growth** if there is a constant $C > 1$ such that $f(n) > C^n$ for all $n \geq 0$.
A group with a finite symmetric set of generators has *polynomial growth rate* if the corresponding *Cayley graph* has polynomial growth. Here the metric ball consists of all elements of the group which can be expressed as products of at most n generators, i.e., it is a closed ball centered in the identity in the **word metric**, cf. Chap. 10.
- **Distance-polynomial graph**
Given a connected graph $G = (V, E)$ of diameter T, for any $2 \leq i \leq T$ denote by G_i the graph (V, E') with $E' = \{e = uv \in E : d_{\text{path}}(u, v) = i\}$. The graph G is called a **distance-polynomial** if the adjacency matrix of any G_i, $2 \leq i \leq T$, is a polynomial in terms of the adjacency matrix of G.
Any **distance-regular** graph is a distance-polynomial.
- **Distance-hereditary graph**
A connected graph is called **distance-hereditary** (Howorka, 1977) if each of its connected induced subgraphs is isometric.
A graph is distance-hereditary if each of its induced paths is isometric. A graph is distance-hereditary, bipartite distance-hereditary, **block graph**, tree if and only if its path metric is a **relaxed tree-like metric** for edge-weights being, respectively, nonzero half-integers, nonzero integers, positive half-integers, positive integers.

A graph is called a **parity graph** if, for any $u, v \in V$, the lengths of all induced $(u - v)$ paths have the same parity. A graph is a parity graph (moreover, distance-hereditary) if and only if every induced subgraph of odd (moreover, any) order of at least five has an even number of Hamiltonian cycles (McKee, 2008).

- **Distance magic graph**
  A graph $G = (V, E)$ is called a **distance magic graph** if it admits a *distance magic labeling*, i.e., a *magic constant* $k > 0$ and a bijection $f : V \to \{1, 2, \dots, |V|\}$ with $\sum_{uv \in E} f(v) = k$ for every $u \in V$. Introduced by Vilfred, 1994, these graphs generalize *magic squares* (such complete n-partite graphs with parts of size n).
  Among trees, cycles and K_n, only P_1, P_3, C_4 are distance magic. The *hypercube graph* $H(m, 2)$ is distance magic if $m = 2, 6$ but not if $m \equiv 0, 1, 3 \pmod 4$.
- **Block graph**
  A graph is called a **block graph** if each of its *blocks* (i.e., a maximal 2-connected induced subgraph) is a complete graph. Any tree is a block graph.
  A graph is a block graph if and only if its path metric is a **tree-like metric** or, equivalently, satisfies the **four-point inequality**.
- **Ptolemaic graph**
  A graph is called **Ptolemaic** if its path metric satisfies the **Ptolemaic inequality**

$$d(x, y)d(u, z) \leq d(x, u)d(y, z) + d(x, z)d(y, u).$$

  A graph is Ptolemaic if and only if it is distance-hereditary and *chordal*, i.e., every cycle of length greater than 3 has a chord. So, any **block graph** is Ptolemaic.
- **k-cocomparability graph**
  A graph $G = (V, E)$ is called (Chang–Ho–Ko, 2003) **k-cocomparability graph** if its vertex-set admits a linear ordering $<$ such that for any three vertices $u < v < w$, $d(u, w) \leq k$ implies $d(u, v) \leq k$ or $d(v, w) \leq k$.
- **Distance-perfect graph**
  Cvetković et al., 2007, observed that any graph of diameter T has at most $k + T^k$ vertices, where k is its **location number** (cf. Chap. 1), i.e., the minimal cardinality of a set of vertices, the path distances from which uniquely determines any vertex. They called a graph **distance-perfect** if it meets this upper bound and proved that such a graph has $T \neq 2$.
- **t-irredundant set**
  A set $S \subset V$ of vertices in a connected graph $G = (V, E)$ is called **t-irredundant** (Hattingh–Henning, 1994) if for any $u \in S$ there exists a vertex $v \in V$ such that, for the path metric d_{path} of G, it holds

$$d_{\text{path}}(v, x) \leq t < d_{\text{path}}(v, V \backslash S) = \min_{u \notin S} d_{\text{path}}(v, u).$$

  The **t-irredundance number** ir_t of G is the smallest cardinality $|S|$ such that S is t-irredundant but $S \cup \{v\}$ is not, for every $v \in V \backslash S$.

The *t-domination number* γ_t and *t-independent number* α_t of G are, respectively, the cardinality of the smallest $(t+1)$-*covering* (by the open balls of the radius $r+1$) and largest $\lceil \frac{t}{2} \rceil$-*packing* of the metric space $(V, d_{\text{path}}(u, v))$; cf. the **radii of metric space** in Chap. 1. Then it holds that $\frac{\gamma_t + 1}{2} \leq ir_t \leq \gamma_t \leq \alpha_t$.
Let B_S denote $\{v \in V : d(v, S) = 1\}$. Then $\max_{S \subset V} |B_S| = |V| - \gamma_1$ and $\max_{S \subset V}(|B_S| - |S|)$ are called the *enclaveless number* and the *differential* of G.

- **r-Locating-dominating set**
  Let $D = (V, E)$ be a digraph and $C \subset V$, and let $B_r^-(v)$ denote the set of all vertices x such that there exists a directed $(x - v)$ path with at most r arcs.
  If $B_r^-(v) \cap C$, $v \in V \setminus C$ (respectively, $v \in V$), are nonempty distinct sets, C is called (Slater, 1984) an **r-locating-dominating set** (respectively, an **r-identifying code**; cf. Chap. 16) of D. Such sets of smallest cardinality are called *optimal*.
- **Locating chromatic number**
  The **locating chromatic number** of a graph $G = (V, E)$ is the minimum number of color classes $C_1, \dots, C_t$ needed to color vertices of G so that any two adjacent vertices have distinct colors and each vertex $u \in V$ has distinct *color code* $(\min_{v \in C_1} d(u, v), \dots, \min_{v \in C_k} d(u, v))$.
- **k-Distant chromatic number**
  The **k-distant chromatic number** of a graph $G = (V, E)$ is the minimum number of colors needed to color vertices of G so that any two vertices at distance at most k have distinct colors, i.e., it is the chromatic number of the **k-power of G**.
- **Distance between edges**
  The **distance between edges** in a connected graph $G = (X, E)$ is the number of vertices in a shortest path between them. So, adjacent edges have distance 1.
  A **distance-k matching** of G is a set of edges no two of which are within distance k. For $k = 1$, it is the usual matching. For $k = 2$, it is also *induced* (or *strong*) matching. A distance-k matching of G is equivalent to an independent set in the **k-power** of the line graph of G. A **distance-k edge-coloring** of G is an edge-coloring such that each color class induces a distance-k matching.
  The **distance-k chromatic index** $\mu_k(G)$ is the least integer t such that there exists a distance-t edge-coloring of G. The **distance-k matching number** $\nu_k(G)$ is the largest integer t such that there exists a distance-t matching in G with t edges. It holds that $\mu_k(G)\nu_k(G) \geq |E|$.
  The **distance between faces** of a plane graph is the number of vertices in a shortest path between them. A **distance-k face-coloring** is a face-coloring such that any two faces at distance at most k have different colors. The **distance-k face chromatic index** is the least integer t such that such coloring exists.
- **Rainbow distance**
  In an edge-colored graph, the **rainbow distance** is (Chartrand and Zhang, 2005) the length of a shortest *rainbow* (i.e., containing no color twice) path.
  In a vertex-colored graph, the **colored distance** is (Dankelmann et al., 2001) the sum of distances between all unordered pairs of vertices having different colors.

- **D-distance graph**
  Given a set D of positive numbers containing 1 and a metric space (X, d), the **D-distance graph** is a graph $G = (V = X, E)$ with the edge-set $E = \{uv : d(u, v) \in D\}$ (cf. **D-chromatic number** in Chap. 1). If (X, d) is path metric of a graph H, then G is called the **distance power** H^D of H.
  Alon–Kupavsky, 2014, call G (in the case $(X, d) = \mathbb{E}^n$, $d = \{1\}$) the *faithful unit-distance graph*, using term *unit-distance graph* for $E \subseteq \{(u, v) : ||u-v||_2 = 1\}$.
  For a positive number t, the *signed distance graph* is (Fiedler, 1969) a signed graph with the vertex-set X in which vertices x, y are joined by a positive edge if $t > d(x, y)$, by a negative edge if $d(x, y) > t$, and not joined if $d(x, y) = t$.
  A D-distance graph is called a **distance graph** (or *unit-distance graph*) if $D = \{1\}$, an *ϵ-unit graph* if $D = [1 - \epsilon, 1 + \epsilon]$, a *unit-neighborhood graph* if $D = (0, 1]$, an *integral-distance graph* if $D = \mathbb{Z}_+$, a *rational-distance graph* if $D = \mathbb{Q}_+$, and a *prime-distance graph* if D is the set of prime numbers (with 1).
  Every finite graph can be represented by a D-distance graph in some $\mathbb{E}^n$. The minimum dimension of such a Euclidean space is called the *D-dimension* of G.
  A *matchstick graph* is a crossingless unit-distance graph in $\mathbb{E}^2$.
- **Distance-number of a graph**
  Given a graph $G = (V, E)$, its *degenerate drawing* is a mapping $f : V \to \mathbb{R}^2$ such that $|f(V)| = |V|$ and $f(uv)$ is an open straight-line segment joining the vertices $f(u)$ and $f(v)$ for any edge $uv \in E$; it is a *drawing* if, moreover, $f(w) \notin f(uv)$ for any $uv \in E$ and $w \in V$.
  The **distance-number** $dn(G)$ **of a graph** G is (Carmi et al., 2008) the minimum number of distinct edge-lengths in a drawing of G.
  The *degenerate distance-number* of G, denoted by $ddn(G)$, is the minimum number of distinct edge-lengths in a degenerated drawing of G. The first of the **Erdös-type distance problems** in Chap. 19 is equivalent to determining $ddn(K_n)$.
- **Dimension of a graph**
  The **dimension** $dim(G)$ **of a graph** G is (Erdös–Harary–Tutte, 1965) the minimum k such that G has a *unit-distance representation* in $\mathbb{R}^k$, i.e., every edge is of length 1. The vertices are mapped to distinct points of $\mathbb{R}^k$, but edges may cross.
  For example, $dim(G) = n - 1, 4, 2$ for $G = K_n, K_{m,n}, C_n$ $(m \geq n \geq 3)$.
- **Bar-and-joint framework**
  A n-dimensional **bar-and-joint framework** is a pair (G, f), where $G = (V, E)$ is a finite graph (no loops and multiple edges) and $f : V \to \mathbb{R}^n$ is a map with $f(u) \neq f(v)$ whenever $uv \in E$. The **framework** is a straight line realization of G in $\mathbb{R}^n$ in which the length of an edge $uv \in E$ is given by $||f(u) - f(v)||_2$.
  The vertices and edges are called *joints* and *bars*, respectively, in terms of Structural Engineering. A **tensegrity structure** (Fuller, 1948) is a mechanically stable bar framework in which bars are either *cables* (tension elements which cannot get further apart), or *struts* (compression elements which cannot get closer together).

A framework (G, f) is *globally rigid* if every framework (G, f'), satisfying $||f(u) - f(v)||_2 = ||f'(u) - f'(v)||_2$ for all $uv \in E$, also satisfy it for all $u, v \in V$. A framework (G, f) is *rigid* if every continuous motion of its vertices which preserves the lengths of all edges, also preserves the distances between all pairs of vertices. The framework (G, f) is *generic* if the set containing the coordinates of all the points $f(v)$ is algebraically independent over the rationals. The graph G is n*-rigid* if every its n-dimensional generic realization is rigid. For generic frameworks, rigidity is equivalent to the stronger property of infinitesimal rigidity.

An *infinitesimal motion* of (G, f) is a map $m : V \to \mathbb{R}^n$ with $(m(u) - m(v))(f(u) - f(v)) = 0$ whenever $uv \in E$. A motion is *trivial* if it can be extended to an isometry of $\mathbb{R}^n$. A framework is an *infinitesimally rigid* if every motion of it is trivial, and it is *isostatic* if, moreover, the deletion of any its edge will cause loss of rigidity. (G, f) is an *elastic framework* if, for any $\epsilon > 0$, there exists a $\delta > 0$ such that for every edge-weighting $w : E \to \mathbb{R}_{>0}$ with $\max_{uv \in E} |w(uv) - ||f(u) - f(v)||_2| \leq \delta$, there exist a framework (G, f') with $\max_{v \in V} ||f(u) - f'(v)||_2 < \epsilon$.

A framework (G, f) with $||f(u) - f(v)||_2 > r$ if $u, v \in V, u \neq c$ and $||f(u), f(v)||_2 \leq R$ if $uv \in E$, for some $0 < r < R$, is called (Doyle–Snell, 1984) a *civilized drawing of a graph*. The random walks on such graphs are recurrent if $n = 1, 2$.

- **Distance constrained labeling**
  Given a sequence $\alpha = (\alpha_1, \ldots, \alpha_k)$ of **distance constraints** $\alpha_1 \geq \cdots \geq \alpha_k > 0$, a λ_α*-labeling* of a graph $G = (V, E)$ is an assignment of labels $f(v)$ from the set $\{0, 1, \ldots, \lambda\}$ of integers to the vertices $v \in V$ such that, for any t with $0 \leq t \leq k$, $|f(v) - f(u)| \geq \alpha_t$ whenever the path distance between u and v is t.
  The *radio frequency assignment problem*, where vertices are transmitters (available channels) and labels represent frequencies of not-interfering channels, consists of minimizing λ. **Distance-two labeling** is the main interesting case $\alpha = (2, 1)$; its *span* is the difference between the largest and smallest labels used.
- **Distance-related graph embedding**
  An *embedding* of the guest graph $G = (V_1, E_1)$ into the host graph $H = (V_2, E_2)$ with $|V_1| \leq |V_2|$, is an injective map from V_1 into V_2.
  The **wire length**, *dilation* and *antidilation* of G in H are

$$\min_f \sum_{(uv) \in E_1} d_H(f(u), f(v)), \ \min_f \max_{(uv) \in E_1} d_H(f(u), f(v)), \ \max_f \min_{(uv) \in E_1} d_H(f(u), f(v)),$$

  respectively, where f is any embedding of G into H. The main **distance-related graph embedding** problems consist of finding or estimating these three parameters.

The *bandwidth* and *antibandwidth* of G is the dilation and antidilation, respectively, of G in a path H with V_1 vertices.

- **Bandwidth of a graph**
Given a graph $G = (V, E)$ with $|V| = n$, its *ordering* is a bijective mapping $f : V \to \{1, \dots, n\}$. Given a number $b > 0$, the *bandwidth problem* for (G, b) is the existence of ordering f with the *stretch* $\max_{uv \in E} |f(u) - f(v)|$ at most b.
The **bandwidth** of G, denoted by $bw(G)$, is the minimum stretch over all f.
The *antibandwidth problem* for G is to find ordering f with maximal $\min_{uv \in E} |f(u) - f(v)|$ (*antibandwidth*).
- **Path distance width of a graph**
Given a connected graph $G = (V, E)$, an ordered partition $V = \cup_{i=1}^{t} L_i$ of its vertices is called a *distance structure* on G if $L_i = \{v \in V : \min_{u \in L_1} d_{\text{path}}(u, v) = i - 1\}$ for $1 \le i \le t$. The structure is *rooted* if $|L_1| = 1$.
The **path distance width** $pwd(G)$ of G is defined (Yamazaki et al., 1999) as $\min \max_{1 \le i \le t} |L_i|$ over all distance structures on G.
An ordered partition $V = \cup_{i=1}^{t} L_i$ is called a *level structure* on G if for each edge uv with $u \in L_i$ and $v \in L_j$, it holds that $|i - j| \le 1$. The *level width* (or *strong pathwidth*) $lw(G)$ is $\min \max_{1 \le i \le t} |L_i|$ over all level structures.
Clearly, $lw(G) \le pdw(G)$. Yamazaki et al., 1999, proved that $pdw(G)$ can be arbitrarily larger than the **bandwidth** $bw(G)$ and $lw(G) \le bw(G) < 2lw(G)$.
- **Tree-length of a graph**
A *tree decomposition* of a graph $G = (V, E)$ is a pair of a tree T with vertex-set W and a family of subsets $\{X_i : i \in W\}$ of V with $\cup_{i \in W} X_i = V$ such that

  1. for every edge $(uv) \in E$, there is a subset X_i containing u, v, and
  2. for every $v \in V$, the set $i \in W : v \in X_i$ induces a connected subtree of T.

  The *chordal graphs* (i.e., ones without induced cycles of length at least 4) are exactly those admitting a tree decomposition where every X_i is a clique.
For tree decomposition, the *tree-length* is $\max_{i \in W} diam(X_i)$ ($diam(X_i)$ is the diameter of the subgraph of G induced by X_i) and *tree-width* is $\max_{i \in W} |X_i| - 1$.
The **tree-length** of G (Dourisboure–Gavoille, 2004) and its **tree-width** (Robertson–Seymour, 1986) are the minima, over all tree decompositions, of above tree-length and tree-width. The *path-length* G is defined taking as trees only paths.
Given a linear ordering $e_1, \dots, e_{|E|}$ of the edges of G, let, for $1 \le i < |E|$, denote by $G_{\le i}$ and $G_{i<}$ the graphs induced by the edges $\{e_1, \dots, e_i\}$ and $\{e_{i+1}, \dots, e_{|E|}\}$, respectively. The *linear-length* is $\max_{1 \le i < |E|} diam(V(G_{\le i}) \cap V(G_{i<}))$. The **linear-length** of G (Umezawa–Yamazaki, 2009) is the minimum of the above linear-length taken over all the linear orderings of its edges.
- **Spatial graph**
A **spatial graph** (or *spatial network*) is a graph $G = (V, E)$, where each vertex v has a spatial position $(v_1, \dots, v_n) \in \mathbb{R}^n$. ($G$ is called a *geometric graph* if it is drawn on $\mathbb{R}^2$ and its edges are straight-line segments.)
The *graph-theoretic dilation* and *geometric dilation* of G are, respectively:

$$\max_{v,u \in V} \frac{d(v,u)}{||v-u||_2} \text{ and } \max_{(vu) \in E} \frac{d(v,u)}{||v-u||_2}.$$

- **Distance Geometry problem**
  Given a weighted finite graph $G = (V, E; w)$, the **Distance Geometry problem** (DGP) is the problem of realizing it as a **spatial graph** $G = (V', E')$, where $x : V \to V'$ is a bijection with $x(v) = (v_1, \dots, v_n) \in \mathbb{R}^n$ for every $v \in V$ and $E' = \{(x(u)x(v)) : (uv) \in E\}$, so that for every edge $(uv) \in E$ it holds that

$$||x(u) - x(u)||_2 = w(uv).$$

  The main application of DGP is the *molecular DGP*: to find the coordinates of the atoms of a given molecular conformation are by exploiting only some of the distances between pairs of atoms found experimentally; cf. [MLLM13].
- **Arc routing problems**
  Given a finite set X, a quasi-distance $d(x, y)$ on it and a set $A \subseteq \{(x, y) : x, y \in X\}$, consider the weighted digraph $D = (X, A)$ with the vertex-set X and arc-weights $d(x, y)$ for all arcs $(x, y) \in A$. For given sets V of vertices and E of arcs, the **arc routing problem** consists of finding a *shortest* (i.e., with minimal sum of weights of its arcs) (V, E)*-tour*, i.e., a circuit in $D = (X, A)$, visiting each vertex in V and each arc in E exactly once or, in a variation, at least once.
  The *Asymmetric Traveling Salesman problem* corresponds to the case $V = X$, $E = \emptyset$; the *Traveling Salesman problem* is the symmetric version of it (usually, each vertex should be visited exactly once). The *Bottleneck Traveling Salesman problem* consists of finding a (V, E)-tour T with smallest $\max_{(x,y) \in T} d(x, y)$.
  The *Windy Postman problem* corresponds to the case $V = \emptyset$, $E = A$, while the Chinese Postman problem is the symmetric version of it.
  The above problems are also considered for general arc- or edge-weights; then, for example, the term *Metric TSP* is used when edge-weights in the Traveling Salesman problem satisfy the triangle inequality, i.e., d is a quasi-semimetric.
- **Steiner distance of a set**
  The **Steiner distance of a set** $S \subset V$ of vertices in a connected graph $G = (V, E)$ is (Chartrand et al., 1989) the minimum *size* (number of edges) of a connected subgraph of G, containing S. Such a subgraph is a tree, and is called a *Steiner tree* for S. Cf. general **Steiner diversity** in **Steiner ratio** (Chap. 1).
  The Steiner distance of the set $S = \{u, v\}$ is the path metric between u and v. The *Steiner* k*-diameter* of G is the maximum Steiner distance of any k-subset of V.
- t**-Spanner**
  A *factor*, i.e., a spanning subgraph, $H = (V, E(H))$ of a connected graph $G = (V, E)$ is called a t**-spanner** (or t*-multiplicative spanner*) of G if, for every $u, v \in V$, the inequality $d^H_{\text{path}}(u, v)/d^G_{\text{path}}(u, v) \leq t$ holds. The value t is called the *stretch factor* (or *dilation*) of H. Cf. **distance-related graph embedding** and **spatial graph**.
  The graph $H = (V, E(H))$ is called a k*-additive spanner* of G if, for every $u, v \in V$, the inequality $d^H_{\text{path}}(u, v) \leq d^G_{\text{path}}(u, v) + k$ holds.

Mulder and Nebeský, 2012, defined, for connected H, the *guide* of (H, G) as the ternary relation $R \subset V \times V \times V$ consisting of ordered triples (u, w, v) such that $uw \in E$ and $d^H_{\text{path}}(u, w) + d^H_{\text{path}}(w, v) = d^H_{\text{path}}(u, v)$. The guide of (G, G) is called the *step* ternary relation; cf. **metric betweenness** in Chap. 1.

- **Optimal realization of metric space**
Given a finite metric space (X, d), a *realization* of it is a weighted graph $G = (V, E; w)$ with $X \subset V$ such that $d(x, y) = d_G(x, y)$ holds for all $x, y \in X$. The realization is **optimal** if it has minimal $\sum_{(uv)\in E} w(uv)$.
- **Proximity graph**
Given a finite subset V of a metric space (X, d), its **proximity graph** is a graph representing neighbor relationships between points of V. Such graphs are used in Computational Geometry and many real-world problems. The main examples are presented below. Cf. **underlying graph of a metric space** in Chap. 1.
A *spanning tree* of V is a set T of $|V|-1$ unordered pairs (x, y) of different points of V forming a tree on V; the *weight* of T is $\sum_{(x,y)\in T} d(x, y)$. A **minimum spanning tree** $MST(V)$ of V is a spanning tree with the minimal weight. Such a tree is unique if the edge-weights are distinct.
A **nearest neighbor graph** is the digraph $NNG(V) = (V, E)$ with vertex-set $V = v_1, \dots, v_{|V|}$ and, for $x, y \in V$, $xy \in E$ if y is the *nearest neighbor* of x, i.e., $d(x, y) = \min_{v_i \in V\setminus\{x\}} d(x, v_i)$ and only v_i with maximal index i is picked. The *k-nearest neighbor graph* arises if k such v_i with maximal indices are picked. The undirect version of $NNG(V)$ is a subgraph of $MST(V)$.
A **relative neighborhood graph** is (Toussaint, 1980) the graph $RNG(V) = (V, E)$ with vertex-set V and, for $x, y \in V$, $xy \in E$ if there is no point $z \in V$ with $\max\{d(x, z), d(y, z)\} < d(x, y)$. Also considered, for $(X, d) = (\mathbb{R}^2, ||x - y||_2)$, the related *Gabriel graph* $GG(V)$ (in general, *β-skeleton*) and *Delaunay triangulation* $DT(V)$; then $NNG(V) \subseteq MST(V) \subseteq RNG(V) \subseteq GG(V) \subseteq DT(V)$.
For any $x \in V$, its *sphere of influence* is the open metric ball $B(x, r_x) = \{z \in X : d(x, z) < r\}$ in (X, d) centered at x with radius $r_x = \min_{z\in V\setminus\{x\}} d(x, z)$. **Sphere of influence graph** is the graph $SIG(V) = (V, E)$ with vertex-set V and, for $x, y \in V$, $xy \in E$ if $B(x, r_x) \cap B(y, r_y) \neq \emptyset$; so, it is a proximity graph and an *intersection graph*. The *closed sphere of influence graph* is the graph $CSIG(V) = (V, E)$ with $xy \in E$ if $\overline{B(x, r_x)} \cap \overline{B(y, r_y)} \neq \emptyset$.

15.3 Distances on Graphs

- **Chartrand–Kubicki–Schultz distance**
The **Chartrand–Kubicki–Schultz distance** (or *ϕ-distance*, 1998) between two connected graphs $G_1 = (V_1, E_1)$ and $G_2 = (V_2, E_2)$ with $|V_1| = |V_2| = n$ is

$$\min\{\sum |d_{G_1}(u, v) - d_{G_2}(\phi(u), \phi(v))|\},$$

where d_{G_1}, d_{G_2} are the path metrics of graphs G_1, G_2, the sum is taken over all unordered pairs u, v of vertices of G_1, and the minimum is taken over all bijections $\phi : V_1 \to V_2$.

- **Subgraph metric**
Let $\mathbb{F} = \{F_1 = (V_1, E_1), F_2 = (V_2, E_2), \dots,\}$ be the set of isomorphism classes of finite graphs. Given a finite graph $G = (V, E)$, denote by $s_i(G)$ the number of *injective homomorphisms* from F_i into G, i.e., the number of injections $\phi : V_i \to V$ with $\phi(x)\phi(y) \in E$ if $xy \in E_i$ divided by the number $\frac{|V|!}{(|V|-|V_i|)!}$ of such injections from F_i with $|V_i| \leq |V|$ into $K_{|V|}$. Set $s(G) = (s_i(G))_{i=1}^{\infty} \in [0,1]^{\infty}$. Let d be the **Cantor metric** (cf. Chap. 18) $d(x, y) = \sum_{i=1}^{\infty} 2^{-i}|x_i - y_i|$ on $[0,1]^{\infty}$ or any metric on $[0,1]^{\infty}$ inducing the *product topology*. Then Bollobás–Riordan, 2007, defined the **subgraph metric** between the graphs G_1 and G_2 as
$$d(s(G_1), s(G_2))$$
and generalized it on *kernels* (or *graphons*), i.e., symmetric measurable functions $k : [0,1] \times [0,1] \to \mathbb{R}_{\geq 0}$, replacing G by k and the above $s_i(G)$ by
$$s_i(k) = \int_{[0,1]^{|V_i|}} \prod_{st \in E_i} k(x_s x_t) \prod_{s=1}^{|V_i|} dx_s.$$

- **Benjamini–Schramm metric**
The rooted graphs (G, o) and (G', o') (where $G = (V, E), G' = (V', E')$ and $o \in V, o' \in V'$) are *isomorphic* is there is a graph-isomorphism of G onto G' taking o to o'. Let X be the set of isomorphism classes of rooted connected locally finite graphs and let $(G, o), (G', o')$ be representatives of two classes.
Let k be the supremum of all radii r, for which rooted **metric balls** $(\overline{B}_G(o, r), o)$ and $(\overline{B}_{G'}(o', r), o')$ (in the usual **path metric**) are isomorphic as rooted graphs. Benjamini and Schramm, 2001, defined the metric 2^{-k} between classes represented by (G, o) and (G', o'). Here $2^{-\infty}$ means 0. Benjamini and Curien, 2011, defined the similar distance $\frac{1}{1+k}$.
- **Rectangle distance on weighted graphs**
Let $G = G(\alpha, \beta)$ be a complete weighted graph on $\{1, \dots, n\}$ with vertex-weights $\alpha_i > 0$, $1 \leq i \leq n$, and edge-weights $\beta_{ij} \in \mathbb{R}$, $1 \leq i < j \leq n$. Denote by $A(G)$ the $n \times n$ matrix $((a_{ij}))$, where $a_{ij} = \frac{\alpha_i \alpha_j \beta_{ij}}{(\sum_{1 \leq i \leq n} \alpha_i)^2}$.
The **rectangle distance** (or *cut distance*) between two weighted graphs $G = G(\alpha, \beta)$ and $G' = G(\alpha', \beta')$ (with vertex-weights (α'_i) and edge-weights (β'_{ij})) is defined (Borgs–Chayes–Lovász–Sós–Vesztergombi, 2007) by
$$\max_{I, J \subset \{1, \dots, n\}} \left| \sum_{i \in I, j \in J} (a_{ij} - a'_{ij}) \right| + \sum_{i=1}^{n} \left| \frac{\alpha_i}{\sum_{1 \leq j \leq n} \alpha_j} - \frac{\alpha'_i}{\sum_{1 \leq j \leq n} \alpha'_j} \right|,$$
where $A(G) = ((a_{ij}))$ and $A(G') = ((a'_{ij}))$.

In the case $(\alpha'_i) = (\alpha_i)$, the rectangle distance is $||A(G) - A(G')||_{cut}$, i.e., the **cut norm metric** (cf. Chap. 12) between matrices $A(G)$ and $A(G')$ and the *rectangle distance* from Frieze–Kannan, 1999. In this case, the l_1- and l_2-metrics between two weighted graphs G and G' are defined as $||A(G) - A(G')||_1$ and $||A(G) - A(G')||_2$, respectively. The subcase $\alpha_i = 1$ for all $1 \leq i \leq n$ corresponds to unweighted vertices. Cf. the **Robinson–Foulds weighted metric**.

Authors generalized the rectangle distance on *kernels* (or *graphons*), i.e., symmetric measurable functions $k : [0,1] \times [0,1] \to \mathbb{R}_{\geq 0}$, using the *cut norm* $||k||_{cut} = \sup_{S,T \subset [0,1]} |\int_{S \times T} k(x,y)dxdy|$.

A map $\phi : [0,1] \to [0,1]$ is *measure-preserving* if, for any measurable subset $A \subset [0,1]$, the measures of A and $\phi^{-1}(A)$ are equal. For a kernel k, define the kernel k^{ϕ} by $k^{\phi}(x,y) = k(\phi(x), \phi(y))$. The **Lovász–Szegedy semimetric** (2007) between kernels k_1 and k_1 is defined by

$$\inf_{\phi} ||k_1^{\phi} - k_2||_{cut},$$

where ϕ ranges over all measure-preserving bijections $[0,1] \to [0,1]$. Cf. **Chartrand–Kubicki–Schultz distance**.

- **Subgraph-supergraph distances**

A *common subgraph* of graphs G_1 and G_2 is a graph which is isomorphic to induced subgraphs of both G_1 and G_2. A *common supergraph* of graphs G_1 and G_2 is a graph which contains induced subgraphs isomorphic to G_1 and G_2.

The **Zelinka distance** d_Z [Zeli75] on the set **G** of all graphs (more exactly, on the set of all equivalence classes of isomorphic graphs) is defined by

$$d_Z = \max\{n(G_1), n(G_2)\} - n(G_1, G_2)$$

for any $G_1, G_2 \in \mathbf{G}$, where $n(G_i)$ is the number of vertices in G_i, $i = 1, 2$, and $n(G_1, G_2)$ is the maximum number of vertices of their common subgraph.

The **Bunke–Shearer metric** (1998) on the set of nonempty graphs is defined by

$$1 - \frac{n(G_1, G_2)}{\max\{n(G_1), n(G_2)\}}.$$

Given any set **M** of graphs, the **common subgraph distance** d_M on **M** is

$$\max\{n(G_1), n(G_2)\} - n(G_1, G_2),$$

and the **common supergraph distance** d_M^* is defined, for any $G_1, G_2 \in \mathbf{M}$, by

$$N(G_1, G_2) - \min\{n(G_1), n(G_2)\},$$

where $n(G_i)$ is the number of vertices in G_i, $i = 1, 2$, while $n(G_1, G_2)$ and $N(G_1, G_2)$ are the maximal order of a common subgraph $G \in \mathbf{M}$ and the minimal order of a common supergraph $H \in \mathbf{M}$, respectively, of G_1 and G_2.

d_M is a metric on **M** if the following condition (i) holds:

(i) if $H \in \mathbf{M}$ is a common supergraph of $G_1, G_2 \in \mathbf{M}$, then there exists a common subgraph $G \in \mathbf{M}$ of G_1 and G_2 with $n(G) \geq n(G_1) + n(G_2) - n(H)$.
d_M^* is a metric on **M** if the following condition (ii) holds:

(ii) if $G \in \mathbf{M}$ is a common subgraph of $G_1, G_2 \in \mathbf{M}$, then there exists a common supergraph $H \in \mathbf{M}$ of G_1 and G_2 with $n(H) \leq n(G_1) + n(G_2) - n(G)$.

One has $d_M \leq d_M^*$ if the condition (i) holds, and $d_M \geq d_M^*$ if (ii) holds.
The distance d_M is a metric on the set **G** of all graphs, the set of all cycle-free graphs, the set of all bipartite graphs, and the set of all trees. The distance d_M^* is a metric on the set **G** of all graphs, the set of all connected graphs, the set of all connected bipartite graphs, and the set of all trees. The Zelinka distance d_Z coincides with d_M and d_M^* on the set **G** of all graphs. On the set **T** of all trees the distances d_M and d_M^* are identical, but different from the Zelinka distance.
The Zelinka distance d_Z is a metric on the set $\mathbf{G}(n)$ of all graphs with n vertices, and is equal to $n - k$ or to $K - n$ for all $G_1, G_2 \in \mathbf{G}(n)$, where k is the maximum number of vertices of a common subgraph of G_1 and G_2, and K is the minimum number of vertices of a common supergraph of G_1 and G_2.
On the set $\mathbf{T}(n)$ of all trees with n vertices the distance d_Z is called the **Zelinka tree distance** (see, for example, [Zeli75]).

- **Fernández–Valiente metric**
Given graphs G and H, let $G_1 = (V_1, E_1)$ and $G_2 = (V_2, E_2)$ be their *maximum common subgraph* and *minimum common supergraph*; cf. **subgraph-supergraph distances**. The **Fernández–Valiente metric** (2001) between G and H is

$$(|V_2| + |E_2|) - (|V_1| + |E_1|).$$

- **Graph edit distance**
The **graph edit distance** (Axenovich–Kézdy–Martin, 2008, and Alon–Stav, 2008) between graphs G and G' on the same labeled vertex-set is defined by

$$d_{ed}(G, G') = |E(G) \Delta E(G')|.$$

It is the minimum number of edge deletions or additions needed to transform G into G', and half of the Hamming distance between their adjacency matrices.
Given a *graph property* (i.e., a family $\mathcal{H}$ of graphs), let $d_{ed}(G, \mathcal{H})$ be $\min\{d_{ed}(G, G') : V(G') = V(G), G' \in \mathcal{H}\}$. Given a number $p \in (0, 1]$, the **edit distance function of a property** $\mathcal{H}$ is (if this limit exists) defined by

$$ed_{\mathcal{H}}(p) = \lim_{n \to \infty} \max \left\{ d_{ed}(G, \mathcal{H}) : |V(G)| = n, |E(G)| = \left\lfloor p \binom{n}{2} \right\rfloor \right\} \left(\binom{n}{2} \right)^{-1}.$$

If $\mathcal{H}$ is *hereditary* (closed under the taking induced subgraphs) and *nontrivial* (contains arbitrarily large graphs), then (Balogh–Martin, 2008) it holds

$$ed_{\mathcal{H}}(p) = \lim_{n\to\infty} \mathbb{E}[d_{ed}(G(n,p),\mathcal{H})]\left(\binom{n}{2}\right)^{-1};$$

$G(n, p)$ is the **random graph** (Chap. 1) on n vertices with edge probability p.
Bunke, 1997, defined the *graph edit distance* between vertex- and edge-labeled graphs G_1 and G_2 as the minimal total cost of matching G_1 and G_2, using deletions, additions and substitutions of vertices and edges. Cf. also **tree, top-down, unit cost** and **restricted edit distance** between rooted trees.
The **Bayesian graph edit distance** between two *relational graphs* (i.e., triples (V, E, A), where V, E, A are the sets of vertices, edges, *vertex-attributes*) is (Myers–Wilson–Hancock, 2000) their graph edit distance with costs defined by probabilities of operations along an editing path seen as a memoryless error process. Cf. **transduction edit distances** (Chap. 11) and **Bayesian distance** (Chap. 14).
The **structural Hamming distance** between two digraphs $G = (X, E)$ and $G' = (X, E')$ is defined (Acid–Campos, 2003) as $SHD(G, G') = |E \Delta E'|$.

- **Edge distance**
  The **edge distance** on the set of all graphs is defined (Baláž et al., 1986) by

$$|E_1| + |E_2| - 2|E_{12}| + ||V_1| - |V_2||$$

  for any graphs $G_1 = (V_1, E_1)$ and $G_2 = (V_2, E_2)$, where $G_{12} = (V_{12}, E_{12})$ is a common subgraph of G_1 and G_2 with maximal number of edges. This distance has many applications in Organic and Medical Chemistry.
- **Contraction distance**
  The **contraction distance** is a distance on the set $\mathbf{G}(n)$ of all graphs with n vertices defined by

$$n - k$$

  for any $G_1, G_2 \in \mathbf{G}(n)$, where k is the maximum number of vertices of a graph which is isomorphic simultaneously to a graph, obtained from each of G_1 and G_2 by a finite number of *edge contractions*. To perform the *contraction* of the edge $uv \in E$ of a graph $G = (V, E)$ means to replace u and v by one vertex that is adjacent to all vertices of $V\backslash\{u, v\}$ which were adjacent to u or to v.
- **Edge move distance**
  The **edge move distance** (Baláž et al., 1986) is a metric on the set $\mathbf{G}(n, m)$ of all graphs with n vertices and m edges, defined, for any $G_1, G_2 \in \mathbf{G}(m, n)$, as the minimum number of *edge moves* necessary for transforming the graph G_1 into the graph G_2. It is equal to $m - k$, where k is the maximum size of a common subgraph of G_1 and G_2.

An *edge move* is one of the *edge transformations*, defined as follows: H can be obtained from G by an edge move if there exist (not necessarily distinct) vertices u, v, w, and x in G such that $uv \in E(G)$, $wx \notin E(G)$, and $H = G - uv + wx$.

- **Edge jump distance**
  The **edge jump distance** is an extended metric (which in general can take the value ∞) on the set $\mathbf{G}(n, m)$ of all graphs with n vertices and m edges defined, for any $G_1, G_2 \in \mathbf{G}(m, n)$, as the minimum number of *edge jumps* necessary for transforming G_1 into G_2.
  An *edge jump* is one of the *edge transformations*, defined as follows: H can be obtained from G by an edge jump if there exist four distinct vertices u, v, w, and x in G, such that $uv \in E(G)$, $wx \notin E(G)$, and $H = G - av + wx$.
- **Edge flipping distance**
  Let $P = \{v_1, \dots, v_n\}$ be a collection of points on the plane. A *triangulation* T of P is a partition of the convex hull of P into a set of triangles such that each triangle has a disjoint interior and the vertices of each triangle are points of P.
  The **edge flipping distance** is a distance on the set of all triangulations of P defined, for any triangulations T and T_1, as the minimum number of edge flippings necessary for transforming T into T_1.
  An edge e of T is called *flippable* if it is the boundary of two triangles t and $t^{'}$ of T, and $C = t \cup t^{'}$ is a convex quadrilateral. The *flipping* e is one of the *edge transformations*, which consists of removing e and replacing it by the other diagonal of C. Edge flipping is an special case of *edge jump*.
  The edge flipping distance can be extended on *pseudo-triangulations*, i.e., partitions of the convex hull of P into a set of disjoint interior *pseudo-triangles* (simply connected subsets of the plane that lie between any three mutually tangent convex sets) whose vertices are given points.
- **Edge rotation distance**
  The **edge rotation distance** (Chartand–Saba–Zou, 1985) is a metric on the set $\mathbf{G}(n, m)$ of graphs with n vertices and m edges, defined, for any G_1, G_2, as the minimum number of *edge rotations* needed for transforming G_1 into G_2.
  An *edge rotation* is one of the *edge transformations*, defined as follows: H can be obtained from G by an edge rotation if there exist distinct vertices u, v, and w in G, such that $uv \in E(G)$, $uw \notin E(G)$, and $H = G - uv + uw$.
- **Tree edge rotation distance**
  The **tree edge rotation distance** is a metric on the set $\mathbf{T}(n)$ of all trees with n vertices defined, for all $T_1, T_2 \in \mathbf{T}(n)$, as the minimum number of *tree edge rotations* necessary for transforming T_1 into T_2. A *tree edge rotation* is an *edge rotation* performed on a tree, and resulting in a tree.
  For $\mathbf{T}(n)$ the tree edge rotation and the edge rotation distances may differ.
- **Edge shift distance**
  The **edge shift distance** (or **edge slide distance**) is a metric (Johnson, 1985) on the set $\mathbf{G}_c(n, m)$ of all connected graphs with n vertices and m edges defined, for any $G_1, G_2 \in \mathbf{G}_c(m, n)$, as the minimum number of *edge shifts* necessary for transforming G_1 into G_2.

An *edge shift* is one of the *edge transformations*, defined as follows: H can be obtained from G by an edge shift if there exist distinct vertices u, v, and w in G such that $uv, vw \in E(G)$, $uw \notin E(G)$, and $H = G - uv + uw$. Edge shift is a special kind of *edge rotation* in the case when the vertices v, w are adjacent in G. The edge shift distance can be defined between any graphs G and H with components $G_i (1 \le i \le k)$ and $H_i (1 \le i \le k)$, respectively, such that G_i and H_i have the same order and the same size.

- **F-rotation distance**
  The **F-rotation distance** is a distance on the set $\mathbf{G}_F(n, m)$ of all graphs with n vertices and m edges, containing a subgraph isomorphic to a given graph F of order at least 2 defined, for all $G_1, G_2 \in \mathbf{G}_F(m, n)$, as the minimum number of *F-rotations* necessary for transforming G_1 into G_2.
  An *F-rotation* is one of the *edge transformations*, defined as follows: let F' be a subgraph of a graph G, isomorphic to F, let u, v, w be three distinct vertices of the graph G such that $u \notin V(F')$, $v, w \in V(F')$, $uv \in E(G)$, and $uw \notin E(G)$; H can be obtained from G by the F-rotation of the edge uv into the position uw if $H = G - uv + uw$.
- **Binary relation distance**
  Let R be a nonreflexive *binary relation* between graphs, i.e., $R \subset \mathbf{G} \times \mathbf{G}$, and there exists $G \in \mathbf{G}$ such that $(G, G) \notin R$.
  The **binary relation distance** is a metric (which can take the value ∞) on the set $\mathbf{G}$ of all graphs defined, for any graphs G_1 and G_2, as the minimum number of *R-transformations* necessary for transforming G_1 into G_2. We say that a graph H can be obtained from a graph G by an *R-transformation* if $(H, G) \in R$.
  An example is the distance between two *triangular embeddings of a complete graph* (i.e., its cellular embeddings in a surface with only 3-gonal faces) defined as the minimal number t such that, up to replacing t faces, the embeddings are isomorphic.
- **Crossing-free transformation metrics**
  Given a subset S of $\mathbb{R}^2$, a *noncrossing spanning tree* of S is a tree whose vertices are points of S, and edges are pairwise noncrossing straight line segments.
  The **crossing-free edge move metric** (see [AAH00]) on the set $\mathbf{T}_S$ of all noncrossing spanning trees of a set S, is defined, for any $T_1, T_2 \in \mathbf{T}_S$, as the minimum number of *crossing-free edge moves* needed to transform T_1 into T_2. Such move is an edge transformation which consists of adding some edge e in $T \in \mathbf{T}_S$ and removing some edge f from the induced cycle so that e and f do not cross.
  The **crossing-free edge slide metric** is a metric on the set $\mathbf{T}_S$ of all *noncrossing spanning trees* of a set S defined, for any $T_1, T_2 \in \mathbf{T}_S$, as the minimum number of *crossing-free edge slides* necessary for transforming T_1 into T_2. Such slide is one of the edge transformations which consists of taking some edge e in $T \in \mathbf{T}_S$ and moving one of its endpoints along some edge adjacent to e in T, without introducing edge crossings and without sweeping across points in S (that gives a new edge f instead of e). The edge slide is a special kind of crossing-free edge

move: the new tree is obtained by closing with f a cycle C of length 3 in T, and removing e from C, in such a way that f avoids the interior of the triangle C.

- **Traveling salesman tours distances**
  The *Traveling Salesman problem* is the problem of finding the shortest tour that visits a set of cities. We will consider only Traveling Salesman problem with undirected links. For an n-city traveling salesman problem, the space $\mathcal{T}_n$ of tours is the set of $\frac{(n-1)!}{2}$ cyclic permutations of the cities $1, 2, \dots, n$.
  The metric D on $\mathcal{T}_n$ is defined in terms of the difference in form: if tours $T, T' \in \mathcal{T}_n$ differ in m links, then $D(T, T') = m$.
  A *k-OPT transformation* of a tour T is obtained by deleting k links from T, and reconnecting. A tour T', obtained from T by a k-OPT transformation, is called a *k-OPT* of T. The distance d on the set $\mathcal{T}_N$ is defined in terms of the 2-OPT transformations: $d(T, T')$ is the minimal i, for which there exists a sequence of i 2-OPT transformations which transforms T to T'. In fact, $d(T, T') \leq D(T, T')$ for any $T, T' \in \mathcal{T}_N$ (see, for example, [MaMo95]). Cf. **arc routing problems**.
- **Orientation distance**
  The **orientation distance** (Chartrand–Erwin–Raines–Zhang, 2001) between two orientations D and D' of a finite graph is the minimum number of arcs of D whose directions must be reversed to produce an orientation isomorphic to D'.
- **Subgraphs distances**
  The standard distance on the set of all subgraphs of a connected graph $G = (V, E)$ is defined by

$$\min\{d_{\text{path}}(u, v) : u \in V(F), v \in V(H)\}$$

  for any subgraphs F, H of G. For any subgraphs F, H of a strongly connected digraph $D = (V, E)$, the standard quasi-distance is defined by

$$\min\{d_{dpath}(u, v) : u \in V(F), v \in V(H)\}.$$

  Using standard operations (rotation, shift, etc.) on the edge-set of a graph, one gets corresponding distances between its edge-induced subgraphs of given size which are subcases of similar distances on the set of all graphs of a given size and order.
  The **edge rotation distance** on the set $\mathbf{S}^k(G)$ of all edge-induced subgraphs with k edges in a connected graph G is defined as the minimum number of *edge rotations* required to transform $F \in \mathbf{S}^k(G)$ into $H \in \mathbf{S}^k(G)$. We say that H can be obtained from F by an *edge rotation* if there exist distinct vertices u, v, and w in G such that $uv \in E(F)$, $uw \in E(G) \setminus E(F)$, and $H = F - uv + uw$.
  The **edge shift distance** on the set $\mathbf{S}^k(G)$ of all edge-induced subgraphs with k edges in a connected graph G is defined as the minimum number of *edge shifts* required to transform $F \in \mathbf{S}^k(G)$ into $H \in \mathbf{S}^k(G)$. We say that H can be obtained from F by an *edge shift* if there exist distinct vertices u, v and w in G such that $uv, vw \in E(F)$, $uw \in E(G) \setminus E(F)$, and $H = F - uv + uw$.

The **edge move distance** on the set $\mathbf{S}^k(G)$ of all edge-induced subgraphs with k edges of a graph G (not necessary connected) is defined as the minimum number of *edge moves* required to transform $F \in \mathbf{S}^k(G)$ into $H \in \mathbf{S}^k(G)$. We say that H can be obtained from F by an *edge move* if there exist (not necessarily distinct) vertices u, v, w, and x in G such that $uv \in E(F)$, $wx \in E(G)\backslash E(F)$, and $H = F - uv + wx$. The edge move distance is a metric on $\mathbf{S}^k(G)$. If F and H have s edges in common, then it is equal to $k - s$.

The **edge jump distance** (which in general can take the value ∞) on the set $\mathbf{S}^k(G)$ of all edge-induced subgraphs with k edges of a graph G (not necessary connected) is defined as the minimum number of *edge jumps* required to transform $F \in \mathbf{S}^k(G)$ into $H \in \mathbf{S}^k(G)$. We say that H can be obtained from F by an *edge jump* if there exist four distinct vertices u, v, w, and x in G such that $uv \in E(F)$, $wx \in E(G)\backslash E(F)$, and $H = F - uv + wx$.

15.4 Distances on Trees

Let T be a *rooted tree*, i.e., a tree with one of its vertices being chosen as the *root*. The *depth* of a vertex v, *depth*(v), is the number of edges on the path from v to the root. A vertex v is called a *parent* of a vertex u, $v = par(u)$, if they are adjacent, and $depth(u) = depth(v) + 1$; in this case u is called a *child* of v. A *leaf* is a vertex without child. Two vertices are *siblings* if they have the same parent.

The *in-degree* of a vertex is the number of its children. $T(v)$ is the subtree of T, rooted at a node $v \in V(T)$. If $w \in V(T(v))$, then v is an *ancestor* of w, and w is a *descendant* of v; $nca(u, v)$ is the *nearest common ancestor* of the vertices u and v.

T is called a *labeled tree* if a symbol from a fixed finite alphabet $\mathcal{A}$ is assigned to each node. T is called an *ordered tree* if a left-to-right order among siblings in T is given. On the set $\mathbb{T}_{rlo}$ of all rooted labeled ordered trees there are three *editing operations*:

- *Relabel*—change the label of a vertex v;
- *Deletion*—delete a nonrooted vertex v with parent v', making the children of v become the children of v'; the children are inserted in the place of v as a subsequence in the left-to-right order of the children of v';
- *Insertion*—the complement of deletion; insert a vertex v as a child of a v' making v the parent of a consecutive subsequence of the children of v'.

For unordered trees above operations (and so, distances) are defined similarly, but the insert and delete operations work on a subset instead of a subsequence.

We assume that there is a *cost function* defined on each editing operation, and the *cost* of a sequence of editing operations is the sum of the costs of these operations.

The *ordered edit distance mapping* is a representation of the editing operations. Formally, the triple (M, T_1, T_2) is an *ordered edit distance mapping* from T_1 to T_2, $T_1, T_2 \in \mathbb{T}_{rlo}$, if $M \subset V(T_1) \times V(T_2)$ and, for any $(v_1, w_1), (v_2, w_2) \in M$, the following conditions hold: $v_1 = v_2$ if and only if $w_1 = w_2$ (*one-to-one condition*),

v_1 is an ancestor of v_2 if and only if w_1 is an ancestor of w_2 (*ancestor condition*), v_1 is to the left of v_2 if and only if w_1 is to the left of w_2 (*sibling condition*).

We say that a vertex v in T_1 and T_2 is *touched by a line* in M if v occurs in some pair in M. Let N_1 and N_2 be the set of vertices in T_1 and T_2, respectively, not touched by any line in M. The *cost* of M is given by $\gamma(M) = \sum_{(v,w)\in M} \gamma(v \to w) + \sum_{v\in N_1} \gamma(v \to \lambda) + \sum_{w\in N_2} \gamma(\lambda \to w)$, where $\gamma(a \to b) = \gamma(a,b)$ is the *cost* of an editing operation $a \to b$ which is a relabel if $a, b \in \mathcal{A}$, a deletion if $b = \lambda$, and an insertion if $a = \lambda$. Here $\lambda \notin \mathcal{A}$ is a special *blank symbol*, and γ is a metric on the set $\mathcal{A} \cup \lambda$ (excepting the value $\gamma(\lambda, \lambda)$).

- **Tree edit distance**
  The **tree edit distance** (see [Tai79]) on the set $\mathbb{T}_{rlo}$ of all rooted labeled ordered trees is defined, for any $T_1, T_2 \in \mathbb{T}_{rlo}$, as the minimum cost of a sequence of editing operations (relabels, insertions, and deletions) turning T_1 into T_2.
  In terms of ordered edit distance mappings, it is equal to $\min_{(M,T_1,T_2)} \gamma(M)$, where the minimum is taken over all such mappings (M, T_1, T_2).
  The **unit cost edit distance** between T_1 and T_2 is the minimum number of three above editing operations turning T_1 into T_2, i.e., it is the tree edit distance with cost 1 of any operation.
- **Selkow distance**
  The **Selkow distance** (or *top-down edit distance*, *degree-*1 *edit distance*) is a distance on the set $\mathbb{T}_{rlo}$ of all rooted labeled ordered trees defined, for any $T_1, T_2 \in \mathbb{T}_{rlo}$, as the minimum cost of a sequence of editing operations (relabels, insertions, and deletions) turning T_1 into T_2 if insertions and deletions are restricted to leaves of the trees (see [Selk77]).
  The root of T_1 must be mapped to the root of T_2, and if a node v is to be deleted (inserted), then any subtree rooted at v is to be deleted (inserted).
  In terms of ordered edit distance mappings, it is equal to $\min_{(M,T_1,T_2)} \gamma(M)$, where the minimum is taken over all such mappings (M, T_1, T_2) such that $(par(v), par(w)) \in M$ if $(v, w) \in M$, where neither v nor w is the root.
- **Restricted edit distance**
  The **restricted edit distance** is a distance on the set $\mathbb{T}_{rlo}$ of all rooted labeled ordered trees defined, for any $T_1, T_2 \in \mathbb{T}_{rlo}$, as the minimum cost of a sequence of editing operations (relabels, insertions, and deletions) turning T_1 into T_2 with the restriction that disjoint subtrees should be mapped to disjoint subtrees.
  In terms of ordered edit distance mappings, it is equal to $\min_{(M,T_1,T_2)} \gamma(M)$, where the minimum is taken over all such mappings (M, T_1, T_2) satisfying the following condition: for all (v_1, w_1), (v_2, w_2), $(v_3, w_3) \in M$, $nca(v_1, v_2)$ is a proper ancestor of v_3 if and only if $nca(w_1, w_2)$ is a proper ancestor of w_3.
  This distance is equivalent to the *structure respecting edit distance* which is defined by $\min_{(M,T_1,T_2)} \gamma(M)$. Here the minimum is taken over all ordered edit distance mappings (M, T_1, T_2), satisfying the following condition: for all (v_1, w_1), (v_2, w_2), $(v_3, w_3) \in M$, such that none of v_1, v_2, and v_3 is an ancestor of the others, $nca(v_1, v_2) = nca(v_1, v_3)$ if and only if $nca(w_1, w_2) = nca(w_1, w_3)$.
  Cf. **constrained edit distance** in Chap. 11.

- **Alignment distance**
  The **alignment distance** (see [JWZ94]) is a distance on the set $\mathbb{T}_{rlo}$ of all rooted labeled ordered trees defined, for any $T_1, T_2 \in \mathbb{T}_{rlo}$, as the minimum *cost* of an *alignment* of T_1 and T_2. It corresponds to a restricted edit distance, where all insertions must be performed before any deletions.
  Thus, one inserts *spaces*, i.e., vertices labeled with a *blank symbol* λ, into T_1 and T_2 so that they become isomorphic when labels are ignored; the resulting trees are overlaid on top of each other giving the *alignment* $T_{\mathcal{A}}$ which is a tree, where each vertex is labeled by a pair of labels. The *cost* of $T_{\mathcal{A}}$ is the sum of the costs of all pairs of opposite labels in $T_{\mathcal{A}}$.
- **Splitting-merging distance**
  The **splitting-merging distance** (see [ChLu85]) is a distance on the set $\mathbb{T}_{rlo}$ of all rooted labeled ordered trees defined, for any $T_1, T_2 \in \mathbb{T}_{rlo}$, as the minimum number of vertex splittings and mergings needed to transform T_1 into T_2.
- **Degree-2 distance**
  The **degree-2 distance** is a metric on the set $\mathbb{T}_l$ of all labeled trees (*labeled free trees*), defined, for any $T_1, T_2 \in \mathbb{T}_l$, as the minimum number of editing operations (relabels, insertions, and deletions) turning T_1 into T_2 if any vertex to be inserted (deleted) has no more than two neighbors. This metric is a natural extension of the **tree edit distance** and the **Selkow distance**.

A *phylogenetic X-tree* is an unordered unrooted tree with the labeled leaf set X and no vertices of degree two. If every interior vertex has degree three, the tree is called *binary*. Let $\mathbb{T}(X)$ denote the set of all phylogenetic X-trees.

- **Robinson–Foulds metric**
  A *cut* $A|B$ of X is a *partition* of X into two subsets A and B (see **cut semimetric**). Removing an edge e from a phylogenetic X-tree induces a cut of the leaf set X which is called the *cut associated with* e.
  The **Robinson–Foulds metric** (or *Bourque metric*, *bipartition distance*) is a metric on the set $\mathbb{T}(X)$, defined, for any phylogenetic X-trees $T_1, T_2 \in \mathbb{T}(X)$, by

$$\frac{1}{2}|\Sigma(T_1)\triangle\Sigma(T_2)| = \frac{1}{2}|\Sigma(T_1) \setminus \Sigma(T_2)| + \frac{1}{2}|\Sigma(T_2) \setminus \Sigma(T_1)|,$$

  where $\Sigma(T)$ is the collection of all cuts of X associated with edges of T.
  The **Robinson–Foulds weighted metric** is a metric on the set $\mathbb{T}(X)$ of all phylogenetic X-trees defined by

$$\sum_{A|B\in\Sigma(T_1)\cup\Sigma(T_2)} |w_1(A|B) - w_2(A|B)|$$

  for all $T_1, T_2 \in \mathbb{T}(X)$, where $w_i = (w(e))_{e\in E(T_i)}$ is the collection of positive weights, associated with the edges of the X-tree T_i, $\Sigma(T_i)$ is the collection of all cuts of X, associated with edges of T_i, and $w_i(A|B)$ is the weight of the edge, corresponding to the cut $A|B$ of X, $i = 1, 2$. Cf. more general **cut norm metric** in Chap. 12 and **rectangle distance on weighted graphs**.

- **μ-metric**
  Given a phylogenetic X-tree T with n leaves and a vertex v in it, let $\mu(v) = (\mu_1(v), \dots, \mu_n(v))$, where $\mu_i(v)$ is the number of different paths from the vertex v to the i-th leaf. Let $\mu(T)$ denote the multiset on the vertex-set of T with $\mu(v)$ being the multiplicity of the vertex v.
  The **μ-metric** (Cardona–Roselló–Valiente, 2008) is a metric on the set $\mathbb{T}(X)$ of all phylogenetic X-trees defined, for all $T_1, T_2 \in \mathbb{T}(X)$, by
  $$\frac{1}{2}|\mu(T_1)\Delta\mu(T_2)|,$$
  where Δ denotes the symmetric difference of multisets.
  Cf. the **metrics between multisets** in Chap. 1 and the **Dodge–Shiode WebX quasi-distance** in Chap. 22.
- **Nearest neighbor interchange metric**
  The **nearest neighbor interchange metric** (or **crossover metric**) on the set $\mathbb{T}(X)$ of all phylogenetic X-trees, is defined, for all $T_1, T_2 \in \mathbb{T}(X)$, as the minimum number of *nearest neighbor interchanges* required to transform T_1 into T_2.
  A *nearest neighbor interchange* consists of swapping two subtrees in a tree that are adjacent to the same internal edge; the remainder of the tree is unchanged.
- **Subtree prune and regraft distance**
  The **subtree prune and regraft distance** is a metric on the set $\mathbb{T}(X)$ of all phylogenetic X-trees defined, for all $T_1, T_2 \in \mathbb{T}(X)$, as the minimum number of *subtree prune and regraft transformations* required to transform T_1 into T_2.
  A *subtree prune and regraft transformation* proceeds in three steps: one selects and removes an edge uv of the tree, thereby dividing the tree into two subtrees T_u (containing u) and T_v (containing v); then one selects and subdivides an edge of T_v, giving a new vertex w; finally, one connects u and w by an edge, and removes all vertices of degree two.
- **Tree bisection-reconnection metric**
  The **tree bisection-reconnection metric** (or **TBR-metric**) on the set $\mathbb{T}(X)$ of all phylogenetic X-trees is defined, for all $T_1, T_2 \in \mathbb{T}(X)$, as the minimum number of *tree bisection and reconnection* transformations required to transform T_1 into T_2.
  A *tree bisection and reconnection transformation* proceeds in three steps: one selects and removes an edge uv of the tree, thereby dividing the tree into two subtrees T_u (containing u) and T_v (containing v); then one selects and subdivides an edge of T_v, giving a new vertex w, and an edge of T_u, giving a new vertex z; finally one connects w and z by an edge, and removes all vertices of degree two.

- **Quartet distance**
The **quartet distance** (see [EMM85]) is a distance of the set $\mathbb{T}_b(X)$ of all binary phylogenetic X-trees defined, for all $T_1, T_2 \in \mathbb{T}_b(X)$, as the number of mismatched *quartets* (from the total number $\binom{n}{4}$ possible quartets) for T_1 and T_2.
This distance is based on the fact that, given four leaves $\{1, 2, 3, 4\}$ of a tree, they can only be combined in a binary subtree in three ways: $(12|34)$, $(13|24)$, or $(14|23)$: the notation $(12|34)$ refers to the binary tree with the leaf set $\{1, 2, 3, 4\}$ in which removing the inner edge yields the trees with the leaf sets $\{1, 2\}$ and $\{3, 4\}$.
- **Triples distance**
The **triples distance** (see [CPQ96]) is a distance of the set $\mathbb{T}_b(X)$ of all binary phylogenetic X-trees defined, for all $T_1, T_2 \in \mathbb{T}_b(X)$, as the number of triples (from the total number $\binom{n}{3}$ possible triples) that differ (for example, by which leaf is the outlier) for T_1 and T_2.
- **Perfect matching distance**
The **perfect matching distance** is a distance on the set $\mathbb{T}_{br}(X)$ of all rooted binary phylogenetic X-trees with the set X of n labeled leaves defined, for any $T_1, T_2 \in \mathbb{T}_{br}(X)$, as the minimum number of interchanges necessary to bring the perfect matching of T_1 to the perfect matching of T_2.
Given a set $A = \{1, \dots, 2k\}$ of $2k$ points, a *perfect matching* of A is a *partition* of A into k pairs. A rooted binary phylogenetic tree with n labeled leaves has a root and $n-2$ internal vertices distinct from the root. It can be identified with a perfect matching on $2n-2$, different from the root, vertices by following construction: label the internal vertices with numbers $n+1, \dots, 2n-2$ by putting the smallest available label as the parent of the pair of labeled children of which one has the smallest label among pairs of labeled children; now a matching is formed by peeling off the children, or sibling pairs, two by two.
- **Tree rotation distance**
The **tree rotation distance** is a distance on the set $\mathbf{T}_n$ of all rooted ordered binary trees with n interior vertices defined, for all $T_1, T_2 \in \mathbf{T}_n$, as the minimum number of *rotations*, required to transform T_1 into T_2.
Given interior edges uv, vv', vv'' and uw of a binary tree, the *rotation* is replacing them by edges uv, uv'', vv' and vw.
There is a bijection between edge flipping operations in triangulations of convex polygons with $n+2$ vertices and rotations in binary trees with n interior vertices.
- **Attributed tree metrics**
An *attributed tree* is a triple (V, E, α), where $T = (V, E)$ is the underlying tree, and α is a function which assigns an *attribute vector* $\alpha(v)$ to every vertex $v \in V$. Given two attributed trees (V_1, E_1, α) and (V_2, E_2, β), consider the set of all *subtree isomorphisms* between them, i.e., the set of all isomorphisms $f : H_1 \to H_2$, $H_1 \subset V_1$, $H_2 \subset V_2$, between their *induced subtrees*.
Given a similarity s on the set of attributes, the similarity between isomorphic induced subtrees is defined as $W_s(f) = \sum_{v \in H_1} s(\alpha(v), \beta(f(v)))$. Let ϕ be the isomorphism with maximal similarity $W_s(\phi) = W(\phi)$.

The following four semimetrics on the set $\mathbf{T}_{att}$ of all attributed trees are used:

$$\max\{|V_1|, |V_2|\} - W(\phi), \quad |V_1| + |V_2| - 2W(\phi) \quad \text{and}$$

$$1 - \frac{W(\phi)}{\max\{|V_1|, |V_2|\}}, \quad 1 - \frac{W(\phi)}{|V_1| + |V_2| - W(\phi)}.$$

They become metrics on the set of equivalences classes of attributed trees: two such trees (V_1, E_1, α) and (V_2, E_2, β) are called *equivalent* if they are *attribute-isomorphic*, i.e., if there exists an isomorphism $g : V_1 \to V_2$ between the trees such that, for any $v \in V_1$, we have $\alpha(v) = \beta(g(v))$. Then $|V_1| = |V_2| = W(g)$.

- **Maximal agreement subtree distance**
  The **maximal agreement subtree distance** (MAST) is a distance of the set $\mathbf{T}$ of all trees defined, for all $T_1, T_2 \in \mathbf{T}$, as the minimum number of leaves removed to obtain a *(greatest) agreement subtree*.
  An *agreement subtree* (or *common pruned tree*) of two trees is an identical subtree that can be obtained from both trees by pruning leaves with the same label.

Chapter 16
Distances in Coding Theory

Coding Theory deals with the design and properties of *error-correcting codes* for the reliable transmission of information across noisy channels in transmission lines and storage devices. The aim of Coding Theory is to find codes which transmit and decode fast, contain many valid code words, and can correct, or at least detect, many errors. These aims are mutually exclusive, however; so, each application has its own good code.

In communications, a *code* is a rule for converting a piece of information (for example, a letter, word, or phrase) into another form or representation, not necessarily of the same sort. *Encoding* is the process by which a source (object) performs this conversion of information into data, which is then sent to a receiver (observer), such as a data processing system. *Decoding* is the reverse process of converting data which has been sent by a source, into information understandable by a receiver.

An *error-correcting code* is a code in which every data signal conforms to specific rules of construction so that departures from this construction in the received signal can generally be automatically detected and corrected. It is used in computer data storage, for example in dynamic RAM, and in data transmission. Error detection is much simpler than error correction, and one or more "check" digits are commonly embedded in credit card numbers in order to detect mistakes. The two main classes of error-correcting codes are *block codes*, and *convolutional codes*.

A *block code* (or *uniform code*) of length n over an alphabet $\mathcal{A}$, usually, over a finite field $\mathbb{F}_q = \{0, \dots, q-1\}$, is a subset $C \subset \mathcal{A}^n$; every vector $x \in C$ is called a *codeword*, and $M = |C|$ is called *size* of the code. Given a metric d on $\mathbb{F}_q^n$ (for example, the **Hamming metric**, **Lee metric**, **Levenstein metric**), the value $d^* = d^*(C) = \min_{x,y \in C, x \neq y} d(x, y)$ is called the **minimum distance** of the code C. The *weight* $w(x)$ of a codeword $x \in C$ is defined as $w(x) = d(x, 0)$. An *(n, M, d^*)-code* is a q-ary block code of length n, size M, and minimum distance d^*. A *binary code* is a code over $\mathbb{F}_2$.

M.M. Deza, E. Deza, *Encyclopedia of Distances*,
DOI 10.1007/978-3-662-44342-2_16

When codewords are chosen such that the distance between them is maximized, the code is called *error-correcting*, since slightly garbled vectors can be recovered by choosing the nearest codeword. A code C is a *t-error-correcting code* (and a *$2t$-error-detecting code*) if $d^*(C) \geq 2t + 1$. In this case each *neighborhood* $U_t(x) = \{y \in C : d(x, y) \leq t\}$ of $x \in C$ is disjoint with $U_t(y)$ for any $y \in C, y \neq x$.

A *perfect code* is a q-ary $(n, M, 2t + 1)$-code for which the M spheres $U_t(x)$ of radius t centered on the codewords fill the whole space $\mathbb{F}_q^n$ completely, without overlapping.

A block code $C \subset \mathbb{F}_q^n$ is called *linear* if C is a vector subspace of $\mathbb{F}_q^n$. An $[n, k]$-code is a k-dimensional linear code $C \subset \mathbb{F}_q^n$ (with the minimum distance d^*); it has size q^k, i.e., it is an (n, q^k, d^*)-code. The *Hamming code* is the linear perfect one-error correcting $(\frac{q^r-1}{q-1}, \frac{q^r-1}{q-1} - r, 3)$-code.

A $k \times n$ matrix G with rows that are basis vectors for a linear $[n, k]$-code C is called a *generator matrix* of C. In *standard form* it can be written as $(1_k|A)$, where 1_k is the $k \times k$ identity matrix. Each *message* (or *information symbol*, *source symbol*) $u = (u_1, \dots, u_k) \in \mathbb{F}_q^k$ can be encoded by multiplying it (on the right) by the generator matrix: $uG \in C$.

The matrix $H = (-A^T|1_{n-k})$ is called the *parity-check matrix* of C. The number $r = n-k$ corresponds to the number of parity check digits in the code, and is called the *redundancy* of the code C. The *information rate* (or *code rate*) of a code C is the number $R = \frac{\log_2 M}{n}$. For a q-ary $[n, k]$-code, $R = \frac{k}{n}\log_2 q$; for a binary $[n, k]$-code, $R = \frac{k}{n}$.

A *convolutional code* is a type of error-correction code in which each k-bit information symbol to be encoded is transformed into an n-bit codeword, where $R = \frac{k}{n}$ is the code rate ($n \geq k$), and the transformation is a function of the last m information symbols, where m is the *constraint length* of the code. Convolutional codes are often used to improve the performance of radio and satellite links.

A *variable length code* is a code with codewords of different lengths.

In contrast to error-correcting codes which are designed only to increase the reliability of data communications, *cryptographic codes* are designed to increase their security. In Cryptography, the sender uses a *key* to encrypt a message before it is sent through an insecure channel, and an authorized receiver at the other end then uses a key to decrypt the received data to a message.

Often, data compression algorithms and error-correcting codes are used in tandem with cryptographic codes to yield communications that are efficient, robust to data transmission errors, and secure to eavesdropping and tampering. Encrypted messages which are, moreover, hidden in text, image, etc., are called *steganographic messages*.

The *encryption/assortment theory of humor* (Flamson–Barrett, 2008) proposes that people signal similarity in locally variable personal features through humor. In a successful joke, both the producer and the receiver share common background information—the key—and the joke is engineered in such a way (via devices such as incongruity) that there is a nonrandom fit between the surface utterance and this information that would only be apparent to a person with access to it. The function of encrypted humor is not secrecy per se, but rather, honestly indexing the presence of shared keys.

16.1 Minimum Distance and Relatives

- **Minimum distance**
  Given a code $C \subset V$, where V is an n-dimensional vector space equipped with a metric d, the **minimum distance** $d^* = d^*(C)$ of the code C is defined by

$$\min_{x,y \in C, x \neq y} d(x, y).$$

  The metric d depends on the nature of the errors for the correction of which the code is intended. For a prescribed correcting capacity it is necessary to use codes with a maximum number of codewords. Such most widely investigated codes are the *q-ary block codes* in the **Hamming metric** $d_H(x, y) = |\{i : x_i \neq y_i, i = 1, \dots, n\}|$.
  For a *linear code* C the minimum distance $d^*(C) = w(C)$, where $w(C) = \min\{w(x) : x \in C\}$ is a *minimum weight* of the code C. As there are $\text{rank}(H) \leq n - k$ independent columns in the parity check matrix H of an $[n, k]$-code C, then $d^*(C) \leq n - k + 1$ (*Singleton upper bound*).
- **Dual distance**
  The **dual distance** $d^\perp$ of a *linear* $[n, k]$-code $C \subset \mathbb{F}_q^n$ is the **minimum distance** of the *dual code* $C^\perp$ of C defined by $C^\perp = \{v \in \mathbb{F}_q^n : \langle v, u \rangle = 0$ for any $u \in C\}$.
  The code $C^\perp$ is a linear $[n, n-k]$-code, and its $(n-k) \times n$ generator matrix is the parity-check matrix of C.
- **Bar product distance**
  Given linear codes C_1 and C_2 of length n with $C_2 \subset C_1$, their *bar product* $C_1|C_2$ is a linear code of length $2n$ defined by $C_1|C_2 = \{x|x + y : x \in C_1, y \in C_2\}$.
  The **bar product distance** between C_1 and C_2 is the minimum distance $d^*(C_1|C_2)$ of their bar product $C_1|C_2$.
- **Design distance**
  A *linear code* is called a *cyclic code* if all cyclic shifts of a codeword also belong to C, i.e., if for any $(a_0, \dots, a_{n-1}) \in C$ the vector $(a_{n-1}, a_0, \dots, a_{n-2}) \in C$. It is convenient to identify a codeword $(a_0, \dots, a_{n-1})$ with the polynomial $c(x) = a_0 + a_1 x + \dots + a_{n-1} x^{n-1}$; then every cyclic $[n, k]$-code can be represented as the principal ideal $\langle g(x) \rangle = \{r(x)g(x) : r(x) \in R_n\}$ of the ring $R_n = \mathbb{F}_q(x)/(x^n - 1)$, generated by the *generator polynomial* $g(x) = g_0 + g_1 x + \dots + x^{n-k}$ of C.

Given an element α of order n in a finite field $\mathbb{F}_{q^s}$, a *Bose-Chaudhuri-Hocquenghem* $[n, k]$*-code* of **design distance** d is a cyclic code of length n, generated by a polynomial $g(x)$ in $\mathbb{F}_q(x)$ of degree $n - k$, that has roots at $\alpha, \alpha^2, \dots, \alpha^{d-1}$. The minimum distance d^* of such a code of odd design distance d is at least d.

A *Reed–Solomon code* is a Bose–Chaudhuri–Hocquenghem code with $s = 1$. The generator polynomial of a Reed–Solomon code of design distance d is $g(x) = (x - \alpha) \dots (x - \alpha^{d-1})$ with degree $n - k = d - 1$; that is, for a Reed–Solomon code the design distance $d = n - k + 1$, and the minimum distance $d^* \geq d$. Since, for a linear $[n, k]$-code, the minimum distance $d^* \leq n - k + 1$ (*Singleton upper bound*), a Reed-Solomon code achieves this bound. Compact disc players use a double-error correcting $(255, 251, 5)$ Reed–Solomon code over $\mathbb{F}_{256}$.

- **Goppa designed minimum distance**
  The **Goppa designed minimum distance** [Gopp71] is a lower bound $d^\star(m)$ for the minimum distance of *one-point geometric Goppa codes* (or *algebraic geometry codes*) $G(m)$. For $G(m)$, associated to the divisors D and mP, $m \in \mathbb{N}$, of a smooth projective absolutely irreducible algebraic curve of genus $g > 0$ over a finite field $\mathbb{F}_q$, one has $d^\star(m) = m + 2 - 2g$ if $2g - 2 < m < n$.
  In fact, for a Goppa code $C(m)$ the structure of the gap sequence at P may allow one to give a better lower bound of the minimum distance (cf. **Feng–Rao distance**).
- **Feng–Rao distance**
  The **Feng–Rao distance** $\delta_{FR}(m)$ is a lower bound for the minimum distance of *one-point geometric Goppa codes* $G(m)$ which is better than the **Goppa designed minimum distance**. The method of Feng and Rao for encoding the code $C(m)$ decodes errors up to half the Feng–Rao distance $\delta_{FR}(m)$, and gives an improvement of the number of errors that one can correct for one-point geometric Goppa codes.
  Formally, the Feng–Rao distance is defined as follows. Let S be a subsemigroup S of $\mathbb{N} \cup \{0\}$ such that the *genus* $g = |\mathbb{N} \cup \{0\} \backslash S|$ of S is finite, and $0 \in S$. The **Feng–Rao distance** on S is a function $\delta_{FR} : S \to \mathbb{N} \cup \{0\}$ such that $\delta_{FR}(m) = \min\{\nu(r) : r \geq m, r \in S\}$, where $\nu(r) = |\{(a, b) \in S^2 : a + b = r\}|$.
  The generalized r**-th Feng–Rao distance** on S is $\delta^r_{FR}(m) = \min\{\nu[m_1, \dots, m_r] : m \leq m_1 < \dots < m_r, m_i \in S\}$, where $\nu[m_1, \dots, m_r] = |\{a \in S : m_i - a \in S \text{ for some } i = 1, \dots, r\}|$. Then $\delta_{FR}(m) = \delta^1_{FR}(m)$. See, for example, [FaMu03].
- **Free distance**
  The **free distance** is the minimum nonzero *Hamming weight* of any codeword in a *convolutional code* or a *variable length code*.
  Formally, the k**-th minimum distance** d^*_k of such code is the smallest Hamming distance between any two initial codeword segments which are k frame long and disagree in the initial frame. The sequence $d^*_1, d^*_2, d^*_3, \dots$ $(d^*_1 \leq d^*_2 \leq d^*_3 \leq \dots)$ is called the *distance profile* of the code. The **free distance** of a convolutional code or a variable length code is $\max_l d^*_l = \lim_{l \to \infty} d^*_l = d^*_\infty$.

- **Effective free distance**
  A *turbo code* is a long *block code* in which there are L input bits, and each of these bits is encoded q times. In the j-th encoding, the L bits are sent through a permutation box P_j, and then encoded via an $[N_j, L]$ block encoder (*code fragment encoder*) which can be thought of as an $L \times N_j$ matrix. The overall turbo code is then a *linear* $[N_1 + \cdots + N_q, L]$-code (see, for example, [BGT93]). The *weight-i input minimum distance* $d^i(C)$ of a turbo code C is the minimum weight among codewords corresponding to input words of weight i. The **effective free distance** of C is its *weight-2 input minimum distance* $d^2(C)$, i.e., the minimum *weight* among codewords corresponding to input words of weight 2.
  Turbo codes were the first practical codes to closely approach the *Shannon limit* (or *channel capacity*), the theoretical limit of maximum information transfer rate over a symmetric memory-less noisy channel. These codes are used in 3G mobile and satellite communications. Another capacity-approaching codes with similar performance are linear LDPC (*low-density parity-check*) codes.
- **Distance distribution**
  Given a code C over a finite metric space (X, d) with the diameter $diam(X, d) = D$, the **distance distribution** of C is a $(D + 1)$-vector $(A_0, \dots, A_D)$, where $A_i = \frac{1}{|C|}|\{(c, c^{'}) \in C^2 : d(c, c^{'}) = i\}|$. That is, one considers $A_i(c)$ as the number of code words at distance i from the codeword c, and takes A_i as the average of $A_i(c)$ over all $c \in C$. $A_0 = 1$ and, if $d^* = d^*(C)$ is the minimum distance of C, then $A_1 = \cdots = A_{d^*-1} = 0$.
  The distance distribution of a code with given parameters is important, in particular, for bounding the probability of decoding error under different decoding procedures from *maximum likelihood* decoding to error detection. It can also be helpful in revealing structural properties of codes and establishing nonexistence of some codes.
- **Unicity distance**
  The **unicity distance** of a cryptosystem (Shannon, 1949) is the minimal length of a cyphertext that is required in order to expect that there exists only one meaningful decryption for it. For classic cryptosystems with fixed key space, the unicity distance is approximated by the formula $H(K)/D$, where $H(K)$ is the *key space entropy* (roughly $\log_2 N$, where N is the number of keys), and D measures the *redundancy* of the plaintext source language in bits per letter.
  A cryptosystem offers perfect secrecy if its unicity distance is infinite. For example, the *one-time pads* offer perfect secrecy; they were used for the "red telephone" between the Kremlin and the White House.
  More generally, **Pe-security distance** of a cryptosystem (Tilburg–Boekee, 1987) is the minimal expected length of cyphertext that is required in order to break the cryptogram with an average error probability of at most Pe.

16.2 Main Coding Distances

- **Arithmetic codes distance**
An *arithmetic code* (or *code with correction of arithmetic errors*) is a finite subset of the set $\mathbb{Z}$ of integers (usually, nonnegative integers). It is intended for the control of the functioning of an *adder* (a module performing addition). When adding numbers represented in the binary number system, a single slip in the functioning of the adder leads to a change in the result by some power of 2, thus, to a single *arithmetic error*. Formally, a single *arithmetic error* on $\mathbb{Z}$ is defined as a transformation of a number $n \in \mathbb{Z}$ to a number $n' = n \pm 2^i$, $i = 1, 2, \dots$.
The **arithmetic codes distance** is a metric on $\mathbb{Z}$ defined, for any $n_1, n_2 \in \mathbb{Z}$, as the minimum number of *arithmetic errors* taking n_1 to n_2. It is $w_2(n_1 - n_2)$, where $w_2(n)$ is the *arithmetic* 2-*weight* of n, i.e., the smallest possible number of nonzero coefficients in representations $n = \sum_{i=0}^{k} e_i 2^i$, where $e_i = 0, \pm 1$, and k is some nonnegative integer. For each n there is a unique such representation with $e_k \neq 0$, $e_i e_{i+1} = 0$ for all $i = 0, \dots, k-1$, which has the smallest number of nonzero coefficients (cf. **arithmetic r-norm metric** in Chap. 12).
- **b-Burst metric**
Given the number $b > 1$ and the set $\mathbb{Z}_m^n = \{0, 1, \dots, m-1\}^n$, each its element $x = (x_1, \dots, x_n)$ can be uniquely represented as

$$(0^{k_1} u_1 v_1^{b-1} 0^{k_2} u_2 v_2^{b-1} \dots),$$

where $u_i \neq 0$, 0^k is the string of $k \geq 0$ zeroes and v^{b-1} is any string of length $b-1$.
The b-**burst metric** between elements x and y of $\mathbb{Z}_m^n$ is (Bridewell and Wolf, 1979) the number of b-tuples uv^{b-1} in $x - y$. It describes the burst errors.
- **Sharma–Kaushik metrics**
Let $q \geq 2$, $m \geq 2$. A *partition* $\{B_0, B_1, \dots B_{q-1}\}$ of $\mathbb{Z}_m$ is called a *Sharma–Kaushik partition* if the following conditions hold:

1. $B_0 = \{0\}$;
2. For any $i \in Z_m$, $i \in B_s$ if and only if $m - i \in B_s$, $s = 1, 2, \dots, q-1$;
3. If $i \in B_s$, $j \in B_t$, and $s > t$, then $\min\{i, m-i\} > \min\{j, m-j\}$;
4. If $s \geq t$, $s, t = 0, 1, \dots, q-1$, then $|B_s| \geq |B_t|$ except for $s = q-1$ in which case $|B_{q-1}| \geq \frac{1}{2}|B_{q-2}|$.

Given a Sharma–Kaushik partition of $\mathbb{Z}_m$, the *Sharma-Kaushik weight* $w_{SK}(x)$ of any element $x \in \mathbb{Z}_m$ is defined by $w_{SK}(x) = i$ if $x \in B_i$, $i \in \{0, 1, \dots, q-1\}$.
The **Sharma–Kaushik metric** [ShKa79] is a metric on $\mathbb{Z}_m$ defined by

$$w_{SK}(x - y).$$

The Sharma–Kaushik metric on Z_m^n is defined by $w_{SK}^n(x - y)$ where, for $x = (x_1, \dots x_n) \in \mathbb{Z}_m^n$, one has $w_{SK}^n(x) = \sum_{i=1}^{n} w_{SK}(x_i)$.

The **Hamming metric** and the **Lee metric** arise from two specific partitions of the above type: $P_H = \{B_0, B_1\}$, where $B_1 = \{1, 2, \ldots, q-1\}$, and $P_L = \{B_0, B_1, \ldots, B_{\lfloor q/2 \rfloor}\}$, where $B_i = \{i, m-i\}$, $i = 1, \ldots, \lfloor \frac{q}{2} \rfloor$.

- **Varshamov metric**

  The **Varshamov metric** between two binary n-vectors $x = (x_1, \ldots, x_n)$ and $y = (y_1, \ldots, y_n)$ from $\mathbb{Z}_2^n = \{0, 1\}^n$ is defined by

$$\max\left(\sum_{i=1}^{n} I_{x_i=1-y_i=0}, \sum_{i=1}^{n} I_{x_i=1-y_i=1}\right).$$

  This metric was introduced by Varshamov, 1965, to describe asymmetric errors.

- **Absolute summation distance**

  The **absolute summation distance** (or *Lee distance*) is the **Lee metric** on the set $\mathbb{Z}_m^n = \{0, 1, \ldots, m-1\}^n$ defined by

$$w_{Lee}(x-y),$$

  where $w_{Lee}(x) = \sum_{i=1}^{n} \min\{x_i, m-x_i\}$ is the *Lee weight* of $x = (x_1, \ldots, x_n) \in \mathbb{Z}_m^n$.

  If $\mathbb{Z}_m^n$ is equipped with the absolute summation distance, then a subset C of $\mathbb{Z}_m^n$ is called a *Lee distance code*. The most important such codes are *negacyclic codes*.

- **Mannheim distance**

  The **Mannheim distance** is a 2D generalization of the **Lee metric**.

  Let $\mathbb{Z}[i] = \{a + bi : a, b \in \mathbb{Z}\}$ be the set of *Gaussian integers*. Let $\pi = a + bi$ $(a > b > 0)$ be a *Gaussian prime*, i.e., either

  (i) $(a + bi)(a - bi) = a^2 + b^2 = p$, where $p \equiv 1 \pmod 4$ is a prime number, or

  (ii) up to an integer, $\pi = p + 0 \cdot i$, where $p \equiv 3 \pmod 4$ is a prime number.

  The **Mannheim distance** is not a metric; it is defined [Hube94a], for any $x, y \in \mathbb{Z}[i]$, as $|x'| + |y'|$, where $x' + y'i = x - y \pmod \pi$, which is defined as $(x-y) - [\frac{(x-y)\bar{\pi}}{\pi\bar{\pi}}]\pi$ in the case (i). Here [.] denotes rounding to the *closest Gaussian integer*, i.e., $[c + di] = [c] + [d]i$ with $[c]$ denoting the rounding to the closest integer.

  In general, the elements of the finite field $\mathbb{F}_p = \{0, 1, \ldots, p-1\}$ for $p \equiv 1 \pmod 4$, $p = a^2 + b^2$, and of the finite field $\mathbb{F}_{p^2}$ for $p \equiv 3 \pmod 4$, $p = a$, can be mapped on a subset of $\mathbb{Z}[i]$ using the complex modulo function $\mu(k) = k - [\frac{k(a-bi)}{p}](a + bi)$, $k = 0, \ldots, p-1$. The set of the selected Gaussian integers $a+bi$ with the minimal *complex modulus* norms $\sqrt{(a+bi)(a-bi)} = \sqrt{a^2 + b^2}$ is called a *constellation*.

  The Mannheim distance between two vectors over $\mathbb{Z}[i]$ is the sum of the Mannheim distances of corresponding components. It was introduced to make 2D QAM-like signals more susceptible to algebraic decoding methods.

For codes over hexagonal signal constellations, a similar metric was introduced over $\mathbb{Z}(\frac{i\sqrt{3}+1}{2})$ in [Hube94b]. Cf. **$\mathbb{Z}(\eta_m)$-related norm metrics** in Chap. 12.

- **Generalized Lee metric**
  Let $\mathbb{F}_{p^m}$ denote the finite field with p^m elements, where p is prime number and $m \geq 1$ is an integer. Let $e_i = (0,\dots,0,1,0,\dots,0)$, $1 \leq i \leq k$, be the standard basis of $\mathbb{Z}^k$. Choose elements $a_i \in \mathbb{F}_{p^m}$, $1 \leq i \leq k$, and the mapping $\phi : \mathbb{Z}^k \to \mathbb{F}_{p^m}$, sending any $x = \sum_{i=1}^k x_i e_i$, $x_i \in \mathbb{Z}^k$, to $\phi(x) = \sum_{i=1}^k a_i x_i$ (mod p), so that ϕ is surjective. So, for each $a \in \mathbb{F}_{p^m}$, there exists $x \in \mathbb{Z}^k$ such that $a = \phi(x)$. For each $a \in \mathbb{F}_{p^m}$, its *k-dimensional Lee weight* is $w_{kL}(a) = \min\{\sum_{i=1}^k |x_i| : x = (x_i) \in \mathbb{Z}, a = \phi(x)\}$.
  The **generalized Lee metric** between vectors (a_j) and (b_j) of $\mathbb{F}^n_{p^m}$ is defined (Nishimura–Hiramatsu, 2008) by
  $$\sum_{j=1}^n w_{kL}(a_j - b_j).$$
  It is the **Lee metric** (or **absolute summation distance**) if $\phi(e_1) = 1$ while $\phi(e_i) = 0$ for $2 \leq i \leq k$. It is the **Mannheim distance** if $k = 2$, $p \equiv 1$ (mod 4), $\phi(e_1) = 1$ while $\phi(e_2) = a$ is a solution in $\mathbb{F}_p$ of the quadratic congruence $x^2 \equiv -1$ (mod p).
- **Poset metric**
  Let $(V_n, \preceq)$ be a *poset* on $V_n = \{1,\dots,n\}$. A subset I of V_n is called *ideal* if $x \in I$ and $y \preceq x$ imply that $y \in I$. If $J \subset V_n$, then $\langle J\rangle$ denotes the smallest ideal of V_n which contains J. Consider the vector space $\mathbb{F}_q^n$ over a finite field $\mathbb{F}_q$. The *P-weight* of an element $x = (x_1,\dots,x_n) \in \mathbb{F}_q^n$ is defined as the cardinality of the smallest ideal of V_n containing the *support* of x: $w_P(x) = |\langle supp(x)\rangle|$, where $supp(x) = \{i : x_i \neq 0\}$.
  The **poset metric** (see [BGL95]) is a metric on $\mathbb{F}_q^n$ defined by
  $$w_P(x - y).$$
  If $\mathbb{F}_q^n$ is equipped with a poset metric, then a subset C of $\mathbb{F}_q^n$ is called a *poset code*. If V_n forms the chain $1 \leq 2 \leq \cdots \leq n$, then the linear code C of dimension k consisting of all vectors $(0,\dots,0,a_{n-k+1},\dots,a_n) \in \mathbb{F}_q^n$ is a perfect poset code with the minimum (poset) metric $d_P^*(C) = n - k + 1$.
  If V_n forms an antichain, then the poset distance coincides with the **Hamming metric**. If V_n consists of finite disjoint union of chains of equal lengths, then the poset distance coincides with the **NRT metric**.
- **Rank metric**
  Let $\mathbb{F}_q$ be a finite field, $\mathbb{K} = \mathbb{F}_{q^m}$ an extension of degree m of $\mathbb{F}_q$, and $\mathbb{K}^n$ a vector space of dimension n over $\mathbb{K}$. For any $a = (a_1,\dots a_n) \in \mathbb{K}^n$ define its *rank*, rank(a), as the dimension of the vector space over $\mathbb{F}_q$, generated by $\{a_1,\dots,a_n\}$.

The **rank metric** (Delsarte, 1978) is a metric on $\mathbb{K}^n$ defined by

$$\text{rank}(a - b).$$

A *constant rank-distance k set* is (Gow et al., 2014) a set U of $n \times n$ matrices over a field $\mathbb{F}$ such that $\text{rank}(A - B) = k$ for all $A, B \in U, A \neq B$ and $\text{rank}(A) = k$ for all $A \in U, A \neq 0$. Such set is called a *partial spread set* if $k = n$; it defines a partial spread in the $(2n - 1)$-dimensional *projective, hermitian polar* or *symplectic polar space*, if U consists of arbitrary, hermitian or symmetric matrices, respectively.

- **Gabidulin–Simonis metrics**
  Let $\mathbb{F}_q^n$ be the vector space over a finite field $\mathbb{F}_q$ and let $F = \{F_i : i \in I\}$ be a finite family of its subsets such that the minimal linear subspace of $\mathbb{F}_q^n$ containing $\cup_{i \in I} F_i$ is $\mathbb{F}_q^n$. Without loss of generality, F can be an antichain of linear subspaces of $\mathbb{F}_q^n$.
  The *F-weight* w_F of a vector $x = (x_1, \dots, x_n) \in \mathbb{F}_q^n$ is the smallest $|J|$ over such subsets $J \subset I$ that x belongs to the minimal linear subspace of $\mathbb{F}_q^n$ containing $\cup_{i \in J} F_i$. A **Gabidulin–Simonis metric** (or F**-distance**, see [GaSi98]) on $\mathbb{F}_q^n$ is defined by

$$w_F(x - y).$$

  The **Hamming metric** corresponds to the case of $F_i, i \in I$, forming the standard basis. The **Vandermonde metric** is F-distance with $F_i, i \in I$, being the columns of a generalized Vandermonde matrix. Among other examples are: the **rank metric** and the *combinatorial metrics* (by Gabidulin, 1984), including the b-**burst metric**.
- **Subspace metric**
  Let $\mathbb{F}_q^n$ be the vector space over a finite field $\mathbb{F}_q$ and let $\mathcal{P}_{n,q}$ be the set of all subspaces of $\mathbb{F}_q^n$. For any subspace $U \in \mathcal{P}_{n,q}$, let $dim(U)$ denote its dimension and let $U^{\perp} = \{v \in \mathbb{F}_q^n : \langle u, v \rangle = 0 \text{ for all } u \in U\}$ be its orthogonal space.
  Let $U + V = \{u + v : u \in U, v \in V\}$, i.e., $U + V$ is the smallest subspace of $\mathbb{F}_q^n$ containing both V and V. Then $dim(U + V) = dim(U) + dim(V) - dim(U \cap V)$. If $U \cap V = \emptyset$, then $U + V$ is a direct sum $U \oplus V$.
  The **subspace metric** between two subspaces U and V from $\mathcal{P}_{n,q}$ is defined by

$$d(U, V) = dim(U + V) - dim(U \cap V) = dim(U) + dim(V) - 2dim(U \cap V).$$

  This metric was introduced by Koetter and Kschischang, 2007, for network coding. It holds $d(U, V) = d(U^{\perp}, V^{\perp})$. Cf. the **lattice valuation metric** in Chap. 10 and distances between subspaces in Chap. 12.

- **NRT metric**
Let $M_{m,n}(\mathbb{F}_q)$ be the set of all $m \times n$ matrices with entries from a finite field $\mathbb{F}_q$ (in general, from any finite alphabet $\mathcal{A} = \{a_1, \dots, a_q\}$). The *NRT norm* $||.||_{RT}$ on $M_{m,n}(\mathbb{F}_q)$ is defined as follows: if $m = 1$ and $a = (\xi_1, \xi_2, \dots, \xi_n) \in M_{1,n}(\mathbb{F}_q)$, then $||0_{1,n}||_{RT} = 0$, and $||a||_{RT} = \max\{i : \xi_i \neq 0\}$ for $a \neq 0_{1,n}$; if $A = (a_1, \dots, a_m)^T \in M_{m,n}(\mathbb{F}_q)$, $a_j \in M_{1,n}(\mathbb{F}_q)$, $1 \leq j \leq m$, then $||A||_{RT} = \sum_{j=1}^{m} ||a_j||_{RT}$.
The **NRT metric** (or *Niederreiter–Rosenbloom–Tsfasman metric*, since introduced by Niederreiter, 1991, and Rosenbloom–Tsfasman, 1997; or *ordered Hamming distance*, in [MaSt99] is a **matrix norm metric** (in fact, an **ultrametric**) on $M_{m,n}(\mathbb{F}_q)$, defined by

$$||A - B||_{RT}.$$

For every matrix code $C \subset M_{m,n}(F_q)$ with q^k elements the minimum NRT distance $d^*_{RT}(C) \leq mn - k + 1$. Codes meeting this bound are called *maximum distance separable codes*.
The most used distance between codewords of a matrix code $C \subset M_{m,n}(F_q)$ is the **Hamming metric** on $M_{m,n}(F_q)$ defined by $||A - B||_H$, where $||A||_H$ is the *Hamming weight* of a matrix $A \in M_{m,n}(\mathbb{F}_q)$, i.e., the number of its nonzero entries.
The **LRTJ-metric** (introduced as *Generalized Lee–Rosenbloom–Tsfasman pseudo-metric* by Jain, 2008) is the **norm metric** for the following generalization of the above norm $||a||_{RT}$ in the case $a \neq 0_{1,n}$:

$$||a||_{LRTJ} = \max_{1 \leq i \leq n} \min\{\xi_i, q - \xi_i\} + \max\{i - 1 : \xi_i \neq 0\}.$$

It is the **Lee metric** for $m = 1$ and the NRT metric for $q = 2, 3$.

- **ACME distance**
The **ACME distance** on a code $C \subset \mathcal{A}^n$ over an alphabet $\mathcal{A}$ is defined by

$$\min\{d_H(x, y), d_I(x, y)\},$$

where d_H is the **Hamming metric**, and d_I is the **swap metric** (cf. Chap. 11), i.e., the minimum number of interchanges of adjacent pairs of symbols, converting x into y.

- **Indel distance**
Let W be the set of all words over an alphabet $\mathcal{A}$. A *deletion* of a letter in a word $\beta = b_1 \dots b_n$ of the length n is a transformation of β into a word $\beta' = b_1 \dots b_{i-1} b_{i+1} \dots b_n$ of the length $n - 1$. An *insertion* of a letter in a word $\beta = b_1 \dots b_n$ of the length n is a transformation of β into a word $\beta'' = b_1 \dots b_i b b_{i+1} \dots b_n$, of the length $n + 1$.

The **indel distance** is a metric on W, defined as the minimum number of deletions and insertions of letters converting α into β. Cf. **indel metric** in Chap. 11.

A *code C with correction of deletions and insertions* is an arbitrary finite subset of W. An example of such a code is the set of words $\beta = b_1 \dots b_n$ of length n over the alphabet $\mathcal{A} = \{0, 1\}$ for which $\sum_{i=1}^{n} ib_i \equiv 0 \pmod{n+1}$. The number of words in this code is equal to $\frac{1}{2(n+1)} \sum_k \phi(k)2^{(n+1)/k}$, where the sum is taken over all odd divisors k of $n + 1$, and ϕ is the *Euler function*.

- **Interval distance**

The **interval distance** (see, for example, [Bata95]) is a metric on a finite group $(G, +, 0)$ defined by

$$w_{int}(x - y),$$

where $w_{int}(x)$ is an *interval weight* on G, i.e., a *group norm* whose values are consecutive nonnegative integers $0, \dots, m$. This distance is used for *group codes* $C \subset G$.

- **Fano metric**

The **Fano metric** is a *decoding metric* with the goal to find the best sequence estimate used for the *Fano algorithm* of *sequential decoding* of *convolutional codes*. In a *convolutional code* each k-bit information symbol to be encoded is transformed into an n-bit codeword, where $R = \frac{k}{n}$ is the code rate ($n \geq k$), and the transformation is a function of the last m information symbols.

The linear time-invariant decoder (*fixed convolutional decoder*) maps an information symbol $u_i \in \{u_1, \dots, u_N\}$, $u_i = (u_{i1}, \dots u_{ik})$, $u_{ij} \in \mathbb{F}_2$, into a codeword $x_i \in \{x_1, \dots, x_N\}$, $x_i = (x_{i1}, \dots, x_{in})$, $x_{ij} \in \mathbb{F}_2$, so one has a code $\{x_1, \dots, x_N\}$ with N codewords which occur with probabilities $\{p(x_1), \dots, p(x_N)\}$. A sequence of l codewords forms a *path* $x = x_{[1,l]} = \{x_1, \dots, x_l\}$ which is transmitted through a *discrete memoryless channel*, resulting in the received sequence $y = y_{[1,l]}$.

The task of a decoder minimizing the sequence error probability is to find a sequence maximizing the joint probability of input and output channel sequences $p(y, x) = p(y|x) \cdot p(x)$. Usually it is sufficient to find a procedure that maximizes $p(y|x)$, and a decoder that always chooses as its estimate one of the sequences that maximizes it or, equivalently, the **Fano metric**, is called a *max-likelihood decoder*.

Roughly, we consider each code as a tree, where each branch represents one codeword. The decoder begins at the first vertex in the tree, and computes the branch metric for each possible branch, determining the best branch to be the one corresponding to the codeword x_j resulting in the largest branch metric, $\mu_F(x_j)$. This branch is added to the path, and the algorithm continues from the new node which represents the sum of the previous node and the number of bits in the current best codeword. Through iterating until a terminal node of the tree is reached, the algorithm traces the most likely path.

In this construction, the **bit Fano metric** is defined by

$$\log_2 \frac{p(y_i|x_i)}{p(y_i)} - R,$$

the **branch Fano metric** is defined by

$$\mu_F(x_j) = \sum_{i=1}^{n} (\log_2 \frac{p(y_i|x_{ji})}{p(y_i)} - R),$$

and the **path Fano metric** is defined by

$$\mu_F(x_{[1,l]}) = \sum_{j=1}^{l} \mu_F(x_j),$$

where $p(y_i|x_{ji})$ are the channel transition probabilities, $p(y_i) = \sum_{x_m} p(x_m) p(y_i|x_m)$ is the probability distribution of the output given the input symbols averaged over all input symbols, and $R = \frac{k}{n}$ is the code rate.
For a hard-decision decoder $p(y_j = 0|x_j = 1) = p(y_j = 1|x_j = 0) = p$, $0 < p < \frac{1}{2}$, the Fano metric for a path $x_{[1,l]}$ can be written as

$$\mu_F(x_{[1,l]}) = -\alpha d_H(y_{[1,l]}, x_{[1,l]}) + \beta \cdot l \cdot n,$$

where $\alpha = -\log_2 \frac{p}{1-p} > 0$, $\beta = 1 - R + \log_2(1-p)$, and d_H is the **Hamming metric**.
The **generalized Fano metric** is defined, for $0 \le w \le 1$, by

$$\mu_F^w(x_{[1,l]}) = \sum_{j=1}^{ln} \left(\log_2 \frac{p(y_j|x_j)^w}{p(y_j)^{1-w}} - wR\right).$$

For $w = 1/2$, it is the Fano metric with a multiplicative constant $1/2$.

- **Metric recursion of a MAP decoding**
  Maximum a posteriori sequence estimation, or *MAP decoding* for *variable length codes*, used the *Viterbi algorithm*, and is based on the **metric recursion**

$$\Lambda_k^{(m)} = \Lambda_{k-1}^{(m)} + \sum_{n=1}^{l_k^{(m)}} x_{k,n}^{(m)} \log_2 \frac{p(y_{k,n}|x_{k,n}^{(m)} = +1)}{p(y_{k,n}|x_{k,n}^{(m)} = -1)} + 2\log_2 p(u_k^{(m)}),$$

  where $\Lambda_k^{(m)}$ is the **branch metric** of branch m at time (level) k, $x_{k,n}$ is the n-th bit of the codeword having $l_k^{(m)}$ bits labeled at each branch, $y_{k,n}$ is the respective received soft-bit, u_k^m is the source symbol of branch m at time k and, assuming statistical independence of the source symbols, the probability $p(u_k^{(m)})$ is equivalent to the probability of the source symbol labeled at branch m, that may be known or estimated. The metric increment is computed for each branch, and the largest value, when using *log-likelihood* values, of each state is used for further recursion. The decoder first computes the metric of all branches, and then the branch sequence with largest metric starting from the final state backward is selected.

- **Distance decoder**
  A graph family A is said (Peleg, 2000) to have an $l(n)$ **distance labeling scheme** if there is a function L_G labeling the vertices of each n-vertex graph $G \in A$ with distinct labels up to $l(n)$ bits, and there exists an algorithm, called a **distance decoder**, that decides the distance $d(u, v)$ between any two vertices $u, v \in X$ in a graph $G \in A$, i.e., $d(u, v) = f(L_G(u), L_G(v))$, polynomial in time in the length of their labels $L(u), L(v)$.
  Cf. **distance constrained labeling** in Chap. 15.
- **Identifying code**
  Let $G = (X, E)$ be a digraph and $C \subset V$, and let $B(v)$ denote the set consisting of v and all of its incoming neighbors in G. If the sets $B(v) \cap C$ are nonempty and distinct, C is called **identifying code** of G. Such sets of smallest cardinality are called (Karpovsky–Chakrabarty–Levitin, 1998) *minimum identifying codes*; denote this cardinality by $M(G)$. An r**-locating-dominating set** (cf. Chap. 15) with $r = 1$ differs from an identifying code only in that $B(v) \cap C$ are not required to be unique *identifying sets* for $v \in C$.
  A *minimum identifying code graph* of order n is a graph $G = (X, E)$ with $X = n$ and $M(G) = \lceil \log 2_2(n + 1) \rceil$ having the minimum number of edges $|E|$.

Chapter 17
Distances and Similarities in Data Analysis

A *data set* is a finite set comprising m sequences $(x_1^j, \ldots, x_n^j)$, $j \in \{1, \ldots, m\}$, of length n. The values $x_i^1, \ldots, x_i^m$ represent an *attribute* S_i.

Among numerical data, **metric data** is any reading at an *interval scale*, measuring the degree of difference between items, or at a *ratio scale* measuring the ratio between a magnitude of a continuous quantity and a unit magnitude of the same kind; with them one have a meter permitting define distances between scale values. **Nonmetric** (or *categorial*, *qualitative*) **data** are collected from *binary* (presence/absence expressed by $1/0$), *ordinal* (numbers expressing rank only), or *nominal* (items are not ordered) *scale*.

Geometric data analysis refer to geometric aspects of image, pattern and shape analysis that treats arbitrary data sets as clouds of points in $\mathbb{R}^n$.

Often data are organized in a **metric database** (especially, **metric tree**), i.e., a database indexed in a metric space. The term *metric indexing* is also used.

Cluster Analysis (or *Classification*, *Taxonomy*, *Pattern Recognition*) consists mainly of partition of data A into a relatively small number of *clusters*, i.e., such sets of objects that (with respect to a selected measure of distance) are at best possible degree, "close" if they belong to the same cluster, "far" if they belong to different clusters, and further subdivision into clusters will impair the above two conditions.

We give three typical examples. In *Information Retrieval* applications, nodes of peer-to-peer database network export data (collection of text documents); each document is characterized by a vector from $\mathbb{R}^n$. An user needs to retrive all documents in the database which are *relevant* to a query object (say, a vector $x \in \mathbb{R}^n$), i.e., belong to the *ball* in $\mathbb{R}^n$, center x, of fixed radius and with a convenient distance function. Such similarity query is called a **metric range query**. In *Record Linkage*, each document (database record) is represented by a term-frequency vector $x \in \mathbb{R}^n$ or a string, and one wants to measure semantic relevancy of syntactically different records. In *Ecology*, let x, y be *species abundance distributions*, obtained by two sample methods (i.e., x_j, y_j are the numbers of individuals of species j,

M.M. Deza, E. Deza, *Encyclopedia of Distances*,
DOI 10.1007/978-3-662-44342-2_17

observed in a corresponding sample); one needs a measure of the distance between x and y, in order to compare two methods.

Once a distance d between objects is selected, it is **intra-distance** or **inter-distance** if the objects are within the same cluster or in two different clusters, respectively.

The **linkage metric**, i.e., a distance between clusters $A = \{a_1, \dots, a_m\}$ and $B = \{b_1, \dots, b_n\}$ is usually one of the following:

average linkage: the average of the distances between the all members of the clusters, i.e., $\frac{\sum_i \sum_j d(a_i, b_j)}{mn}$;

single linkage: the distance $\min_{i,j} d(a_i,_j)$ between the nearest members of the clusters, i.e., the **set-set distance** (cf. Chap. 1);

complete linkage: the distance $\max_{i,j} d(a_i, b_j)$ between the furthest members of the clusters, i.e., the **spanning distance** (cf. Chap. 1);

centroid linkage: the distance between the *centroids* of the clusters, i.e., $||\tilde{a} - \tilde{b}||_2$, where $\tilde{a} = \frac{\sum_i a_i}{m}$, and $\tilde{b} = \frac{\sum_j b_j}{n}$;

Ward linkage: the distance $\sqrt{\frac{mn}{m+n}}||\tilde{a} - \tilde{b}||_2$.

Multidimensional Scaling is a technique developed in the behavioral and Social Sciences for studying the structure of objects or people. Together with Cluster Analysis, it is based on distance methods. But in Multidimensional Scaling, as opposed to Cluster Analysis, one starts only with some $m \times m$ matrix D of distances of the objects and (iteratively) looks for a representation of objects in $\mathbb{R}^n$ with low n, so that their Euclidean distance matrix has minimal square deviation from the original matrix D.

The related *Metric Nearness Problem* (Dhillon–Sra–Tropp, 2003) is to approximate a given finite distance space (X, d) by a metric space (X, d'). Other examples of distance methods in Data Analysis are *distance-based outlier detection* (in Data Mining) and *distance-based redundancy analysis* (in Multivariate Statistics).

There are many **similarities** used in Data Analysis; the choice depends on the nature of data and is not an exact science. We list below the main such similarities and distances.

Given two objects, represented by nonzero vectors $x = (x_1, \dots, x_n)$ and $y = (y_1, \dots, y_n)$ from $\mathbb{R}^n$, the following notation is used in this chapter.

$\sum x_i$ means $\sum_{i=1}^{n} x_i$.

1_F is the *characteristic function* of event F: $1_F = 1$ if F happens, and $1_F = 0$, otherwise.

$||x||_2 = \sqrt{\sum x_i^2}$ is the ordinary Euclidean norm on $\mathbb{R}^n$.

$\overline{x}$ denotes $\frac{\sum x_i}{n}$, i.e., the *mean value* of components of x. So, $\overline{x} = \frac{1}{n}$ if x is a *frequency vector* (*discrete probability distribution*), i.e., all $x_i \geq 0$, and $\sum x_i = 1$; and $\overline{x} = \frac{n+1}{2}$ if x is a *ranking* (*permutation*), i.e., all x_i are different numbers from $\{1, \dots, n\}$.

The *k-th moment* is $\frac{\sum (x_i - \overline{x})^k}{n}$; it is called *variance, skewness, kurtosis* if $k = 2, 3, 4$.

In the binary case $x \in \{0,1\}^n$ (i.e., when x is a binary n-sequence), let $X = \{1 \le i \le n : x_i = 1\}$ and $\overline{X} = \{1 \le i \le n : x_i = 0\}$. Let $|X \cap Y|$, $|X \cup Y|$, $|X \backslash Y|$ and $|X \Delta Y|$ denote the cardinality of the intersection, union, difference and symmetric difference $(X \backslash Y) \cup (Y \backslash X)$ of the sets X and Y, respectively.

17.1 Similarities and Distances for Numerical Data

- **Ruzicka similarity**
  The **Ruzicka similarity** is a similarity on $\mathbb{R}^n$, defined by

$$\frac{\sum \min\{x_i, y_i\}}{\sum \max\{x_i, y_i\}}.$$

  The corresponding **Soergel distance**

$$1 - \frac{\sum \min\{x_i, y_i\}}{\sum \max\{x_i, y_i\}} = \frac{\sum |x_i - y_i|}{\sum \max\{x_i, y_i\}}$$

  coincides on $\mathbb{R}^n_{\ge 0}$ with the **fuzzy polyonucleotide metric** (cf. Chap. 23).
  The **Wave–Edges distance** is defined by

$$\sum \left(1 - \frac{\min\{x_i, y_i\}}{\max\{x_i, y_i\}}\right) = \sum \frac{|x_i - y_i|}{\max\{x_i, y_i\}}.$$

- **Roberts similarity**
  The **Roberts similarity** is a similarity on $\mathbb{R}^n$, defined by

$$\frac{\sum (x_i + y_i) \frac{\min\{x_i, y_i\}}{\max\{x_i, y_i\}}}{\sum (x_i + y_i)}.$$

- **Ellenberg similarity**
  The **Ellenberg similarity** is a similarity on $\mathbb{R}^n$ defined by

$$\frac{\sum (x_i + y_i) 1_{x_i \cdot y_i \neq 0}}{\sum (x_i + y_i)(1 + 1_{x_i y_i = 0})}.$$

  The binary cases of Ellenberg and **Ruzicka similarities** coincide; it is called **Tanimoto similarity** (or **Jaccard similarity of community**, Jaccard, 1908):

$$\frac{|X \cap Y|}{|X \cup Y|}.$$

The **Tanimoto distance** (or **biotope distance** from Chap. 23, **Jaccard distance**) distance on $\{0, 1\}^n$ defined by

$$1 - \frac{|X \cap Y|}{|X \cup Y|} = \frac{|X \Delta Y|}{|X \cup Y|}.$$

- **Gleason similarity**
  The **Gleason similarity** is a similarity on $\mathbb{R}^n$, defined by

$$\frac{\sum(x_i + y_i)1_{x_i \cdot y_i \neq 0}}{\sum(x_i + y_i)}.$$

The binary cases of Cleason, Motyka and Bray–Curtis similarities coincide; it is called **Dice similarity**, 1945 (or *Sørensen similarity*, *Czekanowsky similarity*):

$$\frac{2|X \cap Y|}{|X \cup Y| + |X \cap Y|} = \frac{2|X \cap Y|}{|X| + |Y|}.$$

The **Czekanowsky–Dice distance** (or *nonmetric coefficient*, Bray–Curtis, 1957) is a **near-metric** on $\{0, 1\}^n$ defined by

$$1 - \frac{2|X \cap Y|}{|X| + |Y|} = \frac{|X \Delta Y|}{|X| + |Y|}.$$

- **Intersection distance**
  The **intersection distance** is a distance on $\mathbb{R}^n$, defined by

$$1 - \frac{\sum \min\{x_i, y_i\}}{\min\{\sum x_i, \sum y_i\}}.$$

- **Motyka similarity**
  The **Motyka similarity** is a similarity on $\mathbb{R}^n$, defined by

$$\frac{\sum \min\{x_i, y_i\}}{\sum(x_i + y_i)} = n\frac{\sum \min\{x_i, y_i\}}{\overline{x} + \overline{y}}.$$

- **Bray–Curtis similarity**
  The **Bray–Curtis similarity**, 1957, is a similarity on $\mathbb{R}^n$ defined by

$$\frac{2}{n(\overline{x} + \overline{y})} \sum \min\{x_i, y_j\}.$$

It is called *Renkonen similarity* if x, y are *frequency vectors*.

- **Sørensen distance**
The **Sørensen** (or *Bray–Curtis*) **distance** on $\mathbb{R}^n$ is defined (Sørensen, 1948) by

$$\frac{\sum |x_i - y_i|}{\sum (x_i + y_i)}.$$

- **Canberra distance**
The **Canberra distance** (Lance–Williams, 1967) is a distance on $\mathbb{R}^n$, defined by

$$\sum \frac{|x_i - y_i|}{|x_i| + |y_i|}.$$

- **Kulczynski similarity** 1
The **Kulczynski similarity** 1 is a similarity on $\mathbb{R}^n$ defined by

$$\frac{\sum \min\{x_i, y_i\}}{\sum |x_i - y_i|}.$$

The corresponding distance is

$$\frac{\sum |x_i - y_i|}{\sum \min\{x_i, y_i\}}.$$

- **Kulczynski similarity** 2
The **Kulczynski similarity** 2 is a similarity on $\mathbb{R}^n$ defined by

$$\frac{n}{2}\left(\frac{1}{\overline{x}} + \frac{1}{\overline{y}}\right) \sum \min\{x_i, y_i\}.$$

In the binary case it takes the form

$$\frac{|X \cap Y| \cdot (|X| + |Y|)}{2|X| \cdot |Y|}.$$

- **Baroni–Urbani–Buser similarity**
The **Baroni–Urbani–Buser similarity** is a similarity on $\mathbb{R}^n$ defined by

$$\frac{\sum \min\{x_i, y_i\} + \sqrt{\sum \min\{x_i, y_i\} \sum (\max_{1 \le j \le n} x_j - \max\{x_i, y_i\})}}{\sum \max\{x_i, y_i\} + \sqrt{\sum \min\{x_i, y_i\} \sum (\max_{1 \le j \le n} x_j - \max\{x_i, y_i\})}}.$$

In the binary case it takes the form

$$\frac{|X \cap Y| + \sqrt{|X \cap Y| \cdot |\overline{X \cup Y}|}}{|X \cup Y| + \sqrt{|X \cap Y| \cdot |\overline{X \cup Y}|}}.$$

17.2 Relatives of Euclidean Distance

- **Power (p, r)-distance**
  The **power (p, r)-distance** is a distance on $\mathbb{R}^n$ defined, for $x, y \in \mathbb{R}^n$, by

$$(\sum_{i=1}^{n} |x_i - y_i|^p)^{\frac{1}{r}}.$$

  For $p = r \geq 1$, it is the l_p**-metric**, including the **Euclidean**, **Manhattan** (or *magnitude*) and **Chebyshev** (or *maximum-value*, *dominance*, *template*) **metrics** for $p = 2, 1$ and ∞, respectively.
  The case $(p, r) = (2, 1)$ corresponds to the **squared Euclidean distance**.
  The power (p, r)-distance with $0 < p = r < 1$ is called the **fractional** l_p**-distance** (not a metric since the unit balls are not convex). It is used for "dimensionality-cursed" data, i.e., when there are few observations and the number n of variables is large. The case $0 < p < r = 1$, i.e., of the p-th power of the fractional l_p-distance, corresponds to a metric on $\mathbb{R}^n$.
  The weighted versions $(\sum w_i |x_i - y_i|^p)^{\frac{1}{p}}$ (with nonnegative weights w_i) are also used, for $p = 1, 2$, in applications. Given weights $w_i \geq 0$, the **weighted Manhattan quasi-metric** for $x, y \in \mathbb{R}^n$ is $\sum_{i=1}^{n} d_i$, where every d_i is the quasi-metric defined by $d_i = w_i(x_i - y_i)$ if $x_i > y_i$ and $d_i = W_i(y_i - x_i)$, otherwise.
  The **ordinal distance** on $\mathbb{R}^n$ is defined (Bahari and Van Hamme, 2014) by

$$(\sum_{i=1}^{n} | \sum_{1 \leq j \leq i} (x_j - y_j)|^p)^{\frac{1}{p}}.$$

- **Penrose size distance**
  The **Penrose size distance** is a distance on $\mathbb{R}^n$ defined by

$$\sqrt{n} \sum |x_i - y_i|.$$

  It is proportional to the **Manhattan metric**.
  The **mean character distance** (Czekanowsky, 1909) is defined by $\frac{\sum |x_i - y_i|}{n}$.
  The *Lorentzian distance* is a distance defined by $\sum \ln(1 + |x_i - y_i|)$.

- **Penrose shape distance**
  The **Penrose shape distance** is a distance on $\mathbb{R}^n$ defined by

$$\sqrt{\sum((x_i - \overline{x}) - (y_i - \overline{y}))^2}.$$

  The sum of squares of two above **Penrose distances** is the **squared Euclidean distance**.
- **Effect size**
  Let $\overline{x}, \overline{y}$ be the means of samples x, y and let s^2 be the pooled variance of both samples. The **effect size** (a term used mainly in social sciences) is defined by

$$\frac{\overline{x} - \overline{y}}{s}.$$

  Its symmetric version $\frac{|\overline{x}-\overline{y}|}{s}$ is called *statistical distance* by Johnson–Wichern, 1982, and *standard distance* by Flury–Riedwyl, 1986.
  Cf. the **engineer semimetric** in Chap. 14 and the **ward linkage**.
- **Binary Euclidean distance**
  The **binary Euclidean distance** is a distance on $\mathbb{R}^n$ defined by

$$\sqrt{\sum(1_{x_i>0} - 1_{y_i>0})^2}.$$

- **Mean censored Euclidean distance**
  The **mean censored Euclidean distance** is a distance on $\mathbb{R}^n$ defined by

$$\sqrt{\frac{\sum(x_i - y_i)^2}{\sum 1_{x_i^2+y_i^2\neq 0}}}.$$

- **Normalized l_p-distance**
  The **normalized l_p-distance,** $1 \leq p \leq \infty$, is a distance on $\mathbb{R}^n$ defined by

$$\frac{||x - y||_p}{||x||_p + ||y||_p}.$$

  The only integer value p for which the normalized l_p-distance is a metric, is $p = 2$. Moreover, the distance $\frac{||x-y||_2}{a+b(||x||_2+||y||_2)}$ is a metric for any $a, b > 0$ [Yian91].
- **Clark distance**
  The **Clark distance** (Clark, 1952) is a distance on $\mathbb{R}^n$, defined by

$$\left(\frac{1}{n}\sum\left(\frac{x_i - y_i}{|x_i| + |y_i|}\right)^2\right)^{\frac{1}{2}}.$$

- **Meehl distance**
The **Meehl distance** (or *Meehl index*) is a distance on $\mathbb{R}^n$ defined by

$$\sum_{1\le i\le n-1}(x_i - y_i - x_{i+1} + y_{i+1})^2.$$

- **Hellinger distance**
The **Hellinger distance** is a distance on $\mathbb{R}^n_+$ defined by

$$\sqrt{2\sum\left(\sqrt{\frac{x_i}{\overline{x}}}-\sqrt{\frac{y_i}{\overline{y}}}\right)^2}.$$

Cf. **Hellinger metric** in Chap. 14.
The *Whittaker index of association* is defined by $\frac{1}{2}\sum|\frac{x_i}{\overline{x}}-\frac{y_i}{\overline{y}}|$.

- **Symmetric χ^2-measure**
The **symmetric χ^2-measure** is a distance on $\mathbb{R}^n$ defined by

$$\sum\frac{2}{\overline{x}\cdot\overline{y}}\cdot\frac{(x_i\overline{y}-y_i\overline{x})^2}{x_i+y_i}.$$

- **Symmetric χ^2-distance**
The **symmetric χ^2-distance** (or *chi-distance*) is a distance on $\mathbb{R}^n$ defined by

$$\sqrt{\sum\frac{\overline{x}+\overline{y}}{n(x_i+y_i)}(\frac{x_i}{\overline{x}}-\frac{y_i}{\overline{y}})^2}=\sqrt{\sum\frac{\overline{x}+\overline{y}}{n(\overline{x}\cdot\overline{y})^2}\cdot\frac{(x_i\overline{y}-y_i\overline{x})^2}{x_i+y_i}}.$$

It is a **weighted Euclidean distance**.

- **Weighted Euclidean distance**
The general **quadratic-form distance** on $\mathbb{R}^n$ is defined by

$$\sqrt{(x-y)^T A(x-y)},$$

where A is a real nonsingular symmetric $n\times n$ matrix; cf. **Mahalanobis distance**. The **weighted Euclidean distance** is the case $A=\text{diag}(a_i), a_i\neq 0$, i.e., it is

$$\sqrt{\sum a_i(x_i-y_i)^2}.$$

Some examples are: **pseudo-Euclidean distance** (Chap. 7), **standardized Euclidean distance** and first two metrics (Euclidean $\mathbb{R}^6$-distances) in Sect. 18.3.

- **Mahalanobis distance**
The **Mahalanobis distance** (or **quadratic distance**, or *directionally weighted distance*) is a semimetric on $\mathbb{R}^n$ defined (Mahalanobis, 1936) by

$$||x - y||_A = \sqrt{(x - y)A(x - y)^T},$$

where A is a positive-semidefinite matrix. It is a metric if A is positive-definite. Cf. **Mahalanobis semimetric** in Chap. 14. The square $||x - y||_A^2$ is called *generalized ellipsoid* (or *generalized squared interpoint*) *distance*.
Usually, $A = C^{-1}$, where C is a *covariance matrix* $((Cov(x_i, x_j)))$ of some data points $x, y \in \mathbb{R}^n$ (say, random vectors with the same distribution), or $A = (det(C))^{\frac{1}{n}} C^{-1}$ so that $det(A) = 1$.
Clearly, $||x - y||_I$ is the Euclidean distance. If $C = ((c_{ij}))$ is a diagonal matrix, then $c_{ii} = Var(x_i) = Var(y_i) = \sigma_i^2$ and it holds

$$||x - y||_{C^{-1}} = \sqrt{\sum_i \frac{(x_i - y_i)^2}{\sigma_i^2}}.$$

Such diagonal Mahalanobis distance is called the **standardized Euclidean distance** (or **normalized Euclidean distance**, *scaled Euclidean distance*).
The **maximum scaled difference** (Maxwell–Buddemeier, 2002) is defined by

$$\max_i \frac{(x_i - y_i)^2}{\sigma_i^2}.$$

17.3 Similarities and Distances for Binary Data

Usually, such similarities s range from 0 to 1 or from -1 to 1; the corresponding distances are usually $1 - s$ or $\frac{1-s}{2}$, respectively.

- **Hamann similarity**
  The **Hamann similarity**, 1961, is a similarity on $\{0, 1\}^n$, defined by

$$\frac{2|\overline{X \Delta Y}|}{n} - 1 = \frac{n - 2|X \Delta Y|}{n}.$$

- **Rand similarity**
  The **Rand similarity** (or Sokal–Michener's *simple matching*) is a similarity on $\{0, 1\}^n$ defined by

$$\frac{|\overline{X \Delta Y}|}{n} = 1 - \frac{|X \Delta Y|}{n}.$$

Its square root is called the *Euclidean similarity*. The corresponding metric $\frac{|X \Delta Y|}{n}$ is called the *variance* or *Manhattan similarity*; cf. **Penrose size distance**.

- **Sokal–Sneath similarities**
  The **Sokal-Sneath similarities** 1, 2, 3 are the similarity on $\{0,1\}^n$ defined by

$$\frac{2|\overline{X\Delta Y}|}{n+|\overline{X\Delta Y}|},\quad \frac{|X\cap Y|}{|X\cup Y|+|X\Delta Y|},\quad \frac{|X\Delta Y|}{|\overline{X\Delta Y}|}.$$

- **Russel–Rao similarity**
  The **Russel–Rao similarity** is a similarity on $\{0,1\}^n$, defined by

$$\frac{|X\cap Y|}{n}.$$

- **Simpson similarity**
  The **Simpson similarity** (*overlap similarity*) is a similarity on $\{0,1\}^n$ defined by

$$\frac{|X\cap Y|}{\min\{|X|,|Y|\}}.$$

- **Forbes–Mozley similarity**
  The **Forbes–Mozley similarity** is a similarity on $\{0,1\}^n$ defined by

$$\frac{n|X\cap Y|}{|X||Y|}.$$

- **Braun–Blanquet similarity**
  The **Braun–Blanquet similarity** is a similarity on $\{0,1\}^n$ defined by

$$\frac{|X\cap Y|}{\max\{|X|,|Y|\}}.$$

  The average between it and the **Simpson similarity** is the **Dice similarity**.
- **Roger–Tanimoto similarity**
  The **Roger–Tanimoto similarity**, 1960, is a similarity on $\{0,1\}^n$ defined by

$$\frac{|\overline{X\Delta Y}|}{n+|X\Delta Y|}.$$

- **Faith similarity**
  The **Faith similarity** is a similarity on $\{0,1\}^n$, defined by

$$\frac{|X\cap Y|+|\overline{X\Delta Y}|}{2n}.$$

- **Tversky similarity**
  The **Tversky similarity** is a similarity on $\{0, 1\}^n$, defined by

$$\frac{|X \cap Y|}{a|X\Delta Y| + b|X \cap Y|}.$$

  It becomes the **Tanimoto**, **Dice** and (the binary case of) **Kulczynsky** 1 **similarities** for $(a, b) = (1, 1)$, $(\frac{1}{2}, 1)$ and $(1, 0)$, respectively.
- **Mountford similarity**
  The **Mountford similarity**, 1962, is a similarity on $\{0, 1\}^n$, defined by

$$\frac{2|X \cap Y|}{|X||Y \backslash X| + |Y||X \backslash Y|}.$$

- **Gower–Legendre similarity**
  The **Gower–Legendre similarity** is a similarity on $\{0, 1\}^n$ defined by

$$\frac{|\overline{X\Delta Y}|}{a|X\Delta Y| + |\overline{X\Delta Y}|} = \frac{|\overline{X\Delta Y}|}{n + (a-1)|X\Delta Y|}.$$

- **Anderberg similarity**
  The **Anderberg** (or *Sokal–Sneath* 4 **similarity**) on $\{0, 1\}^n$ is defined by

$$\frac{|X \cap Y|}{4}\left(\frac{1}{|X|} + \frac{1}{|Y|}\right) + \frac{|\overline{X \cup Y}|}{4}\left(\frac{1}{|\overline{X}|} + \frac{1}{|\overline{Y}|}\right).$$

- **Yule similarities**
  The **Yule Q similarity** (Yule, 1900) is a similarity on $\{0, 1\}^n$, defined by

$$\frac{|X \cap Y| \cdot |\overline{X \cup Y}| - |X \backslash Y| \cdot |Y \backslash X|}{|X \cap Y| \cdot |\overline{X \cup Y}| + |X \backslash Y| \cdot |Y \backslash X|}.$$

  The **Yule Y similarity of colligation** (1912) is a similarity on $\{0, 1\}^n$ defined by

$$\frac{\sqrt{|X \cap Y| \cdot |\overline{X \cup Y}|} - \sqrt{|X \backslash Y| \cdot |Y \backslash X|}}{\sqrt{|X \cap Y| \cdot |\overline{X \cup Y}|} + \sqrt{|X \backslash Y| \cdot |Y \backslash X|}}.$$

- **Dispersion similarity**
  The **dispersion similarity** is a similarity on $\{0, 1\}^n$, defined by

$$\frac{|X \cap Y| \cdot |\overline{X \cup Y}| - |X \backslash Y| \cdot |Y \backslash X|}{n^2}.$$

- **Pearson ϕ similarity**
  The **Pearson ϕ similarity** is a similarity on $\{0, 1\}^n$ defined by
$$\frac{|X \cap Y| \cdot |\overline{X \cup Y}| - |X \backslash Y| \cdot |Y \backslash X|}{\sqrt{|X| \cdot |\overline{X}| \cdot |Y| \cdot |\overline{Y}|}}.$$
- **Gower similarity** 2
  The **Gower** 2 (or *Sokal–Sneath* 5) **similarity** on $\{0, 1\}^n$ is defined by
$$\frac{|X \cap Y| \cdot |\overline{X \cup Y}|}{\sqrt{|X| \cdot |\overline{X}| \cdot |Y| \cdot |\overline{Y}|}}.$$
- **Pattern difference**
  The **pattern difference** is a distance on $\{0, 1\}^n$, defined by
$$\frac{4|X \backslash Y| \cdot |Y \backslash X|}{n^2}.$$
- **Q_0-difference**
  The **Q_0-difference** is a distance on $\{0, 1\}^n$, defined by
$$\frac{|X \backslash Y| \cdot |Y \backslash X|}{|X \cap Y| \cdot |\overline{X \cup Y}|}.$$
- **Model distance**
  Let X, Y be two data sets, and let λ_j be the eigenvalues of the symmetrized cross-correlation matrix $C_{X \backslash Y\, Y \backslash X} \times C_{Y \backslash X\, X \backslash Y}$.
  The **model distance** (Todeschini, 2004) is a distance on $\{0, 1\}^n$ defined by
$$\sqrt{|X \backslash Y| + |Y \backslash X| - 2 \sum_j \sqrt{\lambda_j}}.$$
  The **CMD-distance** (or, *canonical measure of distance*, Todeschini et al., 2009) is
$$\sqrt{|X| + |Y| - 2 \sum_j \sqrt{\lambda_j}},$$
  where λ_j are the nonzero eigenvalues of the cross-correlation matrix $C_{X\,Y} \times C_{Y\,X}$.

17.4 Correlation Similarities and Distances

The *covariance* between two real-valued random variables X and Y is $Cov(x, y) = \mathbb{E}[(X - \mathbb{E}[X])(Y - \mathbb{E}[Y])] = \mathbb{E}[XY] - \mathbb{E}[X]\mathbb{E}[Y]$. The *variance* of X is $Var(X) = Cov(X, X)$ and the **Pearson correlation** of X and Y is $Corr(X, Y) = \frac{Cov(X,Y)}{\sqrt{Var(X)Var(Y)}}$; cf. Chap. 14.

Let $(X, Y), (X', Y'), (X'', Y'')$ be independent and identically distributed. The *distance covariance* (Székely, 2005) is the square root of $dCov^2(X, Y) = \mathbb{E}[|X - X'||Y - Y'|] + \mathbb{E}[|X - X'|]\mathbb{E}[|Y - Y'|] - \mathbb{E}[|X - X'||Y - Y''|] - \mathbb{E}[|X - X''||Y - Y'|] = \mathbb{E}[|X - X'||Y - Y'|] + \mathbb{E}[|X - X'|]\mathbb{E}[|Y - Y'|] - 2\mathbb{E}[|X - X'||Y - Y''|]$. It is 0 if and only if X and Y are independent. The **distance correlation** $dCor(X, Y)$ is $\frac{dCov(X,Y)}{\sqrt{dCov(X,X)dCov(Y,Y)}}$.

The vectors x, y below can be seen as *samples* (series of n measurements) of X, Y.

- **Covariance similarity**
  The **covariance similarity** is a similarity on $\mathbb{R}^n$ defined by

$$\frac{\sum(x_i - \overline{x})(y_i - \overline{y})}{n} = \frac{\sum x_i y_i}{n} - \overline{x} \cdot \overline{y}.$$

- **Pearson correlation similarity**
  The **Pearson correlation similarity**, or, by its full name, *Pearson product-moment correlation coefficient*) is a similarity on $\mathbb{R}^n$ defined by

$$s = \frac{\sum(x_i - \overline{x})(y_i - \overline{y})}{\sqrt{\sum(x_j - \overline{x})^2 \sum(y_j - \overline{y})^2}}.$$

  The **Pearson distance** (or **correlation distance**) is defined by

$$1 - s = \frac{1}{2}\sum\left(\frac{x_i - \overline{x}}{\sqrt{\sum(x_j - \overline{x})^2}} - \frac{y_i - \overline{y}}{\sqrt{\sum(y_j - \overline{y})^2}}\right)$$

  A multivariate generalization of the Pearson correlation similarity is the *RV coefficient* (Escoufier, 1973) $RV(X, Y) = \frac{Covv(X,Y)}{\sqrt{Covv(X,X)Covv(Y,Y)}}$, where X, Y are matrices of centered random (column) vectors with covariance matrix $C(X, Y) = \mathbb{E}[X^T Y]$, and $Covv(X, Y)$ is the trace of the matrix $C(X, Y)C(Y, X)$.
- **Cosine similarity**
  The **cosine similarity** (or *Orchini similarity*, *angular similarity*, *normalized dot product*) is the case $\overline{x} = \overline{y} = 0$ of the **Pearson correlation similarity**, i.e., it is

$$\frac{\langle x, y\rangle}{||x||_2 \cdot ||y||_2} = \cos \phi,$$

where ϕ is the angle between vectors x and y. In the binary case, it becomes

$$\frac{|X \cap Y|}{\sqrt{|X| \cdot |Y|}}$$

and is called the **Ochiai–Otsuka similarity**.

In Record Linkage, cosine similarity is called **TF-IDF**; it (or *tf-idf, TFIDF*) are used as an abbreviation of *Frequency–Inverse Document Frequency.*

The **angular semimetric** on $\mathbb{R}^n$ is defined by $\arccos\phi$. The **cosine distance** is $1 - \cos\phi$, and the **Orloci distance** (or *chord distance*) is

$$\sqrt{2(1-\cos\phi)} = \sqrt{\sum\left(\frac{x_i}{||x||_2} - \frac{y_i}{||y||_2}\right)^2}.$$

- **Similarity ratio**

  The **similarity ratio** (or *Kohonen similarity, Kumar–Hassebrook similarity*) is a similarity on $\mathbb{R}^n$ defined by

$$\frac{\langle x, y\rangle}{\langle x, y\rangle + ||x-y||_2^2}.$$

  Its binary case is the **Tanimoto similarity**. Sometimes, the similarity ratio is called the *Tanimoto coefficient* or *extended Jaccard coefficient.*

- **Morisita–Horn similarity**

  The **Morisita–Horn similarity** (Morisita, 1959) is a similarity on $\mathbb{R}^n$ defined by

$$\frac{2\langle x, y\rangle}{||x||_2^2 \cdot \frac{\overline{y}}{\overline{x}} + ||y||_2^2 \cdot \frac{\overline{x}}{\overline{y}}}.$$

- **Spearman rank correlation**

  If the sequences $x, y \in \mathbb{R}^n$ are ranked separately, then the **Pearson correlation similarity** is approximated by the following **Spearman** ρ **rank correlation**:

$$\frac{\sum(a_i - \overline{a})(b_i - \overline{b})}{\sqrt{\sum(a_j - \overline{a})^2 \sum(b_j - \overline{b})^2}} = 1 - \frac{6}{n(n^2-1)}\sum(a_i - b_i)^2,$$

  where $n > 1$ and $a_i = \text{rank}(x_i), b_i = \text{rank}(y_i), a = (a_1, \dots, a_n), b = (b_1, \dots, b_n)$. This approximation is good for such *ordinal* data when it holds $\overline{x} = \overline{y} = \frac{n+1}{2}$.

  The **Spearman footrule** is defined by

$$1 - \frac{3}{n^2-1}\sum|x_i - y_i|.$$

Cf. the **Spearman ρ distance** and **Spearman footrule distance** in Chap. 11. Another correlation similarity for rankings is the **Kendall τ rank correlation**:

$$\frac{2\sum_{1\leq i<j\leq n} sign(x_i - x_j)sign(y_i - y_j)}{n(n-1)}.$$

Cf. the **Kendall τ distance** on permutations in Chap. 11.

- **Global correlation distance**
Let $x \in \mathbb{R}^n$ and (A, d) be a metric space with n points $a_1, \dots, a_n$. For any $d > 0$, the *Moran autocorrelation coefficient* is defined by

$$I(d) = \frac{n\sum_{1\leq i\neq j\leq n} w_{ij}(d)(x_i - \overline{x})(x_j - \overline{x})}{\sum_{1\leq i\neq j\leq n} w_{ij}(d)\sum_{1\leq i\leq n}(x_i - \overline{x})^2},$$

where the weight $w_{ij}(d)$ is 1 if $d(a_i, a_j) = d$ and 0, otherwise. In **spatial analysis**, eventual clustering of (A, d) implies that $I(d)$ decreases with increasing d. $I(d)$ is a global indicator of the presumed **spatial dependence** that evaluate the existence/size of clusters in the spatial arrangement (A, d) of a given variable. The **global correlation distance** is the least value d' for which $I(d) = 0$.

- **Log-likelihood distance**
Given two clusters A and B, their **log-likelihood distance** is the decrease in *log-likelihood* (cf. the **Kullback–Leibler distance** in Chap. 14 and the **log-likelihood ratio quasi-distance** in Chap. 21) as they are combined into one cluster. Simplifying (taking $A, B \subset \mathbb{R}_{>0}$), it is defined by

$$\sum_{x\in A} x\log\frac{x}{|A|} + \sum_{x\in B} x\log\frac{x}{|B|} - \sum_{x\in A\cup B} x\log\frac{x}{|A\cup B|}.$$

- **Spatial analysis**
In Statistics, **spatial analysis** (or *spatial statistics*) includes the formal techniques for studying entities using their topological, geometric, or geographic properties. More restrictively, it refers to *Geostatistics* and *Human Geography*. It considers spatially distributed data as *a priori* dependent one on another.
Spatial dependence is a measure for the degree of associative dependence between independently measured values in an ordered set, determined in samples selected at different positions in a sample space. Cf. **spatial correlation** in Chap. 24. An example of such space-time dynamics: Gould, 1997, showed that $\approx$80 % of the diffusion of HIV in US is highly correlated with the air passenger traffic (origin-destination) matrix for 102 major urban centers.
SADIE (Spatial Analysis by Distance IndicEs) is a methodology (Perry, 1998) to measure the degree of nonrandomness in 2D spatial patterns of populations. Given n sample units $x_i \in \mathbb{R}^2$ with associated counts N_i, the **distance to regularity** is the minimal total Euclidean distance that the individuals in the sample would have to move, from unit to unit, so that all units contained an

identical number of individuals. The **distance to crowding** is the minimal total distance that individuals in the sample must move so that all are congregated in one unit. The *indices of aggregation* are defined by dividing above distances by their mean values. Cf. **Earth Mover's distance** in Chap. 21.

- **Distance sampling**
  Distance sampling is a widely-used group of methods for estimating the density and abundance of biological populations. It is an extension of plot- (or quadrate-based) sampling, where the number of objects at given distance from a point or a segment is counted. Also, *Distance* is the name of a Windows-based computer package that allows to design and analyze distance sampling surveys.
  A standardized survey along a series of lines or points is performed, searching for objects of interest (say, animals, plants or their clusters). *Detection distances* r (perpendicular ones from given lines and radial ones from given points) are measured to each detected object. The *detection function* $g(r)$ (the probability that an object at distance r is detected) is fit then to the observed distances, and this fitted function is used to estimate the proportion of objects missed by the survey. It gives estimates for the density and abundance of objects in the survey area.
- **Cook distance**
  The **Cook distance** is a distance on $\mathbb{R}^n$ giving a statistical measure of deciding if some i-th observation alone affects much regression estimates. It is a normalized **squared Euclidean distance** between estimated parameters from regression models constructed from all data and from data without i-th observation.
  The main similar distances, used in Regression Analysis for detecting influential observations, are *DFITS distance*, *Welsch distance*, and *Hadi distance*.
- **Periodicity p-self-distance**
  Ergun–Muthukrishnan–Sahinalp, 2004, call a data stream $x = (x_1, \dots, x_n)$ *p-periodic approximatively*, for given $1 \le p \le \frac{n}{2}$ and distance function d between p-blocks of x, if the **periodicity p-self-distance** $\sum_{i \neq j} d((x_{jp+1}, \dots, x_{jp+p}), (x_{ip+1}, \dots, x_{ip+p}))$ is below some threshold.
  Above notion of self-distance is different from ones given in Chaps. 1 and 28. Also, the term *self-distance* is used for *round-off error* (or *rounding error*), i.e., the difference between the calculated approximation of a number and its exact value.
- **Distance metric learning**
  Let $x_1, \dots, x_n$ denote the samples in the training set $X \subset \mathbb{R}^m$; here m is the number of features. **Distance metric learning** is an approach for the problem of clustering with side information, when algorithm learns a distance function d prior to clustering and then tries to satisfy some *positive* (or *equivalence*) constraints P and *negative* constraints D. Here S and D are the sets of *similar* (belonging to the same class) and *dissimilar* pairs (x_i, x_j), respectively.
  Usually d is a **Mahalanobis metric** $||x_i - x_j||_A = \sqrt{(x_i - x_j)^T A(x_i - x_j)}$, where A is a positive-semidefinite matrix, i.e., $A = W^T W$ for a matrix W with m columns and $||x_i - x_j||_A^2 = ||Wx_i - Wx_j||^2$. Then, for example, one look for

(Xing et al., 2003) A minimizing $\sum_{(x_i,x_j)\in S} ||x_i - x_j||_A^2$ while $\sum_{(x_i,x_j)\in D} ||x_i - x_j||_A^2 \geq 1$.

- **Heterogeneous distance**

  The following IBL (instance-based learning) setting is used for many real-world applications (neural networks, etc.), where data are incomplete and have both continuous and nominal attributes. Given an $m \times (n+1)$ matrix $((x_{ij}))$, its row $(x_{i0}, x_{i1}, \dots, x_{in})$ denotes an *instance input vector* $x_i = (x_{i1}, \dots, x_{in})$ with output class x_{i0}; the set of m instances represents a training set during learning. For any new input vector $y = (y_1, \dots, y_n)$, the closest (in terms of a selected distance d) instance x_i is sought, in order to *classify* y, i.e., predict its output class as x_{i0}.

  A **heterogeneous distance** $d(x_i, y)$ is defined [WiMa97] by

$$\sqrt{\sum_{j=1}^{n} d_j^2(x_{ij}, y_j)}$$

  with $d_j(x_{ij}, y_j) = 1$ if x_{ij} or y_j is unknown. If the *attribute* (input variable) j is nominal, then $d_j(x_{ij}, y_j)$ is defined, for example, as $1_{x_{ij}\neq y_j}$, or as

$$\sum_a \left| \frac{|\{1 \leq t \leq m : x_{t0} = a, x_{tj} = x_{ij}\}|}{|\{1 \leq t \leq m : x_{tj} = x_{ij}\}|} - \frac{|\{1 \leq t \leq m : x_{t0} = a, x_{tj} = y_j\}|}{|\{1 \leq t \leq m : x_{tj} = y_j\}|} \right|^q$$

  for $q = 1$ or 2; the sum is taken over all output classes, i.e., values a from $\{x_{t0} : 1 \leq t \leq m\}$. For continuous attributes j, the number d_j is taken to be $|x_{ij} - y_j|$ divided by $\max_t x_{tj} - \min_t x_{tj}$, or by $\frac{1}{4}$ of the standard deviation of values x_{tj}, $1 \leq t \leq m$.

Chapter 18
Distances in Systems and Mathematical Engineering

In this chapter we group the main distances used in Systems Theory (such as *Transition Systems*, *Dynamical Systems*, *Cellular Automata*, *Feedback Systems*) and other interdisciplinary branches of Mathematics, Engineering and Theoretical Computer Science (such as, say, *Robot Motion* and *Multi-objective Optimization*).

A *labeled transition system* (LTS) is a triple (S, T, F) where S is a set of *states*, T is a set of *labels* (or *actions*) and $F \subseteq S \times T \times S$ is a ternary relation. Any $(x, t, y) \in F$ represents a t*-labeled transition* from state x to state y. A LTS with $|T| = 1$ corresponds to an *unlabeled transition system*.

A *path* is a sequence $((x_1, t_1, x_2), \dots, (x_i, t_i, x_{i+1}), \dots)$ of transitions; it gives rise to a *trace* $(t_1, \dots, t_i, \dots)$. Two paths are *trace-equivalent* if they have the same traces. The term *trace*, in Computer Science, refers in general to the equivalence classes of strings of a *trace monoid*, wherein certain letters in the string are allowed to commute. It is not related to the *trace* in Linear Algebra.

A LTS is called *deterministic* if for any $x \in S$ and $t \in T$ it holds that $|\{y \in S : (x, t, y) \in F\}| = 1$. Such LTS without output is called a *semiautomaton* (S, T, f) where S is a set of *states*, T is an *input alphabet* and $f : X \times T \to S$ is a *transition function*.

A *deterministic finite-state machine* is a tuple (S, s_0, T, f, S') with S, T, f as above but $0 < |S|, |T| < \infty$, while $s_o \in S$ is an *initial state*, and $S' \subset S$ is the set of *final states*.

The *free monoid* on a set T is a *monoid* (algebraic structure with an associative binary operation and an identity element) T^* whose elements are all the finite sequences $x = x_0, \dots, x_m$ of elements from T. The identity element is the empty string λ, and the operation is string concatenation. The free semigroup on T is $T^+ = T^* \setminus \{\lambda\}$. Let T^ω denote the set of all infinite sequences $x = (x_0, x_1, \dots)$ in T, and let T^∞ denote $T^* \cup T^\omega$.

A finite-state machine is *nondeterministic* if the next possible state is not uniquely determined. A *weighted automaton* is a such machine, say, M equipped

M.M. Deza, E. Deza, *Encyclopedia of Distances*,
DOI 10.1007/978-3-662-44342-2_18

with a cost function $c \geq 0$, over some semiring $(S, \oplus, \otimes)$, on transitions. For a *probabilistic automaton*, the semiring is $(\mathbb{R}_{\geq 0}, +, \times)$ and $0 \leq c \leq 1$.

A **distance automaton** is (Hashiguchi, 1982) a weighted automaton over the *tropical semiring TROP* $= (\mathbb{N} \cup \{\infty\}, \min, +)$. A *run* over a word (string in the *language* of M) $(a_1, \dots, a_k)$ is a sequence $(s_0, \dots, s_k)$ of states. The *run's distance* is the sum $\sum_{i=1}^{k} c(a_i)_{p_{i-1}p_i}$ of costs of involved transitions. The run is *accepting* if s_0 is initial and s_k is a final state. The *distance of a word recognized by M* is the minimum of the distances over the all accepting runs. The *distance of M* is the supremum over the distances of all recognized words. Distance automata are equivalent to finitely generated monoids of matrices over *TROP*: nondeterministic automata recognize the same language as some deterministic ones but with transitions acting on the sets of original states.

18.1 Distances in State Transition and Dynamical Systems

- **Fahrenberg–Legay–Thrane distances**
  Given a *labeled transition system* (LTS) (S, T, F) Fahrenberg–Legay–Thrane, 2011, call T^{∞} the set of *traces* and define a *trace distance* as an extended **hemimetric** (or **quasi-semimetric**) $h : T^{\infty} \times T^{\infty} \to \mathbb{R}_{\geq 0} \cup \{\infty\}$ such that $h(x, y) = \infty$ for any sequences $x, y \in T^{\infty}$ of different length.
  For a given distance d on the set T of labels and a *discount factor* q $(0 < q \leq 1)$, they defined the *pointwise*, *accumulating* and *limit-average trace distance* as, respectively, $PW_{d,q}(x, y) = \sup_i q^{i}d(x_i, y_i)$, $ACC_{d,q}(x, y) = \sum_i q^{i}d(x_i, y_i)$ and $AVG_d = \underline{\lim}_{i \to \infty} \frac{1}{i+1} \sum_{j=0}^{i} d(x_j, y_j)$.
  If d is a **discrete metric**, i.e., $d(t, t') = 1$ whenever $t \neq t'$, then $ACC_{d,1}$ is the **Hamming metric** for finite traces of the same length, and $ACC_{d,q}$ with $q < 1$ and AVG_d are analogs of the Hamming metric for infinite traces.
  Other examples of trace distances are a Cantor-like distance $(1 + \inf\{i : x_i \neq y_i\})^{-1}$ and the *maximum-lead distance*, defined, for $T \subseteq \Sigma \times \mathbb{R}$, by Henzinger–Majumdar–Prabhu, 2005, as $\sup_i |\sum_{j=0}^{i} x_j'' - \sum_{j=0}^{i} y_j''|$ if $x_i' = y_i'$ for all i and ∞, otherwise. Here any $z \in T$ is denoted by (z', z''), where $z' \in \Sigma$ and $z'' \in \mathbb{R}$.
  Fahrenberg–Legay–Thrane, 2011, also define the two following extended simulation hemimetrics between states $x, y \in S$.
  The *accumulating simulation distance* $h_{ac}(x, y)$ and the *pointwise simulation distance* $h_{po}(x, y)$ are the least fixed points, respectively, to the set of equations

$$h_{ac}(x, y) = \max_{t \in T:(x,t,x') \in F} \min_{t' \in T:(y,t',y') \in F} (d(t, t') + qh_{ac}(x', y')) \text{ and}$$

$$h_{po}(x, y) = \max_{t \in T:(x,t,x') \in F} \min_{t' \in T:(y,t',y') \in F} \max\{d(t, t'), h_{po}(x', y')\}.$$

The above hemimetrics generalize the lifting by Alfaro–Faella–Stoelinga, 2004, of the quasi-metric $\max\{x'' - y'', 0\}$ between labels $x, y \in T = \Sigma \times \mathbb{R}$ on an accumulating trace distance and then the lifting of it on the **directed Hausdorff distance** (Chap. 1) between the sets of traces from two given states.
The case $h_{ac}(x, y) = h_{po}(x, y) = 0$ corresponds to the *simulation* of x by y, written $x \preceq y$, i.e., to the existence of a *weighted simulation relation* $R \subseteq S \times S$, i.e., whenever $(x, y) \in R$ and $(x, t, x') \in F$, then $(y, t, y') \in F$ with $(x', y') \in R$.
The case $h_{ac}(x, y) < \infty$ or $h_{po}(x, y) < \infty$ corresponds to the existence of an *unweighted simulation relation* $R \subseteq S \times S$, i.e., whenever $(x, y) \in R$ and $(x, t, x') \in F$, then $(y, t', y') \in F$ with $(x', y') \in R$ and $d(t, t') < \infty$.
The relation $\preceq$ is a pre-order on S. Two states x and y are *similar* if $x \preceq y$ and $y \preceq x$; they are *bisimilar* if, moreover, the simulation R of x by y is the inverse of the simulation of y by x. Similarity is an equivalence relation on S which is coarser than the bisimilarity congruence.
The above trace and similarity system hemimetrics are quantitative generalizations of system relations: trace-equivalence and simulation pre-order, respectively.

- **Cellular automata distances**
  Let $S, |S| \geq 2$, be a finite set (*alphabet*), and let S^∞ be the set of $\mathbb{Z}$-indexed bi-infinite sequences $\{x_i\}_{i=-\infty}^{\infty}$ (*configurations*) of elements of S. A (one-dimensional) *cellular automaton* is a continuous self-map $f : S^\infty \to S^\infty$ that commutes with all *shift* (or *translation*) maps $g : S^\infty \to S^\infty$ defined by $g(x_i) = x_{i+1}$.
  Such cellular automaton form a discrete **dynamical system** with the time set $T = \mathbb{Z}$ (of *cells*, positions of a finite-state machine) on the finite-state space S.
  The main distances between configurations $\{x_i\}_i$ and $\{y_i\}_i$ (see [BFK99]) follow.
  The **Cantor metric** is a metric on S^∞ defined, for $x \neq y$, by

$$2^{-\min\{i \geq 0: |x_i - y_i| + |x_{-i} - y_{-i}| \neq 0\}}.$$

  It corresponds to the case $a = \frac{1}{2}$ of the **generalized Cantor metric** in Chap. 11. The corresponding metric space is compact.
  The **Besicovitch semimetric** is a semimetric on S^∞ defined, for $x \neq y$, by

$$\overline{\lim}_{l \to \infty} \frac{|-l \leq i \leq l : x_i \neq y_i|}{2l + 1}.$$

  Cf. **Besicovitch distance** on measurable functions in Chap. 13. The corresponding semimetric space is complete.
  The **Weyl semimetric** is a semimetric on S^∞, defined by

$$\overline{\lim}_{l \to \infty} \max_{k \in \mathbb{Z}} \frac{|k + 1 \leq i \leq k + l : x_i \neq y_i|}{l}.$$

This and the above semimetric are **translation invariant**, but are neither separable nor locally compact. Cf. **Weyl distance** in Chap. 13.

- **Dynamical system**

A (deterministic) **dynamical system** is a tuple (T, X, f) consisting of a metric space (X, d), called the *state space*, a *time set* T and an *evolution function* $f : T \times X \to X$. Usually, T is a monoid, (X, d) is a manifold locally diffeomorphic to a Banach space, and f is a continuous function.

The system is *discrete* if $T = \mathbb{Z}$ (*cascade*) or if $T = \{0, 1, 2 \dots\}$. It is *real* (or *flow*) if T is an open interval in $\mathbb{R}$, and it is a *cellular automaton* if X is finite and $T = \mathbb{Z}^n$. Dynamical systems are studied in Control Theory in the context of stability; Chaos Theory considers the systems with maximal possible instability.

A discrete dynamical system with $T = \{0, 1, 2 \dots\}$ is defined by a self-map $f : X \to X$. For any $x \in X$, its *orbit* is the sequence $\{f^n(x)\}_n$; here $f^n(x) = f(f^{n-1}(x))$ with $f^0(x) = x$. The orbit of $x \in X$ is called *periodic* if $f^n(x) = x$ for some $n > 0$.

A pair $(x, y) \in X \times X$ is called *proximal* if $\underline{\lim}_{n\to\infty} d(f^n(x), f^n(y)) = 0$, and *distal*, otherwise. The system is called *distal* if any pair (x, y) of distinct points is distal.

The dynamical system is called *expansive* if there exists a constant $D > 0$ such that the inequality $d(f^n(x), f^n(y)) \geq D$ holds for any distinct $x, y \in X$ and some n.

An *attractor* is a closed subset A of X such that there exists an *open neighborhood* U of A with the property that $\lim_{n\to\infty} d(f^n(b), A) = 0$ for every $b \in U$, i.e., A *attracts* all nearby orbits. Here $d(x, A) = \inf_{y\in A} d(x, y)$ is the **point-set distance**.

If for large n and small r there exists a number α such that

$$C(X, n, r) = \frac{|\{(i, j) : d(f^i(x), f^j(x)) \leq r, 1 \leq i, j \leq n\}|}{n^2} \sim r^\alpha,$$

then α is called (Grassberger–Hentschel–Procaccia, 1983) the *correlation dimension*.

- **Melnikov distance**

The evolution of a planar **dynamical system** can be represented in a 3D state space with orthogonal coordinate axes Ox, Ox', Ot. A *homoclinic orbit* (nongeneric orbit that joins a saddle point) can be seen in that space as the intersection with a plane of section $t = const$ of the *stable manifold* (the surface consisting of all trajectories that approach $\gamma_0 = Ot$ asymptotically in forward time) and the *unstable manifold* (the surface consisting of all trajectories that approach Ot in reverse time).

Under a sufficiently small perturbation ϵ which is bounded and smooth enough, Ot persists as a smooth curve $\gamma_\epsilon = \gamma_0 + O(\epsilon)$, and the perturbed system has (not coinciding since $\epsilon > 0$) stable and unstable manifolds (the surfaces consisting of all trajectories that approach γ_ϵ in forward and reverse time, respectively) contained in an $O(\epsilon)$ neighborhood of the unperturbed manifolds.

The **Melnikov distance** is the distance between these manifolds measured along a line normal to the unperturbed manifolds, i.e., a direction that is perpendicular to the unperturbed homoclinic orbit. Cf. Sect. 18.2.

- **Fractal**
  For a metric space, its **topological dimension** does not exceed its **Hausdorff dimension**; cf. Chap. 1. A **fractal** is a metric space for which this inequality is strict. (Originally, Mandelbrot defined a fractal as a point set with noninteger Hausdorff dimension.) For example, the *Cantor set*, seen as a compact metric subspace of $(\mathbb{R}, d(x, y) = |x - y|)$ has the Hausdorff dimension $\frac{\ln 2}{\ln 3}$; cf. another **Cantor metric** in Chap. 11. Another classical fractal, the *Sierpinski carpet* of $[0, 1] \times [0, 1]$, is a **complete geodesic** metric subspace of $(\mathbb{R}^2, d(x, y) = ||x - y||_1)$.
  The term *fractal* is used also, more generally, for a *self-similar* (i.e., roughly, looking similar at any scale) object (usually, a subset of $\mathbb{R}^n$). Cf. **scale invariance**.
- **Scale invariance**
  Scale invariance is a feature of laws or objects which do not change if length scales are multiplied by a common factor.
  Examples of scale invariant phenomena are **fractals** and *power laws*; cf. **scale-free network** in Chap. 22 and *self-similarity* in **long range dependence**. Scale invariance arising from a power law $y = Cx^k$, for a constant C and scale exponent k, amounts to linearity $\log y = \log C + k \log x$ for logarithms.
  Much of scale invariant behavior (and complexity in nature) is explained (Bak–Tang–Wiesenfeld, 1987) by *self-organized cruciality (SOC)* of many **dynamical systems**, i.e., the property to have the critical point of a phase transition as an attractor which can be attained spontaneously without any fine-tuning of control parameters.
  Two moving systems are *dynamically similar* if the motion of one can be made identical to the motion of the other by multiplying all lengths by one scale factor, all forces by another one and all time periods by a third scale factor.
  Dynamic similarity can be formulated in terms of dimensionless parameters as, for example, the **Reynolds number** in Chap. 24.
- **Long range dependence**
  A (second-order stationary) stochastic process X_k, $k \in \mathbb{Z}$, is called **long range dependent** (or *long memory*) if there exist numbers $\alpha, 0 < \alpha < 1$, and $c_\rho > 0$ such that $\lim_{k \to \infty} c_\rho k^\alpha \rho(k) = 1$, where $\rho(k)$ is the autocorrelation function. So, correlations decay very slowly (asymptotically hyperbolic) to zero implying that $\sum_{k \in \mathbb{Z}} |\rho(k)| = \infty$, and that events so far apart are correlated (long memory). If the above sum is finite and the decay is exponential, then the process is *short range*.
  Examples of such processes are the exponential, normal and Poisson processes which are memoryless, and, in physical terms, systems in thermodynamic equilibrium. The above power law decay for correlations as a function of time translates into a power law decay of the Fourier spectrum as a function of frequency f and is called $\frac{1}{f}$ *noise*.

A process has a *self-similarity exponent* (or *Hust parameter*) H if X_k and $t^{-H}X_{tk}$ have the same finite-dimensional distributions for any positive t. The cases $H = \frac{1}{2}$ and $H = 1$ correspond, respectively, to purely random process and to exact self-similarity: the same behavior on all scales. Cf. **fractal**, **scale invariance** and, in Chap. 22, **scale-free network**. The processes with $\frac{1}{2} < H < 1$ are long range dependent with $\alpha = 2(1 - H)$.
Long range dependence corresponds to *heavy-tailed* (or *power law*) distributions. The *distribution function* and *tail* of a nonnegative random variable X are $F(x) = P(X \leq x)$ and $\overline{F(x)} = P(X > x)$. A distribution $F(X)$ is *heavy-tailed* if there exists a number $\alpha, 0 < \alpha < 1$, such that $\lim_{x\to\infty} x^{\alpha}\overline{F(x)} = 1$. Many such distributions occur in the real world (for example, in Physics, Economics, the Internet) in both space (distances) and time (durations). A standard example is the Pareto distribution $\overline{F(x)} = x^{-k}, x \geq 1$, where $k > 0$ is a parameter. Cf. Sect. 18.4 and, in Chap. 29, **distance decay**.
Also, the random-copying model (the cultural analog of genetic drift) of the frequency distributions of various cultural traits (such as of scientific papers citations, first names, dog breeds, pottery decorations) results (Bentley–Hahn–Shennan, 2004) in a power law distribution $y = Cx^{-k}$, where y is the proportion of cultural traits that occur with frequency x in the population, and C and k are parameters.
A general *Lévy flight* is a random walk in which the increments have a power law probability distribution.

- **Lévy walks in human mobility**
  A *jump* is a longest straight line trip from one location to another done without a directional change or pause. Consider a 2D *random walk* (taking successive jumps, each in a random direction) model that involves two distributions: a uniform one for the *turning angle* θ_i and a power law $P(l_i) \sim l_i^{-\alpha}$ for the *jump length* l_i.
  Brownian motion has $\alpha \geq 3$ and *normal diffusion*, i.e., the MSD (mean squared displacement) grows linearly with time t: $MSD \sim t^{\gamma}, \gamma = 1$.
  A *Lévy walk* has $1 < \alpha < 3$. Its jump length is *scale-free*, i.e., lacks an average scale $\overline{l_i}$, and it is *superdiffusive*: $MSD \sim t^{\gamma}, \gamma > 1$. Intuitively, Lévy walks consist of many short jumps and, exceptionally, long jumps eliminating the effect of short ones in average jump lengths.
  Lévy walk dispersal was observed in our Web browsing and image scanning, as well as in foraging animals. It might be an optimal search strategy for finding patches of randomly dispersed unpredictable resources: to cluster, in order to save time and effort, closely located activities and then make many short jumps within the clustered areas and a few long jumps among areas. Scale-free Lévy and Brownian search strategies are effective when resources are abundant.
  Human mobility occurs on many length scales, ranging from walking to air travel. On average, humans spend 1.1 h of their daily time budget traveling. Schafer and Victor, 2000, estimated the average travel distance, per person per year, as 1814, 4382 and 6787 km for 1960, 1990 and 2020, respectively.

Brockmann–Hafnagel–Geisel, 2006, studied long range human traffic via the geographic circulation of money. To track a bill, a user stamps it and enters data (serial number, series and local ZIP code) in a computer. The site www.wheresgeorge.com reports the time and distance between the bill's consecutive sightings. Fifty-seven percent of all ≈465,000 considered bills traveled 50–800 km over 9 months in US. The probability of a bill traversing a distance r (an estimate of the probability of humans moving such a distance) followed, over 10–3,500 km, a power law $P(r) = r^{-1.6}$. Banknote dispersal was fractal, and the bill trajectories resembled Lévy walks.

González–Hidalgo–Barabási, 2008, studied the trajectory of 100,000 anonymized mobile phone users (a random sample of 6 million) over 6 months. The probability of finding a user at a location of *rank* k (by the number of times a user was recorded in the vicinity) was $P(k) \sim \frac{1}{k}$. Forty percent of the time users were found at their first two preferred locations (home, work), while spending remaining time in 5–50 places. Phithakkitnukoon et al., 2011, found that ≈80 % of places visited by mobile phone users are within of their *geo-social radius* (nearest social ties' locations) 20 km.

Jiang–Yin–Zhao, 2009, analyzed people's moving trajectories, obtained from GPS data of 50 taxicabs over 6 months in a large street network. They found a Lévy behavior in walks (both origin-destination and between streets) and attributed it to the fractal property of the underlying street network, not to the goal-directed nature of human movement. Rhee et al., 2009, analyzed ≈1,000 h of GPS traces of walks of 44 participants. They also got Lévy walks.

18.2 Distances in Control Theory

Control Theory deals with influencing the behavior of *dynamical systems*. It considers the feedback loop of a *plant* P (a function representing the object to be controlled, a system) and a *controller* C (a function to design). The output y, measured by a sensor, is fed back to the reference value r.

Then the controller takes the *error* $e = r - y$ to make inputs $u = Ce$. Subject to zero initial conditions, the input and output signals to the plant are related by $y = Pu$, where r, u, v and P, C are functions of the frequency variable s. So, $y = \frac{PC}{1+PC}r$ and $y \approx r$ (i.e., one controls the output by simply setting the reference) if PC is large for any value of s.

If the system is modeled by a system of linear differential equations, then its *transfer function* $\frac{PC}{1+PC}$, relating the output with the input, is a rational function. The plant P is *stable* if it has no poles in the closed right half-plane $\mathbb{C}_+=\{z \in \mathbb{C} : Re(z) \geq 0\}$.

The *robust stabilization problem* is: given a *nominal* plant (a model) P_0 and some metric d on plants, find the open ball of maximal radius which is centered in P_0, such that some controller (rational function) C stabilizes every element of this ball.

The *graph* $G(P)$ of the plant P is the set of all bounded input–output pairs $(u, y = Pu)$. Both u and y belong to the *Hardy space* $H^2(\mathbb{C}_+)$ of the right half-plane; the graph is a closed subspace of $H^2(\mathbb{C}_+) + H^2(\mathbb{C}_+)$. In fact, $G(P)$ is a closed subspace of $H^2(\mathbb{C}^2)$, and $G(P) = f(P) \cdot H^2(\mathbb{C}^2)$ for some function $f(P)$, called the *graph symbol*.

Cf. a **dynamical system** and the **Melnikov distance**.

- **Gap metric**
  The **gap metric** between plants P_1 and P_2 (Zames–El-Sakkary, 1980) is defined by

$$gap(P_1, P_2) = ||\Pi(P_1) - \Pi(P_2)||_2,$$

where $\Pi(P_i)$, $i = 1, 2$, is the orthogonal projection of the graph $G(P_i)$ of P_i seen as a closed subspace of $H^2(\mathbb{C}^2)$. We have

$$gap(P_1, P_2) = \max\{\delta_1(P_1, P_2), \delta_1(P_2, P_1)\},$$

where $\delta_1(P_1, P_2) = \inf_{Q \in H^\infty} ||f(P_1) - f(P_2)Q||_{H^\infty}$, and $f(P)$ is a graph symbol.

Here H^∞ is the space of matrix-valued functions that are analytic and bounded in the open right half-plane $\{s \in \mathbb{C} : \Re s > 0\}$; the H^∞-norm is the maximum singular value of the function over this space.

If A is an $m \times n$ matrix with $m < n$, then its n columns span an n-dimensional subspace, and the matrix B of the orthogonal projection onto the column space of A is $A(A^T A)^{-1} A^T$. If the basis is orthonormal, then $B = AA^T$.

In general, the **gap metric** between two subspaces of the same dimension is the l_2*-norm* of the difference of their orthogonal projections; see also the definition of this distance as an **angle distance between subspaces**.

In applications, when subspaces correspond to autoregressive models, the *Frobenius norm* is used instead of the l_2-norm. Cf. **Frobenius distance** in Chap. 12.

- **Vidyasagar metric**
  The **Vidyasagar metric** (or *graph metric*) between plants P_1 and P_2 is defined by

$$\max\{\delta_2(P_1, P_2), \delta_2(P_2, P_1)\},$$

where $\delta_2(P_1, P_2) = \inf_{||Q|| \leq 1} ||f(P_1) - f(P_2)Q||_{H^\infty}$.

The **behavioral distance** is the gap between *extended* graphs of P_1 and P_2; a term is added to the graph $G(P)$, in order to reflect all possible initial conditions (instead of the usual setup with the initial conditions being zero).

- **Vinnicombe metric**
  The **Vinnicombe metric** (ν*-gap metric*) between plants P_1 and P_2 is defined by

$$\delta_\nu(P_1, P_2) = ||(1 + P_2 P_2^*)^{-\frac{1}{2}}(P_2 - P_1)(1 + P_1^* P_1)^{-\frac{1}{2}}||_\infty$$

if $wno(f^*(P_2)f(P_1)) = 0$, and it is equal to 1, otherwise.
Here $f(P)$ is the graph symbol function of plant P. See [Youn98] for the definition of the *winding number* $wno(f)$ of a rational function f and for a good introduction to Feedback Stabilization.

- **Lanzon–Papageorgiou quasi-distance**
Given a plant P, a *perturbed plant* $\hat{P}$ and an uncertainty structure expressed via a *generalized plant* H, let Δ be the set of all possible perturbations that explain the discrepancy between P and $\hat{P}$. Then **Lanzon–Papageorgiou quasi-distance** (2009) between P and $\hat{P}$ is defined as ∞ if $\Delta = \emptyset$ and $\inf_{\delta\in\Delta} ||\delta||_\infty$, otherwise. This quasi-distance corresponds to the worst-case degradation of the stability margin due to a plant perturbation. For standard uncertainity structures H, it is a metric, but it is only a quasi-metric for multiplicative uncertainity.
- **Distance to uncontrollability**
Linear Control Theory concerns a system of the form $\bar{x} = Ax(t) + Bu(t)$, where, at each time t, $x(t) \in \mathbb{C}^n$ is the *state vector*, $u(t) \in \mathbb{C}^m$ is the *control input vector*, and $A \in \mathbb{C}^{n\times n}$, $B \in \mathbb{C}^{n\times m}$ are the given matrices. The system (matrix pair (A, B)) is called *controllable* if, for any initial and final states $x(0)$ and $x(T)$, there exists $u(t), 0 \leq t \leq T$, that drive the state from $x(0)$ to $x(T)$ within finite time, or, equivalently (Kalman, 1963) the matrix $A - \lambda IB$ has full row rank for all $\lambda \in \mathbb{C}$.
The **distance to uncontrollability** (Paige, 1981, and Eising, 1984) is defined as

$$\min\{||[E, F]|| : (A + E, B + F) \text{ is uncontrollable}\} = \min_{\lambda\in\mathbb{C}} \sigma_n(A - \lambda IB),$$

where $||.||$ is the *spectral* or *Frobenius norm* (cf. Sect. 12.3) and $\sigma_n(A - \lambda IB)$ denotes the n-th largest singular value of the $(n \times (n + m))$-matrix $A - \lambda IB$.
A matrix $A \in \mathbb{C}^{n\times n}$ is *stable* if any its eigenvalue λ has real part $Re(\lambda) < 0$. The **distance to instability** is (Van Loan, 1985) $\min\{||E||\} : A + E$ is unstable$\}$, where $||.||$ is one of two above norms. Cf. **nearness matrix problems** in Chap. 12.

18.3 Motion Planning Distances

Automatic motion planning methods are applied in *Robotics*, *Virtual Reality Systems* and *Computer Aided Design*. A **motion planning metric** is a metric used in automatic motion planning methods.

Let a *robot* be a finite collection of rigid links organized in a kinematic hierarchy. If the robot has n degrees of freedom, this leads to an n-dimensional *manifold* C, called the *configuration space* (or *C-space*) of the robot. The *workspace* W of the robot is the space (usually, $\mathbb{E}^3$) in which the robot moves. Usually, it is modeled as the Euclidean space $\mathbb{E}^3$. A **workspace metric** is a motion planning metric in the workspace $\mathbb{R}^3$.

The *obstacle region* CB is the set of all configurations $q \in C$ that either cause the robot to collide with obstacles B, or cause different links of the robot to collide among themselves. The closure $cl(C_{free})$ of $C_{free} = C \backslash \{CB\}$ is called the *space of collision-free configurations*. A *motion planning algorithm* must find a collision-free path from an initial configuration to a goal configuration.

A **configuration metric** is a motion planning metric on the configuration space C of a robot. Usually, the configuration space C consists of six-tuples $(x, y, z, \alpha, \beta, \gamma)$, where the first three coordinates define the position, and the last three the orientation. The orientation coordinates are the angles in radians.

Intuitively, a good measure of the distance between two configurations is a measure of the workspace region swept by the robot as it moves between them (the **swept volume distance**). However, the computation of such a metric is prohibitively expensive.

The simplest approach has been to consider the C-space as a Cartesian space and to use Euclidean distance or its generalizations. For such **configuration metrics**, one normalizes the orientation coordinates so that they get the same magnitude as the position coordinates. Roughly, one multiplies the orientation coordinates by the maximum x, y or z range of the workspace bounding box. Examples of such metrics are given below.

More generally, the configuration space of a 3D rigid body can be identified with the Lie group $ISO(3)$: $C \cong \mathbb{R}^3 \times \mathbb{R}P^3$. The general form of a matrix in $ISO(3)$ is given by

$$\begin{pmatrix} R & X \\ 0 & 1 \end{pmatrix},$$

where $R \in SO(3) \cong \mathbb{R}P^3$, and $X \in \mathbb{R}^3$.

If X_q and R_q represent the translation and rotation components of the configuration $q = (X_q, R_q) \in ISO(3)$, then a configuration metric between configurations q and r is given by $w_{tr}||X_q - X_r|| + w_{rot} f(R_q, R_r)$, where the **translation distance** $||X_q - X_r||$ is obtained using some norm $||.||$ on $\mathbb{R}^3$, and the **rotation distance** $f(R_q, R_r)$ is a positive scalar function which gives the distance between the rotations $R_q, R_r \in SO(3)$. The rotation distance is scaled relative to the translation distance via the weights w_{tr}, w_{rot}.

There are many other types of metrics used in motion planning methods, in particular, the **Riemannian metrics**, the **Hausdorff metric** and, in Chap. 9, the **separation distance**, the **penetration depth distance** and the **growth distances**.

- **Weighted Euclidean $\mathbb{R}^6$-distance**
  The **weighted Euclidean $\mathbb{R}^6$-distance** is a **configuration metric** on $\mathbb{R}^6$ defined, for any $x, y \in \mathbb{R}^6$, by

$$\left(\sum_{i=1}^{3}|x_i - y_i|^2 + \sum_{i=4}^{6}(w_i|x_i - y_i|)^2\right)^{\frac{1}{2}},$$

where $x = (x_1, \dots, x_6)$, x_1, x_2, x_3 are the position coordinates, x_4, x_5, x_6 are the orientation coordinates, and w_i is the normalization factor. Cf. the general, i.e., in $\mathbb{R}^n$, **weighted Euclidean distance** in Chap. 17.

The *scaled weighted Euclidean* $\mathbb{R}^6$*-distance* is defined, for any $x, y \in \mathbb{R}^6$, by

$$\left(s\sum_{i=1}^{3}|x_i - y_i|^2 + (1-s)\sum_{i=4}^{6}(w_i|x_i - y_i|)^2\right)^{\frac{1}{2}}.$$

This distance changes the relative importance of the position and orientation components through the scale parameter s.

- **Weighted Minkowskian distance**

  The **weighted Minkowskian distance** is a **configuration metric** on $\mathbb{R}^6$ defined, for any $x, y \in \mathbb{R}^6$, by

$$\left(\sum_{i=1}^{3}|x_i - y_i|^p + \sum_{i=4}^{6}(w_i|x_i - y_i|)^p\right)^{\frac{1}{p}}.$$

  It gives the same importance to both position and orientation.

- **Modified Minkowskian distance**

  The **modified Minkowskian distance** is a **configuration metric** on $\mathbb{R}^6$ defined, for any $x, y \in \mathbb{R}^6$, by

$$\left(\sum_{i=1}^{3}|x_i - y_i|^{p_1} + \sum_{i=4}^{6}(w_i|x_i - y_i|)^{p_2}\right)^{\frac{1}{p_3}}.$$

  It distinguishes between position and orientation coordinates using the parameters $p_1 \geq 1$ (for the position) and $p_2 \geq 1$ (for the orientation).

- **Weighted Manhattan distance**

  The **weighted Manhattan distance** is a **configuration metric** on $\mathbb{R}^6$ defined, for any $x, y \in \mathbb{R}^6$, by

$$\sum_{i=1}^{3}|x_i - y_i| + \sum_{i=4}^{6}w_i|x_i - y_i|.$$

- **Robot displacement metric**

  The **robot displacement metric** (or *DISP distance*, Latombe, 1991, and LaValle, 2006) is a **configuration metric** on a configuration space C of a robot defined by

$$\max_{a \in A} ||a(q) - a(r)||$$

for any two configurations $q, r \in C$, where $a(q)$ is the position of the point a in the workspace $\mathbb{R}^3$ when the robot is at configuration q, and $||.||$ is one of the norms on $\mathbb{R}^3$, usually the Euclidean norm. Intuitively, this metric yields the maximum amount in workspace that any part of the robot is displaced when moving from one configuration to another (cf. **bounded box metric**).

- **Euler angle metric**
The **Euler angle metric** is a **rotation metric** on the group $SO(3)$ (for the case of using three—Heading–Elevation–Bank—*Euler angles* to describe the orientation of a rigid body) defined by

$$w_{rot}\sqrt{\Delta(\theta_1, \theta_2)^2 + \Delta(\phi_1, \phi_2)^2 + \Delta(\eta_1, \eta_2)^2}$$

for all $R_1, R_2 \in SO(3)$, given by Euler angles $(\theta_1, \phi_1, \eta_1)$, $(\theta_2, \phi_2, \eta_2)$, respectively, where $\Delta(\theta_1, \theta_2) = \min\{|\theta_1 - \theta_2|, 2\pi - |\theta_1 - \theta_2|\}$, $\theta_i \in [0, 2\pi]$, is the **metric between angles**, and w_{rot} is a scaling factor.

- **Unit quaternions metric**
The **unit quaternions metric** is a **rotation metric** on the *unit quaternion representation* of $SO(3)$, i.e., a representation of $SO(3)$ as the set of points (*unit quaternions*) on the *unit sphere* S^3 in $\mathbb{R}^4$ with identified antipodal points $(q \sim -q)$.
This representation of $SO(3)$ suggested a number of possible metrics on it, for example, the following ones:

1. $\min\{||q - r||, ||q + r||\}$,
2. $|| \ln(q^{-1}r)||$,
3. $w_{rot}(1 - |\lambda|)$,
4. $\arccos |\lambda|$,

where $q = q_1 + q_2 i + q_3 j + q_4 k$, $\sum_{i=1}^{4} q_i^2 = 1$, $||.||$ is a norm on $\mathbb{R}^4$, $\lambda = \langle q, r\rangle = \sum_{i=1}^{4} q_i r_i$, and w_{rot} is a scaling factor.

- **Center of mass metric**
The **center of mass metric** is a **workspace metric**, defined as the Euclidean distance between the *centers of mass* of the robot in the two configurations. The center of mass is approximated by averaging all object vertices.
- **Bounded box metric**
The **bounded box metric** is a **workspace metric** defined as the maximum Euclidean distance between any vertex of the *bounding box* of the robot in one configuration and its corresponding vertex in the other configuration.
The **box metric** in Chap. 4 is unrelated.
- **Pose distance**
A **pose distance** provides a measure of dissimilarity between actions of *agents* (including robots and humans) for Learning by Imitation in Robotics.

In this context, agents are considered as *kinematic chains*, and are represented in the form of a *kinematic tree*, such that every link in the kinematic chain is represented by a unique edge in the corresponding tree.
The configuration of the chain is represented by the *pose* of the corresponding tree which is obtained by an assignment of the pair (n_i, l_i) to every edge e_i. Here n_i is the unit normal, representing the orientation of the corresponding link in the chain, and l_i is the length of the link.
The *pose class* consists of all poses of a given kinematic tree. One of the possible pose distances is a distance on a given pose class which is the sum of measures of dissimilarity for every pair of compatible segments in the two given poses.
Another way is to view a *pose* $D(m)$ in the context of the a precedent and a subsequent frames as a *3D point cloud* $\{D^j(i) : m - a \leq i \leq m + a, j \in J\}$, where J is the joint set. The set $D(m)$ contains $k = |J|(2a + 1)$ points (joint positions) $p_i = (x_i, y_i, z_i)$, $1 \leq i \leq k$. Let $T_{\theta,x,z}$ denote the linear transformation which simultaneously rotates all points of a point cloud about the y axis by an angle $\theta \in [0.2\pi]$ and then shifts the resulting points in the xz plane by a vector $(x, 0, z) \in \mathbb{R}^3$. Then the **3D point cloud distance** (Kover and Gleicher, 2002) between the poses $D(m) = (p_i)_{i\in[1,k]}$ and $D(n) = (q_i)_{i\in[1,k]}$ is defined as

$$\min_{\theta,x,z}\{\sum_{i=1}^{k} ||p_i - T_{\theta,x,z}(q_i)||_2^2\}.$$

Cf. **Procrustes distance** in Chap. 21.

- **Joint angle metric**
For a given frame (or pose) i in an animation, let us define $p_i \in \mathbb{R}^3$ as the global (root) position and $q_{i,k} \in S^3$ as the unit quaternion describing the orientation of a joint k from the joint set J. Cf. **unit quaternions metric** and **3D point cloud distance**. The **joint angle metric** between frames x and y is defined as follows:

$$|p_x - p_y|^2 + \sum_{k\in J} w_k |\log(q_{y,k}^{-1} q_{x,k})|^2.$$

The second term describes the weighted sum of the orientation differences; cf. **weighted Euclidean $\mathbb{R}^6$-distance**. Sometimes, the terms expressing differences in derivatives, such as joint velocity and acceleration, are added.

- **Millibot train metrics**
In *Microbotics* (the field of miniature mobile robots), *nanorobot*, *microrobot*, *millirobot*, *minirobot*, and *small robot* are terms for robots with characteristic dimensions at most one micrometer, mm, cm, dm, and m, respectively.
A *millibot train* is a team of heterogeneous, resource-limited millirobots which can collectively share information. They are able to fuse range information from a variety of different platforms to build a global occupancy map that represents a single collective view of the environment.

In the construction of a **motion planning metric** of millibot trains, one casts a series of random points about a robot and poses each point as a candidate position for movement. The point with the highest overall utility is then selected, and the robot is directed to that point. Thus:
the **free space metric**, determined by free space contours, only allows candidate points that do not drive the robot through obstructions;
the **obstacle avoidance metric** penalizes for moves that get too close to obstacles;
the **frontier metric** rewards for moves that take the robot towards open space;
the **formation metric** rewards for moves that maintain formation;
the **localization metric**, based on the separation angle between one or more localization pairs, rewards for moves that maximize localization (see [GKC04]).
Cf. **collision avoidance distance** and **piano movers distance** in Chap. 19.
A *swarm-bot* can form more complex (more sensors and actuators) and flexible (interconnecting at several angles and with less accuracy) configurations.
The wingspan range of flying robots includes 2.8 cm (quadcopter Lisa/S) and 40 m (Global Hawk). During 2012, a robot Papa Mau (PacX Wave Glider), piloted remotely, swam 16,668 km from San Francisco to Australia.

18.4 MOEA Distances

Most optimization problems have several objectives but, for simplicity, only one of them is optimized, and the others are handled as constraints. *Multi-objective optimization* considers (besides some inequality constraints) an objective vector function $f : X \subset \mathbb{R}^n \to \mathbb{R}^k$ from the *search* (or *genotype*, *decision variables*) space X to the *objective* (or *phenotype*, *decision vectors*) space $f(X) = \{f(x) : x \in X\} \subset \mathbb{R}^k$.

A point $x^* \in X$ is a *Pareto-optimal solution* if, for every other $x \in X$, the decision vector $f(x)$ does not *Pareto-dominate* $f(x^*)$, i.e., $f(x) \leq f(x^*)$. The *Pareto-optimal front* is the set $PF^* = \{f(x) : x \in X^*\}$, where X^* is the set of all Pareto-optimal solutions.

Multi-objective evolutionary algorithms (MOEA) produce, at each generation, an *approximation set* (the found Pareto front PF_{known} approximating the desired Pareto front PF^*) in objective space in which no element Pareto-dominates another element. Examples of **MOEA metrics**, i.e., measures evaluating how close PF_{known} is to PF^*, follow.

- **Generational distance**
  The **generational distance** is defined by

$$\frac{(\sum_{j=1}^{m} d_j^2)^{\frac{1}{2}}}{m},$$

where $m = |PF_{known}|$, and d_j is the Euclidean distance (in the objective space) between $f^j(x)$ (i.e., j-th member of PF_{known}) and the nearest member of PF^*. This distance is zero if and only if $PF_{known} = PF^*$.
The term **generational distance** (or *rate of turnover*) is also used for the minimal number of branches between two positions in any system of ranked descent represented by a hierarchical tree. Examples are: **phylogenetic distance** on a phylogenetic tree (cf. Chap. 23), the number of generations separating a photocopy from the original block print, and the number of generations separating the audience at a memorial from the commemorated event.

- **Spacing**
The **spacing** is defined by

$$\left(\frac{\sum_{j=1}^{m}(\overline{d} - d_j)^2}{m-1}\right)^{\frac{1}{2}},$$

where $m = |PF_{known}|$, d_j is the L_1*-metric* (in the objective space) between $f^j(x)$ (i.e., j-th member of PF_{known}) and the nearest other member of PF_{known}, while $\overline{d}$ is the mean of all d_j.

- **Overall nondominated vector ratio**
The **overall nondominated vector ratio** is defined by

$$\frac{|PF_{known}|}{|PF^*|}.$$

- **Crowding distance**
The **crowding distance** (Deb et al., 2002) is a diversity metric assigned to each Pareto-optimal solution. It is the sum, for all objectives, of the absolute difference of the objective values of two nearest solutions on each side, if they exist.
The *boundary solutions*, i.e., those with the smallest or the highest such value, are assigned an infinite crowding distance.

Part V
Computer-Related Distances

Chapter 19
Distances on Real and Digital Planes

19.1 Metrics on Real Plane

Any L_p**-metric** (as well as any **norm metric** for a given norm $||.||$ on $\mathbb{R}^2$) can be used on the plane $\mathbb{R}^2$, and the most natural is the L_2-metric, i.e., the Euclidean metric $d_E(x, y) = \sqrt{(x_1 - y_1)^2 + (x_2 - y_2)^2}$ which gives the length of the straight line segment $[x, y]$, and is the **intrinsic metric** of the plane.

However, there are other, often "exotic", metrics on $\mathbb{R}^2$. Many of them are used for the construction of *generalized Voronoi diagrams* on $\mathbb{R}^2$ (see, for example, **Moscow metric**, **network metric**, **nice metric**). Some of them are used in Digital Geometry.

- **Erdős-type distance problems**
  Those distance problems were given by Erdős and his collaborators, usually, for the Euclidean metric on $\mathbb{R}^2$, but they are of interest for $\mathbb{R}^n$ and for other metrics on $\mathbb{R}^2$. Examples of such problems are to find out:
  the least number of different distances (or largest occurrence of a given distance) in an m-subset of $\mathbb{R}^2$; the largest cardinality of a subset of $\mathbb{R}^2$ determining at most m distances;
  the minimum diameter of an m-subset of $\mathbb{R}^2$ with only integral distances (or, say, without a pair (d_1, d_2) of distances with $0 < |d_1 - d_2| < 1$);
  the *Erdős-diameter* of a given set S, i.e., the minimum diameter of a rescaled set rS, $r > 0$, in which any two different positive distances differ at least by one;
  the largest cardinality of an *isosceles set* in $\mathbb{R}^2$, i.e., a set of points, any three of which form an isosceles triangle;
  existence of an m-subset of $\mathbb{R}^2$ with, for each $1 \leq i \leq m$, a distance occurring exactly i times (examples are known for $m \leq 8$);
  existence of a dense subset of $\mathbb{R}^2$ with rational distances (Ulam problem);
  existence of $m, m > 7$, noncollinear points of $\mathbb{R}^2$ with integral distances;
  forbidden (not occurring within each part) *distances* of a partition of $\mathbb{R}^2$.

M.M. Deza, E. Deza, *Encyclopedia of Distances*,
DOI 10.1007/978-3-662-44342-2_19

The general **Erdős distinct distances problem**, still open for $n > 2$, is to prove that if $A \subset \mathbb{R}^n$, $|A| = m$ and $d(A)$ denotes the set $\{\sum_{i=1}^{n}(x_i - y_i)^2 : x, y \in A\}$, then $|d(A)| \geq Cm^{\frac{n}{2}}$ for some constant $C > 0$. This problem was generalized for distinct "distances" (cf. Chap. 3) over a finite field. Also, its continuous analog, open **Falconer distance problem** is to prove that if the **Hausdorff dimension** of $A \subset \mathbb{R}^n$ is $> \frac{n}{2}$, then 1-dimensional *Lebesque measure* of $d(A)$ is positive. Related result in Quas, 2009: if the *upper density* of $A \subset \mathbb{R}^n$ is positive, then there is $r_0 > 0$ such that for any $r > r_0$, there are $x, y \in A$ with $d_E(x, y) = r$. The *three-distance theorem* (Sós, 1957): given $a \in (0, 1)$ and $n \in \mathbb{N}$, the points $\{0\}, \{a\}, \{2a\}, \ldots, \{na\}$ (mod 1) on the circle of perimeter 1, partition it into $n + 1$ intervals having at most three lengths, one being the sum of the other two.

- **Distance inequalities in a triangle**

  The multitude of inequalities, involving Euclidean distances between points of $\mathbb{R}^n$, is represented below by some **distance inequalities in a triangle**.
  Let $\triangle ABC$ be a triangle on $\mathbb{R}^2$ with side-lengths $a = d(B, C), b = d(C, A), c = d(A, B)$ and area $\mathcal{A} = \frac{1}{4}\sqrt{(a^2 + b^2 + c^2)^2 - 2(a^4 + b^4 + c^4)}$.
  Let P, P' be two arbitrary interior points in $\triangle ABC$. Denote by D_A, D_B, D_C the distances $d(P, A), d(P, B), d(P, C)$ and by d_A, d_B, d_C the **point-line distances** (cf. Chap. 4) from P to the sides BC, CA, AB opposite to A, B, C. For the point P' define D'_A, D'_B, D'_C and d'_A, d'_B, d'_C similarly.
  The point P is *circumcenter* if $D_A = D_B = D_C$; this distance, $R = \frac{abc}{4\mathcal{A}}$, is circumcircle's radius. The point P is *incenter* if $d_A = d_B = d_C$; this distance, $r = \frac{2\mathcal{A}}{a+b+c}$, is incircle's radius. The *centroid* (the center of mass) is the point G of concurrency of three triangle's medians m_a, m_b, m_c; it holds $d(A, G) = \frac{2}{3}m_a, d(B, G) = \frac{2}{3}m_b, d(C, G) = \frac{2}{3}m_c$. The *symmedian point* is the point of concurrency of three triangle's *symmedians* (reflections of medians at corresponding angle bisectors).
  The *orthocenter* is the point of concurrency of three triangle's altitudes. The centroid is situated on the *Euler line* through the circumcenter and the orthocenter, at $\frac{1}{3}$ of their distance. At $\frac{1}{2}$ of their distance lies the center of the circle going through the midpoints of three sides and the feet of three altitudes.

  - If P and P' are the circumcenter and incenter of $\triangle ABC$, then (Euler, 1765)

    $$d^2(P, P') \geq R(R - 2r)$$

    holds implying $R \geq 2r$ with equality if and only if triangle is equilateral. In fact, the general *Euler's inequality* $R \geq nr$ holds (Klamkin–Tsintsifas, 1979) for the radii R, r of circumscribed and inscribed spheres of an n-simplex.
  - For any P, P', the *Erdős–Mordell* inequality (Mordell–Barrow, 1937) is

    $$D_A + D_B + D_C \geq 2(d_A + d_B + d_C).$$

Liu, 2008, generalized above as follows: for all $x, y, z \geq 0$ it holds

$$\sqrt{D_A D'_A}x^2 + \sqrt{D_B D'_B}y^2 + \sqrt{D_C D'_C}z^2$$
$$\geq 2\left(\sqrt{d_A d'_A}yz + \sqrt{d_B d'_B}xz + \sqrt{d_C d'_C}xy\right).$$

– Lemoine, 1873, proved that

$$\frac{4\mathcal{A}^2}{a^2 + b^2 + c^2} \leq d_A^2 + d_B^2 + d_C^2$$

with equality if and only if P is the symmedian point.

– Posamentier and Salkind, 1996, showed

$$\frac{3}{4}(a+b+c) < m_a+m_b+m_c < a+b+c, \text{ while } \frac{3}{4}(a^2+b^2+c^2) = m_a^2+m_b^2+m_c^2.$$

– Kimberling, 2010, proved that

$$d_A d_B d_C \leq \frac{8\mathcal{A}^3}{27abc}$$

with equality if and only if P is the centroid.
He also gave (together with unique point realizing equality) inequality

$$\frac{(2\mathcal{A})^q}{(a^{\frac{2}{q-1}} + b^{\frac{2}{q-1}} + c^{\frac{2}{q-1}})^{q-1}} \leq d_A^q + d_B^q + d_C^q$$

for any $q < 0$ or $q > 1$. For $0 < q < 1$, the reverse inequality holds.

The side-lengths $d(A, B), d(B, C), d(C, A)$ of a *right triangle* are in arithmetic progression only if their ratio is $3:4:5$. They are in geometric progression only if their ratio is $1 : \sqrt{\varphi} : \varphi$, where φ is the *golden section* $\frac{1+\sqrt{5}}{2}$.

- **City-block metric**
The **city-block metric** is the L_1*-metric* on $\mathbb{R}^2$ defined by

$$||x - y||_1 = |x_1 - y_1| + |x_2 - y_2|.$$

It is also called the **taxicab metric**, **Manhattan metric**, **rectilinear metric**, **right-angle metric**, **4-metric** and, on $\mathbb{Z}^n$, **grid metric**. The *von Neumann neighborhood* of a point is the set of points at a Manhattan distance of 1 from it.

- **Chebyshev metric**
The **Chebyshev metric** (or **chessboard metric**, **king-move metric**, **8-metric**) is the L_∞*-metric* on $\mathbb{R}^2$ defined by

$$||x - y||_\infty = \max\{|x_1 - y_1|, |x_2 - y_2|\}.$$

On $\mathbb{Z}^n$, this metric is called also the **lattice** (or **uniform, sup**) **metric**. A point's *Moore neighborhood* is the set of points at a Chebyshev distance of 1.

- **α-metric**
Given $\alpha \in [0, \frac{\pi}{4}]$, the **$\alpha$-metric** for $x, y \in \mathbb{R}^2$ is defined (Tian, 2005) by

$$d_\alpha(x, y) = \max\{|x_1 - y_1|, |x_2 - y_2|\} - (\sec\alpha - \tan\alpha)\min\{|x_1 - y_1|, |x_2 - y_2|.$$

It is the **city-block metric** if $\alpha = 0$. For $\alpha = \frac{\pi}{4}$, i.e., $\sec\alpha - \tan\alpha = \sqrt{2} - 1$, it is the **Chinese checkers metric** (Chen, 1992). *Chinese checkers* (as well as *Hexagonal chess*, *Masonic chess*, *Sannin shogi*, *Hexshogi*) is a strategy board game with hexagonal cells, while *Tiangular chess*, *Tri-chess*, *Trishogi* have triangular cells. Cf. **hexagonal metric**. Gelişgen and Kaya, 2006, generalized α-metric on $\mathbb{R}^n$.

- **Relative metrics on $\mathbb{R}^2$**
The **(p, q)-relative** and **M-relative** metrics are defined in Chap. 5 on any *Ptolemaic space*. The **(p, q)-relative metric** on $\mathbb{R}^2$ (in general, on $\mathbb{R}^n$) is defined (for x or $y \neq 0$) in the cases $1 \leq p < \infty$ and $p = \infty$, respectively, by

$$\frac{||x - y||_2}{(\frac{1}{2}(||x||_2^p + ||y||_2^p))^{\frac{q}{p}}} \text{ and } \frac{||x - y||_2}{(\max\{||x||_2, ||y||_2\})^q}.$$

Let $f : [0, \infty) \to (0, \infty)$ be a convex increasing function such that $\frac{f(x)}{x}$ is decreasing for $x > 0$. The **M-relative metric** on $\mathbb{R}^2$ (in general, on $\mathbb{R}^n$), is defined by

$$\frac{||x - y||_2}{f(||x||_2) \cdot f(||y||_2)}.$$

In particular, the distance below is a metric if and only if $p \geq 1$:

$$\frac{||x - y||_2}{\sqrt[p]{1 + ||x||_2^p}\sqrt[p]{1 + ||y||_2^p}}.$$

A similar metric on $\mathbb{R}^2\backslash\{0\}$ (in general, on $\mathbb{R}^n\backslash\{0\}$) is defined by $\frac{||x-y||_2}{||x||_2 \cdot ||y||_2}$.

- **MBR metric**
The **MBR metric** (Schönemann, 1982, for bounded response scales in Psychology) is a metric $d((x_1, x_2), (y_1, y_2))$ on $\mathbb{R}^2$, defined by

$$\frac{|x_1 - y_1| + |x_2 - y_2|}{1 + |x_1 - y_1||x_2 - y_2|} = \tanh(\operatorname{arctanh}(|x_1 - y_1|) + \operatorname{arctanh}(|x_2 - y_2|)).$$

- **Moscow metric**
The **Moscow metric** (or **Karlsruhe metric**) is a metric on $\mathbb{R}^2$, defined as the minimum Euclidean length of all *admissible* connecting curves between x and

$y \in \mathbb{R}^2$, where a curve is called *admissible* if it consists only of *radial streets* (segments of straight lines passing through the origin) and *circular avenues* (segments of circles centered at the origin); see, for example, [Klei88]).
If the polar coordinates for points $x, y \in \mathbb{R}^2$ are (r_x, θ_x), (r_y, θ_y), respectively, then the distance between them is equal to $\min\{r_x, r_y\}\Delta(\theta_x - \theta_y) + |r_x - r_y|$ if $0 \le \Delta(\theta_x, \theta_y) < 2$, and is equal to $r_x + r_y$ if $2 \le \Delta(\theta_x, \theta_y) < \pi$, where $\Delta(\theta_x, \theta_y) = \min\{|\theta_x - \theta_y|, 2\pi - |\theta_x - \theta_y|\}, \theta_x, \theta_y \in [0, 2\pi)$, is the **metric between angles**.

- **French Metro metric**
  Given a norm $||.||$ on $\mathbb{R}^2$, the **French Metro metric** is a metric on $\mathbb{R}^2$ defined by

$$||x - y|| \text{ if } x = cy \text{ for some } 0 \neq c \in \mathbb{R} \text{ (i.e., } x_1 y_2 = x_2 y_1),$$

and by

$$||x|| + ||y||, \text{ otherwise.}$$

For the Euclidean norm $||.||_2$, it is called the **Paris metric**, **radial metric**, **hedgehog metric**, or **French railroad metric**, **enhanced SNCF metric**.
In this case it can be defined as the minimum Euclidean length of all *admissible* connecting curves between two given points x and y, where a curve is called *admissible* if it consists only of segments of straight lines passing through the origin.
In graph terms, this metric is similar to the **path metric** of the tree consisting of a point from which radiate several disjoint paths. In the case when only one line radiates from the point, this metric is called the **train metric**.
The Paris metric is an example of an $\mathbb{R}$-**tree** T which is *simplicial*, i.e., its set of points x with $T \backslash \{x\}$ not having exactly two components, is discrete and closed.

- **Lift metric**
  The **lift metric** (or **jungle river metric**, *raspberry picker metric*, *barbed wire metric*) is a metric $d((x_1, x_2), (y_1, y_2))$ on $\mathbb{R}^2$ defined (see, for example, [Brya85]) by

$$|x_1 - y_1| \text{ if } x_2 = y_2,$$

and by

$$|x_1| + |x_2 - y_2| + |y_1| \text{ if } x_2 \neq y_2.$$

It is the minimum Euclidean length of all *admissible* (consisting only of segments of straight lines parallel to the x_1 axis and segments of the x_2 axis) connecting curves between points (x_1, x_2) and (y_1, y_2).
The lift metric is an *nonsimplicial* (cf. **French Metro metric**) $\mathbb{R}$-**tree**.

- **Radar screen metric**
  Given a norm $||.||$ on $\mathbb{R}^2$ (in general, on $\mathbb{R}^n$), the **radar screen metric** is a special case of the t-**truncated metric** (Chap. 4) defined by
  $$\min\{1, ||x - y||\}.$$
- **British Rail metric**
  Given a norm $||.||$ on $\mathbb{R}^2$ (in general, on $\mathbb{R}^n$), the **British Rail metric** is a metric defined as 0 for $x = y$ and, otherwise, by
  $$||x|| + ||y||.$$
  It is also called the **Post Office metric**, **caterpillar metric** and **shuttle metric**.
- **Flower-shop metric**
  Let d be a metric on $\mathbb{R}^2$, and let f be a fixed point (a *flower-shop*) in the plane. The **flower-shop metric** (sometimes called **SNCF metric**) is a metric on $\mathbb{R}^2$ (in general, on any metric space) defined by
  $$d(x, f) + d(f, y)$$
  for $x \neq y$ (and is equal to 0, otherwise). So, a person living at point x, who wants to visit someone else living at point y, first goes to f, to buy some flowers. In the case $d(x, y) = ||x - y||$ and the point f being the origin, it is the **British Rail metric**.
  If $k > 1$ flower-shops $f_1, \dots, f_k$ are available, one buys the flowers, where the detour is a minimum, i.e., the distance between distinct points x, y is equal to $\min_{1 \le i \le k}\{d(x, f_i) + d(f_i, y)\}$.
- **Rickman's rug metric**
  Given a number $\alpha \in (0, 1)$, the **Rickman's rug metric** on $\mathbb{R}^2$ is a 2D case of the **parabolic distance** (Chap. 6) defined by
  $$d((x_1, x_2), (y_1, y_2)) = |x_1 - y_1| + |x_2 - y_2|^{\alpha}.$$
- **Burago–Burago–Ivanov metric**
  The **Burago–Burago–Ivanov metric** [BBI01] is a metric on $\mathbb{R}^2$ defined by
  $$|\,||x||_2 - ||y||_2| + \min\{||x||_2, ||y||_2\} \cdot \sqrt{\angle(x, y)},$$
  where $\angle(x, y)$ is the angle between vectors x and y, and $||.||_2$ is the Euclidean norm on $\mathbb{R}^2$. The corresponding **internal metric** on $\mathbb{R}^2$ is equal to $|\,||x||_2 - ||y||_2|$ if $\angle(x, y) = 0$, and is equal to $||x||_2 + ||y||_2$, otherwise.
- $2n$-**gon metric**
  Given a centrally symmetric regular $2n$-gon K on the plane, the $2n$-**gon metric** is a metric on $\mathbb{R}^2$ defined, for any $x, y \in \mathbb{R}^2$, as the shortest Euclidean length of a polygonal line from x to y with each of its sides parallel to some edge of K.

If K is a square with the vertices $\{(\pm 1, \pm 1)\}$, one obtains the **Manhattan metric**. The Manhattan metric arises also as the **Minkowskian metric** with the unit ball being the *diamond*, i.e., a square with the vertices $\{(1,0), (0,1), (-1,0), (0,-1)\}$.

- **Fixed orientation metric**
  Given a set A, $|A| \geq 2$, of distinct *orientations* (i.e., angles with fixed x axis) on the plane $\mathbb{R}^2$, the A**-distance** (Widmayer–Wu–Wong, 1987) is Euclidean length of the shortest (zig-zag) path of line segments with orientations from A. Any A-distance is a metric; it is called also a **fixed orientation metric**.
  A **fixed orientation metric** with $A = \{\frac{i\pi}{n} : 1 \leq i \leq n\}$ for fixed $n \in [2, \infty]$, is called a **uniform orientation metric**; cf. $2n$**-gon metric** above. It is the L_1-metric, **hexagonal metric**, L_2-metric for $n = 2, 3, \infty$, respectively.
- **Central Park metric**
  The **Central Park metric** is a metric on $\mathbb{R}^2$, defined as the length of a shortest L_1-path (*Manhattan path*) between two points $x, y \in \mathbb{R}^2$ in the presence of a given set of areas which are traversed by a shortest Euclidean path (for example, Central Park in Manhattan).
- **Collision avoidance distance**
  Let $\mathcal{O} = \{O_1, \dots, O_m\}$ be a collection of pairwise disjoint polygons on the Euclidean plane representing a set of obstacles which are neither transparent nor traversable.
  The **collision avoidance distance** (or **piano movers distance**, **shortest path metric with obstacles**) is a metric on the set $\mathbb{R}^2 \backslash \{\mathcal{O}\}$, defined, for any $x, y \in \mathbb{R}^2 \backslash \{\mathcal{O}\}$, as the length of the shortest path among all possible continuous paths, connecting x and y, that do not intersect obstacles $O_i \backslash \partial O_i$ (a path can pass through points on the boundary ∂O_i of O_i), $i = 1, \dots m$.
- **Rectilinear distance with barriers**
  Let $\mathcal{O} = \{O_1, \dots, O_m\}$ be a set of pairwise disjoint open polygonal barriers on $\mathbb{R}^2$. A *rectilinear path* (or *Manhattan path*) P_{xy} from x to y is a collection of horizontal and vertical segments in the plane, joining x and y. The path P_{xy} is called *feasible* if $P_{xy} \cap (\cup_{i=1}^{m} B_i) = \emptyset$.
  The **rectilinear distance with barriers** (or *rectilinear distance in the presence of barriers*) is a metric on $\mathbb{R}^2 \backslash \{\mathcal{O}\}$, defined, for any $x, y \in \mathbb{R}^2 \backslash \{\mathcal{O}\}$, as the length of the shortest *feasible rectilinear path* from x to y.
  The rectilinear distance in the presence of barriers is a restriction of the **Manhattan metric**, and usually it is considered on the set $\{q_1, \dots, q_n\} \subset \mathbb{R}^2$ of n *origin-destination points*: the problem to find such a path arises, for example, in Urban Transportation, or in Plant and Facility Layout (see, for example, [LaLi81]).
- **Link distance**
  Let $P \subset \mathbb{R}^2$. The **polygonal distance** (or **link distance** as defined by Suri, 1986) between any two points of P is the smallest number of edges of a polygonal path in P connecting them if such path exists and ∞, otherwise.

If the path is restricted to be rectilinear, one obtains the *rectilinear link distance*. If each line segment of the path is parallel to one from a set A of fixed orientations, one obtains the *A-oriented link distance*; cf. **fixed orientation metric** above.
If the turning points of the path are constrained to lie on the boundary of P, then the path is called *drp* (diffuse reflection path). The *drp-diameter* of P is the minimum number of *diffuse reflections* (segments in a drp) needed to illuminate any target point from any point light source inside P.

- **Facility layout distances**
A *layout* is a partition of a rectangular plane region into smaller rectangles, called *departments*, by lines parallel to the sides of original rectangle. All interior vertices should be of degree 3, and some of them, at least one on the boundary of each department, are *doors*, i.e., input–output locations.
The problem is to design a convenient notion of distance $d(x, y)$ between departments x and y which minimizes the *cost function* $\sum_{x,y} F(x, y)d(x, y)$, where $F(x, y)$ is some *material flow* between x and y. The main distances used are:
the **centroid distance**, i.e., the shortest Euclidean or **Manhattan distance** between *centroids* (the intersections of the diagonals) of x and y;
the **perimeter distance**, i.e., the shortest rectilinear distance between doors of x and y, but going only along the *walls* (department perimeters).
- **Quickest path metric**
A **quickest path metric** (or **network metric, time metric**) is a metric on $\mathbb{R}^2$ (or on a subset of $\mathbb{R}^2$) in the presence of a given *transportation network*, i.e., a finite graph $G = (V, E)$ with $V \subset \mathbb{R}^2$ and edge-weight function $w(e) > 1$: the vertices and edges are *stations* and *roads*. For any $x, y \in \mathbb{R}^2$, it is the time needed for a *quickest path* (i.e., a path minimizing the travel duration) between them when using, eventually, the network.
Movement takes place, either off the network with unit speed, or along its roads $e \in E$ with fixed speeds $w(e) >> 1$, with respect to a given (usually, Euclidean or **Manhattan**) metric d on the plane. The network G can be accessed or exited only at stations (usual discrete model) or at any point of roads (the continuous model).
The **heavy luggage metric** (Abellanas-Hurtado-Palop, 2005) is a quickest path metric on $\mathbb{R}^2$ in the presence of a network with speed 1 outside of the network and speed ∞ (so, travel time 0) inside of it.
The **airlift metric** is a quickest path metric on $\mathbb{R}^2$ in the presence of an *airports network*, i.e., a planar graph $G = (V, E)$ on n vertices (*airports*) with positive edge weights $(w_e)_{e \in E}$ (*flight durations*). The graph may be entered and exited only at the airports. Movement off the network takes place with unit speed with respect to the Euclidean metric. We assume that going by car takes time equal to the Euclidean distance d, whereas the flight along an edge $e = uv$ of G takes time $w(e) < d(u, v)$. In the simplest case, when there is an airlift between two points $a, b \in \mathbb{R}^2$, the distance between x and y is equal to

$$\min\{d(x,y), d(x,a)+w+d(b,y), d(x,b)+w+d(a,y)\},$$

where w is the flight duration from a to b.

The **city metric** is a quickest path metric on $\mathbb{R}^2$ in the presence of a *city public transportation network*, i.e., a planar straight line graph G with horizontal or vertical edges. G may be composed of many connected components, and may contain cycles.

One can enter/exit G at any point, be it at a vertex or on an edge (it is possible to postulate fixed entry points, too). Once having accessed G, one travels at fixed speed $v > 1$ in one of the available directions. Movement off the network takes place with unit speed with respect to the **Manhattan metric**, as in a large modern-style city with streets arranged in north–south and east–west directions.

A variant of such semimetric is the **subway semimetric** defined [O'Bri03], for $x, y \in \mathbb{R}^2$, as $\min(d(x,y), d(x,L)+d(y,L))$, where d is the Manhattan metric and L is a (*subway*) line.

- **Shantaram metric**
  For any numbers a, b with $0 < b \le 2a \le 2b$, the **Shantaram metric** between two points $x, y \in \mathbb{R}^2$ is $0, a$ or b if x and y coincide in exactly $2, 1$ or 0 coordinates, respectively.
- **Periodic metric**
  A metric d on $\mathbb{R}^2$ is called **periodic** if there exist two linearly independent vectors v and u such that the *translation* by any vector $w = mv+nu$, $m, n \in \mathbb{Z}$, preserves distances, i.e., $d(x,y) = d(x+w, y+w)$ for any $x, y \in \mathbb{R}^2$.
  Cf. **translation invariant metric** in Chap. 5.
- **Nice metric**
  A metric d on $\mathbb{R}^2$ with the following properties is called **nice** (Klein–Wood, 1989):

  1. d induces the Euclidean topology;
  2. The d-circles are bounded with respect to the Euclidean metric;
  3. If $x, y \in \mathbb{R}^2$ and $x \ne y$, then there exists a point $z, z \ne x, z \ne y$, such that $d(x,y) = d(x,z) + d(z,y)$;
  4. If $x, y \in \mathbb{R}^2$, $x \prec y$ (where $\prec$ is a fixed order on $\mathbb{R}^2$, the lexicographic order, for example), $C(x,y) = \{z \in \mathbb{R}^2 : d(x,z) \le d(y,z)\}$, $D(x,y) = \{z \in \mathbb{R}^2 : d(x,z) < d(y,z)\}$, and $\overline{D(x,y)}$ is the closure of $D(x,y)$, then $J(x,y) = C(x,y) \cap \overline{D(x,y)}$ is a curve homeomorphic to $(0,1)$. The intersection of two such curves consists of finitely many connected components.

  Every **norm metric** fulfills 1, 2, and 3 Property 2 means that the metric d is continuous at infinity with respect to the Euclidean metric. Property 4 is to ensure that the boundaries of the correspondent *Voronoi diagrams* are curves, and that not too many intersections exist in a neighborhood of a point, or at infinity.
  A nice metric d has a nice Voronoi diagram: in the Voronoi diagram $V(P, d, \mathbb{R}^2)$ (where $P = \{p_1, \dots, p_k\}$, $k \ge 2$, is the set of *generator points*) each *Voronoi region* $V(p_i)$ is a path-connected set with a nonempty interior, and the system $\{V(p_1), \dots, V(p_k)\}$ forms a *partition* of the plane.

- **Contact quasi-distances**
The **contact quasi-distances** are the following variations of the **distance convex function** (cf. Chap. 1) defined on $\mathbb{R}^2$ (in general, on $\mathbb{R}^n$) for any $x, y \in \mathbb{R}^2$.
Given a set $B \subset \mathbb{R}^2$, the **first contact quasi-distance** d_B is defined by

$$\inf\{\alpha > 0 : y - x \in \alpha B\}$$

(cf. **sensor network distances** in Chap. 29).
Given, moreover, a point $b \in B$ and a set $A \subset \mathbb{R}^2$, the **linear contact quasi-distance** is a **point-set distance** defined by $d_b(x, A) = \inf\{\alpha \geq 0 : \alpha b + x \in A\}$.
The **intercept quasi-distance** is, for a finite set B, defined by $\frac{\sum_{b \in B} d_b(x,y)}{|B|}$.
- **Radar discrimination distance**
The **radar discrimination distance** is a distance on $\mathbb{R}^2$ defined by

$$|\rho_x - \rho_y + \theta_{xy}|$$

if $x, y \in \mathbb{R}^2 \backslash \{0\}$, and by

$$|\rho_x - \rho_y|$$

if $x = 0$ or $y = 0$, where, for each $x \in \mathbb{R}^2$, ρ_x denotes the radial distance of x from $\{0\}$ and, for any $x, y \in R^2 \backslash \{0\}$, θ_{xy} denotes the radian angle between them.
- **Ehrenfeucht–Haussler semimetric**
Let S be a subset of $\mathbb{R}^2$ such that $x_1 \geq x_2 - 1 \geq 0$ for any $x = (x_1, x_2) \in S$.
The **Ehrenfeucht–Haussler semimetric** (see [EhHa88]) on S is defined by

$$\log_2 \left(\left(\frac{x_1}{y_2} + 1 \right) \left(\frac{y_1}{x_2} + 1 \right) \right).$$

- **Toroidal metric**
The **toroidal metric** is a metric on $T = [0, 1) \times [0, 1) = \{x = (x_1, x_2) \in \mathbb{R}^2 : 0 \leq x_1, x_2 < 1\}$ defined for any $x, y \in \mathbb{R}^2$ by

$$\sqrt{t_1^2 + t_2^2},$$

where $t_i = \min\{|x_i - y_i|, |x_i - y_i + 1|\}$ for $i = 1, 2$ (cf. **torus metric**).
- **Circle metric**
The **circle metric** is the **intrinsic metric** on the *unit circle* S^1 in the plane.
As $S^1 = \{(x, y) : x^2 + y^2 = 1\} = \{e^{i\theta} : 0 \leq \theta < 2\pi\}$, it is the length of the shorter of the two arcs joining the points $e^{i\theta}, e^{i\vartheta} \in S^1$, and can be written as

$$\min\{|\theta - \vartheta|, 2\pi - |\theta - \vartheta|\} = \begin{cases} |\vartheta - \theta|, & \text{if } 0 \leq |\vartheta - \theta| \leq \pi, \\ 2\pi - |\vartheta - \theta|, & \text{if } \quad |\vartheta - \theta| > \pi. \end{cases}$$

- **Metric between angles**
  The **metric between angles** Δ is a metric on the set of all angles in the plane $\mathbb{R}^2$ defined for any $\theta, \vartheta \in [0, 2\pi)$ (cf. **circle metric**) by

$$\min\{|\theta - \vartheta|, 2\pi - |\theta - \vartheta|\} = \begin{cases} |\vartheta - \theta|, & \text{if } 0 \le |\vartheta - \theta| \le \pi, \\ 2\pi - |\vartheta - \theta|, & \text{if } \quad |\vartheta - \theta| > \pi. \end{cases}$$

- **Metric between directions**
  On $\mathbb{R}^2$, a *direction* $\hat{l}$ is a class of all straight lines which are parallel to a given straight line $l \subset \mathbb{R}^2$. The **metric between directions** is a metric on the set $\mathcal{L}$ of all directions on the plane defined, for any directions $\hat{l}, \hat{m} \in \mathcal{L}$, as the angle between any two representatives.
- **Angular distance**
  The **angular distance** traveled around a circle is the number $\theta = \frac{l}{r}$ of radians the path subtends, $\theta = \frac{l}{r}$, where l is the path length, and r is the circle('s radius.
- **Circular-railroad quasi-metric**
  The **circular-railroad quasi-metric** on the *unit circle* $S^1 \subset \mathbb{R}^2$ is defined, for any $x, y \in S^1$, as the length of the counterclockwise circular arc from x to y in S^1.
- **Inversive distance**
  The **inversive distance** between two nonintersecting circles in the plane $\mathbb{R}^2$ is defined as the natural logarithm of the ratio of the radii (the larger to the smaller) of two concentric circles into which the given circles can be inverted.
  Let c be the distance between the centers of two nonintersecting circles of radii a and $b < a$. Then their inversive distance is given by

$$\cosh^{-1}\left|\frac{a^2 + b^2 - c^2}{2ab}\right|.$$

  The *circumcircle* and *incircle* of a triangle with *circumradius* R and *inradius* r are at the inversive distance $2\sinh^{-1}(\frac{1}{2}\sqrt{\frac{r}{R}})$.
  Given three noncollinear points, construct three tangent circles such that one is centered at each point and the circles are pairwise tangent to one another. Then there exist exactly two nonintersecting circles, called the *Soddy circles*, that are tangent to all three circles. Their inversive distance is $2\cosh^{-1} 2$.

19.2 Digital Metrics

Here we list special metrics which are used in *Computer Vision* (or *Pattern Recognition, Robot Vision, Digital Geometry*).

A *computer picture* (or *computer image*) is a subset of $\mathbb{Z}^n$ which is called a *digital nD space*. Usually, pictures are represented in the *digital plane* (or *image plane*) $\mathbb{Z}^2$, or in the *digital space* (or *image space*) $\mathbb{Z}^3$. The points of $\mathbb{Z}^2$ and $\mathbb{Z}^3$ are called *pixels* and *voxels*, respectively. An nD *m-quantized space* is a scaling $\frac{1}{m}\mathbb{Z}^n$.

A **digital metric** (see, for example, [RoPf68]) is any metric on a digital nD space. Usually, it should take integer values.

The metrics on $\mathbb{Z}^n$ that are mainly used are the L_1- and L_∞-*metrics*, as well as the L_2-*metric* after rounding to the nearest greater (or lesser) integer. In general, a given list of *neighbors* of a pixel can be seen as a list of permitted *one-step moves* on $\mathbb{Z}^2$. Let us associate a **prime distance**, i.e., a positive weight, to each type of such move.

Many digital metrics can be obtained now as the minimum, over all admissible paths (i.e., sequences of permitted moves), of the sum of corresponding prime distances.

In practice, the subset $(\mathbb{Z}_m)^n = \{0, 1, \dots, m-1\}^n$ is considered instead of the full space $\mathbb{Z}^n$. $(\mathbb{Z}_m)^2$ and $(\mathbb{Z}_m)^3$ are called the *m-grill* and *m-framework*, respectively. The most used metrics on $(\mathbb{Z}_m)^n$ are the **Hamming metric** and the **Lee metric**.

- **Grid metric**
  The **grid metric** is the L_1-*metric* on $\mathbb{Z}^n$. It can be seen as the path metric of an infinite graph: two points of $\mathbb{Z}^n$ are adjacent if their L_1-distance is 1.
  For $n = 2$, this metric is the restriction on $\mathbb{Z}^2$ of the **city-block metric** which is also called the **taxicab** (or **rectilinear**, **Manhattan**, 4-) **metric**.
- **Lattice metric**
  The **lattice metric** is the L_∞-*metric* on $\mathbb{Z}^n$. It can be seen as the path metric of an infinite graph: two points of $\mathbb{Z}^n$ are adjacent if their L_∞-distance is 1. For $\mathbb{Z}^2$, the adjacency corresponds to the king move in chessboard terms, and this graph is called the L_∞-*grid*, while this metric is also called the **chessboard metric**, **king-move metric**, **8-metric**, or **checking distance**.
  This metric is the restriction on $\mathbb{Z}^n$ of the **Chebyshev metric** which is also called the **sup metric**, or **uniform metric**.
- **Hexagonal metric**
  The **hexagonal metric** (or **6-metric**) is a metric on $\mathbb{Z}^2$ with a *unit sphere* (centered at $x \in \mathbb{Z}^2$) defined by $S^1(x) = S^1_{L_1}(x) \cup \{(x_1 - 1, x_2 - 1), (x_1 - 1, x_2 + 1)\}$ for even x_2, and $S^1(x) = S^1_{L_1}(x) \cup \{(x_1 + 1, x_2 - 1), (x_1 + 1, x_2 + 1)\}$ for odd x_2. For any $x = (x_1, x_2), y = (y_1, y_2) \in \mathbb{Z}^2$, this metric $d_6(x, y)$ can be written as

$$\max\left\{|x_2 - y_2|, \frac{|x_2 - y_2|}{2} \pm \left(\frac{x_2 - y_2}{2} + \left\lfloor \frac{x_2 + 1}{2} \right\rfloor - \left\lfloor \frac{y_2 + 1}{2} \right\rfloor - (x_1 - y_1)\right)\right\}.$$

  It is the path metric of the *triangular grid* (or, dually, the minimum number of cell moves of the *hexagonal grid*) on the plane. In *hexagonal coordinates* (h_1, h_2) (in which the h_1- and h_2-axes are parallel to the grid's edges) the hexagonal distance between points $h = (h_1, h_2)$ and $i = (i_1, i_2)$ is

$$d_6(h, i) = \frac{1}{2}(|h_1 - i_1| + |h_2 - i_2| + |(h_1 - i_1) - (h_2 - i_2)|),$$

i.e., $|h_1 - i_1| + |h_2 - i_2|$, if $(h_1 - i_1)(h_2 - i_2) \le 0$, and $\max\{|h_1 - i_1|, |h_2 - i_2|\}$, if otherwise; cf. [LuRo76]. The coordinates (h_1, h_2) of a point x are related to its Cartesian coordinates (x_1, x_2) by $h_1 = x_1 - \lfloor \frac{x_2}{2} \rfloor$, $h_2 = x_2$.

This metric approximates the Euclidean metric better than L_1- or L_∞-*metric*.

The **hexagonal Hausdorff metric** is a metric on the set of all bounded subsets (*pictures*, or *images*) of the hexagonal grid on the plane defined by

$$\inf\{p, q : A \subset B + qH,\ B \subset A + pH\}$$

for any pictures A and B, where pH is the *regular hexagon of size* p (i.e., with $p + 1$ pixels on each edge), centered at the origin and including its interior, and $+$ is the *Minkowski addition*: $A + B = \{x + y : x \in A, y \in B\}$ (cf. **Pompeiu–Hausdorff–Blaschke metric** in Chap. 9). If A is a pixel x, then the distance between x and B is equal to $\sup_{y \in B} d_6(x, y)$, where d_6 is the hexagonal metric.

- **Digital volume metric**

  The **digital volume metric** is a metric on the set K of all bounded subsets (*pictures*, or *images*) of $\mathbb{Z}^2$ (in general, of $\mathbb{Z}^n$) defined by

$$vol(A \triangle B),$$

  where $vol(A) = |A|$, i.e., the number of pixels contained in A, and $A \triangle B$ is the *symmetric difference* between sets A and B.

  This metric is a digital analog of the **Nikodym metric** in Chap. 9.

- **Neighborhood sequence metric**

  On the digital plane $\mathbb{Z}^2$, consider two types of motions: the *city-block motion*, restricting movements only to the horizontal or vertical directions, and the *chessboard motion*, also allowing diagonal movements.

  The use of both these motions is determined by a *neighborhood sequence* $B = \{b(1), b(2), \dots, b(l)\}$, where $b(i) \in \{1, 2\}$ is a particular type of neighborhood, with $b(i) = 1$ signifying unit change in 1 coordinate (*city-block neighborhood*), and $b(i) = 2$ meaning unit change also in 2 coordinates (*chessboard neighborhood*). The sequence B defines the type of motion to be used at every step (see [Das90]).

  The **neighborhood sequence metric** is a metric on $\mathbb{Z}^2$ defined as the length of a shortest path between x and $y \in \mathbb{Z}^2$, determined by a given neighborhood sequence B. It can be written as

$$\max\{d_B^1(u), d_B^2(u)\},$$

  where $u_1 = x_1 - y_1$, $u_2 = x_2 - y_2$, $d_B^1(u) = \max\{|u_1|, |u_2|\}$, $d_B^2(u) = \sum_{j=1}^{l} \lfloor \frac{|u_1| + |u_2| + g(j)}{f(l)} \rfloor$, $f(0) = 0$, $f(i) = \sum_{j=1}^{i} b(j)$, $1 \le i \le l$, $g(j) = f(l) - f(j - 1) - 1$, $1 \le j \le l$.

For $B = \{1\}$ one obtains the **city-block metric**, for $B = \{2\}$ one obtains the **chessboard metric**. The case $B = \{1, 2\}$, i.e., the alternative use of these motions, results in the **octagonal metric**, introduced in [RoPf68].
A proper selection of the B-sequence can make the corresponding metric very close to the Euclidean metric. It is always greater than the chessboard metric, but smaller than the city-block metric.

- ***nD*-neighborhood sequence metric**
The ***nD*-neighborhood sequence metric** is a metric on $\mathbb{Z}^n$, defined as the length of a shortest path between x and $y \in \mathbb{Z}^n$, determined by a given *nD-neighborhood sequence B* (see [Faze99]).
Formally, two points $x, y \in \mathbb{Z}^n$ are called *m-neighbors*, $0 \leq m \leq n$, if $0 \leq |x_i - y_i| \leq 1$, $1 \leq i \leq n$, and $\sum_{i=1}^{n} |x_i - y_i| \leq m$. A finite sequence $B = \{b(1), \dots, b(l)\}$, $b(i) \in \{1, 2, \dots, n\}$, is called an *nD-neighborhood sequence with period l*. For any $x, y \in \mathbb{Z}^n$, a point sequence $x = x^0, x^1, \dots, x^k = y$, where x^i and x^{i+1}, $0 \leq i \leq k - 1$, are r-neighbors, $r = b((i \mod l) + 1)$, is called a *path from x to y determined by B* with length k. The distance between x and y can be written as
$$\max_{1 \leq i \leq n} d_i(u) \text{ with } d_i(x, y) = \sum_{j=1}^{l} \left\lfloor \frac{a_i + g_i(j)}{f_i(l)} \right\rfloor,$$
where $u = (|u_1|, |u_2|, \dots, |u_n|)$ is the nonincreasing ordering of $|u_m|$, $u_m = x_m - y_m$, $m = 1, \dots, n$, that is, $|u_i| \leq |u_j|$ if $i < j$; $a_i = \sum_{j=1}^{n-i+1} u_j$; $b_i(j) = b(j)$ if $b(j) < n - i + 2$, and is $n - i + 1$, otherwise; $f_i(j) = \sum_{k=1}^{j} b_i(k)$ if $1 \leq j \leq l$, and is 0 if $j = 0$; $g_i(j) = f_i(l) - f_i(j - 1) - 1$, $1 \leq j \leq l$.

- **Strand–Nagy distances**
The *face-centered cubic lattice* is $A_3 = \{(a_1, a_2, a_3) \in \mathbb{Z}^3 : a_1 + a_2 + a_3 \equiv 0(\mod 2)\}$, and the *body-centered cubic lattice* is its dual
$$A_3^* = \{(a_1, a_2, a_3) \in \mathbb{Z}^3 : a_1 \equiv a_2 \equiv a_3(\mod 2)\}.$$
Let $L \in \{A_3, A_3^*\}$. For any points $x, y \in L$, let $d_1(x, y) = \sum_{j=1}^{3} |x_j - y_j|$ denote the L_1-metric and $d_\infty(x, y) = \max_{j \in \{1,2,3\}} |x_j - y_j|$ denote the L_∞-metric between them. Two points $x, y \in L$ are called *1-neighbors* if $d_1(x, y) \leq 3$ and $0 < d_\infty(x, y) \leq 1$; they are called *2-neighbors* if $d_1(x, y) \leq 3$ and $1 < d_\infty(x, y) \leq 2$.
Given a sequence $B = \{b(i)\}_{i=1}^{\infty}$ over the alphabet $\{1, 2\}$, a *B-path* in L is a point sequence $x = x^0, x^1, \dots, x^k = y$, where x^i and x^{i+1}, $0 \leq i \leq k - 1$, are 1-neighbors if $b(i) = 1$ and 2-neighbors if $b(i) = 2$.
The **Strand–Nagy distance** between two points $x, y \in L$ (or *B-distance* in Strand and Nagy, 2007) is the length of a shortest B-path between them. For $L = A_3$, it is

$$\min\left\{k : k \geq \max\left\{\frac{d_1(x,y)}{2}, d_\infty(x,y) - |\{1 \leq i \leq k : b(i) = 2\}|\right\}\right\}.$$

The Strand–Nagy distance is a metric, for example, for the periodic sequence $B = (1,2,1,2,1,2,\dots)$ but not for the periodic sequence $B = (2,1,2,1,2,1,\dots)$.

- **Path-generated metric**
  Consider the l_∞*-grid*, i.e., the graph with the vertex-set $\mathbb{Z}^2$, and two vertices being *neighbors* if their l_∞-distance is 1. Let $\mathcal{P}$ be a collection of paths in the l_∞-grid such that, for any $x, y \in \mathbb{Z}^2$, there exists at least one path from $\mathcal{P}$ between x and y, and if $\mathcal{P}$ contains a path Q, then it also contains every path contained in Q. Let $d_\mathcal{P}(x,y)$ be the length of the shortest path from $\mathcal{P}$ between x and $y \in \mathbb{Z}^2$. If $d_\mathcal{P}$ is a metric on $\mathbb{Z}^2$, then it is called a **path-generated metric** (see [Melt91]).
  Let G be one of the sets: $G_1 = \{\uparrow, \rightarrow\}$, $G_{2A} = \{\uparrow, \nearrow\}$, $G_{2B} = \{\uparrow, \nwarrow\}$, $G_{2C} = \{\nearrow, \nwarrow\}$, $G_{2D} = \{\rightarrow, \nwarrow\}$, $G_{3A} = \{\rightarrow, \uparrow, \nearrow\}$, $G_{3B}\{\rightarrow, \uparrow, \nwarrow\}$, $G_{4A} = \{\rightarrow, \nearrow, \nwarrow\}$, $G_{4B}\{\uparrow, \nearrow, \nwarrow, \}$, $G_5 = \{\rightarrow, \uparrow, \nearrow, \nwarrow\}$. Let $\mathcal{P}(G)$ be the set of paths which are obtained by concatenation of paths in G and the corresponding paths in the opposite directions. Any path-generated metric coincides with one of the metrics $d_{\mathcal{P}(G)}$. Moreover, one can obtain the following formulas:

  1. $d_{\mathcal{P}(G_1)}(x,y) = |u_1| + |u_2|$;
  2. $d_{\mathcal{P}(G_{2A})}(x,y) = \max\{|2u_1 - u_2|, |u_2|\}$;
  3. $d_{\mathcal{P}(G_{2B})}(x,y) = \max\{|2u_1 + u_2|, |u_2|\}$;
  4. $d_{\mathcal{P}(G_{2C})}(x,y) = \max\{|2u_2 + u_1|, |u_1|\}$;
  5. $d_{\mathcal{P}(G_{2D})}(x,y) = \max\{|2u_2 - u_1|, |u_1|\}$;
  6. $d_{\mathcal{P}(G_{3A})}(x,y) = \max\{|u_1|, |u_2|, |u_1 - u_2|\}$;
  7. $d_{\mathcal{P}(G_{3B})}(x,y) = \max\{|u_1|, |u_2|, |u_1 + u_2|\}$;
  8. $d_{\mathcal{P}(G_{4A})}(x,y) = \max\{2\lceil(|u_1| - |u_2|)/2\rceil, 0\} + |u_2|$;
  9. $d_{\mathcal{P}(G_{4B})}(x,y) = \max\{2\lceil(|u_2| - |u_1|)/2\rceil, 0\} + |u_1|$;
  10. $d_{\mathcal{P}(G_5)}(x,y) = \max\{|u_1|, |u_2|\}$,

  where $u_1 = x_1 - y_1$, $u_2 = x_2 - y_2$, and $\lceil . \rceil$ is the *ceiling function*: for any real x the number $\lceil x \rceil$ is the least integer greater than or equal to x.
  The metric spaces obtained from G-sets with the same numerical index are isometric. $d_{\mathcal{P}(G_1)}$ is the **city-block metric**, and $d_{\mathcal{P}(G_5)}$ is the **chessboard metric**.
- **Chamfer metric**
  Given numbers α, β with $0 < \alpha \leq \beta < 2\alpha$, the (α, β)*-weighted* l_∞*-grid* is the graph with the vertex-set $\mathbb{Z}^2$, two vertices being adjacent if their l_∞-distance is one, while horizontal/vertical and diagonal edges have *weights* α and β, respectively.
  A **chamfer metric** (or (α, β)*-chamfer metric*, [Borg86]) is the weighted path metric in this graph. For any $x, y \in \mathbb{Z}^2$ it can be written as

  $$\beta m + \alpha(M - m),$$

  where $M = \max\{|u_1|, |u_2|\}$, $m = \min\{|u_1|, |u_2|\}$, $u_1 = x_1 - y_1$, $u_2 = x_2 - y_2$.

If the weights α and β are equal to the Euclidean lengths 1, $\sqrt{2}$ of horizontal/vertical and diagonal edges, respectively, then one obtains the Euclidean length of the shortest chessboard path between x and y. If $\alpha = \beta = 1$, one obtains the **chessboard metric**. The $(3, 4)$-chamfer metric is the most used one for digital images.

A $3D$**-chamfer metric** is the weighted path metric of the graph with the vertex-set $\mathbb{Z}^3$ of *voxels*, two voxels being adjacent if their l_∞-distance is one, while weights α, β, and γ are associated, respectively, to the distance from 6 face neighbors, 12 edge neighbors, and 8 corner neighbors.

- **Weighted cut metric**

  Consider the *weighted* l_∞*-grid*, i.e., the graph with the vertex-set $\mathbb{Z}^2$, two vertices being adjacent if their l_∞-distance is one, and each edge having some positive *weight* (or *cost*). The usual **weighted path metric** between two pixels is the minimal cost of a path connecting them. The **weighted cut metric** between two pixels is the minimal cost (defined now as the sum of costs of crossed edges) of a *cut*, i.e., a plane curve connecting them while avoiding pixels.
- **Knight metric**

  The **knight metric** on $\mathbb{Z}^2$ is the minimum number of moves a chess knight would take to travel from x to $y \in \mathbb{Z}^2$. Its *unit sphere* S^1_{knight}, centered at the origin, contains exactly 8 integral points $\{(\pm 2, \pm 1), (\pm 1, \pm 2)\}$, and can be written as $S^1_{knight} = S^3_{L_1} \cap S^2_{l_\infty}$, where $S^3_{L_1}$ denotes the L_1-sphere of radius 3, and $S^2_{L_\infty}$ denotes the L_∞-sphere of radius 2, both centered at the origin (see [DaCh88]).

  The distance between x and y is 3 if $(M, m) = (1, 0)$, is 4 if $(M, m) = (2, 2)$ and is equal to $\max\{\lceil \frac{M}{2} \rceil, \lceil \frac{M+m}{3} \rceil\} + (M + m) - \max\{\lceil \frac{M}{2} \rceil, \lceil \frac{M+m}{3} \rceil\} \pmod 2$, otherwise, where $M = \max\{|u_1|, |u_2|\}$, $m = \min\{|u_1|, |u_2|\}$, $u_1 = x_1 - y_1$, $u_2 = x_2 - y_2$.
- **Super-knight metric**

  Let $p, q \in \mathbb{N}$. A (p, q)*-super-knight* (or (p, q)*-leaper*, (p, q)*-spider*) is a (variant) chess piece whose move consists of a leap p squares in one orthogonal direction followed by a 90° direction change, and q squares leap to the destination square. Rook, bishop and queen have $q = 0, q = p$ and $q = 0, p$, respectively. Chess-variant terms exist for a $(p, 1)$-leaper with $p = 0, 1, 2, 3, 4$ (*Wazir*, *Ferz*, usual *Knight*, *Camel*, *Giraffe*), and for a $(p, 2)$-leaper with $p = 0, 1, 2, 3$ (*Dabbaba*, usual *Knight*, *Alfil*, *Zebra*).

  A (p, q)**-super-knight metric** (or (p, q)*-leaper metric*) is a metric on $\mathbb{Z}^2$ defined as the minimum number of moves a chess (p, q)-super-knight would take to travel from x to $y \in \mathbb{Z}^2$. Thus, its *unit sphere* $S^1_{p,q}$, centered at the origin, contains exactly 8 integral points $\{(\pm p, \pm q), (\pm q, \pm p)\}$. (See [DaMu90].)

  The **knight metric** is the $(1, 2)$-super-knight metric. The **city-block metric** can be considered as the *Wazir metric*, i.e., $(0, 1)$-super-knight metric.

- **Rook metric**
  The **rook metric** is a metric on $\mathbb{Z}^2$ defined as the minimum number of moves a chess rook would take to travel from x to $y \in \mathbb{Z}^2$. This metric can take only the values $\{0, 1, 2\}$, and coincides with the **Hamming metric** on $\mathbb{Z}^2$.
- **Chess programming distances**
  On a chessboard $\mathbb{Z}_8^2$, *files* are eight columns labeled from a to h and *ranks* are eight rows labeled from 1 to 8. Given two squares, their **file-distance** and **rank-distance** are the absolute differences between the 0 and 7 indices of their files or, respectively, ranks. The **Chebyshev distance** and **Manhattan distance** are the maximum or, respectively, the sum of their file-distance and rank-distance.
  The *center distance* and *corner distance* of a square are its (Chebyshev or Manhattan) distance to closest square among $\{d4, d5, e4, e5\}$ or, respectively, closest corner. For example, the program *Chess 4.x* uses in endgame evaluation $4.7d + 1.6(14 - d')$, where d is the center Manhattan distance of losing king and d' is the Manhattan distance between kings.
  Two kings at rank- and file- distances d_r, d_f, are in *opposition*, which is *direct*, or *diagonal*, or *distant* if $(d_r, d_f) \in \{(0, 2), (2, 0)\}$, or $= (2, 2)$, or their Manhattan distance is even ≥ 6 and no pawns interfere between them.
  Unrelated *cavalry file distance* is the number of files in which it rides.

Chapter 20
Voronoi Diagram Distances

Given a finite set A of objects A_i in a space S, computing the *Voronoi diagram* of A means partitioning the space S into *Voronoi regions* $V(A_i)$ in such a way that $V(A_i)$ contains all points of S that are "closer" to A_i than to any other object A_j in A.

Given a *generator set* $P = \{p_1, \dots p_k\}$, $k \geq 2$, of distinct points (*generators*) from $\mathbb{R}^n$, $n \geq 2$, the ordinary **Voronoi polytope** $V(p_i)$ associated with a generator p_i is defined by

$$V(p_i) = \{x \in \mathbb{R}^n : d_E(x, p_i) \leq d_E(x, p_j) \text{ for any } j \neq i\},$$

where d_E is the Euclidean distance on $\mathbb{R}^n$. The set

$$V(P, d_E, \mathbb{R}^n) = \{V(p_1), \dots, V(p_k)\}$$

is called the *n-dimensional ordinary Voronoi diagram, generated by P*.

The boundaries of (n-dimensional) Voronoi polytopes are called (($n-1$)-dimensional) *Voronoi facets*, the boundaries of Voronoi facets are called *($n-2$)-dimensional Voronoi faces*, ..., the boundaries of 2D Voronoi faces are called *Voronoi edges*, and the boundaries of Voronoi edges are called *Voronoi vertices*.

The ordinary Voronoi diagram can be generalized in the following three ways:

1. The generalization with respect to the generator set $A = \{A_1, \dots, A_k\}$ which can be a set of lines, a set of areas, etc.;
2. The generalization with respect to the space S which can be a sphere (*spherical Voronoi diagram*), a cylinder (*cylindrical Voronoi diagram*), a cone (*conic Voronoi diagram*), a polyhedral surface (*polyhedral Voronoi diagram*), etc.;
3. The generalization with respect to the function d, where $d(x, A_i)$ measures the "distance" from a point $x \in S$ to a generator $A_i \in A$.

M.M. Deza, E. Deza, *Encyclopedia of Distances*,
DOI 10.1007/978-3-662-44342-2_20

This generalized distance function d is called the **Voronoi generation distance** (or *Voronoi distance*, *V-distance*), and allows many more functions than the Euclidean metric on S. If F is a strictly increasing function of a V-distance d, i.e., $F(d(x, A_i)) \leq F(d(x, A_j))$ if and only if $d(x, A_i) \leq d(x, A_j)$, then the generalized Voronoi diagrams $V(A, F(d), S)$ and $V(A, d, S)$ coincide, and one says that the V-distance $F(d)$ is *transformable* to the V-distance d, and that the generalized Voronoi diagram $V(A, F(d), S)$ is a *trivial generalization* of the generalized Voronoi diagram $V(A, d, S)$.

In applications, one often uses for trivial generalizations of the ordinary Voronoi diagram $V(P, d, \mathbb{R}^n)$ the **exponential distance**, the **logarithmic distance**, and the **power distance**. There are generalized Voronoi diagrams $V(P, D, \mathbb{R}^n)$, defined by V-distances, that are not transformable to the Euclidean distance d_E: the **multiplicatively weighted Voronoi distance**, the **additively weighted Voronoi distance**, etc.

The theory of generalized Voronoi diagrams $V(P, D, \mathbb{R}^n)$, where D is a norm metric $||x - p||$ collapses even for the case, when P is a lattice in $\mathbb{R}^n$. But [DeDu13] adapted it for *polyhedral*, i.e., with a polytopal unit ball, norms; $||.||_1$ and $||.||_\infty$ are among them.

For additional information see, for example, [OBS92, Klei89].

20.1 Classical Voronoi Generation Distances

- **Exponential distance**
  The **exponential distance** is the Voronoi generation distance
  $$D_{exp}(x, p_i) = e^{d_E(x, p_i)}$$
  for the trivial generalization $V(P, D_{exp}, \mathbb{R}^n)$ of the ordinary Voronoi diagram $V(P, d_E, \mathbb{R}^n)$, where d_E is the Euclidean distance.
- **Logarithmic distance**
  The **logarithmic distance** is the Voronoi generation distance
  $$D_{\ln}(x, p_i) = \ln d_E(x, p_i)$$
  for the trivial generalization $V(P, D_{\ln}, \mathbb{R}^n)$ of the ordinary Voronoi diagram $V(P, d_E, \mathbb{R}^n)$, where d_E is the Euclidean distance.
- **Power distance**
  The **power distance** is the Voronoi generation distance
  $$D_\alpha(x, p_i) = d_E(x, p_i)^\alpha, \ \alpha > 0,$$
  for the trivial generalization $V(P, D_\alpha, \mathbb{R}^n)$ of the ordinary Voronoi diagram $V(P, d_E, \mathbb{R}^n)$, where d_E is the Euclidean distance.

- **Multiplicatively weighted distance**
The **multiplicatively weighted distance** d_{MW} is the Voronoi generation distance of the generalized Voronoi diagram $V(P, d_{MW}, \mathbb{R}^n)$ (*multiplicatively weighted Voronoi diagram*) defined by

$$d_{MW}(x, p_i) = \frac{1}{w_i} d_E(x, p_i)$$

for any point $x \in \mathbb{R}^n$ and any *generator point* $p_i \in P = \{p_1, \dots, p_k\}$, $k \geq 2$, where $w_i \in w = \{w_i, \dots, w_k\}$ is a given positive *multiplicative weight* of the generator p_i, and d_E is the Euclidean distance.
A *Möbius diagram* (Boissonnat–Karavelas, 2003) is a diagram the **midsets** (bisectors) of which are hyperspheres. It generalizes the Euclidean Voronoi and power diagrams, and it is equivalent to power diagrams in $\mathbb{R}^{n+1}$.
For $\mathbb{R}^2$, the multiplicatively weighted Voronoi diagram is called a *circular Dirichlet tessellation*. An edge in this diagram is a circular arc or a straight line. In the plane $\mathbb{R}^2$, there exists a generalization of the multiplicatively weighted Voronoi diagram, the *crystal Voronoi diagram*, with the same definition of the distance (where w_i is the speed of growth of the crystal p_i), but a different partition of the plane, as the crystals can grow only in an empty area.
- **Additively weighted distance**
The **additively weighted distance** d_{AW} is the Voronoi generation distance of the generalized Voronoi diagram $V(P, d_{AW}, \mathbb{R}^n)$ (*additively weighted Voronoi diagram*) defined by

$$d_{AW}(x, p_i) = d_E(x, p_i) - w_i$$

for any point $x \in \mathbb{R}^n$ and any generator point $p_i \in P = \{p_1, \dots, p_k\}$, $k \geq 2$, where $w_i \in w = \{w_i, \dots, w_k\}$ is a given *additive weight* of the generator p_i, and d_E is the Euclidean distance.
For $\mathbb{R}^2$, the additively weighted Voronoi diagram is called a *hyperbolic Dirichlet tessellation*. An edge in this diagram is a hyperbolic arc or a straight line segment.
- **Additively weighted power distance**
The **additively weighted power distance** d_{PW} is the Voronoi generation distance of the generalized Voronoi diagram $V(P, d_{PW}, \mathbb{R}^n)$ (*additively weighted power Voronoi diagram*) defined by

$$d_{PW}(x, p_i) = d_E^2(x, p_i) - w_i$$

for any point $x \in \mathbb{R}^n$ and any *generator point* $p_i \in P = \{p_1, \dots, p_k\}$, $k \geq 2$, where $w_i \in w = \{w_i, \dots, w_k\}$ is a given *additive weight* of the generator p_i, and d_E is the Euclidean distance.
This diagram can be seen as a Voronoi diagram of circles or as a Voronoi diagram with the *Laguerre geometry*.

The **multiplicatively weighted power distance** $d_{MPW}(x, p_i) = \frac{1}{w_i} d_E^2(x, p_i)$, $w_i > 0$, is transformable to the **multiplicatively weighted distance**, and gives a trivial extension of the multiplicatively weighted Voronoi diagram.

- **Compoundly weighted distance**
  The **compoundly weighted distance** d_{CW} is the Voronoi generation distance of the generalized Voronoi diagram $V(P, d_{CW}, \mathbb{R}^n)$ (*compoundly weighted Voronoi diagram*) defined by

$$d_{CW}(x, p_i) = \frac{1}{w_i} d_E(x, p_i) - v_i$$

  for any point $x \in \mathbb{R}^n$ and any generator point $p_i \in P = \{p_1, \dots, p_k\}$, $k \geq 2$, where $w_i \in w = \{w_i, \dots, w_k\}$ is a given positive *multiplicative weight* of the generator p_i, $v_i \in v = \{v_1, \dots, v_k\}$ is a given *additive weight* of the generator p_i, and d_E is the Euclidean distance.
  An edge in the 2D compoundly weighted Voronoi diagram is a part of a fourth-order polynomial curve, a hyperbolic arc, a circular arc, or a straight line.

20.2 Plane Voronoi Generation Distances

- **Shortest path distance with obstacles**
  Let $\mathcal{O} = \{O_1, \dots, O_m\}$ be a collection of pairwise disjoint polygons on the Euclidean plane, representing a set of obstacles which are neither transparent nor traversable.
  The **shortest path distance with obstacles** d_{sp} is the Voronoi generation distance of the generalized Voronoi diagram $V(P, d_{sp}, \mathbb{R}^2 \backslash \{\mathcal{O}\})$ (*shortest path Voronoi diagram with obstacles*) defined, for any $x, y \in \mathbb{R}^2 \backslash \{\mathcal{O}\}$, as the length of the shortest path among all possible continuous $(x - y)$-paths that do not intersect obstacles $O_i \backslash \partial O_i$ (a path can pass through points on the boundary ∂O_i of O_i), $i = 1, \dots m$.
  The shortest path is constructed with the aid of the *visibility polygon* and the *visibility graph* of $V(P, d_{sp}, \mathbb{R}^2 \backslash \{\mathcal{O}\})$.
- **Visibility shortest path distance**
  Let $\mathcal{O} = \{O_1, \dots, O_m\}$ be a collection of pairwise disjoint line segments $O_l = [a_l, b_l]$ in the Euclidean plane, with $P = \{p_1, \dots, p_k\}$, $k \geq 2$, the set of generator points,

$$VIS(p_i) = \{x \in \mathbb{R}^2 : [x, p_i] \cap]a_l, b_l[= \emptyset \text{ for all } l = 1, \dots, m\}$$

  the *visibility polygon* of the generator p_i, and d_E the Euclidean distance.
  The **visibility shortest path distance** d_{vsp} is the Voronoi generation distance of the generalized Voronoi diagram $V(P, d_{vsp}, \mathbb{R}^2 \backslash \{\mathcal{O}\})$ (*visibility shortest path Voronoi diagram with line obstacles*), defined by

$$d_{vsp}(x, p_i) = \begin{cases} d_E(x, p_i), & \text{if } x \in VIS(p_i), \\ \infty, & \text{otherwise}. \end{cases}$$

- **Network distances**
  A *network* on $\mathbb{R}^2$ is a connected planar geometrical graph $G = (V, E)$ with the set V of vertices and the set E of edges (links).
  Let the generator set $P = \{p_1, \dots, p_k\}$ be a subset of the set $V = \{p_1, \dots, p_l\}$ of vertices of G, and let the set L be given by points of links of G.
  The **network distance** d_{netv} on the set V is the Voronoi generation distance of the *network Voronoi node diagram* $V(P, d_{netv}, V)$ defined as the shortest path along the links of G from $p_i \in V$ to $p_j \in V$. It is the weighted path metric of the graph G, where w_e is the Euclidean length of the link $e \in E$.
  The **network distance** d_{netl} on the set L is the Voronoi generation distance of the *network Voronoi link diagram* $V(P, d_{netl}, L)$ defined as the shortest path along the links from $x \in L$ to $y \in L$.
  The **access network distance** d_{accnet} on $\mathbb{R}^2$ is the Voronoi generation distance of the *network Voronoi area diagram* $V(P, d_{accnet}, \mathbb{R}^2)$ defined by

  $$d_{accnet}(x, y) = d_{netl}(l(x), l(y)) + d_{acc}(x) + d_{acc}(y),$$

  where $d_{acc}(x) = \min_{l \in L} d(x, l) = d_E(x, l(x))$ is the *access distance* of a point x. In fact, $d_{acc}(x)$ is the Euclidean distance from x to the *access point* $l(x) \in L$ of x which is the nearest to x point on the links of G.
- **Airlift distance**
  An *airports network* is an arbitrary planar graph G on n vertices (*airports*) with positive edge weights (*flight durations*). This graph may be entered and exited only at the airports. Once having accessed G, one travels at fixed speed $v > 1$ within the network. Movement off the network takes place with the unit speed with respect to the Euclidean distance.
  The **airlift distance** d_{al} is the Voronoi generation distance of the *airlift Voronoi diagram* $V(P, d_{al}, \mathbb{R}^2)$, defined as the time needed for a *quickest*, i.e., minimizing the travel time, path between x and y in the presence of the airports network G.
- **City distance**
  A *city public transportation network*, like a subway or a bus transportation system, is a planar straight line graph G with horizontal or vertical edges. G may be composed of many connected components, and may contain cycles. One is free to enter G at any point, be it at a vertex or on an edge (it is possible to postulate fixed entry points, too). Once having accessed G, one travels at a fixed speed $v > 1$ in one of the available directions. Movement off the network takes place with the unit speed with respect to the **Manhattan metric**.
  The **city distance** d_{city} is the Voronoi generation distance of the *city Voronoi diagram* $V(P, d_{city}, \mathbb{R}^2)$, defined as the time needed for the *quickest path*, i.e., the one minimizing the travel time, between x and y in the presence of the network G.

The set $P = \{p_1, \dots, p_k\}$, $k \geq 2$, can be seen as a set of some city facilities (say, post offices or hospitals): for some people several facilities of the same kind are equally attractive, and they want to find out which facility is reachable first.

- **Distance in a river**
  The **distance in a river** d_{riv} is the Voronoi generation distance of the generalized Voronoi diagram $V(P, d_{riv}, \mathbb{R}^2)$ (*Voronoi diagram in a river*), defined by
  $$d_{riv}(x, y) = \frac{-\alpha(x_1 - y_1) + \sqrt{(x_1 - y_1)^2 + (1 - \alpha^2)(x_2 - y_2)^2}}{v(1 - \alpha^2)},$$
  where v is the speed of the boat on still water, $w > 0$ is the speed of constant flow in the positive direction of the x_1 axis, and $\alpha = \frac{w}{v}$ $(0 < \alpha < 1)$ is the *relative flow speed.*
- **Boat-sail distance**
  Let $\Omega \subset \mathbb{R}^2$ be a *domain* in the plane (*water surface*), let $f : \Omega \to \mathbb{R}^2$ be a continuous vector field on Ω, representing the velocity of the water flow (*flow field*); let $P = \{p_1, \dots, p_k\} \subset \Omega$, $k \geq 2$, be a set of k points in Ω (*harbors*).
  The **boat-sail distance** [NiSu03] d_{bs} is the Voronoi generation distance of the generalized Voronoi diagram $V(P, d_{bs}, \Omega)$ (*boat-sail Voronoi diagram*) defined by
  $$d_{bs}(x, y) = \inf_{\gamma} \delta(\gamma, x, y)$$
  for all $x, y \in \Omega$, where $\delta(\gamma, x, y) = \int_0^1 \left| F \frac{\gamma'(s)}{|\gamma'(s)|} + f(\gamma(s)) \right|^{-1} ds$ is the time necessary for the boat with the maximum speed F on still water to move from x to y along the curve $\gamma : [0, 1] \to \Omega$, $\gamma(0) = x$, $\gamma(1) = y$, and the infimum is taken over all possible curves γ.
- **Peeper distance**
  Let $S = \{(x_1, x_2) \in \mathbb{R}^2 : x_1 > 0\}$ be the half-plane in $\mathbb{R}^2$, let $P = \{p_1, \dots, p_k\}$, $k \geq 2$, be a set of points contained in the half-plane $\{(x_1, x_2) \in \mathbb{R}^2 : x_1 < 0\}$, and let the *window* be the interval (a, b) with $a = (0, 1)$ and $b = (0, -1)$.
  The **peeper distance** d_{pee} is the Voronoi generation distance of the generalized Voronoi diagram $V(P, d_{pee}, S)$ (*peeper's Voronoi diagram*) defined by
  $$d_{pee}(x, p_i) = \begin{cases} d_E(x, p_i), & \text{if } [x, p] \cap]a, b[\neq \emptyset, \\ \infty, & \text{otherwise}, \end{cases}$$
  where d_E is the Euclidean distance.
- **Snowmobile distance**
  Let $\Omega \subset \mathbb{R}^2$ be a *domain* in the x_1x_2-plane of the space $\mathbb{R}^3$ (a *2D mapping*), and let $\Omega^* = \{(q, h(q)) : q = (x_1(q), x_2(q)) \in \Omega, h(q) \in \mathbb{R}\}$ be the three-dimensional *land surface* associated with the mapping Ω. Let $P = \{p_1, \dots, p_k\} \subset \Omega$, $k \geq 2$, be a set of k points in Ω (*snowmobile stations*).

The **snowmobile distance** d_{sm} is the Voronoi generation distance of the generalized Voronoi diagram $V(P, d_{sm}, \Omega)$ (*snowmobile Voronoi diagram*) defined by

$$d_{sm}(q, r) = \inf_{\gamma} \int_{\gamma} \frac{1}{F\left(1 - \alpha \frac{dh(\gamma(s))}{ds}\right)} ds$$

for any $q, r \in \Omega$, and calculating the minimum time necessary for the snowmobile with the speed F on flat land to move from $(q, h(q))$ to $(r, h(r))$ along the *land path* $\gamma^* : \gamma^*(s) = (\gamma(s), h(\gamma(s)))$ associated with the *domain path* $\gamma : [0, 1] \to \Omega$, $\gamma(0) = q$, $\gamma(1) = r$. Here the infimum is taken over all possible paths γ, and α is a positive constant.
A snowmobile goes uphill more slowly than downhill. The situation is opposite for a forest fire, and it can be modeled using a negative value of α. The resulting distance is called the **forest-fire distance**, and the resulting Voronoi diagram is called the *forest-fire Voronoi diagram*.

- **Skew distance**
Let T be a *tilted plane* in $\mathbb{R}^3$, obtained by rotating the x_1x_2 plane around the x_1 axis through the angle α, $0 < \alpha < \frac{\pi}{2}$, with the coordinate system obtained by taking the coordinate system of the x_1x_2 plane, accordingly rotated. For a point $q \in T$, $q = (x_1(q), x_2(q))$, define the *height* $h(q)$ as its x_3 coordinate in $\mathbb{R}^3$. Thus, $h(q) = x_2(q) \cdot \sin\alpha$. Let $P = \{p_1, \dots, p_k\} \subset T$, $k \geq 2$.
The **skew distance** is the Voronoi generation distance of the generalized Voronoi diagram $V(P, d_{skew}, T)$ (*skew Voronoi diagram*) defined [AACLMP98] by

$$d_{skew}(q, r) = d_E(q, r) + (h(r) - h(q)) = d_E(q, r) + \sin\alpha(x_2(r) - x_2(q))$$

or, more generally, by

$$d_{skew}(q, r) = d_E(q, r) + k(x_2(r) - x_2(q))$$

for all $q, r \in T$, where d_E is the Euclidean distance, and $k \geq 0$ is a constant.

20.3 Other Voronoi Generation Distances

- **Voronoi distance for line segments**
The **Voronoi distance for** (a set of) **line segments** d_{sl} is the Voronoi generation distance of the generalized Voronoi diagram $V(A, d_{sl}, \mathbb{R}^2)$ (*line Voronoi diagram generated by straight line segments*) defined by

$$d_{sl}(x, A_i) = \inf_{y \in A_i} d_E(x, y),$$

where the *generator set* $A = \{A_1, \dots, A_k\}$, $k \geq 2$, is a set of pairwise disjoint straight line segments $A_i = [a_i, b_i]$, and d_E is the ordinary Euclidean distance.

In fact,

$$d_{sl}(x, A_i) = \begin{cases} d_E(x, a_i), & \text{if} \quad x \in R_{a_i}, \\ d_E(x, b_i), & \text{if} \quad x \in R_{b_i}, \\ d_E(x - a_i, \frac{(x-a_i)^T(b_i-a_i)}{d_E^2(a_i,b_i)}(b_i - a_i)), & \text{if} \quad x \in \mathbb{R}^2 \backslash \{R_{a_i} \cup R_{b_i}\}, \end{cases}$$

where $R_{a_i} = \{x \in \mathbb{R}^2 : (b_i - a_i)^T(x - a_i) < 0\}$, $R_{b_i} = \{x \in \mathbb{R}^2 : (a_i - b_i)^T(x - b_i) < 0\}$.

- **Voronoi distance for arcs**
  The **Voronoi distance for** (a set of circle) **arcs** d_{ca} is the Voronoi generation distance of the generalized Voronoi diagram $V(A, d_{ca}, \mathbb{R}^2)$ (*line Voronoi diagram generated by circle arcs*) defined by

$$d_{ca}(x, A_i) = \inf_{y \in A_i} d_E(x, y),$$

where the *generator set* $A = \{A_1, \dots, A_k\}$, $k \geq 2$, is a set of pairwise disjoint circle arcs A_i (less than or equal to a semicircle) with radius r_i centered at x_{c_i}, and d_E is the Euclidean distance. In fact,

$$d_{ca}(x, A_i) = \min\{d_E(x, a_i), d_E(x, b_i), |d_E(x, x_{c_i}) - r_i|\},$$

where a_i and b_i are the endpoints of A_i.

- **Voronoi distance for circles**
  The **Voronoi distance for** (a set of) **circles** d_{cl} is the Voronoi generation distance of a generalized Voronoi diagram $V(A, d_{cl}, \mathbb{R}^2)$ (*line Voronoi diagram generated by circles*) defined by

$$d_{cl}(x, A_i) = \inf_{y \in A_i} d_E(x, y),$$

where the *generator set* $A = \{A_1, \dots, A_k\}$, $k \geq 2$, is a set of pairwise disjoint circles A_i with radius r_i centered at x_{c_i}, and d_E is the Euclidean distance. In fact,

$$d_{cl}(x, A_i) = |d_E(x, x_{c_i}) - r_i|.$$

Examples of above Voronoi distances are $d_{cl}^{\star}(x, A_i) = d_E(x, x_{c_i}) - r_i$ and $d_{cl}^{*}(x, A_i) = d_E^2(x, x_{c_i}) - r_i^2$ (the *Laguerre Voronoi diagram*).

- **Voronoi distance for areas**
  The **Voronoi distance for areas** d_{ar} is the Voronoi generation distance of the generalized Voronoi diagram $V(A, d_{ar}, \mathbb{R}^2)$ (*area Voronoi diagram*) defined by

$$d_{ar}(x, A_i) = \inf_{y \in A_i} d_E(x, y),$$

where $A = \{A_1, \dots, A_k\}$, $k \geq 2$, is a collection of pairwise disjoint connected closed sets (*areas*), and d_E is the ordinary Euclidean distance.

For any generalized generator set $A = \{A_1, \dots, A_k\}$, $k \geq 2$, one can use as the Voronoi generation distance the **Hausdorff distance** from a point x to a set A_i: $d_{Haus}(x, A_i) = \sup_{y \in A_i} d_E(x, y)$, where d_E is the Euclidean distance.

- **Cylindrical distance**
  The **cylindrical distance** d_{cyl} is the **intrinsic metric** on the surface of a cylinder S which is used as the Voronoi generation distance in the *cylindrical Voronoi diagram* $V(P, d_{cyl}, S)$. If the axis of a cylinder with unit radius is placed at the x_3 axis in $\mathbb{R}^3$, the cylindrical distance between any points $x, y \in S$ with the cylindrical coordinates $(1, \theta_x, z_x)$ and $(1, \theta_y, z_y)$ is given by

$$d_{cyl}(x, y) = \begin{cases} \sqrt{(\theta_x - \theta_y)^2 + (z_x - z_y)^2}, & \text{if } \theta_y - \theta_x \leq \pi, \\ \sqrt{(\theta_x + 2\pi - \theta_y)^2 + (z_x - z_y)^2}, & \text{if } \theta_y - \theta_x > \pi. \end{cases}$$

- **Cone distance**
  The **cone distance** d_{con} is the **intrinsic metric** on the surface of a cone S which is used as the Voronoi generation distance in the *conic Voronoi diagram* $V(P, d_{con}, S)$. If the axis of the cone S is placed at the x_3 axis in $\mathbb{R}^3$, and the radius of the circle made by the intersection of the cone S with the x_1x_2 plane is equal to one, then the cone distance between any points $x, y \in S$ is given by

$$d_{con}(x, y) = \begin{cases} \sqrt{r_x^2 + r_y^2 - 2r_x r_y \cos(\theta'_y - \theta'_x)}, \\ \quad \text{if } \theta'_y \leq \theta'_x + \pi \sin(\alpha/2), \\ \sqrt{r_x^2 + r_y^2 - 2r_x r_y \cos(\theta'_x + 2\pi \sin(\alpha/2) - \theta'_y)}, \\ \quad \text{if } \theta'_y > \theta'_x + \pi \sin(\alpha/2), \end{cases}$$

  where (x_1, x_2, x_3) are the Cartesian coordinates of a point x on S, α is the top angle of the cone, θ_x is the counterclockwise angle from the x_1 axis to the ray from the origin to the point $(x_1, x_2, 0)$, $\theta'_x = \theta_x \sin(\alpha/2)$, $r_x = \sqrt{x_1^2 + x_2^2 + (x_3 - \coth(\alpha/2))^2}$ is the straight line distance from the top of the cone to the point (x_1, x_2, x_3).
- **Voronoi distances of order** m
  Given a finite set A of objects in a metric space (S, d), and an integer $m \geq 1$, consider the set of all *m-subsets* M_i of A (i.e., $M_i \subset A$, and $|M_i| = m$). The *Voronoi diagram of order m* of A is a partition of S into *Voronoi regions $V(M_i)$ of m-subsets* of A in such a way that $V(M_i)$ contains all points $s \in S$ which are "closer" to M_i than to any other m-set M_j: $d(s, x) < d(s, y)$ for any $x \in M_i$ and $y \in S \backslash M_i$. This diagram provides first, second, …, m-th closest neighbors of a point in S.
  Such diagrams can be defined in terms of some "distance function" $D(s, M_i)$, in particular, some m-**hemimetric** (cf. Chap. 3) on S. For $M_i = \{a_i, b_i\}$, there were considered the functions $|d(s, a_i) - d(s, b_i)|$, $d(s, a_i) + d(s, b_i)$, $d(s, a_i) \cdot d(s, b_i)$, as well as **2-metrics** $d(s, a_i) + d(s, b_i) + d(a_i, b_i)$ and the area of triangle (s, a_i, b_i).

Chapter 21
Image and Audio Distances

21.1 Image Distances

Image Processing treats signals such as photographs, video, or tomographic output. In particular, *Computer Graphics* consists of image synthesis from some abstract models, while *Computer Vision* extracts some abstract information: say, the 3D description of a scene from video footage of it. From about 2000, analog image processing (by optical devices) gave way to digital processing, and, in particular, digital image editing (for example, processing of images taken by popular digital cameras).

Computer graphics (and our brains) deals with *vector graphics images*, i.e., those represented geometrically by curves, polygons, etc. A *raster graphics image* (or *digital image*, *bitmap*) in $2D$ is a representation of a $2D$ image as a finite set of digital values, called *pixels* (short for picture elements) placed on a square grid $\mathbb{Z}^2$ or a hexagonal grid. Typically, the image raster is a square $2^k \times 2^k$ grid with $k = 8, 9$ or 10.

Video images and *tomographic* or MRI (obtained by cross-sectional slices) images are $3D$ ($2D$ plus time); their digital values are called *voxels* (volume elements). The spacing between two pixels in one slice is referred to as the *interpixel distance*, while the spacing between slices is the *interslice distance*.

A *digital binary image* corresponds to only two values 0,1 with 1 being interpreted as logical "true" and displayed as black; so, such image is identified with the set of black pixels. A *continuous binary image* is a (usually, compact) subset of a **locally compact** metric space (usually, Euclidean space $\mathbb{E}^n$ with $n = 2, 3$).

The *gray-scale images* can be seen as point-weighted binary images. In general, a *fuzzy set* is a point-weighted set with weights (*membership values*); cf. **metrics between fuzzy sets** in Chap. 1. For the gray-scale images, *xyi*-representation is used, where plane coordinates (x, y) indicate shape, while the weight i (short for intensity, i.e., brightness) indicates *texture*. Sometimes, the matrix $((i_{xy}))$ of gray-levels is used.

M.M. Deza, E. Deza, *Encyclopedia of Distances*,
DOI 10.1007/978-3-662-44342-2_21

The *brightness histogram* of a gray-scale image provides the frequency of each brightness value found in that image. If an image has m brightness levels (bins of gray-scale), then there are 2^m different possible intensities. Usually, $m = 8$ and numbers $0, 1, \dots, 255$ represent the intensity range from black to white; other typical values are $m = 10, 12, 14, 16$. Humans can differ between around 10 million different colors but between only 30 different gray-levels; so, color has much higher discriminatory power.

For color images, (RGB)-representation is the better known, where space coordinates R, G, B indicate red, green and blue levels; a $3D$ histogram provides brightness at each point. Among many other $3D$ color models (spaces) are: (CMY) cube (Cyan, Magenta, Yellow colors), (HSL) cone (Hue-color type given as an angle, Saturation in %, Luminosity in %), and (YUV), (YIQ) used, respectively, in PAL, NTSC television. CIE-approved conversion of (RGB) into luminance (luminosity) of gray-level is $0.299R + 0.587G + 0.114B$. The *color histogram* is a feature vector with components representing either the total number of pixels, or the percentage of pixels of a given color in the image.

Images are often represented by *feature vectors*, including color histograms, color moments, textures, shape descriptors, etc. Examples of feature spaces are: *raw intensity* (pixel values), *edges* (boundaries, contours, surfaces), *salient features* (corners, line intersections, points of high curvature), and *statistical features* (moment invariants, centroids). Typical video features are in terms of overlapping frames and motions.

Image Retrieval (similarity search) consists of (as for other data: audio recordings, DNA sequences, text documents, time-series, etc.) finding images whose features have values either mutual similarity, or similarity to a given query or in a given range.

There are two methods to compare images directly: intensity-based (color and texture histograms), and geometry-based (shape representations by *medial axis*, *skeletons*, etc.). The imprecise term *shape* is used for the extent (silhouette) of the object, for its local geometry or geometrical pattern (conspicuous geometric details, points, curves, etc.), or for that pattern modulo a similarity transformation group (translations, rotations, and scalings). The imprecise term *texture* means all that is left after color and shape have been considered, or it is defined via structure and randomness.

The similarity between vector representations of images is measured by the usual practical distances: **l_p-metrics**, **weighted editing metrics**, **Tanimoto distance**, **cosine distance**, **Mahalanobis distance** and its extension, **distance**.

Among probabilistic distances, the following ones are most used: **Bhattacharya 2**, **Hellinger**, **Kullback–Leibler**, **Jeffrey** and (especially, for histograms) **χ^2-**, **Kolmogorov–Smirnov**, **Kuiper distances**.

The main distances applied for compact subsets X and Y of $\mathbb{R}^n$ (usually, $n = 2, 3$) or their digital versions are: **Asplund metric**, **Shephard metric**, **symmetric difference semimetric** $Vol(X \Delta Y)$ (cf. **Nykodym metric**, **area deviation**, **digital volume metric** and their normalizations) and variations of the **Hausdorff distance** (see below).

For Image Processing, the distances below are between "true" and approximated digital images, in order to assess the performance of algorithms. For Image Retrieval, distances are between feature vectors of a query and reference.

- **Color distances**
  The visible spectrum of a typical human eye is about 380–760 nm. It matches the range of wavelengths sustaining photosynthesis; also, at those wavelengths opacity often coincides with impenetrability. A light-adapted eye has its maximum sensitivity at ≈555 nm (540 THz), in the green region of the optical spectrum.
  A *color space* is a 3-parameter description of colors. The need for exactly three parameters comes from the existence of three kinds of receptors (cells on the retina) in the human eye: for short, middle and long wavelengths, corresponding to blue, green, and red. In fact, their respective sensitivity peaks are situated around 570, 543 and 442 nm, while wavelength limits of extreme violet and red are about 700 and 390 nm, respectively. About one of ten women has a 4th type of color receptor. Color blindness is ten times more common in males. People with absent or removed lens of the eye, can see UV (ultraviolet) wavelengths (400–300 nm). The mantis shrimp has 12 types of color receptors including 4 for UV; its species *Gonodactylus smithii* is the only organism known to have optimal polarization vision.
  The CIE (International Commission on Illumination) derived (XYZ) color space in 1931 from the (RGB)-model and measurements of the human eye. In the CIE (XYZ) color space, the values X, Y and Z are also roughly red, green and blue.
  The basic assumption of Colorimetry (Indow, 1991), is that the perceptual color space admits a metric, the true **color distance**. This metric is expected to be almost locally Euclidean, i.e., a **Riemannian metric**. A continuous mapping from the metric space of light stimuli to this metric space is also expected.
  Such a *uniform color scale*, where equal distances in the color space correspond to equal differences in color, is not obtained yet and existing **color distances** are various approximations of it. A first step in this direction was given by *MacAdam ellipses*, i.e., regions on a *chromaticity* (x, y) diagram which contains all colors looking indistinguishable to the average human eye; cf. JND (just-noticeable difference) **video quality metric**. For any $\epsilon > 0$, the **MacAdam metric** in a color space is the metric for which those 25 ellipses are circles of radius ϵ. Here $x = \frac{X}{X+Y+Z}$ and $y = \frac{Y}{X+Y+Z}$ are projective coordinates, and the colors of the chromaticity diagram occupy a region of the real projective plane $\mathbb{R}P^2$.
  The CIE $(L^*a^*b^*)$ (CIELAB) is an adaptation of CIE 1931 (XYZ) color space; it gives a partial linearization of the MacAdam color metric. The parameters L^*, a^*, b^* of the most complete model are derived from L, a, b which are: the luminance L of the color from black $L = 0$ to white $L = 100$, its position a between green $a < 0$ and red $a > 0$, and its position b between green $b < 0$ and yellow $b > 0$.
- **Average color distance**
  For a given $3D$ color space and a list of n colors, let (c_{i1}, c_{i2}, c_{i3}) be the representation of the i-th color of the list in this space. For a color histogram

$x = (x_1, \dots, x_n)$, its *average color* is the vector $(x_{(1)}, x_{(2)}, x_{(3)})$, where $x_{(j)} = \sum_{i=1}^{n} x_i c_{ij}$ (for example, the average red, blue and green values in (RGB)).
The **average color distance** between two color histograms [HSEFN95] is the Euclidean distance of their average colors.

- **Color component distance**
Given an image (as a subset of $\mathbb{R}^2$), let p_i denote the area percentage of this image occupied by the color c_i. A *color component* of the image is a pair (c_i, p_i).
The **color component distance** (Ma–Deng–Manjunath, 1997) between color components (c_i, p_i) and (c_j, p_j) is defined by

$$|p_i - p_j| \cdot d(c_i, c_j),$$

where $d(c_i, c_j)$ is the distance between colors c_i and c_j in a given color space. Mojsilović–Hu–Soljanin, 2002, developed an **Earth Mover's distance**-like modification of this distance.

- **Riemannian color space**
The proposal to measure perceptual dissimilarity of colors by a *Riemannian metric* (cf. Chap. 7) on a strictly convex cone $C \subset \mathbb{R}^3$ comes from von Helmholtz, 1892, and Luneburg, 1947.
Roughly, it was shown in [Resn74] that the only such *GL-homogeneous* cones C (i.e., the group of all orientation preserving linear transformations of $\mathbb{R}^3$, carrying C into itself, acts transitively on C) are either $C_1 = \mathbb{R}_{>0} \times (\mathbb{R}_{>0} \times \mathbb{R}_{>0})$, or $C_2 = \mathbb{R}_{>0} \times C'$, where C' is the set of 2×2 real symmetric matrices with determinant 1. The first factor $\mathbb{R}_{>0}$ can be identified with variation of brightness and the other with the set of lights of a fixed brightness. Let α_i be some positive constants.
The **Stiles color metric** (Stiles, 1946) is the GL-invariant Riemannian metric on $C_1 = \{(x_1, x_2, x_3) \in \mathbb{R}^3 : x_i > 0\}$ given by the *line element*

$$ds^2 = \sum_{i=1}^{3} \alpha_i \left(\frac{dx_i}{x_i}\right)^2.$$

The **Resnikoff color metric** (Resnikoff, 1974) is the GL-invariant Riemannian metric on $C_2 = \{(x, u) : x \in \mathbb{R}_{>0}, u \in C'\}$ given by the *line element*

$$ds^2 = \alpha_1 (\frac{dx}{x})^2 + \alpha_2 ds^2_{C'},$$

where $ds^2_{C'}$, is the **Poincaré metric** (cf. Chap. 6) on C'; so, C_2 with this metric is not isometric to a Euclidean space.

- **Histogram intersection quasi-distance**
Given two color histograms $x = (x_1, \dots, x_n)$ and $y = (y_1, \dots, y_n)$ (with x_i, y_i representing the number of pixels in the bin i), the **histogram intersection quasi-distance** between them (cf. **intersection distance** in Chap. 17) is (Swain–Ballard, 1991) defined by

$$1 - \frac{\sum_{i=1}^{n} \min\{x_i, y_i\}}{\sum_{i=1}^{n} x_i}.$$

For normalized histograms (total sum is 1) the above quasi-distance becomes the usual l_1-metric $\sum_{i=1}^{n} |x_i - y_i|$. The *normalized cross-correlation* (Rosenfeld–Kak, 1982) between x and y is a similarity defined by $\frac{\sum_{i=1}^{n} x_i y_i}{\sum_{i=1}^{n} x_i^2}$.

- **Histogram quadratic distance**
Given two color histograms $x = (x_1, \dots, x_n)$ and $y = (y_1, \dots, y_n)$ (usually, $n = 256$ or $n = 64$) representing the color percentages of two images, their **histogram quadratic distance** (used in IBM's Query By Image Content system) is their **Mahalanobis distance** defined in Chap. 17 by

$$\sqrt{(x-y)^T A(x-y)},$$

where $A = ((a_{ij}))$ is a symmetric positive-definite matrix, and the weight a_{ij} is some, perceptually justified, similarity between colors i and j.
For example (cf. [HSEFN95]), $a_{ij} = 1 - \frac{d_{ij}}{\max_{1 \le p,q \le n} d_{pq}}$, where d_{ij} is the Euclidean distance between 3-vectors representing i and j in some color space.
If (h_i, s_i, v_i) and (h_j, s_j, v_j) are the representations of the colors i and j in the color space (HSV), then $a_{ij} = 1 - \frac{1}{\sqrt{5}}((v_i - v_j)^2 + (s_i \cos h_i - s_j \cos h_j)^2 + (s_i \sin h_i - s_j \sin h_j)^2)^{\frac{1}{2}}$ is used.

- **Histogram diffusion distance**
Given two histogram-based descriptors $x = (x_1, \dots, x_n)$ and $y = (y_1, \dots, y_n)$, their **histogram diffusion distance** (Ling–Okada, 2006) is defined by

$$\int_0^T ||u(t)||_1 dt,$$

where T is a constant, and $u(t)$ is a heat diffusion process with initial condition $u(0) = x - y$. In order to approximate the diffusion, the initial condition is convoluted with a Gaussian window; then the sums of l_1-norms after each convolution approximate the integral.

- **Gray-scale image distances**
Let $f(x)$ and $g(x)$ denote the brightness values of two digital gray-scale images f and g at the pixel $x \in X$, where X is a raster of pixels. Any distance between point-weighted sets (X, f) and (X, g) (for example, the **Earth Mover's distance**) can be applied for measuring distances between f and g. However, the main used distances (called also *errors*) between the images f and g are:

1. The *root-mean-square error* $RMS(f, g) = \left(\frac{1}{|X|} \sum_{x \in X} (f(x) - g(x))^2\right)^{\frac{1}{2}}$ (a variant is to use the l_1-*norm* $|f(x) - g(x)|$ instead of the l_2-*norm*);

2. The **signal-to-noise ratio** $SNR(f, g) = \left(\frac{\sum_{x \in X} g(x)^2}{\sum_{x \in X}(f(x)-g(x))^2}\right)^{\frac{1}{2}}$ (cf. **SNR distance** between sonograms);
3. The *pixel misclassification error rate* $\frac{1}{|X|}|\{x \in X : f(x) \neq g(x)\}|$ (normalized **Hamming distance**);
4. The *frequency root-mean-square error* $\left(\frac{1}{|U|^2}\sum_{u \in U}(F(u) - G(u))^2\right)^{\frac{1}{2}}$, where F and G are the discrete Fourier transforms of f and g, respectively, and U is the frequency domain;
5. The *Sobolev norm of order* δ *error* $\left(\frac{1}{|U|^2}\sum_{u \in U}(1+|\eta_u|^2)^{\delta}(F(u)-G(u))^2\right)^{\frac{1}{2}}$, where $0 < \delta < 1$ is fixed (usually, $\delta = \frac{1}{2}$), and η_u is the $2D$ frequency vector associated with position u in the frequency domain U.

Cf. **metrics between fuzzy sets** in Chap. 1.

- **Image compression L_p-metric**
Given a number r, $0 \leq r < 1$, the **image compression L_p-metric** is the usual **L_p-metric** on $\mathbb{R}^{n^2}_{\geq 0}$ (the set of gray-scale images seen as $n \times n$ matrices) with p being a solution of the equation $r = \frac{p-1}{2p-1} \cdot e^{\frac{p}{2p-1}}$. So, $p = 1, 2$, or ∞ for, respectively, $r = 0, r = \frac{1}{3}e^{\frac{2}{3}} \approx 0.65$, or $r \geq \frac{\sqrt{e}}{2} \approx 0.82$. Here r estimates the *informative* (i.e., filled with nonzeros) part of the image. According to [KKN02], it is the best quality metric to select a lossy compression scheme.
- **Chamfering distances**
The **chamfering distances** are distances approximating Euclidean distance by a weighted path distance on the graph $G = (\mathbb{Z}^2, E)$, where two pixels are neighbors if one can be obtained from another by an *one-step move* on $\mathbb{Z}^2$. The list of permitted moves is given, and a **prime distance**, i.e., a positive weight (cf. Chap. 19), is associated to each type of such move.
An (α, β)**-chamfer metric** corresponds to two permitted moves—with l_1-distance 1 and with l_∞-distance 1 (diagonal moves only)—weighted α and β, respectively.
The main applied cases are $(\alpha, \beta) = (1, 0)$ (the **city-block metric**, or 4**-metric**), (1,1) (the **chessboard metric**, or 8**-metric**), $(1, \sqrt{2})$ (the **Montanari metric**), $(3, 4)$ (the $(3, 4)$**-metric**), $(2, 3)$ (the **Hilditch–Rutovitz metric**), $(5, 7)$ (the **Verwer metric**).
The **Borgefors metric** corresponds to three permitted moves—with l_1-distance 1, with l_∞-distance 1 (diagonal moves only) and knight moves—weighted 5, 7 and 11.
A $3D$**-chamfer metric** (or (α, β, γ)*-chamfer metric*) is the weighted path metric of the infinite graph with the vertex-set $\mathbb{Z}^3$ of voxels, two vertices being adjacent if their l_∞-distance is one, while weights α, β and γ are associated to 6 face, 12 edge and 8 corner neighbors, respectively. If $\alpha = \beta = \gamma = 1$, we obtain l_∞-metric. The $(3, 4, 5)$- and $(1, 2, 3)$-chamfer metrics are the most used ones for digital $3D$ images.

The **Chaudhuri–Murthy–Chaudhuri metric** between sequences $x = (x_1, \dots, x_m)$ and $y = (y_1, \dots, y_n)$ is defined by

$$|x_{i(x,y)} - y_{i(x,y)}| + \frac{1}{1 + \lceil \frac{n}{2} \rceil} \sum_{1 \le i \le n, i \ne i(x,y)} |x_i - y_i|,$$

where the maximum value of $x_i - y_i$ is attained for $i = i(x, y)$. For $n = 2$ it is the $(1, \frac{3}{2})$-chamfer metric.

- **Earth Mover's distance**

  The **Earth Mover's distance** is a discrete form of the **transportation distance** (cf. Chap. 14). Roughly, it is the minimal amount of work needed to transport earth or mass from one position (properly spread in space) to the other (a collection of holes). For any two finite sequences $(x_1, \dots, x_m)$ and $(y_1, \dots, y_n)$ over a metric space (X, d), consider *signatures*, i.e., point-weighted sets $P_1 = (p_1(x_1), \dots, p_1(x_m))$ and $P_2 = (p_2(y_1), \dots, p_2(y_n))$.

  For example, in [RTG00]) signatures represent clustered color or texture content of images: elements of X are centroids of clusters, and $p_1(x_i), p_2(y_j)$ are cardinalities of corresponding clusters. The ground distance d is a **color distance**, say, the Euclidean distance in $3D$ CIE $(L^*a^*b^*)$ color space.

  Let $W_1 = \sum_i p_1(x_i)$ and $W_2 = \sum_j p_2(y_j)$ be the *total weights* of P_1 and P_2, respectively. Then the **Earth Mover's distance** between P_1 and P_2 is defined as

$$\frac{\sum_{i,j} f_{ij}^* d(x_i, y_j)}{\sum_{i,j} f_{ij}^*},$$

  where the $m \times n$ matrix $S^* = ((f_{ij}^*))$ is an *optimal*, i.e., minimizing $\sum_{i,j} f_{ij} d(x_i, y_j)$, *flow*. A *flow* is an $m \times n$ matrix $S = ((f_{ij}))$ with following constraints:

  1. all $f_{ij} \ge 0$;
  2. $\sum_{i,j} f_{ij} = \min\{W_1, W_2\}$;
  3. $\sum_i f_{ij} \le p_2(y_j)$ and $\sum_j f_{ij} \le p_1(x_i)$.

  So, this distance is the average ground distance d that weights travel during an optimal flow. It is not a *bin-to-bin* (component-wise, as L_p-, Kullback–Leibler, χ^2-distances), but a **cross-bin histogram distance**.

  In the case $W_1 = W_2$, the above two inequalities 3. become equalities. Normalizing signatures to $W_1 = W_2 = 1$ (which not changes the distance) allows us to see P_1 and P_2 as probability distributions of random variables, say, X and Y. Then $\sum_{i,j} f_{ij} d(x_i, y_j)$ is $\mathbb{E}_S[d(X, Y)]$, i.e., the Earth Mover's distance coincides, in this case, with the **transportation distance** (Chap. 14).

  For $W_1 \ne W_2$, it is not a metric in general. However, replacing the inequalities 3 in the above definition by equalities

  3′. $\sum_i f_{ij} = p_2(y_j)$ and $\sum_j f_{ij} = \frac{p_1(x_i) W_1}{W_2}$

  produces the Giannopoulos–Veltkamp's **proportional transport semimetric**.

- **Parameterized curves distance**
  The shape can be represented by a parametrized curve on the plane. Usually, such a curve is *simple*, i.e., it has no self-intersections. Let $X = X(x(t))$ and $Y = Y(y(t))$ be two parametrized curves, where the (continuous) parametrization functions $x(t)$ and $y(t)$ on $[0, 1]$ satisfy $x(0) = y(0) = 0$ and $x(1) = y(1) = 1$. The most used **parametrized curves distance** is the minimum, over all monotone increasing parametrizations $x(t)$ and $y(t)$, of the maximal Euclidean distance $d_E(X(x(t)), Y(y(t)))$. It is the Euclidean special case of the **dogkeeper distance** which is, in turn, the **Fréchet metric** for the case of curves.
  Among variations of this distance are dropping the monotonicity condition of the parametrization, or finding the part of one curve to which the other has the smallest such distance [VeHa01].
- **Nonlinear elastic matching distance**
  Consider a digital representation of curves. Let $r \geq 1$ be a constant, and let $A = \{a_1, \dots, a_m\}$, $B = \{b_1, \dots, b_n\}$ be finite ordered sets of consecutive points on two closed curves. For any order-preserving correspondence f between all points of A and all points of B, the *stretch* $s(a_i, b_j)$ *of* $(a_i, f(a_i) = b_j)$ is r if either $f(a_{i-1}) = b_j$ or $f(a_i) = b_{j-1}$, or zero otherwise.
  The **relaxed nonlinear elastic matching distance** is $\min_f \sum(s(a_i, b_j) + d(a_i, b_j))$, where $d(a_i, b_j)$ is the difference between the tangent angles of a_i and b_j. It is a **near-metric** for some r. For $r = 1$, it is called the **nonlinear elastic matching distance**. In general, Younes, 1998, and Mio–Srivastava–Joshi, 2005, introduced *elastic* Riemannian distances between (seen as elastic) plane curves (or enclosed shapes) measuring the minimal cost of elastic reshaping of a curve into another.
- **Turning function distance**
  For a plane polygon P, its *turning function* $T_P(s)$ is the angle between the counterclockwise tangent and the x axis as a function of the arc length s. This function increases with each left hand turn and decreases with right hand turns.
  Given two polygons of equal perimeters, their **turning function distance** is the L_p**-metric** between their turning functions.
- **Size function distance**
  For a shape, seen as a plane graph $G = (V, E)$, and a *measuring function* f on its vertex-set V (for example, the distance from $v \in V$ to the center of mass of V), the *size function* $S_G(x, y)$ is defined, on the points $(x, y) \in \mathbb{R}^2$, as the number of connected components of the restriction of G on vertices $\{v \in V : f(v) \leq y\}$ which contain a point v' with $f(v') \leq x$.
  Given two plane graphs with vertex-sets belonging to a raster $R \subset \mathbb{Z}^2$, their Uras–Verri's **size function distance** is the normalized l_1-distance between their size functions over raster pixels. The **matching distance** (cf. Chap. 1) between the *cornerpoints/cornerlines* multisets of two size functions is also used.
- **Reflection distance**
  For a finite union A of plane curves and each point $x \in \mathbb{R}^2$, let V_A^x denote the union of intervals (x, a), $a \in A$ which are *visible from* x, i.e., $(x, a) \cap A = \emptyset$. Denote by ρ_A^x the area of the set $\{x + v \in V_A^x : x - v \in V_A^x\}$.

The Hagedoorn–Veltkamp's **reflection distance** between finite unions A and B of plane curves is the normalized l_1-distance between the corresponding functions ρ_A^x and ρ_B^x defined by

$$\frac{\int_{\mathbb{R}^2} |\rho_A^x - \rho_B^x| dx}{\int_{\mathbb{R}^2} \max\{\rho_A^x, \rho_B^x\} dx}.$$

- **Distance transform**
Given a metric space $(X = \mathbb{Z}^2, d)$ and a binary digital image $M \subset X$, the **distance transform** is a function $f_M : X \to \mathbb{R}_{\geq 0}$, where $f_M(x) = \inf_{u \in M} d(x, u)$ is the **point-set distance** $d(x, M)$. So, a distance transform can be seen as a gray-scale digital image where each pixel is given a label (a gray-level) which corresponds to the distance to the nearest pixel of the background. Distance transforms, in Image Processing, are also called *distance fields* and **distance maps**; but we reserve the last term only for this notion in any metric space as in Chap. 1.
A *distance transform of a shape* is the distance transform with M being the boundary of the image. For $X = \mathbb{R}^2$, the graph $\{(x, f(x)) : x \in X\}$ of $d(x, M)$ is called the *Voronoi surface* of M.
- **Medial axis and skeleton**
Let (X, d) be a metric space, and let M be a subset of X. The **medial axis** of X is the set $MA(X) = \{x \in X : |\{m \in M : d(x, m) = d(x, M)\}| \geq 2\}$, i.e., all points of X which have in M at least two **elements of best approximation**; cf. **metric projection** in Chap. 1. $MA(X)$ consists of all points of boundaries of *Voronoi regions* of points of M. The *reach* of M is the **set-set distance** (cf. Chap. 1) between M and $MA(X)$.
The *cut locus* of X is the closure $\overline{MA(X)}$ of the medial axis. Cf. **Shankar–Sormani radii** in Chap. 1. The *medial axis transform* $MAT(X)$ is the point-weighted set $MA(X)$ (the restriction of the **distance transform** on $MA(X)$) with $d(x, M)$ being the weight of $x \in X$.
If (as usual in applications) $X \subset \mathbb{R}^n$ and M is the boundary of X, then the **skeleton** $Skel(X)$ of X is the set of the centers of all d-balls inscribed in X and not belonging to any other such ball; so, $Skel(X) = MA(X)$. The skeleton with M being continuous boundary is a limit of *Voronoi diagrams* as the number of the generating points becomes infinite. For $2D$ binary images X, the skeleton is a curve, a single-pixel thin one, in the digital case. The *exoskeleton* of X is the skeleton of the complement of X, i.e., of the background of the image for which X is the foreground.
- **Procrustes distance**
The *shape* of a *form* (configuration of points in $\mathbb{R}^2$), seen as expression of translation-, rotation- and scale-invariant properties of form, can be represented by a sequence of *landmarks*, i.e., specific points on the form, selected accordingly to some rule. Each landmark point a can be seen as an element $(a', a'') \in \mathbb{R}^2$ or an element $a' + a''i \in \mathbb{C}$.

Consider two shapes x and y, represented by their landmark vectors $(x_1, \dots, x_n)$ and $(y_1, \dots, y_n)$ from $\mathbb{C}^n$. Suppose that x and y are corrected for translation by setting $\sum_t x_t = \sum_t y_t = 0$. Then their **Procrustes distance** is defined by

$$\sqrt{\sum_{1 \le t \le n} |x_t - y_t|^2},$$

where two forms are, first, optimally (by least squares criterion) aligned to correct for scale, and their **Kendall shape distance** is defined by

$$\arccos \sqrt{\frac{(\sum_t x_t \overline{y}_t)(\sum_t y_t \overline{x}_t)}{(\sum_t x_t \overline{x}_t)(\sum_t y_t \overline{y}_t)}},$$

where $\overline{\alpha} = a' - a''i$ is the *complex conjugate* of $\alpha = a' + a''i$.
Petitjean, 2002, extended the L_2-Wasserstein distance (cf. Chap. 14) to *colored mixtures*, i.e., ordinary mixtures of random vectors, for which an new axis (the space of colors) has been added. He remarked that the Procrustes distance is an instance of this *colored Wasserstein distance*, when this latter is minimized for a class of affine transformations (rotations and translations).

- **Shape parameters**
Let X be a figure in $\mathbb{R}^2$ with area $A(X)$, perimeter $P(X)$ and convex hull *conv* X. The main **shape parameters** of X are given below.
$D_A(X) = 2\sqrt{\frac{A(X)}{\pi}}$ and $D_P(X) = \frac{P(X)}{\pi}$ are the diameters of circles with area $A(X)$ and with perimeter $P(X)$, respectively.
Feret's diameters $F_x(X), F_y(X), F_{min}(X), F_{\max}(X)$ are the orthogonal projections of X on the x and y axes and such minimal and maximal projections on a line.
Martin's diameter $M(X)$ is the distance between opposite sides of X measured crosswise of it on a line bisecting the figure's area. $M_x(X)$ and $M_y(X)$ are Martin's diameters for horizontal and vertical directions, respectively.
$R_{in}(X)$ and $R_{out}(X)$ are the radii of the largest disc in X and the smallest disc including X. $a(X)$ and $b(X)$ are the lengths of the major and minor semiaxes of the ellipse with area $A(X)$ and perimeter $P(X)$.
Examples of the ratios, describing some shape properties in above terms, follow.
The *area–perimeter ratio* (or *projection sphericity*) and *Petland's projection sphericity ratio* are $ArPe = \frac{4\pi A(X)}{(P(X))^2}$ and $\frac{4A(X)}{\pi(F_{\max}(X))^2}$.
The *circularity shape factor* and *Horton's compactness factor* are $\frac{1}{ArPe}$ and $\frac{1}{\sqrt{ArPe}}$.
Wadell's circularity shape and *drainage-basin circularity shape* ratios are $\frac{D_A(X)}{F_{\max}(X)}$ and $\frac{A(X)}{D_P(X)}$. Both ratios and $ArPe$ are at most 1 with equality only for a disc.
Tickell's ratio is $(\frac{D_A(X)}{D_{out}(X)})^2$. *Cailleux's roundness ratio* is $\frac{2r(X)}{F_{\max}(X)}$, where $r(X)$ is the radius of curvature at a most convex part of the contour of X.

The *rugosity coefficient* and *convexity ratio* (or *solidity*) are $\frac{P(X)}{P(conv\,X)}$ and $\frac{A(X)}{A(conv\,X)}$. Both the solidity and $\frac{P(conv\,X)}{P(X)}$ are at most 1 with equality only for convex sets. The *diameters ratios* are $\frac{MD_x(X)}{F_x(X)}$ and $\frac{MD_y(X)}{F_y(X)}$. The *radii ratio* and *ellipse ratio* are $\frac{R_{in}(X)}{R_{out}(X)}$ and $\frac{a(X)}{b(X)}$. The *Feret's ratio* and *modification ratio* are $\frac{F_{min}(X)}{F_{\max}(X)}$ and $\frac{R_{in}(X)}{F_{\max}(X)}$. The **aspect ratio** in Chap. 1 is the reciprocal of the Feret's ratio.
The *symmetry factor of Blaschke* is $1-\frac{A(X)}{A(S(X))}$, where $S(X) = \frac{1}{2}(X \oplus \{x : -x \in X\})$.

- **Distances from symmetry**
  Many measures of chirality and, in general, given symmetry G of a given set $A \in \mathbb{R}^n$, were proposed. Several examples follow.
  Let A' be the *enantiomorph* (mirror image) of A. Gilat, 1985, proposed to measure *distance from achirality* of A by $\frac{V(A \Delta A')}{V(A)}$; cf. **normalized volume of symmetric difference** in Chap. 9.
  Let shape A be represented by a sequence $(a_1, \dots, a_m)$ of points. Then the **symmetry distance** of A is defined by Zabrodsky–Peleg–Avnir, 1992, as the point-set distance $\inf_b \frac{1}{m} \sum_{i=1}^{m} ||a_i - b_i||_2^2$, where $b = (b_1, \dots, b_n)$ is the L_2-nearest to a representation of a *symmetric* (i.e., invariant to rotation and translation) shape. The *symmetry distance of a function* f with respect to any transformation G is the L_2-distance between f and the nearest function invariant to G.
  If A is a $2D$ object, and it is represented by its *radial function* $R(r)$, then the *distance of A from symmetry G* can be measured (Köhler, 1993) by $\int_0^{2\pi} |G(R(r)) - R(r)|dr$. For a sequence $(a_1, \dots, a_m)$ of points, similar distance is (Köhler, 1999) $\min_p \sum_{i=1}^{m} d_E(a_i, G(p(a_i)))$, where p is any of $m!$ permutations of $(a_1, \dots, a_m)$ and d_E is the Euclidean distance.
- **Tangent distance**
  For any $x \in \mathbb{R}^n$ and a family of *transformations* $t(x, \alpha)$, where $\alpha \in \mathbb{R}^k$ is the vector of k parameters (for example, the scaling factor and rotation angle), the set $M_x = \{t(x, \alpha) : \alpha \in \mathbb{R}^k\} \subset \mathbb{R}^n$ is a manifold of dimension at most k. It is a curve if $k = 1$. The minimum Euclidean distance between manifolds M_x and M_y would be a useful distance since it is invariant with respect to transformations $t(x, \alpha)$.
  However, the computation of such a distance is too difficult in general; so, M_x is approximated by its *tangent subspace at* x: $\{x + \sum_{i=1}^{k} \alpha_k x^i : \alpha \in \mathbb{R}^k\} \subset \mathbb{R}^n$, where the *tangent vectors* x^i, $1 \le i \le k$, spanning it are the partial derivatives of $t(x, \alpha)$ with respect to α. The **one-sided** (or *directed*) **tangent distance** between elements x and y of $\mathbb{R}^n$ is a quasi-distance d defined by

$$\sqrt{\min_{\alpha} ||x + \sum_{i=1}^{k} \alpha_k x^i - y||^2}.$$

The Simard–Le Cun–Denker's **tangent distance** is defined by $\min\{d(x, y), d(y, x)\}$.

- **Pixel distance**
Consider two digital images, seen as binary $m \times n$ matrices $x = ((x_{ij}))$ and $y = ((y_{ij}))$, where a pixel x_{ij} is black or white if it is equal to 1 or 0, respectively. For each pixel x_{ij}, the *fringe distance map to the nearest pixel of opposite color* $D_{BW}(x_{ij})$ is the number of *fringes* expanded from (i, j) (where each fringe is composed by the pixels that are at the same distance from (i, j)) until the first fringe holding a pixel of opposite color is reached.
The **pixel distance** (Smith–Bourgoin–Sims–Voorhees, 1994) is defined by
$$\sum_{1 \le i \le m} \sum_{1 \le j \le n} |x_{ij} - y_{ij}|(D_{BW}(x_{ij}) + D_{BW}(y_{ij})).$$
In a pixel-based device (computer monitor, printer, scanner), the **pixel pitch** (or *dot pitch*) is the spacing between subpixels (dots) of the same color on the inside of a display screen. Closer spacing usually produce a sharper image.
- **Pratt's figure of merit**
In general, a *figure of merit* is a quantity used to characterize the performance of a device, system or method, relative to its alternatives. Given two binary images, seen as nonempty subsets, A and B, of a finite metric space (X, d), their **Pratt's figure of merit** (or *FOM*, Abdou–Pratt, 1979) is a quasi-distance defined by
$$\left(\max\{|A|, |B|\} \sum_{x \in B} \frac{1}{1 + \alpha d(x, A)^2}\right)^{-1},$$
where α is a scaling constant (usually, $\frac{1}{9}$), and $d(x, A) = \min_{y \in A} d(x, y)$ is the **point-set distance**.
Similar quasi-distances are Peli–Malah's *mean error distance* $\frac{1}{|B|} \sum_{x \in B} d(x, A)$, and the *mean square error distance* $\frac{1}{|B|} \sum_{x \in B} d(x, A)^2$.
- **p-th order mean Hausdorff distance**
Given $p \ge 1$ and two binary images, seen as nonempty subsets A and B of a finite metric space (say, a raster of pixels) (X, d), their **p-th order mean Hausdorff distance** is [Badd92] a normalized **L_p-Hausdorff distance**, defined by
$$\left(\frac{1}{|X|} \sum_{x \in X} |d(x, A) - d(x, B)|^p\right)^{\frac{1}{p}},$$
where $d(x, A) = \min_{y \in A} d(x, y)$ is the **point-set distance**. The usual Hausdorff metric is proportional to the ∞-th order mean Hausdorff distance.
Venkatasubraminian's **Σ-Hausdorff distance** $d_{dHaus}(A, B) + d_{dHaus}(B, A)$ is equal to $\sum_{x \in A \cup B} |d(x, A) - d(x, B)|$, i.e., it is a version of L_1-Hausdorff distance.

Another version of the 1st order mean Hausdorff distance is Lindstrom–Turk's *mean geometric error* (1998) between two images (seen as surfaces A and B) defined by

$$\frac{1}{Area(A)+Area(B)}\left(\int_{x\in A} d(x,B)dS + \int_{x\in B} d(x,A)dS\right),$$

where $Area(A)$ denotes the area of A. If the images are seen as finite sets A and B, their *mean geometric error* is defined by

$$\frac{1}{|A|+|B|}\left(\sum_{x\in A} d(x,B) + \sum_{x\in B} d(x,A)\right).$$

- **Modified Hausdorff distance**
  Given two binary images, seen as nonempty subsets A and B of a finite metric space (X,d), their Dubuisson–Jain's **modified Hausdorff distance** (1994) is defined as the maximum of **point-set distances** averaged over A and B:

$$\max\left\{\frac{1}{|A|}\sum_{x\in A} d(x,B), \frac{1}{|B|}\sum_{x\in B} d(x,A)\right\},$$

  while their Eiter–Mannila's **sum of minimal distances** (1997) is defined as

$$\frac{1}{2}\left(\sum_{x\in A} d(x,B) + \sum_{x\in B} d(x,A)\right).$$

- **Partial Hausdorff quasi-distance**
  Given two binary images, seen as subsets $A, B \neq \emptyset$ of a finite metric space (X,d), and integers k,l with $1 \le k \le |A|$, $1 \le l \le |B|$, their Huttenlocher–Rucklidge's **partial (k,l)-Hausdorff quasi-distance** (1992) is defined by

$$\max\{k^{th}_{x\in A} d(x,B), l^{th}_{x\in B} d(x,A)\},$$

  where $k^{th}_{x\in A} d(x,B)$ means the k-th (rather than the largest $|A|$-th ranked one) among $|A|$ distances $d(x,B)$ ranked in increasing order. The case $k = \lfloor\frac{|A|}{2}\rfloor$, $l = \lfloor\frac{|B|}{2}\rfloor$ corresponds to the *modified median Hausdorff quasi-distance*.
- **Bottleneck distance**
  Given two binary images, seen as subsets $A, B \neq \emptyset$ with $|A| = |B| = m$, of a metric space (X,d), their **bottleneck distance** is defined by

$$\min_{f}\max_{x\in A} d(x, f(x)),$$

  where f is any bijective mapping between A and B.

Variations of the above distance are:

1. The **minimum weight matching**: $\min_f \sum_{x \in A} d(x, f(x))$;
2. The **uniform matching**: $\min_f \{\max_{x \in A} d(x, f(x)) - \min_{x \in A} d(x, f(x)\}$;
3. The **minimum deviation matching**: $\min_f \{\max_{x \in A} d(x, f(x)) - \frac{1}{|A|} \sum_{x \in A} d(x, f(x)\}$.

Given an integer t with $1 \leq t \leq |A|$, the t**-bottleneck distance** between A and B [InVe00] is the above minimum but with f being any mapping from A to B such that $|\{x \in A : f(x) = y\}| \leq t$.
The cases $t = 1$ and $t = |A|$ correspond, respectively, to the bottleneck distance and **directed Hausdorff distance** $d_{dHaus}(A, B) = \max_{x \in A} \min_{y \in B} d(x, y)$ (Chap. 1).

- **Hausdorff distance up to** G
Given a group $(G, \cdot, id)$ acting on the Euclidean space $\mathbb{E}^n$, the **Hausdorff distance up to** G between two compact subsets A and B (used in Image Processing) is their **generalized** G**-Hausdorff distance** (cf. Chap. 1), i.e., the minimum of $d_{Haus}(A, g(B))$ over all $g \in G$. Usually, G is the group of all isometries or all translations of $\mathbb{E}^n$.
- **Hyperbolic Hausdorff distance**
For any compact subset A of $\mathbb{R}^n$, denote by $MAT(A)$ its *Blum's medial axis transform*, i.e., the subset of $X = \mathbb{R}^n \times \mathbb{R}_{\geq 0}$, whose elements are all pairs $x = (x', r_x)$ of the centers x' and the radii r_x of the maximal inscribed (in A) balls, in terms of the Euclidean distance d_E in $\mathbb{R}^n$. (Cf. **medial axis and skeleton** transforms for the general case.)
The **hyperbolic Hausdorff distance** [ChSe00] is the **Hausdorff metric** on nonempty compact subsets $MAT(A)$ of the metric space (X, d), where the *hyperbolic distance* d on X is defined, for its elements $x = (x', r_x)$ and $y = (y', r_y)$, by
$$\max\{0, d_E(x', y') - (r_y - r_x)\}.$$
- **Nonlinear Hausdorff metric**
Given two compact subsets A and B of a metric space (X, d), their **nonlinear Hausdorff metric** (or *Szatmári–Rekeczky–Roska wave distance*) is the **Hausdorff distance** $d_{Haus}(A \cap B, (A \cup B)^*)$, where $(A \cup B)^*$ is the subset of $A \cup B$ which forms a closed contiguous region with $A \cap B$, and the distances between points are allowed to be measured only along paths wholly in $A \cup B$.
- **Video quality metrics**
These metrics are between test and reference color video sequences, usually aimed at optimization of encoding/compression/decoding algorithms. Each of them is based on some perceptual model of the human vision system, the simplest ones being RMSE (root-mean-square error) and PSNR (peak signal-to-noise ratio) error measures. The *threshold metrics* estimate the probability of detecting

an *artifact* (i.e., a visible distortion that gets added to a video signal during digital encoding).
Examples are: Sarnoff's JND (just-noticeable differences), Winkler's PDM (perceptual distortion), and Watson's DVQ (digital video quality) metrics. DVQ is an l_p-**metric** between feature vectors representing two video sequences. Some metrics measure special artifacts in the video: the appearance of block structure, blurriness, added "mosquito" noise (ambiguity in the edge direction), texture distortion, etc.

- **Time series video distances**
  The **time series video distances** are objective wavelet-based spatial–temporal **video quality metrics**. A video stream x is processed into a time series $x(t)$ (seen as a curve on coordinate plane) which is then (piecewise linearly) approximated by a set of n contiguous line segments that can be defined by $n + 1$ endpoints (x_i, x_i'), $0 \le i \le n$, in the coordinate plane. In [WoPi99] are given the following (cf. **Meehl distance**) distances between video streams x and y:
  - $Shape(x, y) = \sum_{i=0}^{n-1} |(x_{i+1}' - x_i') - (y_{i+1}' - y_i')|$;
  - $Offset(x, y) = \sum_{i=0}^{n-1} |\frac{x_{i+1}'+x_i'}{2} - \frac{y_{i+1}'+y_i'}{2}|$.
- **Handwriting spatial gap distances**
  Automatic recognition of unconstrained handwritten texts (for example, legal amounts on bank checks or pre-hospital care reports) require measuring the spatial gaps between connected components in order to extract words.
  Three most used ones, among **handwriting spatial gap distances** between two adjacent connected components x and y of text line, are:
  - Seni–Cohen, 1994: the *run-length* (minimum horizontal Euclidean distance) between x and y, and the horizontal distance between their bounding boxes;
  - Mahadevan–Nagabushnam, 1995: Euclidean distance between the convex hulls of x and y, on the line linking hull centroids.

21.2 Audio Distances

Sound is the vibration of gas or air particles that causes pressure variations within our eardrums. *Audio* (speech, music, etc.) *Signal Processing* is the processing of analog (continuous) or, mainly, digital representation of the air pressure waveform of the sound. A *sound spectrogram* (or *sonogram*) is a visual $3D$ representation of an acoustic signal. It is obtained either by a series of bandpass filters (an analog processing), or by application of the *short-time Fourier transform* to the electronic analog of an acoustic wave. Three axes represent time, frequency and *intensity* (acoustic energy). Often this $3D$ curve is reduced to two dimensions by indicating the intensity with more thick lines or more intense gray or color values.

Sound is called *tone* if it is periodic (the lowest *fundamental* frequency plus its multiples, *harmonics* or *overtones*) and *noise*, otherwise. The frequency is measured in *cps* (the number of complete cycles per second) or Hz (Hertz). The range of audible sound frequencies to humans is typically 20 Hz to 20 kHz. A moth *Galleria mellonella* can hear up to 300 kHz, in order to locate predatory bats using ultrasound.

The *power* $P(f)$ of a signal is energy per unit of time; it is proportional to the square of signal's amplitude $A(f)$. *Decibel dB* is the unit used to express the relative strength of two signals. One tenth of 1 dB is *bel*, the original outdated unit.

The amplitude of an audio signal in *dB* is $20 \log_{10} \frac{A(f)}{A(f')} = 10 \log_{10} \frac{P(f)}{P(f')}$, where f' is a reference signal selected to correspond to 0 dB (usually, the threshold of human hearing). The threshold of pain is about 120–140 dB.

Pitch and *loudness* are auditory subjective terms for frequency and amplitude.

The *mel scale* is a perceptual frequency scale, corresponding to the auditory sensation of tone height and based on *mel*, a unit of pitch. It is connected to the acoustic frequency f hertz scale by $Mel(f) = 1127 \ln(1 + \frac{f}{700})$ or, simply, $Mel(f) = 1000 \log_2(1 + \frac{f}{1000}$.

The *Bark scale* (named after Barkhausen) is a psycho-acoustic scale of frequency: it ranges from 0 to 24 *Bark* corresponding to the first 24 critical bands of hearing:

$0, 100, 200, \dots, 1270, 1480, 1720, \dots, 9500, 12000, 15500$ Hz.

Those bands correspond to spatial regions of the basilar membrane (of the inner ear), where oscillations, produced by the sound of given frequency, activate the hair cells and neurons. Our ears are most sensitive in 2,000–5,000 Hz. The Bark scale is connected to the acoustic frequency f kilohertz scale by $Bark(f) = 13 \arctan(0.76f) + 3.5 \arctan(\frac{f}{0.75})^2$.

Terrestrial vertebrates perceive frequency on a logarithmic scale, i.e., pitch perception is better described by frequency ratios than by differences on a linear scale. It is matched by the distribution of cells sensitive to different frequencies in their ears.

Power spectral density $PSD(f)$ of a wave is the power per Hz. It is the Fourier transform of the autocorrelation sequence. So, the power of the signal in the band $(-W, W)$ is given by $\int_{-W}^{W} PSD(f)df$. A *power law noise* has $PSD(f) \sim f^{\alpha}$. The noise is called *violet*, *blue*, *white*, *pink* (or $\frac{1}{f}$), *red* (or *brown(ian)*), *black* (or *silent*) if $\alpha = 2, 1, 0, -1, -2, < -2$. PSD changes by 3α dB per *octave* (i.e., with frequency doubling); it decreases for $\alpha < 0$.

Pink noise occurs in many physical, biological and economic systems; cf. **long range dependence** in Chap. 18. It has equal power in proportionally wide frequency ranges. Humans also process frequencies in a such logarithmic space (approximated by the Bark scale). So, every octave contains the same amount of energy. Thus pink noise is used as a reference signal in Audio Engineering. Steady pink noise (including light music) reduces brain wave complexity and improve sleep quality.

Intensity of speech signal goes up/down within a 3–8 Hz frequency which resonates with the theta rhythm of neocortex. The speakers produce 3–8 syllables per second.

The main way that humans control their *phonation* (speech, song, laughter) is by control over the *vocal tract* (the throat and mouth) shape. This shape, i.e., the cross-sectional profile of the tube from the closure in the *glottis* (the space between the vocal cords) to the opening (lips), is represented by the cross-sectional area function $Area(x)$, where x is the distance to the glottis. The vocal tract acts as a resonator during vowel phonation, because it is kept relatively open. These resonances reinforce the source sound (ongoing flow of lung air) at particular *resonant frequencies* (or *formants*) of the vocal tract, producing peaks in the *spectrum* of the sound.

Each vowel has two characteristic formants, depending on the vertical and horizontal position of the tongue in the mouth. The source sound function is modified by the frequency response function for a given area function. If the vocal tract is approximated as a sequence of concatenated tubes of constant cross-sectional area, then the *area ratio coefficients* are the ratios $\frac{Area(x_{i+1})}{Area(x_i)}$ for consecutive tubes; those coefficients can be computed by *LPC* (linear predictive coding).

The *spectrum* of a sound is the distribution of magnitude (dB) (and sometimes the phases) in frequency (kHz) of the components of the wave. The *spectral envelope* is a smooth contour that connects the spectral peaks. Its estimation is based on either LPC, or FFT (fast Fourier transform) using *real cepstrum*, i.e., the log amplitude spectrum.

FT (Fourier transform) maps time-domain functions into frequency-domain representations. The *complex cepstrum* of the signal $f(t)$ is $FT(\ln(FT(f(t) + 2\pi mi)))$, where m is the integer needed to unwrap the angle or imaginary part of the complex logarithm function. The FFT performs the Fourier transform on the signal and samples the discrete transform output at the desired frequencies usually in the *mel* scale.

Parameter-based distances used in recognition and processing of speech data are usually derived by LPC, modeling the speech spectrum as a linear combination of the previous samples (as in autoregressive processes). Roughly, LPC processes each word of the speech signal in the following 6 steps: filtering, energy normalization, partition into frames, *windowing* (to minimize discontinuities at the borders of frames), obtaining LPC parameters by the autocorrelation method and conversion to the *LPC-derived cepstral coefficients*. LPC assumes that speech is produced by a buzzer at the glottis (with occasionally added hissing and popping sounds), and it removes the formants by filtering.

The majority of distortion measures between sonograms are variations of **squared Euclidean distance** (including a covariance-weighted one, i.e., **Mahalanobis**, distance) and probabilistic distances belonging to following general types: generalized **variational distance**, f**-divergence** and **Chernoff distance**; cf. Chap. 14.

The distances for sound processing below are between vectors x and y representing two signals to compare. For recognition, they are a template reference and input signal, while for noise reduction they are the original (reference) and distorted signal (see, for example, [OASM03]). Often distances are calculated for small segments, between vectors representing short-time spectra, and then averaged.

- **SNR distance**

  Given a sound, let P and A_s denote its average power and RMS (root-mean-square) amplitude. The **signal-to-noise ratio in decibels** is defined by

$$SNR_{dB} = 10\log_{10}\left(\frac{P_{signal}}{P_{noise}}\right) = P_{signal,dB} - P_{noise,dB} = 10\log_{10}\left(\frac{A_{signal}}{A_{noise}}\right)^2.$$

  The *dynamic range* is such ratio between the strongest undistorted and minimum discernable signals. It is roughly 140 dB for human hearing, 40 dB for human speech and 80 dB for a music in a concert hall.

  The *Shannon–Hartley theorem* express the *capacity* (maximal possible information rate) of a channel with additive *colored* (frequency-dependent) Gaussian noise, on the bandwidth B in Hz as $\int_0^B \log_2(1+\frac{P_{signal}(f)}{P_{noise}(f)})df$.

  The **SNR distance** between signals $x = (x_i)$ and $y = (y_i)$ with n frames is

$$10\log_{10}\frac{\sum_{i=1}^{n} x_i^2}{\sum_{i=1}^{n}(x_i - y_i)^2}.$$

  If M is the number of segments, the *segmented SNR* between x and y is defined by

$$\frac{10}{m}\sum_{m=0}^{M-1}\left(\log_{10}\sum_{i=nm+1}^{nm+n}\frac{x_i^2}{(x_i - y_i)^2}\right).$$

  Another measure, used to compare two waveforms x and y in the time-domain, is their **Czekanovski–Dice distance** defined by

$$\frac{1}{n}\sum_{i=1}^{n}\left(1 - \frac{2\min\{x_i, y_i\}}{x_i + y_i}\right).$$

- **Spectral magnitude-phase distortion**

  The **spectral magnitude-phase distortion** between signals $x = x(\omega)$ and $y = y(\omega)$ is defined by

$$\frac{1}{n}\left(\lambda\sum_{i=1}^{n}(|x(w)| - |y(w)|)^2 + (1-\lambda)\sum_{i=1}^{n}(\angle x(w) - \angle y(w))^2\right),$$

  where $|x(w)|, |y(w)|$ are magnitude spectra, and $\angle x(w), \angle y(w)$ are phase spectra of x and y, respectively, while the parameter $\lambda, 0 \leq \lambda \leq 1$, is chosen in order to attach commensurate weights to the magnitude and phase terms. The case $\lambda = 0$ corresponds to the **spectral phase distance**.

  Given a signal $f(t) = ae^{-bt}u(t)$, $a, b > 0$ which has Fourier transform $x(w) = \frac{a}{b+iw}$, its *magnitude* (or *amplitude*) spectrum is $|x| = \frac{a}{\sqrt{b^2+w^2}}$, and its *phase* spectrum (in radians) is $\alpha(x) = \tan^{-1}\frac{w}{b}$, i.e., $x(w) = |x|e^{i\alpha} = |x|(\cos\alpha + i\sin\alpha)$.

The **Fourier distance** and *Fourier phase distance* are $||FFT(x) - FFT(y)||_2$ and $||arg(FFT(x)) - arg(FFT(y))||_2$, where the sums only contain the lower frequency terms of fast Fourier transform in order to reduce noise. The similar *wavelet distance* is based on the *discrete wavelet transform* separating low and high frequencies.

- **Spectral distances**

  Given two discrete spectra $x = (x_i)$ and $y = (y_i)$ with n channel filters, their **Euclidean metric** *EM*, **slope metric** *SM* (Klatt, 1982) and 2**nd differential metric** 2*DM* (Assmann and Summerfield, 1989) are defined, respectively, by

$$\sqrt{\sum_{i=1}^{n}(x_i - y_i)^2}, \sqrt{\sum_{i=1}^{n}(x'_i - y'_i)^2} \text{ and } \sqrt{\sum_{i=1}^{n}(x''_i - y''_i)^2},$$

  where $z'_i = z_{i+1} - z_i$ and $z''_i = \max(2z_i - z_{i+1} - z_{i-1}, 0)$. Comparing, say, the auditory excitation patterns of vowels, *EM* gives equal weight to peaks and troughs although spectral peaks have more perceptual weight. *SM* emphasizes the formant frequencies, while 2*DM* sets to zero the spectral properties other than the formants.

  The **RMS log spectral distance** (or *root-mean-square distance*, *quadratic mean distance*) $LSD(x, y)$ is defined by

$$\sqrt{\frac{1}{n}\sum_{i=1}^{n}(\ln x_i - \ln y_i)^2}.$$

  The corresponding l_1- and l_∞-distances are called *mean absolute distance* and *maximum deviation*. These three distances are related to decibel variations in the log spectral domain by the multiple $\frac{10}{\log 10}$.

  The square of $LSD(x, y)$, via the cepstrum representation $\ln x(\omega) = \sum_{j=-\infty}^{\infty} c_j e^{-j\omega i}$ (where $x(\omega)$ is the *power cepstrum* $|FT(\ln(|FT(f(t))|^2)|^2)$ becomes, in the complex cepstral space, the **cepstral distance**.

  The **log area ratio distance** $LAR(x, y)$ between x and y is defined by

$$\sqrt{\frac{1}{n}\sum_{i=1}^{n} 10(\log_{10} Area(x_i) - \log_{10} Area(y_i))^2},$$

  where $Area(z_i)$ is the cross-sectional area of the i-th segment of the vocal tract.

- **Bark spectral distance**

  Let (x_i) and (y_i) be the *Bark spectra* of x and y, where the i-th component corresponds to the i-th auditory critical band in the Bark scale. The **Bark**

spectral distance (Wang–Sekey–Gersho, 1992) is a perceptual distance, defined by

$$BSD(x, y) = \sum_{i=1}^{n}(x_i - y_i)^2,$$

i.e., it is the **squared Euclidean distance** between the Bark spectra.
A modification of the Bark spectral distance excludes critical bands i on which the loudness distortion $|x_i - y_i|$ is less than the noise masking threshold.

- **Itakura–Saito quasi-distance**
The **Itakura–Saito** (or *maximum likelihood*) **quasi-distance** between LPC-derived spectral envelopes $x = x(\omega)$ and $y = y(\omega)$ is defined (1968) by

$$IS(x, y) = \frac{1}{2\pi}\int_{-\pi}^{\pi}\left(\frac{x(w)}{y(w)} - \ln\frac{x(w)}{y(w)} - 1\right)dw.$$

The **cosh distance** is defined by $IS(x, y) + IS(y, x)$, i.e., is equal to

$$\frac{1}{2\pi}\int_{-\pi}^{\pi}\left(\frac{x(w)}{y(w)} + \frac{y(w)}{x(w)} - 2\right)dw = \frac{1}{2\pi}\int_{-\pi}^{\pi} 2\cosh\left(\ln\frac{x(w)}{y(w)} - 1\right)dw,$$

where $\cosh(t) = \frac{e^t + e^{-t}}{2}$ is the hyperbolic cosine function.

- **Log-likelihood ratio quasi-distance**
The **log-likelihood ratio quasi-distance** between LPC-derived spectral envelopes $x = x(\omega), y = y(\omega)$ is defined (cf. **Kullback–Leibler distance** in Chap. 14) by

$$\frac{1}{2\pi}\int_{-\pi}^{\pi} x(w)\ln\frac{x(w)}{y(w)}dw.$$

The **weighted likelihood ratio distance** between $x(\omega)$ and $y(\omega)$ is defined by

$$\frac{1}{2\pi}\int_{-\pi}^{\pi}\left(\frac{\left(\ln\left(\frac{x(w)}{y(w)}\right) + \frac{y(w)}{x(w)} - 1\right)x(w)}{p_x} + \frac{\left(\ln\left(\frac{y(w)}{x(w)}\right) + \frac{x(w)}{y(w)} - 1\right)y(w)}{p_y}\right)dw,$$

where $P(x)$ and $P(y)$ denote the power of the spectra $x(w)$ and $y(w)$.

- **Cepstral distance**
The **cepstral distance** (or *squared Euclidean cepstrum metric*) $CEP(x, y)$ between the LPC-derived spectral envelopes $x = x(\omega)$ and $y = y(\omega)$ is defined by

$$\frac{1}{2\pi}\int_{-\pi}^{\pi}\left(\ln\frac{x(w)}{y(w)}\right)^2 dw = \frac{1}{2\pi}\int_{-\pi}^{\pi}(\ln x(w) - \ln y(w))^2\, dw$$

$$= \sum_{j=-\infty}^{\infty} (c_j(x) - c_j(y))^2,$$

where $c_j(z) = \frac{1}{2\pi}\int_{-\pi}^{\pi} e^{jwi} \ln |z(w)| dw$ is j-th cepstral (real) coefficient of z derived from the Fourier transform or LPC.

The **quefrency-weighted cepstral distance** (or **Yegnanarayana distance**, *weighted slope distance*) between x and y is defined by

$$\sum_{i=-\infty}^{\infty} i^2(c_i(x) - c_i(y))^2.$$

"Quefrency" and "cepstrum" are anagrams of "frequency" and "spectrum".

The **Martin cepstrum distance** between two ARMs (autoregressive models) is defined, in terms of their cepstra, by

$$\sqrt{\sum_{i=0}^{\infty} i(c_i(x) - c_i(y))^2}.$$

Cf. general **Martin distance** in Chap. 12 and **Martin metric** in Chap. 11.

- **Distances in Musicology**

  Pitch is a subjective correlate of the fundamental frequency. A *note* (or *tone*) is a named pitch. Pitch, seen as extending along a 1D continuum from high to low, is called *pitch height*. But it also varies circularly: a *pitch class* is a set of all pitches that are a whole number of *octaves* (intervals between a frequency and its double) apart. About 10 octaves cover the range of human hearing. In Western music, the most used octave division is the *chromatic scale*: 12 notes $C, C\#, D, D\#, E, F, F\#, G, G\#, A, A\#, B$ drawn usually as *pitch class space*: a circle of *equal temperament*, i.e., divided into 12 equal *semitones* (or *half steps*). The **distance between notes** whose frequencies are f_1, f_2 is $12\log_2(\frac{f_1}{f_2})$ semitones.

  An *interval* is the difference between two pitches. Its *width* is the ratio $\frac{a}{b}$ (with g.c.d$(a,b) = 1$) between their frequencies. The *Benedetti height* of this ratio is ab; *Tenney height* (or **Tenney harmonic distance**) is $\log_2 ab$ and *Kees height* is max (a', b'), where a', b' come from a, b by removing factors of 2. The width of a semitone is $\sqrt[12]{2}$ or 100 *cents*. The width of octave is 2 or 1,200 cents.

  A **pitch distance** (or *melodic distance*) is the size of the section of the pitch-continuum bounded by those two pitches, such as modeled in a given scale. A MIDI (Musical Instrument Digital Interface) number of fundamental frequency f is defined by $p(f) = 69 + 12\log_2 \frac{f}{440}$. The distance between notes, in terms of this *linear pitch space*, becomes the **natural metric** $|p(f_1) - p(f_2)|$ on $\mathbb{R}$. This pitch distance corresponds to physical distance on keyboard instruments, and psychological distance as measured by experiments and musicians.

Using integer notation $0, 1, \ldots, 11$ of pitches, a *pitch interval* $PI(x, y)$ between the pitches x and y is the number of semitones $|x-y|$ that separates them linearly, while a *pitch-interval class* $PIC(x, y)$ is $|x - y| \mod 12$ and an *interval class* $ic(x, y)$ is their **Lee distance** $\min |x - y|, 12 - |x - y|$ on the circle.
In integer notation, the *circle of fifths* is $\{7i \mod 12\}_{0,\ldots,11}$, and its reversal, the *circle of fourths*, is $\{5i \mod 12\}_{0,\ldots,11}$. Neighboring pitches are separated by a *perfect fifth* (interval of five staff positions or seven semitones).
Seven letters of a musical alphabet, C–D–E–F–G–A–B, are called the *natural tones*; they are the names of the white keys on a piano/keyboard, forming an octave. Above sequence and any of its translations is a *major diatonic scale*. A *diatonic scale* is a scale of 7 notes most used in Western music. Its structure is 1–1–0.5–1–1–1–0.5, in terms of interval succession of steps. A **distance model** (in Music) is the alternation of two different intervals to create a nondiatonic musical mode/scale, such as the 1:3 distance model (alternation of semitones and minor thirds).
In tonal music, composition written in home key; it *modulates* (move to other keys) and usually returns. The distance between keys approximates the ease of modulation. Every key is associated with a scale of pitches, usually, major or *minor* diatonic. The **interkey distance** of two keys is 7 minus the number of tones shared by their scales. It is also their distance around the circle of fifths, i.e., the difference in the number of *sharps* (or *flats*) in their signatures. The *relative* (having the same signatures) major and minor key share all seven notes.
A *chord* in music is any set of at least three pitch classes in the same octave that is heard as if sounding simultaneously. Music can be seen as a sequence of chords. *Interval vector* of a given chord c is $V(c) = (c_1, \ldots, c_6)$, where c_i is the number of times i-th interval class (having i or $12-i$ semitones) appears in it. **Intervalic distance** and **Estrada distance** between chord c and c' are (Mathieu, 2002):

$$\sum_{i=1}^{6} |c_i - c_i'| \text{ and } \max |c|, |c'| - |V(c) \cap V(c')| - 1.$$

The **root distance** is the number of fifths between the *roots* (pitches upon which a chord may be built, often by stacking thirds) of the chords. In [RRHD10], a survey of eight **distances between chords** is given: above 3 and those by Chew (2000), Costère (1962), Lerdahl (2001), Paiement et al. (2005) and Yoo et al. (2006).
Alternatively to equal-temperament, *just intonation* is a tuning in which the frequencies of notes are related by ratios of small whole numbers, say, $\frac{3}{2}$ for perfect fifth (G) and $\frac{4}{3}$ for perfect forth (F). The pitches can be arranged in a $2D$ diagram. For an odd number $n > 0$, the *n-limit diagram* contains all rational numbers such that any odd divisor of the numerator or denominator is at most n. Such 5- and 7-limit can be seen as the *hexagonal lattice* $A_2 = \{(a, b, c) \in \mathbb{Z}^3 : a + b + c = 0\}$ and *face-centered cubic lattice* $A_3 = \{(a, b, c) \in \mathbb{Z}^3 :$

$a + b + c \equiv 0 (\text{mod } 2)\}$, respectively, with vector space norms $\sqrt{a^2 + ab + b^2}$ and $\sqrt{a^2 + ab + b^2 + c(a + b + c)}$.

- **Distances between rhythms**

  A rhythm timeline (music pattern) is represented, besides the standard music notation, in the following ways, used in computational music analysis.

  1. By a binary vector $x = (x_1, \dots, x_m)$ of m time intervals (equal in a metric timeline), where $x_i = 1$ denotes a beat, while $x_i = 0$ denotes a rest interval (silence). For example, the five 12/8 metric timelines of Flamenco music are represented by five binary sequences of length 12.
  2. By a *pitch vector* $q = (q_1, \dots, q_n)$ of absolute pitch values q_i and a *pitch difference vector* $p = (p_1, \dots, p_{n-1})$ where $p_i = q_{i+1} - q_i$ represents the number of semitones (positive or negative) from q_i to q_{i+1}.
  3. By an *interonset interval vector* $t = (t_1, \dots, t_n)$ of n time intervals between consecutive onsets.
  4. By a *chronotonic representation* which is a histogram visualizing t as a sequence of squares of sides $t_1, \dots, t_n$; it can be seen as a piecewise linear function.
  5. By a *rhythm difference vector* $r = (r_1, \dots, r_{n-1})$, where $r_i = \frac{t_{i+1}}{t_i}$.

  Examples of general **distances between rhythms** are the Hamming distance, **swap metric** (cf. Chap. 11) and **Earth Mover's distance** between their given vector representations.

  The **Euclidean interval vector distance** is the Euclidean distance between two interonset interval vectors. The Gustafson **chronotonic distance** is a variation of l_1-distance between these vectors using the chronotonic representation.

  Coyle–Shmulevich **interval-ratio distance** is defined by

$$1 - n + \sum_{i=1}^{n-1} \frac{\max\{r_i, r'_i\}}{\min\{r_i, r'_i\}},$$

  where r and r' are rhythm difference vectors of two rhythms (cf. the reciprocal of **Ruzicka similarity** in Chap. 17).

- **Long-distance drumming**

  Long-distance drumming (or *drum telegraphy*) is an early form of long-distance communication which was used by cultures in Africa, New Guinea and the tropical America living in deforested areas. A rhythm could represent an signal, repeat the profile of a spoken utterance or simply be subject to musical laws.

  The *message drums* (or *slit gongs*) were developed from hollow tree trunks. The sound could be understood at $\leq$8 km but usually it was relayed to a next village. Another oldest tools of audio telecommunication were *horns* (tapered sound guides providing an acoustic impedance match between a sound source and free air). Any mode of communication (as by means of drums or horns) for use beyond the range of the articulate voice, is called **distance language**.

Soldier termites of some species drum their heads (11 times per second) on the ground to signal danger. The initial vibrations travel 40 cm, but a chain of soldiers relay the resulting wave, moving 1.3 m/s, over much greater distances.

- **Sonority distance effect**
  People in warm-climate cultures spend more time outdoors and engage, on average, in more distal oral communication. So, such populations have greater *sonority* (audibility) of their phoneme inventory. Munroe et al., 1996 and 2009, observed that speakers in such languages use more simple consonant-vowel syllables, vowels and *sonorant* (say, nasal "n", "m" rather than obstruents as "t", "g") consonants.
  Ember and Ember, 2007, found that number of cold months, as well as the combination of cold climate and sparse vegetation, predicts less sonority. Larger average distance of the baby from its caregivers, as well as higher frequency of premarital and extramarital sex predicts more sonority.
- **Vocal deviation**
  Vocal deviation is (Podos, 2001) the distance of birdsong performance to the upper performance limit. Performance is measured by a variable combining frequency bandwidth and *note repetition rate* (the number of notes per second).
  The **vocal deviation** of a bird is the minimal **point-line distance** (Chap. 4) of data points of its recordings from the (upper-bound regression) line representing performance limit.
- **Acoustics distances**
  The *wavelength* of a sound wave is the distance it travels to complete one cycle. This distance is measured perpendicular to the wavefront in the direction of propagation between one peak of a *sine wave* (sinusoid) and the next corresponding peak. The wavelength of any frequency may be found by dividing the speed of sound (331.4 m/s at sea level) in the medium by the fundamental frequency.
  The *near field* is the part of a sound field (usually within about two wavelengths from the source) where there is no simple relationship between sound level and distance. The *far field* (cf. **Fraunhofer distance** in Chap. 24) is the area beyond the near field boundary. It is comprised of the *reverberant field* and *free field*, where sound intensity decreases as $\frac{1}{d^2}$ with the distance d from the source. This law corresponds to a reduction of $\approx$6 dB in the sound level for each doubling of distance and to halving of loudness (subjective response) for each reduction of $\approx$10 dB.
  The **critical distance** (or *room radius*) is the distance from the source at which the direct sound and reverberant sound (reflected echo produced by the direct sound bouncing off, say, walls, floor, etc.) are equal in amplitude.
  The **pickup distance** of a microphone is the effective distance that it can be used at. For an electric guitar, it is the distance from *pickup* (transducer that captures mechanical vibrations) to strings.
  A directional microphone may be placed farther away from a desired sound source than an omnidirectional one of equal quality; the ratio of distances is called the **distance factor**.

The *proximity effect (audio)* is the anomaly of low frequencies being enhanced when a directional microphone is very close to the source.

Auditory **distance cues** (cf. Chap. 28) are based on differences in loudness, spectrum, direct-to-reverb ratio and binaural ones. The closer sound object is louder, has more bass, high-frequencies, transient detail, dynamic contrast. Also, it appear wider, has more direct sound level over its reflected sound and has greater time delay between the direct sound and its reflections.

The *acoustic metric* is the term used occasionally for some distances between vowels; for example, the Euclidean distance between vectors of formant frequencies of pronounced and intended vowel. Cf. **acoustic metric** in Physics (Chap. 24).

Chapter 22
Distances in Networks

22.1 Scale-Free Networks

A **network** is a graph, directed or undirected, with a positive number (weight) assigned to each of its arcs or edges. Real-world complex networks usually have a gigantic number N of vertices and are sparse, i.e., with relatively few edges.

Interaction networks (Internet, Web, social networks, etc.) tend to be **small-world** [Watt99], i.e., interpolate between regular geometric lattices and random graphs in the following sense. They have a large *clustering coefficient* (the probability that two distinct neighbors of a vertex are neighbors), as lattices in a local neighborhood, while the average path distance between two vertices is small, about $\ln N$, as in a random graph.

A **scale-free network** [Bara01] is a network with probability distribution for a vertex to have degree k being similar to $k^{-\gamma}$, for some constant $\gamma > 0$ which usually belongs to the segment $[2, 3]$. This *power law* implies that very few vertices, called *hubs* (connectors, gateways, super-spreaders), are far more connected than other vertices. The power law (or **long range dependent**, *heavy-tail*) distributions, in space or time, has been observed in many natural phenomena (both physical and sociological).

- **Collaboration distance**
  The **collaboration distance** is the path metric of the *Collaboration graph*, having authors in Mathematical Reviews database as vertices with xy being an edge if authors x and y have a joint publication among the papers from this database.
  The vertex of largest degree (1,416) corresponds to Paul Erdős; the *Erdős number* of a mathematician is his collaboration distance to Paul Erdős. An example of a 3-path: Michel Deza–Paul Erdős–Ernst Gabor Straus–Albert Einstein.

M.M. Deza, E. Deza, *Encyclopedia of Distances*,
DOI 10.1007/978-3-662-44342-2_22

- **Co-starring distance**
  The **co-starring distance** is the path metric of the *Hollywood graph*, having about 250,000 vertices (actors in the Internet Movie database) with xy being an edge if the actors x and y appeared in a feature film together. The vertices of largest degree are Christopher Lee and Kevin Bacon; the trivia game *Six degrees of Kevin Bacon* uses the *Bacon number*, i.e., the co-starring distance to this actor. The *Morphy* and *Shusaku numbers* are the similar measures of a chess or Go player's connection to Paul Morphy and Honinbo Shusaku by way of playing games. *Kasparov number* of a chess-player is the length of a shortest directed path, if any, from him/her to Garry Kasparov; here arc uv means victory of u over v.
  Similar popular examples of such social scale-free networks are graphs of musicians (who played in the same rock band), baseball players (as team-mates), scientific publications (who cite each other), mail exchanges, acquaintances among classmates in a college, business board membership.
  Among other such studied networks are air travel connections, word co-occurrences in human language, US power grid, sensor networks, worm neuronal network, gene co-expression networks, protein interaction networks and metabolic networks (with two substrates forming an edge if a reaction occurs between them via enzymes).
- **WikiDistance**
  In May 2014, Wikipedia had about 30 million articles in 287 languages and 130,000 active editors. English Wikipedia alone had 4.5 million articles, 4 % of estimated number of notable articles needed to cover all human knowledge.
  The **WikiDistance** is the directed path quasi-metric of the *Wikipedia digraph*, having English Wikipedia articles as vertices, with xy being an arc if the article x contains an hyperlink to the article y; cf. http://software.tanos.co.uk/wikidistance and the **Web hyperlink quasi-metric**.
  Gabrilovich–Markovich, 2007, proposed to measure semantic relatedness of two texts by the **cosine distance** (cf. **Web similarity metrics**) between weighted vectors, interpreting texts in terms of affinity with a host of Wikipedia concepts.
  Crandall et al., 2008, considered the social network of Wikipedia editors: two editors are assumed to be connected if one of them posted to the other's discussion page. Brandes et al., 2009, considered the *edit network* of a Wikipedia page, where nodes are the authors of this page and edges correspond to undoing each other edits.
  The *editing depth* of Wikipedia is an indicator of its collaborativeness defined as $D = \frac{e}{a+n} \times (\frac{n}{a})^2$, where e, a, n are the numbers of page edits, articles and *nonarticles* (redirects, talk, user pages). At April 2013, English Wikipedia had $D = 758$.
- **Virtual community distance**
  Largest, in millions of active user accounts in 2012, *virtual communities* (online social networking services) are: *Facebook* (1,000), *Tencent QQ* (712), *Skype* (280), *Google+* (235), *Twitter* (200), *Linkedin* (161).

In 2012, about 30 billion documents were uploaded on Facebook, 300 million tweets sent on Twitter and 24 petabytes of data processed by Google *per day*, while mankind published only ≈5,000 petabytes for the 20,000 years before 2003.

A **virtual community distance** is the path metric of the graph of active users, two of them forming an edge if they are "friends". In Twitter it means that both "follow" each other. In particular, for the **Facebook hop distance** in November 2011, 99.6 % of all pairs of users were connected by paths of length at most 5. The mean distance was 4.74, down from 5.28 in 2008.

The **Twitter friendship distance** in Avril 2010 was 4, 5, 6 among 37 %, 41 %, 13 % of 5.2 billion friendships. The average distance was 4.67 steps. Cf. mean distance 5.2 in Milgram's (1967) theory of *six degrees of separation* on a planetary scale. An example of application: analysing linguistically emotional content of tweets and comments, one can obtain an interaction graph of the targeted region mood.

- **Distance effect in large e-mail networks**

  Takhteyev–Gruzd–Wellman, 2012, considered a sample representing Twitter *tie* (i.e., "follow" relation in both directions) network. They found that distance constrains ties, despite the seeming ease with which they can be formed: 39 % of the ties are shorter than 100 km (within the same regional cluster), ties up to 1,000 km are more frequent than random ones, and ties longer than 5,000 km are rare. Cf. **distance decay** in Chap. 28. But the nonlocal ties are predicted better by the frequency of airline connections than by physical proximity.

  State et al., 2013, started with a graph of a sample about 10 million users of Yahoo! email with an edge between two users whenever they exchanged at least one email message in both directions, during the observation period in 2012. A weighted complete graph of 141 countries was derived, with edge-weight being the rescaled logarithm of the communication density between countries. For each doubling of distance (between each country centroids) and doubling of the number of direct flights, the density decreased by 66 % and increased by 33 %, respectively.

  But the main (besides colonial link and common language) cultural factor, nearly doubling the density, happens to be the common membership in the same civilization from the list produced by Huntigton in a 1993 article *The Clash of Civilizations*: Latin American, Islamic, Orthodox, Sinic, Buddist, Western, African, Hindu, Japonic. For Latin American, Islamic, and Orthodox civilization, this factor increases the density by the factor of 5.4, 3.1 and 2.4, respectively.

- **Network's hidden metric**

  Many social, biologic, and communication networks, including the Internet and Web, are scale-free and *strongly clustered* (many triangular subgraphs). *Greedy routing* is a navigation strategy to do always the locally optimal step with the hope of finding a globally shortest path. Krioukov et al., 2009, found that successful greedy paths are shortest, mostly and asymptotically, in the large complex networks.

  They explain such efficiency by the existence of a **hidden metric space** (V, d) on the set V of nodes, so that a node passes information to the neighbor that is closest in (V, d) to the final destination. Moreover, they suggest that (V, d) is **hyperbolic**, because the nodes are *heterogeneous* (can be classified into groups, subgroups, and so on) implying a tree-like structure of such network.

- **Sexual distance**
  Given a group of people, its *sexual network* is the graph of members two of them forming an edge if they had a sexual contact. The **sexual distance** is the path metric of a sexual network. Such networks of heterosexual individuals are usually scale-free but not small-world since they have no 3-cycles and very few 4-cycles.
  Several sexual networks were mapped in order to trace the spread of sexually infectious diseases. The sexual network of all adults aged 18–35 in Licoma (almost isolated island 18 km^2 on lake Malawi) have a giant connected component containing half of nonisolated vertices, and more than one quarter were connected *robustly*, i.e., by multiple disjoint paths. Also, in the sexual network of students of an Midwestern US high school, 52 % of nonisolated vertices belong to a giant connected component. But this graph contains very few cycles and have large diameter (37).
  A study of persons at risk for HIV (Colorado Springs, 1988–1992) compared their sexual and geographical distance, measured as the actual distance between their residences. The closest (at mean 2.9 km) pairs were HIV-positive persons and their contacts. The most distant (at mean 6.1 km) pairs were prostitutes and their paying partners. The mean distance between all persons in Colorado Springs was 12.4 km compared with 5.4 km between all dyads the study.
  Moslonka-Lefebre et al., 2012, consider weighted sexual networks, where the weight of an edge is the number of sex acts that are actually realized between two individuals per, say, a week. Such model is more consistent with epidemiological data.
  The sexual network for the human race have a giant connected component containing a great many vertices of degree 1 and almost all vertices of larger degree.
- **Subway network core**
  Roth et al., 2012, observed that the world's largest subway networks converge to a similar shape: a core (ring-shaped set of central stations) with quasi-1D/linear branches radiating from it. The average degree of core stations is 2.5; among them $\approx$20 % are transfer stations and >60 % have degree 2.
  The average *radial* (from the geographical barycenter of all stations) distance (in km) to branches stations is about double of such distance to core stations, while the number of branches scales roughly as the square root of the number of stations.
  Cf. **Moscow metric**, **Paris metric** and **subway semimetric** in Chap. 19.
- **Normalized Google distance**
  The number of searches on Google in 2013 was 2.16 trillion.
  The **normalized Google distance** between two search terms x and y is defined (Cilibrasi–Vitanyi, 2005) by

$$\frac{\max\{\log f(x), \log f(y)\} - \log f(x, y)}{\log m - \min\{\log f(x), \log f(y)\}},$$

where m is the total number of web pages searched by Google search engine; $f(x)$ and $f(y)$ are the number of hits for terms x and y, respectively; and $f(x, y)$ is the number of web pages on which both x and y occur.
Cf. **normalized information distance** in Chap. 11.

- **Drift distance**
The **drift distance** is the absolute value of the difference between observed and actual coordinates of a node in a NVE (Networked Virtual Environment).
In models of such large-scale peer-to-peer NVE (for example, Massively Multiplayer Online Games), the users are represented as coordinate points on the plane (*nodes*) which can move at discrete *time-steps*, and each has a visibility range called the *Area of Interest*. NVE creates a synthetic $3D$ world where each user assumes *avatar* (a virtual identity) to interact with other users or computer AI.
The primary metric tool in MMOG and Virtual Worlds is the proximity sensor recording when an avatar is within its specified range.
The term **drift distance** is also used for the current going through a material, in tire production, etc.
- **Betweenness centrality**
For a **geodesic** metric space (X, d) (in particular, for the path metric of a graph), the **stress centrality** of a point $x \in X$ is defined (Shimbel, 1953) by

$$\sum_{y,z \in X,\, y \neq x \neq z} \text{Number of shortest } (y - z) \text{ paths through } x,$$

the **betweenness centrality** of a point $x \in X$ is defined (Freeman, 1977) by

$$g(x) = \sum_{y,z \in X,\, y \neq x \neq z} \frac{\text{Number of shortest } (y - z) \text{ paths through } x}{\text{Number of shortest } (y - z) \text{ paths}},$$

and the **distance-mass function** is a function $M : \mathbb{R}_{\geq 0} \to \mathbb{Q}$ defined by

$$M(a) = \frac{|\{y \in X : d(x, y) + d(y, z) = a \text{ for some } x, y \in X\}|}{|\{(x, z) \in X \times X : d(x, z) = a\}|}.$$

[GOJKK02] estimated that $\frac{M(a)}{a} \approx 4.5$ for the **Internet AS metric**, and ≈ 1 for the **Web hyperlink quasi-metric** for which the shortest paths are almost unique.

- **Distance centrality**
  Given a finite metric space (X, d) (usually, the **path metric** on the graph of a network) and a point $x \in X$, we give here examples of metric functionals used to measure **distance centrality**, i.e., the amount of centrality of the point x in X expressed in terms of its distances $d(x, y)$ to other points.

  1. The **eccentricity** (or *Koenig number*) $\max_{y \in X} d(x, y)$ was given in Chap. 1; Hage-Harary, 1995, considered $\frac{1}{\max_{y \in X} d(x,y)}$.
  2. The **closeness** (Sabidussi, 1966) is the inverse $\frac{1}{\sum_{y \in X} d(x,y)}$ of the *farness*.
  3. Dangalchev, 2006, introduced $\sum_{y \in X, y \neq x} 2^{-d(x,y)}$ which allows the case $d(x, y) = \infty$ (disconnected graphs).
  4. The functions $f_1 = \sum_{y \in X} d(x, y)$ and $f_2 = \sum_{y \in X} d^2(x, y)$; cf. **Fréchet mean** in Chap. 1.

  In *Location Theory* applications, $X' \subset X$ is a set of positions of "clients" and one seeks points $x \in X$ of acceptable facility positions. The appropriate objective function is, say, $\min \max_{y \in X'} d(x, y)$ to locate an emergency service, $\min \sum_{y \in X'} d(x, y)$ for a goods delivering facility and $\max \sum_{y \in X'} d(x, y)$ for a hazardous facility.

22.2 Network-Based Semantic Distances

Among the main lexical networks (such as WordNet, Framenet, Medical Search Headings, Roget's Thesaurus) a semantic lexicon WordNet is the most popular lexical resource used in Natural Language Processing and Computational Linguistics.

WordNet (see http://wordnet.princeton.edu) is an online lexical database in which English nouns, verbs, adjectives and adverbs are organized into *synsets* (synonym sets), each representing one underlying lexical concept.

Two synsets can be linked semantically by one of the following links: upwards x *(hyponym) IS-A* y *(hypernym)* link, downwards x *(meronym) CONTAINS* y *(holonym)* link, or a horizontal link expressing frequent co-occurrence *(antonymy)*, etc. *IS-A* links induce a partial order, called *IS-A taxonomy*. The version 2.0 of WordNet has 80,000 noun concepts and 13,500 verb concepts, organized into 9 and 554 separate *IS-A* hierarchies.

In the resulting DAG (*directed acyclic graph*) of concepts, for any two synsets (or concepts) x and y, let $l(x, y)$ denote the length of the shortest path between them, using only *IS-A* links, and let $LPS(x, y)$ denote their *least common subsumer* (ancestor) by *IS-A* taxonomy. Let $d(x)$ denote the *depth* of x (i.e., its distance from the root in *IS-A* taxonomy) and let $D = \max_{xd}(x)$.

The semantic relatedness of two nouns can be estimated by their **ancestral path distance** (cf. Chaps. 10 and 23), i.e., the length of the shortest *ancestral path* (concatenation of two directed paths from a common ancestor) to them). A list of the other main semantic similarities and distances follows. See also [HRJM13].

- **Length similarities**
  The **path similarity** and **Leacock–Chodorow similarity** between synsets x and y are defined by
  $$\text{path}(x, y) = (l(x, y))^{-1} \text{ and } lch(x, y) = -\ln \frac{l(x, y)}{2D}.$$
  The **conceptual distance** between x and y is defined by $\frac{l(x,y)}{D}$.
- **Wu–Palmer similarity**
  The **Wu–Palmer similarity** between synsets x and y is defined by
  $$wup(x, y) = \frac{2d(LPS(x, y))}{d(x) + d(y)}.$$
- **Resnik similarity**
  The **Resnik similarity** between synsets x and y is defined by
  $$res(x, y) = -\ln p(LPS(x, y)),$$
  where $p(z)$ is the probability of encountering an instance of concept z in a large corpus, and $-\ln p(z)$ is called the *information content* of z.
- **Lin similarity**
  The **Lin similarity** between synsets x and y is defined by
  $$lin(x, y) = \frac{2\ln p(LPS(x, y))}{\ln p(x) + \ln p(y)}.$$
- **Jiang–Conrath distance**
  The **Jiang–Conrath distance** between synsets x and y is defined by
  $$jcn(x, y) = 2\ln p(LPS(x, y)) - (\ln p(x) + \ln p(y)).$$
- **Lesk similarities**
  A *gloss* of a synonym set z is the member of this set giving a definition or explanation of an underlying concept. The **Lesk similarities** are those defined by a function of the overlap of glosses of corresponding concepts; for example, the **gloss overlap** is
  $$\frac{2t(x, y)}{t(x) + t(y)},$$
  where $t(z)$ is the number of words in the synset z, and $t(x, y)$ is the number of common words in x and y.
- **Hirst–St-Onge similarity**
  The **Hirst–St-Onge similarity** between synsets x and y is defined by

$$hso(x, y) = C - L(x, y) - ck,$$

where $L(x, y)$ is the length of a shortest path between x and y using all links, k is the number of changes of direction in that path, and C, c are constants.
The **Hirst–St-Onge distance** is defined by $\frac{L(x,y)}{k}$.

- **Semantic biomedical distances**
The **semantic biomedical distances** are the distances used in biomedical lexical networks. The main clinical terminologies are UMLS (United Medical Language System) and SNOMED (Systematized Nomenclature of Medicine) CT.
An example of such distances used in SNOMED and presented in Melton et al., 2006, is given by the *interpatient distance* between two *medical cases* (sets X and Y of patient data). It is their **Tanimoto distance** (cf. Chap. 1) $\frac{|X \Delta Y|}{|X \cup Y|}$.
The *conceptual distance* between two biomedical concepts in UMLS is (Caviedes and Cimino, 2004) the minimum number of *IS-A* parent links between them in the directed acyclic graph of *IS-A* taxonomy of concepts.
- **Semantic proximity**
For the words in a document, there are short range syntactic relations and long range *semantic correlations*, i.e., meaning correlations between concepts.
The main document networks are Web and bibliographic databases (digital libraries, scientific databases, etc.); the documents in them are related by, respectively, hyperlinks and citation or collaboration.
Also, some semantic tags (keywords) can be attached to the documents in order to index (classify) them: terms selected by author, title words, journal titles, etc.
The **semantic proximity** between two keywords x and y is their **Tanimoto similarity** $\frac{|X \cap Y|}{|X \cup Y|}$, where X and Y are the sets of documents indexed by x and y, respectively. Their **keyword distance** is defined by $\frac{|X \Delta Y|}{|X \cap Y|}$; it is not a metric.
- **Dictionary digraph**
Dictionary digraph (V, E) have the words of a given dictionary as vertices, and arcs $uv \in E$ whenever word u is used to define word v. The *kernel* (V', E') is its subdigraph induced by the vertices with out-degree $\neq 0$. MF (*minimum feedback vertex set*) is a smallest set of vertices, from which any $v \in V$ can be reached.
Picard et al., 2013, found that $|V| \approx 10|V'| \approx 20|MF| \approx 20|V''|$ in such digraphs for four English dictionaries; here (V'', E'') is the *core* (largest strongly connected component) of the kernel. They observed that the words in the kernel V' are learned at a much younger age, and are more concrete, imageable and frequent than the words in $V \setminus V'$. The same is true, but more so, comparing V'' with $V \setminus V''$ and any MF with $V \setminus MF$. Cf. **Swadesh similarity** (Chap. 28).
- **SimRank similarity**
Let D be a directed multigraph representing a cross-referred document corpus (say, a set of citation-related scientific papers, hyperlink-related web pages, etc.) and $I(v)$ be the set of in-neighbors of a vertex v.
SimRank similarity $s(x, y)$ between vertices x and y of D is defined (Jeh and Widom, 2002) as 1 if $x = y$, 0 if $|I(x)||I(y)| = 0$ and, otherwise, as

$$\frac{C}{|I(x)||I(y)|}\sum_{a\in I(x),b\in I(y)} s(a,b),$$

where C is a constant, $0 < C < 1$ (usually, $C = 0.8$ or 0.6 is used).

- **D-separation in Bayesian network**
  A *Bayesian network* is a DAG (digraph with no directed cycles) (V, E) whose vertices represent random variables and arcs represent conditional dependencies; so, the likelihood of each vertex can be calculated from the likelihood of its ancestors. Bayesian networks, including *causal networks*, are used for modeling knowledge.
  A vertex $v \in V$ is called a *collider* of a *trail* (undirected path) t if there are two consecutive arcs $uv, vu' \in E$ on t. A trail t is *active by* a set $Z \subset V$ of vertices if every its collider is or has a descendent in Z, while every other vertex along t is outside of Z. If $X, Y, Z \subset V$ are disjoint sets of vertices, then Z is said (Pearl, 1988) to **d-separate** X from Y if there is no active trail by Z between a vertex in X and a vertex in Y. Such d-separation means that the variable sets, represented by X and Y, are independent conditional on variables, represented by Z, in all probability distributions the DAG (V, E) can represent.
  The minimal set which d-separates vertex v from all other vertices is v's *Markov blanket*; it consists of v's parents, its children, and its children's parents. A *moral graph* of the DAG (V, E), used to find its equivalent undirect form, is the graph (V, E'), where E' consists all arcs from E made undirected plus all missing *marriages* (edges between vertices having a common child).
  Cf. the **Bayesian graph edit distance** in Chap. 15.
- **Forward quasi-distance**
  In a directed network, where edge-weights correspond to a point in time, the **forward quasi-distance** (**backward quasi-distance**) is the length of the shortest directed path, but only among paths on which consecutive edge-weights are increasing (decreasing, respectively).
  The forward quasi-distance is useful in epidemiological networks (disease spreading by contact, or, say, heresy spreading within a church), while the backward quasi-distance is appropriated in P2P (i.e., peer-to-peer) file-sharing networks.
  Berman, 1996, introduced *scheduled network*: a directed network (of, say, airports), in which each edge (say, flight) is labeled by departure and arrival times. Kempe–Kleinberg–Kumar, 2002, defined more general *temporal network*: an edge-weighted graph, in which the weight of an edge is the time at which its endpoints communicated. A path is *time-respecting* if the weights of its edges are nondecreasing. Besides Scheduling and Epidemiology, such networks occur in Distributed Systems (say, dissemination of information using node-to-node communication).
  In order to handle large temporal data on human behavior, Kostakos, 2009, introduced *temporal graph*: an arc-weighted directed graph, where the vertices are instances $a_i t_k$ (person a_i in point t_k of time), and the arcs are $(t_{k+1} - t_k)$-

weighted ones $(a_i t_k, a_i t_{k+1})$ linking time-consecutive pairs and unweighted ones $(a_i t_k, a_j t_k)$ representing a communication (say, e-mail) from a_i to a_j at time t_k. In order to handle *temporally disconnected* (not connected by a time-respecting path) nodes, Tang et al., 2009, defined *time-varying network*: an ordered set $\{D_t\}_{t=1,\dots,T}$ of directed (or not) graphs $D_t = (X, E_t)$, where the arc-sets E_t may change in time and the arcs have temporal duration. As real-world examples, they considered brain cortical and social interaction networks.

22.3 Distances in Internet and Web

Let us consider in detail the graphs of the Web and of its hardware substrate, Internet which are small-world and scale-free.

The *Internet* is the largest WAN (wide area network), spanning the Earth. This publicly available worldwide computer network came from 13-node ARPANET (started in 1969 by US Department of Defense), NSFNet, Usenet, Bitnet, and other networks. In 1995, the National Science Foundation in the US gave up the stewardship of the Internet, and in 2009, US Department of Commerce accepted privatization/internationalization of ICANN, the body responsible for domain names in the Internet.

Its nodes are *routers*, i.e., devices that forward packets of data along networks from one computer to another, using IP (Internet Protocol relating names and numbers), TCP and UDP (for sending data), and (built on top of them) HTTP, Telnet, FTP and many other *protocols* (i.e., technical specifications of data transfer). Routers are located at *gateways*, i.e., places where at least two networks connect.

The links that join the nodes together are various physical connectors, such as telephone wires, optical cables and satellite networks. The Internet uses *packet switching*, i.e., data (fragmented if needed) are forwarded not along a previously established path, but so as to optimize the use of available *bandwidth* (bit rate, in million bits per second) and minimize the *latency* (the time, in milliseconds, needed for a request to arrive).

Each computer linked to the Internet is usually given a unique "address", called its *IP address*. The new Internet Protocol IPv6 has address space $2^{128} \approx 4.4 \times 10^{38}$. The most popular applications supported by the Internet are e-mail, file transfer, Web, and some multimedia as YouTube and Internet TV. In 2012, 144 billions e-mails (68.8 % of which was spam) were sent daily by 2.2 billions users worldwide. In 2015, global IP traffic will reach 1.0 zettabytes (1000^7 bytes) per year.

The *Internet IP graph* has, as the vertex-set, the IP addresses of all computers linked to the Internet; two vertices are adjacent if a router connects them directly, i.e., the passing datagram makes only one *hop*. The Internet also can be partitioned into ASs (administratively Autonomous Systems). Within each AS the intradomain routing is done by IGP (Interior Gateway Protocol), while interdomain routing is done by BGP (Border Gateway Protocol) which assigns an ASN (16-bit number) to each AS. The *Internet AS graph* has ASs (about 42,000 in 2012) as vertices and edges represent the existence of a BGP peer connection between corresponding ASs.

The *World Wide Web* (*WWW* or *Web*, for short) is a major part of Internet content consisting of interconnected documents (resources). It corresponds to HTTP (Hyper Text Transfer Protocol) between browser and server, HTML (Hyper Text Markup Language) of encoding information for a display, and URLs (Uniform Resource Locators), giving unique "addresses" to web pages. The Web was started in 1989 in CERN which gave it for public use in 1993. The *Web digraph* is a virtual network, the nodes of which are *documents* (i.e., static HTML pages or their URLs) which are connected by incoming or outcoming HTML *hyperlinks*, i.e., hypertext links. It was at least 4.64 billion nodes (pages) in the Indexed Web digraph in May 2014.

The number of operating *web sites* (collections of related web pages found at a single address) reached 634 million in 2012 from 18,957 in 1995. In 2012, 54.7 % of websites were in English, followed by 5.9 %, 5.7 % in Russian and German. Along with the Web lies the *Deep* (or *Invisible*) *Web*, i.e., content, which is not indexed by standard search engines. This content (say, unlinked, or having dynamic URL, non-HTML/text, technically limited access, or scripted, requiring registration/login) has (Bergman, 2001) about 3,000 times more pages than Surface Web, where Internet searchers are searching.

There are several hundred thousand *cyber-communities*, i.e., clusters of nodes of the Web digraph, where the link density is greater among members than between members and the rest. The cyber-communities (a customer group, a social network, a concept in a technical paper, etc.) are usually focused around a definite topic and contain a bipartite *hubs-authorities* subgraph, where all hubs (guides and resource lists) point to all authorities (useful and relevant pages on the topic).

Examples of new media, created by the Web are *(we)blogs* (digital diaries posted on the Web), Skype (telephone calls), social sites (as Facebook, Twitter, Linkedin) and Wikipedia (the collaborative encyclopedia). Original Web-as-information-source is often referred as *Web* 1.0, while *Web* 2.0 means present Web-as-participation-platform as, for example, web-based communities, blogs, social-networking (and video-sharing) sites, wikis, hosted services and web applications. For example, with *cloud servers* one can access his data and applications from the Internet rather than having them housed on-site.

Web 3.0 is the third generation of WWW conjectured to include semantic tagging of content. The project Semantic Web by W3C (WWW Consortium) aims at linking to metadata, merging social data and (making all things addressable by the existing naming protocols) transformation of WWW into GGG (Giant Global Graph) of users.

The *Internet of Things* refers to uniquely identifiable objects (things) and their virtual representations in an Internet-like structure. It would encode geographic location and dimensions of 50–100 trillion objects, and be able to follow their movement and send data between them. Every human being is surrounded by 1,000–5,000 objects.

On average, nodes of the Web digraph are of size 10 kilobytes, out-degree 7.2, and probability k^{-2} to have out-degree or in-degree k. A study in [BKMR00] of over 200 million web pages gave, approximately, the largest connected component "core" of 56 million pages, with another 44 million of pages connected to the core (newcomers?), 44 million to which the giant core is connected (corporations?) and 44 million connected to the core only by undirected paths or disconnected from it. For randomly chosen nodes x and y, the probability of the existence of a directed path from x to y was 0.25 and the average length of such a shortest path (if it exists) was 16, while maximal length of a shortest path was over 28 in the core and over 500 in the whole digraph.

A study in [CHKSS07] of Internet AS graphs revealed the following *Medusa structure* of the Internet: "nucleus" (diameter 2 cluster of $\approx$100 nodes), "fractal" ($\approx$15,000 nodes around it), and "tentacles" ($\approx$5,000 nodes in isolated subnetworks communicating with the outside world only via the nucleus).

The distances below are examples of host-to-host **routing metrics**, i.e., values used by routing algorithms in the Internet, in order to compare possible routes. Examples of other such measures are: bandwidth consumption, communication cost, reliability (probability of packet loss). Also, the main computer-related *quality metrics* are mentioned.

- **Distance-vector routing protocol**
  A **distance-vector routing protocol** (DVRP) requires that a router informs its neighbors of topology changes periodically and, in some cases, when a change is detected in the topology of a network. Routers are advertised as vectors of a distance (say, **Internet IP metric**) and direction, given by next hop address and exit interface. Cf. **displacement** in Chap. 24.
  Ad hoc on-demand distance-vector routing is a (both unicast and multicast) routing protocol for mobile and other wireless ad hoc networks. It establishes a route to a destination only on demand and avoids the counting-to-infinity problem of other distance-vector protocols by using sequence numbers on route updates.
  Between nodes of an ad hoc network with end-to-end delay constraints, head-of-line packets compete for access to the shared medium. Each packet with remaining lifetime T and remaining **Internet IP metric** H to its destination, is associated with a ranking function $\gamma(H,T) = \frac{T^\alpha}{H}$, denoting its transmission priority. The number $\alpha \geq 0$ is called **lifetime-distance factor**; it should be optimized in order to minimize the probability of packet loss due to excessive delay.
- **Internet IP metric**
  The **Internet IP metric** (or **hop count**, *RIP metric*, *IP path length*) is the path metric in the *Internet IP graph*, i.e., the minimal number of hops (or, equivalently, routers, represented by their IP addresses) needed to forward a packet of data.
  RIP (a **distance-vector routing protocol** first defined in 1988) imposes a maximum distance of 15 and advertises by 16 nonreachable routes.

- **Internet AS metric**
  The **Internet AS metric** (or *BGP-metric*) is the **path metric** in the *Internet AS graph*, i.e., the minimal number of ISPs (Independent Service Providers), represented by their ASs, needed to forward a packet of data.
- **Geographic distance**
  The **geographic distance** is the **great circle distance** (cf. Chap. 25) on the Earth from the client x (destination) to the server y (source).
  However, for economical reasons, the data often do not follow such geodesics; for example, most data from Japan to Europe transits via US.
- **RTT-distance**
  The **RTT-distance** (or *ping time*) is the round-trip time (to send a packet and receive an acknowledgment back) of transmission between x and y, measured in milliseconds (usually, by the *ping* command).
  See [HFPMC02] for variations of this distance and connections with the above three metrics. Fraigniaud–Lebbar–Viennot, 2008, found that RTT is a C**-inframetric** (Chap. 1) with $C \approx 7$.
- **Synchcronization distance**
  In the Network Time Protocol (NTP), the **synchcronization distance** is the *root dispersion* (maximum error relative to the primary reference source at the root of the synchronization subnet) plus one half the *root delay* (total round-trip delay to the primary reference source at the root of the synchronization subnet).
- **Administrative cost distance**
  The **administrative cost distance** is the nominal number (rating the trustworthiness of a routing information), assigned by the network to the route between x and y. For example, Cisco Systems assigns values $0, 1, \dots, 200, 255$ for the Connected Interface, Static Route, …, Internal BGP, Unknown, respectively.
- **DRP-metrics**
  The DD (Distributed Director) system of Cisco uses (with priorities and weights) the **administrative cost distance**, the **random metric** (selecting a random number for each IP address) and the **DRP** (Direct Response Protocol) metrics. DRP-metrics ask from all DRP-associated routers one of the following distances:

  1. The **DRP-external metric**: the number of BGP (Border Gateway Protocol) hops between the client requesting service and the DRP server agent;
  2. The **DRP-internal metric**: the number of IGP hops between the DRP server agent and the closest border router at the edge of the autonomous system;
  3. The **DRP-server metric**: the number of IGP hops between the DRP server agent and the associated server.

- **Reported distance**
  In a Cisco Systems routing protocol EIGRP, **reported distance** (or *RD*, *advertised distance*) is the total metric along a path to a destination network as advertised by an upstream neighbor. RD is equal to the current lowest total distance through a successor for a neighboring router.

A **feasible distance** is the lowest known distance from a router to a particular destination. This is RD plus the cost to reach the neighboring router from which the RD was sent; so, it is a historically lowest known distance to a particular destination.

- **Network tomography metrics**
Consider a network with fixed routing protocol, i.e., a *strongly connected* digraph $D = (V, E)$ with a unique directed path $T(u, v)$ selected for any pair (u, v) of vertices. The routing protocol is described by a binary *routing matrix* $A = ((a_{ij}))$, where $a_{ij} = 1$ if the arc $e \in E$, indexed i, belongs to the directed path $T(u, v)$, indexed j. The **Hamming distance** between two rows (columns) of A is called the **distance between corresponding arcs** (directed paths) of the network.
Consider two networks with the same digraph, but different routing protocols with routing matrices A and A', respectively. Then a **routing protocol semi-metric** [Vard04] is the smallest Hamming distance between A and a matrix B, obtained from A' by permutations of rows and columns (both matrices are seen as strings).
- **Web hyperlink quasi-metric**
The **Web hyperlink quasi-metric** (or *click count*) is the length of the shortest directed path (if it exists) between two web pages (vertices in the Web digraph), i.e., the minimal number of necessary mouse-clicks in this digraph.
- **Average-clicks Web quasi-distance**
The **average-clicks Web quasi-distance** between two web pages x and y in the Web digraph [YOI03] is the minimum $\sum_{i=1}^{m} \ln p \frac{z_i^+}{\alpha}$ over all directed paths $x = z_0, z_1, \dots, z_m = y$ connecting x and y, where z_i^+ is the out-degree of the page z_i. The parameter α is 1 or 0.85, while p (the average out-degree) is 7 or 6.
- **Dodge–Shiode WebX quasi-distance**
The **Dodge–Shiode WebX quasi-distance** between two web pages x and y of the Web digraph is the number $\frac{1}{h(x,y)}$, where $h(x, y)$ is the number of shortest directed paths connecting x and y.
- **Web similarity metrics**
Web similarity metrics form a family of indicators used to quantify the extent of relatedness (in content, links or/and usage) between two web pages x and y.
Some examples are: topical resemblance in overlap terms, *co-citation* (the number of pages, where both are given as hyperlinks), *bibliographical coupling* (the number of hyperlinks in common) and *co-occurrence frequency* $\min\{P(x|y), P(y|x)\}$, where $P(x|y)$ is the probability that a visitor of the page y will visit the page x.
In particular, **search-centric change metrics** are metrics used by search engines on the Web, in order to measure the degree of change between two versions x and y of a web page. If X and Y are the set of all words (excluding HTML markup) in x and y, respectively, then the **word page distance** is the **Dice distance**

$$\frac{|X \Delta Y|}{|X| + |Y|} = 1 - \frac{2|X \cup Y|}{|X| + |Y|}.$$

If v_x and v_y are weighted vector representations of x and y, then their **cosine page distance** is given (cf. **TF-IDF similarity** in Chap. 17) by

$$1 - \frac{\langle v_x, v_y \rangle}{||v_x||_2 \cdot ||v_y||_2}.$$

- **Web quality control distance function**
  Let P be a query quality parameter and X its domain. For example, P can be query *response time*, or accuracy, relevancy, size of result.
  The **Web quality control distance function** (Chen–Zhu–Wang, 1998) for evaluating the relative goodness of two values, x and y, of parameter P is a function $\rho : X \times X \to \mathbb{R}$ (not a **distance**) such that, for all $x, y, z \in X$:

  1. $\rho(x, y) = 0$ if and only if $x = y$,
  2. $\rho(x, y) > 0$ if and only if $\rho(y, x) < 0$,
  3. if $\rho(x, y) > 0$ and $\rho(y, z) > 0$, then $\rho(x, z) > 0$.

  The inequality $\rho(x, y) > 0$ means that x is better than y; so, it defines a *partial order* (reflexive, antisymmetric and transitive binary relation) on X.
- **Lostness metric**
  Users navigating within hypertext systems often experience *disorientation* (the tendency to lose sense of location and direction in a nonlinear document) and *cognitive overhead* (the additional effort and concentration needed to maintain several tasks/trails at the same time). Smith's **lostness metric** measures it by

$$\left(\frac{d}{t} - 1\right)^2 + \left(\frac{r}{d} - 1\right)^2,$$

  where t is the total number of nodes visited, d is the number of different nodes among them, and r is the number of remaining nodes needed to complete a task.
- **Trust metrics**
  A **trust metric** is, in Computer Security, a measure to evaluate a set of peer certificates resulting in a set of accounts accepted and, in Sociology, a measure of how a member of the group is trusted by the others in the group.
  For example, the UNIX access metric is a combination of only *read*, *write* and *execute* kinds of access to a resource. The much finer *Advogato* trust metric (used in the community of open source developers to rank them) is based on bonds of trust formed when a person issues a certificate about someone else. Other examples are: *Technorati*, *TrustFlow*, Richardson et al., Mui et al., *eBay* trust metrics.
- **Software metrics**
  A **software metric** is a measure of software quality which indicates the complexity, understandability, description, testability and intricacy of code. Managers use mainly **process metrics** which help in monitoring the processes that produce the software (say, the number of times the program failed to rebuild overnight).

An **architectural metric** is a measure of software architecture (development of large software systems) quality which indicates the coupling (interconnectivity of composites), cohesion (intraconnectivity), abstractness, instability, etc.

- **Locality metric**
The **locality metric** is a physical metric measuring globally the locations of the program components, their calls, and the depth of nested calls by

$$\frac{\sum_{i,j} f_{ij} d_{ij}}{\sum_{i,j} f_{ij}},$$

where d_{ij} is a distance between calling components i and j, while f_{ij} is the frequency of calls from i to j. If the program components are of about same size, $d_{ij} = |i - j|$ is taken. In the general case, Zhang–Gorla, 2000, proposed to distinguish *forward* calls which are placed before the called component, and *backward* (other) calls. Define $d_{ij} = d'_i + d''_{ij}$, where d'_i is the number of lines of code between the calling statement and the end of i if call is forward, and between the beginning of i and the call, otherwise, while $d''_{ij} = \sum_{k=i+1}^{j-1} L_k$ if the call is forward, and $d''_{ij} = \sum_{k=j+1}^{i-1} L_k$ otherwise. Here L_k is the number of lines in component k.

- **Reuse distance**
In a computer, the *microprocessor* (or *processor*) is the chip doing all the computations, and the *memory* usually refers to *RAM* (random access memory). A (processor) *cache* stores small amounts of recently used information right next to the processor where it can be accessed much faster than memory. The following distance estimates the cache behavior of programs.
The **reuse distance** (Mattson et al., 1970, and Ding–Zhong, 2003) of a memory location x is the number of distinct memory references between two accesses of x. Each memory reference is counted only once because after access it is moved in the cache. The reuse distance from the current access to the previous one or to the next one is called the *backward* or *forward* reuse distance, respectively.

- **Action at a distance (in Computing)**
In Computing, the **action at a distance** is a class of programming problems in which the state in one part of a program's data structure varies wildly because of difficult-to-identify operations in another part of the program.
In Software Engineering, Holland's *Law of Demeter* is a style guideline: an unit should "talk only to immediate friends" (closely related units) and have limited knowledge about other units; cf. **principle of locality** in Chap. 24.

Part VI
Distances in Natural Sciences

Chapter 23
Distances in Biology

Distances are mainly used in *Biology* to pursue basic classification tasks, for instance, for reconstructing the evolutionary history of organisms in the form of phylogenetic trees. In the classical approach those distances were based on comparative morphology, physiology, mating studies, paleontology and immunodiffusion. The progress of modern *Molecular Biology* also allowed the use of nuclear- and amino-acid sequences to estimate distances between genes, proteins, genomes, organisms, species, etc.

DNA is a sequence of *nucleotides* (or *nuclei acids*) A, T, G, C, and it can be seen as a word over this alphabet of four letters. The (single ring) nucleotides A, G (short for adenine and guanine) are called *purines*, while (double ring) T, C (short for thymine and cytosine) are called *pyrimidines* (in RNA, uracil U replaces T).

Two strands of DNA are held together and in the opposite orientation (forming a double helix) by weak hydrogen bonds between corresponding *base pair* of nucleotides (necessarily, a purine and a pyrimidine) in the strands alignment.

A *transition mutation* is a substitution of a base pair, so that a purine/pyrimidine is replaced by another purine/pyrimidine; say, GC is replaced by AT. A *transversion mutation* is a substitution of a base pair, so that a purine/pyrimidine is replaced by a pyrimidine/purine base pair, or vice versa; say, GC is replaced by TA.

DNA molecules occur (in the nuclei of eukaryote cells) in the form of long chains called *chromosomes*. DNA from one human cell has length/width $\approx$1.8 m/2.4 nm.

Most human cells contain 46 chromosomes (23 pairs, one set of 23 from each parent); the human *gamete* (sperm or egg) is a *haploid*, i.e., contains only one set of 23 chromosomes. The (normal) males and females differ only in the 23rd pair: *XY* for males, and *XX* for females. But a male ant *Mirmecia pilosula* has only 1 chromosome, while a plant *Ophioglossum* has 1,260. A protozoan *Tetrahymena thermophila* occurs in seven different variants (*sexes*) that can reproduce in $\binom{7}{2} = 21$ combinations. A fungus *Cryptococcus neoformans* has two sexes but their ratio is 99.9 %. More than 99 % of multicellular eukaryotes reproduce sexually, while bacteria only reproduce asexually.

M.M. Deza, E. Deza, *Encyclopedia of Distances*,
DOI 10.1007/978-3-662-44342-2_23

A *gene* is a segment of DNA encoding (via *transcription*, information flow to RNA, and then *translation*, information flow from RNA to enzymes) for a protein or an RNA chain. The location of a gene on its chromosome is called the *gene locus*. Different versions (states) of a gene are called its *alleles*.

A *protein* is a large molecule which is a chain of *amino acids*; among them are hormones, catalysts (enzymes), antibodies, etc. The **protein length** is the number of amino acids in the chain; average protein length is around 300.

The *genetic code* is a map, universal to all organisms, of $4^3 = 64$ *codons* (ordered triples of nucleotides) onto 21 messages: 20 *standard* amino acids and stop-signal. It express the *genotype* (information contained in genes, i.e., in DNA) as the *phenotype* (proteins). Some codons have two meanings, one related to protein sequence, and one related to gene control. Slight variations of the code were found for some mitochondria, ciliates, yeasts, etc. The code also was expanded by encoding new amino acids.

Besides genetic and *epigenetic* (not modifying the sequence) changes of DNA, *evolution* (heritable changes) can happen by "protein mutations" (prions) or culturally (via behavior and symbolic communication). Holliger et al., 2012, synthesized (replacing the natural sugar in DNA) a new polymer (*HNA*) capable of replication and evolution.

A *genome* is the entire genetic constitution of a species or of an organism. The human genome is the set of 23 chromosomes consisting of $\approx$3.1 billion base pairs of DNA and organized into $\approx$20,000 genes. But the microscopic flea *Daphnia pulex* has 31,000 genes, and the flower *Paris japonica* genome contains $\approx$150 billion bp. Only $\approx$1.5 % of human DNA are in protein-coding genes, while at least 80 % has some function.

A *hologenome* is the collection of genomes in a *holobiont* (host plus all its symbionts), a possible unit of selection in evolution. The human microbiota consists of $\approx 10^{14}$ (mainly, bacterial and fungal) cells of $\approx$500 species with 3 million distinct genes. But chlorinated water and antibiotics changed this. The most common viral, bacterial and fungal pathogens of humans are genera *Enterovirus*, *Staphylococcus* and *Candida*, respectively. The vast majority of emerging diseases hop into humans from other mammals. The estimated number of mammalian virus species is $58 \times 5{,}500 \approx 320{,}000$.

First known evidence of photosynthetic life, of multicellular organisms and of animals (bilaterians) is dated 3850, 2100 and 560 Ma (million years ago), respectively. During the *Cambrian Explosion* 540–520 Ma, the rate of evolution was 4–5 times faster than in any other era. Discounting viruses, $\approx$ 1.9 million extant species are known: 1,200,000 invertebrates, 290,000 plants, 250,000 bacteria/protists, 70,000 fungi and 60,000 vertebrates, including 5,416 mammals. 7–10 million species are living today. The number of living trees, fishes, ants, viruses are about 4×10^{11}, 3.5×10^{12}, 5×10^{17}, 10^{31}.

About 5,500 species of animals and 29,500 species of plants are protected by (Washington) Convention on International Trade in Endangered Species. But, actually, human–climate–ecosystem interactions already by 2000 significantly altered 75 % of the terrestrial habitats (leading to mass extinction of species) by land use, overharvesting, toxins and invasive species. The tipping point to new geologic epoch, the *Anthropocene*, passed somewhere between the origin of natural language and the invention of steam engine.

About 80 % of species are parasites of others, parasites included; >100 are human-specific ones. The global live biomass is 560 GtC (billion tonnes of organically bound carbon). At least half of it is come from 5×10^{30} prokaryotes. Humans and their main symbionts, domesticated animals and cultivated plants, contribute 0.1, 0.7, 2 GtC.

Ninety-nine percent of species that have ever existed on Earth became extinct. Mean mammalian species' longevity is ≈1 Ma (million years); our direct ancestor, *Homo erectus*, survived from 1.8 to 0.55 Ma ago. Our subspecies is young (0.2–0.4 Ma), 6–7 % of all humans that have ever been born are living today, and their median age is 28.4 years.

The world population was about 1 million 0.05 Ma and 5 million 0.01 Ma ago, after the last glaciation. It grew continuously since the end of the Black Death in 1350, when it was ≈370 million, and reached 3 billion in 1960, 7.2 billion in 2013. But the number of children aged ≤14 leveled off on 1.9 billion, and global population may peak soon.

Gott, 2007: with a 95 % chance, the human race will last anywhere from another 5,000 to 7,800,000 years; the same *doomsday argument* by Carter, 1983, gave only 10,000 years for us. Earth's life was only unicellular 3.8–1.3 Ga (billion years) ago and will be so again in ≈0.8 Ga. But Earth will support some prokaryotes in refuges until mean surface temperature reach 146 °C in 1.6–2.8 Ga. Another Ga life can stay on Mars.

IAM (infinite-alleles model of evolution) assumes that an allele can change from any given state into any other given state. It corresponds to a primary role for *genetic drift* (i.e., random variation in gene frequencies from one generation to another), especially in small populations, over *natural selection* (stepwise mutations). IAM corresponds to low-rate and short-term evolution, while SMM corresponds to high-rate evolution.

SMM (stepwise mutation model of evolution) is more convenient for (recently, most popular) microsatellite data. A *repeat* is a stretch of base pairs that is repeated with a high degree of similarity in the same sequence. *Microsatellites* are highly variable repeating short sequences of DNA; their mutation rate is 1 per 1,000–10,000 replication events, while it is 1 per 1,000,000 for *allozymes* used by IAM. Microsatellite data (for example, for DNA fingerprinting) consist of numbers of repeats of microsatellites for each allele.

Evolution, without design and purpose, has increased the life's size, diversity and maximal complexity. (But organisms can evolve to become simpler and thus multiply faster. For the *Black Queen model*, such evolution pushes microorganisms to lose functions which are performed by another species around.) Evolution has, perhaps, a direction: *convergent gene evolution* (say, bats/dolphins echolocation, primates/crows cognition), increase of energy flow per gram per second (Caisson, 2003), etc.

Natural selection can favor increased evolvability under environmental pressure. Besides natural selection, some species alter their environment through *niche construction*. In general, selection can act at genic, cellular, individual, holobiont and group level. Selection of species and even phyla could happen during rare abrupt extreme events.

Locally and over short time spans, macroevolution, is dominated by biotic factors (competition, predation, etc.) as in the *Red Queen model*. But larger-scale (geographic and temporal) patterns and species diversity are driven largely by extrinsic abiotic factors (climate, landscape, food supply, tectonic events, etc.), as in the *Court Jester model*. The organisms evolve rapidly (sometimes, by macromutations), but most changes cancel each other out. So, in the longer term, the evolution appears slow. It is not simple accumulation of microevolutionary adaptations, but rather nonlinear (or chaotic).

Besides *vertical gene transfer* (reproduction within species), the evolution is affected by *HGT* (horizontal gene transfer), when an organism incorporates genetic material from another one without being its offspring, and *hybridization* (extra-species sexual reproduction). HGT is common among unicellular life and viruses, even across large **taxonomic distance**. It accounts for $\approx$85 % of the prokaryotic protein evolution. HGT happens also in plants and animals, usually, by viruses. 40–50 % of the human genome consists of DNA imported horizontally by viruses. The most taxonomically distant fertile hybrids are (very rare) interfamilial ones, for instance, blue-winged parrot $\times$ cocktatiel, chicken $\times$ guineafowl in birds and (under UV irradiation) carrot with tobacco, rice or barley. In 2012, an RNA-DNA virus hybrid and a *virophage* of a (giant) virus were found.

The *life* is not well defined, say, for viruses. DNA could be only its recent attribute. Neither life can be "anything undergoing evolution", since the unit is this evolution (gene, cell, organism, group, species?) is not clear. Lineweaver, 2012, defined life as a *far-from-equilibrium dissipative system*. For Eigen, life is a type of behavior of matter. An essential feature of life is *autopoiesis* (self-making); for example, the human body replaces 98 % of its atoms every year while maintaining its unique pattern.

Examples of distances, representing general schemes of measurement in Biology, follow.

The term **taxonomic distance** is used for every distance between two *taxa*, i.e., entities or groups which are arranged into a hierarchy (in the form of a tree designed to indicate degrees of relationship).

The *Linnaean taxonomic hierarchy* is arranged in ascending series of ranks: Zoology (Kingdom, Phylum, Class, Order, Family Genus, Species) and Botany (12 ranks). A *phenogram* is a hierarchy expressing *phenetic relationship*, i.e., unweighted overall similarity. A *cladogram* is a strictly genealogical (by ancestry) hierarchy in which no attempt is made to estimate/depict rates or amount of genetic divergence between taxa.

A *phylogenetic tree* is a hierarchy representing a hypothesis of *phylogeny*, i.e., evolutionary relationships within and between taxonomic levels, especially the patterns of lines of descent. The **phenetic distance** is a measure of the difference in phenotype between any two nodes on a phylogenetic tree; see, for example the **biodistances** in Chap. 29.

The **phylogenetic distance** (or **cladistic distance**, **genealogical distance**) between two taxa is the *branch length*, i.e., the minimum number of edges, separating them in a phylogenetic tree. In such edge-weighted tree, the **additive distance** between two taxa is the minimal sum of edge-weights in a path connecting them. The *phylogenetic diversity* is (Faith, 1992) the minimum total length of all the phylogenetic branches required to span a given set of taxa on the phylogenetic tree.

The **evolutionary distance** (or **patristic distance**) between two taxa is a measure of genetic divergence estimating the **temporal remoteness** of their most recent co-ancestor. Their general **immunological distance** is a measure of the strength of antigen–antibody reactions, indicating their evolutionary distance.

23.1 Genetic Distances

The general **genetic distance** between two taxa is a distance between the sets of DNA-related data chosen to represent them. Among the three most popular genetic distances below, the **Nei standard genetic distance** assumes that differences arise due to mutation and genetic drift, while the **Cavalli-Sforza–Edwards chord distance** and the **Reynolds–Weir–Cockerham distance** assume genetic drift only.

A *population* is represented by a vector $x = (x_{ij})$ with $\sum_{j=1}^{n} m_j$ components, where x_{ij} is the frequency of the i-th *allele* (the label for a state of a gene) at the j-th gene locus (the position of a gene on a chromosome), m_j is the number of alleles at the j-th locus, and n is the number of considered loci. Since x_{ij} is the frequency, we have $x_{ij} \geq 0$ and $\sum_{i=1}^{m_j} x_{ij} = 1$. Denote by $\sum$ summation over all i and j.

- **Shared allele distance**

  The **shared allele distance** D_{SA} (Stephens et al., 1992, corrected by Chakraborty–Jin, 1993) between individuals a, b is $1 - SA(a, b)$; for populations x, y it is

$$1 - \frac{\overline{SA(x,y)}}{\overline{SA(x)} + \overline{SA(y)}},$$

  where $SA(a, b)$ denotes the number of shared alleles summed over all n loci and divided by $2n$, while $\overline{SA(x)}$, $\overline{SA(y)}$, and $\overline{SA(x,y)}$ are $SA(a, b)$ averaged over all pairs (a, b) with individuals a, b being in populations x, y, respectively.
- **MHC genetic dissimilarity**

  The **MHC genetic dissimilarity** of two individuals is defined as the number of shared alleles in their MHC (*major histocompatibility complex*).

  MHC is the most gene-dense and fast-evolving region of the mammalian genome. In humans, it is a 3.6 Mb region containing 140 genes on chromosome 6 and called HLA (*human leukocyte antigen system*). HLA has the largest *polymorphism* (allelic diversity) found in the population. Three most diverse loci

(HLA-A, HLA-B,HLA-DRB1) have roughly 1,000, 1,600, 870 known alleles. This diversity is essential for immune function since it broadens the range of *antigens* (proteins bound by MHC and presented to T-cells for destruction); cf. **immunological distance**.
MHC (and related gut microbiota) diversity allows the marking of each individual of a species with a unique body odor permitting kin recognition and mate selection. *MHC-negative assortative mating* (the tendency to select MHC-dissimilar mates) increases MHC variation and so provides progeny with an enhanced immunological surveillance and reduced disease levels.
While about 6 % of the non-African modern human genome is common with other hominins (Neanderthals and Denisovans), the share of such HLA-A alleles is 50 %, 72 %, 90 % for people in Europe, China, Papua New Guinea.

- *Dps* **distance**
The **Thorpe similarity** (proportion of shared alleles) between populations x and y is defined by $\sum \min\{x_{ij}, y_{ij}\}$. The *Dps* **distance** between x and y is defined by

$$-\ln \frac{\sum \min\{x_{ij}, y_{ij}\}}{\sum_{j=1}^{n} m_j}.$$

- **Prevosti–Ocana–Alonso distance**
The **Prevosti–Ocana–Alonso distance** (1975) between populations x and y is defined (cf. **Manhattan metric** in Chap. 19) by

$$\frac{\sum |x_{ij} - y_{ij}|}{2n}.$$

- **Roger distance**
The **Roger distance** D_R (1972) between populations x and y is defined by

$$\frac{1}{\sqrt{2}n} \sum_{j=1}^{n} \sqrt{\sum_{i=1}^{m_j} (x_{ij} - y_{ij})^2}.$$

- **Cavalli-Sforza–Edwards chord distance**
The **Cavalli-Sforza–Edwards chord distance** D_{CH} (1967) between populations x and y (cf. **Hellinger distance** in Chap. 17) is defined by

$$\frac{2\sqrt{2}}{\pi n} \sum_{j=1}^{n} \sqrt{1 - \sum_{i=1}^{m_j} \sqrt{x_{ij} y_{ij}}}.$$

The **Cavalli-Sforza arc distance** between populations x and y is defined by

$$\frac{2}{\pi}\arccos\left(\sum\sqrt{x_{ij}y_{ij}}\right).$$

Cf. **Bhattacharya distance 1** in Chap. 14.

- **Nei–Tajima–Tateno distance**
  The **Nei–Tajima–Tateno distance** D_A (1983) between populations x and y is

$$1-\frac{1}{n}\sum\sqrt{x_{ij}y_{ij}}.$$

  The **Tomiuk–Loeschcke distance** (1998) is $-\ln\frac{1}{n}\sqrt{\sum x_{ij}\sum y_{ij}}$.
  The **Nei standard genetic distance** D_s (1972) between x and y is defined by

$$-\ln\frac{\langle x,y\rangle}{||x||_2\cdot||y||_2}.$$

  Cf. **Bhattacharya distances** in Chap. 14 and **angular semimetric** in Chap. 17.
  Under IAM, D_s increases linearly with time; cf. **temporal remoteness**.
  The **kinship distance** is defined by $-\ln\langle x,y\rangle$. Caballero and Toro, 2002, defined the *molecular kinship coefficient* between x and y as the probability that two randomly sampled alleles from the same locus in them are *identical by state*. Computing it as $\langle x,y\rangle$ and using the analogy with the **coefficient of kinship** defined via *identity by descent*, they proposed several distances adapted to molecular markers (polymorphisms). Cf. **co-ancestry coefficient**.
  The **Nei minimum genetic distance** D_m (1973) between x and y is defined by

$$\frac{1}{2n}\sum(x_{ij}-y_{ij})^2.$$

- **Sangvi χ^2 distance**
  The **Sangvi χ^2 (1953) distance** between populations x and y is defined by

$$\frac{2}{n}\sum\frac{(x_{ij}-y_{ij})^2}{x_{ij}+y_{ij}}.$$

- **Fuzzy set distance**
  The **fuzzy set distance** D_{fs} between populations x and y (Dubois–Prade, 1983; cf. **Tanimoto distance** in Chap. 1) is defined by

$$\frac{\sum 1_{x_{ij}\neq y_{ij}}}{\sum_{j=1}^{n}m_j}.$$

- **Goldstein et al. distance**
The **Goldstein et al. distance** (1995) between populations x and y is

$$(\delta\mu)^2 = \frac{1}{n}\sum(ix_{ij} - iy_{ij})^2.$$

It is the loci-averaged value $(\delta\mu)^2 = (\mu(x)_j - \mu(y)_j)^2$, where $\mu(z)_j = \sum_i iz_{ij}$ is the mean number of repeats of allele at the j-th (microsatellite) locus in population z.
The **Feldman et al. distance** (1997) is $\log(1 - \frac{\sum_i (\delta\mu)_i^2}{M})$, where the summation is over loci and M is the average value of the distance at maximal divergence.
The above two and the next two distances assume high-rate SMM.

- **Average square distance**
The **average square distance** between populations x and y is defined by

$$\frac{1}{n}\sum_{k=1}^{n}\left(\sum_{1\le i<j\le m_j}(i-j)^2 x_{ik}y_{jk}\right).$$

- **Shriver et al. stepwise distance**
The **Shriver et al. stepwise distance** (1995) between populations x and y is

$$D_{SW} = \frac{1}{n}\sum_{k=1}^{n}\sum_{1\le i,j\le m_k}|i-j|(2x_{ik}y_{jk} - x_{ik}x_{jk} - y_{ik}y_{jk}).$$

- **Latter F-statistics distance**
The **Latter F-statistics distance** (1972) between populations x and y is defined by the following F_{ST}-*estimator*:

$$\theta^* = \frac{\sum(x_{ij} - y_{ij})^2}{2(n - \langle x, y\rangle)}.$$

The **Latter distance** D_L (1973) is $-\ln(1 - \theta^*)$.

- **Reynolds–Weir–Cockerham distance**
The **Reynolds–Weir–Cockerham distance** (1983) between populations x and y is defined by

$$D_W = -\ln(1 - \theta),$$

where θ is their **co-ancestry coefficient** estimated as $\frac{a}{a+b}$.
Here a is the variance between populations x and y, and b is the variance within them. If the sample size is large, then D_W is close to the **Latter F-statistics distance**. For short-term evolution (i.e., $\frac{t}{N}$ small), $D_W \approx \frac{t}{2N}$, where

N is the population size, and t is the number of generations; cf. **temporal remoteness**.

- **Co-ancestry coefficient**
The **co-ancestry coefficient** (or *coefficient of kinship*) of two populations (or individuals) x and y is defined (Wright, 1922, and Malécot, 1948) as the probability $\theta(x, y)$ that two alleles, sampled at random from x and y, are *IBD* (or *identical by descent*), i.e., descending from the same ancestral allele.
Two genes can be *IBS* (or *identical by state*), i.e., similar due to random chance. Cf. **Nei standard genetic distance** and **coefficient of relationship**.
An DNA segment, found consistently to be identical in two related people (or populations) is IBD if it is so due to their common ancestry. The total and mean **IBD segment length** of two people g generations since the founding event (i.e., with g meioses on the path of descent) are $\approx \frac{1}{2^g}$-th of total genome length and $\approx \frac{50}{g}$ cM, respectively; cf. the **map distance**. For example, two people are cryptic relatives if those lengths are at least 1,500 and 25 cM.
- **F_{ST}-based distances**
Given a population T of size $|T|$ partitioned into subpopulations $S_1, \dots, S_k$, the *F-statistics* (or *fixation indices*) are the measures

$$F_{IS} = 1 - \frac{H_I}{H_S}, \quad F_{ST} = 1 - \frac{H_S}{H_T}, \quad F_{IT} = 1 - \frac{H_I}{H_T}$$

of the correlation between genes drawn within subpopulations S_i, among them and within the entire T, respectively.
Here H_I, H_S and H_T are the *heterozygosity indices* over (i)ndividuals, (s)ubpopulations and (t)otal T used to compare observed variation in gene frequencies (partitioned into within and between group components) with the expected one in HWE (*Hardy–Weinberg equilibrium*, i.e., an ideal state when allele and genotype frequencies in population remain constant from generation to generation). $H_I = \frac{\sum_{1 \le j \le k} |S_j| H_{obs\, j}}{|T|}$ (where $H_{obs\, j}$ is the observed *heterozygosity*, i.e., proportion of heterozygotes, in subpopulation S_j) is the mean actual heterozygosity in individuals within subpopulations. $H_S = \frac{\sum_{1 \le j \le k} |S_j| H_{exp\, j}}{|T|}$ (where $H_{exp\, j} = 1 - \sum_i p_i^2$ is the expected, assuming HWE, heterozygosity in S_j and p_i is the frequency of the i-th allele of the locus) is the mean expected heterozygosity within subpopulations. $H_T = 1 - \sum_i \bar{p}_i^2$ (where $\bar{p}_i$ is the frequency of the i-th allele averaged over all subpopulations) is the expected, assuming HWE, heterozygosity in T.
The above Nei's (1973) F_{ST} generalizes Wright's (1951) F_{ST}, when there are only two alleles at a locus. This measure is equivalent to the **co-ancestry coefficient** if all the alleles in the population are different. Nei, 1987, generalized F_{ST} to multi-loci as $G_{ST} = 1 - \frac{\bar{H}_S}{\bar{H}_T}$, where H_S and H_T are averaged across all loci.
The above relative measures underestimate the between-population difference if the within-population diversity is high, such as, say, for microsatellites. Slatkin's R_{ST} is an analog of Wright's F_{ST}, adapted for microsatellite loci by assuming

SMM. It is defined by $R_{ST} = \frac{\bar{S}-S_W}{\bar{S}}$, where S_W is the sum over all loci of twice the weighted mean of the within-population variances $var(A)$ and $var(B)$, and $\bar{S}$ is the sum over all loci of twice the variance $var(A \cup B)$ of the combined population.

In fact, S_W and $\bar{S}$ are the **average square distance** within a subpopulation and the entire population. Slatkin (1995) developed R_{ST} using his (1991) SMM-based F-statistics $F_{ST} = \frac{\bar{t}-t_0}{\bar{t}}$, where $\bar{t}$ and t_0 are the average **temporal remoteness** to the closest co-ancestor of any two randomly chosen alleles from the entire population and from the same subpopulation, respectively.

Jost's D_{est} (2008) is an estimator $\frac{k}{k-1}\frac{H_T-H_S}{1-H_S}$ of the actual differentiation based on H's estimated from allele identities rather than ratios of heterozygosity.

The Weir–Cockerham θ_{ST} (1984) is an estimation of F_{ST}, seen as the correlation of pairs of alleles between individuals within a subpopulation and based on partition of variance rather than heterozygosity. The total variance of allele frequency within a population is the sum $a+b+c$ of variances between subpopulations, between individuals within a subpopulation, and between gametes within individuals. Then θ_{ST} is defined, generalizing the **Reynolds–Weir–Cockerham distance**, as $\frac{\sum a}{\sum(a+b+c)}$, where the sum is taken over all alleles and loci.

The **genetic F_{ST}-distance** is the pairwise F_{ST} taking account only of the data for the two subpopulations concerned, not all the data simultaneously. Such a measure is valid only if the breeding system is similar for both populations.

Cavalli-Sforza–Menozzi–Piazza, 1994, evaluated, using 120 blood polymorphisms, the doubled genetic F_{ST}-distance between 42 native human populations and between 9 resulting clusters. The largest such distances between two continents were Africa–Oceania (0.247) and Africa–Americas (0.226), while the shortest distances were Americas–Asia (0.089) and Americas–Europe (0.095). The largest distance in Europe, $F_{ST} = 0.02 - 0.023$, was between Finland and Southern Italy; cf. 0.11 (Europeans–Chinese) and 0.153 (Europeans–Africans (Yoruba)). Mbuti Pygmies (the least "Neanderthal") and Papuans (the most "Denisovan") are the two most divergent living humans with $F_{ST} = 0.377$.

A similar analysis by Atzmon et al., 2010, of seven Jewish groups indicated a common origin and, 100–150 generations ago, the split into Middle Eastern and European clusters. The most distant and differentiated are among Mizrahim: F_{ST} of Iranian Jews to other Jews is 0.016. Ashkenazi Jews have the highest admixture with non-Jews but they are not descendants of converted Khazars or Slavs. The closest by F_{ST} to them are Northern Italians, French, Palestinians, and Druze.

Genetic variation in alleles of genes occurs both within (due to mutations and gene exchange during meiosis) and among (due to natural selection and *genetic drift*, i.e., random gene changes) populations. Human total genetic variation is 0.5 % consisting of 0.1 % in SNPs (single nucleotide polymorphisms), ≈0.4 % in *copy number* (deletion, duplication or more, in a DNA segment, instead of exactly two copies of DNA per cell) and a small variation in repetitive DNA. Each human is born with about 50 new mutations, rarely noticeable. Besides mutations, the main mechanisms of our genetic diversity are migrations and hybridization.

The genetic similarity of humans is 99 % among them, while it is 99 % with Neanderthals and 96–98 % with (having SNP diversity 0.2 %) chimpanzees. After initial division, there was interbreeding with chimpanzees and, later, with Neanderthals (in the Middle East 90,000–65,000 years ago), Denisovans (episodes in East Asia) and archaic Africans (in sub-Saharan Africa ≈35,000 years ago). There are 2 % of archaic genes in sub-Saharan Africans, 2 % of Neanderthal genes in Central Asians and 4 % of them in Europeans and Americans. There are 2.5 % of Neanderthal and 6–8 % Denisovan genes in South Asians and Australo-Melanesians.

Seventy-five to 85 % of human SNP variation 0.1 % is among individuals within any population, 5–10 % between local populations of the same continent, and 6–10 % between large groups from different continents. So, differentiation between continental groups is $F_{ST} \leq 0.1$, less than the threshold 0.25 used to define a *subspecies* (*race*).

- **Temporal remoteness**

  The **temporal remoteness** of most recent common ancestor (or *TMRCA*, *divergence time*, *time to coalescence*) of two taxa is the time (or the number of generations) that has passed since those populations existed as a single one. The *molecular clock hypothesis* estimates that one unit of **Nei standard genetic distance** between two taxa corresponds to 18–20 Ma of their TMRCA.

  A human phylogenetic tree is derived from matrilineal mitochondrial DNA, or patrilineal nonrecombinant part of the Y-chromosome of (usually blood) protein sequences by measuring accumulated mutations. TMRCA is 0.2–0.19 Ma ago along all-female ancestry lines for the *Mitochondrial Eve* and 0.24–0.58 Ma ago along all-male lines for the *Y-chromosomal Adam*.

  The resulting phylogenetic tree is rooted in the common ancestor of chimpanzees and humans, which originated in Africa 8–6 Ma ago. The corresponding **genetic** F_{ST}**-distance** between humans and chimpanzee is about 0.02, i.e., at least 30 million point mutations affecting 80 % of genes.

  Our genus *Homo* had diverged, as a carnivorous scavenger, from the *Australopithecines* (bipedal ape-like using rudimentary stone tools) ≈2.5 Ma ago in East (*Homo habilis*) or South Africa. Then *Homo erectus*, the first global and using fire human species, moved to Eurasia 1.8 Ma ago, followed by Denisovans and, later, Neanderthals, the common ancestor of which split from our line 0.8 Ma ago.

  Archaic Homo sapiens originated 0.5−0.4 Ma ago. They evolved to anatomically modern humans *Homo sapiens sapiens* ≈0.2 Ma ago, as shown by the temporal remoteness of their mitochondrial most recent common ancestor. Then their mitochondrial lineage L3 (among L0, L1, L2, L3) migrated out of (southern or east) Africa 0.125 and 0.065 Ma ago.

  Humans passed via population bottleneck ≈0.074 Ma ago (when Toba supervolcano erupted), followed by a rapid expansion. African-Eurasian divergence happened ≈0.06 Ma ago. Humans arrived ≈0.015 Ma ago in the Americas and

$\approx$2,000 years ago on Madagascar. The last place on Earth (besides the Antarctic and tiny atolls) humans colonized was New Zealand where they arrived $\approx$1,300 years ago.
Savanna living, use of fire, speech and sophisticated hand axes appeared about 1.7, 1.6, 0.6, 0.5 Ma ago. Modern human behavior (language, symbolic thought, *cultural universals*) emerged 0.07–0.05 Ma, i.e., $\approx$3,000 generations, ago.
The main known gene mutations leading to us: improving blood supply to the primate brain, weakening jaw muscle (so skull/brains could expand), speeding up the neuron migration (crucial to intelligence), increasing the production of the salivary enzyme (helping to the emergence of agriculture). Also, noncoding sequence HACNS1 had 16 variations during last 6 Ma; it led to more fine muscle control allowing bipedality and tool use. The gene miR-941, unique to humans, emerged 6–1 Ma ago; it could initiate our advanced brain functions.

- **Pedigree-based distances**
A *cousin* (or *blood relative*) is a relative with whom one shares a common ancestor. A *cousin chart* (or *table of consanguinity*, *family tree*, *pedigree digraph*) is a directed tree, where vertices represent relatives (usually humans, show dogs, race horses or cultivars), and the arc uv means that v is a child of u. So, the in-degree of each vertex is at most two (known parents). Moreover, unoriented edges are added with edge uv meaning *reproductive affinity*, i.e., that u and v are mated.
The **genealogical quasi-distance** (or, in Anthropology, *genealogical distance*, *degree of relative consanguinity*) from the individual x to its relative y is defined (Schneider, 1968) as the number of generations one must go before a common ancestor is found, i.e., it is the length $q(x, y)$ of the shortest directed path to x from a common ancestor of x and y in the family tree. Recently, the value $\min\{q(x, y), q(y, x)\}$ is preferred in English pedigree documents.
An *ancestral path* between the vertices x and y in a family tree (or any acyclic digraph) is a concatenation of two directed paths from a common ancestor to them. The **ancestral path distance** is the length of a shortest ancestral path, i.e., it is $q(x, y) + q(y, x)$. Cf. **genealogical distance** between the vertices x and y (of a phylogenetic tree representing taxa) which is the length of a shortest $(x-y)$-path in the undirected family tree, i.e., also $q(x, y) + q(y, x)$. Cf. ancestral path distance in Chap. 22 and **join semilattice distances** in Chap. 10.
The *ancestral distance of an extant taxon* (Hearn and Huber, 2006) is the time (or the number of speciation events) separating it from its most recent ancestor with at least one extant descendant having an independent trait.
Mycielski and Ulam, 1969, called *genealogical distances* between x and y the value $|S(x)\Delta S(y)|$, where $S(z)$ is the set of ancestors of z in a given family tree, and the **Manhattan metric** between some vector representations of x and y.
Two cousins are *a-removed of degree b* if they are separated by a generations and the minimum number of generations between either cousin and their common ancestor is b. The *direct relatives* are spouses and cousins with $(a, b) = (1, 0), (2, 0), (1, 1), (0, 2)$ and $(0, 1)$, i.e., parents/children, grandpar-

ents/grandchildren, uncles (aunts)/nieces (nephews), first degree cousins and siblings. Clearly, $a = |q(x, y) - q(y, x)|$ and $b = \min\{q(x, y), q(y, x)\}$. Worldwide, $\approx$10 % of marriages are between closer than third degree cousins; the case of third degree cousins results in progeny only slightly more homozygous than the general population.

The above pedigree notions are important also in some family, inheritance and nationality rules. For example, the Roman Catholic Church prohibits marriage of x with a relative y if $q(x, y) + q(y, x) \leq 4$. The closest legally permissible unions are between *double-first cousins*, i.e., those sharing four grandparents (in Muslim populations), or uncle–niece (in South India).

Another example: a *Jew* in Halakha's (Jewish Law) sense is a child born to a Jewish mother or an converted adult. Israel's Law of Return permits independent repatriation to anyone with a nonapostate Jewish grandparent and/or his spouse. In Nazi Germany, a *full Jew* was anyone with three Jewish grandparents, while *part-Jews* of first/second degree were those (not practicing Judaism and not having a Jewish spouse) who had two/one Jewish grandparents.

The *inbreeding coefficient* $F(z)$ of an individual z is the probability of *autozygosity*, i.e., that z received the same ancestral gene from both its parents; so, $F(z)$ is the **co-ancestry coefficient** $\theta(z_1, z_2)$ of its parents z_1, z_2. When pedigree data are available, $\theta(x, y)$ is estimated as $\sum_{z \in Z(x,y)} 0.5^{|P(z)|}(1 + F(z))$, where $Z(x, y)$ is the set of least common ancestors of x and y in the pedigree digraph, and $|P(z)|$ is the number of vertices in the shortest ancestral path between x and y through z. In practice, ancestors z are counted only up to a given number of generations and not all of them are known.

The **coefficient of relationship** between two relatives x and y is the fraction of genome inherited from common ancestors. It is almost 1 for identical twins (they differ due to mutations during development and gene copy number variation) and $\approx \frac{3}{4}$ for *semi-identical twins* inheriting the same genes from only one parent. Otherwise, it is $2\theta(x, y)$, since any progeny have a risk $\frac{1}{2}$ of inheriting identical alleles from both parents. It is $\frac{1}{2}$ for siblings and for parent–offspring.

The **coefficient of relatedness** (or *genetic similarity*) between social partners x, y relative to the population is defined (Hamilton, 1970) by

$$r(x, y) = \frac{cov(g, g')}{cov(g, g)} = \frac{E[(g - E[g])(g' - E[g'])]}{E[(g - E[g])(g - E[g])]},$$

where g, g' are genetic (i.e., heritable) components of the phenotype (for the character of interest) of x, y, respectively, and *cov* denotes a statistical covariance taken over all individuals in the population. This coefficient quantifies the *indirect fitness*, i.e., the component of fitness gained from aiding related individuals.

Fitness is an individual's ability to propagate its genes, i.e., to both survive and reproduce. A measure of it is the average contribution to the gene pool of the next generation that is made by an average individual of the specified genotype or phenotype. The relative *reproductive value* of an individual is the probability that it is the ancestor of a randomly chosen individual in a distant future generation. Fowler–Christakis, 2013: pairs of nonkin friends are, on average, as genetically similar to one another as fourth cousins,

- **Mating distances**
  Individual migration distances are the distances between birthplaces of paired individuals. If the pairs are spouses (gametes) or siblings, we have **marital distance** or **sib distance**, respectively. Also, the **parent–offspring distance** is used to describe gene migration per generation.
  For humans, those distances are measured either in km, or, say, as the number of municipalities crossed by a straight line between municipality midpoints of each pair's birthplaces. The term *marital migration distance* is also used for the distance between premarital town of a person and town of marriage. Cf. **migration distances (in Economics)** in Chap. 28.
  Until the twentieth century, men usually went courting no more than about 8 km from home (the distance they could walk out and back on their day off from work). According to Fox, 80 % of all marriages in history could be between second cousins or closer. Also, young birds, leaving the nest, usually move 4–5 home ranges away; so, they stay within breeding distance of their cousins.
  For a population, **critical mating distances**, are maximum spatial (physical) and genetic (number of genes bearing different alleles) distances allowed for mating; cf. **isolation by distance**. For honey bees, the *mating distance* is the range of queen's nuptial flight from her hive to the drone congregation areas over their hives; it is typically within 7.5 km but can reach 17 km.
- **Lasker distance**
  The **Lasker distance** between two human populations x and y, characterized by surname frequency vectors (x_i) and (y_i), is the number $-\ln 2R_{x,y}$, where $R_{x,y} = \frac{1}{2}\sum_i x_i y_i$ is Lasker's (1977) *coefficient of relationship by isonymy*.
  Surname structure is related to inbreeding and (in patrilineal societies) to random genetic drift, mutation and migration. Surnames can be considered as alleles of one locus, and their distribution can be analyzed by Kimura's theory of neutral mutations; an isonymy points to a common ancestry.
- **Isolation by distance**
  Isolation by distance (or *ibd*, Wright, 1943) is the tendency for most individuals to migrate and find mates between neighbors; so, populations that are a geographically closer are more similar than those that are further apart. It results in a smooth increase in a *cline*, i.e., the gradual change in a character (say, allele frequency, within- or between-population genetic differentiation) or feature (phenotype) with increasing geographic distance. The above distance can be Euclidean or along a great circle, river, or topographic isocline.
  Ibd for humans was studied, for example, via migration patterns and the distribution of surnames (cf. **Lasker distance**). At both continental and global scales, the **genetic F_{ST}-distance** and differentiation in cranial morphology between populations increases with **great circle distance** (cf. Chap. 25).

The geographic distance explains >75 % of the variation between human populations, and this distance from East Africa explains 85 % of the smooth decrease in genetic diversity. Atkinson, 2011, claims that phoneme diversity also declines with distance from Africa. The occurrence of alleles $7R$ and $2R$, linked to risk-taking, of the dopamine-related gene *DRD*4 increases with distance from Africa.

A strong Europe-wide (except Basques, Finns and Sardinians isolates) correlation, based on >300,000 single nucleotide polymorphisms, between geographic and genetic distance was found. South-to-North was the main smooth gradient.

The ibd model explains the emergence of regional differences (races) and new species by restricted gene flow and adaptive variations. *Speciation* (branching of new species from an ancestral population) occurs when subpopulations become reproductively isolated. The dominating mode of speciation is *allopatry* when habitat splits into discontinuous parts by the formation of a physical barrier to gene flow or dispersal. Examples of natural barriers are the Himalayas, Wallace Line, Grand Canyon. All modes, in a continuum from complete (allopatric) to zero (*sympatric*) spatial segregation of diverging groups, occur, mainly, in marine ecosystems.

In spatially extended population, another mode—*topopatric* (or *distance-forced*) speciation can occur via ibd only, without geographic isolation and selection. de Aguiar and Bat-Yam, 2011, gave the conditions for speciation in such population as a function of its density, mutation rate, genome size and **critical mating distances**. They see such speciation as a case of breakdown of unstable uniform distribution, leading to the self-organization of its members into clusters.

Absolute distances between diverged groups can be, say, tens of meters for pathogen resistance to hundreds of kilometers in marine invertebrates.

- **Wright**

  Dispersal neighborhood (DN) is the geographic area within which individuals and genes regularly move and interact. It is estimated as the area within a radius extending two standard deviations from the mean of species's dispersal distribution.

  Richardson et al., 2014, proposed to measure microgeographic adaptive evolution by the **wright** indicating the phenotypic difference between populations relative to the number of species-specific DN's separating them. It is

$$w = \frac{|x_1 - x_2|}{ds_p},$$

  where x_1, x_2 are the means of the genetically determined traits of populations, s_p is the pooled standard deviation of those trait values across populations, and d is the distance in number of DN's separating the two populations.

  The wright is a spatial analog of the *haldane*, a metric for rate of microevolution defined (Gingerich, 1993, and Lynch, 1990) as $\frac{|x_1 - x_2|}{gs_p}$, where g the number of generations separating the populations (or samples of the same populations).

Haldan, 1949, defined the *darwin* as $|\ln x_1 - \ln x_2|$ (or, respectively, $\frac{|\ln x_1 - \ln x_2|}{t}$), where t is the time in Ma separating samples of the same populations).

- **Malécot's distance model**
Genealogy, migration and surname isonymy are used to predict kinship (usually estimated from blood samples). But because of incomplete knowledge on ancestors, pedigree-independent methods for kinship assays utilize the distance-dependent correlations of any parameter influenced by identity in descent: phenotype, gene frequency, or, say, isonymy.
Malécot's distance model (1948, 1959) is expressed by the following *kinship-distance formula* for the mean **coefficient of kinship** between two populations **isolated by distance** d:

$$\theta_d = ae^{-bd}d^c,$$

where $c = 0, \frac{1}{2}$ correspond to one-, two-dimensional migration, b is a function of the *systematic pressure* (joint effect of co-ancestry, selection, mutations and long range migration), and a is the *local kinship* (the correlation between random gametes from the same locality). In fact, the results in 2D for small and moderate distances agree closely with $c = 0$. The model is most successful when the systematic pressure is dominated by migration.
Malécot's model was adapted for the dependency ρ_d of alleles at two loci at distance d (*allelic association, linkage disequilibrium, polymorphism distance*):

$$\rho_d = (1 - L)Me^{-\epsilon d} + L,$$

where d is the distance (say, from a disease gene) between loci along the chromosome (either **genome distance** on the physical scale in kilobases, or **map distance** on the genetic scale in centiMorgans), ϵ is a constant for a specified region, $M \leq 1$ is a parameter expressing mutation rate and L is the parameter predicting association between unlinked loci.
Selection generates long *blocks of linkage disequilibrium* (places in the genome where genetic variations are occurring more often than by chance, as in the genetic drift) across hundreds of kilobases. Using it, Hawks et al., 2007, found that selection in humans much accelerated during the last 40,000 years, driven by exponential population growth and cultural adaptations.
Examples of accelerated (perhaps, under diet and diseases pressures) human evolution and variation include disease resistance, lactose tolerance, skin color, skeletal gracility. A mutation in *microcephalin* (gene MCPH1) appeared 14,000–62,000 years ago and is now carried by 70 % of people but not in sub-Saharan Africa. Distinctive traits of East Asians (about 93 % of Han Chinese and 70 % of Japanese and Tai)—thicker hair shafts, more sweat glands, smaller breasts and specific teeth—are the result of a mutation in gene EDAR that occurred $\approx$35,000 years ago.

The fastest genetic change ever observed in humans is that the ethnic Tibetans split off from the Han Chinese less than 3,000 years ago and since then evolved a unique ability to thrive at high (4,000 m above sea level) altitudes and low oxygen levels. It also come from their gene *EPAS1* found only in Denisovans. An example of quick nongenetic evolution: the average height of a European male at age 21 rose from 167 cm in early 1870s to 178 cm in 1980. Crabtree, 2012, argues that our intellectual and emotional abilities diminish gradually (after a peak 2,000–6,000 years ago) because of weakened control of genetic mutations by natural selection.

Over the past 20,000 years, the average volume of the human brain has decreased by 10 %. Possible reason: our dwindling intelligence (Geary, 2011), or improved brain efficiency (Hawks, 2011), or social self-domestication (Hood, 2014).

23.2 Distances for DNA/RNA and Protein Data

The main way to estimate the genetic distance between DNA, RNA or proteins is to compare their nucleotide or amino acid, sequences, respectively. Besides sequencing, the main techniques used are immunological ones, *annealing* (cf. **hybridization metric**) and *gel electrophoresis* (cf. **read length**).

Distances between nucleotide (DNA/RNA) or protein sequences are usually measured in terms of substitutions, i.e., mutations, between them.

A *DNA sequence* is a sequence $x = (x_1, \dots, x_n)$ over the four-letter alphabet of four nucleotides A, T, C, G (or two-letter alphabet purine/pyrimidine, or 16-letter *dinucleotide* alphabet of ordered nucleotide pairs, etc.). Let $\sum$ denote $\sum_{i=1}^{n}$.

A *protein sequence* is a sequence $x = (x_1, \dots, x_n)$ over a 20-letter alphabet of 20 standard amino acids; $\sum$ again denotes $\sum_{i=1}^{n}$.

A short sequence is called *nullomer* if it do not occur in a given species and *prime* if it has not been found in nature. Hampikian–Andersen, 2007, lists 80 human DNA nullomers of length 11 and many primes: DNA of length 15 and protein of length 5.

For a macromolecule, a *primary structure* is its atomic composition and the chemical bonds connecting atoms. For DNA, RNA or protein, it is specified by its sequence. The *secondary structure* is the 3D form of local segments defined by the hydrogen bonds. The *tertiary structure* is the 3D structure, as defined by atomic positions. The *quaternary structure* describes the arrangement of multiple molecules into larger complexes.

- **Number of DNA differences**

  The **number of DNA differences** between DNA sequences $x = (x_1, \dots, x_n)$ and $y = (y_1, \dots, y_n)$ is the number of mutations, i.e., their **Hamming metric**:

$$\sum 1_{x_i \neq y_i}.$$

- ***p*-distance**

The p-**distance** d_p between DNA sequences $x = (x_1, \dots, x_n)$ and $y = (y_1, \dots, y_n)$ is defined by

$$\frac{\sum 1_{x_i \neq y_i}}{n}.$$

- **Jukes–Cantor nucleotide distance**
  The **Jukes–Cantor nucleotide distance** between DNA sequences x and y is defined, using the p-**distance** d_p with $d_p \leq \frac{3}{4}$, by

$$-\frac{3}{4}\ln\left(1-\frac{4}{3}d_p(x,y)\right).$$

  If the rate of substitution varies with the gamma distribution, and a is the parameter describing the shape of this distribution, then the **gamma distance for the Jukes–Cantor model** is defined by

$$\frac{3a}{4}\left(\left(1-\frac{4}{3}d_p(x,y)\right)^{-1/a}-1\right).$$

- **Tajima–Nei distance**
  The **Tajima–Nei distance** between DNA sequences x and y is defined by

$$-b\ln\left(1-\frac{d_p(x,y)}{b}\right), \quad \text{where}$$

$$b=\frac{1}{2}\left(1-\sum_{j=A,T,C,G}\left(\frac{1_{x_i=y_i=j}}{n}\right)^2+\frac{1}{c}\sum\left(\frac{1_{x_i\neq y_i}}{n}\right)^2\right), \quad \text{and}$$

$$c=\frac{1}{2}\sum_{i,k\in\{A,T,G,C\},j\neq k}\frac{\left(\sum 1_{(x_i,y_i)=(j,k)}\right)^2}{\left(\sum 1_{x_i=y_i=j}\right)\left(\sum 1_{x_i=y_i=k}\right)}.$$

Let $P = \frac{1}{n}|\{1 \leq i \leq n : \{x_i, y_i\} = \{A, G\} \text{ or } \{T, C\}\}|$, and $Q = \frac{1}{n}|\{1 \leq i \leq n : \{x_i, y_i\} = \{A, T\} \text{ or } \{G, C\}\}|$, i.e., P and Q are the frequencies of, respectively, transition and transversion mutations between DNA sequences x and y.
The following four distances are given in terms of P and Q.

- **Jin–Nei gamma distance**
  The **Jin–Nei gamma distance** between DNA sequences is defined by

$$\frac{a}{2}\left((1-2P-Q)^{-1/a}+\frac{1}{2}(1-2Q)^{-1/a}-\frac{3}{2}\right),$$

where the rate of substitution varies with the gamma distribution, and a is the parameter describing the shape of this distribution.

- **Kimura 2-parameter distance**
The **Kimura 2-parameter distance** $K2P$ (Kimura, 1980) between DNA sequences is defined by

$$-\frac{1}{2}\ln(1-2P-Q)-\frac{1}{4}\ln\sqrt{1-2Q}.$$

- **Tamura 3-parameter distance**
The **Tamura 3-parameter distance** between DNA sequences is defined by

$$-b\ln\left(1-\frac{P}{b}-Q\right)-\frac{1}{2}(1-b)\ln(1-2Q),$$

where $f_x = \frac{1}{n}|\{1 \le i \le n : x_i = G \text{ or } C\}|$, $f_y = \frac{1}{n}|\{1 \le i \le n : y_i = G \text{ or } C\}|$, and $b = f_x + f_y - 2f_x f_y$. If $b = \frac{1}{2}$, it is the **Kimura 2-parameter distance**.

- **Tamura–Nei distance**
The **Tamura–Nei distance** between DNA sequences is defined by

$$-\frac{2f_A f_G}{f_R}\ln\left(1-\frac{f_R}{2f_A f_G}P_{AG}-\frac{1}{2f_R}P_{RY}\right)$$
$$-\frac{2f_T f_C}{f_Y}\ln\left(1-\frac{f_Y}{2f_T f_C}P_{TC}-\frac{1}{2f_Y}P_{RY}\right)-$$
$$-2\left(f_R f_Y-\frac{f_A f_G f_Y}{f_R}-\frac{f_T f_C f_R}{f_Y}\right)\ln\left(1-\frac{1}{2f_R f_Y}P_{RY}\right),$$

where $f_j = \frac{1}{2n}\sum(1_{x_i=j}+1_{y_i=j})$ for $j = A, G, T, C$, and $f_R = f_A + f_G$, $f_Y = f_T + f_C$, while $P_{RY} = \frac{1}{n}|\{1 \le i \le n : |\{x_i, y_i\} \cap \{A, G\}| = |\{x_i, y_i\} \cap \{T, C\}| = 1\}|$ (the proportion of transversion differences), $P_{AG} = \frac{1}{n}|\{1 \le i \le n : \{x_i, y_i\} = \{A, G\}\}|$ (the proportion of transitions within purines), and $P_{TC} = \frac{1}{n}|\{1 \le i \le n : \{x_i, y_i\} = \{T, C\}\}|$ (the proportion of transitions within pyrimidines).

- **Lake paralinear distance**
Given two DNA sequences $x = (x_1, \dots, x_n)$ and $y = (y_1, \dots, y_n)$, denote by $det(J)$ the determinant of the 4×4 matrix $J = ((J_{ij}))$, where $J_{ij} = \frac{1}{n}|\{1 \le t \le n : x_t = i, y_t = j\}|$ (joint probability) and indices $i, j = 1, 2, 3, 4$ represent nucleotides A, T, C, G, respectively. Let $f_i(x)$ denote the frequency of the i-th nucleotide in the sequence x (marginal probability), and let $f(x) = f_1(x)f_2(x)f_3(x)f_4(x)$.

The **Lake paralinear distance** (1994) between sequences x and y is defined by

$$-\frac{1}{4}\ln\frac{det(J)}{\sqrt{f(x)f(y)}}.$$

It is a **four-point inequality metric**, and it generalizes trivially for sequences over any alphabet. Related are the **LogDet distance** (Lockhart et al., 1994) $-\frac{1}{4}\ln\det(J)$ and the symmetrization $\frac{1}{2}(d(x,y)+d(y,x))$ of the **Barry–Hartigan quasi-metric** (1987) $d(x,y)=-\frac{1}{4}\ln\frac{det(J)}{\sqrt{f(x)}}$.

- **Eigen–McCaskill–Schuster distance**
The **Eigen–McCaskill–Schuster distance** between DNA sequences $x=(x_1,\dots,x_n)$ and $y=(y_1,\dots,y_n)$ is defined by

$$|\{1\le i\le n:\{x_i,y_i\}\ne\{A,G\},\{T,C\}\}|.$$

It is the number of *transversions*, i.e., positions i with one of x_i, y_i denoting a purine and another one denoting a pyrimidine.

- **Watson–Crick distance**
The **Watson–Crick distance** between DNA sequences $x=(x_1,\dots,x_n)$ and $y=(y_1,\dots,y_n)$ is defined, for $x\ne y$, by

$$|\{1\le i\le n:\{x_i,y_i\}\ne\{A,T\},\{G,C\}\}|$$

It is the **Hamming metric** (**number of DNA differences**) $\sum 1_{x_i\ne\overline{y}_i}$ between x and the *Watson–Crick complement* $\overline{y}=(\overline{y}_1,\dots,\overline{y}_n)$ of y, where $\overline{y}_i=A,T,G,C$ if $y_i=T,A,C,G$, respectively. Let y^* be the reversal $(\overline{y}_n,\dots,\overline{y}_1)$ of $\overline{y}$.
Hybridization is the process of combining complementary single-stranded nucleic acids into a single molecule. *Annealing* is the binding of two strands by the Watson–Crick complementation. *Denaturation* is the reverse process.
A *DNA cube* is any maximal set of DNA n-sequences, such that, for any two x, y of them, it holds that $H(x,y)=\min_{-n\le k\le n}\sum 1_{x_i\ne y^*_{i+k(\bmod n)}}=0$. The **hybridization metric** (Garzon et al., 1997) between DNA cubes A and B is

$$\min_{x\in A,y\in B} H(x,y).$$

- **RNA structural distances**
An *RNA sequence* is a string over the alphabet $\{A,C,G,U\}$ of nucleotides (bases). Inside a cell, such a string folds in 3D space, because of pairing of nucleotide bases (usually, by bonds A–U, G–C and G–U). The *secondary structure* of an RNA is, roughly, the set of helices (or the list of paired bases) making up the RNA. Such structure can be represented as a planar graph and further, as a rooted tree. The *tertiary structure* is the geometric form the RNA takes in space; the secondary structure is its simplified/localized model.

An **RNA structural distance** between two RNA sequences is a distance between their secondary structures. These distances are given in terms of their selected representation. For example, the **tree edit distance** (and other distances on rooted trees given in Chap. 15) are based on the rooted tree representation.

Let an RNA secondary structure be represented by a simple graph (V, E) with vertex-set $V = \{1, \dots, n\}$ such that, for every $1 \leq i \leq n$, $(i, i+1) \notin E$ and $(i, j), (i, k) \in E$ imply $j = k$. Let $E = \{(i_1, j_1), \dots, (i_k, j_k)\}$, and let (ij) denote the transposition of i and j. Then $\pi(G) = \prod_{t=1}^{k}(i_t j_t)$ is an involution.

Let $G = (V, E)$ and $G' = (V, E')$ be such planar graph representations of two RNA secondary structures. The **base pair distance** between G and G' is the number $|E \Delta E'|$, i.e., the **symmetric difference metric** between secondary structures seen as sets of paired bases.

The **Zuker distance** between G and G' is the smallest number k such that, for every edge $(i, j) \in E$, there is an edge $(i', j') \in E'$ with $\max\{|i - i'|, |j - j'|\} \leq k$ and, for every $(k', l') \in E'$, there is an $(k, l) \in E$ with $\max\{|k - k'|, |l - l'|\} \leq k$.

The **Reidys–Stadler–Roselló metric** between G and G' is defined by

$$|E \Delta E'| - 2T,$$

where T is the number of cyclic orbits of length greater than 2 induced by the action on V of the subgroup $\langle \pi(G), \pi(G') \rangle$ of the group Sym_n of permutations on V. It is the number of transpositions needed to represent $\pi(G)\pi(G')$.

Let $I_G = \langle x_i x_j : (x_i, x_j) \in E \rangle$ be the monomial ideal (in the ring of polynomials in the variables $x_1, \dots, x_n$ with coefficients 0, 1), and let $M(I_G)_m$ denote the set of all monomials of total degree $\leq m$ that belong to I_G. For every $m \geq 3$, a Liabrés-Roselló **monomial metric** between $G = (V, E)$ and $G' = (V', E')$ is

$$|M(I_G)_{m-1} \Delta M(I_{G'})_{m-1}|.$$

Chen–Li–Chen, 2010, proposed the following variation of the **directed Hausdorff distance** (cf. Chap. 1) between two intervals $x = [x_1, x_2]$ and $y = [y_1, y_2]$, representing two RNA secondary structures:

$$\max_{a \in x} \min_{b \in y} |a - b| \left(1 - \frac{O(x, y)}{x_2 - x_1 + 1}\right),$$

where $O(x, y) = \min\{x_2, y_2\} - \max\{x_1, y_1\}$, represents the overlap of intervals x and y; it is seen as a negative gap between x and y, if they are disjoint.

- **Fuzzy polynucleotide metric**
The **fuzzy polynucleotide metric** (or **NTV-metric**) is the metric introduced by Nieto, Torres and Valques-Trasande, 2003, on the 12-dimensional unit cube I^{12}. Four nucleotides U, C, A and G of the RNA alphabet being coded as $(1,0,0,0)$, $(0,1,0,0)$, $(0,0,1,0)$ and $(0,0,0,1)$, respectively, 64 possible triplet codons of the genetic code can be seen as vertices of I^{12}.
So, any point $(x_1,\dots,x_{12}) \in I^{12}$ can be seen as a *fuzzy polynucleotide codon* with each x_i expressing the grade of membership of element i, $1 \le i \le 12$, in the fuzzy set x. The vertices of the cube are called the *crisp sets*.
The **NTV-metric** between different points $x, y \in I^{12}$ is defined by
$$\frac{\sum_{1\le i\le 12}|x_i - y_i|}{\sum_{1\le i\le 12}\max\{x_i, y_i\}}.$$
Dress and Lokot showed that $\frac{\sum_{1\le i\le n}|x_i - y_i|}{\sum_{1\le i\le n}\max\{|x_i|,|y_i|\}}$ is a metric on the whole of $\mathbb{R}^n$. On $\mathbb{R}^n_{\ge 0}$ this metric is equal to $1 - s(x,y)$, where $s(x,y) = \frac{\sum_{1\le i\le n}\min\{x_i,y_i\}}{\sum_{1\le i\le n}\max\{x_i,y_i\}}$ is the **Ruzicka similarity** (cf. Chap. 17).
- **Genome rearrangement distances**
The *genomes* of related unichromosomal species or single chromosome organelles (such as small viruses and mitochondria) are represented by the order of genes along chromosomes, i.e., as *permutations* (or *rankings*) of a given set of n homologous genes. If one takes into account the directionality of the genes, a chromosome is described by a *signed permutation*, i.e., by a vector $x = (x_1,\dots,x_n)$, where $|x_i|$ are different numbers $1,\dots,n$, and any x_i can be positive or negative.
The circular genomes are represented by circular (signed) permutations $(x_1,\dots,x_n)$, where $x_{n+1} = x_1$ and so on.
Given a set of considered mutation moves, the corresponding *genomic distance* between two such genomes is the **editing metric** (cf. Chap. 11) with the editing operations being these moves, i.e., the minimal number of moves needed to transform one (signed) permutation into another.
In addition to (and, usually, instead of) local mutation events, such as character indels or replacements in the DNA sequence, the *large* (i.e., happening on a large portion of the chromosome) mutations are considered, and the corresponding genomic editing metrics are called **genome rearrangement distances**. In fact, such rearrangement mutations being rarer, these distances estimate better the true genomic evolutionary distance.
The genome (chromosomal) rearrangements are *inversions* (block reversals), *transpositions* (exchanges of two adjacent blocks), *inverted transpositions* (inversions combined with transpositions) in a permutation, and, for signed permutations, *signed reversals* (sign reversal combined with inversion). The main genome rearrangement distances between two unichromosomal genomes are:
Cayley, **reversal** and **signed reversal** metrics (cf. Chap. 11);

ITT-distance: the minimal number of inversions, transpositions and inverted transpositions needed to transform one of them into another.

Given two circular signed permutations $x = (x_1, \dots, x_n)$ and $y = (y_1, \dots, y_n)$ (so, $x_{n+1} = x_1$, etc.), a *breakpoint* is a number i, $1 \le i \le n$, such that $y_{i+1} \neq x_{j(i)+1}$, where the number $j(i)$, $1 \le j(i) \le n$, is defined by the equality $y_i = x_{j(i)}$. The **breakpoint distance** (Watterson et al., 1982) between genomes, represented by x and y, is the number of breakpoints.

This distance and the **permutation editing metric** (the **Ulam metric** from Chap. 11: the minimal needed number of one-character transpositions) are used for the approximation of genome rearrangement distances.

- **Syntenic distance**

  This is a *genomic distance* between multi-chromosomal genomes, seen as unordered collections of *synteny sets* of genes, where two genes are *syntenic* if they appear in the same chromosome. The **syntenic distance** (Ferretti–Nadeau–Sankoff, 1996) between two such genomes is the minimal number of mutation moves—*translocations* (exchanges of genes between two chromosomes), *fusions* (merging of two chromosomes into one) and *fissions* (splitting of one chromosome into two)—needed to transfer one genome into another. All (input and output) chromosomes of these mutations should be nonempty and not duplicated.

  The above three mutation moves correspond to interchromosomal genome rearrangements which are rarer than intrachromosomal ones; so, they give information about deeper evolutionary history.

- **Genome distance**

  The **genome distance** between two loci on a chromosome is a physical distance: the number of base pairs (bp) separating them on the chromosome.

  In particular, the **intragenic distance** of two neighboring genes is the smallest distance in bp separating them on the chromosome. Sometimes, it is defined as the genome distance between the transcription start sites of those genes.

  Nelson, Hersh and Carrol, 2004, defined the *intergenic distance* of a gene as the amount of noncoding DNA between the gene and its nearest neighbors, i.e., the sum of upstream and downstream distances, where *upstream distance* is the genome distance between the start of a gene's first exon and the boundary of the closest upstream neighboring exon and *downstream distance* is the distance between the end of a gene's last exon and the boundary of the closest downstream neighboring exon. If exons overlap, the intergenic distance is 0.

- **Strand length**

  A single strand of nucleic acid (DNA or RNA sequence) is oriented *downstream*, i.e., from the $5'$ *end* toward the $3'$ *end* (sites terminating at the 5th and 3rd carbon in the sugar-ring; $5'$-phosphate binds covalently to the $3'$-hydroxyl of another nucleotide). So, the structures along it (genes, transcription factors, polymerases) are either downstream or upstream. The **strand length** is the distance from its $5'$ to $3'$ end. Cf. **end-to-end distance** (in Chap. 24) for a general polymer.

For a molecule of *messenger RNA* (mRNA), the **gene length** is the distance from the *cap site* $5'$, where post-translational stability is ensured, to the *polyadenylation site* $3'$, where a poly(A) tail of 50–250 adenines is attached after translation.

- **Map distance**
  The **map distance** between two loci on a genetic map is the recombination frequency expressed as a percentage; it is measured in *centiMorgans* cM (or *map units*), where 1 cM corresponds to a 1 % ($\frac{1}{100}$) chance that a segment of DNA will crossover or recombine within one generation. Genes at map distance 50 cM are unlinked.
  For humans, 1.3 cM corresponds to a **genome distance** of 1 Mb (million bp). In the female this recombination rate (and so map distances) are twice that of the male. In males, the total length of intervals between linked genes is 2,500 cM.
  During meiosis in humans, there is an average of 2–3 crossovers for each pair of homologous chromosomes. The *intermarker meiotic recombination distance* (Dib et al., 1992) counts only meiotic crossovers. *Mitotic crossover* is rare.
- **tRNA interspecies distance**
  Transfer RNA (tRNA) molecules are necessary to translate *codons* (nucleotide triplets) into amino acids; eukaryotes have up to 80 different tRNAs. Two tRNA molecules are called *isoacceptor tRNAs* if they bind the same amino acid.
  The **tRNA interspecies distance** between species m and n is (Xue et al., 2003), averaged for all 20 amino acids, the *tRNA distance for a given amino acid* aa_i which is, averaged for all pairs, the **Jukes–Cantor protein distance** between each isoacceptor tRNAs of aa_i from species m and each isoacceptor tRNAs of the same amino acid from species n.
- **PAM distance**
  There are many notions of similarity/distance (20×20 *scoring matrices*) on the set of 20 standard amino acids, based on genetic codes, physico-chemical properties, secondary structural matching, structural properties (hydrophilicity, polarity, charge, shape, etc.) and observed frequency of mutations. The most frequently used one is the **Dayhoff distance**, based on the 20×20 *Dayhoff PAM250* matrix which expresses the relative mutability of amino acids.
  The **PAM distance** (or **Dayhoff–Eck distance**, *PAM value*) between protein sequences is defined as the minimal number of accepted (i.e., fixed) point mutations per 100 amino acids needed to transform one protein into another.
  1 PAM is a unit of evolution: it corresponds to 1 point mutation per 100 amino acids. PAM values 80, 100, 200, 250 correspond to the distance (in %) 50, 60, 75, 92 between proteins.
- **Genetic code distance**
  The **genetic code distance** (Fitch and Margoliash, 1967) between amino acids x and y is the minimum number of nucleotides that must be changed to obtain x from y. In fact, it is $1, 2$ or 3, since each amino acid corresponds to three bases.
- **Miyata–Miyazawa–Yasanaga distance**
  The **Miyata–Masada–Yasanaga distance** (or *Miyata's biochemical distance*, 1979) between amino acids x, y with polarities p_x, p_y and volumes v_x, v_y, respectively, is defined by

$$\sqrt{\left(\frac{|p_x - p_y|}{\sigma_p}\right)^2 + \left(\frac{|v_x - v_y|}{\sigma_v}\right)^2},$$

where σ_p and σ_v are the standard deviations of $|p_x - p_y|$ and $|v_x - v_y|$.
This distance is derived from the similar *Grantam's chemical distance* (Grantam, 1974) based on polarity, volume and carbon-composition of amino acids.

- **Polar distance (in Biology)**
The following three physico-chemical distances between amino acids x and y were defined in Hughes–Ota–Nei, 1990.
Dividing amino acids into two groups—*polar* (C, D, E, H, K, N, Q, R, S, T, W, Y) and nonpolar (the rest)—the **polar distance** is 1, if x, y belong to different groups, and 0, otherwise. The second polarity distance is the absolute difference between the polarity indices of x and y. Dividing amino acids into three groups—*positive* (H, K, R), *negative* (D, E) and *neutral* (the rest)—the **charge distance** is 1, if x, y belong to different groups, and 0, otherwise.
- **Feng–Wang distance**
Twenty amino acids can be ordered linearly by their *rank-scaled* functions CI, NI of pK_a values for the terminal amino acid groups COOH and NH_3^+, respectively. 17 CI is $1, 2, 3, 4, 5, 6, 7, 7, 8, 9, 10, 11, 12, 13, 14, 14, 15, 15, 16, 17$ for C, H, F, P, N, D, R, Q, K, E, Y, S, M, V, G, A, L, I, W, T, while 18 NI is $1, 2, 3, 4, 5, 5, 6, 7, 8, 9, 10, 10, 11, 12, 13, 14, 15, 16, 17, 18$ for N, K, R, Y, F, Q, S, H, M, W, G, L, V, E, I, A, D, T, P, C.
Given a protein sequence $x = (x_1, \dots, x_m)$, define $x_i < x_j$ if $i < j, CI(x_i) < CI(x_i)$ and $NI(x_i) < NI(x_i)$ hold. Represent the sequence x by the augmented $m \times m$ *Hasse matrix* $((a_{ij}(x)))$, where $a_{ii}(x) = \frac{CI(x_i)+NI(x_i)}{2}$ and, for $i \neq j$, $a_{ij}(x) = -1$ or 1 if $x_i < x_j$ or $x_i \geq x_j$, respectively.
The **Feng–Wang distance** [FeWa08] between protein sequences $x = (x_1, \dots, x_m)$ and $y = (y_1, \dots, y_n)$ is defined by
$$\left|\left|\frac{\lambda(x)}{\sqrt{m}} - \frac{\lambda(y)}{\sqrt{n}}\right|\right|_2,$$
where $\lambda(z)$ denotes the largest eigenvalue of the matrix $((a_{ij}(z)))$.
- **Number of protein differences**
The **number of protein differences** between protein sequences $x = (x_1, \dots, x_m)$ and $y = (y_1, \dots, y_n)$ is just the **Hamming metric** between protein sequences:
$$\sum 1_{x_i \neq y_i}.$$

- **Amino p-distance**
  The **amino p-distance** (or *uncorrected distance*) d_p between protein sequences $x = (x_1, \dots, x_m)$ and $y = (y_1, \dots, y_n)$ is defined by
  $$\frac{\sum 1_{x_i \neq y_i}}{n}.$$
- **Amino Poisson correction distance**
  The **amino Poisson correction distance** between protein sequences x and y is defined, via the **amino p-distance** d_p, by
  $$-\ln(1 - d_p(x, y)).$$
- **Amino gamma distance**
  The **amino gamma distance** (or *Poisson correction gamma distance*) between protein sequences x and y is defined, via the **amino p-distance** d_p, by
  $$a((1 - d_p(x, y))^{-1/a} - 1),$$
  where the substitution rate varies with $i = 1, \dots, n$ according to the gamma distribution with the shape described by the parameter a. For $a = 2.25$ and $a = 0.65$, it estimates the **Dayhoff distance** and **Grishin distances**, respectively. In some applications, this distance with $a = 2.25$ is called simply the **Dayhoff distance**.
- **Jukes–Cantor protein distance**
  The **Jukes–Cantor protein distance** between protein sequences x and y is defined, via the **amino p-distance** d_p, by
  $$-\frac{19}{20} \ln\left(1 - \frac{20}{19} d_p(x, y)\right).$$
- **Kimura protein distance**
  The **Kimura protein distance** between protein sequences x and y is defined, via the **amino p-distance** d_p, by
  $$-\ln\left(1 - d_p(x, y) - \frac{d_p^2(x, y)}{5}\right).$$
- **Grishin distance**
  The **Grishin distance** d between protein sequences x and y can be obtained, via the **amino p-distance** d_p, from the formula
  $$\frac{\ln(1 + 2d(x, y))}{2d(x, y)} = 1 - d_p(x, y).$$

- **k-mer distance**
The **k-mer distance** (Edgar, 2004) between sequences $x = (x_1, \dots, x_m)$ and $y = (y_1, \dots, y_n)$ over a compressed amino acid alphabet is defined by
$$\ln\left(\frac{1}{10} + \frac{\sum_a \min\{x(a), y(a)\}}{\min\{m, n\} - k + 1}\right),$$
where a is any *k-mer* (a word of length k over the alphabet), while $x(a)$ and $y(a)$ are the number of times a occurs in x and y, respectively, as a *block* (contiguous subsequence). Cf. **q-gram similarity** in Chap. 11.
- **Whole genome composition distance**
Let A_k be the set of all $\sum_{i=1}^{k} 4^i$ nonempty words of length at most k over the alphabet of four RNA nucleotides. For an RNA sequence $x = (x_1, \dots, x_n)$ and any $a \in A_k$, let $g_a(x)$ be the number of occurrences of a as a *block* (contiguous subsequence) in x and $f_a(x)$ be $g_a(x)$ divided by the number of blocks of the same length in x.
The **whole genome composition distance** (Wu et al., 2006) between RNA sequences x and y (of two strains of HIV-1 virus) is the Euclidean distance
$$\sqrt{\sum_{a \in A_k} (f_a(x) - f_a(y))^2}.$$
The D^2 *distance* (Torney *et al.*, 1990) is $\sum_{a \in A_k \setminus A_l} (g_a(x) - g_a(y))^2$ for some $l \le k$. The D_2 *statistic* (Lippert *et al.* 2002) is the number of k-word matches of x and y.
- **Additive stem w-distance.**
Given an alphabet $\mathcal{A}$, let $w = w(a, b) > 0$ for $a, b \in \mathcal{A}$, be a weight function on it. The **additive stem w-distance** between two n-sequences $x, y \in \mathcal{A}^n$ is defined (D'yachkov and Voronina, 2008) by
$$D_w(x, y) = \sum_{i=1}^{n-1} (s_i^w(x, x) - s_i^w(x, y)),$$
where $s_i^w(x, y) = w(a, b)$ if $x_i = y_i = a, x_{i+1} = y_{i+1} = b$ and $s_i^w(x, y) = 0$, otherwise. If all $w(a, b) = 1$, then $\sum_{i=1}^{n-1} s_i(x, y)$ is the number of common 2-blocks containing adjacent symbols in the longest common subsequence of x and y; then $D_w(x, y)$ is called a **stem Hamming distance**.
- **ACS-distance.**
Given an alphabet $\mathcal{A}$, the *average common substring length* between sequences $x = (x_1, \dots, x_m)$ and $y = (y_1, \dots, y_n)$ over $\mathcal{A}$ is (Ulitsky et al., 2006) $L(x, y) = \frac{1}{m} \sum_{i=1}^{m} l_i$, where l_i is the length of the longest substring $(x_i, \dots, x_{i-1+l_i})$ which matches a substring of y. So, $L(x, x) = \frac{m+1}{2}$.
The **ACS-distance** is defined by
$$\frac{1}{2}\left(\frac{\log(n)}{L(x, y)} - \frac{\log(m)}{L(x, x)} + \frac{\log(m)}{L(y, x)} - \frac{\log(n)}{L(y, y)}\right).$$

A similar distance was considered (Haubold et al., 2009) replacing the longest common substring by the shortest absent one.

23.3 Distances in Ecology, Biogeography, Ethology

Main distance-related notions in Ecology, Biogeography and Animal Behavior follow.

- **Niche overlap similarities**
  A *niche* is environmental space, while a *biotope* is a geographic space.
  Let $p(x) = (p_1(x), \ldots, p_n(x))$ be a *frequency vector* (i.e., all $p_i(x) \geq 0$ and $\sum_i p_i(x) = 1$) representing an ecological niche of species x, for instance, the proportion of resource i, $i \in \{1, \ldots, n\}$, used by species x.
  Four main **niche overlap similarities** of species x and y are:
  Schoener's D, introduced by Renkonen in 1938:

$$D(x, y) = 1 - \frac{1}{2}\sum_{i=1}^{n} |p_i(x) - p_i(y)|;$$

  cosine similarity (cf. Chap. 17), called in Ecology (from 1973) *Pianka's O*:

$$O(x, y) = \frac{\langle p(x), p(y)\rangle}{||p(x)||_2 \cdot ||p(y)||_2};$$

  Hellinger I (i.e., **fidelity similarity**, cf. Chap. 14) and **Bray–Curtis** (or, since $p(x), p(y)$ are frequency vectors, *Renkonen percentage*) **similarity** (Chap. 17).
- **Ecological distance**
  Let a given species be distributed in subpopulations over a given *landscape*, i.e., a textured mosaic of *patches* (homogeneous areas of land use, such as fields, lakes, forest) and linear *frontiers* (river shores, hedges and road sides). The individuals move across the landscape, preferentially by frontiers, until they reach a different subpopulation or they exceed a maximum **dispersal distance**.
  The **ecological distance** between two subpopulations (patches) x and y is defined (Vuilleumier–Fontanillas, 2007) by

$$\frac{D(x, y) + D(y, x)}{2},$$

  where $D(x, y)$ is the distance an individual covers to reach patch y from patch x, averaged over all successful dispersers from x to y. If no such dispersers exist, $D(x, y)$ is defined as $\min_z(D(x, z) + D(z, x))$.
  Ecotopes are the smallest ecologically distinct features in a landscape mapping.

Effective ecological distance (or *cost-distance*) is the Euclidean distance modified for the effect of landscape and behaviour on the dispersal of an organism between locations in the landscape. Such functional distance can be computed as *least-cost path* using either *cost surface* (matrix between patches assigning degree of permeability depending on hostile habitat or physical barriers), or *resistance surface* accounting for costs (resistance per unit distance) of passing through different landscape elements. Cf. **resistance distance** in Chap. 15. Pinto–Kein, 2009, proposed *least-cost corridors* formed by multiple paths with similar costs, since animals, even birds, rarely move along straight-line paths in a landscape.

- **Biotope distance**
  The *biotopes* here are represented as binary sequences $x = (x_1, \dots, x_n)$, where $x_i = 1$ means the presence of the species i. The **biotope** (or **Tanimoto, Jaccard**; cf. Chap. 17) **distance** between biotopes x and y is defined by

$$\frac{|\{1 \le i \le n : x_i \ne y_i\}|}{|\{1 \le i \le n : x_i + y_i > 0\}|} = \frac{|X \Delta Y|}{|X \cup Y|},$$

  where $X = \{1 \le i \le n : x_i = 1\}$ and $Y = \{1 \le i \le n : y_i = 1\}$.
- **Prototype distance**
  Given a finite metric space (X, d) (usually, a Euclidean space) and a selected, as typical by some criterion, vertex $x_0 \in X$, called the *prototype*, the **prototype distance** of every $x \in X$ is the number $d(x, x_0)$.
  Usually, the elements of X represent phenotypes or morphological traits. The average of $d(x, x_0)$ over $x \in X$ estimates the corresponding *variability*.
- **Critical domain size**
  In Spatial Ecology, the **critical domain size** is (Kierstead and Slobodkin, 1953) the minimal amount of habitat, surrounded by a hostile matrix, required for a population to persist. For example, in the invasion and persistence of algal and insect populations in rivers, such a size is the minimal length of a river (with a given, less than the threshold, flow speed) that supports a population.
- **Island distance effect**
  An *island*, in Biogeography, is any area of habitat surrounded by areas unsuitable for the species on the island: true islands surrounded by ocean, mountains isolated by surrounding lowlands, lakes surrounded by dry land, isolated springs in the desert, grassland or forest fragments surrounded by human-altered landscapes.
  The **island distance effect** is that the number of species found on an island is smaller when the degree of isolation (distance to nearest neighbor and mainland) is larger. Also, organisms with high dispersal, such as plants and birds, are much more common on islands than are poorly dispersing taxa like mammals.
- **Dispersal distance**
  In Biology, the **dispersal distance** is a **range distance** to which a species maintains or expands the distribution of a population. It refers, say, to seed dispersal by pollination and to natal, breeding and migration dispersal. For animals, *natal*

dispersal is permanent emigration from the natal range to a disjoint adult range, and **dispersal distance** is the distance between their barycenters.
When *outcrossing* (gene flow) is used to increase genetic diversity of a plant species, the *optimal outcrossing distance* is the dispersal distance at which seed production is maximized. It is less than the mean pollen dispersal distance.
Plant height matters more than seed mass for its dispersal distance. Unusual way of wind dispersal include tumbleweeds.
Pollen from *Pinus sylvestris* can fly 100 km, but oceanic larvae dispersal is at least one order of magnitude greater than that of pollen-dispersing terrestrial biotas.

- **Long-distance dispersal**
  Long-distance dispersal (or *LDD*) refers to the rare events of biological dispersal on distances an order of magnitude greater than the median **dispersal distance**. For the regional survival of some plants, LDD is more important than local (median-distance) dispersal. The longest recorded distance traveled by a drift seed is 28,000 km by a Mary's bean from the Marshall Islands to Norway.
  LDD emerged in Biogeography as greater factor of biodiversity and species migration patterns than original *vicarience theory* (dispersal via land bridges) based on continental drift. Such relatively recent chance dispersal explain the fast spread of organisms in new habitats, for example, plant pathogens, invasive species and in paleocolonization events, such as the joining of North and South America 3 Ma ago, or Africa and India with Eurasia 30 and 50 Ma ago.
  Human colonization of Madagascar (isolated for 88 Ma) $\approx$2,000 years ago may have resulted from an accidental transoceanic crossing; other animals arrived by rafting from Africa 60–70 Ma ago. LDD followed traders and explorers, especially, in Columbian Exchange after 1492.
  Transoceanic LDD by wind/water currents can explain strong floristic similarities among landmasses in the southern hemisphere. Monkey, rodents, and crocodiles dispersed 50–30 Ma ago to the Americas from Africa via the Atlantic. New fossil primates found in 2012 suggest that anthropoid ancestors originated in Asia and then 40 Ma ago or earlier rafted across the ocean to Africa.
  Free-living microbes occupy every niche but their biodiversity is low, because they are carried by wind thousands of km on dust particles protecting them from UV. Extreme example of such (or via underground rivers, before the continents split) LDD: sunlight-independent bacterium *Desulforudis audaxviator*, living 1–3.3 km deep in South Africa (the only species known to be alone in its ecosystem and radiation-relying), reached deep boreholes in eastern California.
  Some other LDD vehicles are: rafting by water (corals can traverse 40,000 km during their lifetime), migrating birds, human transport, ship ballast water, and extreme climatic events. Snails can travel hundreds of km inside bird guts: 1–10 % of eaten snails survive up to 5 h until being ejected in bird feces.
  Also, cancer invasion (spread from primary tumors invading new tissues) can be thought as an invasive species spread via LDD, followed by localized dispersal.

The most invasive mammal species (besides humans) are: rabbits, black rats, gray squirrels, goats, pigs, deers, mice, cats, red foxes, mongooses. Invasive *Argentine ants* form the largest global insect mega-colony: they do not attack each other.

- **Migration distance (in Biogeography)**

  Migration distance is the distance between regular breeding and foraging areas within seasonal large-scale return movement of birds, fish, insects, turtles, seals, etc.

  The longest such recorded round-trip is 71,000 km pole-to-pole traveled each year by the Arctic tern. The highest migration altitude is 9 km by bar-headed goose. Longest each way migration for a mammal is $\approx$9,800 km, traveled by a humpback whale from the Brazilian coast to Madagascar, and, for an insect, $\approx$4,500 km by desert locust and Monarch butterfly. One of unsolved problems in Biology is: how do the descendants of Monarch butterfly, migrating from Canada to central Mexico for several generations, manage to return to a few small overwintering spots?

  Migration differs from *ranging*, i.e., the movement of an animal beyond its home range which ceases when a suitable new home range (a resource: food, mates, shelter) is found. It differs also from foraging/commuting as occurs, say, for albatrosses or plankton. Wandering albatrosses, having the largest (3.63 m) wingspan, make several-days foraging round trips of up to 3,000 km. Krill, 1–2 cm long, move up to 500 m vertically each night, to feed in the sunlit waters, where plants are abundant, while avoiding being seen by predators. *Mesopelagic* (living 0.2–1 km deep) fish also travel to upper layers at night.

  At the population level, migration involves displacement, while ranging/foraging result only in mixing. Entire species migrate slowly by shifting, because of rapid climate change, their geographical or elevation ranges. Root et al., 2003, claim that butterflies, birds and plants move towards the poles by 6.1 km per decade over the past 100 years. Estimated global mean velocities of change for mean annual temperature and rainfall from 2000 to 2100 are 420 and 220 m per year. During ice ages species move to hotspots, say, volcanoes.

- **Daily distance traveled**

  Daily distance traveled D (m/day) is an important parameter of the energy budget of ranging/foraging mammals.

  The *foraging efficiency* is the ratio $\frac{B}{C}$, where C, B (J/m) are the energy costs of travel and of acquiring energy. Over a day, the expected total net energy return is $D(B - C)$. The *locomotor cost* is the distance traveled per unit energy spent on locomotion. The limb length determines this cost in terrestrial animals but no link with D has been observed. Pontzer, 2011, explains this paradox by high $\frac{B}{C}$ in most taxa: only for $\frac{B}{C} < 10$, would selection for limb length be needed.

  Within species, over a lifetime, increased D is associated with decreased $B - C$, reproductive effort and maintenance. But among species, over evolutionary time, it is associated with a greater number of offspring and their total mass per lifetime.

The mean D traveled by carnivores is four times such distance by herbivores. Also, D and feeding/grooming time are much greater in larger groups of primates. Foraging radius, D and annual travel distance of Neanderthal was $\approx$75 % of that of humans.

- **Collective motion of organisms**

  Organisms aggregate to procure resources (pack-hunting), to find mates (plankton, plants) and to lower predation risk (meerkats, schools of sardines, flocks of starlings). Animals moving in large groups at the same speed and in the same direction, tend to have similar size and to be regularly spaced.

  The near-constant distance which an animal maintains from its immediate neighbors is called the **nearest-neighbor distance** (NDD). When NDD decreases, the mode of movement can change: marching locusts align, ants build bridges, etc.

  Moving in file when perception is limited to one individual (ants, caterpillars in processions up to 300, spiny lobsters in parallel chains of 3–30), animals use tactile cues or just perceive and follow the choice of the preceding individual, such as sheep in mountain path or cows in cattle-handling facilities. Penguins in the huddle move ("traveling wave", like the stop-and-go of cars in a traffic jam) trigger movements in their neighbors as soon as the threshold distance ($\approx$2 cm, i.e., twice the thickness of their compressive feather layer) is formed between two penguins.

  The greatest recorded group of moving animals was a swarm in US, 1875, by 12.5 trillion insects (Rocky Mountain locust, extinct by now) covering 510,000 km^2. A swarm by extant desert locusts in Kenya, 1954, covered 200 km^2. Flights of migratory pest insects occur usually at altitudes up to 1 km, and are downwind; they last for a few hours with displacement up to 400 km. Flocks of *red-billed Quelea* (the most abundant wild bird species) take up to 5 h to fly past. Herring schools occupy up to 4.8 km^3 with density 0.5–1.0 fish per m^3.

  Schools of sardines, anchovy and krill, despite being variable in size, share a ratio $\frac{S}{V} = 3.3\,m^{-1}$ of surface area to volume; it has been interpreted as the optimal trade-off between predator avoidance and resource acquisition.

  The spatiotemporal movement patterns, emerging from such groups, result from interactions between individuals. This local mechanism can be *allelomimesis* ("do what your neighbor does"), social attraction (say, to the center of mass of neighbors), or the threat of cannibalism from behind (in running Mormon crickets), mass mate-searching (in burrow-dwelling crabs). Vicsek, 1995, modelled a swarm as a collection of particles moving with a constant speed but adopting, if perturbation, at each time increment the average motion's direction of the neighbors.

  Migrating birds tend to fly in a V, J, or W shaped formation. In energy-saving V-formation (or *skein*), they sync their flapping to hook the former bird's updraft. The birds flying at the front and the tips are rotated.

  Most spectacular are aerial displays of flocks of starlings highly variable in shape. Scale-free behavioral correlation was observed: regardless of flock size, the correlations of a bird's orientation and velocity with the other birds did not vary and was near-instantaneous. Cf. *SOC* in **scale invariance** (Chap. 18).

Silverberg et al., 2013, discovered self-organized emergent behavior in *moshing* (when 100–100,000 fans at heavy metal concert form circles and then run together with abandon, bouncing off one another). In fact, the speed distribution of people closely matches that of molecules in a 2D gas at equilibrium and moshing corresponds to domination of the model's parameters by noise.
Such emerging, when their number increases, collective behavior can be seen as a *critical phase transition*; it was observed also for simple automatons.
Besides animals, collective directed motion occurs also in cellular populations. Some aggregated bacterial populations (say, foraging swarms of billions of *Paenibacillus vortex*) can migrate rapidly and coordinately over a surface. A *grex* is an slug-like aggregate 2–4 mm long of up to 100,000 amoebas formed when they are under stress. It moves as a unit, only forward, 1 mm/h.
In a multicellular organism, collective cell migration occurs (usually by *chemotaxis*: response to chemical concentration) throughout embryonic development, wound healing (fibroplasts and epithelial cells), immune response (leukocytes), and cancerous tumor invasion. Similarly to migration of songbirds, cancerous cells prepare for metastatic travel by gathering proteins near their leading edges. During development, some cells migrate to very long distances. For example, newborn neurons in the adult brain can traverse $\frac{2}{3}$ of its length.

- **Distances in Animal Behavior**
The first such distance was derived by Hediger for zoos; his *interanimal distance* is the maximum species-specific distance at which conspecifics approach each other. In 1955, he defined *flight distance* (run boundary), *critical distance* (attack boundary), *personal distance* (at which members of noncontact species feel comfortable) and *social distance* (at which within-species groups tolerate each other).
The exact such distances are highly context dependent. An example: a tamer manipulate a semi-tamed lion moving in and out of its critical zone.
For humans, flight and critical distances have been, with few exceptions, eliminated. So, Hall adapted above space boundaries; cf. his **distances between people** in Chap. 28. The main **distances in Animal Behavior** follow.
The **individual distance**: the distance which an animal attempts to maintain between itself and other animals. It ranges between "proximity" and "far apart" (for example, ≤ 8 m and ≥ 61 m in elephant social calls). Bell et al., 2012, found that gaining and maintaining a preferred interanimal distance, accounts for much of the variability in dodging by rats and field crickets.
The **group distance**: the distance which a group of animals attempts to maintain between it and other groups. Cf. the **nearest-neighbor distance**.
The **alert distance**: the distance from the disturbance source (say, a predator or a dominating conspecific) when the animal changes its behavior (say, turns towards as *perception advertisement*) in response to an approaching threat.
The **flight initiation distance** (or FID, *escape distance*): the distance from the disturbing stimulus when escape begins. FID, corrected for the distance to refuge, is a measure of animal's boldness.

The **reaction distance**: the distance at which the animal reacts to the appearance of prey; *catching distance*: the distance at which the predator can strike a prey.
The **detection distance**: the maximal distance from the observer at which the individual or cluster of them is seen or heard. For example, it is 2,000 m for an eagle searching for displaying sage-grouse, 200 m for a male-searching female sage-grouse and 1,450 m for a sage-grouse scanning for a flying eagle.
The **social recognition distance**: the distance over which a contact call can be identified as belonging to a family.
In the main nonresource-based mating system, *lek mating*, females in estrous visit a congregation of displaying males, the *lek* and mate preferentially with males of higher **lekking distance rank**, i.e., relative distance from male territory (the median of his positions) to the center of the lek. High-ranking individuals have smaller, centrally located (so, less far to travel and more secure) home ranges.
The **sleeping distance** of a mating pair: for example, it is no further than 0.5 m in Arctic blue fox, but more than 2–2.5 m one month after copulation.
The **distance-to-shore**: the distance to the coastline used to study clustering of whale strandings (by distorted echo-location, anomalies of magnetic field, etc.).
The **gape distance**: the width of the widely opened mouth of a vertebrate.

- **Animal depth/distance perception**
Many animals, including humans, have two eyes with overlapping visual fields that use parallax (cf. **parallax distance** in Chap. 26) for depth perception and distance estimation. Some animals (for example, pigeons) use *motion parallax* in which they move head to gain different viewpoints. Another example: the velocity of the mantis's head movement is kept constant during peering. So, the distance to the target (prey) is inversely proportional to the velocity of the retinal image.
All animals have a binocular region (growing as eyes become more forward-facing) which allows for vision through the clutter, as long as the width of the objects causing clutter is less than the **interpupillary distance** d.
Changizi–Shimojo, 2008, suggested that the degree of binocular convergence is selected to maximize how much the animal can see. Most animals exist in non-cluttered environments or surroundings where the cluttering objects are bigger in size than d. They tend to have sideways-facing eyes allowing panoramic vision. But humans and other large mammals evolved in leafy environments like forests and their forward-facing eyes (and smaller distance d) maximize ability to see.
Larvae of the sunburst diving beetle (*Thermonectus marmoratus*) have 6 pairs of eyes. Four eyes of two frontal pairs (used to scan potential prey) have bifocal lenses and at least two 1D-retinas: distant and close-up. The two focused images produced by the lens sit at different distances and vertically separated.
- **Distance-related animal settings**
Spatial fidelity zones specific to individuals (say, at a given distance from a colony center, or within a particular zone of the total foraging area) have been observed for some social insect species, molluscan communities, birds, etc.
Home range is the area where an animal (or a group) lives and travels within. Within it, the area of intensive and exclusive use by resident animals is the *core*

area. The *distance between range centroids* of two individuals (or groups) is a parameter used in studies of spatially based animal social systems. Cf. **dispersal distance**.

An animal is **territorial** if it consistently occupies, marks and defends a *territory*, i.e., an area with a focused resource (say, a nest, den, mating site or sufficient food resources). Territories are held by an individual, a mated pair, or a group. An *extraterritorial foray* is the movement of a territorial animal into a conspecific's territory. *Dear enemy recognition* is the stronger response of a territorial animal to strangers than to its neighbors from adjacent territories.

The **defense region** is the region that a male must defend in a mating competition to monopolize a female. It can be 1D (burrow, tunnel), 2D (dry land), bounded 3D (tree, coral reef), or open 3D (air, water). Puts, 2010, claims that 1D and 2D (as for humans) mating environments favor the evolution of contests.

The reliability of a threat display in animal contests is maintained by the *proximity risk*, i.e., the display is credible only within a certain distance of the opponent. This threshold distance is related to weaponry and the species-specific fighting technique. Here, greater formidability and dominance can be reached solely behaviorally; for example, an elephant's musth status overrides its body size and tusks.

The **landscape of fear** of a foraging animal is defined by the spatial variation of presumed predation risk. Its horizontal and vertical components correspond to terrestrial and aerial predators. It include clearness of sightlines (to spot predators), shrubs/trees/edge cover and the interplay of the distances to food and shelter. For example, small fish stay close to the coral reef when grazing seaweed; this creates "grazing halos" of bare sand, visible from space, around all reefs. Similar natural features are "fairy rings" of green eelgrass (up to 1,500 m in width, off Denmark's coast), of mushrooms (10–600 m) and of barren sand (2–15 m, in Africa).

The **domain of danger** (DOD, or **Voronoi polygon**, cf. Chap. 20) of an animal, risking predation, in aggregation is the area closer to it than to any other group member. *Selfish herd theory* (Hamilton, 1971) posits that a cover-seeking dominant animal tends to minimize its DOD by occupying the center, thus reducing its risk by placing another individual between itself and a predator or parasite. Moreover, some fish bite a group member, when exposed to a searching predator.

During traveling, dominant animals are closer to the front of the herd. During foraging, their trajectories are shorter, more direct and more aligned both with their nearest neighbors and with the whole herd.

Distance senses include sight, hearing, and smell (they can be in stereo), while *contact senses* include taste, the senses of pressure, thermoception, and *internal senses* include the sense of balance and muscle stretch. The buzzard can see small rodents from a height of 4.6 km. The spotted hyena hears noises from predators feeding on carcasses over distances of up to 10 km. The silkmoth detects pheromones up to 11 km distant. The grizzly bear smells food from up to 29 km away.

An example of unexplained **distance prediction** by animals is given (Vannini et al., 2008) by snails *Cerithidea decollata* migrating up and down mangrove shores in synchrony with tidal phases. In the absence of visual cues and chemical marks, snails cluster just above the high water line, and the distance from the ground correlates better with the incoming tide level than with previous ones.

Ants initially wander randomly and upon finding food return to their colony while laying down pheromone trails. So, when one ant finds a **shortest path** to a food source, other (and eventually all) ants are likely to follow it. Inspired by this idea, the *ant colony optimization algorithm* (ACO) is a probabilistic technique for finding shortest paths through graphs; cf. **arc routing problems** in Chap. 15. Also, ants routinely find the maximal distance from all entrances to dispose of dead bodies.

The **distance effect avoidance** is the observed selection of some good distant source of interest over a poor but nearer one in the same direction. For example, females at a chorusing lek of anurians or arthropods may use the lower pitch of a bigger or better distant male's call to select it over a weaker but louder call nearby. High-quality males help them by timing their calls to precede or follow those of inferior males. Ant colonies can (Franks et al., 2007) select a good distant nest over a poorer one in the way, even when it is 9 times closer. Ants compensate for the distance effect by increasing recruitment latencies and quorum thresholds at nearby poor nests.

In land locomotion, animals crawl, walk, run, hop, climb or crawl, slither, undulate. In fluids (water, air) animals swim and fly by beating flagella, tails, wings, undulating their bodies, or actuating pumps. Some animals can switch the medium. Fish *Exocoetidae* can spend 45 s in flight gliding up to 200 m at altitudes of up to 6 m; using waves, it can span distances up to 400 m. Some squids fly in shoals covering up to 50 m at 6 m above the water. Squirrels *Petauristinae*, snakes *Chrysopelea* and lemurs *Dermoptera* can glide with small loss of height up to 200, 100 and 70 m, respectively. The deepest dive for a flying bird is 210 m by a thick-billed murre.

Flying and swimming animals can move through volumes with six degrees of freedom: three translational (left/right, forwards/backwards, up/down) and three rotational (pitch, roll, yaw). Surface-constrained animals have only three degrees: left/right, forwards/backwards and yaw; moving in 3D, they have higher place field resolution in the horizontal plane and showed a preference for movement in the horizontal. In terms the number of body lengths per second, a mite *Paratarsotomus macropalpis* is the fastest known running animal; cf. its 322 body lengths with 16 for the cheetah.

Navigating animals use an egocentric orientation mechanism and simple panoramic views, within which proximal objects dominate because their image on the retina change significantly with displacement. Animals rely on the spatial arrangement of the objects/landmarks across the scene rather than on their individual identification and geometric cues. Humans and, perhaps, chimpanzees and capuchin monkeys, possess, in addition, an allocentric reference system, centered on objects/features of the environment, and a more flexible geometric representation of space, with true distance and direction, i.e., closer to an abstract mental map.

Gaze monitoring and *pointing*: four great apes, canids and ravens follow another's head and eye orientation into distant space, even behind an obstacle. Moreover, bonobos and chimpanzees take barrier opacity into consideration. African elephants can use communicative intent of human pointing as a cue to find food. Horses can use their facial expressions (direction of eyes and ears) to "talk" to other horses.

Great apes, dolphins, elephants and magpies recognise themselves in mirrors. *Metacognition* (cognitive self-awareness) was found in great apes, dolphins and rhesus monkeys. A basic *Theory of Mind* (ability to attribute mental states), mental time travel, meta-tool use and empathy are expected in primates and corvides. Mammals, birds and octopuses possess neurological substrates generating consciousness. Chimpanzees are the only known non-human animals with a system of intentional communication. But shared intentionality and cumulative culture seems to be uniquely human.

- **Animal communication**

  Only humans, songbirds, hummingbirds, parrots, cetaceans and bats have complex, learned vocalisation. Conceptual generalizations (bottlenose dolphins can transmit up to 9 km identity information independent of the caller's voice/location), syntax (alarm calls of some monkeys and songs of Bengal finches are built as "word sequences") and meta-communication ("play face" and tail-to-the-right signals in dogs that the subsequent aggressive signal is a play) have been observed.

  Matters of relevance *at a distance* (a distant food source or shelter) can be communicated by body language. For example, honeybees dancing convey the polar coordinates (distance D to the goal and angle between the direction towards it and to that of the Sun's azimuth) of locations of interest. The mean number of waggings of bee's *waggle phases* increases with D. Also, wolves, before a hunt, howl to rally the pack, become tense and have their tails pointing straight. Dogs express their spatial needs by body language and vocalizations. Stiffness, pilorection, aggressive barking/lunging are *distance-increasing*, while play bow, tail wagging to the right, "positive" barking/lunging are *distance-decreasing signals*.

  A **distance pheromone** is a soluble (for example, in the urine) and/or evaporable substance emitted by an animal, as a chemosensory cue, in order to send a message (on alarm, sex, food trail, recognition, etc.) to other members of the same species. In contrast, a *contact pheromone* is such an insoluble nonevaporable substance; it coats the animal's body and is a contact cue. The *action radius* of a distance pheromone is its attraction (or repulsion) range, the maximum distance over which animals can be shown to direct their movement to (or from) a source. In species, such as carnivores occurring at low densities or having large home ranges, individuals are widely spaced and communicate via chemical broadcast signaling at *latrines*, i.e., collections of scent marks (feces, urine or glandular secretions), or via visually conspicuous landmarks of the boundary such as scratches and middens. Herrings communicate by farting. Shelter-dwelling caterpillars ballistically eject faecal pellets great distances (7–39 times their body length) at great speeds, in order to remove olfactory chemical cues for natural enemies.

The **communication distance** is the maximal distance at which the receiver can still get the signal. Animals can vary the signal amplitude and visual display with *receiver distance* in order to ensure signal transmission.

For example, baleen whales have been observed calling more loudly to each other in order to compensate for human-generated noise in modern oceans.

Another example of *distance-dependent communication* is the protective coloration of some aposematic animals: it switches from *conspicuousness* (signaling nonedibility) to *crypsis* (camouflage) with increasing distance from a predator.

Examples of interspecies communication of nonhuman animals, other than predator–prey signaling, are: eavesdropping, heterospecific alarm calls and cooperative hunting.

The main modes of animal communication are infrasound (<20 Hz), sound, ultrasound (>20 kHz), vision (light), chemical (odor), tactile, seismic and electrical. Infrasound, low-pitched sound (as territorial calls) and light in air can be long-distance. Some frogs, spiders, insects, small mammals have vibrotactile sense.

A blue whale infrasound could (prior to noise pollution caused by ships) travel over 6,000 km through the ocean water using the **SOFAR channel** (Chap. 25).

Most elephant communication is in the form of infrasonic rumbles which may be heard by conspecifics 5–10 km away. Also, they drum their soles on the ground, and resulting seismic waves can be detected as far as 16–32 km.

Many animals hear infrasound generated by earthquakes, tsunami and hurricanes before they strike. Elephants can hear storms 160–240 km away.

High-frequency sounds attenuate more rapidly with distance, more directional and vulnerable to scattering. But ultrasounds are used by bats (echo-location) and arthropods. Rodents use them to communicate to nearby receivers without alerting predators and competitors. Some anurans shift to ultrasound signals in the presence of continuous background noise (such as waterfall, human traffic).

Animals, including frogs, insects, birds and whales, increase the minimum frequency, amplitude or **signal-to-noise ratio** (Chap. 21) in the presence of antropogenic noise.

- **Plant long-distance communication**

  Long-distance signaling was observed from roots and mature leaves, exposed to an environmental stress, to newly developing leaves of a higher plant.

  This communication is done cell-to-cell through the plant vascular transpiration system. In this system, macromolecules (except for water, ions and hormones) carry nutrients and signals, via *phloem* and *xylem* tissues, only in one direction: from lower mature regions to shoots. The identity of long-distance signals in plants is still unknown but the existence of information macromolecules is expected. Large-scale RNA-based communication between a parasitic plant and its host was found.

Besides the above vascular signaling, plants communicate chemically with each other or with mutualistic animals (pollinators, bodyguards, etc.). For example, plants respond to attack by herbivores or pathogens with the release of volatile compounds, informing neighboring plants and attracting predators of attackers. Some 80 % of plants are colonized by ectosymbiotic fungi that form a network of fine white threads, *mycorrhizae*, which take in water and minerals from the soil, and hand some over to the plant in exchange for nutrients. A mycorrhizal network can take over an entire forest and tie together plants of different species. Plants use this network as a signaling and kin (or host) detection system too. They assist neighbors or kin in deterring pests, attracting pollinators and nutrient uptake.

- **Internodal distance**
  A *node* on a plant stem is a joint where a leaf is attached. The **internodal distance** (or *internode length*) is the distance between two consecutive nodes.
  A *ramet* is an independent member of a clone. The **interramet distance** (or *propagule dispersal distance*) is the internodal distance in plant clonal species.
- **Insecticide distance effect**
  The main means of pest (termites, ants, etc.) control are chemical liquid insecticides and repellents. The efficiency of an insecticide can be measured by its **all dead distance**, i.e., the maximum distance from the standard toxicant source within which no targeted insects are found alive after a fixed period.
  The **insecticide distance effect** is that the toxicant is spread through the colony because insects groom and feed each other. The toxicant should act slowly in order to maximize this effect and minimize secondary repellency created by the presence of dying, dead and decaying insects. Nearly all animals, when they die, emit the same stench of fatty acids which acts as repellent and it is universal.
- **Body size rules**
  Body size, measured as mass or length, is one of the most important traits of an organism. Food webs, describing "who eat whom" (cf. **trophic distance**), are *nearly interval*, i.e., the species can be ordered so that almost all the resources of each consumer are adjacent in the order. Zook et al., 2011, found that ordering by body size is the best proxy to produce this near-interval ordering.
  The lower limit (10 kg and 2 g) to body size is set by the size of offspring for marine and by energetic limitations for terrestrial mammals. The largest known sizes for them are 190 and 16 t, but the upper limit is still unclear,
  According to Payne et al., 2008, the maximum size of the Earth's organisms increased by 16 orders of magnitude over the last 3.5 billion years. Seventy-five percent of the increase happened in two great leaps (about 1,900 and 600–400 Ma ago: the appearance of eukaryotic cells and multi-cellularity) due to leaps in the oxygen level, and each time it increased about million times.
  Smith et al., 2010: the maximum size of mammals increased (from 2 g to 190 t) near-exponentially after the C–T (Cretaceous–Paleogene) extinction of the nonavian dinosaurs 65.5 Ma ago; on each continent, it leveled off within 25 Ma. Kurbel, 2013, claims that after this C–T event, homeothermic animals (mammals and birds) radiated globally from northern Asia and became dominant.

The maximum size of insects also followed O_2 level 350 – 150 Ma ago, reaching 71 cm. Then it dropped (while O_2 went up) with evolution of birds and 65 Ma ago with their specialization and evolution of bats. Larsson–Dececchi, 2013, explain the origin of birds by a change of body-to-limb length ratio in Maniraptoran dinosaurs: the hind legs shrank, while for elimbs got long enough to work as an airfoil. From 230–220 to 163 Ma ago, theropods shrinked ($\approx$0.5% of mass) to first birds.

Evans et al., 2012, claim that an increase in size (100, 1000, 5000 times) of land and marine mammals took 1.6, 5.1, 10 and 1, 1.3, 5 million generations, respectively. Mouse-sized mammals evolved into elephant-sized ones during 24 million generations, but decreasing in size occurred about 30 times faster.

Clauset and Erwin, 2008: 60 Ma of mammalian body size evolution can be explained by simple diffusion model of a trade-off between the short-term selective advantages (**Cope's rule**, common among mammals: a slight within-lineage drift toward larger masses) and long-term selective risks of increased size. The size has costs as well as benefits; for example, reversals to unicellularity occurred at least five times in cyanobacteria. It favors the individual but renders the clade more susceptible to extinction via, for example, dietary specialization. Large size enhances reproductive success, the ability to avoid predators and capture prey, and improves thermal efficiency. In large carnivores, bigger species dominate better over smaller competitors. Predator–prey mass ratio is typically around 10. But, for example, cookiecutter shark, only 0.5–1 m in length, preys on all larger animals in ocean, and the larvae of beetle *Epomis* preys on amphibians. By mean body size (67 kg now and 50 kg in the Stone Age) humans are a small *megafauna* ($\geq$44 kg) species. A rapid average decline of $\approx$20 % in size-related traits was observed in human-harvested species. One of main human effects on nature is the decline of the apex consumers (top predators and large plant eaters).

Given below are the other main rules of large-scale Ecology involving body size.

Foster's (or *island*) **rule** is a principle that members of a species get smaller or bigger depending on the resources available in the environment. Damuth, 1993: there is an optimum mammal body size $\approx$1 kg for energy acquisition, and so island species should, in the absence of the usual competitors and predators, evolve to it.

Insular dwarfism is an evolutionary trend of the reduction in size of large mammals when their gene pool is limited to a very small environment (say, islands). One explanation is that food decline activates processes where only the smaller of the animals survive since they need fewer resources and reproduce faster.

Island gigantism is a form of natural selection where the size of animals isolated on an island increases dramatically over generations due the removal of constraints.

Abyssal gigantism is a tendency of deep-sea species to be larger than their shallow-water counterparts. For example, the colossal squid and the king-of-herrings (giant oarfish) can reach 14 and 17 m in length. It can be adaptation for scarcer food resources (delaying sexual maturity results in greater size), greater pressure and lower temperature.

The ratio $\frac{S}{V}$ (surface area to volume) is the main compactness measure for 3D shapes in Biology. Higher $\frac{S}{V}$ permits smaller cells to gather nutrients and reproduce very rapidly. Also, smaller animals in hot and dry climates lose heat better through the skin and cool the body. But lower $\frac{S}{V}$ (and so, larger size) improves temperature control: slower heat loss or gain. **Bergmann's rule** is a principle that, within a species, the body size increases with colder climate. For example, Northern Europeans on average are taller than Southern ones.

Also, 1 °C of warming reduces the adult body mass of cold-blooded organisms by 2.5 % on average. For warm-blooded animals, **Allen's rule** holds: those from colder climates have shorter limbs than the equivalent ones from warmer climates.

Rensch's rule is that males are the larger sex in big-bodied species (such as humans) and the smaller sex in small-bodied species (such as spiders). It holds for plants also. Often, natural selection on females to maximize fecundity results in female-biased sexual size dimorphism, whereas sexual selection for large males promotes male-biased dimorphism. The males in some cichlid fish are up to 60 times larger than that of the females, while tremoctopus females may reach 2 m versus the males, at most a few cm long.

Size-assortative mating (positive correlation between male and female size among couples) has been found in crustaceans, insects, birds, reptiles, fishes and humans, for which it is a part of *homophily* (tendency to associate and bond with similar others). Humans have by far the largest, among apes, penises and breasts.

An **allometric law** is a relation between the size of an organism and the size of any of its parts or attributes; say, eye, brain and body sizes are closely correlated in vertebrates. Examples of related power laws are, in terms of animal's body mass M (or, assuming constant density of biomass, of body size) are proportionalities of metabolic rate to $M^{0.75}$ (**Kleiber's law**) and of life span to $M^{0.25}$. Niklas–Enquist, 2001, proposed length-biomass scaling to $M^{0.25}$ for primary producers. Muller et al., 2013, claim that animal's dry matter intake (in kg per day) is $0.026M^{0.885}$.

A cellular organism (for example, bacteria) of linear size S has, roughly, internal metabolic activity proportional to cell volume (so, to S^3) and flux of nutrient and energy dissipation proportional to cell envelope area (so, to S^2). Hence, this size S is close to their ratio. For viral particles, there is no metabolism, and their size is, roughly, proportional to the 3rd root of the genome size.

Cognitive and behavioral capacities do not correlate either with body or brain size, nor with their ratio, which is, say, $\frac{1}{7}, \frac{1}{40}, \frac{1}{2496}$ for small (0.06 mg) ant, human and shark. The *encephalization quotient* is the ratio of actual to predicted brain mass for a given size animal; it is the record 7.4–7.8 for humans. The number of neurons is $302, 85 \times 10^9, 2 \times 10^{11}$ in a nematode, human and elephant. Fish with smaller brain have more offspring. Echinoderms (say, starfish) lack a brain entirely.

Bromage et al., 2012, found a strong correlation between body mass and *RI* (*repeat interval*), i.e., the number of days between adjacent *striae of Retzius* in primate's enamel. RI is also represented by the *lamellae* (increments in bone). RI is an integer within $[1, 11]$; the mean RI is 8–9 in humans. RI (>1) also correlates with all metabolic rates and common life history traits except estrous cyclicity.

- **Size spectrum**
  The term **size spectrum** is used generally when comparing objects of a given class, say, shoes or phones. But mainly, it is (Sheldon–Parsons, 1967) the relationship between body size of individuals and their abundance or biomass, regardless of their species, in a given (aquatic or soil) **size-based food web**.
  For a population, the main considered sizes (lengths or masses) are: maximal, *asymptotic* (which individuals would reach if they were to grow indefnitely), of *maximal yield* (with highest biomass) and average in maturity. Example of corresponding *size-spectrum models*: Andersen and Beyer, 2006, derived proportionality of the number of individuals of given species and size to their asymptotic size raised to the power -2.05.
- **Trophic distance**
  Given an ecosystem, its *ecological network* is a digraph in which species are (biomass- or abundance-weighted) vertices with two of them being connected by arc or edge if there is a trophic or, respectively, symbiotic interaction. A *community food web* (or *ecological pyramid*) is a such digraph with only trophic arcs.
  The **trophic distance** from resource u to consumer v is the length of a shortest *food chain* (directed $(u - v)$ path) if it exists,
  The *trophic level* of a vertex v is 1 if it is a primary resource (usually, *producer* as plants, algae, phytoplankton) and 1 plus the trophic level of its principal diet, otherwise. The *fractional tropic level* of v is (Pauly–Palomares, 2005) 1 plus the weighted average (using stomach contents) trophic level of all its food items.
  The *mean trophic level* for fishery overall catch should be preserved to avoid *fishing down the food web*, when fisheries in a given ecosystem deplete the large predatory fish and end up with small fish and invertebrates.
  In a **size-based food web**, the layers are defined by body-size class rather than by trophic level. Community-based *predator–prey body mass ratios* (PPMR) and *transfer efficiency* (TE) are key parameters in such webs. In marine food webs, typically, PPMR$\in [100, 3000]$ and TE$\in [0.1, 0.13]$, i.e., 10–13 % of prey biomass is converted into predator production.
  An *energy* and *functional food webs* are weighted digraphs where arcs correspond to energy flow and interaction strength. Consumers at each level convert to tissue about 10 % of their food's chemical energy.

23.4 Other Biological Distances

Here we collect the main examples of other notions of distance and distance-related models used in Biology.

- **Immunologic distance**
  An *antigen* (or immunogen, pathogen) is any molecule eliciting an immune response. Once it gets into the body, the immune system either neutralizes its

pathogenic effect or destroys the infected cells. The most important cells in this response are white blood cells: *T-cells* and *B-cells* responsible for the production and secretion of *antibodies* (specific proteins that bind to the antigen).
When an antibody strongly matches an antigen, the corresponding B-cell is stimulated to divide, produce clones of itself that then produce more antibodies, and then differentiate into a plasma or memory cell. A secreted antibody binds to an antigen, and antigen–antibody complexes are removed.
A mammal (usually a rabbit) when injected with an antigen will produce immunoglobulins (antibodies) specific for this antigen. Then *antiserum* (blood serum containing antibodies) is purified from the mammal's serum. The produced antiserum is used to pass on passive immunity to many diseases.
Immunological distance procedures (*immunodiffusion* and, the mainly used now, *micro-complement fixation*) measure the relative strengths of the immunological responses to antigens from different taxa. This strength is dependent upon the similarity of the proteins, and the dissimilarity of the proteins is related to the evolutionary distance between the taxa concerned.
The *index of dissimilarity* $id(x, y)$ between two taxa x and y is the factor $\frac{r(x,x)}{r(x,y)}$ by which the *heterologous* (reacting with an antibody not induced by it) antigen concentration must be raised to produce a reaction as strong as that to the *homologous* (reacting with its specific antibody) antigen.
The **immunological distance** between two taxa is given by

$$100(\log id(x, y) + \log id(y, x)).$$

It can be 0 for two closely related species. It is not symmetric in general.
Earlier immunodiffusion procedures compared the amount of precipitate when heterologous bloods were added in similar amounts as homologous ones, or compared with the highest dilution giving a positive reaction.
The name of the applied antigen (target protein) can be used to specify immunological distance, say, albumin, transferring lysozyme distances. Proponents of the *molecular clock hypothesis* estimate that one unit of albumin distance between two taxa corresponds to $\approx$0.54 Ma of their divergence time, and that one unit of **Nei standard genetic distance** corresponds to 18–20 Ma.
Adams and Boots, 2006, call the *immunological distance* between two immunologically similar pathogen strains (actually, serotypes of dengue virus) their *cross-immunity*, i.e., 1 minus the probability that primary infection with one strain prevents secondary infection with the other. Lee and Chen, 2004, define the *antigenic distance* between two influenza viruses to be the reciprocal of their *antigenic relatedness* which is (presented as a percentage) the geometric mean $\sqrt{\frac{r(x,y)}{r(x,x)}\frac{r(y,x)}{r(y,y)}}$ of two ratios between the heterologous and homologous antibody *titers*.
An antiserum *titer* is a measurement of concentration of antibodies found in a serum. Titers are expressed in their highest positive dilution.

- **Metabolic distance**
Enzymes are proteins that *catalyze* (increase the rates of) chemical reactions.
The **metabolic distance** (or *pathway distance*) between enzymes is the minimum number of metabolic steps separating two enzymes in the metabolic pathways.
- **Pharmacological distance**
The *protein kinases* are enzymes which transmit signals and control cells using transfer of *phosphate groups* from high-energy donor molecules to specific target proteins. So, many drug molecules (against cancer, inflammation, etc.) are kinase inhibitors (blockers). Designed drugs should be *specific* (say, not to bind to $\geq 95\,\%$ of other proteins), in order to avoid toxic side-effects.
Given a set $\{a_1, \dots, a_n\}$ of drugs in use, the *affinity vector* of kinase x is defined as $(-\ln B_1(x), \dots, -\ln B_n(x))$, where $B_i(x)$ is the *binding constant* for the reaction of x with drug a_i, and $B_i(x) = 1$ if no interaction was observed. The binding constants are the average of several experiments where the concentration of binding kinase is measured at equilibrium. The **pharmacological distance** (Fabian et al., 2005) between kinases x and y is the Euclidean distance $(\sum_{i=1}^{n}(\ln B_i(x) - \ln B_i(y))^2)^{\frac{1}{2}}$ between their affinity vectors.
The *secondary structure* of a protein is given by the hydrogen bonds between its residues. A *dehydron* in a solvable protein is a hydrogen bond which is solvent-accessible. The *dehydron matrix* of kinase x with residue-set $\{R_1, \dots, R_m\}$ is the $m \times m$ matrix $((D_{ij}(x)))$, where $D_{ij}(x)$ is 1 if residues R_i and R_j are paired by a dehydron, and is 0, otherwise. The **packing distance** (Maddipati–Fernándes, 2006) between kinases x and y is the Hamming distance $\sum_{1 \leq i,j \leq m} |D_{ij}(x) - D_{ij}(y)|$ between their dehydron matrices; cf. **base pair distance** among **RNA structural distances**. The *environmental distance* (Chen, Zhang and Fernándes, 2007) between kinases is a normalized variation of their packing distance.
Besides hydrogen bonding, residues in protein helices adopt backbone dihedral angles. So, the secondary structure of a protein much depends on its sequence of dihedral angles defining the backbone. Wang and Zheng, 2007, presented a variation of **Lempel–Ziv distance** between two such sequences.
- **Global distance test**
The *secondary structures* of proteins are mainly composed of the α-helices, β-sheets and loops. Protein *tertiary structure* refers to the 3D structure of a single protein molecule. The α and β structures are folded into a compact globule.
The **global distance test** (GDT) is a measure of similarity between two (model and experimental) proteins x and y with identical *primary structures* (amino acid sequences) but different tertiary structures. GDT is calculated as the largest set of amino acid residues' α carbon atoms in x falling within a defined **cutoff distance** (cf. Chap. 29) d_0 of their position y.
For proteins, in order for this set to define all intermolecular stabilizing (relevant short range) interactions, $d_0 = 0.5$ nm is usually sufficient. Sometimes, $d_0 = 0.6$ nm, in order to include contacts acting through another atom.

- **Migration distance (in Biomotility)**
  The **migration** (or *penetration*) **distance**, in cattle reproduction and human infertility diagnosis, is the distance in mm traveled by the vanguard spermatozoon during sperm displacement *in vitro* through a capillary tube filled with homologous cervical mucus or a gel mimicking it. Sperm swim 1–4 mm/min. 90 % of human sperm swim forward with small side-to-side movements, while $\approx$5 % swim in a faster-paced helical pattern and the remaining $\approx$5 % swim in a hyper-helical manner, where the sperm are more active but less directional.
  Such measurements, under different specifications (duration, temperature, etc.) of incubation, estimate the ability of spermatozoa to colonize the oviduct *in vivo*.
  In general, the term **migration distance** is used in biological measurements of directional motility using controlled migration; for example, determining the molecular weight of an unknown protein via its migration distance through a gel, or comparing the migration distance of mast cells in different peptide media.
- **Penetration distance**
  The **penetration distance** is a general term used in (especially, biological) measurements for the distance from the given surface to the point where the concentration of the penetrating substance (say, a drug) in the medium (say, a tissue) had dropped to the given level. Several examples follow.
  During penetration of a macromolecular drug into the tumor interstitium, *tumor interstitial penetration* is the distance that the drug carrier moved away from the source at a vascular surface; it is measured in 3D to the nearest vascular surface.
  During the intraperitoneal delivery of cisplatin and heat to tumor metastases in tissues adjacent to the peritoneal cavity, the *penetration distance* is the depth to which the drug diffuses directly from the cavity into tissues. Specifically, it is the distance beyond which such delivery is not preferable to intravenous delivery.
  It can be the distance from the cavity surface into the tissues within which drug concentration is, for example, (a) greater, at a given time point, than that in control cells distant from the cavity, or (b) is much higher than in equivalent intravenous delivery, or (c) has a first peak approaching its plateau value within 1 % deviation.
  The *penetration distance* of a drug in the brain is the distance from the probe surface to the point where the concentration is roughly half its far field value.
  The *penetration distance* of chemicals into wood is the distance between the point of application and the 5 mm cut section in which the contaminant concentration is at least 3 % of the total.
  The *forest edge-effect penetration distance* is the distance to the point where invertebrate abundance ceased to differ from forest interior abundance.
  Cf. **penetration depth distance** in Chap. 9, **penetration depth** in Chap. 24 and **distance sampling** in Chap. 17.
- **Capillary diffusion distance**
  One of the diffusion processes is *osmosis*, i.e., the net movement of water through a permeable membrane to a region of lower solvent potential. In the respiratory system (the alveoli of mammalian lungs), oxygen O_2 diffuses into the blood and carbon dioxide CO_2 diffuses out.

The **capillary diffusion distance** is, similarly to **penetration distance**, a general term used in biological measurements for the distance, from the capillary blood through the tissues to the mitochondria, to the point where the concentration of oxygen has dropped to the given low level.
This distance is measured as the average distance from the capillary wall to the mitochondria, or the distance between the closest capillary endothelial cell to the epidermis, or in percentage terms, say, the distance where a given percentage (95 % for maximal, 50 % for average) of the fiber area is served by a capillary.
Another practical example: the *effective diffusion distance* of nitric oxide NO in microcirculation *in vivo* is the distance within which N concentration is greater than the equilibrium dissociation constant of the target enzyme for oxide action.
Cf. the **immunological distance** for immunodiffusion and, in Chap. 29, the **diffusion tensor distance** among **distances in Medicine**.

- **Főrster distance**
  FRET (fluorescence resonance energy transfer; Főrster, 1948) is a distance-dependent quantum mechanical property of a *fluorophore* (molecule component causing its fluorescence) resulting in direct nonradiative energy transfer between the electronic excited states of two dye molecules, the donor fluorophore and a suitable acceptor fluorophore, via a dipole. In FRET microscopy, fluorescent proteins are used as noninvasive probes in living cells since they fuse genetically to proteins of interest.
  The efficiency of FRET transfer depends on the square of the donor electric field magnitude, and this field decays as the inverse sixth power of the intermolecular separation (the physical donor–acceptor distance). The distance at which this energy transfer is 50 % efficient, i.e., 50 % of excited donors are deactivated by FRET, is called the **Főrster distance** of these two fluorophores.
  Measurable FRET occurs only if the donor–acceptor distance is less than ≈ 10 nm, the mutual orientation of the molecules is favorable, and the spectral overlap of the donor emission with acceptor absorption is sufficient.
- **Gendron–Lemieux–Major distance**
  The **Gendron–Lemieux–Major distance** (2001) between two base-base interactions, represented by 4×4 *homogeneous transformation matrices* X, Y, is

$$\frac{S(XY^{-1}) + S(X^{-1}Y)}{2},$$

  where $S(M) = \sqrt{l^2 + (\theta/\alpha)^2}$, l is the translation length, θ is the rotation angle, and α is a scaling factor between the translation and rotation contributions.
- **Spike train distances**
  A human brain has 85×10^9 *neurons* (nerve cells) each communicating with an average 1,000 other neurons dozens of times per second. Most neurons are capable of making 10^4–10^6 individual microconnections. One human brain, using $\approx 10^{15}$ synapses, produces $\approx 6.4 \times 10^{18}$ nerve impulses per second.
  The neuronal response to a stimulus is a continuous time series. It can be reduced, by a threshold criterion, to a simpler discrete series of *spikes* (short electrical

pulses). A *spike train* is a sequence $x = (t_1, \dots, t_s)$ of s events (neuronal spikes, or heart beats, etc.) listing absolute spike times or interspike time intervals. The main **distances between spike trains** $x = x_1, \dots, x_m$ and $y = y_1, \dots, y_n$ follow.

1. The **spike count distance** is defined by

$$\frac{|n-m|}{\max\{m,n\}}.$$

2. The **firing rate distance** is defined by

$$\sum_{1 \le i \le s} (x'_i - y'_i)^2,$$

   where $x' = x'_1, \dots, x'_s$ is the sequence of local firing rates of train $x = x_1, \dots, x_m$ partitioned in s time intervals of length T_{rate}.
3. Let $\tau_{ij} = \frac{1}{2}\min\{x_{i+1} - x_i, x_i - x_{i-1}, y_{i+1} - y_i, y_i - y_{i-1}\}$ and $c(x|y) = \sum_{i=1}^{m}\sum_{j=1}^{n} J_{ij}$, where $J_{ij} = 1$ if $0 < x_i - y_i \le \tau_{ij}$, $= \frac{1}{2}$ if $x_i = y_i$ and $= 0$, otherwise. The **event synchronization distance** (Quiroga et al., 2002) is defined by

$$1 - \frac{c(x|y) + c(y|x)}{\sqrt{mn}}.$$

4. Let $x_{isi}(t) = \min\{x_i : x_i > t\} - \max\{x_i : x_i < t\}$ for $x_1 < t < x_m$, let $I(t) = \frac{x_{isi}(t)}{y_{isi}(t)} - 1$ if $x_{isi}(t) \le x_{isi}(t)$ and $I(t) = 1 - \frac{y_{isi}(t)}{x_{isi}(t)}$, otherwise. The time-weighted and spike-weighted **ISI distances** (Kreuz et al., 2007) are

$$\int_0^T |I(t)|dt \text{ and } \sum_{i=1}^{m} |I(x_i)|.$$

5. Various *information distances* were applied to spike trains: the **Kullback–Leibler distance**, and the **Chernoff distance** (cf. Chap. 14). Also, if x and y are mapped into binary sequences, the **Lempel–Ziv distance** and a version of the **normalized information distance** (cf. Chap. 11) are used.
6. The **Victor–Purpura distance** (1996) is a cost-based **editing metric** (i.e., the minimal cost of transforming x into y) defined by the following operations with their associated costs: insert a spike (cost 1), delete a spike (cost 1), shift a spike by time t (cost qt); here $q > 0$ is a parameter. The **fuzzy Hamming distance** (cf. Chap. 11), introduced in 2001, identifies cost functions of shift preserving the triangle inequality.
7. The **van Rossum distance**, 2001, is defined by

$$\sqrt{\int_0^\infty (f_t(x) - f_t(y))^2 dt},$$

where x is convoluted with $h(t) = \frac{1}{\tau}e^{-t/\tau}$ and $\tau \approx 12$ ms (best); $f_t(x) = \sum_0^m h(t - x_i)$. This and above distances are the most commonly used metrics.

8. Given two sets of spike trains labeled by neurons firing them, the **Aronov et al. distance** (2003) between them is a cost-based **editing metric** (i.e., the minimal cost of transforming one into the other) defined by the following operations: insert or delete a spike (cost 1), shift a spike by time t (cost qt), relabel a spike (cost k), where $q, k > 0$ are parameters.

- **Bursting distances**
  Bursts refers to the periods in a spike train when the spike frequency is relatively high, separated by periods when it is relatively low or spikes are absent.
  Given neurons $x_1, \dots, x_n$ and SBEs (synchronized bursting events) $Y_1, \dots, Y_m$ with similar patterns of neuronal activity, let C^{ij} denote the cross-correlation between the activity of a neuron in Y_i and Y_j maximized over neurons, and let C_{ij} denote the correlation between neurons x_i and x_j averaged over SBEs.
  Baruchi and Ben-Jacob, 2004, defined the **interSBE distance** between Y_i and Y_j and the **interneuron distance** between x_i and x_j by $\frac{1}{m}(\sum_{s=1}^m (C^{is} - C^{js})^2)^{\frac{1}{2}}$ and $\frac{1}{n}(\sum_{s=1}^n (C_{is} - C_{js})^2)^{\frac{1}{2}}$, respectively.
- **Long-distance neural connection**
  Unlike Computing, neural systems are not exclusively optimized for minimal global wiring, but for a variety of factors including the minimization of processing steps. Kaiser and Hilgetag, 2006, showed that, due to the existence of long-distance projections, the total wiring among 95 primate (Macaque) cortical areas could be decreased by 32 %, and the wiring of neuronal networks in the nematode *C. elegans* could be reduced by 48 % on the global level. For example, >10 % of the primate cortical projections connect components separated by >40 mm, while 69 mm is the maximal possible distance. For the global *C. elegans* network, some connections are almost as long as the entire organism.
  The *global workspace theory* (Baars, 1988, 1997, 2003) posits that consciousness arises when neural representations of external stimuli are made available widespread to global areas of the brain and not restricted to the originating local areas. Dehaene et al., 2006, showed that distant areas of the brain are connected to each other and these connections are especially dense in the prefrontal, cingulate and parietal regions of the cortex which are involved in planning, reasoning and short-memory. These long-distance and long-lasting connections may be the architecture linking the separate regions/processes together during a single global conscious state.
  In autism there are more local connections and more local processing, while the psychosis/schizophrenia spectrum is marked by more long-distance connections.

About 5 %, 10 %, 6.7 % of variation in individual intelligence is predicted by activity level in LPFC (lateral prefrontal cortex), by the strength of neural pathways connecting left LPFC to the rest of the brain and by overall brain size.

- **Long-distance cell communication**
  Human cell size is within [4–135] μm; typically, 10 μm. In *gap junctions*, the intercellular spacing is reduced from 25–450 nm to a gap of 1–3 nm, bridged by hollow tubes. Animal cells may communicate locally, either directly through gap junctions, or by cell–cell recognition (in immune cells), or (*paracrine signaling*) using messenger molecules that travel, by diffusion, only short distances. Mammal's, astrocytes form, via gap junctions, a network of neurons and vasculature. Neurons may use interferon signals transmitted over great distances to fend off viral infection.
  In *synaptic signaling*, the electrical signal along a neuron's axon triggers the release of a neurotransmitter to diffuse across the synapse through a gap junction. Signal transmission through the nervous system is a long-distance signaling. Slower long-distance signaling is done by hormones transported in the blood. A hormone reaches all parts of the body, but only target cells have receptors for it.
  Another means of long-distance cell communication, via TNTs (*tunneling nanotubes*), was found in 1999. TNTs are membrane tubes, 50–200 nm thick with length up to several cell diameters. Cells can send out several TNTs, creating a network lasting hours. TNTs can carry cellular components and pathogens (HIV and prions). Also, electrical signals can spread bidirectionally between TNT-connected cells (over distances 10–70 μm) through interposed gap junctions.
  Some bacteria gain energy by oxidizing H_2S via electron transfer, hundreds of cell-lengths away. Thousands of *Desulfobulbus* form cm-long conductive chains, transporting electrons from H_2S-rich marine sediment to the upper O_2-rich one.
- **Length constant**
  In an excitable cell (nerve or muscle), the **length constant** is the distance over which a nonpropagating, passively conducted electrical signal decays to $\frac{1}{e}$ (36.8 %) of its maximum.
  During a measurement, the **conduction distance** between two positions on a cell is the distance between the first recording electrode for each position.
- **Ontogenetic depth**
  The **ontogenetic depth** (or *egg-adult distance*) is (Nelson, 2003) the number of cell divisions, from the unicellular state (fertilized egg) to the adult metazoan capable of reproduction (production of viable gametes).
  The **mitotic length** is the number of intervening mitoses, from the normal (neither immortal nor malignant) cells in the immature precursor stage to their progeny in a state of *mitotic death* (terminal differentiation) and phenotypic maturity.
- **Interspot distance**
  A *DNA microarray* is a technology consisting of an arrayed series of thousands of *features* (microscopic spots of DNA oligonucleotides, each containing picomoles of a specific DNA sequence) that are used as probes to hybridize a *target*

(cRNA sample) under high-stringency conditions. Probe-target hybridization is quantified by fluorescence-based detection of fluorophore-labeled targets to determine the relative abundance of nucleic acid sequences in the target.
The **interspot distance** is the **spacing distance** (Chap. 29) between features. Typical values are 375, 750, 1,500 μm ($1\ \mu m = 10^{-6}$ m).

- **Read length**
In gene sequencing, automated sequencers transform electropherograms (obtained by *electrophoresis* using fluorescent dyes) into a four-color chromatogram where peaks represent each of the DNA bases A, T, C, G. Chromosomes stained by some dyes show a 2D pattern of traverse bands of light and heavy staining.
The **read length** is the length, in the number of bases, of the sequence obtained from an individual clone chosen. Computers then assemble those short blocks into long continuous stretches which are analyzed for errors, gene-coding regions, etc.
- **Action at a distance along DNA/RNA**
An **action at a distance along DNA/RNA** happens when an event at one location on a molecule affects an event at a distant (say, more than 2,500 base pairs) location on the same molecule.
Many genes are regulated by distant (up to a million bp away and, possibly, located on another chromosome) or short (30–200 bp) regions of DNA, *enhancers*. Enhancers increase the probability of such a gene to be transcribed in a manner independent of distance and position (the same or opposite strand of DNA) relative to the transcription initiation site (the promoter).
DNA supercoiling is the twisting of a DNA double helix around its axis, once every 10.4 bp of sequence (forming circles and figures of eight) because it has been bent, overwound or underwound. Such folding puts a long range enhancer, which is far from a regulated gene in **genome distance**, geometrically closer to the promoter.
The *genomic radius of regulatory activity* of a genome is the genome distance of the most distant known enhancer from the corresponding promoter; in the human genome it is $\approx 10^6$ bp (for the enhancer of SSH, *Sonic Hedgehog* gene).
There is evidence that genomes are organized into enhancer–promoter loops. But the long range enhancer function is not fully understood yet.
Similarly, some viral RNA elements interact across thousands of intervening nucleotides to control translation, genomic RNA synthesis and mRNA transcription.
Genes are controlled either locally (from the same molecule) by specialized *cis* regulators, or at a distance by *trans* regulators. Comparing genes in key brain regions of human and primates, the most drastic changes were found in *trans*-controlled genes.
- **Length variation in 5-HTTLPR**
5-HTTLPR is a repeat polymorphic region in *SLC6A4*, the gene (on chromosome 17) coding for SERT (serotonin transporter) protein. This polymorphism has

short (14 repeats) and *long* (16 repeats) variations. So, an individual can have short/short, short/long, or long/long genotypes at this location in the DNA.
A short/short allele leads to less transcription for SLC6A4, and its carriers are more attuned and responsive to their environment; so, social support is more important for their well-being. They have less gray matter, more neurons and a larger thalamus. Whereas $\frac{2}{3}$ of East Asians have the short/short variant, only $\frac{1}{5}$ of Americans and Western Europeans have it.
Other gene variants of central neurotransmitter systems—dopamine receptor (*DRD4 7R*), dopamine/serotonin breaking enzyme (*MAOA VNTR*) and μ-opioid receptor (*OPRM1 A118G*)—are also associated with novelty-seeking, plasticity and social sensitivity. They appeared < 0.08 Ma ago and spread into 20–50 % of the population. They generate anxiety and aggression, but could be selected for extending behavioral range and boosting resilience at the group level.

- **Telomere length**
The *telomeres* are the caps of repetitive DNA sequences ($(TTAGGG)_n$ in vertebrates cells) at both ends of each linear chromosome in the cell nucleus. They are long stretches of noncoding DNA protecting coding DNA. The number n of $TTAGGG$ repeats is called the **telomere length** (TL); it is $\approx$2,000 in humans. TL is a robust indicator of biological age and a prognostic marker of disease risk. A limit of life - about 120 years - can be defined by TL in blood stem cells.
Every time a normal cell divides, its telomeres shorten and eventually they are so short that cell stops dividing, self-destructs, or tries to self-replicate and creates cancer. The *Hayflick limit* is the maximal number of divisions beneath which a normal cell will stop dividing, because of shortened telomeres or DNA damage, and die; for humans it is about 52.
Human telomeres are 3–20 kilobases in length, and they lose $\approx$100 bp, i.e., 16 repeats, at each mitosis (i.e., every 20–180 min). But telomere length can increase: by transfer of repeats between telomers or by action of enzyme *telomerase*. In humans, telomerase acts only in germ, stem or proliferating tumor cells.
Hydras, lobsters, planarian flatworms, trees maintain telomere lengths. Also, bacterial colonies and *Turritopsis dohrnii*, whose medusa form can revert to the polyp stage, are *biologically immortal*, i.e., there is no aging (sustained increase of mortality rate with age) since the Hayflick limit does not apply. Animals with negligible aging die mainly because of growth: they lose agility to get food. The oldest living animals are some sponges and black corals: 2,000–10,000 years. The oldest known cell line is 11,000 years-old canine transmissible venereal tumor.
Phenoptsis is genetically programmed death of organism. It acts quickly in *semelparous* (capable of only single reproduction) species, say, Pacific salmon, cicada, mayflies, annual plants and some bamboo, arachnids, squids. Extreme examples: the male praying mantis ejaculates only after being decapitated by the female, and the *Adactyllidium* tick larvae kill their mother eating her from the inside out.

Aging (or *catabiosis*) is slow phenoptsis in other, i.e., *iteroparous*, species. The telomere shortening is one of the main mechanisms of aging. Vascular decease, osteoarthritis, cancer and menopause are other means of human phenoptsis. Mortality rate of people with cancer behave as if the cancer had aged them by 15 years.

- **Gerontologic distance**

  The **gerontologic distance** between individuals of ages x and y from a population with survival fraction distributions $S_1(t)$ and $S_2(t)$, respectively, is defined by

$$\left| \ln \frac{S_2(y)}{S_1(x)} \right| .$$

  A function $S(t)$ can be either an empirical distribution, or a parametric one based on modeling. The main survival functions $S(t)$ are: $\frac{N(t)}{N(0)}$ (where $N(t)$ is the number of survivors, from an initial population $N(0)$, at time t), e^{kt} (exponential model), $e^{\frac{a}{b}(1-e^{bt})}$ (Gompertz model), and $e^{-\frac{at^{b+1}}{b+1}}$ (Weibull model); here a and b are, respectively, age independent and age dependent mortality rate coefficients. But *late-life mortality deceleration* was observed for humans and fruit flies: the probability that organism's somatic cells become senescent tends to be independent of its age in the long-time limit. The 1-year probability of death at advanced age asymptotically approaches 44 % for women and 54 % for men. Such a plateau is typical for many Markov processes. Human *species-specific life span* (age at which death rates of different populations converge) is close to 95 years.

  Since the 1960s mortality rates among those over 80 years have decreased by 1.5 % per year. But the age of super-long livers is linked to their genes rather than their lifestyle; at least 100 genes are linked to longevity.

  Distances are used in Human Gerontology also to model the link between geographical distance and contact between adult children and their elderly parents.

  Aging/death are adaptive species-specific trade-offs with reproduction. But the *Akela effect* (long post-reproductory period with intergenerational transfers) was observed, besides humans, in toothed whales and some elephants, primates, birds.

- **Distance to death**

  Eighty percent of the persons who die in any one year are age 65 or older. Elderly persons think and talk readily about death, but perceived temporal nearness of it is not quantified by ≈50 % of them. Still this proximity determines one's attitude on it.

  Gerstorf et al., 2008: relative to age-related decline, mortality-related one (i.e., **distance to death**) in reported life satisfaction account for more variance in the change of subjective well-being. At a point about 4 years before death of an old,

i.e., 70+ years, person, this decline showed a two-fold increase (three-fold for the *oldest old*, i.e., 85+ years) in steepness relative to the preterminal phase.
Bosworth et al., 1999: distance to death explains much of the variance in intellectual performance (verbal meaning, psychomotor speed, spatial and reasoning abilities) associated with age. Higher baseline intelligence test scores are associated with reduced risk of mortality and reduced effects of impending death on cognition.
The *terminal drop hypothesis* (Riegel–Riegel, 1972) states that death is preceded by a decrease in cognitive (especially, verbal) functioning over an ≈ 5 years period.
The *cascade model* (Birren–Cunningham, 1985) posits primary (normal), secondary (disease-related) and tertiary (distance to death) aging, which influence 3rd, 2–3rd and 1–3rd, respectively, classes of intellectual function: *crystallized abilities* (to think logically and solve problems knowledge-independently), *fluid abilities* (to use skills, knowledge and experience) and perceptual speed.
Borjigin et al., 2013, observed neural correlates of heightened conscious processing at near-death: a surge 30 s of coherence and connectivity in the dying rat's brain.
Micromort and *microlife* are the units of risk: 10^{-6} probability of death and half an hour ($\approx 10^{-6}$-th of 57 years) change of life expectancy, respectively.

- **Distance running model**
  Bipedality is a key behavior of hominins which appeared 6–4.2 Ma ago. It allowed australopithecines to see approaching danger further off, to walk long distances and to use hands for gathering food. Our genus Homo emerged ≈ 2.5 Ma ago.
  The distance running model anthropogenesis, proposed in [BrLi04], claims that our capacity to run long distances in the savanna arised, prior to the invention of the spear, as adaptation for persistence hunting (by running prey to exhaustion) and scavenging (allowing to compete for widely dispersed carcasses).
  This model specifies how endurance running defined the human body form, producing balanced head, low/wide shoulders, narrow chest, short forearms and heels, large hip, etc. Even now, a good athlete can run at 20 km/h for several hours which is comparable to endurance specialists as, say, zebras and antelopes. By sweating we can dissipate body heat faster than any other large mammal and reach large sustainable distance. The capacity of humans to travel vast distances using little energy contributed also to the evolution of their complex social networks.
- **Distance coercion model**
  The **distance coercion model** [OkBi08] of the origin of uniquely human kinship-independent cooperation see all complex symbolic speech, cognitive virtuosity, transmission of fitness-relevant information, etc. as elements and effects of this cooperation catalyzed by advances in lethal projectile weapons.
  The model argues that such cooperation can arise only as a result of the pursuit of individual self-interest by animals who can project "death from a distance".

Among rare organisms able to project coercive threat remotely, humans are the most efficient on long distances, say, to kill adult conspecifics up to 18–27 m by throwing a spear and up to 91 m by a bow. The chimpanzee and Neanderthal also could throw objects but not with human's precision.

The model posits that this capacity, permitting to repel predators and scavenge their kills in the African savanna, briefly preceded the emergence of brain expansion and social support. Comparing with Neanderthals, evidence of a huge number of injuries suggests that their hunting involved dangerously close contact with large prey animals; they used conventional spears rather than true projectile weapons.

Throwing and language capacities enabled humans to survive rapid climatic and environmental changes, to spread and to become the dominant large-scale species on the planet. Historical increases in social cooperation could be associated with prior acquisition of a new coercive technology; for instance, the bow and agricultural civilizations, gunpowder weaponry and the modern state.

Humans are most efficient enforcers of cooperation (even relying mainly on indirect cues): our cognitive abilities expanded the range of situations in which cooperation can be favored. Also, while the *strong reciprocity* (generous third-party enforcement) is prevalent in large societies, Marlowe et al., 2012, claim that motivated by the basic emotion of anger, humans-special tendency to retaliate on their own behalf, even at a cost, is sufficient to explain the origin of human cooperation.

- **Distance model of altruism**

  In Evolutionary Ecology, altruism is explained by kin selection, reciprocity, sexual selection, etc. The cooperation between nonrelatives was a driving force in some major transitions (say, from symbiotic bacteria to mitochondria, eukaryotes or multicellular organisms). Individual selection, including *social selection* in which fitness is influenced by the behaviors of others, interacts with group selection.

  The **distance model of altruism** [Koel00] claims that altruists spread locally, i.e., with small interaction distance and offspring dispersal distance, while the egoists invest in increasing of those distances. The intermediate behaviors are not maintained, and evolution will lead to a stable bimodal spatial pattern.

- **Distance grooming model of language**

  In primates, being groomed produces mildly narcotic effects, because it stimulates the production of the body's natural opiates, the *endogenous opioid peptides*.

  Language, according to Dunbar, 1993, evolved in archaic *Homo sapiens* as more distance/time efficient replacement of social grooming. Their brain size expanded (as 2.5 Ma ago with *Homo habilis*) 0.5 Ma ago from 900 cm^3 in *Homo erectus* to 1,300 cm^3, and they lived in large groups (over 120 individuals) requiring cohesion. Language allowed them to produce the reinforcing, social-bonding effects of grooming at a distance and to use more efficiently the time available for social interaction.

Language achieves this through information transfer, gossip and emotional means (say, laughter, facial expression, Duchenne smile). Many primate species extensively use contact calls such as the long-distance *pant-hoot* call of chimpanzees. Dunbar interprets such calls as a grooming-at-a-distance from which language evolved. But gestures are far more likely precursor of language than vocalizations.
He observed the link between group and brain sizes in primates and deduced that human social networks tend to be structured in layers: 5 intimates (support clique), 15 best friends (sympathy group), active network of "persons" (50 good friends and 150 friends), 500 acquaintances, 1500 "people I recognize". One need to be in contact every week, month, half-year, year with groups 1–4, respectively, A natural group size (*Dunbar's number*) is 150 for humans and 50 for chimpanzees.
Dunbar explain above sizes by cognitive and time constraints on the number of relationships ego can maintain at a given level of intensity. The clique size correlates with the highest achievable order of *intentionality recursion*, in which mind states are reflexively attributed to others. 0-order means responses to stimuli (as bacteria and computers); 1st order: belief about the real or imagined world (as most organisms with brains); 2nd order: belief about the mental state of others; i-th order: as, for $i = 5$, in the sentence "I *think* that you *believe* that I *suppose* that we *understand* that Jane *wants*...". We operate usually at 3rd and sometimes at 4th or 5th order. Language is essential for 4th order recursion.

Chapter 24
Distances in Physics and Chemistry

24.1 Distances in Physics

Physics studies the behavior and properties of matter in a wide variety of contexts, ranging from the submicroscopic particles from which all ordinary matter is made (*Particle Physics*) to the behavior of the material Universe as a whole (*Cosmology*).

Physical forces which act at a distance (i.e., a push or pull which acts without "physical contact") are nuclear and molecular attraction and, beyond the atomic level, gravity (completed, perhaps, by anti-gravity), static electricity, and magnetism. Last two forces can be both push and pull, depending on the charges of involved bodies. The nucleon–nucleon interaction (or *residual strong force*) is attractive but becomes repulsive at very small distances keeping the nucleons apart. Dark matter is attractive while dark energy is repulsive (if they exist).

Distances on a relatively small scale are treated in this chapter, while large distances (as in Astronomy and Cosmology) are the subject of Chaps. 25 and 26.

The distances having physical meaning range from 1.6×10^{-35} m (**Planck length**) to 8.8×10^{26} m (estimated size of the observable Universe). We can see things of about 10^{-4} to 10^{21} m and measure them within $[10^{-18}, 10^{27}]$ m. The smallest measurable distance, time and weight are 10^{-18} m (by LHC), 10^{-17} sec and 10^{-24} g.

The Theory of Relativity, Quantum Theory and Newtonian laws permit us to describe and predict the behavior of physical systems in the range 10^{-15} to 10^{12} m, i.e., from proton to Solar System. Weakened description is still possible up to 10^{25} m.

The world appears Euclidean at distances less than about 10^{25} m (if gravitational fields are not too strong). Relativity and Quantum Theory effects, governing Physics on very large and small scales, are already accounted for in technology, say, of GPS satellites and nanocrystals of solar cells.

M.M. Deza, E. Deza, *Encyclopedia of Distances*,
DOI 10.1007/978-3-662-44342-2_24

- **Moment**
  In Physics and Engineering, **moment** is the product of a quantity (usually, force) and a distance (or a power of it) to some point associated with that quantity.
- **Momentum**
  In classical mechanics, **momentum** $\mathbf{p} = (p_x, p_y, p_z)$ is the product $m\mathbf{v}$ of the mass m and velocity vector $\mathbf{v} = (v_x, v_y, v_z)$ of an object.
  In relativistic 4D mechanics, *momentum-energy* $(\frac{E}{c}, p_x, p_y, p_z)$, where c is the speed of light and $E = mc^2$ is energy, is compared with space-time (ct, x, y, z).
- **Displacement**
  In Mechanics, a *displacement* (or *relative position*) *vector* of a moving particle from its initial position P_i to the final position P_f, is the vector $\overrightarrow{P_i P_f} = \overrightarrow{0(P_f - P_i)}$, where O is a reference point (usually the origin of a coordinate system).
  A **displacement** is the length $||P_f - P_i||_2$ of this vector, i.e., the Euclidean distance from P_i to P_f. It is never greater than the distance traveled by a particle.
- **Acceleration distance**
  The **acceleration distance** is the minimum distance at which an object (or, say, flow, flame), accelerating in given conditions, reaches a given speed.
- **Mechanic distance**
  The **mechanic distance** is the position of a particle as a function of time t.
  For a particle, moving linearly with initial position x_0 and initial speed v_0, which is acted upon by a constant acceleration a, it and the speed are given by

$$x(t) = x_0 + v_0 t + \frac{1}{2}at^2 \quad \text{and} \quad v(t) = v_0 + at.$$

  So, the **acceleration distance** fallen under uniform acceleration a, in order to reach a speed v, is $\frac{v^2}{2a}$. A body is *free falling* if it is falling subject only to acceleration g by gravity; the **free fall distance** (distance fallen by it) is $y(t) = \frac{1}{2}gt^2$.
- **Terminal distance**
  The **terminal distance** is the distance of an object, moving linearly in a resistive medium, from an initial position to a stop.
  If object's initial position and speed are x_0, v_0, and the drag per unit mass in the medium is proportional to speed with constant of proportionality β, then the position and speed of a body are given by

$$x(t) = x_0 + \frac{v_0}{\beta}(1 - e^{-\beta t}) \quad \text{and} \quad v(t) = x'(t) = v_0 e^{-\beta t}.$$

  The speed decreases to 0, and the body reaches a *maximum terminal distance*

$$x_{terminal} = \lim_{t \to \infty} x(t) = x_0 + \frac{v_0}{\beta}.$$

For a body, moving from initial position (x_0, y_0) and speed (v_{x_0}, v_{y_0}), the position $(x(t), y(t))$ is $x(t) = x_0 + \frac{v_{x0}}{\beta}(1 - e^{-\beta t})$, $y(t) = (y_0 + \frac{v_{y0}}{\beta} - \frac{g}{\beta^2}) + \frac{v_{y0}\beta - g}{\beta^2}e^{-\beta t}$. The horizontal motion ceases at a maximum terminal distance $x_{terminal} = x_0 + \frac{v_{x0}}{\beta}$.

- **Ballistics distances**
  Ballistics is the study of the motion of *projectiles*, i.e., bodies which are propelled (or thrown) with some initial velocity, and then allowed to be acted upon by the forces of gravity and possible drag.
  The **trajectory**, **range** and **height** of a projectile are its parabolic path, total horizontal distance traveled and maximum upward distance reached. If projectile is launched on flat ground at velocity v and angle θ to the horizontal, then at the time t of motion, its horizontal and vertical positions are

$$x(t) = vt\cos\theta \quad \text{and} \quad y(t) = vt\cos\theta - \frac{1}{2}gt^2.$$

  So, the range, realized by the *time of flight* $t_{of} = \frac{2v\sin\theta}{g}$, and height are

$$x_{\max} = x(t_{of}) = \frac{v^2\sin 2\theta}{g} \quad \text{and} \quad y_{\max} = y(\frac{1}{2}t_{of}) = \frac{v\sin^2\theta}{2g},$$

  which are maximized when $\theta = \pi/4$ and $\theta = \pi/2$, respectively.
  The *bullet drop* is the height it loses, because of gravity, between leaving the rifle and reaching the target. In order to ensure that the "zero" (point at which the bullet's path intersects with the *LOS*, line of sight, to the target) will be at a specific range, the shooter should set (using a *sight*, device mounted on the rifle) the *bore angle* between the rifle bore and the LOS. A properly adjusted rifle barrel and sight are said to be *zeroed* (or *sighted-in*). The shooter zeroes rifle at a standard *zero range* and then adjustments are made for other ranges.
  The **point-blank range** is the distance at which the bullet is expected to strike a target of a given size without adjusting the elevation of the firearm.
- **Interaction distance**
  The **impact parameter** is the perpendicular distance between the velocity vector of a projectile and the center of the object it is approaching.
  The **interaction distance** between two particles is the farthest distance of their approach at which it is discernible that they will not pass at the impact parameter, i.e., their distance of closest approach if they had continued to move in their original direction at their original speed.
  The *coefficient of restitution* (COR) of colliding objects A, B is the ratio of speeds after and before an impact, taken along its line. The collision is *inelastic* if COR < 1. COR^2 is the ratio of *rebound* and *drop* distances if A bounces off stationary B.

- **Mean free path (length)**
  The **mean free path (length)** of a particle (photon, atom or molecule) in a medium measures its probability to undergo a situation of a given kind K; it is the average of an exponential distribution of distances until the situation K occurs. In particular, this average distance d is called:

  nuclear collusion length if K is a nuclear reaction;
  interaction length if K is an interaction which is neither elastic, nor quasi-elastic;
  scattering length if K is a scattering event;
  attenuation length (or *absorption length*) if K means that the probability $P(d)$, that a particle has not been absorbed, drops to $\frac{1}{e} \approx 0.368$, cf. **Beer–Lambert law**;
  radiation length (or *cascade unit*) if K means that the energy of (high energy electromagnetic-interacting) relativistic charged particles drops by the factor $\frac{1}{e}$;
  free streaming length if K means that particles become nonrelativistic.

  In Gamma-ray Radiography, the *mean free path* of a beam of photons is the average distance a photon travels between collisions with atoms of the target material. It is $\frac{1}{\alpha\rho}$, where α is the material *opacity* and ρ is its density.
- **Neutron scattering length**
  In Physics, *scattering* is the random deviation or reflection of a beam of radiation or a stream of particles by the particles in the medium.
  In Neutron Interferometry, the **scattering length** a is the zero-energy limit of the scattering amplitude $f = -\frac{\sin\delta}{k}$. Since the *total scattering cross-section* (the likelihood of particle interactions) is $4\pi|f|^2$, it can be seen as the radius of a hard sphere from which a point neutron is scattered.
  The spin-independent part of the scattering length is the *coherent scattering length*. In order to expand the scattering formalism to absorption, the scattering length is made complex $a = a'ia''$.
  Thomson scattering length is the *classical electron radius* $\approx 2.818 \times 10^{-15}$ m.
- **Inelastic mean free path**
  In Electron Microscopy, the **inelastic mean free path** (or IMFP) is the average total distance that an electron traverses between events of inelastic scattering, while the **effective attenuation length** (or EAL) is an experimental parameter reflecting the average net distance traveled.
  The EAL is the thickness in the material through which electron can pass with probability $\frac{1}{e}$ that it survives without inelastic scattering. It is about 20 % less than the IMFP due to the elastic scatterings which deflect the electron trajectories.
  Both are smaller than the total electron range which may be 10–100 times greater.
- **Sampling distance**
  In Electron Spectroscopy for chemical analysis, the **sampling distance** is the lateral distance between areas to be measured for characterizing a *surface*, i.e., the volume from which the photo-electrons can escape.

- **Debye screening distance**
The **Debye screening distance** (or *Debye length, Debye–Hückel length*) is the distance over which a local electric field affects the distribution of mobile charge carriers (for example, electrons) present in the material (plasmas and other conductors).
Its order increases with decreasing concentration of free charge carriers, from 10^{-4} m in gas discharge to 10^5 m in intergalactic medium.
- **Range of a charged particle**
The **range of a charged particle**, passing through a medium and ionizing, is the distance to the point where its energy drops to almost zero.
- **Gyroradius**
The **gyroradius** (or *cyclotron radius, Larmor radius*) is the radius of the circular orbit of a charged particle in the presence of a uniform magnetic field.
- **Radius of gyration**
The **radius of gyration** of a body about a given axis is the distance from this axis to the centre of gyration. It is the RMS (square root of the mean of the squares) of the distances from the axis of rotation to all the points in the body.
- **Inverse-square distance laws**
Any law stating that some physical quantity is inversely proportional to the square of the distance from the source that quantity.
Newton's **law of universal gravitation** (checked above 6×10^{-5} m): the gravitational attraction between two point-like masses m_1, m_2 at distance d is

$$G\frac{m_1 m_2}{d^2},$$

where $G = 6.67384(80) \times 10^{-11}\,\mathrm{m}^3\,\mathrm{kg}^{-1}\,\mathrm{s}^{-2}$ is the Newton *gravitational constant*.
The existence of extra dimensions, postulated by M-theory, will be checked by LHC (Large Hadron Collider at CERN, near Geneva) based on the inverse proportionality of the gravitational attraction in nD space to the $(n-1)$-th degree of the distance between objects; if the Universe has a 4th dimension, LHC will find out the inverse proportionality to the cube of the small interparticle distance.
Coulomb law: the force of attraction or repulsion between two point-like objects with charges e_1, e_2 at distance d is given by

$$k_e\frac{e_1 e_2}{d^2},$$

where k_e is the *Coulomb constant* depending upon the medium that the charged objects are immersed in. The gravitational and electrostatic forces of two bodies with *Planck mass* m_P and $e_1 = e_2 = 1$ have equal strength.
The *intensity* (power per unit area in the direction of propagation) of a spherical wavefront (light, sound, etc.) radiating from a point source decreases (assuming that there are no losses caused by absorption or scattering) inversely proportional

to the square d^2 of the distance from the source (cf. **distance decay** in Chap. 29). However, for a radio wave, it decrease like $\frac{1}{d}$.

- **Range of fundamental forces**
  The fundamental forces (or interactions) are gravity and electromagnetic, weak nuclear and strong nuclear forces. The **range** of a force is considered *short* if it decays (approaches 0) exponentially as the distance d increases.
  Both electromagnetic force and gravity are forces of infinite range which obey **inverse-square distance laws**. The shorter the range, the higher the energy. Both weak and strong forces are very short range (about 10^{-17} and 10^{-15} m, respectively) which is limited by the uncertainty principle.
  At subatomic distances, Quantum Field Theory describes electromagnetic, weak and strong interactions with the same formalism but different constants. Quantum Electrodynamics describes electromagnetism via photon exchanges between charged particles and Quantum Chromodynamics describe strong interactions via gluon exchanges between quarks. Strong interaction force grows stronger with the distance. Three forces almost coincide at very large energy, but at large distances they are irrelevant compared with gravity. The number of fundamental particles increases on smaller distance scales. But at macroscopic scales, those particles can collectively create emerging phenomena, say, superconductivity.
  General Relativity has been probed from submillimeter up to Solar System scales but at cosmological scale it require the presence of *dark matter* and *dark energy*. Maxwell's electromagnetism has been probed from atomic distances up to 1.3 AU (order of the **coherence lengths** of the magnetic fields dragged by the solar wind) but it does not explain magnetic fields found in galaxies, clusters and voids. The main hypothesis: at cosmological scale the repulsive force of putative *dark energy*, due to *vacuum energy* (or *cosmological constant*) overtakes gravity; cf. the **metric expansion of space** in Chap. 26. Dark energy is the only substance known to act both on subatomic and cosmological scale. Its effect is measured only on a scale larger than superclusters. Khoury–Weltman, 2004, in order to explain dark energy, conjectured *fifth force* with range depending on density of matter in its environment, say, 1 mm in Earth's vicinity and 10^7 light-years in cosmos.
  An alternative to dark energy: possible, in String Theory, modifications of gravity at ultra large distances (i.e., small curvatures) due to some specific compactification of, say, δ extra dimensions; their size is expected to be $\sim 10^{16}, 1, 10^{10}$ mm for $\delta = 1, 2, 5$, respectively. Another alternative is extended, via dropping the *Lorenz condition*, Maxwell's theory of electromagnetism by Beltrán and Maroto, 2011.
  It allow the propagation, in addition to usual photons, the *longitudinal* (wave in which electric field points along direction of motion) and *temporal* (wave of pure electric potential) modes of light, produced from quantum fluctuations during inflation. Wavelengths of longitudinal electric waves are longer than the longest (millions of km) observed ones but still less than observable Universe; they generate magnetic fields from subgalactic up to the present **Hubble radius**.

Wavelengths of temporal waves are many orders of magnitude larger than observable Universe; they may explain the actual quantity of dark energy in the Universe.

- **EM radiation wavelength range**
  The *wavelength* is the distance $\lambda = \frac{c}{f}$ the wave travels to complete one cycle.
  Electromagnetic (EM) **radiation wavelength range** is infinite and continuous in principle. The limits of short and long waves are the vicinity of the **Planck length** and the size of Universe, respectively.
  The wavelengths are: <0.01 nm for gamma rays, 0.01–10 nm for X-rays, 100–400 nm for ultraviolet, 400–780 nm for visible light, 0.78–1,000 μm for infrared (in lasers), 1–330 mm for microwave, 0.33–3,000 m for radio frequency radiation, >3 km for low frequency, and ∞ for static field.
  Besides gamma rays, X-rays and far ultraviolet, the EM radiation is *nonionizing*, i.e., passing through matter, it only *excites* electrons: moves them to a higher energy state, instead of removing them completely from an atom or molecule.
- **Fraunhofer distance**
  The Maxwell equations, governing the field strength decay, can be approximated as d^{-3}, d^{-2} and d^{-1} for three regions surrounding an radiating antenna: the *reactive near field*, the *radiating near field* (or *Fresnel region*) and the *far field* (or *Fraunhofer region*). The Fresnel region begins about at $0.62(\frac{D^3}{\lambda})^{\frac{1}{2}}$, where D is the largest dimension of the antenna and λ is the wavelength. The propagating waves start to dominate here, but only in the far field the distribution of the diffracted energy no longer change with distance.
  The **Fraunhofer** (or *far field*, *Rayleigh*) **distance** is $\frac{2D^2}{\lambda}$, the distance where the far field is considered to begin. Cf. the **acoustic distances** in Chap. 21.
  In Optics, beam divergence is defined by its *radius*, i.e., for a Gaussian beam, the distance from the beam propagation axis where intensity drops to $\frac{1}{e^2} \approx 13.5\,\%$ of the maximum. The beam's *waist* (or *focus*) is the position on its axis where the radius is at its minimum. The *imaginary-distance BPM* (Jevick–Hermansson, 1989) refers to beam propagating the (complex electric) field along the imaginary axis.
  The beam's **Rayleigh length** (or *Rayleigh range*) Rl is the distance from the waist (in the propagation direction) where the radius increases from w_0 to $\sqrt{2}w_0$, i.e., the beam propagates without diverging much. For Gaussian beams, $Rl = \frac{\pi w_0^2}{\lambda}$, where λ is the vacuum wavelength divided by the refractive index of the material. The Rayleigh length divides the *near-field* and *mid-field*; it is the distance from the waist at which the wavefront curvature is at a maximum. The divergence really starts in the *far field* where the beam radius is at least $10Rl$. The *confocal parameter* (or *depth of focus*) of the beam is $2Rl$. Cf. the **lens distances** in Chap. 28.

- **Half-value layer**
  Ionizing radiation consists of highly-energetic particles or waves (especially, X-rays, gamma rays and far ultraviolet light) which are progressively absorbed during propagation through the surrounding medium, via *ionization*, i.e., removing an electron from some of its atoms or molecules. The **half-value layer** is the depth within a material where half of the incident radiation is absorbed.
  A basic rule of protection against ionizing radiation exposure: multipliyng the distance from its source by a *distance factor* r decreases this exposure to $\frac{1}{r^2}$ of it.
  In Maxwell Render light simulation software, the *attenuation distance* (or *transparency*) is the thickness of object that absorbs 50 % of light energy.
- **Compton wavelength**
  Compton scattering is the scattering of (X-ray or gamma ray) photons by electrons. It results in a decrease in energy (so, increase in wavelength) of the photon.
  Compton wavelength $\lambda_C(m)$ and **reduced Compton wavelength** $\overline{\lambda}_C(m)$ of a particle with rest mass m (where c is the speed of light, $\hbar$ is the *reduced Planck constant* $\frac{h}{2\pi}$ and l_P, m_P are Planck length and mass) are defined by

$$\frac{1}{2\pi}\lambda_C(m) = \overline{\lambda}_C(m) = \frac{\hbar}{mc} = \frac{m_P}{m}l_P.$$

- **Radiation attenuation with distance**
  Radiation is the process by which energy is emitted from a source and propagated through the surrounding medium. Radiant energy described in wave terms includes sound and electromagnetic radiation, such as light, X-rays and gamma rays. The incident radiation partially changes its direction, gets absorbed, and the remainder transmitted. The change of direction is *reflection*, *diffraction*, or *scattering* if the direction of the outgoing radiation is reversed, split into separate rays, or randomized (diffused), respectively. Scattering occurs in nonhomogeneous media.
  In Physics, *attenuation* is any process in which the flux density, power amplitude or intensity of a wave, beam or signal decreases with increasing distance from the energy source, as a result of absorption of energy and scattering out of the beam by the transmitting medium. It comes in addition to the divergence of flux caused by distance alone as described by the **inverse-square distance laws**.
  Attenuation of light is caused mainly by scattering and absorption of photons. The primary causes of attenuation in matter are the *photoelectric effect* (emission of electrons), *Compton scattering* (cf. **Compton wavelength**) and *pair production* (creation of an elementary particle and its antiparticle from a high-energy photon).
  In Physics, *absorption* is a process in which atoms, molecules, or ions enter some bulk phase - gas, liquid or solid material; in *adsorption*, the molecules are taken up by the surface, not by the volume. *Absorption of EM radiation* is the process by which the energy of a photon is taken up (and destroyed) by, for example, an

atom whose valence electrons make the transition between two electronic energy levels. The absorbed energy may be re-emitted or transformed into heat.
Attenuation is measured in units of decibels (dB) or *nepers* ($\approx$8.7 dB) per length unit of the medium and is represented by the medium *attenuation coefficient* α. When possible, specific absorption or scattering coefficient is used instead.
Attenuation (or *loss*) *of signal* is the reduction of its strength during transmission. In Signal Propagation, attenuation of a propagating EM wave is called the *path loss*. Path loss may be due to free-space loss, refraction, diffraction, reflection, absorption, aperture-medium coupling loss, etc. of antennas. Path loss in decibels is $L = 10n \log_{10} d + C$, where n is the path loss exponent, d is the transmitter-receiver distance in m, and C is a constant accounting for system losses.
The *free-space path loss* (FSPL) is the loss in signal strength of an EM wave that would result from a line-of-sight path through free space, with no obstacles to cause reflection or diffraction. FSPL is $(\frac{4\pi d}{\lambda})^2$, where d is the distance from the transmitter and λ is the signal wavelength (both in m), i.e., in dB it is $10\log_{10}(FSPL) = 20\log_{10} d + 20\log_{10} f - 147.56$, where f is the frequency in Hz.

- **Beer–Lambert law**
The **Beer–Lambert law** is an empirical relationship for the *absorbance Ab* of a substance when a radiation beam of given frequency goes through it:

$$Ab = \alpha d = -\log_a T,$$

where $a = e$ or (for liquids) 10, d is the *path length* (distance the beam travels through the medium), $T = \frac{I_d}{I_0}$ is the *transmittance* (I_d and I_0 are the intensity of the transmitted and incident radiation), and α is the medium *opacity* (or *linear attenuation coefficient, absorption coefficient*); α is the fraction of radiation lost to absorption and/or scattering per unit length of the medium.
The *extinction coefficient* is $\frac{\lambda_w}{4\pi}\alpha$, where λ_w is the same frequency wavelength in a vacuum. In Chemistry, α is given as ϵC, where C is the absorber *concentration*, and ϵ is the *molar extinction coefficient*.
The **optical depth** is $\tau = -\ln \frac{I_d}{I_0}$, measured along the true (slant) optical path.
The **penetration depth** (or **attenuation length**, *mean free path, optical extinction length*) is the thickness d in the medium where the intensity I_d has decreased to $\frac{1}{e}$ of I_0; so, it is $\frac{1}{\alpha}$. Cf. **half-value layer**.
Also, in Helioseismology, the (meridional flow) *penetration depth* is the distance from the base of the solar convection zone to the location of the first reversal of the meridional velocity. In an information network, the *message penetration distance* is the maximum distance from the event message traverses in the valid routing region.
The **skin depth** is the thickness d where the amplitude A_d of a propagating wave (say, alternating current in a conductor) has decreased to $\frac{1}{e}$ of its initial value A_0; it is twice the penetration depth. The *propagation constant* is $\gamma = -\ln \frac{A_d}{A_0}$.

The Beer–Lambert law can describe also the attenuation of solar or stellar radiation. The main components of the atmospheric light attenuation are: absorption and scattering by aerosols, Rayleigh scattering (from molecular oxygen O_2 and nitrogen Ni_2) and (only absorption) by carbon dioxide CO_2, O_2, nitrogen dioxide NiO_2, water vapor, ozone O_3. Cf. **atmospheric visibility distances** in Chap. 25.

The sea is nearly opaque to light: less than 1 % penetrates 100 m deep. Cf. **distances in Oceanography** in Chap. 25. In Oceanography, attenuation of light is the decrease in its intensity with depth due to absorption (by water molecules) and scattering (by suspended fine particles). The transparency of the water in oceans and lakes is measured by the *Secchi depth* d_S at which the reflectance equals the intensity of light backscattered from the water. Then $\alpha = \frac{10d_S}{17}$ is used as the *light attenuation coefficient* in the Beer–Lambert law $\alpha d = -\ln \frac{I_d}{I_0}$, in order to estimate I_d, the intensity of light at depth d, from I_0, its intensity at the surface.

In Astronomy, attenuation of EM radiation is called *extinction* (or *reddening*). It arises from the absorption and scattering by the interstellar medium, the Earth's atmosphere and dust around an observed object.

The *photosphere* of a star is the surface where its optical depth is $\frac{2}{3}$. energy emitted. The *optical depth of a planetary ring* is the proportion of light blocked by the ring when it lies between the source and the observer.

- **Arago distance**

  The *Arago point* is a *neutral point* (where the degree of polarization of skylight goes to zero) located $\approx 20°$ directly above the *antisolar point* (the point on the celestial sphere that lies directly opposite the sun from the observer) in relatively clear air and at higher elevations in turbid air.

  So, the **Arago distance**, i.e., the angular distance from the antisolar point to the Arago point, is a measure of *atmospheric turbidity* (effect of aerosols in reducing the transmission of direct solar radiation).

  Another useful measure of turbidity is *aerosol optical depth*, i.e., the **optical depth** due to extinction by the aerosol component of the atmosphere.

- **Sound attenuation with distance**

  Vibrations propagate through elastic solids and liquids, including the Earth, and consist of *elastic* (or *seismic*, *body*) *waves* and *surface* (occurring since it acts as an solid–gas interface) *waves*. Elastic waves are: primary (P) wave moving in the propagation direction of the wave and *secondary* (S) wave moving in this direction and perpendicular to it. Surface waves are: the *Love* wave moving perpendicular to the direction of the wave and the *Rayleigh* (R) wave moving in the direction of the wave and circularly within the vertical surface perpendicular to it. The attenuation of P- and S-waves is proportional to $\frac{1}{d^2}$ or $\frac{1}{d}$, when propagated by the surface or inside of an infinite elastic body. For the R-wave, it is proportional to $\frac{1}{\sqrt{d}}$.

  Sound propagates through gas (say, air) as a P-wave and attenuates over a distance, at a rate of $\frac{1}{d^2}$. The *far field* (cf. **Fraunhofer distance**) is the part of a

sound field in which sound pressure (if it is the same in all directions) decreases according to the *inverse distance law* $\frac{1}{d}$; but sound intensity decreases as $\frac{1}{d^2}$.
In natural media, further weakening occurs from *attenuation*, i.e., *scattering* (reflection of the sound in other directions) and *absorption* (conversion of the sound energy to heat). Cf. **critical distance** among **acoustics distances** in Chap. 21.
The **sound extinction distance** is the distance over which its intensity falls to $\frac{1}{e}$ of its original value. For sonic boom intensities (say, supersonic flights), the lateral *extinction distance* is the distance where in 99 % of cases the sound intensity is lower than 0.1–0.2 mbar (10–20 pascals) of atmospheric pressure. Cf. earthquake *extinction length* in **distances in Seismology** (Chap. 25).
Water is transparent to sound. Sound energy is absorbed (due to viscosity) and ≈6 % of it is scattered (due to water inhomogeneities). Absorbed less, low frequency sounds can propagate over large distances along lines of minimum sound speed. High frequency waves attenuate more rapidly. So, low frequency waves are dominant further from the source (say, a musical band or earthquake).
Attenuation of ultrasound waves with frequency f MHz at a given distance r cm is $\alpha f r$ decibels, where α is the *attenuation coefficient* of the medium. It is used in Ultrasound Biomicroscopy; in a homogeneous medium (so, without scattering) α is 0.0022, 0.18, 0.85, 20, 41 for water, blood, brain, bone, lung, respectively.

- **Lighting distance**
  Sound travels through air at 330–350 m per second (depending on altitude, relative humidity, pressure, etc.), while the speed of light is $c \approx 300 \times 10^6$ m/s. So, the **lighting distance** (of a lightning bolt from an observer) in km is $\approx\frac{1}{3}$ of the delay, in seconds, between observer's seeing it and hearing thunder.
- **Optical distance**
  The **optical distance** (or *optical path length*) is a distance dn traveled by light, where d is the physical distance in a medium and $n = \frac{c}{v}$ is the medium's *refractive index* (c, v are the speeds of an EM wave in a vacuum and in the medium). By *Fermat's principle* light follows the shortest optical path. Cf. **optical depth**.
  The **light extinction distance** is the distance where light propagating through a given medium reaches its *steady-state speed*, i.e., a characteristic speed that it can maintain indefinitely. It is proportional to $\frac{1}{\rho\lambda}$, where ρ is the density of the medium and λ is the wavelength, and it is very small for most common media.
- **Edge perimeter distance**
  In semiconductor technology, the **edge perimeter distance** is the distance from the edge of a *wafer* (thin slice with parallel faces cut from a semiconductor crystal) in a wafer carrier to the top face of the wafer carrier.
- **Proximity effects**
  In Electronic Engineering, an alternating current flowing through an electric conductor induces (via the associated magnetic field) eddy currents within the conductor. The *electromagnetic proximity effect* is the "current crowding" which occurs when such currents are flowing through several nearby conductors such as

within a wire. It increases the alternating current resistance (so, electrical losses) and generates undesirable heating.
In Nanotechnology, the *quantum* $\frac{1}{f}$ *proximity effect* is that the $\frac{1}{f}$ fundamental noise in a semiconductor sample is increased by the presence of another similar current-carrying sample placed in the close vicinity.
The *superconducting proximity effect* is the propagation of superconductivity through a NS (normal-superconductor) interface, i.e., a very thin layer of "normal" metal behaves like a superconductor (that is, with no resistance) when placed between two thicker superconductor slices.
In Lithography, if a material is exposed to an electronic beam, some molecular chains break and many electron scattering events occur. Any pattern written by the beam on the material can be distorted by this *E-beam proximity effect.*
In LECD (localized electrochemical deposition) technique for fabrication of miniature devices, the electrode (anode) is placed close to the tip of a fabricated structure (cathode). Voltage is applied and the structure is grown by deposition. The *LECD proximity effect*: at small cathode-anode distances, migration overcomes diffusion, the deposition rate increases greatly and the products are porous.
In Atomic Physics, the **proximity effect** refers to the intramolecular interaction between two (or more) functional groups (in terms of group contributions models of a molecule) that affects their properties and those of the groups located nearby. Cf. also *proximity effect (audio)* among **acoustics distances** in Chap. 21.
The term *proximity effect* is also used more abstractly, to describe some undesirable proximity phenomena. For example, the *proximity effect in the production of chromosome aberrations* (when ionizing radiation breaks double-stranded DNA) is that DNA strands can misrejoin if separated by less than $\frac{1}{3}$ of the diameter of a cell nucleus. The *proximity effect in innovation process* is the tendency to the geographic agglomeration of innovation activity.

- **Hopping distance**
  Hopping is atomic-scale long range dynamics that controls diffusivity and conductivity. For example, oxidation of DNA (loss of an electron) generates a radical cation which can migrate a long (more that 20 nm) distance, called the **hopping distance**, from site to site before it is trapped by reaction with water.
- **Atomic jump distance**
  In the solid state the atoms are about closely packed on a rigid lattice. The atoms of some elements (carbon, hydrogen, nitrogen), being too small to replace the atoms of metallic elements on the lattice, are located in the interstices between metal atoms and they diffuse by squeezing between the host atoms.
  Interstitial diffusion is the only mechanism by which atoms can be transported through a solid substance while, in a gas or liquid, mass transport is possible by both diffusion and the flow of fluid (for example, convection currents).
  The **jump distance** is the distance an atom is moved through the lattice in a given direction by one exchange of its position with an adjacent lattice site.
  Some crystals can jump 1,000 times their own length under light, since light energy rearranges atoms and builds strain, which is then explosively released.

The **mean square diffusion distance** d_t from the starting point which a molecule will have diffused in time t, satisfies $d_t^2 = r^2 N = r^2 \nu t = 2nDt$, where r is the jump distance, N is the number of jumps (equal to νt assuming a fixed jump rate ν), $n = 1, 2, 3$ for $1, 2, 3$-dimensional diffusion, and $D = \frac{\nu r^2}{2n}$ is the *diffusivity* in cm^2/s.
In diffusion alloy bonding, a **characteristic diffusion distance** is the distance between the joint interface and the site wherein the concentration of the diffusing substance (say, aluminum in high carbon-steel) falls to zero up to a given error.

- **Diffusion length**
Diffusion is a process of spontaneous spreading of matter, heat, momentum, or light: particles move to lower chemical potential implying concentration change. In Microfluidics, the **diffusion length** is the distance from the point of initial mixing to the complete mixing point where the equilibrium composition is reached.
In semiconductors, electron–hole pairs are generated and recombine. The (*minority carrier*) **diffusion length** of a material is the average distance a minority carrier can move from the point of generation until it recombines with majority carriers. Also, in electron transport by diffusion, the *diffusion length* is the distance over which concentration of free charge carriers injected into semiconductor falls to $\frac{1}{e}$ of its original value.
Cf. **jump distance** and, in Chap. 23, **capillary diffusion distance**.
- **Thermal diffusion length**
The heat propagation into material is represented by the **thermal diffusion length**, i.e., the propagation distance of the thermal wave producing an attenuation of the peak temperature to about 0.1 of the maximum surface value.
For lasers with femtosecond pulse duration, it is so small that the beam's energy, not being absorbed by laser-induced plasma, is fully deposited into the target.
The propagation of the laser-generated shock wave is approximated as *blast wave* (instantaneous, massless point explosion). The **expansion distance** is the distance between the surface of the target and the position of a blast wave; it depends on the energy converted into the plasma state.
- **Thermal entrance length**
In heat transfer at a boundary (surface) within a fluid, the **thermal entrance length** is the distance required for the *Nusselt number* (ratio of convective to conductive heat transfer across normal to the boundary) associated with the pipe flow to decrease to within 5 % of its value for a fully developed heat flow.
- **Distance-to-spot ratio**
The **distance-to-spot ratio** of an infrared temperature sensing device is the ratio of the distance to the object and the diameter of the t° measurement area.
- **Bjerrum length**
The **Bjerrum length** is the separation at which the electrostatic interaction between two elementary charges is comparable in magnitude to the thermal energy scale, $k_B T$, where k_B is the Boltzmann constant and T is the temperature in kelvin.

- **Lagrangian radius**
  The **Lagrangian radius** of the particle is the distance from the explosion center to a particle at the moment the shock front passes through it. Cf. in Chap. 25 unrelated *Lagrangian radii* in the item **radii of a star system**.
- **Reynolds number**
  For an object of a **characteristic length** (Chap. 29) l, flowing with mean relative velocity v in a *fluid* (liquid or gas) of the density ρ and dynamic viscosity μ, the **Reynolds number** is the ratio $Re = \frac{\rho v l}{\mu}$ of inertial forces to viscous forces.
  The flow is smooth (or *laminar*) if Re is low (viscous forces dominate), rough (or *turbulent*) if Re is high (usually $Re \geq 10^5$) and *transitional* in between. In a *Stokes flow* (laminar flow with very low Re), the inertial forces are negligible.
  The *law of the wall* (von Kármán, 1930) states that the average velocity of a turbulent flow close to the *wall* (boundary of the fluid region) is proportional to $\ln y^+$, where the **wall distance** $y^+ = \frac{u'}{\nu} y$ is the distance y to the wall, made dimensionless with the friction velocity u' at the wall and fluid's kinematic viscosity ν.
  In swimming, Re is 10^{-5}, 4×10^{-3}, 10^{-1}–10, 5×10^4 and 3×10^8 for bacterium, spermatozoa, small zooplankton, large fish and whale, respectively. In flying, Re is 30–4×10^4 for insects, 10^3–10^5 for birds, 1.6×10^6 for a glider and 2×10^9 for Boeing 747. Blood flow has $Re = 2 \times 10^{-3}, 140, 500$ and 3.4×10^3 in capillary, vein, artery and aorta, respectively. Re is a *dimensionless parameter*, i.e., the units of measurement in it cancel out. Examples of other such flow parameters follow.
  The *Mach number* Ma is a ratio of the speed of flow to the speed of sound in a fluid. Ma is ratio of inertia to *compressibility* (volume change as a response to a pressure). The flow is *subsonic*, *supersonic*, *transonic* or *hypersonic* if $Ma < 1$, $Ma > 1$, $0.8 \leq Ma \leq 1.5$ or $Ma \geq 5$, respectively. Ma governs *compressible* (i.e., those with $Ma > 0.3$) flows. The *Froude number* $Fr = \frac{v}{gl}$, where g is Earth gravity, is the ratio of the inertia to gravitation; it governs open-channel flows.
  The *lift* L and *drag* D are perpendicular and, respectively, parallel (to the oncoming flow direction) components of the force fluid flowing past the surface of a body exerts on it. In a flight without wind, the *lift-to-drag* ratio $\frac{L}{D}$ is the horizontal distance traveled divided by the altitude lost. $\frac{L}{D}$ is $4, 17, 20, 37$ for cruising house sparrow, Boeing 747, albatross, Lockheed U-2, respectively. Küchemann, 1978, found that the maximal (so, range-maximizing) $\frac{L}{D}$ for high Ma is $\approx 4 + \frac{12}{Ma}$.
- **Turbulence length scales**
  Turbulence is the time dependent chaotic behavior of fluid flows. The turbulent field consists of the superposition of interacting *eddies* (coherent patterns of velocity, vorticity and pressure) of different length scales. The kinetic energy cascades from the eddies of largest scales down to the smallest ones generated from the larger ones through the nonlinear process of vortex stretching.
  The **turbulence length scales** are measures of the eddy scale sizes in turbulent flow. Such standard length scales for largest, smallest and intermediate eddy sizes are called **integral length scale**, **Kolmogorov microscale** and

Taylor microscale, respectively. The corresponding ranges are called *energy-containing*, *dissipation* and *inertial range*.
Integral length scale measures the largest separation distance over which components of the eddy velocities at two distinct points are correlated; it depends usually of the flow geometry. For example, the largest integral scale of pipe flow is the pipe diameter. For atmospheric turbulence, this length can reach several hundreds km. On intermediate Taylor microscale, turbulence kinetic energy is neither generated nor destroyed but is transferred from larger to smaller scales.
At the smallest scale, the dynamics of the small eddies become independent of the large-scale eddies, and the rate at which energy is supplied is equal to the rate at which it is dissipated into heat by viscosity. The Kolmogorov length microscale is given by $\tau = (\frac{\nu^3}{\epsilon})^{\frac{1}{4}}$, where ϵ is the average rate of energy dissipation per unit mass and ν is the kinematic viscosity of the fluid. This microscale describe the smallest scales of turbulence before viscosity dominates. Similarly, the *Batchelor scale* (usually, smaller) describes the smallest length of fluctuations in scalar concentration before molecular diffusion dominates. *Quantum turbulence* is the chaotic motion of quantum fluids (say, *superfluids*) at high flow rates and close to absolute zero.
Turbulence is well described by the *Navier–Stokes equations*. Clay Mathematics Institute list the investigation, whether those equations in 3D always have a nonsingular solution, among the six US$1,000,000-valued open problems.

- **Meter of water equivalent**
The **meter of water equivalent** (mwe) of a material is the thickness of that material that provides the equivalent shielding of 1 m of water.
Also, the *mass balance* of a glacier is reported in mwe as the ratio of the volume (of water that would be obtained from melting the snow/ice) and area; it gives the change of thickness in water depth.
Unrelated *centimeter of water* (cmH_2O) is the pressure $\approx$98.1 Pa (pascals) exerted by a column of water of 1 cm in height at 4 °C at acceleration g. A similar *manometric* unit of pressure is the *millimeter of mercury* (mmHg) $\approx$1 Torr $\approx$133.3 Pa.
- **Hydraulic diameter**
For flow in a (in general, noncircular) pipe or tube, the **hydraulic diameter** is $\frac{4A}{P}$, where A is the cross-sectional area and P is the *wetted perimeter*, i.e., the perimeter of all channel walls that are in direct contact with the flow. So, in open liquid flow, the length exposed to air is not included in P. The hydraulic diameter of a circular tube is equal to its inside diameter.
The **hydraulic radius** is (nonstandardly) defined as $\frac{1}{4}$ of the hydraulic diameter.
- **Hydrodynamic radius**
The **hydrodynamic radius** (or *Stokes radius*, *Stokes–Einstein radius*) of a molecule, undergoing diffusion in a *solution* (homogeneous mixture of two or more substances), is the hypothetical radius of a hard sphere which diffuses with the same rate as the molecule. Cf. the **characteristic diameters** in Chap. 29.

- **Wigner–Seitz radius**
  The **Wigner–Seitz radius** is the radius of a sphere whose volume is equal to the mean volume $\frac{V}{N} = \frac{1}{n}$ per particle in a solid; n is the particle density. So, it is $(\frac{4}{3\pi n})^{\frac{1}{3}}$, an estimation of the mean interparticle distance.
- **Chromatographic migration distances**
  In thin layer Chromatography, the **solvent migration distance** is the distance d_{sl} traveled by the front line of the liquid or gas entering a chromatographic bed for *elution* (the process of using a solvent to extract an absorbed substance from a solid medium). The **migration distance of substance** is distance d_{sb} traveled by the center of a spot. The *retardation* and *retention factors* are $\frac{d_{sl}}{d_{sb}}$ and $\frac{d_{sb}}{d_{sl}} - 1$.
  The **retention distance** is a measure of equal-spreading of the spots on the chromatographic plate, defined via retention factors of sorted compounds.
- **Droplet radii**
  Let A be a small liquid droplet in equilibrium with a *supersaturated vapor*, i.e., a vapor which will begin to condense in the presence of nucleation centers.
  Let ρ_l, ρ_v be the liquid and vapor density, respectively, and let p_l, p_v be the liquid and vapor pressure. Let γ and γ_0 be the actual value at the surface of tension and planar limit value of surface tension.
  The **capillarity radius** R_c of A is defined by the *Young–Laplace equation* $\frac{\gamma_0}{R_c} = \frac{1}{2}(p_l - p_v)$.
  The **surface of tension radius** (or **Kelvin–Laplace radius**, *equilibrium radius of curvature*) R_s is defined by $\frac{\gamma}{R_s} = \frac{1}{2}(p_l - p_v)$. The reciprocal of R_s is the mean curvature $H = \frac{1}{2}(k_1 + k_2)$ (cf. Chap. 8) of the *Gibbs surface of tension* for which the Young–Laplace equation holds exactly for all droplet radii.
  The **equimolar radius** (or *Gibbs adsorption radius*) R_e of A is the radius of a ball of *equimolar* (i.e., with the same *molar concentration*) volume. Roughly, this ball has uniform density ρ_l in the cubic cell of density ρ_v.
  The **Tolman length** and the **excess equimolar radius** of the droplet A are $\delta = R_e - R_s$ and $\tau = R_e - R_c$, respectively.
  On the other hand, the **cloud drop effective radius** is a weighted mean of the size distribution of cloud droplets.
- **Dephasing length**
  Intense laser pulses traveling through plasma can generate, for example, a *wake* (the region of turbulence around a solid body moving relative to a liquid, caused by its flow around the body) or X-rays. The **dephasing length** is the distance after which the electrons outrun the wake, or (for a given mismatch in speed of pulses and X-rays) laser and X-rays slip out of phase.
- **Healing length**
  A *Bose–Einstein condensate* (BEC) is a state of dilute gas of weakly interacting bosons confined in an external potential, and cooled to temperatures near absolute zero (0 K, i.e., −273.15 °C), so that a large fraction of them occupy the lowest quantum state of the potential, and quantum effects become apparent on a macroscopic scale. Examples of BEC are *superconductors* (materials loosing all electrical resistance if cooled below critical temperature), *superfluids*

(liquid states with no viscosity) and *supersolids* (spatially ordered materials with superfluid properties). A BEC at 4.5×10^{-9} K was obtained by Leanhardt et al. in 2003.

The **healing length** of BEC is the width of the bounding region over which the probability density of the condensate drops to zero. For a superfluid, say, it is a length over which the wave function can vary while still minimizing energy.

- **Coupling length**

  In optical fiber devices mode coupling occurs during transmission by multimode fibers (mainly because of random bending of the fiber axis). Between two modes, a and b, the **coupling length** l_c is the length for which the complete power transfer cycle (from a to b and back) take place, and the **beating length** z is the length along which the modes accumulate a 2π phase difference. The resonant coupling effect is *adiabatic* (no heat is transferred) if and only if $l_c > z$.

  Furuya–Suematsu–Tokiwa, 1978, define the *coupling length* of modes a and b as the length of transmission at which the ratio $\frac{I_a}{I_b}$ of mode intensities reach e^2.

- **Localization length**

  Generally, the **localization length** is the average distance between two obstacles in a given scale. The localization scaling theory of metal-insulator transitions predicts that, in zero magnetic field, electronic wave functions are always localized in disordered 2D systems over a length scale called the **localization length**.

- **Thermodynamic length**

  Thermodynamic length (Weinhold, 1975) is a Riemannian metric defined on a manifold of equilibrium states of a thermodynamic system.

  It is a path function that measures the distance along a path in the state space. Cf. the **thermodynamic metrics** in Chap. 7.

- **Magnetic length**

  The **magnetic length** (or *effective magnetic length*) is the distance between the effective magnetic poles of a magnet.

  The *magnetic correlation length* is a magnetic-field dependent **correlation length**.

- **Correlation length**

  The **correlation length** (or *correlation radius*) is the distance from a point beyond which there is no further correlation of a physical property associated with that point. It is used mainly in statistical mechanics as a measure of the order in a system for phase transitions (fluid, ferromagnetic, nematic).

  For example, in a spin system at high temperature, the correlation length is $-\frac{\ln d \cdot C(d)}{d}$ where d is the distance between spins and $C(d)$ is the correlation function.

  In particular, the *percolation correlation length* is an average distance between two sites belonging to the same cluster, while the *thermal correlation length* is an average diameter of spin clusters in thermal equilibrium at a given temperature. In second-order phase transitions, the correlation length diverges at the *critical point*.

In wireless communication systems with multiple antennas, **spatial correlation** is a correlation between a signal's direction and the average received signal gain.

- **Long range order**
  A physical system has **long range order** if remote portions of the same sample exhibit correlated behavior. For example, in crystals and some liquids, the positions of an atom and its neighbors define the positions of all other atoms.
  Examples of long range ordered states are: superfluidity and, in solids, magnetism, charge density wave, superconductivity. Most strongly correlated systems develop long range order in their ground state.
  Short range refers to the finite **correlation length**, say, to the first- or second-nearest neighbors of an atom.
  The system has **long range order**, *quasi-long range order* or is *disordered* if the corresponding correlation function decays at large distances to a constant, to 0 polynomially, or to 0 exponentially. Cf. **long range dependency** in Chap. 18.
- **Spatial coherence length**
  The **spatial coherence length** is the propagation distance from a coherent source to the farthest point where an electromagnetic wave still maintains a specific degree of coherence. This notion is used in Telecommunication Engineering (usually, for the optical regime) and in synchrotron X-ray Optics (the advanced characteristics of synchrotron sources provide highly coherent X-rays).
  The spatial coherence length is about 20 cm, 100 m and 100 km for helium–neon, semiconductor and fiber lasers, respectively. Cf. *temporal coherence length* which describes the correlation between signals observed at different moments of time.
  For vortex-loop phase transitions (superconductors, superfluid, etc.), **coherence length** is the diameter of the largest thermally excited loop. Besides coherence length, the second **characteristic length** (cf. Chap. 29) in a superconductor is its **penetration depth**. If the ratio of these values (the *Ginzburg–Landau parameter*) is $<\sqrt{2}$, then the phase transition to superconductivity is of second-order.
- **Decoherence length**
  In disordered media, the **decoherence length** is the propagation distance of a wave from a coherent source to the point beyond which the phase is irreversibly destroyed (for example, by a coupling with noisy environment).
- **Critical radius**
  Critical radius is the minimum size that must be formed by atoms or molecules clustering together (in a gas, liquid or solid) before a new-phase inclusion (a bubble, a droplet, or a solid particle) is stable and begins to grow.
- **Binding energy**
  The **binding energy** of a system is the mechanical energy required to separate its parts so that their relative distances become infinite. For example, the binding energy of an electron or proton is the energy needed to remove it from the atom or the nucleus, respectively, to an infinite distance.
  In Astrophysics, *gravitational binding energy* of a celestial body is the energy required to disassemble it into dust and gas, while the lower *gravitational*

potential energy is needed to separate two bodies to infinite distance, keeping each intact.

- **Metric theory of gravity**

A **metric theory of gravity** assumes the existence of a symmetric metric (seen as a property of space-time itself) to which matter and nongravitational fields respond. Such theories differ by the types of additional gravitational fields, say, by dependency or not on the location and/or velocity of the local systems. General Relativity is one such theory; it contains only one gravitational field, the space-time metric itself, and it is governed by Einstein's partial differential equations. It has been found empirically that, besides Nordström's 1913 *conformally-flat scalar theory*, every other metric theory of gravity introduces auxiliary gravitational fields.

A **bimetric theory of gravity** is (Rosen, 1973) a metric theory of gravity in which two, instead of one, metric tensors are used for, say, effective Riemannian and background Minkowski space-times. But usually, rather two frames (not two metric tensors) are considered. Cf. **multimetric** in Chap. 3.

The *Brans–Dicke theory* is a metric theory of gravity, in which $\frac{1}{G}$, where G is the gravitational constant, is replaced by a scalar field. Another direct competitor of General Relativity is affine *Einstein–Cartan–Sciama–Kibble theory* relaxing the assumption that the metric be torsion-free and interpreting spin as affine torsion. It supposes (Sakharov, 1967) *Induced Gravity* with space-time background emerging as a mean field approximation of underlying microscopic degrees of freedom. Such quantum gravity is implied by a *World Crystal model* of **quantum space-time**.

Østvang, 2001, proposed a quasi-metric framework for relativistic gravity.

Classical physics adequately describes gravity only for masses of 10^{-23}–10^{30} kg.

- **Schwarzschild radius**

The **Schwarzschild radius** of a mass m is the radius $r_s(m) = \frac{2Gm}{c^2} = \frac{2m}{m_P} l_P$ of a sphere S such that, if m is compressed within S, it will become a *Schwarzschild* (i.e., uncharged and with angular momentum zero) *black hole*, and so, the escape speed from the surface of S would be the speed c of light.

For such hole, the radii of *photon sphere* (where photons are forced by gravity to travel in circular orbits), of *marginally bound orbit* (where a test particle starts to be gravitationally bound) and of *marginally stable orbit* (smallest circular orbit for material, usually the inner edge of the accretion cloud) are $\frac{3}{2}r_g$, $2r_g$ and $3r_g$.

A typical (stellar) black hole has mass $\approx 6\, M_{Sun}$ (where $M_{Sun} = 2 \times 10^{30}$ kg is the solar mass), diameter ≈ 18 km, temperature $\approx 10^{-8}$ K and lifetime $\approx 2 \times 10^{68}$ years. The black holes in our galaxy and in the galaxy NGC 4889 (largest known black hole at 2013) are supermassive: 4×10^6 and 21×10^9 suns. The radius of Sgr A^*, "our" black hole, is at most 12.5 light-hours (45 AU) since, otherwise, the star S2 would be ripped apart by hole's tidal forces.

Most black holes do not exceed 0.1 % of the mass of their host galaxies, but the one in NGC 1277 reached 17×10^9 suns, i.e., 14 % of the mass of this galaxy. The smallest known black holes, XTE J1650-500 and IGR J17091-3624, have mass 3.8 and less than 3 suns. The transition point separating neutron stars and black

holes is expected within 1.7–2.7M_{Sun}. *Neutron stars* are composed of the densest known form of matter. The radius of J0348-0432, the largest known (2.04 M_{Sun}) neutron star is $\approx$10 km, i.e., only about twice its Schwarzschild radius.
The "mini" black hole would be a hypothetical *Planck particle* with mass $\sqrt{\pi}m_P$, for which $r_s(m) = \lambda_C(m)$ (cf. **Compton wavelength**), and radius $r_s(\sqrt{\pi}m_P) = 2\sqrt{\pi}l_P$. Cf. *planckeon* in **quantum space-time**; it should have radius $l_P = \bar{\lambda}_C(m_P)$ and mass m_P, for which $\bar{\lambda}_C(m) = \frac{1}{2}r_s(m)$.
A *quasar* (quasi-stellar radio source) is a compact region in the center of a massive galaxy surrounding its central supermassive black hole; its size is 10–10,000 times the Schwarzschild radius of the black hole.
The Schwarzschild radius of observable Universe is $\approx$10 Gly. So, Pathria and Good, 1972, then Poplawski, from 2010, proposed that the observable Universe is the interior of a black hole existing inside a larger universe, or multiverse.

- **Jeans length**
The **Jeans length** (or *acoustic instability scale*) is (Jeans, 1902) the length scale $L_J = v_s t_g = \frac{v_s}{\sqrt{\rho G}}$ of a cloud (usually, of interstellar dust) where thermal energy causing the cloud to expand, is counteracted by self-gravity causing it to collapse. Here v_s, t_g, ρ are the speed of sound, gravitational free fall time and enclosed mass density. So, L_j is also the distance a sound wave would travel in the collapse time.
The *Jeans mass* is the mass contained in a sphere of Jeans length diameter.
- **Acoustic metric**
In Acoustics and Fluid Dynamics, the **acoustic metric** (or **sonic metric**) is a characteristic of sound-carrying properties of a given medium: air, water, etc.
In General Relativity and Quantum Gravity, it is a characteristic of signal-carrying in a given *analog model* (with respect to Condensed Matter Physics) where, for example, the propagation of scalar fields in curved *space-time* is modeled (see, for example, [BLV05]) as the propagation of sound in a moving fluid, or slow light in a moving fluid dielectric, or *superfluid* (quasi-particles in quantum fluid).
The passage of a signal through an acoustic metric modifies the metric; for example, the motion of sound in air moves air and modifies the local speed of the sound. Such "effective" (i.e., recognized by its "effects") **Lorentz metric** (cf. Chap. 26) governs, instead of the background metric, the propagation of fluctuations: the particles associated to the perturbations follow geodesics of that metric.
In fact, if a fluid is barotropic and inviscid, and the flow is irrotational, then the propagation of sound is described by an **acoustic metric** which depends on the density ρ of flow, velocity $\mathbf{v}$ of flow and local speed s of sound in the fluid. It can be given by the *acoustic tensor*

$$g = g(t, \mathbf{x}) = \frac{\rho}{s} \begin{pmatrix} -(s^2 - v^2) & \vdots & -\mathbf{v}^T \\ \cdots & & \cdots \\ -\mathbf{v} & \vdots & 1_3 \end{pmatrix},$$

where 1_3 is the 3×3 identity matrix, and $v = ||\mathbf{v}||$. The *acoustic line element* is

$$ds^2 = \frac{\rho}{s}\left(-(s^2 - v^2)dt^2 - 2\mathbf{v}d\mathbf{x}dt + (d\mathbf{x})^2\right) = \frac{\rho}{s}\left(-s^2dt^2 + (d\mathbf{x} - \mathbf{v}dt)^2\right).$$

The signature of this metric is (3, 1), i.e., it is a **Lorentz metric**. If the speed of the fluid becomes supersonic, then the sound waves will be unable to come back, i.e., there exists a *mute hole*, the acoustic analog of a *black hole*.
The **optical metrics** are also used in analog gravity and effective metric techniques; they correspond to the representation of a gravitational field by an equivalent optical medium with magnetic permittivity equal to electric one.

- **Aichelburg–Sexl metric**
In Quantum Gravity, the **Aichelburg–Sexl metric** (Aichelburg and Sexl, 1971) is a 4D metric created by a relativistic particle (having an energy of the order of the Planck mass) of momentum p along the x axis, described by its *line element*

$$ds^2 = dudv - d\rho^2 - \rho^2 d\phi^2 + 8p \ln \frac{\rho}{\rho_0} \delta(u)du^2,$$

where $u = t - x, v = t + x$ are null coordinates, ρ and ϕ are standard polar coordinates, $\rho = \sqrt{y^2 + z^2}$, and ρ_0 is an arbitrary scale constant.
This metric admits an nD generalization (de Vega and Sánchez, 1989), given by

$$ds^2 = dudv - (dX^i)^2 + f_n(\rho)\delta(u)du^2,$$

where X^i are the traverse coordinates, $\rho = \sqrt{\sum_{1 \le i \le n-2} (X^i)^2}$, $f_n(\rho) = K(\frac{\rho}{\rho_0})^{4-n}$, $k = \frac{8\pi^{2-0.5n}}{n-4}\Gamma(0.5n - 1)GP$, $n > 4$, $f_4 = 8GP \ln \frac{\rho}{\rho_0}$, P is the particle's momentum.

- **Quantum space-time**
Quantum space-time is a generalization of the usual space-time in which some variables that ordinarily commute are assumed not to commute, form a different Lie algebra, and, as a result, some variables may become discrete. For example, noncommutative field theory supposes that, on sufficiently small (quantum) distances, the spatial coordinates do not commute, i.e., it is impossible to measure exactly the position of a particle with respect to more than one axis. Any noncommutative algebra with ≥ 4 generators could be interpreted as a quantum space-time.

At Planck scale $l_P \approx 1.6 \times 10^{-35}$ m, "quantum foam" (Wheeler, 1950) is expected: violent warping and turbulence of space-time, which loses the smooth continuous structure (apparent macroscopically) of a *Riemannian manifold*, to become discrete, fractal, nondifferentiable.

Many models of *granular space* were proposed. *Planckeon* is (Markov, 1965) a hypothetical "grain of space" of size l_P and Planck rest mass m_P. In the *World Crystal model*, quantum space-time is a lattice with spacing of the order l_P, and matter creates defects generating curvature and all effects of General Relativity.

A **quantum metric** is a general term used for a metric expected to describe the space-time at quantum scales. Cf. **Rieffel metric space**, **Fubini–Study distance** (Chap. 7), **quantum graph** (Chap. 15), statistical geometry of fuzzy lumps [ReRo01], quantization of the **semimetric cone** (Chap. 1) in [IKP90].

Loop Quantum Gravity (LQG), *String Theory*, *Causal Sets* and *Black Hole Thermodynamics*, predict a quantum space-time at Planck scale. LQG predict, moreover, that its geometry (area, volume) is quantized via **spin networks** (Chap. 15). Analyses of gamma ray bursts rule out quantum graininess at $>10^{-48}$ m.

- **Distances between quantum states**

  A **distance between quantum states** is a metric which is preferably preserved by unitary operations, monotone under quantum operations, stable under addition of systems and having clear operational interpretation.

  The pure states correspond to the rays in the Hilbert space of wave functions. Every mixed state can be purified in a larger Hilbert space. The mixed quantum states are represented by *density operators* (i.e., positive operators of unit trace) in the complex projective space over the infinite-dimensional Hilbert space. Let X denote the set of all density operators in this Hilbert space. For two given quantum states, represented by $x, y \in X$, we mention the following main distances on X.

  The **trace distance** is a metric on density matrices defined by

$$T(x,y) = \frac{1}{2}||x-y||_{tr} = \frac{1}{2}\mathrm{Tr}\sqrt{(x-y)^*(x-y))} = \frac{1}{2}\mathrm{Tr}\sqrt{(x-y)^2} = \frac{1}{2}\sum_i |\lambda_i|,$$

  where λ_i are eigenvalues of the Hermitian matrix $x - y$. It is the maximum probability that a quantum measurement will distinguish x from y. Cf. the **trace norm metric** $||x - y||_{tr}$ in Chap. 12. When matrices x and y commute, i.e., are diagonal in the same basis, $T(x, y)$ coincides with **variational distance** in Chap. 14.

  The **quantum fidelity similarity** is defined (Jorza, 1994) by

$$F(x,y) = (\mathrm{Tr}(\sqrt{x}y\sqrt{x}))^2 = (||\sqrt{x}\sqrt{y}||_{tr})^2.$$

  When the states x and y are *classical*, i.e., they commute, $\sqrt{F(x,y)}$ is the classical **fidelity similarity** $\rho(P_1, P_2) = \sum_z \sqrt{p_1(z)p_2(z)}$ from Chap. 14.

When x and y are pure states, $F(x, y)$ is called *transition probability* and $\sqrt{F(x, y)} = |\langle x', y' \rangle|$ (where x', y' are the unit vectors representing x, y) is called *overlap*. In general, $F(x, y)$ is the maximum overlap between purifications of x and y. Useful lower and upper bounds for $F(x, y)$ are

$$\mathrm{Tr}(xy) + \sqrt{2((\mathrm{Tr}(xy))^2 - \mathrm{Tr}(xyxy))} \ (\textit{subfidelity}) \text{ and}$$

$$\mathrm{Tr}(xy) + \sqrt{(\mathrm{Tr}(x))^2 - \mathrm{Tr}(x^2))(\mathrm{Tr}(y))^2 - \mathrm{Tr}(y^2))} \ (\textit{super-fidelity}).$$

The **Bures–Uhlmann distance** is $\sqrt{2(1 - \sqrt{F(x, y)})}$. The **Bures length** (or *Bures angle*) $\arccos \sqrt{F(x, y)}$; it is the minimal such distance between purifications of x and y. Cf. the **Bures metric** and **Fubini–Study distance** in Chap. 7. In general, the Riemannian **monotone metrics** in Chap. 7 generalize the **Fisher information metric** on the class of probability densities (*classical* or commutative case) to the class of density matrices (quantum or noncommutative case).
The distances based on the *Shannon entropy* $H(p) = -\sum_i p_i \log p_i$ are generalized on quantum setting via the *von Neumann entropy* $S(x) = -\mathrm{Tr}(x \log x)$.
The **sine distance** (Rastegin, 2006) is a metric defined by

$$\sin \min_{x', y'} (\arccos(|\langle x', y' \rangle|)) = \sqrt{1 - F(x, y)},$$

where x', y' are purifications of x, y. It holds $1 - \sqrt{F(x, y)} \leq T(x, y) \leq \sqrt{1 - F(x, y)}$.
Examples of other known metrics generalized to the class of density matrices are the **Hilbert–Schmidt norm metric**, **Sobolev metric** (cf. Chap. 13) and **Monge–Kantorovich metric** (cf. Chap. 21).

- **Action at a distance (in Physics)**
An **action at a distance** is the interaction, without known mediator, of two objects separated in space. Einstein used the term *spooky action at a distance* for quantum mechanical interaction (like *entanglement* and *quantum nonlocality*) which is instantaneous, regardless of distance. His **principle of locality** is: distant objects cannot have direct influence on one another, an object is influenced directly only by its immediate surroundings.
Alice–Bob distance is the distance between two entangled particles, "Alice" and "Bob". Quantum Theory predicts that the correlations based on quantum entanglement should be maintained over arbitrary Alice–Bob distances. But a *strong nonlocality*, i.e., a measurable action at a distance (a superluminal propagation of real, physical information) never was observed and is not expected. Salart et al., 2008, estimated that such signal should be 10,000 times faster than light.
At 2012, some quantum information—the polarization property of a photon—to its mate in an entangled pair of photons, was teleported over 143 km. Lee et al., 2011, teleported, preserving superposition states, wave packets of light up to a bandwidth of 10 MHz. Wallraff et al., 2013 designed two atom-like systems,

which formed at $\approx$2 cm a type of weakly bound molecule, due to the exchange of photons.

Two-particle entanglement occurs in any temperature. Nuclear spins of ions can encode *qubits* (units of quantum information). Simmons et al., 2013, found that the superposition states (spins) of about 37 % of phosphorus ions in a sample (of silicon doped by P) survived 39 min at 25 °C and 3 h at −269 °C.

"Mental action at a distance" (say, telepathy, clairvoyance, distant anticipation, psychokinesis) is controversial because it challenge classical concepts of time/causality as well as space/distance.

The term *short range interaction* is used for the transmission of action at a distance by a material medium from point to point with a certain velocity dependent on properties of this medium. In Information Storage, the term *near-field interaction* is used for very short distance interaction using scanning probe techniques. *Near-field communication* is a set of standards-based technologies enabling short range ($\leq$4 cm) wireless communication between electronic devices.

- **Macroscale entanglement/superposition**

  Quantum superposition is the addition of the amplitudes of wave-functions, occurring when an object simultaneously "possesses" two or more values for an observable quantity, say, the position or energy of a particle. If the system interacts with its environment in a *thermodynamically irreversible way* (say, the quantity is measured), then *quantum decoherence* occurs: the state randomly collapses onto one of those values. But it can happen also without any influence from the outside world.

  Superposition and *entanglement* (nonlocal correlation which cannot be described by classical communication or common causes) were observed at atomic scale. Entangling in time (a pair of photons that never existed at the same time) was observed as well. With increasing duration and size/complexity of objects, these quantum effects are lost: decoherence, due to many interactions at the molecular level, occurs. To find out this threshold, if any, is a hot research topic.

  Nimmrichter and Hornberger, 2013 assign the *macroscopicity* μ to a quantum state, if the *equivalent* (in terms of ruling out even a minimal modification of Quantum Mechanics, which would predict a failure of the superposition principle on the macroscale) superposition state of a single electron last for 10^{τ} seconds. The record score so far is $\mu \approx 12$ and 24 looks reachable. But the Schrödinger's cat (seen as a 4-kg sphere of water) in a superposition, where it sits in two positions spaced 10 cm apart for 1 s, would score unconceivable 57.

  Szarek–Aubrun–Ye, 2013, found a threshold $k_0 \approx \frac{N}{5}$ such that two subsystems of k particles each of a system of N identical particles in a random pure state, typically share entanglement if $k > k_0$, and typically do not share it if $k < k_0$.

  "Warm" quantum coherence was observed in plant photosynthesis, animal magnetoreception, our sense of smell and microtubules inside brain neurons. Hameroff and Penrose, 2014: EEG rhythms (brain waves) and consciousness derive from quantum vibrations in microtubules, i.e., on the quantum-realm scale ($\sim$100 nm) rather than, or in addition to, the larger scale of neurons (4–100 μm).

- **Entanglement distance**
  The **entanglement distance** is the maximal distance between two entangled electrons in a *degenerate* electron gas beyond which all entanglement is observed to vanish. *Degenerate matter* (say, a white dwarf star) is matter having so high density that the main contribution to its pressure arises from the *Pauli exclusion principle*: no two identical fermions may occupy the same quantum state together.
- **Tunneling distance**
  Quantum Tunneling is the quantum mechanical phenomenon where a particle tunnels through a barrier that it classically could not surmount.
  For example, in *STM* (Scanning Tunneling Microscope), electron tunneling current and a net electric current from a metal tip of STM to a conducting surface result from overlap of electron wavefunctions of tip and sample, if they are brought close enough together and an electric voltage is applied between them.
  The tip-sample current depends exponentially (about $exp(-d^{0.5})$) on their distance d, called **tunneling distance**. Formally, d is the sum of the radii of the electron delocalization regions in the donor and the acceptor atoms.
  By keeping the current constant while scanning the tip over the surface and measuring its height, the contours of the surface can be mapped out. The tunneling distance is longer (<1 nm) in aqueous solution than in vacuum (<0.3 nm).

24.2 Distances in Chemistry and Crystallography

Main chemical substances are ionic (held together by ionic bonds), metallic (giant close packed structures held together by metallic bonds), giant covalent (as diamond and graphite), or molecular (small covalent). Molecules are made of a fixed number of atoms joined together by covalent bonds; they range from small (single-atom molecules in the noble gases) to very large ones (as in polymers, proteins or DNA).

The largest known (55 tons and 12, 4 m in diameter) crystal is a selenite found in Naica Mine, Mexico. The largest stable synthetic molecule is PG_5 with a diameter of 10 nm and a mass equal to 2×10^8 hydrogen atoms.

The **interatomic distance** of two atoms is the distance (in angstroms or picometers, where $1\,Å = 10^{-10}\,m = 10\,pm$) between their nuclei. The bond between helium atoms in molecules He_2 is the longest (54.6 Å) and weakest known; it is 0.75 Å in H_2.

- **Atomic radius**
  Quantum Mechanics implies that an atom is not a ball having an exactly defined boundary. Hence, **atomic radius** is defined as the distance from the atomic nucleus to the outermost stable electron orbital in a atom that is at equilibrium.

Atomic radii represent the sizes of isolated, electrically neutral atoms, unaffected by bonding.
Atomic radii are estimated from **bond distances** if the atoms of the element form bonds; otherwise (like the noble gases), only **van der Waals radii** are used.
The atomic radii of elements increase as one moves down the column (or to the left) in the Periodic Table of Elements. internuclear distance, Re is the equilibrium internuclear distance (bond length)

- **Bond distance**
  The **bond distance** (or *bond length*) is the equilibrium *internuclear distance* of two bonded atoms. For example, typical bond distances for carbon-carbon bonds in an organic molecule are 0.15, 0.13 and 0.12 pm (picometers 10^{-9} m) for single, double and triple bonds, respectively. The atomic nuclei repel each other; the **equilibrium distance** between two atoms in a molecule is the internuclear distance at the minimum of the electronic (or potential) energy surface.
  Depending on the type of bonding of the element, its atomic radius is called *covalent* or *metallic*. The *metallic radius* is one half of the **metallic distance**, i.e., the closest internuclear distance in a *metallic crystal* (lattice of metallic element). *Covalent radii* of atoms of elements that form covalent bonds are inferred from bond distances between pairs of covalently-bonded atoms. If the two atoms are of the same kind, then their covalent radius is one half of their bond distance. Covalent radii for other elements is inferred by combining the radii of those that bond with bond distances between pairs of atoms of different kind.
- **van der Waals contact distance**
  Intermolecular distance data are interpreted by viewing atoms as hard spheres. The spheres of two neighboring nonbonded atoms (in touching molecules or atoms) are supposed to just touch. So, their interatomic distance, called the **van der Waals contact distance**, is the sum of radii, called **van der Waals radii** (of *effective sizes*), of their hard spheres.
  The van der Waals contact distance corresponds to a "weak bond", when repulsion forces of electronic shells exceed London (attractive electrostatic) forces.
- **Molecular RMS radius**
  The **molecular RMS radius** (cf. **radius of gyration** in Sect. 24.1) is the root-mean-square distance of a molecule's atoms from their common center of gravity:

$$\sqrt{\frac{\sum_{1 \le i \le n} d_{0i}^2}{n+1}} = \sqrt{\frac{\sum_i \sum_j d_{ij}^2}{(n+1)^2}},$$

  where n is the number of atoms in the molecule, d_{0i} is the Euclidean distance of the i-th atom from the center of gravity of the molecule (in a specified conformation), and d_{ij} is the Euclidean distance between the i-th and j-th atoms.

The **mean molecular radius** is the number $\frac{\sum_i r_i}{n}$, where n is the number of atoms, and r_i is the Euclidean distance of the i-th atom from the centroid $\frac{\sum_j x_{ij}}{n}$ of the molecule (here x_{ij} is the i-th Cartesian coordinate of the j-th atom).

- **Molecular sizes**
  There are various descriptions of the **molecular sizes**; examples as follows.
  The *kinetic diameter* of a molecule (most applicable to transport phenomena) is its smallest effective dimension.
  The **effective diameter** of a molecule is the general extent of the electron cloud surrounding it as calculated in any of several ways.
  Sometimes, it is defined as diameter of the sphere containing 98 % of the total electron density; then its half is close to the experimental **van der Waals radius**.
  The **effective molecular radius** is the size a molecule displays in solution. For liquids and solids it is usually defined via packing density.
  For a gas, molecular sizes can be estimated from the intermolecular separation, speed, mean free path and collision rate of gas molecules.
  For example, in the model of kinetic theory of gases, assuming that molecules interact like hard spheres, the **molecular diameter** d is $\sqrt{\frac{m}{\pi\sqrt{2}\rho}}$, where m is the mass of molecule, l is mean free path and ρ is density.
- **Range of molecular forces**
  Molecular forces (or interactions) are the following electromagnetic forces: ionic bonds (charges), hydrogen bonds (dipolar), dipole-dipole interactions, London forces (the attraction part of van der Waals forces) and steric repulsion (the repulsion part of van der Waals forces). If the distance (between two molecules or atoms) is d, then (experimental observation) the potential energy function P relates inversely to d^n with $n = 1, 3, 3, 6, 12$ for the above five forces, respectively.
  The **range** (or the *radius*) of an interaction is considered *short* if P approaches 0 rapidly as d increases. It is also called *short* if it is at most 3 Å; so, only the range of steric repulsion is short (cf. **range of fundamental forces**).
  An example: for polyelectrolyte solutions, the long range ionic solvent–water force competes with the shorter range water–water (hydrogen bonding) force.
  In protein molecule, the range of London van der Waals force is $\approx$5 Å, and the range of hydrophobic effect is up to 12 Å, while the length of hydrogen bond is $\approx$3 Å, and the length of *peptide bond* is $\approx$1.5 Å.
- **Chemical distance**
  Various chemical systems (single molecules, their fragments, crystals, polymers, clusters) are well represented by graphs where vertices (say, atoms, molecules acting as monomers, molecular fragments) are linked by, say, chemical bonding, van der Waals interactions, hydrogen bonding, reactions path.
  In Organic Chemistry, a *molecular graph* $G = (V, E)$ is a graph representing a given molecule, so that the vertices $v \in V$ are atoms and the edges $e \in E$ correspond to electron pair bonds. The *Wiener number* of a molecule is one half of the sum of all pairwise distances between vertices of its molecular graph. The

Wiener polarity index is the number of unordered vertex pairs at distance 3 in this graph. Cf. **Wiener-like distance indices** in Chap. 1.
The (bonds and electrons) *BE-matrix* of a molecule is the $|V| \times |V|$ matrix $((e_{ij}))$, where e_{ii} is the number of free unshared valence electrons of the atom A_i and, for $i \neq j$, $e_{ij} = e_{ji} = 1$ if there is a bond between atoms A_i and A_j, and $= 0$, otherwise.
Given two *stoichiometric* (i.e., with the same number of atoms) molecules x and y, their **Dugundji–Ugi chemical distance** is the **Hamming metric**

$$\sum_{1 \leq i,j \leq |V|} |e_{ij}(x) - e_{ij}(y)|,$$

and their **Pospichal–Kvasnička chemical distance** is

$$\min_{\pi} \sum_{1 \leq i,j \leq |V|} |e_{ij}(x) - e_{\pi(i)\pi(j)}(y)|,$$

where π is any permutation of the atoms. The above distance is equal to $|E(x)| + |E(y)| - 2|E(x, y)|$, where $E(x, y)$ is the edge-set of the maximum common subgraph of the molecular graphs $G(x)$ and $G(y)$. Cf. **Zelinka distance** in Chap. 15.
The **Pospichal–Kvasnička reaction distance**, assigned to a molecular transformation $x \to y$, is the minimum number of *elementary transformations* needed to transform $G(x)$ onto $G(y)$.

- **Molecular similarities**
  Given two 3D molecules x and y characterized by some structural (shape or electronic) property P, their similarities are called **molecular similarities**.
  The main electronic similarities correspond to some correlation similarities from Chap. 17. For example, the *Carbó similarity* (Carbó–Leyda–Arnau, 1980) is the **cosine similarity** (cf. Chap. 17) defined by

$$\frac{\langle f(x), f(y) \rangle}{||f(x)||_2 \cdot ||f(y)||_2},$$

where the *electron density function* $f(z)$ of a molecule z is the volumic integral $\int P(z)dv$ over the whole space.
The *Hodgkin–Richards similarity* (1991) is defined (cf. the **Morisita–Horn similarity** in Chap. 17) by

$$\frac{2\langle f(x), f(y) \rangle}{||f(x)||_2^2 + ||f(y)||_2^2},$$

where $f(z)$ is the electrostatic potential or electrostatic field of a molecule z. Petitjean, 1995, proposed to use the distance $V(x \cup y) - V(x \cap y)$, where the *volume* $V(z)$ of a molecule z is the union of *van der Waals spheres* of its atoms. Cf. **van der Waals contact distance** and, in Chap. 9, **Nikodym metric** $V(x\Delta y)$.

- **End-to-end distance**

  A *polymer* is a large macromolecule composed of repeating structural units connected by covalent chemical bonds.

  For a coiled polymer, the **end-to-end distance** (or *displacement length*) is the distance between the ends of the polymer chain. The maximal possible such distance (i.e., when the polymer is stretched out) is called *contour length*.

  The root-mean-square end-to-end distance of ideal linear or randomly branched polymer scales as $n^{0.5}$ or, respectively, $n^{0.25}$ if n is the number of monomers. For a polymer chain following a random walk in 3D, it is also 6 times **molecular RMS radius**. The **strand length** in Chap. 23 is the end-to-end distance for a special linear polymer, single-stranded RNA or DNA.
- **Persistence length**

  The **persistence length** of a polymer chain is the length over which correlations in the direction of the tangent are lost.

  The molecule behaves as a flexible elastic rod for shorter segments, while for much longer ones it can only be described statistically, like a 3D random walk. Cf. **correlation length**.

  Twice the persistence length is the *Kuhn length*, i.e., the length of hypothetical segments which can be thought of as if they are freely jointed with each other in order to form given polymer chain.
- **Bend radius**

  In Polymer Tubing, the **bend radius** of a tube is the distance from the center of an imaginary circle on which the arc of the bent tube falls to a point on that arc.
- **Intermicellar distance**

  Micelle is an electrically charged particle built up from polymeric molecules or ions and occurring in certain colloidal electrolytic solutions like soaps and detergents. This term is also used for a submicroscopic aggregation of molecules, such as a droplet in a colloidal system, and for a coherent strand or structure in a fiber.

  The **intermicellar distance** is the average distance between micelles.
- **Interionic distance**

  An *ion* is an atom that has a positive or negative electrical charge. The **interionic distance** is the distance between the centers of two adjacent (bonded) ions. **Ionic radii** are inferred from ionic bond distances in real molecules and crystals.

  The ion radii of *cations* (positive ions, for example, sodium Na^+) are smaller than the atomic radii of the atoms they come from, while *anions* (negative ions, for example, chlorine Cl^-) are larger than their atoms.
- **Repeat distance**

  Given a periodic layered structure, its **repeat distance** is the period, i.e., the spacing between layers (say, lattice planes, bilayers in a liquid-crystal system, or graphite sheets along the unit cell's hexagonal axis).

A crystal lattice, the unit cell in it and cell spacing are called also a *repeat pattern*, *basic repeat unit* and *cell repeat distance* (or *lattice spacing*, *interplaner distance*).
The *repeat distance* in a polymer is the ratio of the unit cell length along its axis of propagation to the number of monomeric units this length covers.

- **Metric symmetry**
  The full crystal symmetry is given by its *space group*.
  The **metric symmetry** of the crystal lattice is its symmetry without taking into account the arrangement of the atoms in the unit cell.
  In between lies the *Laue group* giving equivalence of different reflexions, i.e., the symmetry of the crystal diffraction pattern. In other words, it is the symmetry in the *reciprocal space* (taking into account the reflex intensities).
  The Laue symmetry can be lower than the metric symmetry (for example, an orthorhombic unit cell with $a = b$ is metrically tetragonal) but never higher.
  There are seven *crystal systems*—*triclinic*, *monoclinic*, *orthorhombic*, *tetragonal*, trigonal, *hexagonal*, and *cubic* (or *isometric*). Taken together with possible *lattice centerings*, there are 14 *Bravais lattices*.
- **Homometric structures**
  Two structures of identical atoms are **homometric** if they are characterized by the same multiset of interatomic distances; cf. **distance list** in Chap. 1.
  Homometric crystal structures produce identical X-ray diffraction patterns.
  In Music, two rhythms with the same multiset of intervals are called **homometric**.
- **Dislocation distances**
  In Crystallography, a *dislocation* is a defect extending through a crystal for some distance (**dislocation path length**) along a *dislocation line*. It either forms a complete loop within the crystal or ends at a surface or other dislocation.
  The **mean free path** of a dislocation is (Gao et al., 2007), in 2D, the average distance between its origin and the nearest particle or, in 3D, the maximum radius of a dislocation loop before it reaches a particle in the slip plane.
  The **pinning distance** is the distance between two endpoints of a mobile dislocation, where one of the endpoints has to be within the volume. It is a characteristic length for the dislocation microstructure.
  The *Burgers vector* of a dislocation is a crystal vector denoting the direction and magnitude of the atomic displacement that occurs within a crystal when a dislocation moves through the lattice. A dislocation is called *edge*, *screw* or *mixed* if the angle between its line vector and the Burgers vector is 90°, 0° or otherwise, respectively. The **edge dislocation width** is the distance over which the magnitude of the displacement of the atoms from their perfect crystal position is greater than $\frac{1}{4}$ of the magnitude of the Burgers vector.
  The *dislocation density* ρ is the total length of dislocation lines per unit volume; typically, it is 10 km per cm^3 but can reach 10^6 km per cm^3 in a heavily deformed

metal. The **average distance** between dislocations depends on their arrangement; it is $\rho^{-\frac{1}{2}}$ for a quadratic array of parallel dislocations. If the average distance decreases, dislocations start to cancel each other's motion.
The **spacing dislocation distance** is the minimum distance between two dislocations which can coexist on separate planes without recombining spontaneously.

- **Dynamical diffraction distances**
Diffraction is the apparent bending of propagating waves around obstacles of about the wavelength size. Diffraction from a 3D periodic structure such as an atomic crystal is called *Bragg diffraction*. It is a convolution of the simultaneous scattering of the probe beam (light as X-rays, or matter waves such as electrons or neutrons) by the sample and interference (superposition of reflections from crystal planes).
The *Bragg Law*, modeling diffraction as reflexion from crystal planes of atoms, states that waves (with wavelength λ scattered under angle θ from planes at spacing d) interfere only if they remain in phase, i.e., $\frac{2d\sin\theta}{\lambda}$ is an integer.
The decay of intensity with depth traversed in the crystal occurs by *dynamical extinction*, redistributing energy within the wave field, and by *photoelectric absorption* (a loss of energy from the wave field to the atoms of the crystal).
The former *kinematic* theory works for imperfect crystals and estimates absorption. The *dynamical* (multiple diffraction) theory is used to model the *perfect* (no disruptions in the periodicity) crystals. It considers the incident and diffracted wave fronts as coupled/interacting parts of a wave field and the periodically varying electrical susceptibility of the medium so as to satisfy the Maxwell equations.
Dynamical theory distinguishes two cases: Laue (or *transmission*) and Bragg (or *reflexion*) case, when the reflected wave is directed toward the inside and outside of the crystal. The wave field is represented by its *dispersion surface*. The inverse of the diameter of this surface is called (Autier, 2001) the **Pendellösung distance** Λ_L in the Laue case and the **extinction distance** Λ_B in the Bragg case.
At the exit face of the crystal, the wave splits into two single waves with different directions: incident 0-beam and diffracted H-beam. With increasing thickness of the crystal, the wave leaving it will first appear mainly in the 0-beam, then entirely in the H-beam at thickness $\frac{\Lambda_L}{2}$, and then it will oscillate between these beams with a period Λ_L, called the **Pendellösung length**; cf. similar **coupling length**.
The wave amplitude (and the intensity of the diffracted beam) is transferred back-and-forth once, i.e., the physical distance acquires a phase change of 2π. Pendellösung oscillations happen also in Bragg case, but with very rapidly decaying amplitudes, and *Pendellösung fringes* are visible only for θ close to 0^0 or 45^0.
Diffraction that involves multiple scattering events is called *extinction* since it reduces the observed integrated diffracted intensity. Extinction is very significant

for perfect crystals and is then called *primary extinction*. In the Bragg case, the **primary extinction length** (James, 1964) is the inverse of the *extinction factor* (maximum extinction coefficient for the middle of the range of total reflection):

$$\frac{\pi V \cos\theta}{\lambda r_e |F| C},$$

where F, C (valued 1 or $\cos 2\theta$) are the structure and polarization factors, V is the volume of unit cell, $r_e \approx 2.81794 \times 10^{-15}$m is the *classical electron radius* and λ is the X-ray beam wavelength. The diffracted intensity with sufficiently large thickness no longer increases significantly with increased thickness.
The **extinction length** of an electron or neutron diffraction is $\frac{\pi V \cos\theta}{\lambda |F|}$. Half of it gives the number of atom planes needed to reduce the beam to 0 intensity.
The **X-ray penetration depth** (or *attenuation length*, *mean free path*, *extinction distance*) is (Wolfstieg, 1976) the depth into the material where the intensity of the diffracted beam has decreased e-fold. Cf. **penetration depth**.
In Gullity, 1956, *X-ray penetration depth* is the depth z such that $\frac{I_z}{I_\infty} = 1 - \frac{1}{e}$, where I_∞, I_z are the total diffraction intensities given from the whole specimen and, respectively, the range between the surface and the depth, z, from it.

- **X-ray absorption length**
  The *absorption edge* is a sharp discontinuity in the absorption spectrum of X-rays by an element that occurs when the energy of the photon is just above the **binding energy** of an electron in a specific shell of the atom.
  The **X-ray absorption length** of a crystal is the thickness s of the sample such that the intensity of the X-rays incident upon it at an energy 50 eV above the absorption edge is attenuated e-fold.
  For an X-ray laser, the *extinction length* is the thickness needed to fully reflect the beam; usually, it is a few microns while the absorption length is much larger.
  In Segmüller, 1968, the *absorption length* is $\frac{\sin\theta}{\mu}$, where μ is the linear absorption coefficient, and the beam enters the crystal at an angle θ.
- **Diamond-cutting distances**
  Diamond is the hardest natural gem and the only gemstone composed of a single element—carbon. Diamond takes a fine polish, which makes its surfaces highly reflective. Color in diamond (the rarest being red) is caused by structural irregularities, or trace elements. Diamonds are graded according to carat weight, clarity, color and cut. Diamonds are cut to maximize the play of light within the stone. Their beauty comes from a combination of *fire* (rainbow flash from within) and *brilliance* (burst of sparkling light). Both are a direct result of the cut.
  A faceted stone can be divided into an upper (*crown*) and lower (*pavilion*) section. The perimeter, where both parts meet, is referred to as the *girdle*. The **depth of a gemstone** is measured from the *table* (highest crown facet) to the *culet* (tip of the pavilion). On a round brilliant diamond, the **depth percentage** represents the ratio of the table-culet distance to the average girdle diameter.
  Normally, the table is the largest surface on a gemstone. On a round brilliant-cut diamond it forms an octagon, but some cutting styles do not have a table.

The **table percentage** of a diamond represents the ratio of table width to overall stone width. A beautiful, well-cut stone will normally have a table percentage 53–64 %. A stone's *luster* (appearance of the surface dependent upon its reflecting qualities) is directly affected by its depth and table percentages.

Chapter 25
Distances in Earth Science and Astronomy

25.1 Distances in Geography

- **Spatial scale**
  In Geography, **spatial scales** are shorthand terms for distances, sizes and areas. For example, micro, meso, macro, mega may refer to local (0.001–1), regional (1–100), continental (100–10,000), global (>10,000) km, respectively.
- **Earth radii**
  The Earth's maximal and minimal *radii* (the center-surface distances) are 6,384 km (the Chimborazo's summit) and 6,353 km (the Arctic Ocean's floor). An object, moved from the 2nd spot to the 1st, will loose $\approx 1\,\%$ of its weight.
  In the ellipsoidal model, the Earth's *equatorial radius* (*semi-major axis*) a, is 6,378 km and the *polar radius* (*semi-minor axis*) b, is 6,357 km. The equatorial and polar *radii of curvature* are $\frac{b^2}{a}$ and $\frac{a^2}{b}$. The *mean radius* is $\frac{2a+b}{3} = 6{,}371$ km. The Earth's *authalic* and *volumetric radius* (the radii of the spheres with the same surface area and volume, respectively, as the Earth's ellipsoid) are 6,371 and 6,371 km; cf. the **characteristic diameters** in Chap. 29.
  In Telecommunications, the *effective Earth radius* is the radius of a sphere for which the distance to the radio horizon, assuming rectilinear propagation, is the same as that for the Earth with an assumed uniform vertical gradient of atmospheric refractive index. For the standard atmosphere, this radius is $\frac{4}{3}$ that of the Earth.
- **Great circle distance**
  The **great circle distance** (or **orthodromic distance**, *air line*) is the shortest distance between points x and y on the Earth's surface measured along a path on this surface. It is the length of the *great circle* arc, passing through x and y, in the spherical model of the planet. Cf. **spherical metric** in Chap. 6.

M.M. Deza, E. Deza, *Encyclopedia of Distances*,
DOI 10.1007/978-3-662-44342-2_25

Let δ_1, ϕ_1 be the latitude and the longitude of x, and δ_2, ϕ_2 be those of y; let r be the Earth's radius. Here $2r^2 = a^2 + b^2$, where a and b are the equatorial and polar radii of the Earth. Then the great circle distance is equal to

$$r \arccos(\sin \delta_1 \sin \delta_2 + \cos \delta_1 \cos \delta_2 \cos(\phi_1 - \phi_2)).$$

In the spherical coordinates (θ, ϕ), where ϕ is the azimuthal angle and θ is the colatitude, the great circle distance between $x = (\theta_1, \phi_1)$ and $y = (\theta_2, \phi_2)$ is

$$r \arccos\left(\cos \theta_1 \cos \theta_2 + \sin \theta_1 \sin \theta_2 \cos(\phi_1 - \phi_2)\right).$$

For $\phi_1 = \phi_2$, the formula above reduces to $r|\theta_1 - \theta_2|$.
The **tunnel distance** between points x and y is the length of the line segment through 3D space connecting them. For a spherical Earth, this line is the chord of the great circle between the points.
The Earth resembles a flattened spheroid with extreme values for the radius of curvature of 6,335.4 km at the equator and 6,399.6 km at the poles. The *spheroidal distance* between points x and y is their distance in this spheroidal model.
The *geoid* (the shape the Earth would have if it was entirely covered by water and influenced by gravity alone) looks like a lumpy potato; cf. **potato radius**.

- **Loxodromic distance**
  A *loxodrome* (or, *rhumb line*) is a curve on the Earth's surface that crosses each meridian at the same angle. It is the path taken by a ship or plane maintaining a constant compass direction; it is a straight line on the Mercator projection.
  The **loxodromic distance** is a distance between two points on the Earth's surface on the rhumb line joining them. It is never shorter than the great circle distance.
  The **nautical distance** is the length in nautical miles of the rhumb line joining any two places on the Earth's surface. One nautical mile is equal to 1,852 m.
- **Continental shelf distance**
  Article 76 of the United Nations Convention on the Law of the Sea (1999) defined the *continental shelf of a coastal state* (its sovereignty domain) as the seabed and subsoil of the submarine areas that extend beyond its *territorial sea* as the natural prolongation of its land territory to the outer edge of the continental margin. It postulated that the **continental shelf distance**, i.e., the **range distance** from the baselines from which the breadth of the territorial sea is measured to above the other edge, should be within 200–350 nautical miles (370–648 km), and gave rules of its (almost) exact determination.
  Territorial sea is a belt of coastal waters extending at most 22 km. Next 370 km form the *exclusive economic zone*; first 22 km of it form the *contiguous zone*.
  Article 47 of the same convention postulated that, for an archipelagic state, the ratio of the area of its waters (sovereignty domain) to the area of its land, including atolls, should be between 1 to 1 and 9 to 1, and elaborated case-by-case rules.

There is no defined bottom underground and upper airspace limit for sovereignty. The international waters/seabed and celestial bodies are the common heritage of mankind for the signatories of the Law of the Sea and Outer Space (1967) treaties. Among divided islands, the largest one (785,753 km^2) is New Guinea and the smallest one is Koiluoto (200 m ×110 m, shared by Finland and Russia.

- **Port-to-port distance**
  The **port-to-port distance** is the shortest great circle distance between two ports that does not intersect any land contours.
  Officially published **distance between ports** represent the shortest navigable route or longer routes using favorable currents and/or avoiding some dangers to navigation. Reciprocal distances between two ports may differ.
- **Airway distance**
  An *airway* is a designated route in the air. Low altitude (or *victor*) airways are those below 5,500 m AMSL (above mean sea level). High altitudes (or *jet*) airways are those above 5,500 m AMSL. **Airway distance** is the actual (as opposed to straight line) distance flown by the aircraft between two points, after deviations required by air traffic control and navigation along published routes.
  The *stage length* is the distance of a nonstop leg of an itinerary. *Radar altitude* is the height with respect to the terrain below.
- **Point-to-point transit**
  Point-to-point transit is a route structure (common among low-fare airlines) where a plane, bus or train travels directly to a destination, rather than going through a central hub as in a *spoke-hub network*.
  A **point-to-point telecommunication** is a connection restricted to two endpoints as opposed to a *point-to-multipoint link* used in *hub and switch circuits*; cf. **flower-shop metric** in Chap. 19.
- **Lighthouse distance**
  The **lighthouse distance** is the distance from which the light from the lighthouse is first seen from of a sailboat. This distance (in feet) is $\approx 1.17(\sqrt{h_e} + \sqrt{h_l})$, where h_l is the lighthouse's height above tide level and h_e is the observer's eye level above sea. For $h_l = 0$, it estimates the **distance to horizon**.
- **Optical horizon**
  Optical (or, say, *neutrino, gravitational wave*) **horizon** is the farthest distance that any photon (respectively, neitrino and gravitational wave) can freely streem.
- **Distance to horizon**
  The *horizon* is the locus of points at which line of sight is tangent to the surface of the planet. At a height h above the surface of a spherical planet of radius R without atmosphere, the line-of-sight distance to the horizon is $d = \sqrt{(R+h)^2 - R^2}$, and the arc length distance to it along the curved planet's surface is $R\cos^{-1}(\frac{R}{R+h})$.
  Taking the equatorial radius 6,378 km of the Earth as a typical value, gives $d \approx 357\sqrt{h}$ m for small $\frac{h}{R}$. Allowing for refraction, gives roughly $d \approx 386\sqrt{h}$ m.

The *middle distance* is halfway between the observer and the horizon.

- **Radio distances**

  Marconi's law, 1897, claims that the maximum signalling distance of an antenna in meters is cH^2, where H is antenna's height and c is a constant.

  The *electrical length* is the length of a transmission medium or antenna element expressed as the number of wavelengths of the signal propagating in the medium. In coaxial cables and optical fibers, it is ≈ 1.5 times the physical length.

  The **electrical distance** is the distance between two points, expressed in terms of the duration of travel of an electromagnetic wave in free space between the two points. The *light microsecond*, $\approx$300 m, is a convenient unit of electrical distance.

  The main modes of electromagnetic wave (radio, light, X-rays, etc.) propagation are *direct wave* (line-of-sight), *surface wave* (interacting with the Earth's surface and following its curvature) and *skywave* (relying on refraction in the ionosphere).

  The **line-of-sight distance** is the distance which radio signals travel, from one antenna to another, by a *line-of-sight path*, where both antennas are visible to one another, and there are no metallic obstructions.

  The *radio horizon* is the locus of points in telecommunications at which direct rays from an antenna are tangential to the surface of the Earth. The **horizon distance** is the distance on the Earth's surface reached by a direct wave; due to ionospheric refraction or tropospheric events, it is sometimes greater than the distance to the visible horizon. In television, the **horizon distance** is the distance of the farthest point on the Earth's surface visible from a transmitting antenna.

  The **skip distance** is the shortest distance that permits a radio signal (of given frequency) to travel as a skywave from the transmitter to the receiver by reflection (hop) in the ionosphere.

  If two radio frequencies are used (for instance, 12.5 and 25 kHz in maritime communication), the **interoperability distance** and **adjacent channel separation distance** are the range within which all receivers work with all transmitters and, respectively, the minimal distance which should separate adjacent tunes for narrow-band transmitters and wide-band receivers, in order to avoid interference.

  DX is amateur radio slang (and Morse code) for distance; **DXing** is a distant radio exchange (amplifiers required). Specifically, *DX* can mean **distance unknown**, short for DXing and a far-away station that is hard to hear.

  Radio waves need 138 ms to go round the world and $\approx$2.57 s to be reflected from the Moon. **Long delayed echoes** (LDEs) are radio echoes which return to the sender later than 2.7 s after transmission; it is a rare and not explained phenomenon.

- **Ground sample distance**

  In Remote Sensing of the surfaces of terrestrial objects of the Solar System, including the Earth, the **ground sample distance** (or *GSD*, *ground sampling distance*, *ground-projected sample interval*) is the spacing of areas represented by each pixel in a digital photo of the ground from air or space.

For example, in an image with GSD 22 m, provided by UK-DMC2 (a British Earth imaging satellite), each pixel represents a ground area of 22^2 m^2.

- **Map's distance**
  The **map's distance** is the distance between two points on the map (not to be confused with **map distance** from Chap. 23). The length of a curved line feature on a map is usually measured by an *opisometer* (or *curvimeter*).
  The **horizontal distance** is determined by multiplying the map's distance by the numerical scale of the map.
  Map resolution is the size of the smallest feature that can be represented on a surface; more generally, it is the accuracy at which the location and shape of map features can be depicted for a given map scale.
- **Equidistant map**
  An **equidistant map** is a map projection of Earth having a well-defined nontrivial set of *standard lines*, i.e., lines (straight or not) with constant scale and length proportional to corresponding lines on the Earth. Some examples are:
  Sanson–Flamsteed equatorial map: all parallels are straight lines;
  cylindric equidistant map: the vertical lines and equator are straight lines;
  an *azimuthal equidistant map* preserves distances along any line through the central point; a *Werner cordiform map* preserves, moreover, distances along any arc centered at that point.
  Maurer–Close (or *doubly equidistant*) *map* preserves distances from two central points. If those points are identical, the map is azimuthal equidistant.
  A *gnomonic map* displays all great circles as straight lines; so, it preserves the shortest route between two locations.
- **Distance cartogram**
  A **distance** (or *linear*) **cartogram** is a diagram or abstract map in which distances are distorted proportionally to the value of some thematic variable. Typically, it shows the relative travel times and directions from vertices in a network.
- **Tolerance distance**
  In GIS (computer-based Geographic Information System), the **tolerance distance** is the maximal distance between points which must be established so that gaps and overshoots can be corrected (lines snapped together) as long as they fall within it.
- **Space syntax**
  Space syntax is a set of theories and techniques (cf. Hiller–Hanson, 1984) for the analysis of spatial configurations complementing Transport Engineering and geographic accessibility analysis in a GIS (Geographical Information System).
  It breaks down space into components, analyzed as networks of choices, and then represents it by maps and graphs describing the relative connectivity and integration of parts. The basic notions of space syntax are, for a given space:

  isovist (or visibility polygon), i.e., the field of view from any fixed point;
  axial line, i.e., the longest line of sight and access through open space;
  convex space, i.e., the maximal inscribed convex polygon (all points within it are visible to all other points within it).

These components are used to quantify how easily a space is navigable, for the design of settings where way-finding is important such as museums, airports, hospitals. Space syntax has also been applied to predict the correlation between spatial layouts and social effects such as crime, traffic flow, sales per unit area, etc.

- **Defensible space**
  In landscape use, **defensible space** refers to the 30 m zone surrounding a structure that has been maintained/designed to reduce fire danger. The first 9 m ($\approx$30 feet) is where vegetation is kept to a minimum combustible mass. The remaining area 9–30 m is the *reduced fuel zone*, where fuels and vegetation should be separated (by thinning, pruning, etc.) vertically and/or horizontally.
- **Sanitation distances**
  The **drinking distance** of a dwelling is its distance from the closest source of water.
  A *latrine* is a communal facility containing (usually many) toilets. It should be at most 50 m away from dwellings to be served and at least 50 m away from communal food-storage and preparation. A latrine should be $\geq$30 m from water-storage and treatment facilities, as well as from surface water and shallow groundwater sources. A septic tank should be $\geq$15 m from a water supply well.
  The **vertical separation distance** is the distance between the bottom of the drain field of a sewage septic system and the underlying water table. This separation distance allows pathogens (disease-causing bacteria, viruses, or protozoa) in the effluent to be removed by the soil before it comes in contact with the groundwater.
- **Setback distance**
  In land use, a **setback** (or *buffer*) **distance** is the minimum horizontal distance at which a building or other structure must legally be from property lines, or the street, or a watercourse, or any other place which needs protection. Setbacks may also allow for public utilities to access the buildings, and for access to utility meters. Cf. also **buffer distance** and **clearance distance** in Chap. 29.
- **Shy distance**
  Shy distance is the space left between vehicles (or pedestrians and vehicles) as they pass each other.
- **Distance-based numbering**
  The **distance-based exit number** is a number assigned to a road junction, usually an exit from a freeway, expressing in miles (or km) the distance from the beginning of the highway to the exit. A *milestone* (or *kilometer sign*) is one of a series of numbered markers placed along a road at regular intervals.
  The *Kilometer Zero* (or *km* 0) is a particular location (usually in the nation's capital city), from which distances are traditionally measured. For France, it is on the square facing the entrance of Notre Dame cathedral in Paris. For Russia, it is in a short passage connecting Red Square with Manege Square in Moscow.
  Distance-based house addressing is the system when buildings and blocks are numbered according to the distance, i.e., the number of increments (feet or division of miles), from a given baseline. For example, the number $67W430$

in Naperville, US, can express that the house is 67 miles west of downtown Chicago. One of the GIS-inspired guidelines: to use the address $n = \frac{d}{10} + 100$, where d is the distance in feet of the house from the reference point; roughly, $d = \frac{n}{500}$ miles.

Metes and bounds is a traditional system of land description (in Real Estate and town boundary determination) by courses and distances. *Metes* is a boundary defined by the measurement of each straight run specified as **displacement**, i.e., by the distance and direction. *Bounds* refers to a general boundary description in terms of local geography (along some watercourse, public road, wall, etc.). The boundaries are described in a running prose style, all the way around the parcel of land in sequence.

Surveying is the technique of determining the terrestrial and spatial position of points and the distances and angles between them; cf., for example, *Surveyor's Chain measures* among **imperial length measures** in Chap. 27.

- **Driveway distances**

  A *driveway* is a private road giving access from a public way. The main **driveway distances** follow.

  The *throat length* is the distance between the street and the end of the driveway inside the land development. It should be 200–250 feet (about 61–76 m or 15 car lengths) for shopping centers and 25–28, 9–15 m for small developments with or without signalized access.

  The optimal one-way *driveway width* is 4.5–5 m. Driveways entering a roadway at traffic signals should have two outbound lanes (for right and left turns) at least 7 m and an inbound lane at least 4.5 m wide. The normal width of residential driveways is 4.5–7.5 m.

  The *turn radius* is the extent that the edge of a commercial driveway is "rounded" to permit easier entry/exit by turning vehicles. In urban settings, it is 8–15 m.

- **Road sight distances**

  In Transportation Engineering, the *normal visual acuity* is the ability of a person to recognize a letter (or an object) of size 25 mm from a distance of 12 m.

  The **visibility distance** of a traffic control device is the maximum distance at which one can see it, while its **legibility distance** is the distance from which the driver can discern the intended message in order to have time to take the necessary action. For a safety sign, the **distance factor** is the ratio of the observation distance to the size of the symbol or text.

  The **clear sight distance** is the length of highway visible to a driver. The **safe sight distance** is the necessary sight distance needed to a driver in order to accomplish a fixed task. The main safe distances, used in Road Design, are:

  - the **stopping sight distance**—-to stop the vehicle before reaching an unexpected obstacle;
  - the *maneuver sight distance*—-to drive around an unexpected small obstacle;
  - the *road view sight distance*—-to anticipate the alignment (eventually curved and horizontal/vertical) of the road (for instance, choosing a speed);
  - the *passing sight distance*—-to overtake safely (the distance the opposing vehicle travels during the overtaking maneuver).

The *safe overtaking distance* is the sum of four distances: the passing sight distance, the *perception-reaction distance* (between decision and action), the distance physically needed for overtaking and the buffer safety distance.
Also, adequate sight distances are required locally: at intersections and in order to process information on traffic signs. A warning "objects in mirror are closer than they appear" can be required on vehicle's passenger side mirrors.
In a traffic flow, the *spacing* (or *distance headway*) is the distance between corresponding points (front to front) of consecutive vehicles moving in the same lane, while the *clearance* is the spacing minus the length of the leading vehicle. The corresponding time measures are *headway* and *gap*.

- **Road travel distance**
The **road travel** (or *road, driving, wheel, actual*) **distance** between two locations (say, cities) of a region is the length of the shortest road connecting them.
Some GISs (Geographic Information Systems) approximate road distances as the l_p-metric with $p \approx 1.7$ or as a linear function of **great circle distances**; in the US the *distance factor* (multiplier) is $\approx$1.15 in an east–west direction and $\approx$1.21 in the north–south direction. Several relevant notions of distance follow.
The **GPS navigation distance**: the distance directed by GPS (Global Positioning System, cf. **radio distance measurement** in Chap. 29) navigation devices. But this shortest route, from the GPS system point of view, is not always the best, for instance, when it directs a large truck to drive through a tiny village; cf. the Talmudic little boy's paradox among **distance-related quotes** in Chap. 28.
The **official distance**: the officially recognized (by, say, an employer or an insurance company) driving distance between two locations that will be used for travel or mileage reimbursement. Distance data (shortest paths between locations) are taken from a large web map service (say, MapQuest, Google, Yahoo or Bing) which uses a variation of the *Dijkstra algorithm*; cf. **Steiner ratio** in Chap. 1.
The **distance between zip codes** (in general, postal or telephone area codes) is the estimated driving distance (or driving time) between two corresponding locations.
Time-distance and **cost-distance** are time and cost measures of how far apart places are. The **journey length** is a general notion of distance used as a reference in transport studies. It can refer to, say, the average distance traveled per person by some mode of transport (walk, cycle, car, bus, rail, taxi) or a statutory vehicle distance as in the evaluation of aircraft fuel consumption.
An *odometer* is an instrument that indicates distance traveled by a vehicle. A *hubometer* is such device mounted on the axle of a vehicle, while a *trip meter* is an electronic device recording such distance in any particular journey.
Distance-based (or *mileage-based, per-mile*) *pricing* means that vehicle charges are based on the amount a vehicle is driven during a time period.

- **Horizontal distance**
  The **horizontal distance** (or **ground distance**) is the distance on a true level plane between two points, such as scaled off the map (it does not take into account the relief between two points). There are two types of horizontal distance: **straight line distance** (the length of the straight line segment between two points as scaled off the map), and **distance of travel** (the length of the shortest path between two points as scaled off the map, in the presence of roads, rivers, etc.).
  The *thalweg* (valley way) of a river or valley is the deepest inline within it.
  The *stream gradient* is the slope measured (say, in m/km) by the ratio of drop in a stream per unit distance; the *relief ratio* is such average drop. The *gradient of a road* is the ratio of the vertical to the horizontal distance, measured in m/km or as slop tangent of the angle of the elevation. The *pitch* (or *slope*, *incline*) of a roof is the ratio of the rise to the roof span, expressed in cm/m.
- **Slope distance**
  The **slope distance** (or **slant distance**) is the inclined distance (as opposed to the true horizontal or vertical distance) between two points.
  In Engineering, the **rollout distance** is the distance that a boulder or rock took to finally reach its resting point after rolling down a slope. The *release height* is the height at which a boulder or rock was released in relation to a slope.
  Naismith's rule in mountaineering: eight units of walking flat distance are time-equivalent to one unit of climb on a typical decline 12°.
  Craeme et al., 2014, claim that the cost for an organism of mass M kg to walk uphill, gaining 100 m in altitude, is $2.94M$ kJ.
  Walking uphill, humans and animals minimize metabolic energy expenditure; so, at critical slopes, they shift to zigzag walking. Langmuir's hiking handbook advises one to do it at 25°. Llobera and Sluckin, 2007, explain switchbacks in hill trails by the need to zigzag in order to maintain the critical slope, $\approx 16°$ uphill and $\approx 12.4°$ downhill. Skiing and sailing against the wind also require zigzagging.
  The west face of Mount Thor, in the Canadian Arctic, is the Earth's greatest *vertical drop*: a uninterrupted wall 1,250 m, with an average angle of 105°. The world record for the longest *rappel* (slope descent using ropes), 33 days, was set here in 2006. The world's highest unclimbed mountain is Gangkhar Puensum (7,570 m) on the Bhutan–Tibet border. The most dangerous by fatality rate mountains are: Annapurna and K2, 10th and 2nd highest ones: 8,091 and 8,611.
- **Vertical distance**
  The **vertical distance** of a location is its height above or depth below a fixed reference, say, the Earth's surface, mean sea level (MSL) or its model. On other planets, the elevations of solid surface are measured relative to the mean datum.
  The terms *elevation* (or *geometric height*), *altitude* (or *geopotential height*) and *depth* are used for points/planes on the ground, in the air and below the surface, respectively. *AMSL*, *AGL*, *AAE* and (in Broadcasting) *HAAT* mean height above MSL, ground level, nearest aerodrome and average (surrounding) terrain, respectively. The *height of an aircraft* is its AGL, i.e., AMSL plus elevation of the ground.

The **orthometric height** is the vertical distance of an object above the *geoid*, i.e., a surface of a constant potential which is the best approximation, in a least-square sense, of the global mean sea level.
The average and maximal land heights are 840 and 8,848 m (Mount Everest), while the average and maximal depths of the ocean are 3,730 and 10,911 m (Challenger Deep in Mariana Trench). The surface's points closest (6,353 km) and farthest (6,384 km) from the Earth's center are the bottom of the Arctic Ocean and the summit of the Andean volcano Chimborazo (6,268 m).

- **Prominence**
In Topography, **prominence** (or *autonomous height, relative height, shoulder drop*) is a measure of the stature of a summit of a hill or mountain. The prominence of a peak is the minimum height of climb to the summit on any route from a higher peak (called the *parent peak*), or from sea level if there is no higher peak. The lowest point on that route is the *col*. So, the prominence of any island or continental highpoint is equal to its elevation above sea level.
The highest mountains of the two largest isolated landmasses, Afro-Eurasia (Mount Everest) and the Americas (Aconcagua), have the most prominent peaks, 8,848 and 6,962 m. But from its ocean base, the elevation of the Hawaiian volcano Mauna Kea (4,205 m) is 10,203 m, and the mountain with the highest (5,486 m) elevation from its land base is Mount McKinley (6,193 m) in Alaska.
The *topographic isolation* of a summit is the great circle distance to the nearest point of equal elevation; for Everest, it is 40,008 km (Earth's circumference between the poles). *Spire measure* (or *ORS*, short for *omnidirectional relief and steepness*) is a rough measure of the visual "impressiveness" of a peak. It averages out how high and steep a peak is in all directions above local terrain.
- **Special parallels and meridians**
A network of *parallels* and *meridians* (lines of **latitude** and **longitude**, cf. Chap. 25) provides a locational system on Earth, using North Pole, South Pole (parallels 90°N and 90°S), rotation axis, and *equatorial plane* (an imaginary plane passing through Earth halfway between the poles and perpendicular to rotation axis).
The *equator* is the imaginary midline, where the equatorial plane intersects Earth's surface. It is the parallel of 0° latitude separating North and South hemispheres.
The *Prime meridian* and the *Date line* are internationally agreed at 0° and (with some bends, so as not to cross any land) $\approx$180° longitude; they form a great circle separating the Eastern and Western hemispheres. The point 0°, 0° is located in the Atlantic Ocean $\approx$614 km south of Accra, Ghana.
A degree of latitude varies from 110.567 km apart at the equator to 111.699 km at the poles; each minute ($\frac{1}{60}$-th of a degree) is $\approx$1 mile. A degree of longitude shrinks from 111.321 km at the equator to 0 at the poles.
The *circle of illumination* is the great circle that divides Earth between a light half and a dark half. The *land hemisphere* is the hemisphere containing the largest possible area, $\approx\frac{7}{8}$, of land. It is centered on 47°13′N 1°32′W (in the city of Nantes, France). The other half is the *water hemisphere*.

The *tropic of Cancer* and the *tropic of Capricorn* are parallels at which the Sun is directly overhead at the northern and the southern summer solstice, respectively. Their positions depend on the Earth's axial tilt. The region between them, centered on the equator, is the *tropics*. The regions around 25–30°N and S are *subtropics*, and the regions 30–35°N and S are *horse latitudes* (or *subtropical highs*). *Equatorial* and *polar* regions are within a few degrees of the equator or a pole. The *Arctic circle* and the *Antarctic circle* are parallels at which the Sun does not appear above the horizon at the northern and the southern winter solstice.
The longest land, continuous land, continuous sea latitudes are 48°24′53N (10,726 km France–Ukraine–Kazakhstan–China), 78°35′S (7,958 km Antarctica), 55°59′S (22,471 km). The longest land, continuous land, continuous sea longitudes are 22°E (13,035 km Russia–China–Tailand), 99°1′30″E (7,590 km), 34°45′45″W (15,986 km). The longest continuous land and sea distances along a great circle are 13,573 km (Liberia–Suez Canal–China) and ≈32,000 km.
Many parallels and meridians, often named and/or approximated, represent political boundaries. For example, 49°N latitude is (much of, from British Columbia to Manitoba) the border between Canada and US, 38°N is the boundary between North and South Korea, and 60°S is the northern boundary of Antarctica in the Antarctic Treaty. 2°20′14.025″E longitude is the *Paris meridian* (historic rival of the Prime meridian through Greenwich), 52.5°E is the official meridian of Iran, and ≈70°E was agreed in 1941 by Nazi Germany and the Empire of Japan as division of their spheres of interest in Asia. The *Brandt Line* (Brandt, 1980), represents socio-economic and political divide between the "rich North" and the "poor South". It encircles the world at ≈30°N latitude, passing between North and Central America, north of Africa and the Middle East, then going north so as to exclude China, Mongolia, Korea and going south so as to include Australia.

- **Remotest places on Earth**
  In medieval geographies, *ultima Thule* was any distant place located beyond the borders of the known world. Eratosthenes (c. 276–195 BC), measuring the *oikoumene* (inhabited world), put its northern limit in a mythical island Thule.
  The remotest island is uninhabited Bouvet island in the South Atlantic Ocean. Its nearest (1,600 km) land is Antarctica and nearest inhabited land is Tristan da Cunha, the remotest inhabited archipelago.
  Among other remotest (i.e., lacking normal transportation links) places on the Earth are: Kergelen (France), Pitcairn (UK), Svalbard (Norway) archipelagos, Easter (Chili), Foula (UK), Macquarie (Australia) islands, Motuo (China) county, McMurdo Station (Antarctica-US), La Rinconada (Peru, at an altitude of 5,100 km) towns, Alert (Canada, 800 km below the North Pole) village.
  The *continental pole of inaccessibility* (*Point Nocean*), the point on land farthest (2,514 km) from any ocean, lies in the Xinjiang, China, around 45°22′N 88°11′E.
  The *oceanic pole of inaccessibility* (*Point Nemo*), the point farthest (2,690 km) from any land, lies in the South Pacific Ocean at 48°52.6′S 123°23.6′W.

The *northern pole of inaccessibility* (84°03′N 174°51′W) is the point on the Arctic Ocean pack ice, 661 km from the North Pole, farthest (1,094 km) from any land mass. The *southern pole of inaccessibility* (82°06′S 54°58′E) is the point on the Antarctic, 878 km from the South Pole, farthest (1,300 km) from the ocean.

For a country, accessibility to its coast from its interior is measured by the ratio of coastline length in meters to land area in km^2. This ratio is the highest (10,100) for Tokelau and the lowest nonzero (0.016) for the Democratic Republic of the Congo. Canada has the longest (202,080 km) coastline.

The largest *antipodal* (diametrically opposite) land masses are the Malay Archipelago–Amazon Basin, and east China + Mongolia, antipodal to Chile + Argentina. Capitals close to being antipodes are: Buenos Aires–Beijing, Madrid–Wellington, Lima-Bangkok, Quito–Singapore, Montevideo–Seoul.

Politically unaccessed areas include isolated people (as Sentinelese and $\approx$100 tribes in dense forests) and unclaimed areas (antarctic Marie Byrd Land, Bir Tawil).

Counting as different only population centers at >1,000 km, the *point of minimum aggregate travel* (or *geometric median*, cf. **Fréchet mean** in Chap. 1) of the world's population lies around Afghanistan-Kashmir. This point is closest, 5,200 km of the mean great circle distance, to all humans, and its antipodal point is the farthest from mankind. But the closest, 5,600 km, point to the world's entire wealth (measured in GNP) lies in southern Scandinavia.

In terms of altitude, the number of people decreases faster than exponentially with increasing elevation (Cohen–Small, 1998). Within 100 m of sea level, lies 15.6 % of all inhabited land but 33.5 % of the world population live there. Altitude of residence (hypoxia?) is a risk factor for psychological distress and suicide in bipolar disorder.

- **Latitudinal distance effect**

  Diamond, 1997, explained the larger spread of crops and domestic animals along an east–west, rather than north–south, axis by the greater longitudinal similarity of climates and soil types.

  Ramachandran and Rosenberg, 2006, confirmed that genetic differentiation increases (and so, cultural interaction decreases) more with *latitudinal distance* in the Americas than with *longitudinal distance* in Eurasia. Randler, 2008: within the same time zone, people in the east get up and go to bed earlier than people in the west.

  Turchin–Adams–Hall, 2006, observed that $\approx$80 % of land-based, contiguous historical empires are wider in the east–west compared to the north–south directions. Three main exceptions—Egypt (New Kingdom), Inca, Khmer—obey a more general rule of expansion within an ecological zone.

  Taylor et al., 2014: polyandry in species is more common in northern latitudes. The *latitudinal biodiversity gradient* refers to the decrease in in both terrestrial and marine biodiversity, that occurs, the past 30 Ma, from the equator to the poles for most fauna and flora. Mace and Pagel, 1995 and 2004, found the same gradient for the density (number per range) of language groups and cultural variability.

Around 60 % of the world's languages are found in the great belts of equatorial forest. Papua New Guinea (14 % of languages), sub-Saharan Africa and India have the largest linguistic diversity. The number of phonemes in a language decrease, but the number of color terms increase, from the equator to the poles.

25.2 Distances in Geophysics

- **Atmospheric visibility distances**
  Atmospheric extinction (or *attenuation*) is a decrease in the amount of light going in the initial direction due to *absorption* (stopping) and scattering (direction change) by particles with diameter 0.002–100 μm or gas molecules. The dominant processes responsible for it are *Rayleigh scattering* (by particles smaller than the wavelength of the incident light) and absorption by dust, ozone O_3 and water. For example, mountains in the distance look blue due to the Rayleigh scattering effect.
  In extremely clean air in the Arctic or mountainous areas, the visibility can reach 70–100 km. But it is often reduced by air pollution and high humidity: haze (in dry air) or mist (moist air). *Haze* is an atmospheric condition where dust, smoke and other dry particles (from farming, traffic, industry, fires, etc.) obscure the sky. The World Meteorological Organization classifies the horizontal obscuration into the categories of fog (a cloud in contact with the ground), ice fog, steam fog, mist, haze, smoke, volcanic ash, dust, sand and snow. Fog and mist are composed mainly of water droplets, haze and smoke can be of smaller particle size.
  Visibility of less than 100 m is usually reported as zero. The international definition of *fog*, *mist* and *haze* is a visibility of <1 km, 1–2 km and 2–5 km.
  In the air pollution literature, **visibility** is the distance at which the contrast of a visual target against the background (usually, the sky) is equal to the threshold contrast value for the human eye, necessary for object identification, while **visual range** is the distance at which the target is just visible. Visibility can be smaller than the visual range since it requires recognition of the object.
  Visibility is usually characterized by either visual range or by the *extinction coefficient* (attenuation of light per unit distance due to four components: scattering and absorption by gases and particles in the atmosphere). It has units of inverse length and, under certain conditions, is inversely related to the visual range.
  Meteorological range (or *standard visibility*, *standard visual range*) is an instrumental daytime measurement of the (daytime sensory) visual range of a target. It is the furthest distance at which a black object silhouetted against a sky would be visible assuming a 2 % threshold value for an object to be distinguished from the background. Numerically, it is $\ln 50$ divided by the extinction coefficient.

In Meteorology, **visibility** is the distance at which an object or light can be clearly discerned with the unaided eye under any particular circumstances. It is the same in darkness as in daylight for the same air. **Visual range** is defined as the greatest distance in a given direction at which it is just possible to see and identify with the unaided eye in the daytime, a prominent dark object against the sky at the horizon, and at night, a known, unfocused, moderately intense light source.
The International Civil Aviation Organization defines the **nighttime visual range** as the greatest distance at which lights of 1,000 candelas can be seen and identified against an unlit background. Daytime and nighttime ranges measure the atmospheric attenuation of contrast and flux density, respectively.
In Aviation Meteorology, the **runway visual range** is the maximum distance along a runway at which the runway markings are visible to a pilot after touchdown. It is measured assuming constant contrast and luminance thresholds.
Oblique visual range (or *slant visibility*) is the greatest distance at which a target can be perceived when viewed along a line of sight inclined to the horizontal.

- **Atmosphere distances**
The **atmosphere distances** are the altitudes above Earth's surface (mean sea level) which indicate approximately the following specific (in terms of temperature, gravity, electromagnetism, etc.) layers of its atmosphere.
Below 1–2 km: *planetary boundary layer*, where winds are directly retarded by surface friction. The reminder of the atmosphere: the *free atmosphere*.
From 8 km: the death zone for human climbers (lack of oxygen).
From the *Armstrong line* (18.900–19.350 km) water boils at 37 °C (low pressure) and a pressure suit is needed.
Below 7–20 km (over the poles and equator, respectively): the *troposphere* in which temperature decreases with height (the weather and clouds occur here).
Above the troposphere to $\approx$51 km: the *stratosphere*, where the temperature increases with height (the ozone layer is at 19–48 km). The *tropopause* (its boundary with the troposphere) occurs at a pressure $\approx$0.1 bar; it is observed also on Jupiter, Saturn, Uranus, Neptune and expected on any thick-atmosphere exoplanet.
Above the stratosphere to 80–85 km: the *mesosphere*, in which temperature again decreases with height. Above the mesosphere to 500–1,000 km: the *thermosphere*, where the temperature again increases with height.
20–100 km: *near space* (or *upper atmosphere*), above airliners but below satellites.
100 km: the *Kármán line* prescribed by Fédération Aéronautique International as the boundary separating Aeronautics and Astronautics, near and outer space.
Above the thermosphere to $\approx$190,000 km : the *exosphere*, where molecules are still gravitationally bound but they can escape into space. Below the exosphere: the *homosphere*, where atmosphere has relatively uniform composition since turbulence causes a continuous mixing. The reminder of the atmosphere: the *heterosphere*.

From 50–80 to 2,000 km: the *ionosphere*, an electrically conducting region. From ≈160 km upwards: the *anacoustic zone*, where distances between air particles are so great that sound can no longer propagate; high-frequency sounds disappear first.
Up to 6–10 Earth radii on the sunward side: the *magnetosphere*, where Earth's magnetic field still dominates that of the solar wind. *Geospace* is the region from the beginning of ionosphere to the end of magnetosphere.
The altitude of the International Space Station is 278–460 km. 35,786 km: the altitude of geostationary (communication and weather) satellites. For observation and science satellites, it is 480–770 km and 4,800–9,700 km, respectively. Geocentric orbits with altitudes up to 2,000 km, 2,000–35,786 km and more than 35,786 km are called low, medium and high Earth orbits, respectively.
From 320,000 km: Moon's (at 356,000–406,700 km) gravity exceeds Earth's.
1,496,000 km = 0.011 AU: Earth's **Hill radius**, where Sun's gravity Earth's.

- **Wind distances**
Examples of wind-related distances follow.
Monin-Obukhov length: a rough measure of the height over the ground, where mechanically produced (by vertical wind shear) turbulence becomes smaller than the buoyant production of turbulent energy (dissipative effect of negative buoyancy). In the daytime over land, it is usually 1–50 m.
The **aerodynamic roughness length** (or *roughness length*) z_0 is the height at which a wind profile assumes zero velocity.
The **wind daily run** is the distance that results by integrating the wind speed, measured at a point, over 24 h. The fastest recorded wind speed near Earth's surface was 318 mph (i.e., 511.76 km/h) in Oklahoma, US, in 1999.
Rossby radius of deformation is the distance that cold pools of air can spread under the influence of the *Coriolis force*, i.e., the apparent deflection of moving objects when they are viewed from a rotating reference frame. It is the length scale at which effects, caused by Earth's rotation and the inertia of the mass experiencing the effect, become as important as buoyancy or gravity wave effects in the evolution of the flow about some disturbance.
The *aerial plankton* carried aloft by winds or convection, consists of bacteria, fungi, spores, pollen and small invertebrates. Even in the upper troposphere (8–15 km altitude), viable bacteria cells represent ≈20 % of 0.25–1 μm-sized particles.
The *jet streams* are fast flowing, narrow air currents found in the atmosphere. The strongest jet streams are, both west-to-east and in each hemisphere, the *Polar jet*, at 7–12 km above sea level, and the weaker *Subtropical jet* at 10–16 km.
The *atmospheric rivers* are narrow (a few hundred km across but several thousand km long) corridors of atmospheric water vapor transport over mid-latitude ocean regions. They account for over 90 % of such global meridional daily transport.
A *teleconnection* refers to climate anomalies being related to each other at large, say, thousands of km distances. For example, teleconnection between sea level pressure at Tahiti and Darwin, Australia, defines ENSO (El Niño Southern Oscillation).

- **Distances in Oceanography**
  Sea-surface height (SSH) is the height of the ocean's surface. **Decay distance**: the distance through which ocean waves travel after leaving the generating area. **Wavelength** is the distance between the troughs at the bottom of consecutive waves. A wave's *height* and *amplitude* are its trough-crest and rest-crest distances.
  The **significant wave height** (SWH) is the mean height $H_{1/3}$ of the highest third of waves. More modern and very close value is four times the root-mean-square of the surface elevation. Unusual *rogue waves* are defined as those with height $> 2H_{1/3}$; the tallest recorded one (by ship USS Ramapo in 1933) was 34 m. A wave up to $11H_{\frac{1}{3}}$ is possible. Large internal waves occur at thermoline and salwater-freshwater interface. A hundred-year wave is a statistically projected water wave, the height of which, on average, is met or exceeded once in a 100 years for a given location.
  The maximum horizontal distance inland and height reached there by tsunami waters are called the **run-up** (or *inundation*) **distance** and **run-up height**. It was 1,100 and 524 m for the 1958 Lituya Bay, Alaska, megatsunami, the largest recorded.
  Deep water (or *short*, *Stokesian*) **wave**: a surface ocean wave that is traveling in water depth greater than one-quarter of its wavelength; the velocity of deep water waves is independent of the depth. **Shallow water** (or *long*, *Lagrangian*) **wave**: a surface ocean wave of length 25 or more times larger than the water depth.
  Littoral (or *intertidal*): the zone between high and low water marks. Sometimes, *littoral* refers to the zone between the shore and water depths of $\approx$200 m.
  Oceanographic (or *thermal*) **equator**: the zone of maximum sea surface temperature located near (generally, north) the geographic equator. Sometimes, it is defined more specifically as the zone within which the sea surface temperature exceeds 28 °C. Below about 500 m, all of the world's oceans are at about 1.1 °C.
  Standard depth: a depth below the sea surface at which water properties should be measured and reported (in m): 0, 10, 20, 30, 50, 75, 100, 150, 200, 250, 300, 400, 500, 600, 800, 1,000, 1,200, 1,500, 2,000, 2,500, 3,000, 4,000, $\ldots$, 9,000, 10,000.
  Charted depth: the recorded vertical distance from the *lowest astronomical tide* (LAT, the lowest low water that can be expected in normal circumstances) to the seabed. **Drying height**: the vertical distance of the seabed that is exposed by the tide, above sea level at LAT. Actual depth of water is height of tide + charted depth or height of tide—drying height. **Tidal range**: the difference between the heights of high water and low water at any particular place. The empirical *rule of twelfths* assigns 6 h for it and subdivides the range into $1, 2, 3, 3, 2, 1$ twelfths per hour.
  The **thermocline**, **halocline** and **pycnocline**: the layers where the water temperature, salinity and density, respectively, change rapidly with depth.
  Depth of no motion: a reference depth in a body of water at which it is assumed that the horizontal velocities are practically zero. On a horizontal scale, *ocean fronts* are the boundaries between water masses with different properties.

Plankton (viruses, bacteria, phytoplankton, zooplankton and small pelagic larvae) aggregate at the clines, depth of no motion and persistent ocean fronts. $\approx$ 75 % of the water column's biomass consist of plankton organized in thin (< 3–4 m) layers 1–12 km in horizontal extent. Standard proxies for phyto- and zooplankton abundance are chlorofill-a imagery and sound attenuation. Giant (up to 130,000 km^2) bacterial mats float in the oxygen minimum zone off Chili and Peru.

Depth of the effective sunlight penetration: the depth at which $\approx$1 % of solar energy penetrates; in general, it does not exceed 100 m. The ocean is opaque to electromagnetic radiation with a small window in the visible spectrum. But it is transparent to acoustic transmission.

Depth of compensation: the depth at which illuminance has diminished to the extent that oxygen production through photosynthesis and oxygen consumption through respiration by plants are equal. The maximum depth for photosynthesis depends on plants and weather. Within the *epipelagic zone* there is enough light for photosynthesis, and thus plants and animals are largely concentrated here.

Below the *mesopelagic zone* lies the **aphotic zone** which is not exposed to sunlight. Organisms there depend on "marine snow" (a continuous shower of mostly organic detritus, decaying creatures and feces, falling from above) and chemosynthesis. The *deep sea* is the layer in the ocean below thermocline, at the depth 1,800 m or more.

The **pelagic zone** consists of all the sea other than that near the coast or the sea floor, while the *benthic zone* is the ecological region at the very bottom of the sea.

The ocean is divided into the following horizontal layers from the top down.

- From the surface down 200 m: *epipelagic* (sunlit zone);
- 200–1,000 m: *mesopelagic* (twilight zone);
- 1,000–4,000 m: *bathypelagic* (dark zone);
- 4,000–6,000 m: *abyssopelagic* (abyss);
- below 6,000 m: *hadalpelagic* (trenches).

Fast-flowing floods of turbid water form *abyssal channels* in sea floor.

The **deep sound** (or **SOFAR**, i.e., SOund Fixing And Ranging) **channel** is a layer of ocean water where the speed of sound is at a minimum ($\approx$1,480 m/s), because water pressure, temperature and salinity cause a minimum of water density. Sound waves of low frequency, caught and bent here, can travel hundreds of km. In low and middle latitudes, the SOFAR channel axis lies 600–1,200 m below the sea surface; it is deepest in the subtropics and comes to the surface in high latitudes.

The **SLD** (sonic layer depth) is the depth of maximum sound speed above this axis. The **best depth** for a submarine to avoid detection is SLD plus 100 m.

Mixing length: the distance at which an *eddy* (a circular movement of water) maintains its identity until it mixes. An eddy can reach 500 km across and persist for months. Cf. the **mean free path** and **diffusion length** in Chap. 24.

Mixed layer depth: the depth of the bottom of the *mixed layer*, i.e., a nearly isothermal surface layer of 40–150 m depth where water is mixed through wave action or thermohaline convection.

Depth of exponential mixing or **depth of homogeneous mixing** refers to a surface turbulent mixing layer in which the distribution of a constituent decreases exponentially, or is constant, respectively, with height.

Depth of frictional resistance: the depth at which the wind-induced current direction is 180° from that of the true wind.

The *fetch* (or *fetch length*): the horizontal distance along open water over which wave-generating wind or waves have traveled uninterrupted. In an enclosed body of water, the *fetch* is the distance between the points of minimum and maximum water-surface elevation. In Meteorology, the *fetch* is the distance upstream of a measurement site, receptor site, or region of interest that is relatively uniform.

The total volume of Earth's water is $\approx$1.39 billion km^3 (0.2 % Earth's mass) of which $\approx$96 % is liquid. For each 1 °C increase, in temperature, the sea level could rise by 5–20 m. Global sea level rose at 1.7 mm per year over 1950–2009.

- **River length**

  Meaning of **river length**, i.e., the distance between source and mouth, depend on their definitions, *anabranches* (multiple channels), map scale, etc.

  The **maximal river length** is the length of the longest continuous river channel in a given river system, regardless of name. Here, a river's "true source" is considered to be the source of whichever tributary is farthest from the mouth.

  The world's longest rivers are Nile-Kagera ($\approx$6,650–6,853 km), Amazon-Ucayali-Apurmac ($\approx$6,400–6,992 km) and Yangtze ($\approx$6,300–6,418 km).
- **Soil distances**

  Soil is composed of particles of broken rock that have been altered by chemical and environmental processes that include weathering and erosion. It is a mixture of mineral and organic constituents that are in solid, gaseous and aqueous states.

  A *soil horizon* is a specific layer in the land area that is parallel to the soil surface and possesses physical characteristics which differ from the layers above and beneath. Each soil type usually has three to four horizons.

  - *A Horizon* (or *topsoil*): the upper layer (usually 5–20 cm) with most organic matter accumulation and soil life.
  - *B Horizon* (or *subsoil*): the deeper layer accumulating by *illuviation* (action of rainwater), iron, clay, aluminum and organic compounds.
  - *C Horizon*: the layer which is little affected by soil forming processes.
  - *R Horizon*: the layer of partially weathered bedrock at the base of the soil profile.

  The *pedosphere* is the outermost layer of the Earth that is composed of soil and subject to soil formation processes. It lies below the vegetative cover of the biosphere and above the groundwater and *lithosphere* (outermost shell of the Earth).

Larger *Critical Zone* includes vegetation, the pedosphere, groundwater aquifer systems and ends in the bedrock where the biosphere and *hydrosphere* (combined mass of Earth's water) cease to make significant changes to the chemistry.
The *water table* (or *phreatic surface*) is the level at which the groundwater pressure is equal to atmospheric pressure.
The *cryosphere* is the part of the hydrosphere describing the Earth's ice: sea/lake/river ice, snow cover, glaciers, ice caps, ice sheets and frozen ground including permafrost. The Bentley Subglacial Trench in Antarctica is the world's deepest, 2,555 m, ice.
The Earth is now in a warm phase of the 5th (Quaternary) major Ice Age. This Age started 2.58 Ma ago and the last glacial expansion ended $\approx$11,500 years ago with the start of the Holocene. Next one is expected within coming 1,500–10,000 years unless anthropogenic global warming will delay it. The planet has warmed by only 0.74 °C since the early 1900s.

- **Frost line (in Earth Science)**
  The **frost line** (or *freezing depth*) is the depth to which the groundwater in soil is expected to freeze. In polar locations with year-round permafrost, the *thaw depth* is the depth to which the permafrost is expected to thaw each summer.
  In tropical regions, *frost line* may refer to the vertical geographic elevation below which frost does not occur. The *climatic snow line* is the point above which snow and ice cover the ground throughout the year; seasonally, snow occurs much lower. Cf. **frost line (in Astrophysics)**.
- **Moho distance**
  The Earth's *oceanic crust* (or *sima* for Si and Mg in basaltic rocks) is the surface, 5–10 km thick, of the ocean basins. The *continental crust* (or *sial* for Si and Al) is the layer of granitic rocks, 20–90 km thick, forming continents and continental shelves. The *Moho interface* (or *Mohorovicić seismic discontinuity*) is the boundary between the crust and the mantle, where the velocity of seismic P-waves increases. The **Moho distance** is the crustal thickness, i.e., the distance from a surface's point to the *Moho interface* beneath it.
  The world's lowest sea-drilled point was 10,680 m-deep (in the Gulf of Mexico) under 1,259 m of water. The Japanese research vessel *Chikyu*, aiming to the Moho interface, drilled 7,740 m below the sea level and 2,466 m below the seafloor. Cf. the lowest point on dry land (the shore of Dead Sea: 418 m), deepest cave (Krubera, Caucasus: 2,191 m), deepest mine (Mponeng gold mine, South Africa: about 4 km) and deepest drill (Kola Superdeep Borehole: 12,262 m). The temperature rises usually by 1° every 33 m.
  The *Curie depth* is the depth (usually 10–50 km) at which the temperature reaches the *Curie point* at which rocks lose their ferromagnetic properties.
  The Earth's mantle extends from the Moho interface to the mantle-core boundary at a depth of $\approx$2,890 km. The liquid outer core of radius 3,480 km contains a solid inner core (expanding $\approx$0.5 mm per year) of radius 1,220 km. The mantle is divided into the upper and the lower mantle at about 660 km. Main other seismic boundaries are at about 60–90 km (Hales discontinuity), 50–150 km (Gutenberg discontinuity), 220 km (Lehmann discontinuity), 410 km, 520 km, and 710 km.

The *lithosphere* comprises the crust and the rigid portion of the upper mantle that behaves elastically on large time scales. Its thickness is the depth of the isotherm $\approx$1,000° of the transition between brittle and viscous behavior. The lithosphere is broken into *tectonic plates* which float on the more plastic part of the mantle, the *asthenosphere*, 100–200 km deep.
The Eurasian and African plates are moving at the rates of 2 and 2.15 cm per year. The maximum earthquakes occur on the boundaries of the major moving plates. The eastern part of Indo-Australian plate is moving north 5.6 cm per year while the western part (India) is moving (north-east) only 3.7 cm per year due to impediment by Himalayas. The sequence of rare intraplate earthquakes in April 2012 off the coast of Sumatra, may eventually split this plate in two.

- **Distances in Seismology**
The Earth's crust is broken into tectonic plates that move around (at some cm per year) driven by the thermal convection of the deeper mantle and by gravity. At their boundaries, plates stick most of the time and then slip suddenly.
An *earthquake*, i.e., a sudden (several seconds) motion or trembling in the Earth, caused by the abrupt release of slowly accumulated strain, was, from 1906, seen mainly as a rupture (the sudden appearance, nucleation and propagation of a new crack or fault) due to elastic rebound. However, from 1966, it is seen within the framework of slippage along a pre-existing fault or plate interface, as the result of stick-slip frictional instability. One of most important parameters controlling fault instability is the **slip-weakening distance** D_c over which a fault weakens during its seismogenic motion. The coefficient of friction degrades linearly with slip until D_c is reached; then it stays constant.
So, an earthquake happens when dynamic friction becomes less than static friction. The advancing boundary of the slip region is called the *rupture front*. The standard approach assumes that the fault is a definite surface of tangential displacement discontinuity, embedded in a liner elastic crust.
Most earthquakes occur at near-vertical faults but a magnitude 6.0 earthquake at Kohat, Pakistan, in 1992, moved a 80 km^2 swath of land 30 cm horizontally. Almost all (81 and 17 %) world's largest earthquakes occur along the *Ring of Fire* (circum-Pacific seismic belt) and the *Alpide belt* (from Java to Sumatra through the Himalayas, the Mediterranean, and out into the Atlantic).
90 % of earthquakes are of tectonic origin, but they can also be caused by volcanic eruption, nuclear explosion and work in a large dam, well or mine. Earthquakes can be measured by **focal depth**, speed of slip, intensity (modified Mercali scale of earthquake effects), magnitude, acceleration (main destruction factor), etc.
The Richter scale of magnitude is computed from the amplitude and frequency of shock waves received by a seismograph, adjusted to account for the **epicentral distance**. An increase of 1.0 of this magnitude corresponds to an increase of 10 times in amplitude of the waves and $\approx$31 times in energy; the largest recorded value is 9.5 (Chile, 1960). Asteroid's impact in Yucatan 66 Ma ago was 12.55.

An earthquake first releases energy in the form of shock *pressure waves* that move quickly through the ground with an up-and-down motion. Next come shear S-waves which move along the surface, causing much damage: *Love waves* in a side-to-side fashion, followed by *Rayleigh waves* which have a rolling motion. The earthquake *extinction length* is the distance over which the S-wave energy is decreased by $\frac{1}{e}$.

Distance attenuation models (cf. **distance decay** in Chap. 29), used in Earthquake Engineering for buildings and bridges, postulate acceleration decay with an increase of some **site-source distance**, i.e., the distance between seismological stations and the crucial (for the given model) "central" point of the earthquake.

The simplest model is the *hypocenter* (or focus), i.e., the point inside the Earth from which an earthquake originates (the waves first emanate, the seismic rupture or slip begins). The *epicenter* is the point of the Earth's surface directly above the hypocenter. This terminology is also used for other catastrophes, such as an impact or explosion of a nuclear weapon, meteorite or comet but, for an explosion in the air, the term *hypocenter* refers to the point on the Earth's surface directly below the burst. A list of the main Seismology distances follows.

The **focal depth**: the distance between the hypocenter and epicenter. Earthquake is *shallow-*, *mid-* or *deep-focus* if it is <70, 70–300 or 300–700 km.

The **hypocentral distance**: the distance from the station to the hypocenter.

The **epicentral distance** (or **earthquake distance**): the **great circle distance** from the station to the epicenter.

The **Joyner–Boore distance** (1981): the distance from the station to the closest point, located over the *rupture surface* (rupturing portion of the fault plane).

The **rupture distance**: the distance from the station to the closest point on the rupture surface. The **seismogenic depth distance**: the distance from the station to the closest point of the rupture surface within the *seismogenic zone*, i.e., the depth range where the earthquake may occur; usually at depth 8–12 km.

The **crossover distance**: the distance on a seismic refraction survey time-distance chart at which the travel times of the direct and refracted waves are the same.

Also used are the distances from the station to:

- the center of static energy release and the center of static deformation of the fault plane;
- the surface point of maximal macroseismic intensity, i.e., of maximal ground acceleration (it can be different from the epicenter);
- the epicenter such that the reflection of body waves from the *Moho interface* (the crust-mantle boundary) contribute more to ground motion than directly arriving shear waves (it is called the *critical Moho distance*);
- the line extending the *fault trace* (top edge of the rupture) in both directions;
- the sources of noise and disturbances: oceans, lakes, rivers, railroads, buildings.

The **space-time link distance** between two earthquakes x and y is defined by

$$\sqrt{d^2(x, y) + C|t_x - t_y|^2},$$

where $d(x, y)$ is the distance between their epicenters or hypocenters, $|t_x - t_y|$ is the time lag, and C is a scaling constant needed to connect distance and time.
The *earthquake distance effect*: at greater distances from its center, the perception of an earthquake weaken and lower frequency shaking dominates it. Many animals hear infrasound of imminent earthquakes and feel primary P-waves.
Another space-time measure for catastrophic events is **distance between landfalls** for hurricanes hitting a US state. It is (Landreneau, 2003) the length of state's coastline divided by the number of hurricanes which have affected it from 1899.

- **Plume height**
  In a volcanic eruption, **plume height** refers to the highest point the eruptive cloud reaches before it flattens out and begins to drift downwind.
  The Volcanic Explosivity Index (VEI) is a scale measuring known eruptions by their volume of ejecta and plume height from VEI 0 (1,000 m^3, $<$ 100 m) to *mega-colossal* (or *super-volcano*) VEI 8 (1,000 km^3, $>$ 50 km).
- **Weather distance records**
  For a tornado, maximum width of damage, highest elevation, longest path: 4,000 m, 3,650 m, 472 km. The longest transport of a surviving human and of an object: 398 m and 359 km (personal check).
  Longest path of a tropical cyclone: 13,500 km; highest storm surge: 13 m.
  Largest snowflake and hail diameter: 38 and 20 cm. Longest lighting bolt: 190 km.
  Greatest minute, hour, day, month, year rainfall: 31.2 mm, 0.3 m, 1.82 m, 9.3 m, 26.47 m. Lowest and highest mean annual precipitation: 0.762 mm and 11.872 m.
- **Extent of Earth's biosphere**
  Life has adapted to every (except, perhaps, ocean vent locales >130 °C) ecological niche possessing liquid water and a source of free energy (say, sunlight, plate tectonics, water-rock chemistry). The main physical factors are temperature and pressure; their range for known active life as $[-20°, 122°]$C and $[5 \times 10^{-2}, 1.3 \times 10^3]$ bar. But the range, say, $[-30°, 135°]$C looks possible. The acidity/alkalinity range of known life is $[1, 11]$ on the pH scale, from acidic hot springs to soda lakes.
  In Jones–Lineweaver, 2010, the depth 5–10 km of the 122 °C isotherm and the altitude 10–15 km (a *tropopause* boundary of the vertical movement of particles) are the boundaries of active life. In Nussinov–Lysenko, 1991, the boundaries of biosphere are *Moho interface* (say, −30 km) and *Kármán line* (100 km).
  For humans, the typical bounds for main physiologic factors are: core temperature 35–38 °C, serum pH 7,35–7.45, plasma *osmolality* 270–290 mOsm/kg, fasting plasma glucose 3.3–5.6 mmol/l and serum calcium 2.2–2.6 mmol/l. But

there are permanent human habitations at mean annual temperatures of 34.4 °C, −19.7 °C and at an altitude of 5.1 km. Birds usually fly at altitudes 0.65–1.8 km but a vulture collided, at 11.3 km, with an aircraft. Deepest multicellular life are worms found at depths up to 3.6 km in gold mines and at a depth of 7.7 km in the Japan Trench.

Microbes, supported by chemosynthesis, have been found in cores drilled 5.3 km, in hydrothermal vents at 11 km depth and below 400 m of basalt rock +265 m of sediment +2.6 km ocean. Such *deep biosphere* (1–10 % of the world's biomass and the Earth vastest) is expected below the surface of continents and the bottom of the ocean. The same 19 deep-rock bacteria found to be similar worldwide.

The ranges for latent life (*cryptobiosis*: reversible state of low or undetectable metabolism) are much larger. Fungi and bacterial spores were found at an altitude 18–41 km. Examples of survival limits follow.

Some frogs, turtles and snakes survive the winter by freezing solid. A brine shrimp *Artemia* tolerates salt amounts of 25 %. Tardigrades, in cryptobiosis, survive −272 °C, 151 °C (a few minutes), pressure 6,000 bar, radiation 6,200 g (gray) and 120 years without water. Fly's larva *Polypedilium vanderplanki* dehydrates, in dry period, to 3 % water content, and it can survive −270 °C, 102 °C, radiation 7,000 g and 18 months in outer space vacuum. A parasitic leach *Ozobranchus jantseanus* survives −196 °C (24 h) and −90 °C during 32 months. Archea *Thermococcus gammatolerans* survive 30,000 gray of gamma rays. A bacterium survived 30 months on the Moon. Bacteria growing under hypergravity 403,627 g were cultivated.

Deinococcus radiodurans can survive extreme cold, dehydration, vacuum, radiation and acid; it has been listed by Guinness as the world's toughest bacterium. A bacteria *Tersicoccus phoenicis* has only been found in two spacecraft assembly clean rooms and is resistant to the methods used to clean such facilities.

Millions of years old (nondormant, just slow metabolizing) microbs, reproducing only every 10,000 years, were found in ocean floor. Bacterial spores were revived after 34,000 years of stasis; it was claimed also for 40 Ma old spores. A 1,300 years old lotus seed and 2,000 year old seed from extinct Judean date palm were germinated. *Silene stenophylla* was grown from 31,800 years old fruit.

Among the proponents of *panspermia* (the hypothesis that life, via extremophile bacteria and crystallized viruses surviving in space, propagates throughout the Universe) Yang et al., 2009, expect microbe density to be 10^{-3}–10^{-2} cells/m^3 at altitude 100 km and 10^{-6}–10^{-4} at 500 km. A large amount is expected at the altitude of the ISS (278–460 km). Napier–Wickramasinghe, 2010, claim that 10^{14}–10^{16} microorganisms ($\approx$10 tonnes) per year are ejected from Earth at survivable temperatures. Organics preserved in cometary amorphous ice and meteorite-formed glass can be transported from one planet to another.

A total of 7.5×10^{15} terrestrial microbes could reach the Moon per year, and the Solar System could be surrounded by an expanding biosphere of radius > 5 parsecs containing 10^{19}–10^{21} microbes. Wainwright et al., 2010, point out that no ubiquitous ultrasmall bacteria (passing through 0.1–0.2 μ filters) were found but large *Bacillus* and eukaryotes (5–100 μ fungal spores) have been isolated from the stratosphere. So, some viable but nonculturable microbes could be incoming from space. Hoover, 2011, found microfossils similar to filamentous prokaryotes in CI1 (Alais, Ivuna and Orgueil) and CM2 (Murchison and Murray) meteorites. Life on Mars, if any, is expected to be of the same origin (and, perhaps, earlier) as that on Earth, but it would have to be under at least 1 m of soil/rock to survive. Impact of icy comets crashing into Earth billions of years ago could have produced a variety of prebiotic or life-building compounds, including amino acids.

Interstellar panspermia, when the Sun passes a star-forming cloud, and even intergalactic panspermia, when galaxies collide, are debated. But on a cosmic scale, even enthusiasts of panspermia see it as a local, "a few megaparsec", phenomenon.

25.3 Distances in Astronomy

A *celestial object* (or *celestial body*) is a term describing astronomical objects such as stars and planets. The *celestial sphere* is the projection of celestial objects into their apparent positions in the sky as viewed from the Earth. The *celestial equator* is the projection of the Earth's equator onto the celestial sphere. The *celestial poles* are the projections of Earth's North and South Poles onto the celestial sphere. The *hour circle* of a celestial object is the great circle of the celestial sphere, passing through the object and the celestial poles.

The *ecliptic* is the intersection of the plane, containing the Earth's orbit, with the celestial sphere: seen from the Earth, it is the path that the Sun appears to follow over the course of a year. The *vernal equinox point* (or the *First point in Aries*) is one of the two points on the celestial sphere, where the equator intersects the ecliptic: it is the position of the Sun at the time of the vernal equinox.

In Astronomy, the *horizon* is the horizontal plane through the eyes of the observer. The *horizontal coordinate system* is a celestial coordinate system using the observer's local horizon as the *fundamental plane*, the locus of points having an altitude of 0°. The horizon is the line separating Earth from sky; it divides the sky into the upper hemisphere that the observer can see, and the lower hemisphere that he cannot. The pole of the upper hemisphere (the point of the sky directly overhead) is called the *zenith*; the pole of the lower hemisphere is called the *nadir*.

In general, an **astronomical distance** is a distance from one celestial body to another measured in light-years (ly), parsecs (pc), or astronomical units (au). The average distance between stars (in a galaxy like our own) is several ly; it is ≈ 6.57

ly in the solar neighborhood. The average distance between galaxies (in a cluster) is only about 20 times their diameter, i.e., several megaparsecs (Mpc). The separation between clusters of galaxies is typically of order 10 Mpc.

The large structures are groups of galaxies, clusters, galaxy clouds (or groups of clusters), superclusters, and supercluster complexes (or galaxy filaments, great walls). The Universe appears as a collection of giant bubble-like voids separated by great walls, with the superclusters appearing as relatively dense nodes. In the Universe, the average density of stars is about 1.4 per 100 billion cubic light-years, i.e., the average distance between them is about 4,150 light-years. The mean density of visible matter (i.e., galaxies) in the Universe is estimated as $\sim 10^{-31}$ g/cm^3, while 1 g/cm^3 is the mass density of water.

- **Latitude**
  In spherical coordinates (r, θ, ϕ), the **latitude** is the **angular distance** δ from the xy plane (*fundamental plane*) to a point, measured from the origin; $\delta = 90° - \theta$, where θ is the **colatitude**.
  In a *geographic coordinate system* (or *earth-mapping coordinate system*), the **latitude** is the angular distance from the Earth's equator to an object, measured from the center of the Earth. Latitude is measured in degrees, from $-90°$ (South Pole) to $+90°$ (North Pole). *Parallels* are the lines of constant latitude. The **colatitude** is the angular distance from the Earth's North Pole to an object.
  The **celestial latitude** is an object's latitude (measured in degrees) on the celestial sphere from the intersection of the fundamental plane with the celestial sphere in a given *celestial coordinate system.* In the *equatorial coordinate system* the fundamental plane is the plane of the Earth's equator; in the *ecliptic coordinate system* the fundamental plane is the plane of the ecliptic; in the *galactic coordinate system* the fundamental plane is the plane of the Milky Way; in the *horizontal coordinate system* the fundamental plane is the observer's horizon.
  Geomagnetic latitude is a parameter analogous to geographic latitude, except that bearing is with respect to the magnetic poles. The intersection between the magnetic and rotation axes of the Earth is located $\approx$500 km North from its centre.
- **Longitude**
  In spherical coordinates (r, θ, ϕ), the **longitude** is the **angular distance** ϕ in the xy plane from the x axis to the intersection of a great circle, that passes through the point, with the xy plane.
  In a *geographic coordinate system* (or *Earth-mapping coordinate system*), the **longitude** is the angular distance measured eastward along the Earth's equator from the *Greenwich meridian* (or *Prime meridian*) to the intersection of the meridian that passes through the object. Longitude is measured in degrees, from 0° to 360°. A *meridian* is a great circle, passing through Earth's North and South Poles; the meridians are the lines of constant longitude.
  The **celestial longitude** is the longitude of a celestial object (measured in units of time) on the celestial sphere measured eastward, along the intersection of the fundamental plane with the celestial sphere in a given *celestial coordinate system,* from the chosen point. In the *equatorial coordinate system* the fundamental plane

is the plane of the Earth's equator; in the *ecliptic coordinate system* it is the plane of the ecliptic; in the *galactic coordinate system* it is the plane of the Milky Way; and in the *horizontal coordinate system* it is the observer's horizon.

- **Declination**
  In the *equatorial* (or *geocentric*) *coordinate system*, the **declination** δ is the **celestial latitude** of a celestial object on the celestial sphere, measured from the celestial equator. Declination is measured in degrees, from $-90°$ to $+90°$.
- **Right ascension**
  In the *equatorial* (or *geocentric*) *coordinate system*, fixed to the stars, the **right ascension** RA is the **celestial longitude** of a celestial object on the celestial sphere, measured eastward along the celestial equator from the First point in Aries to the intersection of the hour circle of the celestial object. RA is measured in units of time with 1 h approximately equal to 15°.
  The time needed for one complete cycle of the precession of the equinoxes is called a *Platonic* (or *Great*) *year*); it is 257–258 centuries and slightly decreases. This cycle is important in Astrology. Also, it is close to the Maya calendar's longest cycle—5 *Great Periods* of 5,125 years; cf. **distance numbers** in Chap. 29.
  The time (225–250 million Earth years) it takes the Solar System to revolve once around the center of the Milky Way (*Solar circle*) is called the *Galactic year*.
- **Hour angle**
  In the *equatorial* (or *geocentric*) *coordinate system*, fixed to the Earth, the **hour angle** HA is the **celestial longitude** of a celestial object on the celestial sphere, measured along the celestial equator from the observer's meridian to the intersection of the hour circle of the celestial object.
  HA gives the time elapsed since the celestial object's last transit at the observer's meridian (for $HA > 0$), or the time until the next transit (for $HA < 0$).
- **Polar distance (in Geography)**
  In the *equatorial* (or *geocentric*) *coordinate system*, the **polar distance** (or *codeclination*) PD is the **colatitude** of a celestial object, i.e., the **angular distance** from the celestial pole to a celestial object on the celestial sphere. Similarly as the **declination** δ, it is measured from the celestial equator: $PD = 90° \pm \delta$. An object on the celestial equator has $PD = 90°$.
- **Ecliptic latitude**
  In the *ecliptic coordinate system*, the **ecliptic latitude** is the **celestial latitude** (in degrees) of a celestial object on the celestial sphere from the ecliptic.
  The object's **ecliptic longitude** is its **celestial longitude** on the celestial sphere measured eastward along the ecliptic from the First point in Aries.
- **Zenith distance**
  In the *horizontal* (or *Alt/Az*) *coordinate system*, the **zenith distance** (or *North polar distance*, *zenith angle*) ZA is the object's **colatitude**, measured from the zenith.

- **Altitude**
  In the *horizontal* (or *Alt/Az*) *coordinate system*, the **altitude** ALT is the **celestial latitude** of an object from the horizon. It is the complement of the **zenith distance** ZA: $ALT = 90° - ZA$. Altitude is measured in degrees.
- **Azimuth**
  In the *horizontal* (or *Alt/Az*) *coordinate system*, the **azimuth** is the **celestial longitude** of an object, measured eastward along the horizon from the North point. Azimuth is measured in degrees, from 0° to 360°.
- **Elliptic orbit distance**
  The **elliptic orbit distance** is the distance from a mass m which a satellite body has in an elliptic orbit about the mass M at the focus. This distance is given by

$$r(\theta) = \frac{a(1-e^2)}{1+e\cos\theta},$$

  where a is the *semi-major axis* (half of the major diameter), e is the *eccentricity* $\frac{c}{a}$ (where c is half the distance between the foci), and θ is the orbital angle.
  The **periapsis distance** and **apoapsis distance** are the closest and farthest distances $r_- = r(0) = a(1-e)$ and $r_+ = r(\pi) = a(1+e)$. An *anomaly* (in Astronomy) is a quantity measured with respect to an apsis, usually the periapsis. The **orbital distance** is mean $r(\theta)$ over the *eccentric anomaly*, i.e., $\frac{1}{2}(r_+ + r_-) = a$, while such mean distance over the **true anomaly** (the **angular distance** of a point in an orbit past the point of **periapsis**) is the *semi-minor axis* $b = a\sqrt{1-e^2}$.
  For *orbital period* T, Newton made precise the 3rd Kepler's law ($T^2 \sim a^3$) by

$$T^2 = \frac{4\pi^2}{G(M+m)}a^3.$$

  The *near-Earth objects* are asteroids, comets, spacecraft and large meteoroids whose *perigee* is closer than 1.3 AU; the largest such asteroids are 1036 Ganymede and 433 Eros, about 34 km across. The *perigee* and *apogee* are the points at periapsis and apoapsis distances of an elliptical orbit around the Earth, while the *perihelion* and *aphelion* are such points around the Sun.
  The *periastron* and *apastron* of a double star are the closest and farthest points of the smaller star to its primary.
- **Minimum orbit intersection distance**
  The **minimum orbit intersection distance** (MOID) between two bodies is the distance between the closest points of their gravitational *Kepler orbits* (ellipse, parabola, hyperbola or straight line).
  An asteroid or comet is a *potentially hazardous object* (PHO) if its Earth MOID is less than 0.05 AU and its diameter is at least 150 m. Impact with a PHO occurs on average around once per 10,000 years. The only known asteroid whose hazard could be above the background is 1950 DA (of mean diameter 1.2 km) which can,

with probability $\frac{1}{300}$, hit Earth on March 16, 2880. The closest known geocentric distance for a comet was 0.0151 AU (Lexell's comet on July 1, 1770).

- **Impact distances**
After an impact event, the falling debris forms an *ejecta blanket*, i.e., a generally symmetrical apron of ejecta that surrounds crater. About half the volume of ejecta falls within 2 radii from the center of the crater, and over 90 % falls within ≈5 radii. Beyond it, the debris are discontinuous and are called *distal ejecta*.
Main parameter of an impact crater is the ratio of *rim-to-floor depth* d to the *rim-to-rim diameter* D. The *simple craters* are small with $\frac{1}{7} \leq \frac{d}{D} \leq \frac{1}{5}$ and a smooth bowl shape. If $D > D_0$, where the *transitional diameter* D_0 scales as the inverse power of the planet's surface gravity, the initially steep crater walls collapse gravitationally downward and inward, forming a complex structure. On Earth, $2 \leq D_0 \leq 4$ km depending on target rock properties; on the Moon, $15 \leq D_0 \leq 20$ km.
The largest known (diameter of 300 km) and old (2,023 Ma ago) *astrobleme* (meteorite impact crater) is Vredefort Dome, 120 km south–west of Johannesburg. It was the world's greatest known single energy release event and largest asteroid known to have impacted the Earth (≈10 km). The diameter of MAPCIS crater in Australia is 600 km, but it is not confirmed impact crater.
Sometimes, the term *impact distance* is used more generally as a **setback distance** from some possible hazard (say, explosion, toxic chemical release, odor from swine facilities) or from the action of some equipment (say, laser, homogenizer); Cf. **standoff distance** and **protective action distance** in Chap. 29.
- **Elongation**
Elongation (or *digression*) is the angular distance in longitude of a celestial body from another around which it revolves (usually a planet from the Sun).
- **Lunar distance**
The **lunar distance** is the **angular distance** between the Moon and another celestial object.
In Astronomy, *new moon* (or *dark moon*) is a lunar phase that occurs at the moment of conjunction in **ecliptic longitude** with the Sun. If, moreover, the Sun, Moon, and Earth are aligned exactly, a solar eclipse occurs. *Full moon* occurs when the Moon is on the opposite side of the Earth from the Sun. If, moreover, the Sun, Earth, and Moon are aligned exactly, a lunar eclipse occurs.
A *supermoon* (or *perigee-syzygy of the Earth–Moon–Sun system*) is the near-coincidence of a full moon or a new moon with the closest approach the Moon makes to the Earth on its orbit, resulting in its largest apparent size.
- **Sun–Earth–Moon distances**
The Sun, Earth and Moon have masses 1.99×10^{30}, 5.97×10^{24}, 7.36×10^{22} kg and equatorial radii 695,500, 6,378, 1,738 km, respectively.
Earth's axial tilt varies 22.1–24.5° about every 41,000 years, its rotation occurs about every 19,000 years and eccentricity cycles 0.003–0.058 about every 0.1 Ma.

The Earth and the Moon are at a mean distance of 1 AU $\approx$1.496 × 10^8 km from the Sun. This distance increases at the present rate $\approx$15 cm per year.
The Moon, at distance 0.0026 AU ($\approx$60 Earth radii $R_\oplus$), is within the **Hill radius** (1,496,000 km) of the Earth, but well outside of the **Roche radius** (9,496 km).
Asimov argued that the Earth–Moon system is a double planet because their diameter and mass ratios ($\approx$4:1 and $\approx$81:1) are smallest for a planet in the Solar System. Also, the Sun's gravitational effect on the Moon is more than twice that of Earth's. But the *barycenter* (common center of mass) of the Earth and Moon lies well inside the Earth, $\approx\frac{3}{4}$ of its radius.
The Moon has a greater tidal influence on the Earth than the Sun. Because of tidal forces, the Moon is receding from the Earth at $\approx$3.8 cm per year. So, Earth's rotation is slowing, and Earth's day increases by $\approx$23 s every million years (excluding glacial rebounds). At present rate, the Moon's orbital distance will reach, $\approx$1 Ga from now, $67R_\oplus$, and Earth's axial tilt will become chaotic.

- **Opposition distance**
  A *syzygy* is a straight line configuration of three celestial bodies A, B, C. Then, as seen from A, B and C are in *conjunction*, and the passage of B in front of C is called *occultation* if the apparent size of B is larger, and *transit*, otherwise. *Appulse* is the closest approach of B and C as seen from A.
  If B and C are planets orbiting the star A, then C said to be in *opposition* to A, and the distance between B and C (roughly, their closest approach) is called their **opposition distance**. It can vary at different oppositions.
  The closest possible distance between Earth and a planet is 38 million km: the minimal opposition distance with Venus. The closest known distance between two stars is 80,000 km in the binary HM Cancri; their orbital period is 5.4 min. The orbital period of exoplanet OPH 11b around OPH 11 is 1,000–3,000 years. The largest and smallest known orbits of a planet around a single star are $\sim$ 650 AU (by HD 106906b) and 0.006 AU (by Kepler-70b). The closest known approach between planets is 0.0016 AU $\approx$240,000 km (by Kepler-70b and Kepler-70c).
- **Planetary aspects**
  In Astrology, an *aspect* is an angle (measured by the **angular distance** of ecliptic longitude, as viewed from Earth) the planets make to each other and other selected points in the *horoscope*, i.e., a chart representing the apparent positions and selected angles of the celestial bodies at the time of an event, say, a person's birth. Astrology claims a link between aspects and events in the human world.
  Major aspects are 1–10° (*conjunction*) and 90° (*square*), 180° (*opposition*) for which an *orb* (error) of 5–10° is usually allowed. Then follow 120 ± 4° (*trine*), 60 ± 4° (*sextile*) and (with orb 2°) 150° (*quincunx*), 45°, 135°, 72°, 144°. Other aspects are based on the division of the zodiac circle by 7, 9, 10, 11, 14, 16 or 24.

- **Primary-satellite distances**
  Consider two celestial bodies: a *primary* M and a smaller one m (a satellite, orbiting around M, or a secondary star, or a comet passing by).
  Let ρ_M, ρ_m and R_M, R_m be the densities and radii of M and m. The **Roche radius** (or *Roche limit, tidal radius*) of the pair (M, m) is the maximal distance between them within which m will disintegrate due to the tidal forces of M exceeding the gravitational self-attraction of m. This distance is $\approx 1.26 R_M \sqrt[3]{\frac{\rho_M}{\rho_m}}$ or $\approx 2.423 R_M \sqrt[3]{\frac{\rho_M}{\rho_m}}$ if m is rigid or fluid. The *Roche lobe* of a star is the region of space around the star within which orbiting material is gravitationally bound to it.
  The *tidal locking radius* of M is the distance at which the axial and orbital rotations of m become synchronized, i.e., the same side of m always faces M. The Moon is tidally locked by the Earth. Pluto and Charon are mutually tidally locked.
  Let $d(m, M)$ denote the **mean distance** between m and M, i.e., the arithmetic mean of their maximum and minimum distances; let S_m and S_M denote the masses of m and M. The barycenter of (M, m) is the point (in a focus of their elliptical orbits) where M and m balance and orbit each other. The distance from M to the barycenter is $d(m, M)\frac{S_m}{S_m+S_M}$. For the (*Earth, Moon*) system, it is 4,670 km (1,710 km below the Earth's surface). Pluto and Charon, the largest of its five moons, form rather a *binary system* since their barycenter lies outside of either body.
  The *Hill sphere* of a body is the region in which it dominates the attraction of satellites. The **Hill radius** of m in the presence of M is $\approx d(m, M)\sqrt[3]{\frac{S_m}{3S_M}}$; within it m can have its own satellites. The Earth's Hill radius is 0.01 AU; in the Solar System, Neptune has the largest Hill radius, 0.775 AU.
  The pair (M, m) can be characterized by five **Lagrange points** L_i, $1 \leq i \leq 5$, where a third, much smaller body (say, a spacecraft) will be relatively stable because its centrifugal force is equal to the combined gravitational attraction of M and m. These points are:
  L_1, L_2, L_3 lying on the line through the centers of M and m, so that $d(L_3, m) = 2d(M, m)$, $d(M, L_2) = d(M, L_1) + d(L_1, m) + d(m, L_2)$, $d(L_1, m) = d(m, L_2)$ (the satellite SOHO is at the point L_1 of the Sun–Earth system, where the view of the Sun is uninterrupted; the satellites WMAP and Planck are at L_2); L_4 and L_5 lying on the orbit of m around M and forming equilateral triangles with the centers of M and m. (These points are more stable; each of them forms with M and m a partial solution of the unsolved gravitational 3-*body problem*. Objects orbiting at L_4 and L_5 are called *Trojans* of Greek or Trojan *camp*, respectively. The Moon was created 4.5 Ga ago by impact of a Mars-sized Trojan planetoid on the Earth. The first known Sun–Earth Trojan asteroid is 2010 TK7, $\approx$300 m across.)

Other instances of the circular restricted 3-body problem are provided by planet–*co-orbital moons* and star–planet–*quasi-satellite* systems. *Co-orbital moons* are natural satellites that orbit at a very similar distance from their parent planet. Only Saturn's system is known to have them; it has three sets.

Orbital resonance occurs when the bodies orbital periods are in a close-to-integer ratio. For example, Pluto–Neptune are in a 2:3 ratio and Jupiter's moons Ganymede–Europa–Io are in a 1:2:4 ratio. Earth and Venus are in a *quasi-resonance* only 0.032 % away from 8:13. A *quasi-satellite* is an object in a 1:1 orbital resonance with its planet that stays close to the planet over many orbital periods. The largest of four known Earth's quasi-satellites is 3753 Cruithne, $\approx$5 km across.

The most tenuously linked long-distance binary in the Solar System is 2001 QW322: two icy bodies ($\approx$ 130 km in diameter) in the Kuiper belt, at mean distance $> 10^5$ km, orbiting each other at 3 km/h.

The elliptic restricted 3-body problem treats the *circumbinary* (orbiting two stars) planets such as Kepler-16b. A planet PH1 was found in a quadruple (binary pair) star system Kepler-64. Systems with up to 7 stars are known.

- **Dynamical spacing**

Let M, m_1, m_2 be the masses of a star and two adjacent planets orbiting it with semi-major axes a_1 and a_2. The *mutual Hill radius* of two planets is

$$R_H = \frac{a_1 + a_2}{2}\sqrt[3]{\frac{m_1 + m_2}{3M}}$$

and their **dynamical spacing** is (Gladman, 1993; Chambers et al., 1996)

$$\Delta = \frac{|a_2 - a_1|}{R_H}.$$

Fang–Margot, 2013, claim that on average $\Delta = 21.7$, and $\Delta < 10$ leads to instability in giga-year time span. In the Solar System, $\Delta > 26$ for terrestrial planets.

- **Titius–Bode law**

The **Titius–Bode law**, 1766, is an empirical rule approximating the mean distance d_i of i-th planet from the Sun (its orbital *semi-major axis*) by $\frac{3k+4}{10}$ AU.

Here 1 AU $\approx 1.5 \times 10^8$ km and $k_1 = 0 = 2^{-\infty}$ (for Mercury), $k_i = 2^{i-2}$ for $i \geq 2$, i.e., Venus, Earth, Mars, Ceres (the largest one in the Asteroid belt, $\approx\frac{1}{3}$ of its mass), Jupiter, Saturn, Uranus, Pluto. (But Neptune does not fit in the law while Pluto fits Neptune's spot $k = 2^7$.) The best fit for the form (Wurm, 1787) $d_i = AC^{i-2} + B$ is given by $C \approx 1.925, A \approx 0.334, B = 0.382$. Cf. **elliptic orbit distance**.

In the Solar System, the period ratios between adjacent orbits scatter around the dominant 5 : 2 ratio; it is 3 : 2 for Earth–Venus and 2 : 1 for Mars–Earth.

A *generalized Titius–Bode relation* $d_i = AC^i$ for some A, C fits even better for many other exoplanet systems showing such preference towards near mean motion resonance; cf. **dynamical spacing**. It helps to locate undetected exoplanets.
Hamano et al., 2013, claim that between 108 (as Venus) and 150 (as Earth) million km from the Sun, there is a critical distance explaining their difference. Venus has a similar size and bulk composition to those of Earth, but it lacks water. Earth solidified from its molten magma state within several million years, trapping water in rock and under its hard surface, while Venus got more of the Sun's heat and remained in molten state for $\approx$100 Ma giving time for any water to escape.

- **Planetary distance ladder**
The scale of interstellar-medium dust, *chondrules* (round grains found in stony meteorites, the oldest solid material in the Solar System), *boulders* (rock with grain size of diameter $\geq$256 mm), *planetesimals* (kilometer-sized solid objects in protoplanetary disks) and *protoplanets* (internally melted Moon-to-Mars-sized planetary embryos) is 10^{-6}, 10^{-3}, 10^0, 10^3 and 10^6 m.
In the Solar System's protoplanetary gas/dust disk, the binary electrostatic coagulation of dust/ice grains resulted in the creation, of planetesimals. Then gravity took over the accretion process. The growth was *runaway* (when $T_1 < T_2$, for growth time scales of the first and second most-massive bodies) at first and then (with $T_1 > T_2$ at some *transition radius*) it became *oligarchic*. A few tens of protoplanets were formed and then, by giant impacts, they were transformed into Earth and the other rocky planets. The process took $\approx$90 Ma from $\approx$4.57 to $\approx$4.48 Ga ago.
The free-floating planet with lowest mass known, six Jupiters, is PSO J318.5-22.
- **Potato radius**
The basic shape-types of objects in the Universe are: an irregular dust, rounded "potatoes" (asteroids, icy moons), spheres (planets, stars, black holes), disks (Saturn's rings, galactic disks) and halos (elliptic galaxies, globular star clusters). At mean radius $R <$ few km, objects (dust, crystals, life forms) have irregular shape dominated by nonmass-dependent electronic forces. Solid objects with $R >$ 200–300 km are gravity-dominated spheres. If both energy E and angular momentum L are exported (by some dissipative processes), the object, if large enough, collapses into a sphere. If only E is exported, the shape is a disk. If neither E, nor L is exported, the shape is a *halo*, i.e., the body is spheroidal.
If R ($R >$ few km) increases, there is a smooth size-dependent transition to more and more rounded potatoes until $\approx$200–300 km, where gravity begins to dominate. Ignoring surface tension, erosion and impact fragmentation, the potato shape comes mainly from a compromise between electronic forces and gravity. It also depends on the density and the yield strength of the (rocky or icy) material.
Lineweaver and Norman, 2010, define the **potato radius** R_{pot} as this potato-to-sphere transition radius. They derived $R_{pot} = 300$ km for asteroids (Vesta, Pallas, Ceres have $R = 265, 275, 475$ km, respectively) and $R_{pot} = 200$

km for icy moons (Hyperion, Mimas, Miranda have $R = 140, 198, 235$ km, respectively).

In 2006, the IAU (International Astronomical Union) defined a *planet* as a orbiting body which has sufficient mass for its self-gravity to overcome rigid body forces so that it assumes a *hydrostatic equilibrium* (nearly round) shape and cleared the neighborhood around its orbit. If the body has not cleared its neighborhood, it is called a *dwarf planet*. The potato radius, at which self-gravity makes internal overburden pressures equal to the yield strengths of the material, marks the boundary of hydrostatic equilibrium used in above IAU definition.

Buchhave et al., 2014: planets smaller than 1.7 Earths are likely to be completely rocky, while those larger than 3.9 Earths are probably gas giants.

- **Frost line (in Astrophysics)**

  In Astrophysics, by analogy with **frost line (in Earth Science)**, the **frost** (or *snow*, *ice*) **line** is the distance from a star (or a nebula's protostar) where hydrogen compounds such as water, ammonia, methane condense into ice grains. It separates an inner region of rocky objects from an outer region of icy objects. Water and methane condensate at 180 and 40 K, respectively. Sun's water-frost and methane-frost lines are roughly at 2.7 and 48 AU, i.e., in the Asteroid belt (between the orbits of Mars and Jupiter) and the Kuiper belt (at 30–55 AU). On the other hand, inside of ≈ 0.1 AU, rocky grains cannot exist: dust evaporates.

  Martin and Livio, 2012, claim that a giant planet like Jupiter should be in the right location outside of the frost line to produce an asteroid belt of the appropriate size, offering the potential for life on a nearby rocky planet like Earth.

- **Solar distances**

  Following a supernova explosion 4,570 Ma ago in our galactic neighborhood, the Sun was formed 4,567 Ma ago by rapid gravitational collapse of a fragment (about 1 parsec across) of a giant (about 20 parsecs) hydrogen molecular cloud. The mean distance of the Sun from Earth is 1 AU $\approx 1.496 \times 10^8$ km. The mean distance of the Sun from the Milky Way core is 27,200 light-years.

  The Sun is more massive than 95 % of nearby stars and its orbit around the Galaxy is less eccentric than ≈ 93 % of similar (i.e., of spectral types F, G, K) stars within 40 parsecs. The Sun's mass (99.86 % of the Solar System) is 1.988×10^{30} kg.

  The Sun's radius is 6.955×10^5 km; it is measured from its center to the edge of the *photosphere* (≈ 500 km thick layer below which the Sun is opaque to visible light). The Sun will expand ≈ 256 times in 5.4–8 Ga and then become a white dwarf.

  The Sun does not have a definite boundary, but it has a well-defined interior structure: the *core* extending from the center to ≈ 0.2 solar radii, the *radiative zone* at ≈ 0.2–0.8 solar radii, where thermal radiation is sufficient to transfer the intense heat of the core outward, the *tachocline* (transition layer) and the *convection zone*, where thermal columns carry hot material to the surface (photosphere) of the Sun.

The principal zones of the *solar atmosphere* (the parts above the photosphere) are: temperature minimum, chromosphere, transition region, corona, and heliosphere.
The *chromosphere*, a $\approx$3,000 km deep layer, is more visually transparent. The *corona* is a highly rarefied region continually varying in size/shape; it is visible only during a total solar eclipse. The chromosphere-corona region is much hotter than the Sun's surface. Extending further, the corona becomes the *solar wind*, a very thin gas of charged particles that travels through the Solar System.
The *heliosphere* is the teardrop-shaped region around the Sun created by the solar wind and filled with solar magnetic fields and outward-moving gas. It extends from $\approx$20 solar radii (0.1 AU) outward 86–100 AU past the orbit of Pluto to the *heliopause*, its outermost edge, where the interstellar medium and solar wind pressures balance. The interstellar medium and solar wind are moving supersonically in opposite directions, towards and away from the Sun. The points, $\approx$80 and $\approx$230 AU from the Sun, where the solar wind and interstellar medium become subsonic, are the *termination shock* and *bow shock*, respectively.
The *tidal truncation radius* (100,000–200,000 AU, say, $\approx$2 ly from the Sun) is the outer limit of the *Oort cloud*. It is the boundary of the Solar System, i.e., Sun's *Hill/Roche sphere*, where its gravity is overtaken by the galactic tidal force.

- **Dyson radius**
  The **Dyson radius** of a star is the radius of a hypothetical *Dyson sphere* around it, i.e., a megastructure (say, a system of orbiting star-powered satellites) meant to completely encompass a star and capture a large part of its energy output. The solar energy, available at distance d (measured in AU) from the Sun, is $\frac{1366}{d^2}$ watts/m^2. The inner surface of the sphere is intended to be used as a habitat.
  For example, at Dyson radius 300×10^6 km from the Sun a continuous structure with ambient temperature 20 °C (on the inner surface) and efficiency 3 % of power generation (by a heat flux to −3 °C on the outer surface) is conceivable.
- **Star's radii**
  The **corotation radius** of a star is the distance from it where the centrifugal force on a particle corotating with it balances the gravitational attraction, i.e., the accretion disk rotates at the same angular velocity as the star.
  The **Bondi-Hoyle accretion radius** is the radius where star's gravitational energy is larger than the kinetic energy and, so, at which material is bound to star.
  The **Hayashi radius** (or *Hayashi limit*) of a star is its maximum radius for a given mass. A star within *hydrostatic equilibrium* (where the inward force of gravity is matched by the outward pressure of the gas) cannot exceed this radius.
  The **Eddington radius** (or *Eddington limit*) of a star is the radius where the gravitational force inwards equals the continuum radiation force outwards, assuming hydrostatic equilibrium and spherical symmetry. A star exceeding it would initiate a very intense continuum driven stellar wind from its outer layers.
  The largest and smallest known stars, the red hypergiant UY Scuti and red dwarf OGLE-TR-122b, have respective radii 1708 ± 192 and 0.12 solar radii.

- **Galactocentric distance**
  A star's **galactocentric distance** (or *galactocentric radius*) is its **range distance** from the galactic center; it may also refer to a distance between two galaxies. The *galactic anticenter* is the point lying opposite, for an observer on Earth, this center.
  The Sun's present galactocentric distance is nearly fixed $\approx$8.4 kiloparsec, i.e., 27,400 light-years, but it may have been 2.5–5 kpc in the past. *Einasto's law*, 1963, claims that the density $\rho(r)$ of a spherical stellar system (say, a galaxy or its halo) varies as $\exp(-Ar^{\alpha})$, where r is the distance from the center.
- **M31–M33 bridge**
  Braun and Thilker, 2004, discovered that the distance 782,000 light-years between Andromeda (M31) and Triangulum (M33) galaxies is spanned by a link consisting of about 500 million Sun's masses of ionized hydrogen.
  A third of all baryonic matter is in stars and galaxies; another $\frac{1}{3}$ is diffuse and thought to be in filamentary networks spread through space. Remaining $\frac{1}{3}$, called *warm-hot intergalactic medium* (WHIM), is expected to be of intermediate density. The **M31–M33 bridge** consists of WHIM, the first evidence of this medium. Such WHIM bridges are likely remnants of collisions between galaxies.
- **Radii of a star system**
  Given a star system (say, a galaxy or a globular cluster), its **half-light radius** (or **effective radius**) hr is the distance from the core within which half the total luminosity from the system, assumed to be circularly symmetric, is received. The **core radius** cr is the distance from the core at which the apparent surface luminosity has dropped by half; so, $cr \leq hr$. In general, **isophotal radius** is the size attributed to the system corresponding to a particular level of surface brightness.
  The **half-mass radius** $r_{0.5}$ is the radius from the core that contains half the total mass of the system. In general, the *Lagrangian radii* are the distances from the center at which various percentages of the total mass are enclosed.
  The **tidal radius** of a globular cluster is the distance from its center at which the external gravitation of the galaxy has more influence over the stars in the cluster than does the cluster itself.
  Unlike a star cluster, all galaxies are filled with and surrounded by a halo of dark matter acting as a sort of glue within and between galaxies. Thin tendrils of dark matter connect nodes of galaxy clusters, creating a cosmic web.
  The **virial radius** R_{vir} of a galaxy is the radius centered on it containing matter at 200 (sometimes, $18\pi^2 \approx 178$ or 130) times the critical (or, mean) density of the Universe. The mass within R_{vir} is a measure of the total mass inside a dark matter halo. Kravtsov, 2011, claim that $r_{0.5} \approx 0.015 R_{vir}$. Also, Harris, 2013, explains speed anomalies of Earth's satellites by 0.005–008 % increase of its mass due to a dark matter's disk, 191 km thick and 70,000 km across, around the equator.

- **Habitable zone radii**
A maximally Earth-like mean temperature is expected at the distance $\sqrt{\frac{L_{star}}{L_{sun}}}$ AU from a star, where L is the total radiant energy.
The **habitable zone radii** of a star are the minimal and maximal orbital radii r, R such that liquid water may exist on a *terrestrial* (i.e., primarily composed of silicate rocks or metals) planet orbiting within this range, so that life, constructed from carbon and reliant on liquid water, could develop there in a similar way as on the early Earth. For Sun, $[r, R]$ is $[0.99, 1.70]$ AU; it includes Earth and Mars. For best candidates—orange dwarf stars—HZ is within $[0.5, 1]$ AU.
The **Kasting distance** (or *habitable zone distance*) of an exoplanet, at distance d from its star, is an index defined by

$$HZD(d) = \frac{2d - (R + r)}{R - r}.$$

So, $-1 \leq HZD(d) \leq 1$ correspond to $r \leq d \leq R$.
The above notion of surface habitability is modeled from temperature/humidity; the edges r, R of HZ are determined by loss of water and, respectively, by the maximum greenhouse provided by a CO_2 atmosphere. Among known exoplanets of 2–10 Earth's mass, the best candidates for Earth-like habitability are GJ 667Cc, HD-85512b, Kepler-22b (22, 36, 600 ly away) orbiting, respectively, stars Gliese 667C, Gliese 370, Kepler-22 in the constellations Scorpius, Vela, Cygnus of our galaxy.
Habitable zones are 10–14 times wider for subsurface life. Protected inside a warm mineral-rich rocks, it can be much more typical than Earth's surface life.
Petigra et al., 2013: *eta-Earth* of our galaxy, i.e., fraction of Sun-like stars with an *Earth-size* (1–2 Earth's radii) planet orbiting in habitable zones, is 22 ± 8 %, depending on the definition of HZ. There could be 40 billion habitable Earth-size planets in the Milky Way with the nearest one being within 12 ly. They found 603 *Earths* 2.0, i.e., rocky planets with an atmosphere circling in HZ of a Sun-like star in 200–400 days and receiving 0.25–4 of stellar energy coming to Earth.
But for Heller–Armstrong, 2014, planet or moon habitability should not be defined by the stellar HZ, since, for example, tidal heating can render terrestrial or icy worlds habitable, even more than Earth, beyond it. They expect such *superhabitable* objects around orange dwarfs, including Alpha Centauri B, third-nearest star.
Our *galactic habitable zone* is (Lineweaver et al., 2004) a slowly expanding region between 7 and 9 kpc of **galactocentric distance**; so, the minimal and maximal radii are 22,000 and 28,000 ly. They used 4 prerequisites for complex life: the presence of a host star, enough heavy elements to form terrestrial planets, sufficient time (4 ± 1 Ga) for biological evolution and an supernovae-free environment.
- **Earth similarity index**
The **Earth similarity index** of a planet P is (Schulze-Makuch et al., 2011):

$$ESI(P) = \prod_{i=1}^{n}(1 - |\frac{x_i(P) - x_i(E)}{x_i(P) + x_i(E)}|)^{\frac{w_i}{n}},$$

where $x_i(P)$ is a planetary parameter (including surface temperature, escape velocity, mean radius, bulk density), $x_i(E)$ is the reference value for Earth (i.e., 14.85° C, 1,1,1), w_i is a weight (5.58, 0.70, 0.57, 1.07) and n is the number of parameters. ESI(P)= 1, 0.84, 0.83, 0.78, 0.64 for Earth, GJ-667Cc, Kepler-62e and Venus, Mars. Many exomoons and unconfirmed NASA Kepler candidates rank within [0.76, 0.90]. Terrestrial, but only simple extremophilic, life might be possible if ESI(P)> 0.6, while plants/animals may require > 0.8.
The same authors proposed a *planetary habitability index* based on the presence of a stable substrate, atmosphere, magnetic field, available energy, appropriate chemistry and the potential for holding a liquid solvent, such as 100-km deep ocean beneath the surface of Jupiter's moon Europa and hydrocarbon lakes on Saturn's moon Titan. Unicellular life has been found in the most adverse conditions on Earth. So, the presence of extremophiles on Mars and, with very different biochemistry, on Europa and Titan is plausible. For primary producers (plants), Earth was more habitable 500 Ma ago, with less seasonal ice and deserts. Observing oxygen in a planet's atmosphere will indicate photosynthetic life since the photosynthesis is the only known process able to release O_2 in any real quantity. But the importance of oxygen and carbon can be a peculiarity of Earth life. For Oze et al., 2012, low (<40) hydrogen/methane ratio indicate that life is likely present. Also, infrared, or heat, radiation can indicate an alien civilization.

- **SETI detection ranges**

  SETI (Search for Extra Terrestrial Intelligence) involves using radio telescopes to search for a possible alien radio transmission. The recorded signals are mostly random noise but in 1977 a strong signal (called WOW!) was received at $\leq$10 kHz of the frequency $\approx$1420.406 MHz (21 cm) of the hydrogen line. Also, a puzzling radio source SHGb02+14a was observed three times in 2003 at $\approx$1420 MHz.

  SETI detection ranges are the maximal distances over which detection is still possible using given frequency, antenna dish size, receiver bandwidth, etc. They are low for broadband signals from Earth (from 0.007 AU for AM radio up to 5.4 AU for EM radio) but reach 720 light-years for the S-Band of the world's largest (with dish's diameter 305 m) single-aperture radio telescope at Arecibo.

  SETI searches in the microwave window 1–10 GHz (the part of the radio spectrum that can pass through the atmosphere), especially around the "waterhole" 1,420–1,666 MHz (21–18 cm) between hydrogen, H, and hydroxyl, OH.

  All known signals with spectral width <5 Hz arise from artificial sources; so, such extraterrestrial signal will indicate an intelligent civilization. SETI searched those narrow band signals in L-band (1.1–1.9 GHz) from 86 stars in the Kepler field of view hosting most life-promising exoplanets, but not found none. Tarter et al., 2013, deduce from it that the number of *Kardashev type II* (using all energy from their star; our total power consumption today is $\approx$0.01 % of the sunlight

falling on Earth) civilizations in the Milky Way loud in L-band, is less than 1 in a million per sun-like star. The volume V of our galaxy is about $\pi(50000^2)1000 \approx 7.9 \times 10^{12}$ (ly)^3. If N civilizations are distributed there uniformly with spacing d, then $d^3 = \frac{V}{N}$.
Active SETI (or *METI*) consists of sending radio or optical signals into space hoping that they will be picked up by an alien intelligence. The first radio signals from Earth to reach space were produced around 1940 but TV and radio signals decompose into static within 1–2 ly. In 1974 Arecibo telescope sent an elaborate radio signal aimed at the star cluster M13 located 25,000 ly away.
About the perceived risk of revealing the location of the Earth to an alien civilization, METI enthusiasts reply that an advanced civilization within a radius of 100 ly already knows of our existence due to electromagnetic signals leaking from TV, radio and radar. But now, with digital transmissions replacing analogue ones and virtually no radiation escaping into outer space, the Earth become electronically invisible to aliens. Still, a civilization even slightly more advanced than ours could detect the lights of our big cities from up to 500 light years away, using its sun as a gravitational lens. Also, some life (plants, lichens, algae, bacterial mats) can be recognized by its light signature from space.
Besides radio signals and light, nonmicrobial alien life can be discovered by analyzing the output of methane or oxygen in the atmosphere of exoplanets.

- **Voyager 1 distance**
The Voyager 1 is a 722-kg robotic space probe launched by NASA in 1977; it has power to operate its radio transmitters until 2025 but only 68 kB of memory. It is currently the farthest man-made object from Earth, the first probe to leave the Solar System (in 2013) and the fastest probe (moving at $\approx$17 km/s or 3.6 AU/year). As of May 2014, **Voyager 1 distance** from Earth was 1.9×10^{10} km $\approx$127 while for Voyager 2, it was $\approx$104.7 AU $\approx$14.5 light-hours.
The NASA *Stardust* spacecraft (1999–2006) achieved the longest distance (30 AU) traveled by a return mission and the farthest distance (2.7 AU) solar powered spacecraft has traveled from the Sun. Amino acid glycine was found in its comet sample.
The Earth–Moon distance ($\approx$1.28 light-seconds) can be covered, with current technologies, in $\approx$8 h. The distance from Earth to other planets ranges from 3 light-minutes to $\approx$4 light-hours. At Voyager 1's current rate, a journey to Proxima Centauri (the nearest known star, 4.24 ly away) would take 72,000 years.
Interstellar travel will be possible only with new technology, say, beamed-light sails, hydrogen-fuelled ramjet, nuclear pulse propulsion, warp drive, wormholes. Human spaceflight beyond the close neighborhood in the Solar System looks, as now, unlikely, because of duration, cost and health threat due to microgravity, radiation and isolation. Also, long (more than a month) sojourns in space produce potentially serious brain anomalies and severe eyesight problems. Still, the project *Inspiration Mars* will send a crew of two for a 501 days fly-by mission to Mars, using its next closest approach (57.6 million km) in 2018; cf. **opposition distance**.

- **Earth in space**
  The Earth, spinning 0.5 km/s, orbits the Sun at 30 km/s. The Sun orbits the galactic center at 219 km/s and it moves at 16.5 km/s, with respect to the motion of its galactic neighborhood, towards Vega, a star in the constellation Lyra.
  The *Local Bubble* is a cavity, 300–800 ly across (with hydrogen density 0.05 atoms per cm^3, one tenth of the galactic density) in the *Local* (or *Orion-Cygnus*) *Arm* of the Milky Way. The Solar System has been traveling through this Bubble for the last 5–10 Ma and is located now close to its inner rim, about half-way along the Arm's length. From 0.044–0.15 Ma ago and for another 0.01–0.02 Ma, the Sun is traversing the *Local Interstellar Cloud* 30 ly across at 23 km/s.
  The Milky Way (0.1 Mly across and 1 kly thick) and Andromeda galaxies are 2.5 Mly apart and are approaching at 100–140 km/s. In $4 + 1.3 + 0.1$ Ga (3 consecutive collisions) they will merge to form the *Milkomeda*, new elliptical galaxy in which our Solar System would remain intact but Sun's **galactocentric distance** will be 0.16 Mly. Their stars will not collide but central black holes will merge.
  Our *Local Group* (LG) is a *poor* (small and not centered) cluster, 10 Mly across, consisting of Andromeda (M31), Milky Way (MW), Triangulum and about 50 small galaxies. It lies in the outskirts (on a small filament connecting the Fornax and Virgo clusters) of our small Local Supercluster (LSC), 110 Mly across and with a mass 10^{15} suns. The number of galaxies per unit volume, in the LSC, falls off with the square of the distance from its center, near the Virgo cluster.
  The LSC belongs to the Pisces-Cetus supercluster complex, 1 Gly long and 150 Mly (46 Mpc) wide; its mass is 10^{18} suns. Fairall, 1994, proposed to unite the LSC and (the nearest) Centaurus superclusters via the zone, obscured by the Milky Way, with the Fornax Wall, creating *the Centaurus Great Wall.*
  The *Extended Local Group* is the LG plus the "nearby" (3.9 Mpc) Maffei and Sculptor groups. It belongs to our *Local Filament* (LF, or *Coma-Sculptor Cloud*), a branch of the Fornax-Virgo filament of the LSC.
  The LF bounds the *Local Void* (LV), extending 60 Mpc from the edge of the LG. The *Local Sheet* (LS) is all LF's matter within 7 Mpc. The Milky Way and Andromeda are encircled by 12 large galaxies arranged in a ring about 24-Mly across.
  With respect to the CMBR (cosmic microwave background radiation) filling the Universe almost uniformly, the Solar System, Milky Way, and LG velocities are 369, 600, 627 km/s. *Peculiar velocities* V_{pec} are the deviations from the Hubble expansion, i.e., $V_{pec} = V_{obs} - H_0 d$, where V_{obs} is the observed velocity, d is the distance and H_0 is the Hubble constant, ≈ 72 km/s for every Mpc. The Hubble flow, dominating at large distances, is negated by gravity at smaller distances; for example, its recession velocity is <1 mm/s at the edge of the Solar System.
  According to Tully et al., 2007, the Local Sheet is moving as a unit with low internal dispersion; the LG moves at only 66 km/s with respect to the LS. The bulk flow of the LS is sharply discontinuous from the flows of other

nearby structures. The vector of this flow has, with respect to the CMBR, amplitude 631 km/s. It can be decomposed into a vector sum of three quasi-orthogonal components: *local* (259 km/s away from the center of the Local Void), *intermediate* (185 km/s to the Virgo cluster) and *large* (455 km/s towards the *Great Attractor* (GrAt)).

All matter within 4.6 Mpc moves away from the Local Void at 268 km/s. It will collide, in $\approx$10 Ga, with the nearest adjacent filament, the Leo Spur. The Local Sheet moves toward the Virgo cluster, at the distance 17 Mpc. All matter within 50 Mpc moves at 600 km/s towards overdensities at 200 Mly (GrAt dominated by the Norma cluster) and 600 Mly (Shapley supercluster, roughly behind GrAt).

Chapter 26
Distances in Cosmology and Theory of Relativity

26.1 Distances in Cosmology

The *Universe* is defined as the whole space-time continuum in which we exist, together with all the energy and matter within it.

Cosmology is the study of the large-scale structure of the Universe. Specific cosmological questions of interest include the *isotropy* of the Universe (on the largest scales, the Universe looks the same in all directions, i.e., is invariant to rotations), the *homogeneousness* of the Universe (any measurable property of the Universe is the same everywhere, i.e., it is invariant to translations), the density of the Universe, the equality of matter and antimatter, and the origin of density fluctuations in galaxies.

Hubble, 1929, discovered that all galaxies have a positive *redshift*, i.e., all galaxies, except for a few nearby galaxies like Andromeda, are receding from the Milky Way. By the Copernican principle (that we are not at a special place in the Universe), we deduce that all galaxies are receding from each other, i.e., we live in an expanding Universe, and the further a galaxy is away from us, the faster it is moving away (this is now called the *Hubble law*). The *Hubble flow* is the general outward movement of galaxies and clusters of galaxies resulting from the expansion of the Universe. It occurs radially away from the observer, and obeys the Hubble law. The gravitation in galaxies can overcome this expansion, but the clusters and superclusters (largest gravitationally bound objects) only slow the rate of their expansion.

In Cosmology, the prevailing scientific theory about the early development and shape of the Universe is the *Big Bang Theory*. The observation that galaxies appear to be receding from each other, combined with the General Theory of Relativity, leads to the construction that, as one goes back in time, the Universe becomes increasingly hot and dense, then leads to a gravitational singularity, at which all distances become zero, and temperatures and pressures become infinite.

M.M. Deza, E. Deza, *Encyclopedia of Distances*,
DOI 10.1007/978-3-662-44342-2_26

The term *Big Bang* is used to refer to a hypothesized point in time when the observed expansion of the Universe began. Based on measurements of this expansion, it is currently believed that the Universe has an age of $\approx$13.82 Ga (billion years).

In Cosmology (or, more exactly, *Cosmography*, the measurement of the Universe) there are many ways to specify the distance between two points, because in the expanding Universe, the distances between comoving objects are constantly changing, and Earth-bound observers look back in time as they look out in distance. The unifying aspect is that all distance measures somehow measure the separation between events on *radial null trajectories*, i.e., trajectories of photons which terminate at the observer. In general, the **cosmological distance** is a distance far beyond the boundaries of our Galaxy.

The geometry of the Universe is determined by several *cosmological parameters*: the *cosmic scale factor* a, the *Hubble constant* H, the *density* ρ and the *critical density* ρ_{crit} (the density required for the Universe to stop expansion and, eventually, collapse back onto itself), the *cosmological constant* Λ, the *curvature* k *of the Universe*. Many of these quantities are related under the assumptions of a given *cosmological model*. The most common cosmological models are the closed and open *Friedmann–Lemaître cosmological models* and the *Einstein–de Sitter cosmological model*.

This model assumes a homogeneous, isotropic, constant curvature Universe with zero cosmological constant Λ and pressure p. For constant mass M of the Universe, $H^2 = \frac{8}{3}\pi G\rho$, $t = \frac{2}{3}H^{-1}$, $a = \frac{1}{R_C}(\frac{9GM}{2})^{\frac{1}{3}}t^{\frac{2}{3}}$, where $G = 6.67 \times 10^{-11}\,\mathrm{m^3\,kg^{-1}\,s^{-2}}$ is the *gravitational constant*, $R_C = |k|^{-\frac{1}{2}}$ is the *radius of curvature*, and t is the age of the Universe.

The **scale factor** $a = a(t)$ is an expansion parameter, relating the size of the Universe $R = R(t)$ at time t to its size $R_0 = R(t_0)$ at time t_0 by $R = aR_0$.

The *Hubble constant* H is the constant of proportionality between the speed of expansion v and the size of the Universe R, i.e., $v = HR$. This equality is just the *Hubble law* with the Hubble constant $H = \frac{a'(t)}{a(t)}$. This is a linear redshift-distance relationship, where redshift is interpreted as recessional velocity v, typically expressed in km/s.

The current value of the Hubble constant is $H_0 = 71 \pm 4\,\mathrm{km}\,s^{-1}\,\mathrm{Mpc}^{-1}$, where the subscript 0 refers to the present epoch because H changes with time. The *Hubble time* and the **Hubble distance** are defined by $t_H = \frac{1}{H_0} \approx 13.82$ Ga and $D_H = \frac{c}{H_0} \approx 4.24$ Gpc. The *Hubble volume* (or *Hubble sphere*) is the region of the Universe surrounding an observer beyond which the recessional velocity exceeds the speed c of light, i.e., any object beyond *particle horizon* (4.4×10^{26} m $=$ 47 light-Ga), is receding (due to the expansion of the Universe itself) at a rate greater than c.

The volume of observable Universe is the volume $\approx 4.1 \times 10^{34}$ cubic light-years, or $\approx 3.4 \times 10^{80}\,\mathrm{m}^3$, of Universe with a comoving size of $\frac{c}{H_0}$, i.e., a sphere with radius $\approx$14 Gpc (about 3 times larger than that of Hubble volume). It has mass $\approx 1.6 \times 10^{53}$ kg and contains $\approx 10^{23}$ stars (in at least 8×10^{10} galaxies) and $\approx 10^{80}$ atoms.

The mass density ρ (ρ_0 in the present epoch) and the value of the cosmological constant Λ are dynamical properties of the Universe; today $\rho \sim 9.4 \times 10^{-27}$ kg m^{-3} and $\Lambda \sim 10^{-52}$ m^{-2}. They can be made into dimensionless parameters Ω_M and Ω_Λ by $\Omega_M = \frac{8\pi G\rho_0}{3H_0^3}$, $\Omega_\Lambda = \frac{\Lambda}{3H_0^3}$. A third parameter $\Omega_R = 1 - \Omega_M - \Omega_\Lambda$ measures the "curvature of space". These parameters determine the geometry of the Universe if it is homogeneous, isotropic, and matter-dominated.

The velocity of a galaxy is measured by the *Doppler effect*, i.e., the fact that light emitted from a source is shifted in wavelength by the motion of the source. (The Doppler shift is reversed in some *metamaterials*: a light source moving toward an observer appears to reduce its frequency.) A relativistic form of the Doppler shift exists for objects traveling very quickly, and is given by $\frac{\lambda_{obser}}{\lambda_{emit}} = \sqrt{\frac{c+v}{c-v}}$, where λ_{emit} is the emitted wavelength, and λ_{obser} is the shifted (observed) wavelength. The change in wavelength with respect to the source at rest is called the *redshift* (if moving away), and is denoted by the letter z. The relativistic redshift z for a particle is given by $z = \frac{\Delta\lambda_{obser}}{\lambda_{emit}} = \frac{\lambda_{obser}}{\lambda_{emit}} - 1 = \sqrt{\frac{c+v}{c-v}} - 1$.

The cosmological redshift is directly related to the scale factor $a = a(t)$: $z+1 = \frac{a(t_o)}{a(t_e)}$. Here $a(t_o)$ is the value of the scale factor at the time the light from the object is observed, and $a(t_e)$ is its value at the time it was emitted. It is usually chosen $a(t_o) = 1$, where t_o is the present age of the Universe.

- **Metric expansion of space**

  The **metric expansion of space** is the averaged increase of measured distances between objects in the Universe with time.

  It is not a motion of space and not a motion into pre-existing space. Only distances expand (and contract). The expansion has no center: all distances increase by the same factor, and every observer sees the same expanding cosmos. The observed *Hubble law* quantifies expansion from an observer. Expansion rate between two points in free space 1 m apart is 2.2×10^{-18} m/s.

  The mean distances between widely separated galaxies increase by $\approx$1 % every 140 million years. **FLRW metric** models, at large (superclusters of galaxies) scale, this expansion. On the scales of galaxies, there is no expansion since the metric of the local Universe has been altered by the presence of the mass of the galaxy. Full expansion, at the Hubble rate $\approx$7,000 km/s, commences only at distances $\approx$100 Mpc. Superclusters are expanding but remain *gravitationally bound*, i.e., their expansion rate is decelerated.

  Expansion is thought to start due to cosmic inflation and then, due mainly to inertia. Its rate decelerated about 12 Ga ago due to gravity and then, from about 6 Ga ago when putative dark energy took over, accelerated. Now, for every megaparsec of distance from the observer, the rate of expansion increases by about 74 km/s. When the Universe doubles in volume, the dark energy doubles too. In 10^{11} years our galaxy will be the only one left in the observable Universe.

  The Universe was *radiation-dominated* with the **scale factor** $a(t) \sim t^{\frac{1}{2}}$ first $\approx$70,000 years, then *matter-dominated* with $a(t) \sim t^{\frac{2}{3}}$ until $\approx$4.5 Ma ago, then *dark-energy-dominated* with $a(t) \sim \exp(Ht)$ and the Hubble constant

$H = \sqrt{\frac{8\pi G\rho}{3}} = \sqrt{\frac{\Lambda}{3}}$. In fact, its expansion caused the matter surpass the radiation in energy density and further, when matter and radiation dropped to low concentrations, the repulsive *dark energy* (or *vacuum energy*) overtook the gravity of matter.

The most commonly accepted scenario for the future is the *Big Freeze*: continued expansion results in a universe that asymptotically approaches 0 K and the *Heat Death*, a state of maximum entropy in which everything is evenly distributed. Caldwell, 2003, claimed that the scale factor a will became infinite in the finite future, resulting in *Big Rip*, final singularity in which all distances diverge to ∞.

- **Zero-gravity radius**

  For a cluster of mass M, its **zero-gravity** (or *zero-velocity*, *turnover*) **radius** R_V is (Sandage, 1986, and Chernin–Teerikorpi–Baryshev, 2006) the distance r from the cluster's barycenter, where the radial force $\frac{GM}{r^2}$ of the point mass M gravity become equal to the radial force ($G2\rho_V \frac{4\pi}{3} r^3$ divided by r^2) of vacuum antigravity. So,

  $$R_V^3 = \frac{3M}{8\pi\rho_V}.$$

  Here G is the *gravitational constant* and $\rho_V \approx 7 \times 10^{-30}$ g/cm^3 is the constant density of dark energy inferred from global observations of supernovae 1a.

  The **Einstein–Straus radius** R_M is the radius besides which expansion rate reach the global level. It is estimated that $\frac{R_M}{R_V}$ is 1.5–1.7 if the ratio of local and global density of dark energy is 0.1–1. If above ration is 1, then $R_M = R_V(1 + z_V)$, where $z_V \approx 0.7$ is the global zero-acceleration redshift.

  For the *Local Group* (LG), containing Milky Way and of mass $2 - 3.5 \times 10^{12}$ suns, above model corresponds to observed $R_V = 1.3 - 1.55$ Mpc and $R_M = 2.2 - 2.6$ Mpc. The Virgo cluster, dominating Local supercluster, contains over 1,000 galaxies in a volume slightly larger than LG; its mass is $\approx 10^{15}$ suns and $R_V = 10.3$ Mpc.

- **Hubble distance**

  The **Hubble distance** (or **cosmic light horizon**, **Hubble radius**) is an increasing maximum distance $D_H = ct_H$ that a light signal could have traveled since the Big Bang, the beginning of the Universe. Here c is the speed of light and t_H is the *Hubble time* (or *age of the Universe*). It holds $t_H = \frac{1}{H_0}$, where H_0 is the *Hubble constant* which is estimated as $71 \pm 4\,\text{km}\,\text{s}^{-1}\,\text{Mpc}^{-1}$ at present. So, at present, $t_H \approx 4.32 \times 10^{17}$ s ≈ 13.82 Ga, and $D_H = \frac{c}{H_0} \approx 13.82$ billion light-years $\approx 1.31 \times 10^{26}$ m ≈ 4.24 Gpc, i.e., 4.6×10^{61} **Planck lengths**.

  But we are observing now, due to the space expansion, objects much farther away than a static distance 13.82 Gly.

  For small $\frac{v}{c}$ or small distance d in the expanding Universe, the velocity is proportional to the distance, and all distance measures, for example, **angular diameter distance**, **luminosity distance**, etc., converge. In the linear approximation, this

reduces to $d \approx zD_H$. But for large $\frac{v}{c}$, the relativistic **Lorentz length contraction** $L = L_0\sqrt{1-(\frac{v}{c})^2}$, where L_0 is a **proper length**, of an object traveling at velocity v relative to an observer, become noticeable to that observer.
Above Hubble radius was measured (by the Wilkinson Microwave Anisotropy Probe) as a **light travel distance** to the source of *cosmic background radiation*. Other estimations: 13.1 Gly (calibrating the distances to supernovae of a standard brightness), 14.3 Gly (measuring radio galaxies of a standard size) and 14.5 Gly (basing on the abundance ratio of uranium/thorium chondritic meteorites, [Dau05]).

- **Cosmic sound horizon**
Cosmic background radiation (CMB) is thermal radiation (strongest in the microwave region of the radio spectrum) filling the observable Universe almost uniformly. It originated $t_r \approx 380{,}000$ years after the Big Bang (or at a redshift of $z = 1{,}100$), at *recombination*, when the Universe (ionized plasma of electrons and *baryons*, i.e., protons and neutrons) cooled to below 3,000 K. (Now, the Universe's temperature is $\approx$1,100 times cooler and its size is $\approx$1,100 times larger.)
The electrons and protons start to form neutral hydrogen atoms, allowing photons (trapped before by Thomson scattering) to travel freely. During next $\approx$100,000 years radiation decoupled from the matter and the Universe became transparent.
The plasma of photons and baryons can be seen as a single fluid. The gravitational collapse around "seeds" (point-like overdensities produced during inflation) into dark matter hierarchical halos was opposed by outward radiation pressure from the heat of photon-matter interactions. This competition created longitudinal (acoustic) oscillations in the photon-baryon fluid, analogic to sound waves, created in air by pressure differences, or to ripples in a pond.
At recombination, the only remaining force on baryons is gravitation, and the pattern of oscillations (configuration of baryons and, at the centers of perturbations, dark matter) became frozen into the CMB. Baryon radiative cooling into gas and stars let this pattern of seeds to grow into structure of the Universe.
More matter existed at the centers and edges of these waves, leading eventually to more galaxies there. Today, we detect the sound waves (regular, periodic fluctuations in the density of the visible baryonic matter) via the primary CMB anisotropies.
These *baryon acoustic oscillations* (BAO) started at $t = 0$ (post-inflation) and stopped at $t = t_r$ (recombination). The **cosmic sound horizon** is the distance sound waves could have traveled. At recombination, it was $\approx c_s t_r \sim 100$ kpc, approximating the speed c_s of sound as $\frac{c}{\sqrt{3}}$.
Expanding by factor $1 + z = 1100$, it is 120–150 Mpc today. It is a standard ruler; an excess of galaxy pairs separated by this horizon was confirmed. Cf. **cosmological distance ladder** and, in Chap. 24, **acoustic metric**.
- **GZK-horizon**
Greisen and Kuzmin–Zatsepin, 1966, computed that a cosmic ray with kinetic energy over *GZK-limit* (5×10^{19} eV) traveling from its distant, over

GZK-horizon (50 Mpc $\approx$163 Mly) source, will be absorbed (due to slowing interaction with photons of the CMB and associated mean path) and so never observed on Earth.
Several cosmic rays apparently exceeding *GZK-limit* were observed; this *GZK-paradox* is still unexplained.

- **Comoving distance**
The standard Big Bang model uses *comoving coordinates*, where the spatial reference frame is attached to the average positions of galaxies. With this set of coordinates, both the time and expansion of the Universe can be ignored and the shape of space is seen as a spatial hypersurface at constant cosmological time.
The **comoving** (or *cosmological*) **distance** is a distance (denoted χ or d_{comov}) in comoving coordinates between two points in space at a single cosmological time, i.e., the distance between two *nearby* (close in redshift z) objects, which remains constant with epoch if these objects are moving with the Hubble flow.
The (cosmological) **proper distance** d_{proper} is a distance between two nearby events in the frame in which they occur at the same time. It is the distance measured by a ruler at the time t_o of observation. It holds

$$d_{comov}(x,y) = d_{proper}(x,y) \cdot \frac{a(t_o)}{a(t_e)} = d_{proper}(x,y) \cdot (1+z),$$

where $a(t)$ is the **scale factor**. In the time t_o, i.e., at the present, $a = a(t_o) = 1$, and $d_{comov} = d_{proper}$, In general, $d_{proper}(t) = a(t)d_{comov}$, for a cosmological time t.
The total **line-of-sight comoving distance** D_C from us to a distant object is computed by integrating the infinitesimal $d_{comov}(x,y)$ contributions between nearby events along the time ray from the time t_e, when the light from the object was emitted, to the time t_o, when the object is observed:

$$D_C = \int_{t_e}^{t_o} \frac{cdt}{a(t)}.$$

In terms of redshift, D_C from us to a distant object is computed by integrating the infinitesimal $d_{comov}(x,y)$ contributions between nearby events along the radial ray from $z = 0$ to the object: $D_C = D_H \int_0^z \frac{dz}{E(z)}$, where D_H is the **Hubble distance**, and $E(z) = (\Omega_M(1+z)^3 + \Omega_R(1+z)^2 + \Omega_\Lambda)^{\frac{1}{2}}$.
In a sense, the comoving distance is the fundamental distance measure in Cosmology, since all other distances can simply be derived in terms of it.

- **Proper motion distance**
The **proper motion distance** (or **transverse comoving distance**, *contemporary angular diameter distance*) D_M is a distance from us to a distant object defined as the ratio of the actual transverse velocity (in distance over time) of the object to its *proper motion* (in radians per unit time). It is given by

$$D_M = \begin{cases} D_H \frac{1}{\sqrt{\Omega_R}} \sinh(\sqrt{\Omega_R} D_C / D_H), & \text{for } \Omega_R > 0, \\ D_C, & \text{for } \Omega_R = 0, \\ D_H \frac{1}{\sqrt{|\Omega_R|}} \sin(\sqrt{|\Omega_R|} D_C / D_H), & \text{for } \Omega_R < 0, \end{cases}$$

where D_H is the **Hubble distance**, and D_C is the **line-of-sight comoving distance**. For $\Omega_\Lambda = 0$, there is an analytic solution (z is the *redshift*):

$$D_M = D_H \frac{2(2 - \Omega_M(1-z) - (2 - \Omega_M)\sqrt{1 + \Omega_M z})}{\Omega_M^2 (1+z)}.$$

The proper motion distance D_M coincides with the line-of-sight comoving distance D_C if and only if the curvature of the Universe is equal to zero. The **comoving distance** between two events at the same redshift or distance, but separated in the sky by some angle $\delta\theta$, is equal to $D_M \delta\theta$.
The distance D_M is related to the **luminosity distance** D_L and the **angular diameter distance** D_A by $D_M = (1+z)^{-1} D_L = (1+z) D_A$.

- **Luminosity distance**
The **luminosity distance** D_L is a distance from us to a distant object defined by the relationship between the observed flux S and emitted luminosity L:

$$D_L = \sqrt{\frac{L}{4\pi S}}.$$

This distance is related to the **proper motion distance** D_M and t the **angular diameter distance** by $D_L = (1+z) D_M = (1+z)^2 D_A$, where z is the *redshift*. The luminosity distance does take into account the fact that the observed luminosity is attenuated by two factors, the relativistic redshift and the Doppler shift of emission, each of which contributes an $(1+z)$ attenuation: $L_{obser} = \frac{L_{emiss}}{(1+z)^2}$.
The *corrected luminosity distance* D'_L is defined by $D'_L = \frac{D_L}{1+z}$.

- **Distance modulus**
The **distance modulus** is $DM = 5 \ln(\frac{D_L}{10pc})$, where D_L is the **luminosity distance**. The distance modulus is the difference between the *absolute magnitude* (the brightness that star would appear to have if it was at a distance of 10 parsec) and *apparent magnitude* (the actual brightness) of an astronomical object.
Distance moduli are most commonly used when expressing the distances to other galaxies. For example, the Andromeda Galaxy's DM is 24.5, and the Virgo cluster has DM equal to 31.7. For a much smaller object (planet, comet or asteroid), the *absolute magnitude* is its apparent visual magnitude at zero phase angle and at unit (1 AU) heliocentric and geocentric distances. The brightest (with peak apparent magnitude −7.5) recorded stellar event was the supernova in 1006.

- **Angular diameter distance**
The **angular diameter distance** (or *angular size distance*) D_A is a distance from us to a distant object defined as the ratio of an object's physical transverse size to its angular size (in radians). It is used to convert angular separations in telescope images into proper separations at the source. It is special for not increasing indefinitely as $z \to \infty$; it turns over at $z \sim 1$, and so more distant objects actually appear larger in angular size. D_A is related to the **proper motion distance** D_M and the **luminosity distance** D_L by $D_A = \frac{D_M}{1+z} = \frac{D_L}{(1+z)^2}$, where z is the *redshift*.
The *distance duality* $\frac{D_L(z)}{D_A(z)} = (1+z)^2$ links D_L, based on the apparent luminosity of standard candles (for example, supernovae) and D_A, based on the *apparent size* ("visual diameter" measured as an angle) of standard rulers (for example, **cosmic sound horizon**). It holds for any general **metric theory of gravity** (cf. Chap. 24) in any background in which photons travel on unique null geodesics.
If the angular diameter distance is based on the representation of object diameter as angle × distance, the **area distance** is defined similarly according the representation of object area as solid angle × distance^2.
- **Einstein radius**
General Relativity predicts *gravitational lensing*, i.e., deformation of the light from a *source* (a galaxy or star) in the presence of a *gravitational lens*, i.e., a body of large mass M (another galaxy, or a black hole) bending it.
If the source S, lens L and observer O are all aligned, the gravitational deflection is symmetric around the lens. The **Einstein radius** is the radius of the resulting *Einstein* (or *Chwolson*) *ring*. In radians it is

$$\sqrt{M\frac{4G}{c^2}\frac{D(L,S)}{D(O,L)D(O,S)}},$$

where $D(O, L)$ and $D(O, S)$ are the **angular diameter distances** of the lens and source, while $D(L, S)$ is the angular diameter distance between them.
- **Light travel distance**
The **light travel** (or *look-back*) **distance** is a distance from us to a distant object, defined by $D_{lt} = cD_t$, where D_t is the difference $t_o - t_e$ between the time, when the object was observed, and the time, when the light from it was emitted. The *look-back time* D_t is a **proper time**, but D_{lt} is not a **proper distance**.
D_{lt} is not a very useful distance, because it is hard to determine t_e, the age of the Universe at the time of emission of the light which we see. Cf. **Hubble radius**.
- **Parallax distance**
Given an object O viewed along two different lines of sight, its *parallax* is the angle $p = AOB$ between its directions of view from the two ends of a baseline AB. If $AO \approx BO$ and p, AB are small, the distance $A0$ can be easily estimated.
Animals use their two eyes (*stereoopsis*) or two positions of moving head (*motion parallax*) as points A, B. Cf. **animal depth/distance perception** in Chap. 23.

In Astronomy, the **parallax distance** is a distance D_P from us to a distant object (say, a star) defined from measuring of *stellar parallaxes*, i.e., its apparent changes of position in the sky caused by the motion of the observer on the Earth. Usually, it is the *annual parallax*, i.e., p is the angle Earth–star–Sun (in arsec) subtended at a star by the mean radius of the Earth's orbit around the Sun, and this distance (in parsecs, corresponding to $p = 1$ arcsec) is given by $D_P = \frac{1}{p}$.

- **Kinematic distance**
  The **kinematic distance** is the distance to a galactic source, which is determined from differential rotation of the galaxy: the *radial velocity* of a source directly corresponds to its **galactocentric distance**. But the *kinematic distance ambiguity* arises since, in our inner galaxy, any given galactocentric distance corresponds to two distances along the line of sight, *near* and *far* kinematic distances.
  This problem is solved, for some galactic regions, by measurement of their absorption spectra, if there is an interstellar cloud between the region and observer.
- **Radar distance**
  The **radar** (or *target*) **distance** D_R is a distance from us to a distant object, measured by a *radar*, i.e., a high frequency radio pulse sent out for a short interval of time. When it encounters a conducting object, sufficient energy is reflected back to allow radar to detect it. Since radio waves travel in air at close to their speed c (of light) in vacuum, one can calculate the distance D_R of the detected object from the round-trip time t between the transmitted and received pulses as

$$D_R = \frac{1}{2}ct.$$

  In general, *Einstein protocol* is to measure the distance between two objects A and B as $\frac{1}{2}c(t_3 - t_1)$. Here a light pulse is sent from A to B at time t_1 (measured in A), received at time t_2 (measured in B) and immediately sent back to A with a return time t_3 (measured with A).
- **Cosmological distance ladder**
  For measuring distances to astronomical objects, one uses a kind of "ladder" of different methods; each method applies only for a limited distance, and each method which applies for a larger distance builds on the data of the preceding methods.
  The starting point is knowing the distance from the Earth to the Sun; this distance is called one *astronomical unit* (*AU*), and is roughly 150 million km. Distances in the inner Solar System are measured by bouncing radar signals off planets or asteroids, and measuring the time until the echo is received.
  The next step in the ladder consists of simple geometrical methods; with them, one can go to a few hundred ly. The distance to nearby stars can be determined by their *parallaxes*: using Earth's orbit as a baseline, the distances to stars are measured by triangulation. This is accurate to about 1 % at 50 ly, 10 % at 500 ly. Using data acquired by the geometrical methods, and adding *photometry* (measurements of the brightness) and spectroscopy, one gets the next step in the ladder

for stars so far away that their parallaxes are not measurable yet. The *distance-luminosity relation* is that the light intensity from a star is inversely proportional to the square of its distance; cf. **distance modulus**.

The distance to the stellar cluster *Pleiades* is thought to be 135 parsecs. But satellite Hipparcos gave, by measuring the parallax of stars in the cluster, only 118 parsecs. This *Hipparcos anomaly* is a major unsolved problem in Astronomy.

For even larger distances, are used *standard candles*, i.e., several types of cosmological objects, for which one can determine their absolute brightness without knowing their distances. *Primary standard candles* are the *Cepheid* variable stars. They periodically change their size and temperature. There is a relationship between the brightness of these pulsating stars and the period of their oscillations, and this relationship can be used to determine their absolute brightness. Cepheids can be identified as far as in the Virgo cluster (60 Mly).

Secondary standard candles are supernovae 1a (having equal peak brightness), red giant branch stars, active galactic nuclei and entire galaxies. Main other techniques to estimate the **angular diameter distance** to galaxies are *gravitational lensing* (cf. **Einstein radius**) and using *baryon acoustic oscillations* matter clustering (cf. **cosmic sound horizon**) as a standard ruler.

For very large distances (hundreds of Mly or several Gly), the cosmological redshift and the Hubble law are used. A complication is that it is not clear what is meant by "distance" here, and there are several types of distances used here: **luminosity distance**, **proper motion distance**, **angular diameter distance**, etc.

Depending on the situation, there is a large variety of special techniques to measure distances in Cosmology, such as **light echo distance**, **Bondi radar distance**, **RR Lyrae distance** and *secular*, *statistical*, *expansion*, *spectroscopic* parallax distances. For example, NASA's Chandra X-ray Observatory measures since 2000 the distance to a distant source via the delay of the halo of scattering material (interstellar dust grains) between it and the Earth.

26.2 Distances in Theory of Relativity

The *Minkowski space-time* (or *Minkowski space*, *Lorentz space-time*, *flat space-time*) is the usual geometric setting for the Einstein Special Theory of Relativity. In this setting the three ordinary dimensions of space are combined with a single dimension of time to form a 4D *space-time* $\mathbb{R}^{1,3}$ in the absence of gravity. See, for example, [Wein72] for details.

Vectors in $\mathbb{R}^{1,3}$ are called *4-vectors* (or *events*). They can be written as (ct, x, y, z), where the first component is the unidirectional *time-like dimension* (c is the speed of light in vacuum, and t is the time), while the other three components are bidirectional *spatial dimensions*. Formally, c is a conversion factor from time to space.

In fact, c is the speed of gravitational waves and any massless particle: the *photon* (carrier of electromagnetism), the *gluon* (carrier of the strong force) and the *graviton* (theoretical carrier of gravity). It is the highest possible speed for any physical interaction in nature and the only speed independent of its source and the motion of an observer.

In the *spherical coordinates*, the events can be written as (ct, r, θ, ϕ), where r $(0 \leq r < \infty)$ is the *radius* from a point to the origin, ϕ $(0 \leq \phi < 2\pi)$ is the azimuthal angle in the xy plane from the x axis (*longitude*), and θ $(0 \leq \theta \leq \pi)$ is the polar angle from the z axis (*colatitude*). 4-vectors are classified according to the sign of their squared *norm*:

$$||v||^2 = \langle v, v \rangle = c^2t^2 - x^2 - y^2 - z^2.$$

They are said to be *time-like*, *space-like*, and *light-like* (*isotropic*) if their squared norms are positive, negative, or equal to zero, respectively. The set of all light-like 4-vectors forms the *light cone*. If the coordinate origin is singled out, the space can be broken up into three domains: domains of *absolute future* and *absolute past*, falling within the light cone, whose points are joined to the origin by time-like vectors with positive or negative value of time coordinate, respectively, and the domain of *absolute elsewhere*, falling outside of the light cone, whose points are joined to the origin by space-like vectors.

A *world line* of an object is the sequence of events that marks its time history. A world line is a *time-like* curve tracing out the path of a single point in the Minkowski space-time, i.e., at any point its *tangent vector* is a time-like 4-vector. All world lines fall within the light cone, i.e., the curves whose tangent vectors are light-like 4-vectors correspond to the motion of light and other particles of zero rest mass.

World lines of particles at constant speed (equivalently, of free falling particles) are called *geodesics*. In Minkowski space they are straight lines. A geodesic in Minkowski space which joins two given events x and y, is the longest curve among all world lines which join these two events. This follows from the **Einstein time triangle inequality** (cf. **inverse triangle inequality** and, in Chap. 5, **reverse triangle inequality**):

$$||x + y|| \geq ||x|| + ||y||,$$

according to which a time-like broken line joining two events is shorter than the single time-like geodesic joining them, i.e., the **proper time** of the particle moving freely from x to y is greater than the **proper time** of any other particle whose world line joins these events. It holds also in Minkowski space extended to any number of spatial dimensions, assuming null or time-like vectors in the same time direction. It is called *twin paradox*.

The *space-time* is a 4D *manifold* which is the usual mathematical setting for the Einstein General Theory of Relativity, which is the generalization of Special Relativity to include gravitation. Here the three spatial components with a single time-like component form a 4D space-time in the presence of gravity. Gravity is equivalent to the geometric properties of space-time, and in the presence of gravity the geometry of space-time is curved. (Bean, 2009, found evidence that over extragalactic distances gravity exerts a greater pull on time than on space.) So, the space-time is a 4D curved manifold for which the tangent space to any point is the Minkowski space, i.e., it is a *pseudo-Riemannian manifold*—a manifold, equipped with a nondegenerate indefinite metric (called **pseudo-Riemannian metric** in Chap. 7) of signature (1,3).

In the General Theory of Relativity, gravity is described by the properties of the local geometry of space-time. In particular, the gravitational field can be built out of a **metric tensor**, a quantity describing geometrical properties space-time such as distance, area, and angle. Matter is described by its *stress-energy tensor*, a quantity which contains the density and pressure of matter. The strength of coupling between matter and gravity is determined by the gravitational constant G.

The *Einstein field equation* is an equation in the General Theory of Relativity, that describes how matter creates gravity and, conversely, how gravity affects matter. A solution of the Einstein field equation is a certain **Einstein metric** appropriated for the given mass and pressure distribution of the matter.

A *black hole* is an astrophysical object that is theorized to be created from the collapse of a neutron or "quark" star. The gravitational forces are so strong in a black hole that they overcome neutron degeneracy pressure and, roughly, collapse to a *singularity* (point of infinite density and space-time curvature). Even light cannot escape the gravitational pull of a black hole within the black hole's its *event horizon*.

Uncharged black holes are called *Schwarzschild* or *Kerr black holes* if their angular momentum is zero or not, respectively. Charged black holes are called *Kerr–Newman* or *Reissner–Nordström black holes* if they are spinning or not, respectively.

Universe and black hole both have singularities—in time and space, respectively. *Naked* (not surrounded by a black hole) singularities were not observed but they might exist also. **Kerr metric** and **Reissner–Nordström metric below** admit such case. Also, a *kugelblitz* is a putative black hole formed from energy as opposed to mass.

Experimentally, General Relativity is still untested for strong fields (such as near neutron-star surfaces or black-hole horizons) or over distances on a galactic scale and larger. Neither Newton law of gravitation was tested below 6×10^{-5} m.

Putative *gravitational waves* (fluctuations in the curvature of space-time propagating as a wave, predicted by Einstein), have been detected in 2014. Also predicted *frame-dragging effect* (the spinning Earth pulls space-time around with it) is under probe. The *geodetic effect*, confirming that space-time acts on matter, was found.

- **Minkowski metric**

  The **Minkowski metric** is a **pseudo-Riemannian metric**, defined on the *Minkowski space* $\mathbb{R}^{1,3}$, i.e., a 4D real vector space which is considered as the *pseudo-Euclidean space* of signature $(1, 3)$. It is defined by its **metric tensor**

$$((g_{ij})) = \begin{pmatrix} 1 & 0 & 0 & 0 \\ 0 & -1 & 0 & 0 \\ 0 & 0 & -1 & 0 \\ 0 & 0 & 0 & -1 \end{pmatrix}.$$

The *line element* ds^2 of this metric are given by

$$ds^2 = c^2dt^2 - dx^2 - dy^2 - dz^2.$$

In *spherical coordinates* (ct, r, θ, ϕ), one has $ds^2 = c^2dt^2 - dr^2 - r^2d\theta^2 - r^2\sin^2\theta d\phi^2$.

The pseudo-Euclidean space $\mathbb{R}^{1,3}$ of signature $(1, 3)$ with the *line element*

$$ds^2 = -c^2dt^2 + dx^2 + dy^2 + dz^2$$

can also be used as a space-time model of the Special Theory of Relativity.

Above notion of *space-time* (Minkowski, 1908) was the first application of geometry to a nonlength-like quantity. But there were some precursors of such union of space and time. Lagrange, 1797, observed that with time as a 4th coordinate, "one can regard mechanics as 4-dimensional geometry". Schopenhauer wrote in *On the Fourfold Root of the Principle of Sufficient Reason* (1813): "... it is only by the combination of Time and Space that the representation of coexistence arises." Poe wrote "Space and Duration are one" in *Eureka: A Prose Poem* (1848).

Wells wrote on the first page of *The Time Machine* (1895): 'Clearly,' the Time Traveler proceeded, 'any real body must have extension in four directions: it must have Length, Breadth, Thickness, and Duration ... There is no difference between Time and any of the tree dimensions of Space except that our consciousness moves along it". *Quechua*, the language of Inca and eight to ten million modern speakers, have a single concept, *pacha*, for the location in time and space.

- **Proper distance and time**

  In Relativistic Physics, **proper distance** and **proper time** between any two events are true physical distance and time difference: the spatial distance between them when the events are simultaneous and the temporal distance between them when the events occur at the same spatial location. They are the *invariant* (with respect to Lorentz transformations, describing a transition to a coordinate system associated with a moving body) intervals of a space-like path or pair of space-like separated events, and, respectively, of a time-like path or pair of time-like separated events.

  In General Relativity, **proper time** is the pseudo-Riemannian arc length of world lines in 4D-spacetime. In particular, in SR (Special Relativity), it is

  $$\tau = \int_P \sqrt{dt^2 - c^{-2}(dx^2 + dy^2 + dz^2)},$$

  where t and x, y, z are time and spatial coordinates, while P is the path of the clock in space-time. In the subcase of inertial motion, it become

  $$\Delta\tau = \sqrt{(\Delta t)^2 - c^{-2}((\Delta x)^2 + (\Delta y)^2 + (\Delta z)^2)},$$

  where Δ means "the change in" between two events. Cf. the **kinematic metric**.

In SR, the **proper distance** between two space-like separated events is

$$\Delta\sigma = \sqrt{(\Delta x)^2 + (\Delta y)^2 + (\Delta z)^2 - c^2(\Delta t)^2}.$$

- **Proper length**
  In Special Theory of Relativity, the **proper** (or *rest*) **length** between two space-like separated events is the distance between them, such as measured in an inertial frame of reference in which the events are simultaneous. In contrast to invariant **proper distance**, such simultaneity depends on the observer.
  In a flat space-time, the proper length between two events is the proper length of a straight path between them. General Relativity consider the curved space-times in which may be more than one straight path (geodesic) between two events.
  So, the general **proper length** is defined as the path integral $\int_P \sqrt{-g_{ij}dx^i dx^j}$, where g_{ij} is the metric tensor for the space-time with signature $(1, 3)$, along the shortest curve joining the endpoints of the space-like path P at the same time.
- **Affine space-time distance**
  Given a space-time (M^4, g), there is a unique affine parametrization $s \to \gamma(s)$ for each light ray (i.e., light-like geodesic) through the observation event p_o, such that $\gamma(0) = p_o$ and $g(\frac{d\gamma}{dt}, U_o) = 1$, where U_o is the 4-velocity of the observer at p_o (i.e., a vector with $g(U_o, U_o) = -1$). In this case, the **affine space-time distance** is the *affine parameter* s, viewed as a distance measure.
  This distance is monotone increasing along each ray, and it coincides, in a small neighborhood of p_o, with the Euclidean distance in the rest system of U_o.
- **Lorentz metric**
  A **Lorentz metric** is a **pseudo-Riemannian metric** (i.e., **nondegenerate indefinite metric**, cf. Chap. 7) of signature $(1, p)$.
  The curved space-time of the General Theory of Relativity can be modeled as a *Lorentzian manifold* (a manifold equipped with a Lorentz metric) of signature $(1, p)$. The *Minkowski space* $\mathbb{R}^{1,p}$ with the flat **Minkowski metric** is a model of it, in the same way as Riemannian manifolds can be modeled on Euclidean space.
  Given a rectifiable non-space-like curve $\gamma : [0, 1] \to M$ in the space-time M, the *length* of the curve is defined as $l(\gamma) = \int_0^1 \sqrt{-\langle \frac{d\gamma}{dt}, \frac{d\gamma}{dt}\rangle}dt$. For a space-like curve, we set $l(\gamma) = 0$ and define the **Lorentz distance** between two points $p, q \in M$ as

$$\sup_{\gamma\in\Gamma} l(\gamma),$$

  if $p \prec q$, i.e., if the set Γ of *future directed* non-space-like curves from p to q is nonempty; otherwise, this distance is 0.
  The **Lorentz–Minkowski distance** is a **pseudo-Euclidean distance** (Chap. 7) $\sqrt{D(x, y)}$, where $D(x, y) = |x_1 - y_1|^2 - \sum_{2\le i\le n} |x_i - y_i|^2$. The points are called

time-, *space-*, *null-separated* if $D(x, y)$ is more, less or equal to 0, respectively, i.e., if they can be joined by a time-like, space-like or null path.

- **Distances on causal sets**

  Causal Set Theory is a fundamentally discrete approach to quantum gravity. A *causal set* (or *causet*) is a partially ordered set $(X, \preceq)$, which is *locally finite*, i.e., the *interval* $(x, y) = \{z \in X : x \prec z \prec y\}$ is finite for any $x, y \in X$. A *link* is a pair $x, y \in X$ such that $x \prec y$ and $(x, y) = \emptyset$. A *chain* is a subcauset such that $x \prec y$ or $y \prec x$ for any two its elements x, y.

  Given $x, y \in X$ with $x \prec y$, their **time-like distance** $d_t(x, y)$ is (Brightwell–Gregory, 1991) the *length* (number of links) in any *geodesic* between them, i.e., longest chain between and including x and y. Given two unrelated elements $x, y \in X$, their **naive space-like distance** is defined (Brightwell–Gregory, 1991) as

  $$d_{ns}(x, y) = \min_{u,v \in X : u \prec (x,y) \prec v} d_t(u, v).$$

  Rideout–Wallden, 2013, modified $d_{ns}(x, y)$, replacing the minimum above with an average over suitably selected minimumizing pairs.

  The elements of X can be seen as events in a discrete space-time, where the partial order represent causal relationship. In a causet embedded in Minkowski space-time, the distance $d_t(x, y)$ is proportional to the **proper time**. A related discrete space-time is a random poset obtained by sampling from a compact domain in a space-time manifold. Cf. also **D-separation in Bayesian network** in Chap. 22.

- **Kinematic metric**

  Given a set X, a **kinematic metric** (or *abstract Lorentzian distance*) is (Pimenov, 1970) a function $\tau : X \times X \to \mathbb{R}_{\geq 0} \cup \{\infty\}$ such that, for all $x, y, z \in X$:

  1. $\tau(x, x) = 0$;
  2. $\tau(x, y) > 0$ implies $\tau(y, x) = 0$ (*antisymmetry*);
  3. $\tau(x, y), \tau(y, z) > 0$ implies $\tau(x, z) > \tau(x, y) + \tau(y, z)$ (**inverse triangle inequality** or *anti-triangle inequality*).

  The *space-time* set X consists of *events* $x = (x_0, x_1)$ where, usually, $x_0 \in \mathbb{R}$ is the *time* and $x_1 \in \mathbb{R}^3$ is the *spatial location* of the event x. The inequality $\tau(x, y) > 0$ means *causality*, i.e., x can influence y; usually, it is equivalent to $y_0 > x_0$ and the value $\tau(x, y) > 0$ can be seen as the largest (since it depends on the speed) **proper time** of moving from x to y.

  If the gravity is negligible, then $\tau(x, y) > 0$ implies $y_0 - x_0 \geq ||y_1 - x_1||_2$, and $\tau_p(x, y) = ((y_0 - x_0)^p - ||y_1 - x_1||_2^p))^{\frac{1}{p}}$ (as defined by Busemann, 1967) is a real number. For $p \approx 2$ it is consistent with Special Relativity observations.

  A kinematic metric is not our usual distance metric; also it is not related to the **kinematic distance** in Astronomy.

  But Zapata, 2013, proved that $\sup_z \max(|\tau(x, z) - \tau(y, z)|, |\tau(z, x) - \tau(z, y)|)$ is a continuous metric on a compact part of space-time and it generates the same topology as a nonphysical coordinate-dependent Euclidean distance $\sqrt{\sum_{i=2}^{4}|x_i - y_i|^2}$.

- **Galilean distance**
The **Galilean distance** is a distance on $\mathbb{R}^n$ defined by

$$|x_1 - y_1| \text{ if } x_1 \neq y_1,$$

and by

$$\sqrt{(x_2 - y_2)^2 + \ldots + (x_n - y_n)^2} \text{ if } x_1 = y_1.$$

The space $\mathbb{R}^n$, equipped with the Galilean distance, is called *Galilean space*. For $n = 4$, it is a setting for the space-time of classical mechanics according to Galilei–Newton in which the distance between two events taking place at the points p and q at the moments of time t_1 and t_2 is defined as the time interval $|t_1 - t_2|$, while if $t_1 = t_2$, it is defined as the distance between the points p and q.

- **Einstein metric**
In the General Theory of Relativity, describing how space-time is curved by matter, the **Einstein metric** is a solution to the Einstein field equation

$$R_{ij} - \frac{g_{ij}R}{2} + \Lambda g_{ij} = \frac{8\pi G}{c^4}T_{ij},$$

i.e., a **metric tensor** $((g_{ij}))$ of signature $(1, 3)$, appropriated for the given mass and pressure distribution of the matter. Here $E_{ij} = R_{ij} - \frac{g_{ij}R}{2} + \Lambda g_{ij}$ is the *Einstein curvature tensor*, R_{ij} is the *Ricci curvature tensor*, R is the *Ricci scalar*, and T_{ij} is a *stress-energy tensor*. *Empty space* (*vacuum*) is the case of $R_{ij} = 0$.
Einstein introduced in 1917 the *cosmological constant* Λ to counteract the effects of gravity on ordinary matter and keep the Universe *static*, i.e., with **scale factor** always being 1. He put $\Lambda = \frac{4\pi G\rho}{c^2}$. The static Einstein metric for a homogeneous and isotropic Universe is given by the *line element*

$$ds^2 = -dt^2 + \frac{dr^2}{(1 - kr^2)} + r^2(d\theta^2 + \sin^2\theta d\phi^2),$$

where k is the curvature of the space-time. The radius of this curvature is $\frac{c}{\sqrt{4\pi G\rho}}$ and numerically it is of the order 10 Gly. Einstein from 1922 call this model his "biggest blunder" but Λ was reintroduced in modern dynamic models as dark energy.

- **de Sitter metric**
The **de Sitter metric** is a maximally symmetric vacuum solution to the Einstein field equation with a positive cosmological constant Λ, given by the *line element*

$$ds^2 = dt^2 + e^{2\sqrt{\frac{\Lambda}{3}}t}(dr^2 + r^2d\theta^2 + r^2\sin^2\theta d\phi^2).$$

Expansion of Universe is accelerating at the rate consistent with $\Lambda \sim 10^{-123}$, but Hartie–Hawking–Hertog, 2012, gave a quantum model of it with $\Lambda < 0$.
The most symmetric solutions to the Einstein field equation in a vacuum for $\Lambda = 0$ and $\Lambda < 0$ are the flat **Minkowski metric** and the **anti de Sitter metric**.
The n-dimensional *de Sitter space* dS_n and *anti de Sitter space* AdS_n are Lorentzian manifold analogs of elliptic and hyperbolic space, respectively.
In order to explain the *hierarchy problem* (why the weak nuclear force is 10^{32} times stronger than gravity?), Randall and Sundram, 1999, proposed that Universe is 5D anti de Sitter space AdS_5 with elementary particles, except for the graviton, being on a $(3+1)$-D brane or branes. This **Randall–Sundrum metric** is $ds^2 = e^{-2ky} g_{ab} dx^a dy^b + dy^2$, where k is of order the Planck scale ($\sim 10^{-35}$ m) and x^a, y are coordinates in 4D and extra-dimension.

- **BTZ metric**
The **BTZ metric** (Banados, Teitelboim and Zanelli, 2001) is a black hole solution for $(2+1)$-dimensional gravity with a negative cosmological constant Λ.
There are no such solutions with $\Lambda = 0$. BTZ black holes without any electric charge are locally isometric to *anti de Sitter space*.
This metric is given by the *line element*

$$ds^2 = -k^2(r^2 - R^2)dt^2 + \frac{1}{k^2(r^2 - R^2)}dr^2 + r^2 d\theta^2,$$

where R is the black hole radius, in the absence of charge and angular momentum.

- **Schwarzschild metric**
The **Schwarzschild metric** is a vacuum solution to the *Einstein field equation* around a spherically symmetric mass distribution; this metric represents the Universe around a black hole of a given mass, from which no energy can be extracted.
It was found by Schwarzschild, 1915, only a few months after the publication of the Einstein field equation, and was the first exact solution of this equation.
The *line element* of this metric is given by

$$ds^2 = \left(1 - \frac{r_g}{r}\right) c^2 dt^2 - \frac{1}{\left(1 - \frac{r_g}{r}\right)} dr^2 - r^2(d\theta^2 + \sin^2\theta d\phi^2),$$

where $r_g = \frac{2Gm}{c^2}$ is the *Schwarzschild radius* and m is the mass of the black hole. This solution is only valid for radii larger than r_g, as at $r = r_g$ there is a coordinate singularity. This problem can be removed by a transformation to a different choice of space-time coordinates, called *Kruskal–Szekeres coordinates*.
As $r \to +\infty$, the Schwarzschild metric approaches the **Minkowski metric**.

- **Kottler metric**
The **Kottler metric** is the unique spherically symmetric vacuum solution to the Einstein field equation with a cosmological constant Λ. It is given by

$$ds^2 = -\left(1 - \frac{2m}{r} - \frac{\Lambda r^2}{3}\right) dt^2 + \left(1 - \frac{2m}{r} - \frac{\Lambda r^2}{3}\right)^{-1} dr^2 + r^2(d\theta^2 + \sin^2\theta d\phi^2).$$

It is called also the **Schwarzschild–de Sitter metric** for $\Lambda > 0$ and **Schwarzschild–anti de Sitter metric** for $\Lambda < 0$. Cf. **Delaunay metric** in Chap. 7.

- **Reissner–Nordström metric**
  The **Reissner–Nordström metric** is a vacuum solution to the Einstein field equation around a spherically symmetric mass distribution in the presence of a charge; it represents the Universe around a charged black hole. This metric is given by

$$ds^2 = \left(1 - \frac{2m}{r} + \frac{e^2}{r^2}\right) dt^2 - \left(1 - \frac{2m}{r} + \frac{e^2}{r^2}\right)^{-1} dr^2 - r^2(d\theta^2 + \sin^2\theta d\phi^2),$$

  where m is the mass of the hole, e is the charge ($e < m$), and we have used units with the speed c of light and the *gravitational constant* G equal to one.
- **Kerr metric**
  The **Kerr metric** (or **Kerr–Schild metric**) is an exact solution to the Einstein field equation for empty space (vacuum) around an axially symmetric, rotating mass distribution, This metric represents the Universe around a rotating black hole. Its *line element* is given (in *Boyer–Lindquist form*) by

$$ds^2 = \rho^2 \left(\frac{dr^2}{\Delta} + d\theta^2\right) + (r^2 + a^2)\sin^2\theta d\phi^2 - dt^2 + \frac{2mr}{\rho^2}(a \sin^2\theta d\phi - dt)^2,$$

  where $\rho^2 = r^2 + a^2\cos^2\theta$ and $\Delta = r^2 - 2mr + a^2$. Here m is the mass of the black hole and a is the angular velocity as measured by a distant observer.
  The **Schwarzschild metric** is the Kerr metric with $a = 0$. A black hole is rotating if radiation processes are observed outside its *Schwarzschild radius* (the event horizon radius as dependent on the mass only) but inside its *Kerr radius* (where the rotational kinetic energy is comparable with the rest energy). For the Earth, those radii are about 1 cm and 3 m, respectively.
  In 2013, the spin of a black hole was directly measured for the first time: the central black hole of the galaxy NGC 1365 rotates at 84 % of the speed c of light.
- **Kerr–Newman metric**
  The **Kerr–Newman metric** is an exact, unique and complete solution to the Einstein field equation for empty space (vacuum) around an axially symmetric, rotating mass distribution in the presence of a charge, This metric represents the Universe around a rotating charged black hole. Its *line element* is given by

$$ds^2 = -\frac{\Delta}{\rho^2}(dt - a\sin^2\theta d\phi)^2 + \frac{\sin^2\theta}{\rho^2}((r^2 + a^2)d\phi - a dt)^2 + \frac{\rho^2}{\Delta}dr^2 + \rho^2 d\theta^2,$$

where $\rho^2 = r^2 + a^2 \cos^2\theta$ and $\Delta = r^2 - 2mr + a^2 + e^2$. Here m is the mass of the black hole, e is the charge, and a is the angular velocity.
The Kerr–Newman metric becomes the **Kerr metric** if the charge is 0 and the **Reissner–Nordström metric** if the angular momentum is 0.

- **Ozsváth–Schücking metric**
The **Ozsváth–Schücking metric** (1962) is a rotating vacuum solution to the field equations having in Cartesian coordinates the form

$$ds^2 = -2[(x^2 - y^2)\cos(2t) - 2xy\sin(2t)]dt^2 + dx^2 + dy^2 - 2dtdz.$$

- **Static isotropic metric**
The **static isotropic metric** is the most general solution to the Einstein field equation for empty space (vacuum); this metric can represent a static isotropic gravitational field. The *line element* of this metric is given by

$$ds^2 = B(r)dt^2 - A(r)dr^2 - r^2(d\theta^2 + \sin^2\theta d\phi^2),$$

where $B(r)$ and $A(r)$ are arbitrary functions.

- **Eddington–Robertson metric**
The **Eddington–Robertson metric** is a generalization of the **Schwarzschild metric** which allows that the mass m, the *gravitational constant* G, and the density ρ are altered by unknown dimensionless parameters α, β, and γ (all equal to 1 in the *Einstein field equation*). The *line element* of this metric is given by

$$ds^2 = \left(1 - 2\alpha\frac{mG}{r} + 2(\beta - \alpha\gamma)\left(\frac{mG}{r}\right)^2 + \dots\right)dt^2 - \left(1 + 2\gamma\frac{mG}{r} + \dots\right)dr^2 - r^2(d\theta^2 + \sin^2\theta d\phi^2).$$

- **Janis–Newman–Wincour metric**
The **Janis–Newman–Wincour metric** is the most general spherically symmetric static and asymptotically flat solution to the Einstein field equation coupled to a massless scalar field. It is given by the *line element*

$$ds^2 = -\left(1 - \frac{2m}{\gamma r}\right)^{\gamma} dt^2 + \left(1 - \frac{2m}{\gamma r}\right)^{-\gamma} dr^2 + \left(1 - \frac{2m}{\gamma r}\right)^{1-\gamma} r^2(d\theta^2 + \sin^2\theta d\phi^2),$$

where m and γ are constants. For $\gamma = 1$ one obtains the **Schwarzschild metric**. In this case the scalar field vanishes.

- **FLRW metric**
  The **FLRW** (Friedmann–Lemaître–Robertson–Walker) **metric** is a exact solution to the Einstein field equation for a simply connected, homogeneous, isotropic expanding (or contracting) Universe filled with a constant density and negligible pressure. This metric represents a matter-dominated Universe filled with a *dust* (pressure-free matter); it models the **metric expansion of space**.
  Its *line element* is usually written in the *spherical coordinates* (ct, r, θ, ϕ):
  $$ds^2 = c^2dt^2 - a(t)^2\left(\frac{dr^2}{1-kr^2} + r^2(d\theta^2 + \sin^2\theta d\phi^2)\right),$$
  where $a(t)$ is the **scale factor** and k is the *curvature* of the space-time.
- **Vaidya metric**
  The **Vaidya metric** is a inhomogeneous solution to Einstein field equation describing a spherically symmetric space-time composed purely of radially propagating radiation. It has been used to describe the radiation emitted by a shining star, by a collapsing star and by evaporating black hole.
  The Vaidya metric is a nonstatic generalization of the **Schwarzschild metric** and the radiation limit of the **LTB metric**. Let $M(u)$ be the mass parameter; the *line element* of this metric (Vaidya, 1953) is given by
  $$ds^2 = -[1 - 2\frac{M(u)}{r}]du^2 + 2dudr + r^2(d\theta^2 + \sin^2\theta d\phi^2).$$
- **LTB metric**
  The **LTB** Lemaître–Tolman–Bondi) **metric** is a solution to the Einstein field equation describing a spherical (finite or infinite) cloud of *dust* (pressure-free matter) that is expanding or collapsing under gravity.
  The LTB metric describes an inhomogeneous space-time expected on very large (Gpc) scale. It generalizes the **FLRW metric** and the **Schwarzschild metric**.
  The *line element* of this metric in the *spherical coordinates* is:
  $$ds^2 = dt^2 - \frac{(R')^2}{1+2E}dr^2 - R^2(d\theta^2 + \sin^2\theta d\phi^2),$$
  where $R = R(t, r)$, $R' = \frac{\partial R}{\partial r}$, $E = E(r)$. The shell $r = r_0$ at a time $t = t_0$ has an area $4\pi R^2(r_0, t_0)$, and the areal radius R evolves with time as $\frac{\partial R}{\partial t} = 2E + \frac{2M}{R}$, where $M = M(r)$ is the gravitational mass within the comoving sphere at radius r.
- **Kantowski–Sachs metric**
  The **Kantowski–Sachs metric** is a solution to the Einstein field equation, given by the *line element*
  $$ds^2 = -dt^2 + a(t)^2dz^2 + b(t)^2(d\theta^2 + \sin\theta d\phi^2),$$

where the functions $a(t)$ and $b(t)$ are determined by the Einstein equation. It is the only homogeneous model without a 3D transitive subgroup.
In particular, the Kantowski–Sachs metric with the *line element*

$$ds^2 = -dt^2 + e^{2\sqrt{\Lambda}t}dz^2 + \frac{1}{\Lambda}(d\theta^2 + \sin^2\theta d\phi^2)$$

describes an anisotropic Universe with two spherical dimensions having a fixed size during the cosmic evolution and exponentially expanding 3rd dimension.

- **Bianchi metrics**
The **Bianchi metrics** are solutions to the Einstein field equation for cosmological models that have spatially homogeneous sections, invariant under the action of a 3D Lie group, i.e., they are real 4D metrics with a 3D isometry group, transitive on 3-surfaces. Using the Bianchi classification of 3D Lie algebras over Killing vector fields, we obtain the nine types of Bianchi metrics.
Each Bianchi model B defines a transitive group G_B on some 3D simply connected manifold M; so, the pair (M, G) (where G is the maximal group acting on X and containing G_B) is one of eight Thurston *model geometries* if M/G' is compact for a discrete subgroup G' of G. In particular, Bianchi type IX corresponds to the geometry S^3. Only the model geometry $S^2 \times \mathbb{R}$ is not realized in this way.
The Bianchi type I metric is a solution to the Einstein field equation for an anisotropic homogeneous Universe, given by the *line element*

$$ds^2 = -dt^2 + a(t)^2dx^2 + b(t)^2dy^2 + c(t)^2dz^2,$$

where the functions $a(t)$, $b(t)$, and $c(t)$ are determined by the Einstein equation. It corresponds to flat spatial sections, i.e., is a generalization of the **FLRW metric**.
The Bianchi type IX metric, or **Mixmaster metric** (Misner, 1969), exhibits chaotic dynamic behavior near its curvature singularities.

- **Kasner metric**
The **Kasner metric** is a Bianchi type I metric, which is a vacuum solution to the Einstein field equation for an anisotropic homogeneous Universe, given by

$$ds^2 = -dt^2 + t^{2p_1}dx^2 + t^{2p_2}dy^2 + t^{2p_3}dz^2,$$

where $p_1 + p_2 + p_3 = p_1^2 + p_2^2 + p_3^2 = 1$.
The equal-time slices of Kasner space-time are spatially flat, but space contracts in one dimension (i with $p_i < 0$), while expanding in the other two. The volume of the spatial slices is proportional to t; so, $t \to 0$ can describe either a Big Bang or a *Big Crunch*, depending on the sense of t.

- **GCSS metric**
A **GCSS** (i.e., **general cylindrically symmetric stationary**) **metric** is a solution to the *Einstein field equation*, given by the *line element*

$$ds^2 = -f\,dt^2 + 2k\,dt d\phi + e^{\mu}(dr^2 + dz^2) + l d\phi^2,$$

where the space-time is divided into two regions: the interior, with $0 \le r \le R$, to a cylindrical surface of radius R centered along z, and the exterior, with $R \le r < \infty$. Here f, k, μ and l are functions only of r, and $-\infty < t, z < \infty, 0 \le \phi \le 2\pi$; the hypersurfaces $\phi = 0$ and $\phi = 2\pi$ are identical.

- **Lewis metric**
The **Lewis metric** is a **cylindrically symmetric stationary metric** which is a solution to the *Einstein field equation* for empty space (vacuum) in the exterior of a cylindrical surface. The *line element* of this metric has the form

$$ds^2 = -f\,dt^2 + 2k\,dt d\phi - e^{\mu}(dr^2 + dz^2) + l\,d\phi^2,$$

where $f = ar^{-n+1} - \frac{c^2}{n^2 a} r^{n+1}$, $k = -Af$, $l = \frac{r^2}{f} - A^2 f$, $e^{\mu} = r^{\frac{1}{2}(n^2-1)}$ with $A = \frac{cr^{n+1}}{naf} + b$. The constants n, a, b, and c can be either real or complex, the corresponding solutions belong to the *Weyl class* or *Lewis class*, respectively. In the last case, such metrics form a subclass of the *Kasner type metrics*.

- **van Stockum dust metric**
The **van Stockum dust metric** is a stationary cylindrically symmetric solution to the Einstein field equation for empty space (vacuum) with a rigidly rotating infinitely long dust cylinder. The *line element* of this metric for the interior of the cylinder is given (in comoving, i.e., corotating, coordinates) by

$$ds^2 = -dt^2 + 2ar^2 dt d\phi + e^{-a^2 r^2}(dr^2 + dz^2) + r^2(1 - a^2 r^2) d\phi^2,$$

where $0 \le r \le R$, R is the radius of the cylinder, and a is the angular velocity of the dust particles. There are three vacuum exterior solutions (i.e., **Lewis metrics**) that can be matched to the interior solution, depending on the mass per unit length of the interior (the *low mass case*, the *null case*, and the *ultrarelativistic case*). Under some conditions (for example, if $ar > 1$), the existence of *closed time-like curves* (and, hence, time-travel) is allowed.

- **Levi-Civita metric**
The **Levi-Civita metric** is a static cylindrically symmetric vacuum solution to the Einstein field equation, with the *line element*, given (in the Weyl form) by

$$ds^2 = -r^{4\sigma} dt^2 + r^{4\sigma(2\sigma-1)}(dr^2 + dz^2) + C^{-2} r^{2-4\sigma} d\phi,$$

where the constant C refers to the deficit angle, and σ is a parameter.
In the case $\sigma = -\frac{1}{2}$, $C = 1$ this metric can be transformed either into the *Taub's plane symmetric metric*, or into the *Robinson–Trautman metric*.

- **Weyl–Papapetrou metric**
The **Weyl–Papapetrou metric** is a stationary axially symmetric solution to the Einstein field equation, given by the *line element*

$$ds^2 = Fdt^2 - e^{\mu}(dz^2 + dr^2) - Ld\phi^2 - 2Kd\phi dt,$$

where F, K, L and μ are functions only of r and z, $LF + K^2 = r^2$, $\infty < t, z < \infty$, $0 \le r < \infty$, and $0 \le \phi \le 2\pi$; the hypersurfaces $\phi = 0$ and $\phi - 2\pi$ are identical.

- **Bonnor dust metric**
  The **Bonnor dust metric** is a solution to the *Einstein field equation* which is an axially symmetric metric describing a cloud of rigidly rotating dust particles moving along circular geodesics about the z axis in hypersurfaces of $z =$ *constant*. The *line element* of this metric is given by

$$ds^2 = dt^2 + (r^2 - n^2)d\phi^2 + 2ndtd\phi + e^{\mu}(dr^2 + dz^2),$$

  where, in Bonnor comoving (i.e., corotating) coordinates, $n = \frac{2hr^2}{R^3}$, $\mu = \frac{h^2r^2(r^2-8z^2)}{2R^8}$, $R^2 = r^2 + z^2$, and h is a rotation parameter.
  As $R \to \infty$, the metric coefficients tend to Minkowski values.
- **Weyl metric**
  The **Weyl metric** is a general static axially symmetric vacuum solution to the Einstein field equation given, in Weyl canonical coordinates, by the *line element*

$$ds^2 = e^{2\lambda}dt^2 - e^{-2\lambda}\left(e^{2\mu}(dr^2 + dz^2) + r^2d\phi^2\right),$$

  where λ and μ are functions only of r and z such that $\frac{\partial^2\lambda}{\partial r^2} + \frac{1}{r}\cdot\frac{\partial\lambda}{\partial r} + \frac{\partial^2\lambda}{\partial z^2} = 0$, $\frac{\partial\mu}{\partial r} = r(\frac{\partial\lambda^2}{\partial r} - \frac{\partial\lambda^2}{\partial z})$, and $\frac{\partial\mu}{\partial z} = 2r\frac{\partial\lambda}{\partial r}\frac{\partial\lambda}{\partial z}$.
- **Zipoy–Voorhees metric**
  The **Zipoy–Voorhees metric** (or γ-*metric*) is a **Weyl metric**, obtained for $e^{2\lambda} = \left(\frac{R_1+R_2-2m}{R_1+R_2+2m}\right)^{\gamma}$, $e^{2\mu} = \left(\frac{(R_1+R_2+2m)(R_1+R_2-2m)}{4R_1R_2}\right)^{\gamma^2}$, where $R_1^2 = r^2 + (z-m)^2$, $R_2^2 = r^2 + (z+m)^2$. Here λ corresponds to the Newtonian potential of a line segment of mass density $\gamma/2$ and length $2m$, symmetrically distributed along the z axis.
  The case $\gamma = 1$ corresponds to the **Schwartzschild metric**, the cases $\gamma > 1$ ($\gamma < 1$) correspond to an oblate (prolate) spheroid, and for $\gamma = 0$ one obtains the flat Minkowski space-time.
- **Straight spinning string metric**
  The **straight spinning string metric** is given by the *line element*

$$ds^2 = -(dt - ad\phi)^2 + dz^2 + dr^2 + k^2r^2d\phi^2,$$

  where a and $k > 0$ are constants. It describes the space-time around a straight spinning string. The constant k is related to the string's mass-per-length μ by $k = 1 - 4\mu$, and the constant a is a measure of the string's spin. For $a = 0$ and $k = 1$, one obtains the **Minkowski metric** in cylindrical coordinates.

- **Tomimatsu–Sato metric**
  A **Tomimatsu–Sato metric** [ToSa73] is one of the metrics from an infinite family of spinning mass solutions to the Einstein field equation, each of which has the form $\xi = U/W$, where U and W are some polynomials.
  The simplest solution has $U = p^2(x^4-1) + q^2(y^4-1) - 2ipqxy(x^2-y^2)$, $W = 2px(x^2-1) - 2iqy(1-y^2)$, where $p^2+q^2=1$. The *line element* for it is given by

$$ds^2 = \Sigma^{-1}\left((\alpha dt + \beta d\phi)^2 - r^2(\gamma dt + \delta d\phi)^2\right) - \frac{\Sigma}{p^4(x^2-y^2)^4}(dz^2+dr^2),$$

  where $\alpha = p^2(x^2-1)^2 + q^2(1-y^2)^2$, $\beta = -\frac{2q}{p}W(p^2(x^2-1)(x^2-y^2) + 2(px+1)W)$, $\gamma = -2pq(x^2-y^2)$, $\delta = \alpha + 4((x^2-1)+(x^2+1)(px+1))$, $\Sigma = \alpha\delta - \beta\gamma = |U+W|^2$.
- **Gödel metric**
  The **Gödel metric** is an exact solution to the Einstein field equation with cosmological constant for a rotating Universe, given by the *line element*

$$ds^2 = -(dt^2 + C(r)d\phi)^2 + D^2(r)d\phi^2 + dr^2 + dz^2,$$

  where (t, r, ϕ, z) are the usual *cylindrical coordinates.*
  The *Gödel Universe* is homogeneous if $C(r) = \frac{4\Omega}{m^2}\sinh^2\left(\frac{mr}{2}\right)$, $D(r) = \frac{1}{m}\sinh(mr)$, where m and Ω are constants. The Gödel Universe is singularity-free. But there are *closed time-like curves* through every event, and hence time-travel is possible here. The condition required to avoid such curves is $m^2 > 4\Omega^2$.
- **Conformally stationary metric**
  The **conformally stationary metrics** are models for gravitational fields that are time-independent up to an overall conformal factor. If some global regularity conditions are satisfied, the space-time must be a product $\mathbb{R} \times M^3$ with a (Hausdorff and paracompact) 3-manifold M^3, and the *line element* of the metric is given by

$$ds^2 = e^{2f(t,x)}(-(dt + \sum_{\mu}\phi_\mu(x)dx_\mu)^2 + \sum_{\mu,\nu} g_{\mu\nu}(x)dx_\mu dx_\nu),$$

  where $\mu, \nu = 1, 2, 3$. The conformal factor e^{2f} does not affect the light-like geodesics apart from their parametrization, i.e., the paths of light rays are completely determined by the Riemannian metric $g = \sum_{\mu,\nu} g_{\mu\nu}(x)dx_\mu dx_\nu$ and the one-form $\phi = \sum_\mu \phi_\mu(x)dx_\mu$ which both live on M^3.
  In this case, the function f is called the *redshift potential*, the metric g is called the **Fermat metric**. For a static space-time, the geodesics in the Fermat metric are the projections of the null geodesics of space-time.

In particular, the **spherically symmetric and static metrics**, including models for nonrotating stars and black holes, *wormholes*, *monopoles*, *naked singularities*, and (boson or fermion) stars, are given by the *line element*

$$ds^2 = e^{2f(r)}(-dt^2 + S(r)^2dr^2 + R(r)^2(d\theta^2 + \sin^2\theta d\phi^2)).$$

Here, the one-form ϕ vanishes, and the Fermat metric g has the special form

$$g = S(r)^2dr^2 + R(r)^2(d\theta^2 + \sin^2\theta d\phi^2).$$

For example, the conformal factor $e^{2f(r)}$ of the **Schwartzschild metric** is equal to $1 - \frac{2m}{r}$, and the corresponding Fermat metric has the form

$$g = \left(1 - \frac{2m}{r}\right)^{-2}\left(1 - \frac{2m}{r}\right)^{-1} r^2(d\theta^2 + \sin\theta d\phi^2).$$

- **pp-wave metric**
  The **pp-wave metric** is an exact solution to the Einstein field equation, in which radiation moves at the speed c of light. The *line element* of this metric is given (in Brinkmann coordinates) by

  $$ds^2 = H(u, x, y)du^2 + 2dudv + dx^2 + dy^2,$$

  where H is any smooth function. The term "pp" stands for *plane-fronted waves with parallel propagation* introduced by Ehlers–Kundt, 1962.
  The most important class of particularly symmetric pp-waves are the **plane wave metrics**, in which H is quadratic. The *wave of death*, for example, is a *gravitational* (i.e., the space-time curvature fluctuates) plane wave exhibiting a strong nonscalar null curvature singularity which propagates through an initially flat space-time, progressively destroying the Universe.
  Examples of axisymmetric pp-waves include the *Aichelburg–Sexl ultraboost* modeling the motion past a spherically symmetric gravitating object at nearly c, and the *Bonnor beam* modeling the gravitational field of an infinitely long beam of incoherent electromagnetic radiation. The Aichelburg–Sexl wave is obtained by boosting the Schwarzschild solution to the speed c at fixed energy, i.e., it describes a Schwarzschild black hole moving at c. Cf. **Aichelburg–Sexl metric** (Chap. 24).
- **Bonnor beam metric**
  The **Bonnor beam metric** is an exact solution to the Einstein field equation which models an infinitely long, straight beam of light. It is an **pp-wave metric**. The interior part of the solution (in the uniform plane wave interior region which is shaped like the world tube of a solid cylinder) is defined by the *line element*

  $$ds^2 = -8\pi mr^2du^2 - 2dudv + dr^2 + r^2d\theta^2,$$

where $-\infty < u, v < \infty$, $0 < r < r_0$, and $-\pi < \theta < \pi$. This is a null dust solution and can be interpreted as incoherent electromagnetic radiation.
The exterior part of the solution is defined by

$$ds^2 = -8\pi m r_0^2(1 + 2\log(r/r_0))du^2 - 2dudv + dr^2 + r^2 d\theta^2,$$

where $-\infty < u, v < \infty$, $r_0 < r < \infty$, and $-\pi < \theta < \pi$.

- **Plane wave metric**
The **plane wave metric** is a vacuum solution to the Einstein field equation, given by the *line element*

$$ds^2 = 2dwdu + 2f(u)(x^2 + y^2)du^2 - dx^2 - dy^2.$$

It is conformally flat, and describes a pure radiation field. The space-time with this metric is called the *plane gravitational wave*. It is an **pp-wave metric**.

- **Wils metric**
The **Wils metric** is a solution to the *Einstein field equation*, given by

$$ds^2 = 2xdwdu - 2wdudx + \left(2f(u)x(x^2 + y^2) - w^2\right) du^2 - dx^2 - dy^2.$$

It is conformally flat, and describes a pure radiation field which is not a *plane wave*.

- **Koutras–McIntosh metric**
The **Koutras–McIntosh metric** is a solution to the Einstein field equation, given by the *line element*

$$ds^2 = 2(ax + b)dwdu - 2awdudx + \left(2f(u)(ax + b)(x^2 + y^2) - a^2w^2\right) du^2 - dx^2 - dy^2.$$

It is conformally flat and describes a pure radiation field which, in general, is not a *plane wave*. It gives the **plane wave metric** for $a = 0, b = 1$, and the **Wils metric** for $a = 1, b = 0$.

- **Edgar–Ludwig metric**
The **Edgar–Ludwig metric** is a solution to the *Einstein field equation*, given by

$$\begin{aligned} ds^2 &= 2(ax + b)dwdu - 2awdudx \\ &\quad + \left(2f(u)(ax + b)(g(u)y + h(u) + x^2 + y^2) - a^2w^2\right) du^2 - dx^2 - dy^2. \end{aligned}$$

This metric is a generalization of the **Koutras–McIntosh metric**. It is the most general metric which describes a conformally flat pure radiation (or null fluid) field which, in general, is not a *plane wave*. If plane waves are excluded, it has the form

$$ds^2 = 2xdwdu - 2wdudx + \left(2f(u)x(g(u)y + h(u) + x^2 + y^2) - w^2\right) du^2 - dx^2 - dy^2.$$

- **Bondi radiating metric**
The **Bondi radiating metric** describes the asymptotic form of a radiating solution to the Einstein field equation, given by the *line element*

$$ds^2 = -\left(\frac{V}{r}e^{2\beta} - U^2r^2e^{2\gamma}\right) du^2 - 2e^{2\beta}dudr - 2Ur^2e^{2\gamma}dud\theta$$
$$+ r^2(e^{2\gamma}d\theta^2 + e^{-2\gamma}\sin^2\theta d\phi^2),$$

where u is the retarded time, r is the **luminosity distance**, $0 \leq \theta \leq \pi, 0 < \phi < 2\pi$, and U, V, β, γ are functions of u, r, and θ.
- **Taub–NUT de Sitter metric**
The **Taub–NUT de Sitter metric** (cf. **de Sitter metric**) is a positive-definite (i.e., Riemannian) solution to the Einstein field equation with a cosmological constant Λ, given by the *line element*

$$ds^2 = \frac{r^2 - L^2}{4\Delta}dr^2 + \frac{L^2\Delta}{r^2 - L^2}(d\psi + \cos\theta d\phi)^2 + \frac{r^2 - L^2}{4}(d\theta^2 + \sin^2\theta d\phi^2),$$

where $\Delta = r^2 - 2Mr + L^2 + \frac{\Lambda}{4}(L^4 + 2L^2r^2 - \frac{1}{3}r^4)$, L and M are parameters, and θ, ϕ, ψ are the *Euler angles*.
If $\Lambda = 0$, one obtains the **Taub–NUT metric** (cf. Chap. 7).
- **Eguchi–Hanson de Sitter metric**
The **Eguchi–Hanson de Sitter metric** (cf. **de Sitter metric**) is a positive-definite (i.e., Riemannian) solution to the Einstein field equation with a cosmological constant Λ, given by the *line element*

$$ds^2 = \left(1 - \frac{a^4}{r^4} - \frac{\Lambda r^2}{6}\right)^{-1} dr^2 + \frac{r^2}{4}\left(1 - \frac{a^4}{r^4} - \frac{\Lambda r^2}{6}\right)(d\psi + \cos\theta d\phi)^2 +$$
$$+ \frac{r^2}{4}(d\theta^2 + \sin^2\theta d\phi^2),$$

where a is a parameter, and θ, ϕ, ψ are the *Euler angles*.
If $\Lambda = 0$, one obtains the **Eguchi–Hanson metric**.
- **Barriola–Vilenkin monopole metric**
The **Barriola–Vilenkin monopole metric** is given by the *line element*

$$ds^2 = -dt^2 + dr^2 + k^2r^2(d\theta^2 + \sin^2\theta d\phi^2),$$

with a constant $k < 1$. There is a deficit solid angle and a singularity at $r = 0$; the plane $t = constant$, $\theta = \frac{\pi}{2}$ has the geometry of a cone.

This metric is an example of a conical singularity; it can be used as a model for *monopole*, i.e., a hypothetical isolated magnetic poles. It has been theorized that such things might exist in the form of tiny particles similar to electrons or protons, formed from topological defects in a similar manner to cosmic strings. Cf. **Gibbons–Manton metric** in Chap. 7.

- **Bertotti–Robinson metric**

  The **Bertotti–Robinson metric** is a solution to the Einstein field equation in a universe with a uniform magnetic field. The *line element* of this metric is

$$ds^2 = Q^2(-dt^2 + \sin^2 t\,dw^2 + d\theta^2 + \sin^2\theta\,d\phi^2),$$

  where Q is a constant, $t \in [0, \pi]$, $w \in (-\infty, +\infty)$, $\theta \in [0, \pi]$, and $\phi \in [0, 2\pi]$.

- **Wormhole metric**

  A *wormhole* is a hypothetical region of space-time containing a *world tube* (the time evolution of a closed surface) that cannot be continuously deformed to a *world line* (the time evolution of a point).

  Wormhole metric is a theoretical distortion of space-time in a region of the Universe that would link one location or time with another, through a "shortcut", i.e., a path that is shorter in distance or duration than would otherwise be expected. A wormhole geometry can only appear as a solution to the Einstein equations if the *stress-energy tensor* of matter violates the *null energy condition* at least in a neighborhood of the wormhole throat.

  Einstein–Rosen bridge (1935) is a nontraversable wormhole formed from either black hole or spherically symmetric vacuum regions; it possesses a singularity and impenetrable event horizon. Traversable wormholes, as well as *warp drive* (faster-than-light propulsion system) and *time machines*, permitting journeys into the past, require bending of space-time by *exotic matter* (negative mass or energy).

  Whereas the curvature of space produced by the attractive gravitational field of ordinary matter acts like a converging lens, negative energy acts like a diverging lens. The negative mass required for engineering, say, a wormhole of throat diameter 4.5 m, as in Stargate's inner ring (from TV franchise *Stargate*), is $\approx -3 \times 10^{27}$ kg. But oscillating warp and tweaking wormhole's geometry (White, 2012) can greatly reduce it. Also, negative energy can be created (Butcher, 2014) at the centre of a wormhole if its throat is orders of magnitude longer than its mouth.

  Lorentzian wormholes, not requiring exotic matter to exist, were proposed, using higher-dimensional extensions of Einstein's theory of gravity, by Bronnikov–Kim, 2002, and Kanti–Kleihaus–Kunz, 2011. Still, only atomic-scale wormholes would be practical to build, using them solely for superluminal information transmission.

  Lorentzian wormholes can be seen as maximally entangled states of two black holes in a *Einstein–Podolsky–Rosen correlation*, i.e., *nonclassical* one (it cannot be approximated by convex combinations of product states). Maldacena–Susskind, 2013: any two entangled quantum subsystems (cf. Chap.24) are connected by a such wormhole. For Sonner, 2013, gravity might emerge from quantum entanglement.

- **Morris–Thorne metric**
  The **Morris–Thorne metric** (Morris–Thorne, 1988) is a traversable **wormhole metric** which is a solution to the Einstein field equation with the *line element*

$$ds^2 = e^{\frac{2\Phi(w)}{c^2}} c^2 dt^2 - dw^2 - r(w)^2(d\theta^2 + \sin^2\theta d\phi^2),$$

  where $w \in [-\infty, +\infty]$, r is a function of w that reaches some minimal value above zero at some finite value of w, and $\Phi(w)$ is a gravitational potential allowed by the space-time geometry. It is the most general static and spherically symmetric metric able to describe a stable and traversable wormhole.
  Morris–Thorne–Yurtsever, 1988, stated that two closely spaced (10^{-9}–10^{-10} m) concentric thin charged hollow spheres the size of 1 AU can create negative energy, required for engineering this wormhole, using the quantum *Casimir effect*.
- **Alcubierre metric**
  The **Alcubierre metric** (Alcubierre, 1994) is a **wormhole metric** which is a solution to the Einstein field equation, representing *warp drive space-time* where the existence of *closed time-like curves* is allowed. The Alcubierre construction corresponds to a *warp* (i.e., faster than light) drive in that it causes space-time to contract in front of a spaceship bubble and expand behind, thus providing the spaceship with a velocity that can be much greater than the speed of light relative to distant objects, while the spaceship never locally travels faster than light.
  In this case, only the relativistic principle that a space-traveler may move with any velocity up to, but not including or exceeding, the speed of light, is violated.
  The *line element* of this metric has the form

$$ds^2 = -dt^2 + (dx - vf(r)dt)^2 + dy^2 + dz^2,$$

  where $v = \frac{dx_s(t)}{dt}$ is the apparent velocity of the warp drive spaceship, $x_s(t)$ is spaceship trajectory along the coordinate x, the radial coordinate is $r = ((x - x_s(t))^2 + y^2 + z^2)^{\frac{1}{2}}$, and $f(r)$ an arbitrary function subject to the conditions that $f = 1$ at $r = 0$ (the location of the spaceship) and $f(\infty) = 0$.
  Another warp drive space-time was proposed by Krasnikov, 1995. **Krasnikov metric** in the 2D subspace t, x is given by the *line element*

$$ds^2 = -dt^2 + (1 - k(x,t))dxdt + k(x,t)dx^2,$$

  where $k(x,t) = 1-(2-\delta)\theta_\epsilon(t-x)(\theta_\epsilon(x) - \theta_\epsilon(x+\epsilon-D))$, D is the distance to travel, θ_ϵ is a smooth monotonic function satisfying $\theta_\epsilon(z) = 1$ at $z > \epsilon$, $\theta_\epsilon(z) = 0$ at $z < 0$ and δ, ϵ are arbitrary small positive parameters.

- **Misner metric**
  The **Misner metric** (Misner, 1960) is a metric, representing two black holes, instantaneously at rest, whose throats are connected by a *wormhole*. The *line element* of this metric has the form

$$ds^2 = -dt^2 + \psi^4(dx^2 + dy^2 + dz^2),$$

  where the *conformal factor* ψ is given by

$$\psi = \sum_{n=-N}^{N} \frac{1}{\sinh(\mu_0 n)} \frac{1}{\sqrt{x^2 + y^2 + (z + \coth(\mu_0 n))^2}}.$$

  The parameter μ_0 is a measure of the ratio of mass to separation of the throats (equivalently, a measure of the distance of a loop in the surface, passing through one throat and out of the other). The summation limit N tends to infinity.
  The topology of the *Misner space-time* is that of a pair of asymptotically flat sheets connected by a number of wormholes. In the simplest case, it can be seen as a 2D space $\mathbb{R} \times S^1$, in which light progressively tilts as one moves forward in time, and has *closed time-like curves* (so, time-travel is possible) after a certain point.
- **Rotating C-metric**
  The **rotating C-metric** is a solution to the *Einstein–Maxwell equations*, describing two oppositely charged black holes, uniformly accelerating in opposite directions. The *line element* of this metric has the form

$$ds^2 = A^{-2}(x + y)^{-2}\left(\frac{dy^2}{F(y)} + \frac{dx^2}{G(x)} + k^{-2}G(X)d\phi^2 - k^2A^2F(y)dt^2\right),$$

  where $F(y) = -1 + y^2 - 2mAy^3 + e^2A^2y^4$, $G(x) = 1 - x^2 - 2mAx^3 - e^2A^2x^4$, m, e, and A are parameters related to the mass, charge and acceleration of the black holes, and k is a constant fixed by regularity conditions.
  This metric should not be confused with the **C-metric** from Chap. 11.
- **Myers–Perry metric**
  The **Myers–Perry metric** describes a 5D rotating black hole. Its *line element* is

$$ds^2 = -dt^2 + \frac{2m}{\rho^2}(dt - a\sin^2\theta d\phi - b\cos^2\theta d\psi)^2 +$$

$$+\frac{\rho^2}{R^2}dr^2 + \rho^2 d\theta^2 + (r^2 + a^2)\sin^2\theta d\phi^2 + (r^2 + b^2)\cos^2\theta d\psi^2,$$

  where $\rho^2 = r^2 + a^2\cos^2\theta + b^2\sin^2\theta$, and $R^2 = \frac{(r^2+a^2)(r^2+b^2)-2mr^2}{r^2}$. Above black hole is asymptotically flat and has an event horizon with S^3 topology.

Emparan and Reall, 2001, using the possibility of rotation in several independent rotation planes, found a 5D *black ring*, i.e., asymptotically flat black hole solution with the event horizon's topology of $S^1 \times S^2$.

- **Ponce de León metric**
  The **Ponce de León metric** (1988) is a 5D metric, given by the *line element*

$$ds^2 = l^2 dt^2 - (t/t_0)^2 p l^{\frac{2p}{p-1}} (dx^2 + dy^2 + dz^2) - \frac{t^2}{(p-1)^2} dl^2,$$

  where l is the 5th (space-like) coordinate. This metric represents a 5D apparent vacuum. It is not flat but embed the flat 4D **FLRW metric**.
- **Kaluza–Klein metric**
  The **Kaluza–Klein metric** is a metric in the *Kaluza–Klein model* of 5D space-time which seeks to unify classical gravity and electromagnetism.
  Kaluza, 1921 (but sent to Einstein in 1919), found that, if the Einstein theory of pure gravitation is extended to a 5D space-time, the Einstein field equation can be split into an ordinary 4D gravitation tensor field, plus an extra vector field which is equivalent to the Maxwell equation for the electromagnetic field, plus an extra scalar field known as the *dilation* (or *radion*).
  Klein, 1926, assumed that the 5th dimension (i.e., 4th spatial dimension) is curled up in a circle of an unobservable size, below 10^{-20} m. Almost all modern higher-dimensional unified theories are based on Kaluza–Klein approach.
  An alternative proposal is that the extra dimension(s) is extended, and the matter is trapped in a 4D submanifold. In a model of a such large extra dimension, the 5D metric of a universe can be written in Gaussian normal coordinates as

$$ds^2 = -(dx_5)^2 + \lambda^2(x_5) \sum_{\alpha,\beta} \eta_{\alpha\beta} dx_\alpha dx_\beta,$$

  where $\eta_{\alpha\beta}$ is the 4D **metric tensor** and $\lambda^2(x_5)$ is any function of the 5th coordinate.
  In particular, the *STM* (space-time-matter) *theory* (Wesson and Ponce de León, 1992) relate the 5th coordinate to mass via either $x_5 = \frac{Gm}{c^2}$ or $x_5 = \frac{h}{mc}$, where G is the *Newton gravitational constant* and h is the *Planck constant*.
  The **Ponce de León metric** is a STM solution. In STM (or *induced matter*) theory, the 4D curvature arises not due to the distribution of matter in the Universe (as claims Relativity Theory) but because the Universe is embedded in some higher-dimensional vacuum manifold M, and all the matter in our world can be thought of as being manifestations of the geometrical properties of M.
  Wesson and Seahra, 2005, claim that the Universe may be a 5D black hole. Life is not excluded since in 5D there is no physical plughole and the "tidal" forces are negligible. Suitable manifolds for such STM theory are given by two isometric solutions of the 5D vacuum field equations: *Liu–Mashhoon–Wesson metric* and *Fukui–Seahra–Wesson metric*; both embed 4D **FLRW metric**.

- **Carmeli metric**
  The **Carmeli metric** (Carmeli, 1996) is given by the *line element*

$$ds^2 = dx^2 + dy^2 + dz^2 - \tau^2 dv^2,$$

  where $\tau = \frac{1}{H}$ is the inverse of *Hubble constant* and v is the *cosmological recession velocity*. So, comparing with the **Minkowski metric**, it has τ and velocity v, instead of c and time t. This metric was used in *Carmeli's Relativity Theory* which is intended to be better than General Relativity on cosmological scale.
  The Carmeli metric produces the *Tulli–Fisher type relation* in spiral galaxies: 4th power of the rotation speed is proportional to the mass of galaxy; it obviate the need for dark matter. This metric predicts also cosmic acceleration.
  Including *icdt* component of the **Minkowski metric**, gives the **Kaluza–Klein–Carmeli metric** (Harnett, 2004) defined by

$$ds^2 = dx^2 + dy^2 + dz^2 - c^2 dt^2 - \tau^2 dv^2.$$

- **Prasad metric**
  A *de Sitter Universe* can be seen as the sum of the external and internal space. The internal space has a negative constant curvature $-\frac{1}{r^2}$ and can be characterized by the symmetry group $SO_{3,2}$. The **Prasad metric** of this space is given, in hyperspherical coordinates, by the *line element*

$$ds^2 = r^2 \cos^2 t (d\chi^2 + \sinh^2 \chi (d\theta^2 + \sin^2 \theta d\phi^2)) - r^2 dt^2.$$

  The value $\sin \chi$ is called *adimensional normalized radius* of the de Sitter Universe.
  The external space has constant curvature $\frac{1}{R^2}$ and can be characterized by the symmetry group $SO_{4,1}$. Its metric has the *line element* of the form

$$ds^2 = R^2 \cosh^2 t (d\chi^2 + \sin^2 \chi (d\theta^2 + \sin^2 \theta d\phi^2)) - R^2 dt^2.$$

Part VII
Real-World Distances

Chapter 27
Length Measures and Scales

The term *length* has many meanings: distance, extent, linear measure, span, reach, end, limit, etc.; for example, the length of a train, a meeting, a book, a trip, a shirt, a vowel, a proof. The *length* of an object is its linear extent, while the *height* is the vertical extent, and *width* (or *breadth*) is the side-to-side distance at 90° to the length, wideness. The *depth* is the distance downward, distance inward, deepness, profundity, drop.

The ancient Greek mathematicians saw all numbers as lengths (of segments), areas or volumes. In Mathematics, a **length function** is a function $l : G \to \mathbb{R}_{\geq 0}$ on a group $(G, +, 0)$ such that $l(0) = 0$ and $l(g) = l(-g)$, $l(g + g') \leq l(g) + l(g')$ for $g, g' \in G$.

In Engineering and Physics, "length" usually means "distance". **Unit distance** is a distance taken as a convenient unit of length in a given context.

In this chapter we consider length only as a measure of physical distance. We give selected information on the most important length units and present, in length terms, a list of interesting physical objects.

27.1 Length Scales

The main length measure systems are: Metric, Imperial (British and American), Japanese, Thai, Chinese Imperial, Old Russian, Ancient Roman, Ancient Greek, Biblical, Astronomical, Nautical, and Typographical.

There are many other specialized length scales; for example, to measure cloth, shoe size, gauges (such as interior diameters of shotguns, wires, jewelry rings), sizes for abrasive grit, sheet metal thickness, etc.

Many units express relative or inverse distances. Quantities measured in **reciprocal length** include: radius of curvature, density of a linear feature in an area (say, km per km^2), magnitude of vectors (in Crystallography and Spectroscopy),

M.M. Deza, E. Deza, *Encyclopedia of Distances*,
DOI 10.1007/978-3-662-44342-2_27

optical power (cf. **lens distances** in Chap. 29), absorption (or attenuation) coefficient (cf. Chap. 24), gain (in Laser Physics). Common units used for this measurement include inverse meter m^{-1} (called *diopter* in Optics), cm^{-1} and cycles per unit length (for spatial frequency)

Some units express nonlength quantities in length terms. For example, the *denudation rate* (wearing down of the Earth's surface) is measured in cm per 1,000 years. Cf. also **meter of water equivalent** in Chap. 24.

- **International Metric System**
  The **International Metric System** (or SI, short for *Système International*), also known as MKSA (meter–kilogram–second–ampere), is a modernized version of the system of units, established by the *Treaty of the Meter* from 20 May 1875, which provides a logical and interconnected framework for all measurements in science, industry and commerce. The system is built on a foundation consisting of the following seven *SI base units*, assumed to be mutually independent:
  (1) length: **meter** (m); it is equal to the distance traveled by light in a vacuum in 1/299,792,458 of a second; (2) time: *second* (s); (3) mass: *kilogram* (kg); (4) temperature: *kelvin* (K); (5) electric current: *ampere* (A); (6) luminous intensity: *candela* (cd); (7) amount of substance: *mole* (mol).
  Meter is defined as a **proper length** (the length of the object in its rest frame, cf. Sect. 26.2). So, it is well defined only over short distances where relativistic effects are negligible (cf. **Lorentz length contraction** in Sect. 26.1), and all cosmic distances, given in meters, are approximations.
  Originally, on March 26, 1791, the *mètre* (French for meter) was defined as $\frac{1}{10000000}$ of the distance from the North Pole to the equator along the Dunkirk–Barcelona meridian. The name *mètre* was derived from the Greek *metron* (measure). In 1799 the standard of *mètre* became a meter-long platinum–iridium bar kept in Sèvres, a town outside Paris, for people to come and compare their rulers with. (The metric system, introduced in 1793, was so unpopular that Napoleon was forced to abandon it and France returned to the *mètre* only in 1837.) In 1960–1983, the **meter** was defined in terms of wavelengths.
  The initial metric unit of mass, the *gram*, was defined as the mass of one cubic centimeter of water at its temperature of maximum density. A *metric ton* (or *metric tonne*, *tonne*) is a unit of mass equal to 1,000 kg; this non-SI unit is used instead of the SI term *megagram* (10^6 g). For capacity, the *litre* (liter) was defined as the volume of a cubic decimeter.
- **Metrication**
  The **metrication** is an ongoing (especially, in US, UK and Caribbean countries) process of conversion to the **International Metric System**, SI. Only US, Liberia and Myanmar have not fully switched to SI. For example, US uses only miles for road distance signs (milestones). Altitudes in aviation are usually described in feet, and resolutions of output devices are specified in *dpi* (dots per inch). In shipping, nautical miles and knots are used; both are accepted for use with SI.
  Hard metric means designing in the metric measures from the start and conformation, where appropriate, to internationally recognized sizes and designs.

Soft metric means multiplying an inch-pound number by a metric conversion factor and rounding it to an appropriate level of precision; so, the soft converted products do not change size. The *American Metric System* consists of converting traditional units to embrace the uniform base 10 used by the Metric System.
Such SI-Imperial hybrid units, used in soft metrication, are, for example, *kiloyard* (914.4 m), *kilofoot* (304.8 m), *mil* or *milli-inch* (25.4 μm), and *min* or *microinch* (25.4 nm). The *metric inch* (2.5 cm $\approx$1 inch) and *metric foot* (30 cm) were used in some Soviet computers when building from American blueprints.
In athletics and skating, races of 1,500 or 1,600 m are often called *metric miles*. Examples of traditional units adapted to the meter are Chinese *li* = 500 m = 1,500 *chi* (Chinese feet), Thai *wa* = 2 m = 4 *sok*, Vietnamese *xich* = 1 m = 1,000 *ly*.

- **Meter, in Poetry and Music**
In Poetry, *meter* (or *cadence*) is a measure of rhythmic quality, the regular linguistic sound patterns of a verse or line in it. The meter of a verse is the number of lines, the number of syllables in each line and their arrangement as sequences of *feet*. Each foot is a specific sequence of syllable types—such as unstressed/stressed or long/short. Fussell, 1965, define four types of meter: syllabic, accentual, accentual-syllabic and quantitative, where patterns are based on *syllable weight* (number and/or duration of segments in the rhyme) rather than stress.
Hypermeter is part of a verse with an extra syllable; *metromania* is a mania for writing verses and *metrophobia* is a fear/hatred of poetry.
In Music, *meter* (or *metre*) is the regular rhythmic patterns of a musical line, the division of a composition into parts of equal time, and the subdivision of them. It is derived from the poetic meter of song. Different tonal preferences in voiced speech are reflected in music; it explains why Eastern and Western music differ.
Metrical rhythm is where each time value is a multiple or fraction of a fixed unit (*beat*) and normal accents re-occur regularly providing systematic grouping (*measures*). *Isometre* is the use of a *pulse* (unbroken series of periodically occurring short stimuli) without a regular meter, and *polymetre* is the use of two or more different meters simultaneously, whereas *multimetre* is using them in succession.
A rhythmic pattern/unit is either *intrametric* (confirming the pulses on the metric level), or *contrametric* (syncopated, not following the beat/meter), or *extrametric* (irregular with respect to the metric structure of the piece). Rhythms/chords with the same multiset of intervals/distances are called *homometric*.
A temporal pattern is *metrically represented* if it can be subdivided into equal time intervals. A *metronome* is any device that produces regular, metrical ticks (beats); *metronomy*: measurement of time by a metronome or, in general, an instrument.

- **Meter-related terms**
  We present this large family of terms by the following examples (besides the unit of length and use in Poetry and Music).
  Metrograph: a device attached to a locomotive to record its speed and the number and duration of its stops. Cf. unrelated *metrography* in Medicine (Chap. 29).
  Metrogon: a high resolution, low-distortion, extra-wide field photographic lens design used extensively in aerial photography.
  The names of various measuring instruments contain *meter* at the end, say, *ammeter*, *gas meter*, *multimeter* (or *volt-ohm meter*).
  Metrosophy: a cosmology based on strict number correspondences.
  Metrology: the science of, or a system of, weights and measures.
  A **metric meterstick** is a rough rule of thumb for comprehending a metric unit; for example, 5 cm is the side of a matchbox, and 1 km is $\approx$10 minutes' walk.
  Metering: an equivalent term for a *measurement* (assignment of numbers to objects or events); *micrometry*: measurement under the microscope; *hypsometry*: measurement of heights; *telemetry*: technology that allows remote measurement; *archeometry*: the science of exact measuring referring to the remote past.
  Hedonimetry: the study of happiness as a measurable economic asset; *psychometry*: alleged psychic power enabling one to divine facts by handling objects.
  Psychometrics: the theory and technique of psychological measurement; *psychrometrics*: the determination of physical and thermodynamic properties of gas-vapor mixtures; *biometrics*: the identification of humans by their characteristics or traits; *cliometrics*: the systematic application of econometric techniques and other formal or mathematical methods to the study of history.
  Metric, as a nonmathematical term, is a standard unit of measure (for example, *font metrics* refer to numeric values relating to size and space in the font) or, more generally, part of a system of parameters; cf. **quality metrics** in Chap. 29.
  Antimetric matrix: a square matrix A with $A = -A^T$; an *antimetric electrical network* is one that exhibits antisymmetrical electrical properties.
  Isometropia: equality of refraction in both eyes; *hypermetropia* is farsightedness.
  Isometric particle: a virus which (at the stage of virion capsid) has icosahedral symmetry. *Isometric process*: a thermodynamic process at constant volume.
  Metrohedry: overlap in 3D of the lattices of twin domains in a crystal.
  Multimetric crystallography: to consider (Janner, 1991), in addition to the Euclidean metric tensor, *pseudo-Euclidean tensors* (hyperbolic rotations) attached to the same basis; cf. **pseudo-Euclidean distance** in Chap. 7 and **multimetric** in Chap. 3.
  Metria: a genus of moths of the *Noctuidae* family.
  Metrio: Greek coffee with one teaspoon of sugar (medium sweet). In Anthropology, *metriocranic* means having a skull that is moderately high compared with its width, with a breadth-height index 92–98.
  Metroid: the name of a series of video games produced by Nintendo and *metroids* are a fictional species of parasitic alien creatures from those games.

Examples of companies with a meter-related name are: Metron, Metric Inc., MetaMetrics Inc., Metric Engineering, Panametric, Prometric, Unmetric, World Wide Metric. *Metric* is also a Canadian New Wave rock band.

- **Metric length measures**
  kilometer (km) $= 1{,}000\,\text{m} = 10^3$ m;
  meter (m) $= 10$ dm $= 10^0$ m;
  decimeter (dm) $= 10$ cm $= 10^{-1}$ m;
  centimeter (cm) $= 10$ mm $= 10^{-2}$ m;
  millimeter (mm) $= 1{,}000\,\mu\text{m} = 10^{-3}$ m;
  micrometer (or non-SI *micron*; μm) $= 1{,}000$ nm $= 10^{-6}$ m;
  nanometer (or non-SI 10 *angstroms* Å; nm) $= 1{,}000$ pm $= 10^{-9}$;
  picometer (pm) $= 1{,}000$ fm $= 10^{-12}$ m;
  femtometer (or non-SI *fermi*; fm) $= 1{,}000$ attometers $= 10^{-15}$ m.
  The numbers 10^{3t} ($t = -8, \dots, -1, 1, \dots, 8$) are given by *metric prefixes*: yocto-(y), zepto-(z), atto-(a), femto-(f), pico-(p), nano-(n), micro-(μ), milli-(m), kilo-(k), mega-(M), giga-(G), tera-(T), peta-(P), exa-(E), zetta-(Z), yotta-(Y), respectively, while 10^t ($t = -2, -1, 1, 2$) are given by centi-(c), deci-(d), deca-(da), hecto-(h).
  In computers, a *bit* (binary digit) is the basic unit of information, a *byte* (or *octet*) is 8 bits, and 10^{3t} bytes for $t = 1, \dots, 8$ are kilo-(KB), mega-(MB), giga-(GB), ..., yottabyte (YB), respectively. Sometimes (because of $2^{10} = 1{,}024 \approx 10^3$) the binary terms kibi-(KiB), mebi-(MiB), gibibyte (GiB), etc., are used for 2^{10t} bytes.
- **Imperial length measures**
  The **Imperial length measures** (as slightly adjusted by a treaty in 1959) are:
  (land) *league* = 3 international miles;
  (international) *mile* = 5,280 feet = 1,609.344 m;
  (US survey) *mile* = 5,280 US feet $\approx$1,609.347 m;
  data (or *tactical*) *mile* = 6,000 feet = 1,828.8 m and *radar mile* = 12.204 μs (time it takes a radar pulse to travel one data mile forth and back);
  (international) *yard* = 0.9144 m = 3 *feet* = $\frac{1}{2}$ *fathom*;
  (international) *foot* = 0.3048 m = 12 *inches*;
  (international) *inch* = 2.54 cm = 12 *lines*;
  (a unit of measure of height of equipment) *rack unit* = $\frac{7}{4}$ inch;
  (a unit of measure in advertising space) *agate line* $\approx \frac{1}{14}$ inch;
  (a unit of computer mouse movement) *mickey* = $\frac{1}{200}$ inch;
  mil (British *thou*) = $\frac{1}{1{,}000}$ inch; *mil* is also an angular measure $\frac{\pi}{3{,}200} \approx 0.001$ radian.
  In addition, *Surveyor's Chain measures* are: *furlong* = 10 chains = $\frac{1}{8}$ mile; *chain* = 100 links = 66 feet; *rope* = 20 feet; *rod* (or *pole*) = 16.5 feet; *link* = 7.92 inches. Mile, furlong and fathom come from the slightly shorter Greco-Roman milos (milliare), stadion and orguia, mentioned in the New Testament.
  For measuring cloth, old measures are used: *bolt* = 40 yards; *goad* = $\frac{3}{2}$ yard; *ell* = $\frac{5}{4}$ yard = 45 inches; *quarter* = $\frac{1}{4}$ yard; *finger* = $\frac{1}{8}$ yard; *nail* = $\frac{1}{16}$ yard.

Other old English units of length: *barleycorn* = $\frac{1}{3}$ inch; *digit* = $\frac{3}{4}$ inches and *palm, hand, shaftment, span, cubit* = 3, 4, 6, 9, 18 inches, respectively.

- **Cubit**
  The **cubit**, originally the length of the forearm from the elbow to the tip of the middle finger, was the ordinary unit of length in the ancient Near East which varied among cultures and with time. It is the oldest recorded measure of length. The cubit was used, in the temples of Ancient Egypt from at least 2,700BC, as follows: 1 *ordinary Egyptian cubit* = 6 *palms* = 24 *digits* = 45 cm (18 inches), and 1 *royal Egyptian cubit* = 7 *palms* = 28 *digits* ≈52.6 cm. Relevant Sumerian measures were: 1 *ku* = 30 *shusi* = 25 *uban* = 50 cm, and 1 *kus* = 36 *shusi*.
  Biblical measures of length are the *cubit* and its multiples by 4, $\frac{1}{2}$, $\frac{1}{6}$, $\frac{1}{24}$ called *fathom, span, palm, digit*, respectively. But the length of this cubit is unknown; it is estimated now as ≈44.5 cm (as Roman *cubitus*) for the common cubit, used in commerce, and 51–56 cm for the sacred one, used for building.
  The *Talmudic cubit* is 48–57.6 cm. The *pyramid cubit* (25.025 inches ≈63.567 cm), derived in Newton's Biblical studies, is supposed to be the basic one in the dimensions of the Great Pyramid and far-reaching numeric relations on them.
  Thom, 1955, claim that the *megalithic yard*, 82.966 cm, was the basic unit used for stone circles in Britain and Brittany c. 3,500BC. Butler–Knight, 2006, derived this unit as $1/(360 \times 366^2)$-th of 40,075 km (the Earth's equatorial circumference), linking it to the putative Megalithic 366-degree circle and Minoan 366-day year. Such a "366 geometry" is a part of the pseudoscientific metrology
- **Nautical length units**
  The main **nautical length units** (also used in aerial navigation) are:
  sea league = 3 sea (nautical) miles;
  nautical mile = 1,852 m (originally defined as 1 min of arc of latitude);
  geographical mile ≈1855.32 m (the average distance on the Earth's surface, represented by 1 min of arc along the Earth's equator);
  (international) *short cable length* = $\frac{1}{10}$ nautical mile
  (US customary) *cable length* = 120 fathoms = 720 feet = 219.456 m;
  fathom = 6 feet = 1.8288 m.
- **Preferred design sizes**
  Objects are often manufactured in a series of sizes of increasing magnitude. In Industrial Design, *preferred numbers* are standard guidelines for choosing such product sizes within given constraints of functionality, usability, compatibility, safety or cost. **Preferred design sizes** are such lengths, diameters and distances. Four basic *Renard's series* of preferred numbers divide the interval from 10 to 100 into 5, 10, 20, or 40 steps, with the factor between two consecutive numbers being constant (before rounding): the 5, 10, 20, or 40th root of 10. Since the **International Metric System** (SI) is decimally-oriented, the International Organization for Standardization (ISO) adopted Renard's series as the main

preferred numbers for use in setting metric sizes. But, for example, the ratio between adjacent terms (i.e., notes) in the Western musical scale is 12th root of 2.

In the widely used ISO paper size system, the height-to-width ratio of all pages is the *Lichtenberg ratio*, i.e., $\sqrt{2}$. The system consists of formats An, Bn and (used for envelopes) Cn with $0 \le n \le 10$, having widths $2^{-\frac{1}{4}-\frac{n}{2}}$, $2^{-\frac{n}{2}}$ and $2^{-\frac{1}{8}-\frac{n}{2}}$, respectively. The above measures are in m; so, the area of An is 2^{-n} m^2. They are rounded and expressed usually in mm; for example, format A4 is 210×297 and format B7 (used also for EU and US passports) is 88×125.

- **Typographical length units**
  PostScript point $= \frac{1}{72}$ inch $=$ 100 gutenbergs $=$ 0.3527777778 mm;
  TeX point (or *printer's point*) $= \frac{1}{72.27}$ inch $=$ 0.3514598035 mm;
  ATA point (or *Anglo-Saxon point*) $= \frac{1}{72.272}$ inch $=$ 0.3514598 mm;
  point (Didot, European) $=$ 0.37593985 mm, *cicero* $=$ 12 points Didot;
  pica (Postscript, TeX or ATA) $=$ 12 points in the corresponding system;
  twip $= \frac{1}{20}$ of a point in the corresponding system.
  In display systems, *twip* is $\frac{1}{1440}$ inch, and *himetric* is 0.01 mm.
- **Astronomical length units**
  The **Hubble distance** (cf. Chap. 26) or *Hubble length* is $D_H = \frac{c}{H_0} \approx 1.31 \times 10^{26}$ m $\approx$4.237 Gpc $\approx$ 13.82 Gly (used to measure distances $d > \frac{1}{2}$ Mpc in terms of redshift z: $d = zD_H$ if $z \le 1$, and $d = \frac{(z+1)^2-1}{(z+1)^2+1}D_H$, otherwise).
  gigaparsec (Gpc) $=$ 10^3 megaparsec (Mpc) $=$ 10^6 kiloparsec (kpc) $=$ 10^9 parsecs;
  hubble (or light-gigayear, light-Ga, Gly) $=$ 10^3 million light-years (Mly);
  siriometer $=$ 10^6 AU $\approx$15.813 ly (about twice the Earth-Sirius distance);
  parsec (pc) $= \frac{648000}{\pi} = \cot(\frac{1}{3600}) \approx$ 206,265 AU $\approx$3.262 light-years $=$ 3.08568×10^{16} m (the distance from an imaginary star, when the lines drawn from it to the Earth and Sun form the maximum angle of one arcsecond; SI-accepted);
  light-year (ly, the distance light travels in vacuum in 365.25 days) $\approx 9.46053 \times 10^{15}$ m $\approx 5.2595 \times 10^5$ light-minutes $\approx \pi \times 10^7$ light-seconds $\approx$0.3066 parsec;
  spat (used formerly) $=$ 10^{12} m $=$ 10^3 gigameters $\approx$6.6846 AU;
  astronomical unit (AU) $=$ 149,597,870.69$\pm$0.03 km $\approx$499 light-seconds (mean Earth-Sun distance; used to measure distances within the Solar System; SI-accepted);
  light-second $\approx 2.998 \times 10^8$ m (the Earth–Moon distance is $\approx$1.28 light-seconds); radii of Moon, Earth, Jupiter and Sun: 1,737, 6,371, 69,911 and 695,510 km;
  picoparsec $\approx$30.86 km; cf. other funny units such as *sheppey* 1.4 km (closest distance at which sheep remain picturesque), *beard-second* 5 nm (distance that a beard grows in a second), *microcentury* $\approx$52.5 min (length of lectures), *nanocentury* $\approx \pi$ sec.

- **Natural length units**
  Natural units are units of measurement based only on physical constants, for example, the speed c of light, gravitational constant G, reduced Planck constant $\hbar$, *Boltzmann constant* k_B, *Coulomb's constant* k_e, proton's *elementary charge* e, *fine-structure constant* $\alpha = \frac{e^2 k_e}{\hbar c} \approx \frac{1}{137}$ and masses m_e, m_p of electron and proton.
  Planck length (smallest measurable length) is $l_P = \sqrt{\frac{\hbar G}{c^3}} \approx 1.6162 \times 10^{-35}$ m. (The *Stoney length*, used formerly, is $\sqrt{\alpha} l_P$.) l_P is the **reduced Compton wavelength** $\bar{\lambda}_C(m) = \frac{m_P}{m} l_P$, and also half of the **Schwarzschild radius** $r_s(m) = 2Gmc^{-2}$ (Chap. 24) for $m = m_P$ *(Planck mass)* $= \sqrt{\frac{\hbar c}{G}} \approx 2.18 \times 10^{-8}$ kg $\approx$22 mg.
  The remaining base Planck units are *Planck time* $t_p = \frac{l_P}{c} \approx 5.39 \times 10^{-44}$ s, *Planck temperature* $T_P = \frac{m_P c^2}{k_B} \approx 1.42 \times 10^{32}$ K, and *Planck charge* $q_P = \sqrt{\frac{\hbar c}{k_e}} = \frac{e}{\sqrt{\alpha}} \approx 1.88 \times 10^{-18}$ C. The *Planck area* A_P is l_P^2, *Planck energy* E_P is $m_P c^2 \approx 1.22 \times 10^{28}$ eV $\approx$500 kWh, and *Planck density* ρ_P is $m_P l_P^{-3} \approx 5.16 \times 10^{96}$ kg/m^3. Only black holes exceed ρ_P; some theories (for example, *Landau poles*) allow to exceed T_P.
  The Planck units come by a normalization of the *geometrized units* for the expressing SI units second, kilogram, kelvin and coulomb as $c, \frac{G}{c^2}, \frac{Gk_B}{c^4}$ and $\frac{\sqrt{Gk_e}}{c^2}$ m, respectively.
  The length unit of Quantum Chromodynamics (or strong interactions) is $\bar{\lambda}_C(m_p) \approx 2.103 \times 10^{-16}$ m. The majority of lengths, used in experiments on nuclear fundamental forces, are integer multiples of $\lambda_C(m_p) = 2\pi\bar{\lambda}_C(m_p) \approx$ 1.32 fm.
  X unit $\approx 1.002 \times 10^{-13}$ m $\approx$0.1 pm measures wavelengths of X- and gamma rays. The atomic unit of length is the **Bohr radius** (or *bohr*) $\alpha_0 \approx 5.292 \times 10^{-11}$ m $\approx$53 pm = 0.53 Å, the most probable distance between the proton and electron in a hydrogen atom. It is $\alpha^{-1}\bar{\lambda}_C(m_e) = \alpha^{-2} r_e$, where $r_e \approx 2.818 \times 10^{-15}$ m is the **Thomson scattering length** (Chap. 24), i.e., the *classical electron radius*.
  In the units of Particle Physics, 1/eV = 10^{-9}/GeV is $\frac{\hbar c}{\text{eV}} = \frac{E_P}{\text{eV}} l_P \approx 1.97 \times 10^{-7}$ m.
- **Length scales in Physics**
  In Physics, a **length scale** (or **distance scale**) is a distance range determined with the precision of a few orders of magnitude, within which given phenomena are consistently described by a theory. Roughly, the scales $<10^{-15}$, 10^{-15}–10^{-6}, 10^{-6}–10^{6} and $>10^{6}$ m are called *subatomic*, *atomic to cellular* (microscopic), *human* (macroscopic) and *astronomical*, respectively.
  Bacteria (and human ovums) are roughly on the geometrical mean (10^{-4} m) of Nature's hierarchy of sizes. Dawkins, 2006, used term **middle world** for our realm between two counterintuitive extreme levels of existence: the microscopic world of quarks/atoms and the Universe at the galactic/universal level. The limit scales correspond to the **Planck length** l_P and **Hubble distance** $\approx 4.6 \times 10^{61}\ l_P$.

In terms of their constituents, Chemistry (molecules, atoms), Nuclear (say, proton, electron, photon), Hadronic (exited states) and Standard Model (quarks and leptons) are applicable at scales $\geq 10^{-10}$, $\geq 10^{-14}$, $\geq 10^{-15}$ and $\geq 10^{-18}$ m.
At the *meso-* (or *nanoscopic*) *scale*, 10^{-9}–10^{-7} m, materials and phenomena can be described continuously and statistically, and average macroscopic properties (say, temperature and entropy) are relevant. At the *atomic scale*, $\sim 10^{-10}$ m = 1 Å, the atoms should be seen as separated. The *electroweak scale*, $\sim 10^{-18}$ m (100–1,000 GeV, in terms of energy) will be probed by the LHC (Large Hadron Collider). The *Planck scale* (Quantum Gravity), $\sim 10^{-35}$ m ($\sim 10^{19}$ GeV) is not yet accessible.
Both, uncertainty principle from Quantum Mechanics and gravitational collapse (black hole formation) from classical General Relativity, indicate some minimum length of order the Planck length l_P where the notion of distance loses operational meaning. Also, *doubly special relativity* adds minimum length and maximum energy scales to observer-independent maximum velocity c.
At short distances, classical geometry is replaced by "quantum geometry" described by 2D conformal field theory (CFD). As two points are getting closer together, the vacuum fluctuations of the gravitational field make the distance between them fluctuate randomly, and its mean value tends to a limit, of the order of l_P. So, no two events in space-time can ever occur closer together.
In String Theory, space-time geometry is not fundamental and, perhaps, it only emerges at larger distance scales. The *Maldacena duality* is the conjectured equivalence between an M-theory defined on a ("large, relativistic") space, and a (quantum, without gravity) CFD defined on its (lower dimensional) conformal boundary.
The Big Bang paradigm supposes a minimal length scale and a smooth distribution (homogeneous and isotropic) at a large scale. For Vilenkin et al., 2011, the main theories admitting "before the Big Bang" (*cyclical universe*, *eternal inflation*, *multiverse*, *cosmic egg*) still require a beginning. For Hartle–Hawkings, 1983, time emerged continuously from space after the Universe was at the age t_P.

- **Glashow's snake**

  Uroboros, the snake that bites in its own tail, is an ancient symbol representing the fundamental in different cultures: Universe, eternal life, integration of the opposite, self-creation, etc. **Glashow's snake** is a sketch of the cosmic uroboros by Glashow, 1982, arraying four fundamental forces and the distance scales over which they dominate (62 orders of magnitude from the *Planck scale* $\sim 10^{-35}$ m to the *cosmological scale* $\sim 10^{26}$ m) in clock-like form around the serpent. The dominating forces are:

  1. gravity: in the *macrocosmos* from cosmic to planetary distances;
  2. electromagnetism: from mountains to atoms (say, within $[10^{-10}, 2 \times 10^5]$ m);
  3. weak and strong forces: in the *microcosmos* inside the atom (say, $<10^{-12}$ m).

  No objects are known within $[10^{-14}, 10^{-10}]$ m (the largest nucleus and smallest atom). As distances decrease and energies increase, the last three forces become

equivalent around the length 10^{-28} m. Then gravity is included (super-unification happens) linking the largest and smallest: the snake swallows its tail.
Also, a symmetry between small and large distances, called *T-duality*, claims: two superstring theories are *T-dual* if one compactified on a space of large volume is equivalent to the other compactified on a space of small volume.
Cosmic inflation (expansion by a factor of at least 10^{78} in volume, to the size of a grain of sand, from 10^{-36} to $\approx 10^{-32}$ second after the Big Bang) may have created the large scale of the Universe out of quantum-scale fluctuations. Strong and weak forces describe both atomic nuclei and energy generation in stars. Cf. **range of fundamental forces** in Chap. 24.
In *Conformal Cyclic Cosmology* (Penrose, 2010), the Universe is a sequence of *aeons* (space-times with **FLRW metrics** g_i), where the future time-like singularity of each aeon is the Big Bang singularity of the next. In an eaon's beginning and end, distance and time do not exist; only *conformal* (preserving angles) geometry holds. Any eaon is attached to the next one by a conformal rescaling $g_{i+1} = \Omega^2 g_i$.

27.2 Orders of Magnitude for Length

In this section we present a selection of such orders of length, expressed in meters.

$1.616252(81) \times 10^{-35}$: Planck length;

10^{-34}: length of a putative *string* in *M-theory* which supposes that all forces and elementary particles arise by vibration of such strings (but there is no even agreement that there are smallest fundamental objects);

1.01×10^{-25}: *Schwarzschild radius* ($\frac{2Gm}{c^2}$: the value below which mass m collapses into a black hole) of an average (68 kg) human;

$10^{-24} = 1$ **yoctometer**: effective cross-section radius of 1 MeV neutrinos is 2×10^{-23};

10^{-22}: a certain quantum roughness starts to show up, while the space appears completely smooth at the scale of 10^{-14};

$10^{-21} = 1$ **zeptometer**: *preons*, hypothetical components of quarks/leptons;

$10^{-18} = 1$ **attometer**: size of up quark and down quarks; sizes of strange, charm and bottom quarks are 4×10^{-19}, 10^{-19} and 3×10^{-20};

$10^{-15} = 1$ **femtometer** (or *fermi*);

1.75×10^{-15} and 1.5×10^{-14}: diameter of the smallest (H, hydrogen) and largest (U, uranium-234) nucleus;

1.68×10^{-15}: diameter of proton, range of the weak nuclear force;

$10^{-12} = 1$ **picometer**: distance between atomic nuclei in a white dwarf star;

10^{-11}: wavelength of the hardest (shortest) X-rays and longest gamma rays;

0.62 and 5.2 Å: diameter of the smallest (helium) and largest (caesium) atom;

$10^{-10} = 1$ Å (angstrom): diameter of a typical atom;

0.74 Å and 1,000 Å: diameter of the smallest (H_2) and largest (a SiO_2) molecule;

1.54 Å: length of a typical covalent bond (C-C);
3.4 Å: distance between base pairs in a DNA molecule;
$10^{-9} = 1$ **nanometer**: diameter of typical molecule;
10^{-8}: wavelength of softest X-rays and most extreme ultraviolet;
1.1×10^{-8}: diameter of prion (smallest self-replicating biological entity);
in 2012;
9×10^{-8}: human immunodeficiency virus, HIV; in general, capsid sizes of known viruses range from 1.7×10^{-8} (*Porsine circovirus*) to 1.5×10^{-6} (*pithovirus sibericum*);
10^{-7}: size of chromosomes and largest particle fitting through a surgical mask;
2×10^{-7}: limit of resolution of the light microscope;
3.8–7.6 $\times 10^{-7}$: wavelength of visible (to humans) light;
$10^{-6} = 1$ **micrometer** (or *micron*);
10^{-6}–10^{-5}: diameter of a typical bacterium; known (nondormant) bacteria range from 2–3 $\times 10^{-7}$ (*Mycoplasma genitalium*) to 7.5×10^{-4} (*Thiomargarita Namibiensis*);
8.5×10^{-6}: size of *Ostreococcus*, the smallest free-living eukaryotic unicellular organism, while the length of a nerve cell of the Colossal Squid can reach 12 m;
10^{-5}: typical size of (a fog, mist, or cloud) water droplet;
10^{-5}, 1.5×10^{-5}, and 2×10^{-5}: widths of cotton, silk, and wool fibers;
2×10^{-4}: approximately, the lower limit for the human eye to discern an object;
5×10^{-4}: diameter of a human ovum, grain of salt;
$10^{-3} = 1$ **millimeter**;
5×10^{-3}: length of average red ant; in general, insects range from 1.39×10^{-4} (*Dicopomorpha echmepterygis*, the smallest animal) to 5.67×10^{-1} (*Phobaeticus chani*);
7.7×10^{-3}, $5 \times 10^{q-2}$ and 9.2×10^{-2}: length of the smallest ones: vertebrate (frog *Paedophryne amauensis*), warm-blooded vertebrate (bee hummingbird *Mellisuga helenae*) and primate (lemur *Microcebus berthae*);
8.9×10^{-3}: Schwarzschild radius of the Earth;
$10^{-2} = 1$ **centimeter**;
5.8×10^{-2}: length of uncoiled sperm of the fruit fly *Drosophila bifurca* (it is 20 fly's bodylengths and the longest sperm cell of any known organism);
$10^{-1} = 1$ **decimeter**: wavelength of the lowest microwave and highest UHF radio frequency, 3 GHz;
1 **meter**: wavelength of the lowest UHF and highest VHF radio frequency, 300 MHz;
1.5: average ground level of the Maldives above sea level;
2.77–3.44: wavelength of the broadcast radio FM band, 108–87 MHz;
5.5 and ≈ 3: height of the tallest animal (giraffe) and extinct primate *Gigantopithecus*;
$10 = 1$ **decameter**: wavelength of the lowest VHF and highest shortwave radio frequency, 30 MHz;

$20, 33, 37, 55$: lengths of the longest animals (tapeworm *Diphyllobothrium Klebanovski*, blue whale, lion's mane jellyfish, bootlace worm *Lineus longissimus*);
99.6: height of the world's tallest flowering plant, a tasmanian *Eucalyptus* Centurion (after 100 m, the distribution of the products of photosynthesis become impossible);
$100 = 1$ **hectometer**: wavelength of the lowest HF (high radio frequency) and highest MF (medium radio frequency), 3 MHz;
115.5: height of the world's tallest living tree, a californian *sequoia* Hyperion;
$139, 324, 541, 830$ and 8.5: heights of the Great Pyramid of Giza, Eiffel Tower in Paris, One World Trade Center in New York, Burj Khalifa skyscraper in Dubai and 11,000 years-old Tower of Jericho;
187–555: wavelength of the broadcast radio AM band, 1,600–540 kHz;
340: distance which sound travels in air in one second;
$10^3 = 1$ **kilometer**;
2.954×10^3: Schwarzschild radius of the Sun;
$10^4 = 1$ *miriameter* (used formerly): *scandinavian mile* (Norwegian/Swedish *mil*);
8,848 and 10,911: the highest (Mount Everest) and deepest (Mariana Trench) points on the Earth's surface;
$5 \times 10^4 = 50$ km: the maximal distance at which the light of a match can be seen (at least 10 photons arrive on the retina during 0.1 s);
$1.11 \times 10^5 = 111$ km: one degree of latitude on the Earth;
1.5×10^4–1.5×10^7: wavelengths of sound audible to humans (20 Hz to 20 kHz);
1.37×10^5 and 1.9×10^6: length of the world's longest tunnel, Delaware Aqueduct, New York, and of longest street, Jounge Street, Ontario;
2×10^5: wavelength of a typical tsunami;
$10^6 = 1$ **megameter**, thickness of Earth's atmosphere;
2.22×10^6: diameter of Typhoon Tip (northwest Pacific Ocean, 1979), the most intense tropical cyclone on record;
2.33×10^6: diameter of the plutoid Eris, the largest (together with Pluto itself) dwarf planet, at 67.67 AU from the Sun; the smallest dwarf planet is Ceres (the largest asteroid in the Asteroid belt) of diameter 9.42×10^5 and at 2.77 AU;
3.48×10^6: diameter of the Moon;
9.3×10^6 and 2.1×10^7: length of Trans-Siberian Railway and China's Great Wall;
1.28×10^7 and 4.01×10^7: Earth's equatorial diameter and length of the equator;
4.5×10^7: distance from which Earth's good-looking photograph, *The Blue Marbre*, was taken in 1972 by the Apollo 17 mission. Other famous Earth's images are *Earthrise* (1968, by the Apollo 8), *Pale Blue Dot* (0.12 pixel against the space's vastness; 1990, by Voyager 1) and one from Saturn's neighborhood (2013, by NASA's Cassini).
1.4×10^8: mean diameter of Jupiter;
1.67×10^8: diameter of OGLE-TR-122b, the smallest known star;
$\approx 3 \times 10^8$ (299,792.458 km): distance traveled by light in one second;

3.84×10^8: Moon's orbital distance from the Earth;
4.002×10^8: the farthest distance a human has ever been from Earth (Apollo 13 mission, 1970, passed over the far side of the Moon);
$10^9 = 1$ **gigameter**;
1.39×10^9: Sun's diameter and orbital distance of a planet with 3, 3-h "year";
6.37×10^9: distance at which Earth's gravity becomes $\frac{1}{1{,}000{,}000}$ of that on its surface;
5.83×10^{10}: orbital distance of Mercury from the Sun;
1.496×10^{11} (1 astronomical unit, AU): mean Earth-Sun distance;
$\approx$2.8 AU (near the middle of the Asteroid belt): Sun's water **frost line** (the distance where it is cold enough, $\approx -123\,°C$, for water to condense into ice), separating terrestrial and jovian planets; it is the radius of the inner Solar System;
5.7×10^{11}: length of the longest observed comet tail (Hyakutake, 1996); the Great Comet of 1997 (Hale–Bopp) has biggest known nucleus (>60 km);
$10^{12} = 1$ **terameter** (formerly, *spat*);
15.8 AU: diameter of the largest known star, red supergiant UY Scuti;
30.1 AU: radius of the outer Solar System (orbital distance of Neptune);
50 AU: distance from the Sun to the *Kuiper cliff*, the abrupt outer boundary of the *Kuiper belt* (the region of trans-Neptunian objects around Sun). Only three Solar System objects (dwarf planets) are known to have a perihelion of greater than 50 AU: 2012 VP113 (80 AU), 90377 Sedna (76 AU) and 2004 XR190 (51 AU);
937 AU: aphelion of Sedna, the farthest known Solar System object (its orbital period is about 11,400 years);
$10^{15} = 1$ **petameter**;
1.1 light-year $\approx 10^{16}$: the closest passage (in 1.36 Ma) of Gliese-710, a star expected to perturb dangerously the *Oort cloud* of long-period comets;
50,000–100,000 AU: distance from the Sun to the boundaries of the Oort cloud;
1.3 parsec $\approx 4 \times 10^{16} \approx 4.24$ ly: distance to Proxima Centauri, the nearest star;
$\approx 6.15 \times 10^{17}$: radius of humanity's radio bubble, caused by high-power TV broadcasts leaking through the atmosphere into outer space;
$10^{18} = 1$ **exameter**;
$1.57 \times 10^{18} \approx 50.9$ pc, $\approx$250 pc, 46 pc: distances to supernova 1987A, to *pulsar* (a rapidly rotating neutron star) Geminga (remains of a supernova 0.3 Ma ago which created the Local Bubble), to IK Pegasi B, the nearest known supernova candidate;
$2.59 \times 10^{20} \approx 8.4$ kpc $\approx$27,400 ly: distance from the Sun to the geometric center of our Milky Way galaxy (in Sagittarius A^*, a putative supermassive black hole);
12.9 kpc and 52.8 kpc: distances to the closest (Canis Major Dwarf) and the largest (Large Magellanic Cloud) of 26 satellite galaxies of the Milky Way;
$9.46 \times 10^{20} \approx 30.66$ kpc $\approx 10^5$ ly: diameter of the Milky Way. The largest known galaxy, C 1101, at the center of the cluster Abell 2029, is $\approx$6 Mly across;
$10^{21} = 1$ **zettameter**;
$2.23 \times 10^{22} = 725$ kpc $= 2.54$ Mly: distance to Andromeda (M31), the closest (and approaching at 100–140 km/s) large galaxy; also, it is the farthest naked eye object;

$5.7 \times 10^{23} = 59$ Mly: distance to Virgo, the nearest (and approaching) major cluster;
$10^{24} = 1$ **yottameter**;
$2 \times 10^{24} = 60$ Mpc $= 110$ Mly: diameter of the Local (or Virgo) supercluster; *Cold Spot Supervoid*, the largest known;
4 Gly: length of the wall *U1.27* of quasars, the largest known superstructure;
12.7 Gly: distance to the *quasar* (very active distant galactic nucleus) CFHQS J2329-0301 ($z = 6.43$, while 6.5 is the "Wall of Invisibility" for visible light);
13.14 Gly ($z \approx 9.4$): the most distant gamma ray burst observed, GRB 090429B (possibly, the farthest object, ever seen in the Universe);
13.3 Gly: distance to the farthest and earliest ($\approx$420 Ma after Big Bang) known galaxy MACS0647-JD ($z \approx 10.7$). The formation of the first stars (at the end of the "Dark Age", when matter consisted of clouds of cold hydrogen) corresponds to $z \approx 20$ when the Universe was $\approx$200 Ma old;
$1.3 \times 10^{26} = 13.82$ Gly $= 4.24$ Gpc: **Hubble radius** of the Universe measured as the **light travel distance** to the source of CMB radiation;
$4.4 \times 10^{26} = 47$ Gly $= 14.4$ Gpc: *particle horizon* (present radius of the Universe measured as a **comoving distance**); it is larger than the Hubble radius, since the Universe is expanding). It is $\approx 2\,\%$ larger than the radius of the *visible universe* including only signals emitted later than $\approx$380,000 years after the Big Bang;
The size of whole Universe can be now much larger than the size of the observable one, even infinite, if its curvature is 0. If the Universe is finite but unbounded or if it is nonsimply connected, then it can be smaller than the observable one.
Projecting into the future: the scale of the Universe will be 10^{31} in 10^{14} years (last red dwarf stars die) and 10^{37} in 10^{20} years (stars have left galaxies). If protons decay, their half-life is $\geq 10^{35}$ years; their estimated number in the Universe is 10^{77};
The Universe, in the current *Heat Death* scenario, achieves beyond 10^{1000} years such a low-energy state that quantum events become major macroscopic phenomena, and space-time loses its meaning again, as below the Planck time/length;
The hypothesis of parallel universes estimates that one can find another identical copy of our Universe within the distance $10^{10^{118}}$ m.

Chapter 28
Distances in Applied Social Sciences

In this chapter we present selected distances used in real-world applications of Human Sciences. In this and the next chapter, the expression of distances ranges from numeric (say, in m) to ordinal (as a degree assigned according to some rule) and nominal.

Depending on the context, the distances are either practical ones, used in daily life and work outside of science, or uncountable ones, used figuratively, say, as metaphors for remoteness (being apart, being unknown, coldness of manner, etc.).

28.1 Distances in Perception and Psychology

- **Distance ceptor**
  A **distance ceptor** is a nerve mechanism of one of the organs of special sense whereby the subject is brought into relation with his distant environment.
- **Oliva et al. perception distance**
  Let $\{s_1, \dots, s_n\}$ be the set of stimuli, and let q_{ij} be the conditional probability that a subject will perceive a stimulus s_j, when the stimulus s_i was shown; so, $q_{ij} \geq 0$, and $\sum_{j=1}^{n} q_{ij} = 1$. Let q_i be the probability of presenting the stimulus s_i.
  The **Oliva et al. perception distance** [OSLM04] between stimuli s_i and s_j is

$$\frac{1}{q_i + q_j} \sum_{k=1}^{n} \left| \frac{q_{ik}}{q_i} - \frac{q_{jk}}{q_j} \right|.$$

- **Visual space**
  Visual space refers to a stable perception of the environment provided by vision, while **haptic space** (or *tactile space*) and **auditory space** refer to such internal representation provided by the senses of pressure perception and audition. The

M.M. Deza, E. Deza, *Encyclopedia of Distances*,
DOI 10.1007/978-3-662-44342-2_28

geometry of these spaces and the eventual mappings between them are unknown. But Lewin et al., 2012, found that sensitivity to touch is heritable, and linked to hearing. The main observed kinds of distortion of vision and haptic spaces versus physical space follow; the first three were observed for auditory space also.

- *Distance-alleys*: lines with corresponding points perceived as equidistant, are, actually, some hyperbolic curves. Usually, the parallel-alleys are lying within the distance-alleys and, for visual space, their difference is small at >1.5 m.
- *Oblique effects*: performance of certain tasks is worse when the orientation of stimuli is oblique rather than horizontal or vertical.
- *Equidistant circles*: the **egocentric distance** is direction-dependent; the points perceived as equidistant from the subject lie on egg-like curves, not on circles.

These effects and **size-distance invariance hypothesis** should be incorporated in a good model of visual space. In a visual space the distance d and direction are defined from the self, i.e., as the **egocentric distance**. There is evidence that visual space is almost affine and, if it admits a metric d, then d is a **projective metric**, i.e., $d(x,y) + d(y,z) = d(x,z)$ for any perceptually collinear points x, y, z.
The main models for visual space are a Riemannian space of constant negative curvature (cf. **Riemannian color space** in Chap. 21), a general Riemannian/Finsler space, or an affinely connected (so, not metric, in general) space [CKK03].
An *affine connection* is a linear map sending two vector fields into a third one. The expansion of perceived depth on near and its contraction at far distances hints that the mapping between visual and physical space is not affine.
Amedi et al., 2002, observed the convergence of visual and tactile shape processing in the human lateral occipital complex. The *vOICe technology* (OIC for "Oh I see!") explores cross-modal binding for inducing visual sensations through sound (mental imagery and artificial synesthesia). Some blind people "see" by echolocation. The cane extends peri-hand space of blind users and, in general, extrapersonal or far space can remap as peripersonal or near space when using tools.

- **Length-related illusions**
The most common optical illusions distort size or length. For example, in the *Müller–Lyer illusion*, one of two lines of equal length appear shorter because of the way the arrows on their ends are oriented. Pigeons and parrots also are susceptible to it. Segall et al., 1963, found that the mean fractional misperception varies cross-culturally from 1.4 to 20.3 % with maximum for Europeans. Also, urban residents and younger subjects are much more susceptible to this illusion.
In the *Luckiech–Sander illusion* (1922), the diagonal bisecting the larger, left-hand parallelogram appears to be longer than the diagonal bisecting the smaller, right-hand parallelogram, but is in fact of the same length.
The perspective created in *Ponzo illusion* (1911) increases the perceived distance and so, compliant with **Emmert's size-distance law**, perceived size increases.
The *Moon illusion* (mentioned in clay tablets at Nineveh in the seventh century BC) is that the Moon, despite the constancy of its visual angle ($\approx 0.52°$), at the

horizon may appear to be about twice the zenith Moon. This illusion (and similar *Sun illusion*) could be cognitive: the zenith moon is perceived as approaching. (Plug, 1989, claim that the *distance to the sky*, assumed unconsciously, is about 10–40 m cross-culturally and independent of the consciously perceived distance.)
The *Ebbenhouse illusion*: the diameter of the circle, surrounded by smaller circles, appears to be larger than one of the same circle nearby, surrounded by larger circles.
In *vista paradox* (Walker–Rupich–Powell, 1989), a large distant object viewed through a window appears to both shrink in size and recede in distance as the observer approaches; a similar framing effect works in the *coffee cup illusion* (Senders, 1966). In the *Pulfrich depth illusion* (1922), lateral motion of an object is interpreted as having a depth component.
An *isometric illusion* (or *ambiguous figure*) is a shape that can be built of same-length (i.e., isometric) lines, while relative direction between its components are not clearly indicated. The *Necker Cube* is an example.
The *Charpentier size-weight illusion* (1891): the larger of two graspable/liftable objects of equal mass is misperceived to be less heavy than the smaller.

- **Size-distance invariance hypothesis**
  The SDIH (**size-distance invariance hypothesis**) by Gilinsky, 1951, is that $\frac{S'}{D'} = C\frac{S}{D}$ holds, where S, D are the physical and S', D' are perceived size and distance of visual stimulus, while C is an observer constant. A simplified formula is $\frac{S'}{D'} = 2\tan\frac{\alpha}{2}$, where α is the angular size of the stimulus.
  A version of SDIH is the **Emmert's size-distance law**: $S' = CD'$. This law accounts for *size constancy*: object's size is perceived to remain constant despite changes in the retinal image (more distant objects appear smaller because of perspective). The Müller–Lyer and Ponzo illusions are examples of size constancy.
  The Moon and Ebbenhouse illusions are called **size-distance paradoxes** since they unbalance SDIH. They are misperceptions of visual angle and examples of **distance constancy**: distance is perceived constant despite changes in the retinal image.
  If an observer's head translates smoothly through a distance K as he views a stationary target point at pivot distance D_p, then the point will appear to move through a displacement W' when it is perceived to be at a distance D'. The **apparent distance/pivot distance hypothesis** (Gogel, 1982): it holds $\frac{D'}{D_p} + \frac{W'}{K} = 1$.
  The **size-distance centration** is the overestimation of the size of objects located near the focus of attention and underestimation of it at the periphery.
  Hubbard and Baiard, 1988, gave to subjects name and size S of a familiar object and asked imaged distances d_F, d_O, d_V. Here the object mentally looks to be of the indicated size at the *first-sight distance* d_F. The object become, while mentally walking (zooming), too big to be seen fully with zoom-in at the *overflow distance* d_O, and too small to be identified with zoom-out at the *vanishing point distance* d_V. Consistently with SDIH, d_F was linearly related to S. For d_O and d_V, the relation were the power functions with exponents about 0.9 and 0.7. The

time needed to imagine d_O increased slower than linearly with the *scan distance* $d_O - d_F$.
Konkle and Oliva, 2011, found that the real-world objects have a consistent visual size at which they are drawn, imagined, and preferentially viewed. This size is proportional to the logarithm of the object's assumed size, and is characterized by the ratio of the object and the frame of space around it. This size is also related to the first-sight distance d_F and to the typical distance of viewing and interaction. A car at a typical viewing distance of 9.15 m subtends a visual angle of 30°, whereas a raisin held at an arm's length subtends 1°. Cf. the **optimal eye-to-eye distance** and, in Chap. 29, the *TV viewing distance* in the **vision distances**.
Similarly, Palmer et al., 1981, found that in goodness judgments of photographs of objects, the $\frac{3}{4}$ *perspective* (or 2.5 *view*, *pseudo-3D*), in which the front, side, and top surfaces are visually present, were usually ranked highest. Cf. the *axonometric projection* in the **representation of distance in Painting**.

- **Egocentric distance**
  The **egocentric distance** is the perceived absolute distance from the self (observer or listener) to an object or a stimulus; cf. **subjective distance**. Usually, such visual distance underestimates the actual physical distance to far objects, and overestimates it for near objects. Such distortion decreases in a lateral direction.
  In Visual Perception, the *action space* of a subject is 1–30 m; the smaller and larger spaces are called the *personal space* and *vista space*, respectively.
  The **exocentric distance** is the perceived relative distance between objects.
- **Distance cues**
  The **distance cues** are cues used to estimate the **egocentric distance**.
  For a listener at a fixed location, the main auditory distance cues include: *intensity*, *direct-to-reverberant energy ratio* (in the presence of sound reflecting surfaces), *spectrum* and *binaural differences*; cf. **acoustics distances** in Chap. 21.
  For an observer, the main visual distance cues include:

  - *relative size*, *relative brightness*, *light and shade*;
  - *height in the visual field* (in the case of flat surfaces lying below the level of the eye, the more distant parts appear higher);
  - *interposition* (when one object partially occludes another from view);
  - *binocular disparities*, *convergence* (depending on the angle of the optical axes of the eyes), *accommodation* (the state of focus of the eyes);
  - *aerial perspective* (distant objects become bluer and paler), *distance hazing* (distant objects become decreased in contrast, more fuzzy);
  - *motion perspective* (stationary objects appear to a moving observer to glide past).

  Examples of the techniques which use the above distance cues to create an optical illusion for the viewer, are:

- *distance fog*: a 3*D* computer graphics technique such that objects farther from the camera are progressively more blurred (obscured by haze). It is used, for example, to disguise the too-short **draw distance**, i.e., the maximal distance in a 3D scene that is still drawn by the rendering engine;
- *forced perspective*: a technique to make objects appear either far away, or nearer depending on their positions relative to the camera and to each other.
- *lead room*: space left in the direction the subject is facing or moving.

- **Subjective distance**
  The **subjective distance** (or *cognitive distance*) is a mental representation of actual distance molded by an individual's social, cultural and general life experiences; cf. **egocentric distance**. Cognitive distance errors occur either because information about two points is not coded/stored in the same branch of memory, or because of errors in retrieval of this information.
  For example, the length of a route with many turns and landmarks is usually overestimated. In general, the filled or divided space (distance or area) appears greater than the empty or undivided one. Also, affective signals of threat and disgust increase and decrease, respectively, perceived proximity.
  Human mental maps, used to find out distance and direction, rely mainly, instead of geometric realities, on real landscape understanding, via webs of landmarks. Ellard, 2009, suggests that this loss of natural navigation skills, coupled with the unique ability to imagine themselves in another location, may have given modern humans the freedom to create a reality of their own.
- **Geographic distance biases**
  Sources of distance knowledge are either symbolic (maps, road signs, verbal directions) or directly perceived ones during locomotion: environmental features (visually-perceived turns, landmarks, intersections, etc.), travel time/effort.
  They relate mainly to the perception and cognition of **environmental distances**, i.e., those that cannot be perceived in entirety from a single point of view but can still be apprehended through direct travel experience.
  Examples of **geographic distance biases** (subjective distance judgments) are:

- observers are quicker to respond to locations preceded by locations that were either close in distance or were in the same region;
- distances are overestimated when they are near to a reference point; for example, intercity distances from coastal cities are exaggerated;
- subjective distances are often asymmetrical as the perspective varies with the reference object: a small village is considered to be close to a big city while the big city is likely to be seen as far away from it;
- traveled routes segmented by features are subjectively longer than unsegmented routes; moreover, longer segments are relatively underestimated;
- increasing the number of pathway features encountered and recalled by subjects leads to increased distance estimates;
- structural features (such as turns and opaque barriers) breaking a pathway into separate vistas strongly increase subjective distance (suggesting that distance

knowledge may result from a process of summing vista distances) (turns are often memorized as straight lines or right angles);
- *Chicago–Rome illusion*: belief that some European cities are located far to the south of their actual location; in fact, Chicago and Rome are at the same latitude (42°), as are Philadelphia and Madrid (40°), etc.;
- *Miami–Lima illusion*: belief that US east coast cities are located to the east of the west coast cities of South America; in fact, Miami is 3° west of Lima.

Such illusions could be perceptually based mental representations that have been distorted through normalization and/or conceptual nonspatial plausible reasoning.
Thorndyke and Hayes-Roth, 1982, compared distance judgments of people with navigation- and map-derived spatial knowledge. Navigation-derived route distance estimates were more accurate than Euclidean judgments, and this difference diminished with increased exploration. The reverse was true for map subjects, and no improvement was observed in the map learning.
Turner–Turner, 1997, made a similar experiment in a plane virtual building. Route distances were much underestimated but exploration-derived Euclidean judgments were good; repeated exposure did not help. The authors suggest that exploration of virtual environments is similar to navigation in the real world but with a restricted field of view, as in tunnels, caves or wearing a helmet, watching TV.
Krishna et al., 2008, compared spatial judgments of *self-focused* ("Western") and *relationship-focused* ("Eastern") people. The former ones were more likely to misjudge distance (when multiple features should be considered), to pay attention to only focal aspects of stimuli and ignore the context and background information.

- **Psychogeography**
  Psychogeography is (Debord, 1955) the study of the precise laws and specific effects of the geographical environment, consciously organized or not, on the emotions and behavior of individuals. An example of related notions is a *desire path* (or *social trail*), i.e., a path developed by erosion caused by animal or human footfall, usually the shortest or easiest route between an origin and destination.
  Also, the psychoanalytic study of spatial representation within the unconscious construction of the social and physical world is called *psychogeography*. In general, *depth psychology* refer to unconscious-accounting approaches to therapy and research.
- **Psychological Size and Distance Scale**
  The CID (*Comfortable Interpersonal Distance*) scale by Duke and Nowicky, 1972, consists of a center point 0 and eight equal lines emanating from it. Subjects are asked to imagine themselves on the point 0 and to respond to descriptions of imaginary persons by placing a mark at the point on a line at which they would like the imagined person to stop, that is, the point at which they would no longer feel comfortable. CID is then measured in mm from 0.

The GIPSDS (**Psychological Size and Distance Scale**) by Grashma and Ichiyama, 1986, is a 22-item rating scale assessing interpersonal status and affect. Subjects draw circles, representing the drawer and other significant persons, so that the radii of the circles and the distances between them indicate the thoughts and feelings about their relationship. These distances and radii, measured in mm, represent the **psychological distance** and status, respectively. Cf. related questionnaire on http://www.surveymonkey.com/s.aspx?sm=Nd8c_2fazsxMZfK9ryhvzPlw_3d_3d.

- **Visual Analogue Scales**
  In Psychophysics and Medicine, a **Visual Analogue Scale** (or *VAS*) is a self-report device used to measure the magnitude of internal states such as pain and mood (depression, anxiety, sadness, anger, fatigue, etc.) which range across a continuum and cannot be measured directly. Usually, VAS is a horizontal (or vertical, for Chinese subjects) 10 cm line anchored by word descriptors at each end.
  The *VAS score* is the distance, measured in mm, from the left hand (or lower) end of the line to the point marked by the subject. The VAS tries to produce *ratio data*, i.e., ordered data with a constant scale and a natural zero.
  Amongst scales used for pain-rating, the VAS is more sensitive than the simpler verbal scale (six descriptive or activity tolerance levels), the Wong–Baker facial scale (six grimaces) and the numerical scale (levels $0, 1, 2, \dots, 10$). Also, it is simpler and less intrusive than questionnaires for measuring internal states.
- **Psychological distance**
  CLT (*construal level theory*) in Liberman–Trope, 2003, defines **psychological distance** from an event or object as a common meaning of spatial ("where"), temporal ("when"), social ("who") and hypotheticality ("whether") distance from it.
  Expanding spatial, temporal, social and hypotheticality horizons in human evolution, history and child development is enabled by our capacity for *mental construals*, i.e., abstract mental representations. Any event or object can be represented at *lower-level* (concrete, contextualized, secondary) or *higher-level* (abstract, more schematic, primary) construal.
  More abstract construals lead to think of more distant (spatially, temporally, socially, hypothetically) objects and vice versa. People construe events at greater, say, temporal distance in terms of their abstract, central, goal-related features and pro-arguments, while nearer events are treated situation-specifically at a lower level of counter-arguments. Examples are: greater moral concern over a distant future event, more likely victim's forgiveness of the earlier transgression, more intense affective consumer's reaction when a positive outcome is just missed.
  CLT implied that the four dimensions are functionally similar. For example, increase of distance in only one dimension leads to greater moral concern. Zhang and Wang, 2008, observed that stimulating people to consider spatial distance influences their judgments along three other dimensions, but the reverse is not true.

It is consistent with a claim by Boroditsky, 2000, that the human cognitive system is structured around only concepts emerging directly out of experience, and that other concepts are then built in a metaphorical way. Williams and Bargh, 2008, also claim that psychological distance is a derivative of spatial distance. Spatial concepts such as "near/far" are present at 3–4 months of age since the relevant information is readily available to the senses, whereas abstract concepts related to internal states are more difficult to understand. Also, spatial relations between oneself, one's caretakers and potential predators have primary adaptive significance.

- **Time-distance relation (in Psychology)**

  People often talk about time using spatial linguistic metaphors (a long vacation, a short concert) but much less talk about space in terms of time. This bidirectional but asymmetric relation suggests that spatial representations are primary, and are later co-opted for other uses such as time.

  Casasanto and Boroditsky, 2008, showed that people, in tasks not involving any linguistic stimuli or responses, are unable to ignore irrelevant spatial information when making judgments about duration, but not the converse. So, the metaphorical space-time relationship observed in language also exists in our more basic representations of distance and duration. Mentally representing time as a linear spatial path may enable us to conceptualize abstract (as moving a meeting forward, pushing a deadline back) and impossible (as time-travel) temporal events.

  In Psychology, the *Kappa effect* is that among two journeys of the same duration, one covering more distance appears to take longer, and the *Tau effect* is that among two equidistant journeys, one taking more time to complete appears to have covered more distance. Jones–Huang, 1982, see them as effects of *imputed velocity* (subjects impute uniform motion to discontinuous displays) on judgments of both time and space, rather than direct effect of time (distance) on distance (time) judgment.

  Fleet–Hallet–Jepson, 1985, found spatiotemporal inseparability in early visual processing by retinal cells. Maruya–Sato, 2002, reported a new illusion (the time difference of two motion stimuli is converted in the illusory spatial offset) indicating interchangeability of space and time in early visual processing. Simner–Mayo–Spiller, 2009, tested ten individuals with time-space synesthesia. The differences appear at the level of higher processing because of different representations: space is represented in retinotopic maps within the visual system, while time is processed in the cerebellum, basal ganglia and cortical structures. Evidence from lesion and human functional brain imaging/interference studies point towards the posterior parietal cortex as the main site where spatial and temporal information converge and interact with each other. Cf. also **spatial–temporal reasoning**.

  In human-computer interaction, *Fitts's law* claims that the average time taken to position a mouse cursor over an on-screen target is $a + b\log_2(1 + \frac{D}{W})$, where D is the distance to the center of the target, W is the width (along the axis of motion) of the target and a, b represent the start/stop time and device's speed.

People in the West construct mental timelines going from the left; those with damaged right side of their brain have trouble imagining past, i.e., timeline's left side. Núñez, 2012, found that our spatial representation of time is not innate but learned. The Aymara of the Andes place the past in front and the future behind. The Pormpuraaw of Australia place the past in the east and the future in the west. Some Mandarin speaker have the past above and future below.
For the Yupno of Papua New Guinea, past and future are arranged as a nonlinear 3D bent shape: the past downhill and the future uphill of the local river. Inside of their homes, Yupno point towards the door when talking about the past, and away from the door to indicate future. Yupno also have a native counting system and number concepts but they ignore the number-line concept. They place numbers on the line but only in a categorical manner, ignoring line's extension.

- **Symbolic distance effect**
  In Psychology, the brain compares two concepts (or objects) with higher accuracy and faster reaction time if they differ more on the relevant dimension. For example, the performance of subjects when comparing a pair of positive numbers (x, y) decreases for smaller $|x - y|$ (*behavioral numerical distance effect*).
  The related *magnitude effect* is that performance decreases for larger $\min\{x, y\}$. For example, it is more difficult to measure a longer distance (say, 100 m) to the nearest mm than a short distance (say, 1 cm). Those effects are valid also for congenitally blind people; they learn spatial relation via tactile input (interpreting, say, numerical distance by placing pegs in a peg board).
  A current explanation is that there exists a mental line of numbers which is oriented from left to right (as $2, 3, 4$) and nonlinear (more mental space for smaller numbers). So, close numbers are easier to confuse since they are represented on the mental line at adjacent and not always precise locations. Possible mental lines, explaining such confusion, are *linear-scalar* (the psychological distance $d(a, a+1)$ between adjacent values is constant but the amount of noise increases as ka) or *logarithmic* (amount of noise is constant but $d(a, a + 1)$ decreases logarithmically).
  Related *SNARC* (spatial-numerical association of response codes) *effect* is that smaller (or larger) numbers are responded to more easily with responses toward a left (or, respectively, right) location. Also, smaller numbers promote a left-oriented gaze-direction whereas the opposite is true for higher numbers. Similar spatial-musical association SMARC and a mental line of pitches were observed.
- **Law of proximity**
  Gestalt psychology is a theory of mind and brain of the Berlin School, in which the brain is holistic, parallel and self-organizing. Perceptual organization is composed of grouping and segregation. The visual grouping of discrete elements is determined by proximity, similarity, common fate, good continuation, closure (Wertheimer, 1923), and, more recently, common region, connectedness or synchrony.
  In particular, the **law of proximity** is that spatial or temporal proximity of elements may induce the mind to perceive a collective or totality.

- **Emotional distance**
  The **emotional distance** is the degree of emotional detachment (toward a person, group or events), aloofness, indifference by personal withdrawal, reserve.
  The Bogardus Social Distance Scale (cf. **social distance**) measures the distance between two groups by averaged emotional distance of their members.
  Spatial empathy is the awareness that an individual has to the proximity, activities, and comfort of people surrounding him.
  The *propinquity effect* is the tendency for people to get emotionally involved with those who have higher *propinquity* (physical/psychological proximity) with them, i.e., whom they encounter often. Walmsley, 1978, proposed that emotional involvement decreases as $d^{-\frac{1}{2}}$ with increasing **subjective distance** d.
- **Psychical distance**
  Psychical (or *psychic*) **distance** is a term having no commonly accepted definition. In several dictionaries, it is a synonym for the **emotional distance**. This term was introduced in [Bull12] to define what was called later the **aesthetic distance** (cf. the **antinomy of distance**) as a degree of the emotional involvement that a person, interacting with an aesthetic artifact or event, feels towards it.
  In Marketing, the psychic distance mean the level of attraction or detachment to a particular country resulting from the degree of uncertainty felt about it.
- **Distancing**
  Distancing (from the verb *to distance*, i.e., to move away from or to leave behind) is any behavior or attitude causing to be or appearing to be at a distance.
  Uncountable noon **distantness** (or **farness**) is the state or quality of being distant, remote, far-off, way in the distance. Archaic meaning: distant parts or regions.
  Distancy, **farawayness**, **distaunce** are rare/obsolete synonyms for distance, while **indistancy** is either nearness, or lack (or want) of distance (or separation).
  Self-distance is the ability to critically reflect on yourself and your relations from an external perspective; not to confound with mathematical notions of **self-distance** in Chaps. 1 and 17.
  Outdistancing means to outrun, especially in a long-distance race, or, in general, to surpass by a wide margin, especially through superior skill or endurance.
  In Martial Arts, **distancing** is the selection of an appropriate *combat range*, i.e., distance from the adversary. For other examples of spatial distancing; cf. **distances between people** and, in Chap. 29, **safe distancing** from a risk factor.
  Social distancing during pandemic refers to focused measures to increase the physical distance between individuals, or activity restrictions, such as increasing distance between student desks, canceling sports activities, and closing schools.
  In *Mediation* (a form of alternative dispute resolution), **distancing** is the impartial and nonemotive attitude of the mediator versus the disputants and outcome.
  In Psychoanalysis, **distancing** is the tendency to put persons and events at a distance. It concerns both the patient and the psychoanalyst.
  In Developmental Psychology, **distancing** (Werner–Kaplan, 1964, for deaf-blind patients) is the process of establishing the individuality of a subject as an essential phase (prior to symbolic cognition and linguistic communication)

in learning to treat symbols and referential language. For Sigel (1970, for preschool children), **distancing** is the process of the development of cognitive representation: cognitive demands by the teacher or the parent help to generate a child's representational competence. **Distancing from role identities** is the first step of 7th (individualistic) of nine stages of ego development in Loevinger, 1976.
In the books by Kantor, **distancing** refers to APD (Avoidant Personality Disorder): fear of intimacy and commitment in confirmed bachelors, "femmes fatales", etc. **Associational distancing** refers to individual's dissociation with those in the group inconsistent with his desired social identity.
The **distancing language** is phrasing used by a person to avoid thinking about the subject or content of his own statement (for example, referring to death).
Distancing by scare quotes is placing quotation marks around an item (single word or phrase) to indicate that the item does not signify its literal or conventional meaning. The purpose could be to distance the writer from the quoted content, to alert the reader that the item is used in an unusual way, or to represent the writer's concise paraphrasing. **Neutral distancing** convey a neutral writer's attitude, while distancing him from an item's terminology, in order to call attention to a neologism, jargon, a slang usage, etc; sometimes italics are used for it.
Cf. **technology-related distancing**, **antinomy of distance**, **distanciation**.

28.2 Distances in Economics and Human Geography

- **Technology distances**
The **technological distance** between two firms is a distance (usually, χ^2- or **cosine distance**) between their *patent portfolios*, i.e., vectors of the number of patents granted in (usually, 36) technological subcategories. Other measures are based on the number of patent citations, co-authorship networks, etc.
Granstrand's **cognitive distance** between two firms is the **Steinhaus distance** $\frac{\mu(A\Delta B)}{\mu(A\cup B)} = 1 - \frac{\mu(A\cap B)}{\mu(A\cup B)}$ between their technological profiles (sets of ideas) A and B seen as subsets of a *measure space* $(\Omega, \mathcal{A}, \mu)$.
Olsson, 2000, defined the metric space (I, d) of all ideas (as in human thinking), $I \subset \mathbb{R}^n_+$, with some **intellectual distance** d. The closed, bounded, connected *knowledge set* $A_t \subset I$ extends with time t. New elements are, normally, convex combinations of previous ones: *innovations* within gradual technological progress. Exceptionally, breakthroughs (Kuhn's paradigm shifts) occur.
The similar notion of *thought space* (of ideas/knowledge and relationships among them in thinking) was used by Sumi et al., 1997, for computer-aided thinking with text; they proposed a system of mapping text-objects into metric spaces.
Introduced by Patel, 1965, the **economic distance** between two countries is the time (in years) for a lagging country to catch up to the same per capita income level as the present one of an advanced country. Introduced by Fukuchi–Satoh,

1999, the **technology distance** between countries is the time (in years) when a lagging country realizes a similar technological structure as the advanced one has now. The basic assumption of the *Convergence Hypothesis* is that the technology distance between two countries is smaller than the economic one.

- **Production Economics distances**
  In quantitative Economics, a *technology* is modeled as a set of pairs (x, y), where $x \in \mathbb{R}^m_+$ is an *input* vector, $y \in \mathbb{R}^m_+$ is an *output* vector, and x can produce y. Such a set T should satisfy standard economical regularity conditions.
  The **directional distance function** of input/output x, y toward a (projected and evaluated) direction $(-d_x, d_y) \in \mathbb{R}^m_- \times \mathbb{R}^m_+$ is (Chambers–Chung–Färe, 1996)
  $$\sup\{k \geq 0 : ((x - kd_x), (y + kd_y)) \in T\}.$$
  For $d_x = x, d_y = y$, it is a scaled version of the **Shephard input distance function** (Shephard, 1953 and 1970) $\sup\{k \geq 0 : (x, \frac{y}{k}) \in T\}$.
  The *frontier* $f_s(x)$ is the maximum feasible output of a given input x in a given system (or year) s. The **distance to frontier** (Färe–Crosskopf–Lovell, 1994) of a production point (x, y), where $y = g_s(x))$, is $\frac{g_s(x)}{f_s(x)}$.
  The *Malmquist index* measuring the change in TFP (total factor productivity) between periods s, s' (or comparing to another unit in the same period) is $\frac{g'_s(x)}{f_s(x)}$.
  The *distance to frontier* is the inverse of TFP in a given industry (or of GDP per worker in a given country) relative to the existing maximum (the frontier, usually, US). In general, the term *distance-to-target* is used for the deviation in percentage of the actual value from the planned one.
  Consider a *production set* $T \subset \mathbb{R}^{n_1} \times \mathbb{R}^{n_2}$ (input, output). The measure of the technical efficiency, given by Briec–Lemaire, 1999, is the point-set distance $\inf_{y \in we(T)} ||x - y||$ (in a given norm $||.||$ on $\mathbb{R}^{n_1+n_2}$) from $x \in T$ to the *weakly efficient set* $we(T)$. It is the set of minimal elements of the poset $(T, \preceq)$ where the *partial order* $\preceq$ ($t_1 \preceq t_2$ if and only if $t_2 - t_1 \in K$) is induced by the cone $K = int(\mathbb{R}^{n_1}_{>0} \times \mathbb{R}^{n_2}_{>0}) + \{0\}$.
- **Distance to default**
  A *call option* is a financial contract in which the buyer gets, for a fee, the right to buy an agreed quantity of some commodity or financial instrument from the seller at a certain time (the expiration date) for a certain price (the *strike price*).
  Let us see a firm's equity E as a call option on the firm's assets A, with the total *liabilities* (debt) L being the strike price, i.e., $E = \max(0, A - L)$ with $A < L$ meaning the firm's default. Applying Black–Sholes, 1973, and Merton, 1974, option pricing formulas, the **distance to default** t periods ahead is defined by
  $$D2D_t = \frac{\ln \frac{A_t}{D} + t(\mu_A - \frac{1}{2}\sigma_A^2)}{\sigma_A \sqrt{t}},$$
  where μ_A is the rate of growth of A and σ_A is its *volatility* (standard deviation of yearly logarithmic returns). A Morningstar's credit score is $cs = \frac{7}{2}(D2D+SS)+$

$8BR + CC \times \max(D2D, SS, BR)$, where SS, BR and CC are the solvency, business risk and cash flow cushion scores. The resulting credit rating AAA, AA, A, BBB etc., corresponds to cs within $[16, 23)$, $[23, 61)$, $[61, 96)$, etc.

- **Action distance**
  The **action distance** is the distance between the set of information generated by the Active Business Intelligence system and the set of actions appropriate to a specific business situation. Action distance is the measure of the effort required to understand information and to effect action based on that information. It could be the physical distance between information displayed and action controlled.
- **Effective trade distance**
  There is large border effect of political boundaries on the volume of trade and on relative prices. The border introduces costs related to tariffs, market regulations, differences in product packages and languages.
  Engel–Rogers, 1996, showed that the dispersion of prices within a country is orders of magnitude smaller than across countries, and estimated that the US-Canadian border was equivalent to a distance of 120,000 km. McCallum, 1995, found that inter-provincial trade within Canada was, on average, 22 times larger that the trade of any province with any State from US. Cf. **impact of distance on trade**.
  Borraz et al., 2012, showed that the "online border" in E-commerce is equivalent to the average distance from the online warehouse to the offline stores.
  [HeMa02] defined the **effective trade distance** between countries x and y with populations $x_1, \dots, x_m$ and $y_1, \dots, y_n$ of their main agglomerations as

$$\left(\sum_{1 \le i \le m} \frac{x_i}{\sum_{1 \le t \le m} x_t} \sum_{1 \le j \le n} \frac{y_j}{\sum_{1 \le t \le n} y_t} d_{ij}^r \right)^{\frac{1}{r}},$$

  where d_{ij} is the bilateral distance (in km) of the corresponding agglomerations x_i, y_i, and r measures the sensitivity of trade flows to d_{ij}.
  As an **internal distance of a country**, measuring the average distance between producers and consumers, Head and Mayer [HeMa02] proposed $0.67\sqrt{\frac{area}{\pi}}$.
- **Impact of distance on trade**
  Bilateral trade decreases with distance; this effect slightly increased over the last century. Webb, 2007, claims that an average distance of trade in 1962 of 4,790 km changed only to 4,938 km in 2000.
  The relationship between shipments and distance, found in Hillberry–Hummels, 2008, is highly nonlinear: at the beginning, there is a sharp reduction in value with distance; but, once a distance-threshold is achieved the negative effect vanishes.
  An example of used measures is the average distance traveled by heavy trucks between actual origins and destinations in their deliveries of commodities.
  Frankel–Rose, 2000, estimated impact of certain distance variables on trade, for example, $+340, +200, +80, +0.8, -0.2, -1.1$ % for common currency, com-

mon language, common border, economic size (1 % GDP increase), physical size (1 % increase), physical distance (1 % increase), respectively.
Using the **gravity models** with 16 *CAGE* (cultural, administrative, geographic, economic) distances between countries, Ghemawat, 2004, developed *CAGE Distance Framework* for managers considering international strategies. His distances are *cultural* (different languages, ethnicities, religions, social norms), *administrative* (absence of shared monetary or political association, institutional weakness), *geographic* (physical remoteness, different climates, lack of common border or waterway access, weak transportation or communication links) and *economic* (difference in consumer incomes, cost and quality of natural, financial, human resources).
Most affected industries are: meat, cereals, tobacco (by linguistic ties), gold, electricity, textile (by preferential trading agreements), electricity, gas, live animals (by physical remoteness). The wealth difference decreases trade in metals, fertilisers, meat, but increases trade in coffee, tea, animal oils, office machines.

- **Long-distance trade routes**
Examples of such early historic routes are the *Amber Road* (from northern Africa to the Baltic Sea), *Via Maris* (from Egypt to modern day Iran, Iraq, Turkey, Syria), the route *from the Varangians to the Greeks* (from Scandinavia across Kievan Rus' to the Byzantine Empire), the *Incense Road* (from Mediterranean ports across the Levant and Egypt through Arabia to India), Roman-Indian routes, Trans-Saharan trade, *Grand Trunk Road* (from Calcutta to Peshawar) and the *Ancient Tea Route* (from Yunnan to India via Burma, to Tibet and to central China).
The *Silk Road* was, from the second century BC, a network of trade routes connecting East, South, and Western Asia with the Mediterranean world, North/Northeast Africa and Europe. Extending 6,500 km, it enabled traders to transport goods, slaves and luxuries such as silk, other fine fabrics, perfumes, spices, medicines, jewels, as well as the spreading of knowledge, ideas, cultures, plants, animals and diseases. But the Silk Road became unsafe and collapsed in the tenth century after the fall of the Tang Dynasty of China, the destruction of the Khazar Khaganate and, later, the Turkic invasions of Persia and the Middle East.
During fifth to tenth centuries, the *Radhanites* (medieval Jewish merchants) dominated trade between the Christian and Islamic worlds, covering much of Europe, North Africa, Middle East, Central Asia and parts of India and China. They carried commodities combining small bulk and high demand (spices, perfumes, jewelry, silk). The *Maritime Republics* (mercantile Italian city-states, especially Genoa, Venice, Pisa, Amalfi) dominated long-distance trade during tenth to thirteenth centuries. The spice trade from Asia to Europe became, via new sea routes, a Portuguese monopoly (fifteenth to seventeenth centuries) replaced by the Dutch, and soon after the English and the French. During thirteenth to seventeenth centuries, the *Hanseatic League* v(an alliance of trading cities and their guilds) dominated trade along the coast of Northern Europe.

- **Relational proximity**
  Economic Geography considers to nongeographical types of proximity (organizational, institutional, cognitive, etc.). In particular, **relational proximity** (or trust-based interaction between actors) is an inclusive concept of the benefits derived from spatially localized sets of economic activities. It generates relational capital through the dynamic exchange of locally produced knowledge.
  The five dimensions of relational proximity are proximity: of contact (directness), through time (continuity, stability), in diversity (multiplicity, scope), in mutual respect and involvement (parity), of purpose (commonality).
  Individuals are close to each other in a relational sense when they share the same interaction structure, make transactions or realize exchanges. They are *cognitively close* if they share the same conventions and have common values and representations (including knowledge and technological capabilities).
  Bouba-Olga and Grossetti, 2007, divide socio-economic proximity into relational one (role of social networks) and *mediation proximity* (role of newspapers, directories, Internet, agencies, etc.). Tranos and Nijkamp, 2013: physical distance and relational proximities have a significant impact on Internet's infrastructure.
- **Migration distance (in Economics)**
  The **migration distance**, in Economic Geography, is the distance between the geographical centers of the municipalities of origin and destination.
  Ravenstein's 2nd and 3rd laws of migration (1880) are that the majority of migrants move a short distance, while those move longer distances tend to choose big-city destinations. About 80 % of migrants move within their own country.
  Migration tends to be an act of aspiration; it generally improves migrant's wealth and lifestyle. *Existential migrants* refer to voluntary noneconomic expatriates with "existential wanderlust". Madison, 2006, defines them as seeking greater possibilities for self-actualising, exploring foreign cultures in order to assess own identity, and ultimately grappling with issues of home/belonging in the world generally.
- **Commuting distance**
  The **commuting distance** is the distance (or travel time) separating work and residence when they are located in separated places (say, municipalities).
- **Consumer access distance**
  Consumer access distance is a distance measure between the consumer's residence and the nearest provider where he can get specific goods or services (say, a store, market or a health service). For example, *food miles* refers to the distance food is transported from the time of its production until it reaches consumers.
  Measures of geographic access and spatial behavior include distance measures (**map's distance**, **road travel distance**, perceived travel time, etc.), **distance decay** (decreased access with increasing distance) effects, transportation availability and *activity space* (the area of $\approx\frac{2}{3}$ of the consumer's routine activities).
  For example, by US Medicare standards, consumers in urban, suburban, rural areas should have a pharmacy within $2, 5, 15$ miles, respectively. The patients residing outside of a 15-miles radius of their hospital are called *distant patients*.

Food grown within 100 miles of its point of purchase or consumption is *local food.*
Similar studies for retailers revealed that the negative effect of distance on store choice behavior was (for all categories of retailers) much larger when this behavior was measured as "frequency" than when it was measured as "budget share".

- **Distance decay (in Spatial Interaction)**
In general, **distance decay** or the **distance effect** (cf. Chap. 29) is the attenuation of a pattern or process with distance. In Spatial Interaction, **distance decay** is the mathematical representation of the inverse ratio between the quantity of obtained substance and the distance from its source.
This decay measures the effect of distance on accessibility and number of interactions between locations. For example, it can reflect a reduction in demand due to the increasing travel cost. The quality of streets and shops, height of buildings and price of land decrease as distance from the center of a city increases.
The *bid-rent distance decay* induces, via the cost of overcoming distance, a class-based spatial arrangement around a city: with increasing distance (and so decreasing rent) commercial, industrial, residential and agricultural areas follow.
In location planning for a service facility (fire station, retail store, transportation terminal, etc.), the main concerns are *coverage standard* (the maximum distance, or travel time, a user is willing to overcome to utilize it) and distance decay (demand for service decays with distance).
An example of related *size effect*: doubling the size of a city leads usually to a 15 % decrease of resource use (energy, roadway amount, etc.) per capita, a rise of $\approx$15 % in socio-economic well-being (income, wealth, the number of colleges, etc.), but also in crime, disease and average walking speed. Bettencourt et al., 2007, observed that "social currencies" (information, innovation, wealth) typically scale superlinearly with city size, while basic needs (water and household energy consumption) scale linearly and transportation/distribution infrastructures scale sublinearly.
Distance decay is related to **friction of distance** which posits that in Geography, the *absolute distance* (say, in km) requires some amount of effort, money, time and/or energy to overcome. The corresponding cost is called *relative distance*; it describes the amount of social, cultural, or economic connectivity between two places.

- **Gravity models**
The general **gravity model** for social interaction is given by the *gravity equation*

$$F_{ij} = a\frac{M_i M_j}{D_{ij}^b},$$

where F_{ij} is the "flow" (or "gravitational attraction", *interaction, mass-distance function*) from location i to location j (alternatively, between those locations),

D_{ij} is the "distance" between i and j, M_i and M_j are the relevant economic "masses" of i and j, and a, b are parameters. Cf. Newton's **law of universal gravitation** in Chap. 24, where $b = 2$. The first instances were formulated by Reilly (1929), Stewart (1948), Isard (1956) and Tinbergen (1962).

If F_{ij} is a monetary flow (say, export values), then M is GDP (gross domestic product), and D_{ij} is the distance (usually the **great circle distance** between the centers of countries i and j). For trade, the true distances are different and selected by economic considerations. But the distance is a proxy for transportation cost, the time elapsed during shipment, cultural distance, and the costs of synchronization, communication, transaction. The **distance effect on trade** is measured by the parameter b; it is 0.94 in Head, 2003, and 0.6 in Leamer–Levinsohn, 1994.

If F_{ij} is a people (travel or migration) or message flow, then M is the population size, and D_{ij} is the travel or communication cost (distance, time, money).

If F_{ij} is the force of attraction from location i to location j (say, for a consumer, or for a criminal), then, usually $b = 2$. Reilly's *law of retail gravitation* is that, given a choice between two cities of sizes M_i, M_j and at distances D_i, D_j, a consumer tends to travel further to reach the larger city with the equilibrium point defined by

$$\frac{M_i}{D_i^2} = \frac{M_j}{D_j^2}.$$

- **Nearness principle**

  The **nearness principle** (or Zipf's *least effort principle*, in Psychology) is the following basic geographical heuristic: given a choice, a person will select the route requiring the least expenditure of effort, i.e., path of least resistance.

  This principle is used, for example, in transportation planning and locating of serial criminals: they tend to commit their crimes fairly close to where they live.

  The **first law of geography** (Tobler, 1970) is: "Everything is related to everything else, but near things are more related than distant things".
- **Distances in Criminology**

  Geographic profiling (or *geoforensic analysis*) aims to identify the spatial behavior (target selection and likely *offender's heaven*, i.e., the residence or workplace) of a serial criminal as it relates to the spatial distribution of linked crime sites.

  The **offender's buffer zone** is an area surrounding the offender's heaven, from which little or no criminal activity will be observed; usually, such a zone occurs for premeditated personal offenses. The primary streets and network arterials that lead into the buffer zone tend to intersect near the estimated offender's heaven. A 1 km buffer zone was found for UK serial rapists. Most personal offenses occur within about 2 km from the offender's heaven, while property thefts occur further away.

  Given n crime sites $(x_i, y_i), 1 \leq i \leq n$ (where x_i and y_i are the latitude and longitude of the i-th site), the *Newton–Swoope Model* predicts the offender's heaven to be within the circle around the point $(\frac{\sum_i x_i}{n}, \frac{\sum_i y_i}{n})$ with the search radius being

$$\sqrt{\frac{\max|x_{i_1} - x_{i_2}| \cdot \max|y_{i_1} - y_{i_2}|}{\pi(n-1)^2}},$$

where the maxima are over (i_1, i_2), $1 \le i_1 < i_2 \le n$. The *Ganter–Gregory Circle Model* predicts the offender's heaven to be within a circle around the first offense crime site with diameter the maximum distance between crime sites.
The *centrographic models* estimate the offender's heaven as a *center*, i.e., a point from which a given function of travel distances to all crime sites is minimized; the distances are the Euclidean distance, the Manhattan distance, the **wheel distance** (i.e., the actual travel path), perceived travel time, etc. Many of these models are the reverse of Location Theory models aiming to maximize the placement of distribution facilities in order to minimize travel costs. These models (*Voronoi polygons*, etc.) are based on the **nearness principle** (*least effort principle*).
The **journey-to-crime decay function** is a graphical **distance curve** used to represent how the number of offenses committed by an offender decreases as the distance from his/her residence increases. Such functions are variations of the center of gravity functions; cf. **gravity models**.
For detection of criminal, terrorist and other hidden networks, there are many data-mining techniques which extract latent associations (distances and **near-metrics** between people) from graphs of their co-occurrence in relevant documents, events, etc. In, say, drug cartel networks, better to remove *betweeners* (not well-connected bridges between groups, as paid police) instead of hubs (kingpins).
Electronic tagging consist of a device attached to a criminal or vehicle, allowing their whereabouts to be monitored using GPS. An *ankle monitor* (or *tether*) is a such tracking device that individuals under house arrest or parole are often required to wear. The *range of a tether* (or *inclusion zone*, 10–50 m) depends usually on the gravity of the crime; it is set by the offender's probation officer.

- **Drop distance**
  In judicial hanging, the **drop distance** is the distance the executed is allowed to fall. In order to reduce the prisoner's physical suffering (to about a third of a second), this distance is pre-determined, depending on his/her weight, by special *drop tables*. For example, the (US state) Delaware protocol prescribes, in pounds/feet, about 252, 183 and 152 cm for at most 55, 77 and at least 100 kg. Unrelated *hanging distance* is the minimum (horizontal) distance needed for hanging a hammock.
  In Biosystems Engineering, a ventilation jet *drop distance* is defined as the horizontal distance from an air inlet to the point where the jet reaches the occupational zone. In Aviation, an airlift *drop distance* (or *drop height*) is the vertical distance between the aircraft and the drop zone over which the airdrop is executed.

- **Distance telecommunication**
  Distance telecommunication is the transmission of signals over a distance for the purpose of communication. In modern times, this process almost always involves the use of electromagnetic waves by transmitters and receivers.
  Nonelectronic visual signals were sent by fires, beacons, smoke signals, then by mail, pigeon post, hydraulic semaphores, heliographs and, from the fifteenth century, by maritime flags, semaphore lines and signal lamps.
  Audio signals were sent by drums, horns (cf. **long-distance drumming** in Chap. 21) and, from nineteenth century, by telegraph, telephone, and radio.
  Advanced electrical/electronic signals are sent by television, videophone, fiber optical telecommunications, computer networking, analog cellular mobile phones, SMTP email, Internet and satellite phones.
- **Distance supervision**
  Distance supervision refers to the use of interactive distance technology (landline and cell phones, Email, chat, text messages to cell phone and instant messages, video teleconferencing, Web pages) for live (say, work, training, psychological umbrella, mental health worker, administrative) supervision.
  Such supervision requires tolerance for ambiguity when interacting in an environment that is devoid of nonverbal information.
- **Distance education**
  Distance education is the process of providing instruction when students and instructors are separated by physical distance, and technology is used to bridge the gap. *Distance learning* and *distance* (or *online*) degrees are the desired outcomes. A *semi-distance program* is one combining online and residential courses.
  The **transactional distance** (Moore, 1993) is a perceived degree of separation during interaction between students and teachers, and within each group. This distance decreases with *dialog* (a purposeful positive interaction meant to improve the understanding of the student), with larger autonomy of the learner, and with lesser predetermined structure of the instructional program.
  Vygotsky's *zone of proximal development* is the distance between the actual developmental level as determined by independent problem solving and the level of potential development as determined through problem solving under adult guidance, or in collaboration with more capable peers.
- **Distance selling**
  Distance selling, as opposed to face-to-face selling in shops, covers goods or services sold without face-to-face contact between supplier and consumer but through *distance communication means*: press adverts with order forms, catalog sales, telephone, tele-shopping, e-commerce (via Internet), m-commerce (via mobile phone). Examples of the relevant legislation are Consumer Protection (Distance Selling) Directive 97/7/EC and Regulations 2000 in EU.
  The main provisions are: clear prior information before the purchase, its confirmation in a durable medium, delivery within 30 days, "cooling-off" period of seven working days during which the consumer can cancel the contract without any reason and penalty. Exemptions are: *Distance marketing* (financial services

sold at distance), business-to-business contracts and some purchases (say, of land, or at an auction, or from vending machines).

- **Approximative human-scale distances**
  An **arm's length** is a distance (about 0.7 m, i.e., within **personal distance**) sufficient to exclude intimacy, i.e., discouraging familiarity or conflict; its analogs are: Italian *braccio*, Turkish *pik*, and Old Russian *sazhen*.
  The **reach distance** is the difference between the maximum reach and arm's length distance. The **striking distance** is a short, easily reachable distance.
  The **whiffing** (or **spitting**, **poking**) **distance** is a very close distance.
  A **stone's throw** is a distance of about 25 fathoms (46 m).
  The **hailing** (or **shouting**, **calling**) **distance** is the distance within which the human voice can be heard. **Far cry**: distance estimated in audibility's terms.
  The **walking distance** is the context-depending distance normally reachable by walking. In Japan, its standard unit is 80 m, i.e., 1 min walking time. Some UK high schools define 2 and 3 miles as the statutory walking distance for children younger and older than 11 years. Typical urban walking distance to transit is 400 m. *Pace out* means to measure distance by *pacing* (walking with even steps).
  The **acceptable commute distance**, in Real Estate, is the distance that can be covered in an acceptable travel time and increases with better connectivity.
- **Optimal eye-to-eye distance**
  The **optimal eye-to-eye distance** between two persons was measured for some types of interaction. For example, such an optimal viewing distance between a baby and its mother's face, with respect to the immature motor and visual systems of the newborn, is 20–30 cm. During the first weeks of life the accommodation system does not yet function and the lens of the newborn is locked at the **focal distance** of about 19 cm. At ages 8–14 months, babies are judging distances well; they fear a distance with mother (separation anxiety stage) and strangers. Also, left-sided cradling/holding preference have been found in humans and great apes.
- **Language style matching**
  During conversation, texting, emailing, and other forms of interactive communication, people unconsciously mimic their partners' language use patterns.
  The *LSM* (language style matching) *score* of a dyad $(1, 2)$ of persons, with respect to a *function word type* k, $1 \le k \le 9$, is $LSM_i = 2\frac{\min(l_{1k}, l_{2k})}{l_{1k}+l_{2k}}$, where l_{ik} $(i = 1, 2)$ is the percentage of person i's words of type k. Each dyad's total *LSM* is the mean of its LSM_i across the nine types of *function words*: auxiliary verbs (say, *am*, *will*, *have*), articles, common adverbs (say, *hardly*, *often*), personal pronouns, indefinite pronouns, prepositions, negations, conjunctions (say, *and*, *but*) and quantifiers.
  LSM is high within the first 15–30 s of any interaction and is generally unconscious. Women use conjunctions at much higher rates.
  LSM predicts successful hostage negotiations (Taylor–Thomas, 2008), task group cohesiveness (Gonzales–Hancock–Pennebaker, 2010), and the formation and persistence of romantic relationships (Ireland et al., 2011).

However (Manson et al., 2013), the probability of diad's cooperation in a post-conversation one-shot prisoner's dilemma, is positively related, instead of LSM, to the convergence of their speech rates (mean syllable duration).

- **Distances between people**

  In [Hall69], four interpersonal bodily distances were introduced: the *intimate distance* for lovers, childrens, pets (from touching to 46 cm), the *personal-casual distance* for conversations among friends (46–120 cm), the *social-consultative distance* for conversations among acquaintances (1.2–3.7 m), and the *public distance* for public speaking (over 3.7 m). To each of those *proxemics distances*, there corresponds an intimacy/confidence degree and appropriated sound level.

  The distance which is appropriate for a given social situation depends on culture, gender and personal preference. For example, under Islamic law, proximity (being in the same room or secluded place) between a man and a woman is permitted only in the presence of their *mahram* (a spouse or anybody from the same sex or a pre-puberty person from the opposite sex). For an average westerner, personal space is about 70 cm in front, 40 cm behind and 60 cm on either side.

  In interaction between strangers, the interpersonal distance between women is smaller than between a woman and a man. The bonding hormone oxytocin discourages partnered (but not a single) men from getting close to a female stranger; they, if were given oxytocin, stayed 10–15 cm farther from the attractive woman.

  An example of other cues of nonverbal communication is given by angles of vision which individuals maintain while talking. The **people angular distance in a posture** is the spatial orientation, measured in degrees, of an individual's shoulders relative to those of another; the position of a speaker's upper body in relation to a listener's (for example, facing or angled away). Speaker uses 45° open position in order to make listener feel comfortable and direct body point in order to exert pressure. In general, to approach men directly from the front or women from behind is rude. Also, this distance reveals how one feels about people nearby: the upper body unwittingly angles away from disliked persons and during disagreement.

  Eye-contact decreases with spatial proximity. Persons stand closer to those whose eyes are shut. The *Steinzor effect* is the finding that members of leaderless discussion groups seated in circles, are most apt to address remarks to or to get responses from persons seated opposite or nearly opposite them, while in the presence of a strong leader, it happens with persons seated alongside or nearly alongside.

  Distancing behavior of people can be measured, for example, by the *stop distance* (when the subject stops an approach since she/he begins to feel uncomfortable), or by the *quotient of approach*, i.e., the percentage of moves made that reduce the interpersonal distance to all moves made.

  Humans and monkeys with amygdala lesions have much smaller than average preferred interpersonal distance. *Peripersonal*, i.e., within reach of any limb of an individual, *space* is located dorsally in the parietal lobe whereas *extrapersonal* (outside his reach) *space* is located ventrally in the temporal lobe.

- **Death of Distance**

  Death of Distance is the title of the influential book [Cair01] arguing that the telecommunication revolution (the Internet, mobile telephones, digital TV, etc.) initiated the "death of distance" implying fundamental changes: three-shift work, lower taxes, prominence of English, outsourcing, new ways of government control and citizens communication, but also management-at-a-distance and concentration of elites within the "latte belt". Physical distance (and so, Economic Geography) do not matter; we all live in a "global village". Friedman, 2005, announced: "The world is flat". Gates, 2006, claimed: "With the Internet having connected the world together, someone's opportunity is not determined by geography". The proportion of long-distance relationships in foreign relations increased.

  Similarly (see [Ferg03]), steam-powered ships and the telegraph (as railroads previously and cars later) led, via falling transportation/communication costs, to the "annihilation of distance" in the nineteenth and twentieth centuries. Heine wrote in 1843: "Space is killed by the railways, and we are left with time alone". Further in the past, archaeological evidence points out the appearance of long-distance trade ($\approx$0.14 Ma ago), and the innovation of projectile weapons and traps ($\approx$0.04 Ma ago) which allowed humans to kill large game (and other humans) from a safe distance.

  But already Orwell, 1944, dismissed as "shallowly optimistic" the the phrases "airplane and radio have abolished distance" and "all parts of the world are now interdependent". Heidegger wrote in 1950: "All distances in time and space are shrinking...The peak of this abolition of every possibility of remoteness is reached by television..." but "The frank abolition of all distances brings no nearness". Edgerton, 2006, claims that new technologies foster self-sufficiency and isolation instead.

  Modern technology eclipsed distance only in that the time to reach a destination has (usually) shrunk. Distances still matter for, say, a company's strategy on the emerging markets (cf. **impact of distance on trade**) and for political legitimacy. "Tyranny of distance" still affects small island states in the Pacific.

  Partridge et al., 2007, report that proximity to higher-tiered urban centers (with their higher-order services, urban amenities, higher-paying jobs, lower-cost products) increasingly favors local job growth. Increased access to services and knowledge exchange requires more face-to-face interaction and so, an increase in the role of distance. Economic and innovation activity are highly localized spatially and tend to agglomerate more. Also, the social influence of individuals, measured by the frequency of memorable interactions, is heavily determined by distance. Goldenberg–Levy, 2009, show that the IT (Information Technology) revolution which occurred in the 1990s, increased local social interactions (as email, Facebook communication, baby name diffusion) to a greater degree than long-distance ones.

In military affairs, Boulding, 1965, and Bandow, 2004, argued that twentieth century technology reduced the value of proximity for the projection of military power because of "a very substantial diminution in the cost of transportation of armed forces" and "an enormous increase in the range of the deadly projectile". It was used as partial justification for the withdrawal of US forces from overseas bases in 2004. But Webb, 2007, counter-argues that any easing of transport is countered by increased strain put upon its modes since both sides will take advantage of the falling costs to send more supplies. Also, the greatest movement of logistics continues to be conducted by sea, with little improvement in speed since 1900.

- **Technology-related distancing**
  The *Moral Distancing Hypothesis* postulates that technology increases the propensity for unethical conduct by creating a **moral distance** between an act and the moral responsibility for it.
  Print technologies divided people into separate communication systems and distanced them from face-to-face response, sound and touch. TV involved audile-tactile senses and made distance less inhibiting, but it exacerbated *cognitive distancing*: story and image are biased against space/place and time/memory.
  This distancing has not diminished with computers but interactivity has increased. In terms of Hunter: technology only re-articulates *communication distance*, because it also must be regarded as the space between understanding and not. The collapsing of spatial barriers diminishes economic but not social and cognitive distance.
  The *Psychological Distancing Model* in [Well86] relates the immediacy of communication to the number of information channels: sensory modalities decrease progressively as one moves from face-to-face to telephone, videophone, and e-mail. Skype communication is rated higher than phone since it creates a sense of co-presence. People phone with bad news but text with good news.
  Online settings tend to filter out social and relational cues. The lack of instant feedback (since e-mail communication is asynchronous) and low bandwidth limit visual/aural cues. For example, moral and cognitive effects of distancing in online education are not known at present. Also, the shift from face-to-face to online communication can diminish social skills, lead to smaller and more fragmented networks and so, increase feeling of isolation and alienation. But it can be only a bias, based on traditional spatiotemporal assumptions that farness translates into an increase in mediation and results in blurring of the communication.
  Virtual distance is the perceived distance between individuals when their primary way of communication is not face-to-face. The main markers of virtual distance are physical, operational and affinity distances.
  Mejias, 2005, define *epistemological distance* and *ontological distance* between things as the difference, respectively, in degree of knowledge justification and in ability of subjects to act upon things. He argue that we should strive towards ontological nearness, using modern information and communication technologies to manipulate temporal/spatial and epistemological distances to attain this goal.

Mejias, 2007, see some new advantages in "uniform distancelessness", deplored by Heidegger. *Networked proximity* (nearness mediated through new technology) provides shift from physical proximity to informational availability as the main measure of social relevance. It facilitates new kinds of spatially unbound community, and these emerging forms of sociality could be no less meaningful than the older ones. Networked sociality reconfigures distance rather than eliminates it.

28.3 Distances in Sociology and Language

- **Sociometric distance**
  The **sociometric distance** refers to some measurable degree of mutual or social perception, acceptance, and understanding. Hypothetically, greater sociometric distance is associated with more inaccuracy in evaluating a relationship.
- **Social distance**
  In Sociology, the **social distance** is the extent to which individuals or groups are removed or excluded from participating in one another's lives; a degree of understanding and intimacy which characterize personal and social relations generally. This notion was originated by Simmel in 1903; in his view, the social forms are the stable outcomes of distances interposed between subject and object. For example (Mulgan, 1991), the centers of global cities are socially closer to each other than to their own peripheries. In general, the notion of social distance is conceptualized in affective, normative or interactive way, i.e., in terms of sympathy the members of a group feel for another group, norms to define in- and outsider, or the frequency/intensity of interactions between two groups.
  The *Social Distance Scale* by Bogardus, 1947, offers the following response items: would marry, would have as a guest in my household, would have as next door neighbor, would have in neighborhood, would keep in the same town, would keep out of my town, would exile, would kill; cf. **emotional distance**. The responses for each (say, ethnic/racial) group are averaged across all respondents which yields (say, racial) distance quotient ranging from 1.00 to 8.00.
  Dodd and Nehnevasja, 1954, attached distances of 10^t m, $0 \leq t \leq 7$, to eight levels of the Bogardus scale. Many studies on conflicts in ex-Yugoslavia consider *ethnic distance* defined via some modified Bogardus scale, i.e., in terms of acceptance of a particular relation with an abstract person from the other group. Caselli and Coleman, 2012, defined *ethnic distance* as the cost to be born by a member of one group to successfully pass himself as a member of the other group.
  An example of relevant models: Akerlof [Aker97] defines an *agent* x as a pair (x_1, x_2) of numbers, where x_1 represents the initial, i.e., inherited, social position, and the position expected to be acquired, x_2. The agent x chooses the value x_2 so as to maximize

$$f(x_1) + \sum_{y \neq x} \frac{e}{(h + |x_1 - y_1|)(g + |x_2 - y_1|)},$$

where e, h, g are parameters, $f(x_1)$ represents the intrinsic value of x, and $|x_1 - y_1|$, $|x_2 - y_1|$ are the inherited and acquired *social distances* of x from any agent y (with the social position y_1) of the given society.
Hoffman, Cabe and Smith, 1996, define *social distance* as the degree of reciprocity that subjects believe exists within a social interaction.

- **Rummel sociocultural distances**
  [Rumm76] defined the main sociocultural distances between two persons as follows.

  - **Personal distance**: one at which people begin to encroach on each other's territory of personal space.
  - **Psychological distance**: perceived difference in motivation, temperaments, abilities, moods, and states (subsuming **intellectual distance**).
  - **Interests-distance**: perceived difference in wants, means, and goals (including **ideological distance** on socio-political programs).
  - **Affine distance**: degree of sympathy, liking or affection between the two.
  - **Social attributes distance**: differences in income, education, race, sex, etc.
  - **Status-distance**: differences in wealth, power, and prestige (including **power distance**).
  - **Class-distance**: degree to which one person is in general authoritatively superordinate to the other.
  - **Cultural distance**: differences in meanings, values and norms reflected in differences in philosophy-religion, science, ethics, law, language, and fine arts.

- **Cultural distance**
  The **cultural distance between countries** $x = (x_1, \dots, x_5)$ and $y = (y_1, \dots, y_5)$ (usually, US) is derived (in [KoSi88]) as the following composite index

$$\sum_{i=1}^{5} \frac{(x_i - y_i)^2}{5V_i},$$

where V_i is the variance of the index i, and the five indices represent [Hofs80]:

1. Power distance (preferences for equality);
2. Uncertainty avoidance (risk aversion);
3. Individualism versus collectivism;
4. Masculinity versus femininity (gender specialization);
5. Confucian dynamism (long-term versus short-term orientation).

The above **power distance** measures the extent to which the less powerful members of institutions and organizations within a country expect and accept that power is distributed unequally, i.e., how much a culture has respect for authority. For example, Latin Europe and Japan fall in the middle range.

But for Shenkar–Luo–Yeheskel, 2008, above distance is merely a measure of how much a country strayed from the core culture of the multinational enterprise. They propose instead (especially, as a regional construct) the *cultural friction* linking goal incongruity and the nature of cultural interaction.

In order to explain multinational enterprise behavior, Kostova, 1999, introduced the *institutional distance* between its home and host countries as the difference in their regulative, cognitive, and normative institutions. Habib–Zurawicki, 2002, consider effects of the *corruption distance*, i.e., such difference in corruption levels.

Wirsing, 1973, defined *social distance* as a "symbolic gap" between rulers and ruled designed to set apart the political elite from the public. It consists of reinforced and validated ideologies (a formal constitution, a historical myth, etc.). Davis, 1999, theorized social movements (in Latin America) in terms of their shared distance from the state: geographically, institutionally, socially (class position and income level) and culturally.

The Inglehart–Welzel *cultural map of the world* represent countries as points on $\mathbb{R}^2$, in which the two dimensions (*traditional/secular-rational* and *survival/self-expression*) explain >70 % of cross-national variance in 10 indicators.

- **Political distance**

  A finite metric space $(X = \{x_1, \dots, x_n\}, d)$ can be seen as a *political space* with the points and distance seen as positions (policy proposals) and some **ideological distance**, respectively. Usually, (X, d) is a subspace of $([0,1]^m, ||x - y||_2)$.

  Let $\{v_1, \dots, v_n\}$ be the vote shares of all candidates $\{c_1, \dots, c_n\}$ of an election or, say, allocated seat shares of all competing parties; so, $\sum_{i=1}^{n} v_i = 1$. The *index of political diversity* (Ortuño-Ortin and Weber, 2008) is $\sum_{1 \le i < j \le n} v_i v_j d(x_i, x_j)$.

  The **mean minimum political distance**, cf. http://wiki.electorama.com/wiki, is (the case $m = 1$ of) $\sum_{i=1}^{n} v_i \min_{j \in E} d(x_i, y)$, where $E = \{1 \le i \le n : c_i \text{ is elected}\}$. Cf. **distance-rationalizable voting rule** in Chap. 11.

- **Surname distance model**

  A **surname distance model** was used in [COR05] in order to estimate the preference transmission from parents to children by comparing, for 47 provinces of mainland Spain, the 47×47 distance matrices for **surname distance** with those of **consumption distance** and **cultural distance**.

  The distances were l_1-distances $\sum_i |x_i - y_i|$ between the frequency vectors (x_i), (y_i) of provinces x, y, where z_i is, for the province z, either the frequency of the i-th surname (**surname distance**), or the budget share of the i-th product (**consumption distance**), or the population rate for the i-th cultural issue, say, rate of weddings, newspaper readership, etc. (**cultural distance**), respectively.

  Other (matrices of) distances considered there are:

  - *geographical distance* (in km, between the capitals of two provinces);
  - *income distance* $|m(x) - m(y)|$, where $m(z)$ is mean income in the province z;

- *climatic distance* $\sum_{1 \le i \le 12} |x_i - y_i|$, where z_i is the average temperature in the province z during the i-th month;
- *migration distance* $\sum_{1 \le i \le 47} |x_i - y_i|$, where z_i is the percentage of people (living in the province z) born in the province i.

Strong *vertical preference transmission*, i.e., correlation between surname and consumption distances, was detected only for food items.

- **Distance as a metaphor**
  Lakoff and Núñez, 2000, claim that mathematics emerged via conceptual metaphors grounded in the human body, its motion through space and time, and in human sense perceptions: change is motion, arithmetic is motion along a path, etc.
  For them, the mathematical idea of distance comes from the activity of measuring, and the corresponding technique consists of rational numbers and metric spaces. The idea of proximity/connection comes from connecting and corresponds to topological space. The idea of subtraction mathematizes the ordinary idea of distance.
- **Metaphoric distance**
  A **metaphoric distance** is any notion in which a degree of similarity between two difficult-to-compare things is expressed using spatial notion of distance as an implicit bidirectional and understandable metaphor. Some examples are:
  Internet and Web bring people closer: proximity in subjective space is at-handiness;
  professional distance: teacher–student, therapist–patient, manager–employee;
  financial distance: degree of separation in couple's money/property arrangements;
  competitive distance (incomparability) between two airline product offerings;
  metaphoric distance that a creative thinker takes from the problem, i.e., degree of intuitivity, required to evolve/reshape concepts into new ideas.
  The **distance-similarity metaphor** (Montello et al., 2003) is a design principle, where relatedness in nonspatial data is projected onto distance, so that semantically similar documents are placed closer to one another in an information space. It is the inverse of the Tobler's **first law of geography**; cf. **nearness principle**. This metaphor is used in Data Mining, Pattern Recognition and **Spatialization**.
  Comparing the linguistic metaphor *proximity→similarity* with its mental counterpart, Casasanto (2008), found that stimuli (pairs of words or pictures) presented closer together on the computer screen were rated more similar in conceptual judgments of abstract entities or unseen object properties but, less similar in perceptual judgments of visual appearance of faces and objects.
- **Spatial cognition**
  Spatial cognition concerns the knowledge about spatial properties of objects and events: location, size, distance, direction, separation/connection, shape, pattern, and movement. For instance, it consider navigation (locomotion and way-finding) and orientation during it: recognition of landmarks and *path*

integration (an internal measuring/computing process of integrating information about movement).
Spatial cognition addresses also our (spatial) understanding of the World Wide Web and computer-simulated virtual reality.
Men surpass women on tests of spatial relations, mental rotation and targeting, while women have better fine motor skills and spatial memory for immobile objects and their location. Such selection should come from a division of labor in Pleistocene groups: hunting of mobile prey for men and gathering of immobile plant foods for women. Women's brains are 10–15 % smaller than men's, but their frontal lobe (decision-making, problem-solving), limbic cortex (emotion regulation) and hippocampus (spatial memory) are proportionally larger, while the parietal cortex (spatial perception) and amygdala (emotional memory) are smaller. Men's brains contain stronger front-to-rear connections (suggesting greater synergy between perception and action) while those of women are better connected from left to right (facilitating emotional processing and the ability to infer others' intentions).
One of the *cultural universals* (traits common to all human cultures) is that men on average travel greater distances over their lifetime. They are less likely than women to migrate within the country of their birth but more likely to emigrate.

- **Size representation**
  Konkle and Oliva, 2012, found that object representations is differentiated along the ventral temporal cortex by their real-world size. Both big and small objects activated most of temporal cortex but fMRI voxels with a big- or small-object preference were consistently found along its medial or, respectively, lateral parts. These parts overlapped with the regions known to be active when identifying spaces to interact with (say, streets, elevators, cars, chairs) or, respectively, processing information on tools, ones we usually pick up.
  Different-sized objects have different action demands and typical interaction distances. Big/small preferences are object-based rather than retinotopic or conceptual. They may derive from systematic biases, say, eccentricity biases and size-dependent biases in the perceptual input and in functional requirements for action. For example, over the viewing experience, in the lifetime or over evolutionary time, the smaller objects tend to be rounder, while larger objects tend to extend more peripherally on the retina. Cf. the **size-distance invariance hypothesis** and, in Chap. 29, **neurons with spatial firing properties**.
- **Spatialization**
  Spatialization (Lefebvre, 1991) refer to the spatial forms that social activities and material things, phenomena or processes take on. It includes cognitive maps, cartography, everyday practice and imagination of possible spatial worlds.
  One of the debated definitions of consciousness: it is a notion of self in space and an ability to make decisions based on previous experience and the current situation. Self-awareness permits to evaluate the distance that separates one from his objectives and to adjust his behavior in order to approach those aims.
  We usually give the upper face or upper torso as *egocenter* (spatial seat of self).
  The term *spatialization* is also used for information display of nonspatial data.

- **Spatial reasoning**
  Spatial reasoning is the domain of spatial knowledge representation: spatial relations between spatial entities and reasoning based on these entities and relations.
  As a modality of human thought, spatial reasoning is a process of forming ideas through the spatial relationships between objects (as in Geometry), while verbal reasoning is the process of forming ideas by assembling symbols into meaningful sequences (as in Language, Algebra, Programming). *Spatial intelligence* is the ability to comprehend 2D and 3D images and shapes.
  Luria, 1973, called the ability to derive the abstract concepts from spatially organized heteromodal information, the *quasi-spatial synthesis*.
  Spatial–temporal reasoning (or *spatial ability*) is the capacity to visualize spatial patterns, to manipulate them mentally over a time-ordered sequence of spatial transformations and to draw conclusions about them from limited information.
  Specifically, *spatial visualization ability* is the ability to manipulate mentally 2D and 3D figures. *Spatial skills* is the ability to locate objects in a 3D-world using sight or touch. *Spatial acuity* is the ability to discriminate two closely-separated points or shapes (say, two similar polygons with different numbers of sides).
  Visual thinking (or *visual/spatial learning*, *picture thinking*) is the common (about 60 % of the general population) phenomenon of thinking through visual processing. Spatial–temporal reasoning is prominent among visual thinkers, as well as among *kinesthetic learners* (who learn through body mapping and physical patterning) and *logical thinkers* (mathematical/systems thinking) who think in patterns and relationships and may work without this being pictorially.
  In Computer Science, spatial–temporal reasoning aims at describing, using abstract relation algebras, the common-sense background knowledge on which human perspective of physical reality is based. It provides rather inexpensive reasoning about entities located in space and time.
- **Spatial language**
  Spatial language consists of natural-language spatial relations used to indicate where things are, and so to identify or refer to them. It usually expresses imprecise and context-dependent information about space.
  Among spatial relations there are *topological* (such as on, to, in, inside, at), *path-related* (such as across, through, along, around), *distance-related* and more complex ones (such as right/left, between, opposite, back of, south of, surround).
  A **distance relation** is a spatial relation which specifies how far the object is away from the reference object: near, far, close, etc.
  The **distance concept of proximity** (Pribbenow, 1992) is the area around the RO (reference object) in which it can be used for localization of the LO (local object), so that there is visual access from RO and noninterruption of the spatial region between objects, while LO is less directly related to a different object. Such proximity can differ with physical distance as, for example, in "The Morning Star is to the left of the church". The area around RO, in which a particular relation

is accepted as a valid description of the distance between objects, is called the *acceptance area*.
Pribbenow, 1991, proposed five distance distinctions: *inclusion* (acceptance area restricted to projection of RO), *contact/adjacency* (immediate neighborhood of RO), *proximity*, *geodistance* (surroundings of RO) and *remoteness* (the complement of the proximal region around RO).
For Jackendorff–Landau, 1992, there are three degrees of distance distinctions in English: interior of RO (*in*, *inside*), exterior but in contact (*on*, *against*), proximate (*near*), plus corresponding negatives, such as *outside*, *off of*, *far from*. A spatial reference system is mainly egocentric, or relative (such as *right*, *back*) for the languages spoken in industrialized societies, while the languages spoken in small scale societies rely rather on an allocentric, or absolute set of coordinates.
Semantics of spatial language is considered in Spatial Cognition, Linguistics, Cognitive Psychology, Anatomy, Robotics, Artificial Intelligence and Computer Vision. Cognitively based common-sense spatial ontology and metric details of spatial language are modeled for eventual interaction between Geographic Information Systems and users. An example of far-reaching applications is Grove's *clean space*, a neuro-linguistic psychotherapy based on the spatial metaphors produced by (or extracted from) the client on his present and desired "space" (state).

- **Language distance from English**
  Such measures are based either on a typology (comparing formal similarities between languages), or language trees, or performance (mutual intelligibility and learnability of languages). For example, Rutheford, 1983, defined **distance from English** as the number of differences from English in the following three-way typological classification: subject/verb/object order, topic-prominence/subject-prominence and pragmatic word-order/grammatical word-order. It gives distances $1, 2, 3$ for Spanish, Arabic/Mandarin, Japanese/Korean.
  Borland, 1983, compared several languages of immigrants by their acquisition of four areas of English syntax: copula, predicate complementation, negation and articles. The resulting ranking was English, Spanish, Russian, Arabic, Vietnamese.
  Elder–Davies, 1998, used ranking based on the following three main types of languages: isolating, analytic or root (as Chinese, Vietnamese), inflecting, synthetic or fusional (as Arabic, Latin, Greek), agglutinating (as Turkish, Japanese). It gave ranks $1, 2, 4, 5$ for Romance, Slavic, Vietnamese/Khmer, Japanese/Korean, respectively, and 3 for Chinese, Arabic, Indonesian, Malay.
  The **language distance index** (Chiswick–Miller, 1998) is the inverse of the *language score* of the average speaking proficiency, after 24 weeks of instruction, of English speakers learning this language. This score was measured at regular intervals by increments of 0.25; it ranges from 1 (hardest to learn) to 3 (easiest to learn). The score was, for example, $1.00, 1.25, 1.50, 1.75, 2.00, 2.25, 2.50, 2.75, 3.00$ for Japanese, Cantonese, Mandarin, Hindi, Hebrew, Russian, French, Dutch, Afrikaans.

In addition to the above distances, based on syntax, and **linguistic distance**, based on pronunciation, see the lexical semantic distances in Chap. 22.
Cf. **clarity similarity** in Chap. 14, **distances between rhythms** in Chap. 21, **Lasker distance** in Chap. 23 and **surname distance model** in Chap. 28.
Translations of the English noun *distance*, for example, into French, Italian, German, Swedish, Spanish, Interlingua, Esperanto are: distance, distanza, distanz, distans, distancia, distantia, distanco.
The word *distance* has Nr. 625 in the list (Wiktionary:Frequency lists/PG/2006/04) of the common English words in the books found on Project Gutenberg. It has Nrs. 835, 1035, 2404 in contemporary poetry, fiction, TV/movie and overall Nrs. 1513 (written), 1546 (spoken). It comes from Latin *distantia* (distance, farness, difference) and *distans*, present participle of *distare*: *di* (apart) + *stare* (to stand).
The **longest English word** (noncoined and nontechnical) is *antidisestablishmentarianism* (28 letters). Examples of funny distance-related words in *Urban Dictionary* (Web-based dictionary of slang in English) are: *distading* (start and give up on many goals in quick succession), *distarnated* (having no friends and being hated by everyone), *distanitus* (illness one suffer from spotting a person which looks really good from across the room but is a butterface at 5 feet distance), *distance* (space provided when someone is *dissing*, i.e., show disrespect for, someone else).

- **Editex distance**

  The main phonetic encoding algorithms are (based on English language pronunciation) *Soundex*, *Phonix* and *Phonex*, converting words into one-letter three-digit codes. The letter is the first one in the word and the three digits are derived using an assignment of numbers to other word letters. Soundex and Phonex assign:
  0 to *a, e, h, i, o, u, w, y*; 1 to *b, p, f, v*; 2 to *c, g, j, k, q, s, x, z*; 3 to *d, t*; 4 to *l*; 5 to *m, n*; 6 to *r*.
  Phonix assigns the same numbers, except for 7 (instead of 1) to *f* and *v*, and 8 (instead of 2) to *s, x, z*.
  The **Editex distance** (Zobel–Dart, 1996) between two words x and y is a cost-based **editing metric** (i.e., the minimal cost of transforming x into y by substitution, deletion and insertion of letters). For substitutions, the costs are 0 if two letters are the same, 1 if they are in the same letter group, and 2, otherwise.
  The *syllabic alignment distance* (Gong–Chan, 2006) between two words x and y is another cost-based **editing metric**. It is based on Phonix, the identification of syllable starting characters and seven edit operations.

- **Phone distances**

  A *phone* is a sound segment having distinct acoustic properties, and is the basic sound unit. A *phoneme* is a minimal distinctive feature/unit in the language (a set of phones which are perceived as equivalent to each other in a given language).
  The number of phonemes (consonants) range, among about 6,000 languages spoken now, from 11 (6) in Rotokas to 112 (77) in Taa (languages spoken by about 4,000 people in Papua New Guinea and Botswana, respectively).
  The main classes of the **phone distances** (between two phones x and y) are:

– *Spectrogram-based distances* which are physical-acoustic distortion measures between the sound spectrograms of x and y;
– *Feature-based phone distances* which are usually the **Manhattan distance** $\sum_i |x_i - y_i|$ between vectors (x_i) and (y_i) representing phones x and y with respect to a given inventory of phonetic features (for example, nasality, stricture, palatalization, rounding, syllabicity).

The **Laver consonant distance** refers to the improbability of confusing 22 consonants among ≈ 50 phonemes of English, developed by Laver, 1994, from subjective auditory impressions. (Chen–Wang–Jia–Dang, 2013, considered similar perception distance between two types of Chinese initials.) The smallest distance, 15 %, is between phonemes [*p*] and [*k*], the largest one, 95 %, is, for example, between [*p*] and [*z*]. Laver also proposed a quasi-distance based on the likelihood that one consonant will be misheard as another by an automatic speech-recognition system.

Each vowel can be represented by a pair (F_1, F_2) of resonant frequencies of the vocal tract (first and second formants). For example, [*u*], [*a*], [*i*] are represented by (350, 800), (850, 1150), (350, 1700) in *mels* (cf. Chap. 21). The International Phonetic Alphabet identifies 7 levels of *height* (F_1) and 5 levels of *backness* (F_2).

Chang et al., 2013, produced English language map of the brain; they found the set of neurons in the sensorimotor cortex which controls muscles (in the tongue, lips, jaw, larynx) and fires in unique combination for each sound.

- **Phonetic word distance**

  The **phonetic word** (or *pronunciation*, *Levenstein phonological*) **distance** between two words x and y seen as strings of phones is the **Levenstein metric** with costs (cf. Chap. 11): the minimal cost of transforming x into y by substitution, deletion and insertion of phones. Given a **phone distance** $r(u, v)$ on the International Phonetic Alphabet with the additional phone 0 (silence), the cost of substitution of phone u by v is $r(u, v)$, while $r(u, 0)$ is the cost of insertion or deletion of u.

  Levenstein orthographic distance (or *LPD*) is the same measure, but operating on letters instead of phonemes. Words with larger mean LPD to (but smaller mean frequency of) its 20 closest neighbors are easier to recognize.

  The average adult has a vocabulary of about 40,000–50,000 words.

- **Linguistic distance**

  The **linguistic distance** between two languages is a term loosely used to describe their difference. The mutual intelligibility of the two languages depends on the degree of their lexical, phonetical, morphological, and syntactical similarity.

  The **lexical similarity** is the percentage of *common* (similar in form and meaning) words in their standardized wordlists. English was evaluated to have a lexical similarity of 60 %, 27 %, 24 % with German, French and Russian, respectively. Cf. **language distance index**, **language distance effect**, **Swadesh similarity**.

  Specifically, the **linguistic (dialectal) distance** between language varieties X and Y is the mean, for a fixed sample S of notions, **phonetic word distance** between

cognate (i.e., having the same meaning) words s_X and s_Y, representing the same notion $s \in S$ in X and Y, respectively.
One example of a *dialect continuum* (as *ring species* in Biology) is Dutch-German: their mutual intelligibility is small but a chain of dialects connects them.

- **Swadesh similarity**
The *Swadesh word list* of a language (Swadesh, 1940–1950) is a list of vocabulary with (usually, 100) basic words which are acquired by the native speakers in early childhood and supposed to change very slowly over time. The **Swadesh similarity** between two languages is the percentage of *cognate* (having similar meaning and sound) words in their Swadesh lists. *Glottochronology* is a method of assessing the temporal divergence of two languages based on this similarity.
The first 12 items of the original final Swadesh list: *I, you, we, this, that, who?, what?, all, many, one, two*. Cf. the first 12 most frequently used English words: *the, of, and, a, to, in, is, you, that, it, he, was* in all printed material and *I, the, and, to, a, of, that, in, it, my, is, you* across both spoken and written texts.
Acerbi et al., 2013: the frequency of emotional words declined in English-language books over twentieth century, but the use of fear-related words increased from 1980. The *half-life* of a word is the number of years after which it has a 50 % probability of having been replaced by a new noncognate word; roughly, it is within 750–20,000 years, say, 9,000, 3,200, 1,900. *stab, bird, we*.
Pagel et al., 2013, suggest existence of a *proto-Eurasian mother tongue*. For example, they list 15,000 years old words cognate in at least 4 Eurasiatic language families: thou, I, not, that, we, to give, who, this, what, man/male, ye, old, mother, to hear, hand, fire, to pull, black, to flow, bark, ashes, to spit, worm.
- **Language distance effect**
In Foreign Language Learning, Corder, 1981, conjectured the existence of the following **language distance effect**: where the mother tongue (L1) is structurally similar to the target language, the learner will pass more rapidly along the developmental continuum (or some parts of it) than where it differs; moreover, all previous learned languages have a facilitating effect.
Ringbom, 1987, added: the influence of the L1 is stronger at early stages of learning, at lower levels of proficiency and in more communicative tasks. But such correlation could be indirect. For example, the written form of modern Chinese does not vary among the regions of China, but the spoken languages differ sharply, while spoken German and Yiddish are close but have different alphabets.
- **Long-distance dependence (in Language)**
In Language, **long-distance dependence** (or *syntactic binding*) is a construction, including wh-questions (such as "Who do you think he likes"), topicalizations (such as "Mary, he likes"), *easy*-adjectives (such as "Mary is easy to talk to"), relative clauses (such as "I saw the woman who I think he likes")—which permits an element in one position (*filler*) to fulfill the grammatical role associated with another nonadjacent position (*gap*). The *filler-gap distance*, in terms of

the number of intervening clauses or words between them in a sentence, can be arbitrary large. Cf. **long range dependence** in Chap. 18.
An *anaphora* is a subsequent reference to an entity already introduced in discourse. In order to be interpreted, anaphora must get its content from an antecedent in the sentence which in English is usually syntactically local as in "Mary excused *herself*". A **long-distance anaphora** is an anaphora with antecedent outside of its local domain, as in "The players told us stories about *each other*". Its resolution (finding what it refers to) is a hard problem of machine translation.
The **anaphoric distance** is (Ariel, 1990) the number of words between an anaphora and its antecedent. The **referential distance** (or **textual distance**) is (Givón, 1983) the amount of clauses between them. In general, each text can be represented as a tree in which *discourse units* (normally, clauses) are vertices and *rhetorical relations* (sequence, joint, cause, elaboration, etc.) are edges.
The **rhetorical distance** between two discourse units is (Fox, 1987) the minimal number of "sequence"-edges on a path between them.

28.4 Distances in Philosophy, Religion and Art

- **Zeno's distance dichotomy paradox**
This paradox by the pre-Socratic Greek philosopher Zeno of Elea claims that it is impossible to cover any distance, because half the distance must be traversed first, then half the remaining distance, then again half of what remains, and so on.
The paradoxical conclusion is that travel over any finite distance can neither be completed nor begun, and so all motion must be an illusion.
But, in fact, dividing a finite distance into an infinite series of small distances and then adding the all together gives back the finite distance one started with.
- **Space (in Philosophy)**
The present Newton–Einstein notion of **space** was preceded by Democritus's (c. 460–370 BC) *Void* (the infinite container of objects), Plato's (c. 424–348 BC) *Khora* (an interval between being and nonbeing in which Forms materialize) and Aristotle's (380–322 BC) *Cosmos* (a finite system of relations between objects). Cf. **Minkowski metric** (Chap. 26) for the origin of the space-time concept.
Like the Hindu doctrines of Vedanta, Spinoza (1632–77) saw our Universe as a mode under two (among an infinity of) attributes, *Thought* and *Extension*, of *God* (unique absolutely infinite, eternal, self-caused substance, without personality and consciousness). These parallel (but without causal interaction) attributes define how substance can be understood: to be composed of thoughts and physically extended in space, i.e., to have breadth and depth. So, the Universe is deterministic.
For Newton (1642–1727) space was absolute: it existed permanently and independently of whether there is any matter in it. It is a framework of creation,

stage setting within which physical phenomena occur. For Leibniz's (1646–1716) space was a collection of relations between objects, given by their distance and direction from one another, i.e., an idealized abstraction from the relations between individual entities or their possible locations which must therefore be discrete.

For Kant (1724–1804) space is not substance or relation, but a part of an unavoidable systematic framework used by the humans to organize their experiences. Disagreement continues between philosophers over whether space is an entity, a relationship between entities, or part of a conceptual framework.

In *biocentic cosmology* (Lanza, 2007), build on quantum physics, space is a form of our animal understanding and does not have an observer-independent reality, while time is the process by which we perceive changes in the Universe. Also, space-time could be not fundamental, but emerging from a deeper quantum reality.

Free space refers to a perfect vacuum, devoid of all particles; it is excluded by the uncertainty principle. The quantum vacuum is devoid of atoms but contains subatomic short-lived particles—photons, gravitons, etc.

A *parameter space* is the set of values of parameters in a mathematical model.

In Mathematics and Physics, the *phase space* (Gibbs, 1901) is a space in which all possible states of the system are represented as unique points; cf. Chap. 18.

- **Kristeva nonmetric space**

  Kristeva, 1980, considered the basic psychoanalytic distinction (by Freud) between pre-Oedipal and Oedipal aspects of personality development. Narcissistic identification and maternal dependency, anarchic component drives, polymorphic erotogenicism, and primary processes characterize the pre-Oedipal. Paternal competition and identification, specific drives, phallic erotogenicism, and secondary processes characterize Oedipal aspects.

  Kristeva describes the pre-Oedipal feminine phase by an enveloping, amorphous, "nonmetric" space (Plato's *khora*) that both nourishes and threatens; it also defines and limits self-identity. She characterizes the Oedipal male phase by a metric space (Aristotle's *topos*); the self and the self-to-space are more precise and well defined in topos. Kristeva insists also on the fact that the semiotic process is rooted in feminine libidinal, pre-Oedipal energy which needs channeling for social cohesion.

  Deleuze–Guattari, 1980, divide *multiplicities* (networks, manifolds, spaces) into *striated* (metric, hierarchical, centered, numerical) and *smooth* ("nonmetric, rhizomic, those that occupy space without counting it and can be explored only by legwork").

  The above French post-structuralists use the metaphor of *nonmetric* in line with a systematic (but generating controversy) use of topological terms by the psychoanalyst Lacan. In particular, he sought the space J (of *Jouissance*, i.e., sexual relations) as a metric space and used metaphorically the *Heine–Borel theorem* (that closed and bounded subspaces of Euclidean spaces are their only compact subspaces).

Back to Mathematics, when a notion, theorem or algorithm is extended from metric to general distance space, the latter is called **nonmetric space**.

- **Emerson distance between persons**
  We call the **Emerson distance between persons** the separation between "gods", which was required by an American poet and philosopher Ralph Waldo Emerson (1803–82) in his Essay 16 *Manners*: "Let the incommunicable objects of nature and the metaphysical isolation of man teach us independence... We should meet each morning, as from foreign countries, and spending the day together, should depart at night, as into foreign countries. In all things I would have the island of a man inviolate. Let us sit apart as the gods, talking from peak to peak all round Olympus... Lovers should guard their strangeness... Every natural function can be dignified by deliberation and privacy." At the end of his 1836 book *Nature*, Emerson also wrote: "Every spirit builds itself a house; and beyond its house, a world; and beyond its world a heaven... Build, therefore, your own world."
  Similar dignified separation is mentioned in quotes from the Russian philosopher Mikhail Bakhtin (1895–1975): "The feeling of respect creates a distance, both in relation to the other person, and in relation to one's own self" and the Bohemian-Austrian poet Rainer Maria Rilke (1875–1926) wrote: "Once the realization is accepted that even between the closest human beings infinite distances continue, a wonderful living side by side can grow, if they succeed in loving the distance between them which makes it possible for each to see the other whole against the sky."
- **Nietzsche's Ariadne distance**
  The German philosopher Friedrich Wilhelm Nietzsche (1844–1900), treated distance in a sensual/erotic way. In "On the Genealogy of Morals" (1887) he wrote:
  "The pathos of nobility and distance... the fundamental total feeling on the part of a higher ruling nature in relation to a lower nature, to a 'beneath'—that is the origin of the opposition between 'good' and 'bad.'"
  His Zarathustra favors *fernstenliebe* (love of the farthest) over Christian love of the neighbor. Moreover, *fernstenliebe* is to love neither objects, nor ends—but rather, distance/endlessness itself, which makes all distances recur and perpetuate themselves.
  The courtly troubadours of the twelfth century valued eroticization of the unattainable object, while for German romanticism (for example, Novalis, Schopenhauer, Wagner) there can be no satisfaction in erotic relations, or in life itself, as long as distance remains. In Wagner's opera, Tristan laments: "Blessed nearness, tedious distance."
  Kuzma, 2013, claims that Nietzsche, by the early 1880s, "rehabilitated erotic distance", in response to its denigration and the consummatory idealism and passive nihilism of the German romantic tradition. This rehabilitation of courtly love culminated in the concept of an absolute separation and eternal recurrence.
  According to Kuzma, Ariadne in '*Thus Spoke Zarathustra* (1883–1885) is not only the symbol of the human soul and life, but Nietzsche's privileged name for absolute, infinite spatially and eternal distance itself, for an eternity conceived in

the absence of every end, any possible object to attain and every Other to love. To desire Ariadne, is to desire the incessant prolongation of longing in the absence of all fulfilment. Zarathustra does not seek rest, consummation, and release, but affirms a sort of metaphysical *coitus reservatus*, the eternal prolongation of boundless and unresolved desire, implying "voluptuousness of the future" and "love of fate".
The eternal recurrence requires spatial or temporal infinity. Nietzsche, in his posthumous notes, posits finite matter and infinite cyclical time.

- **Heidegger's de-severance distance**
The German philosopher Martin Heidegger (1889–1976), sought space in terms of limit and event of placing, not merely a location. He wrote: "space is something that has been spaced, or made room for, and that which is let into its bounds".
His main notion, *Dasein* (Being there), means Being-in-the-world, as opposed to the Cartesian abstract agent, a subject, or the objective world alone. Dasein is revealed by projection into, and engagement with, a personal world, one's environment. It is *ontically* (in factual existence) closest to itself yet ontologically farthest.
For Heidegger, Dasein dwells spatially in the world, but in the *equipmental space* (functional places, defined by Dasein-centered totalities of involvements) rather than in physical, Cartesian space, and this spatiality is characterized by *de-severance*, where "*de-severing* amounts to making the farness vanish—that is, making the remoteness of something disappear, bringing it close". Not only reducing physical distance, de-severance is the *reach* of Dasein's skilled practical activity.
An entity is *nearby* if it is readily available for some such activity, and *far away* if it is not. Nearness comes into being when the thing is examined. We reach it through things; it is nearness that makes the thingness of the thing appear. Cf. *Heidegger's Topology* (MIT Press, 2007) by Malpas. The following quotes (of 1924, 1954, 1966, 1971) illustrate **Heidegger's de-severance distance**:
"Man, as existing transcendence abounding in and surpassing toward possibilities, is a creature of distance. Only through the primordial distances he establishes toward all being in his transcendence does a true nearness to things flourish in him."
"Longing is the agony of the nearness of the distant."
"Then thinking would be coming-into-the-nearness-of distance."
"What is this uniformity in which everything is neither far nor near – is, as it were, without distance? Everything gets lumped together into uniform distancelessness."
Cf. the **technology-related distancing** and **death of distance** in Sect. 28.2.
French philosopher/writer Maurice Blanchot (1907–2003) considered Nietzsche, Heidegger and Lévinas via their metaphorics of distance. For example, he wrote: "A distance is synonymous with extreme non-coincidence."

"Far and near are dimensions of what escapes presence as well as absence under attraction of [impersonal] 'it'. It draws away, draws close, the same ghostly affirmation, the same premises of non-presence."
"To the proximity of the most distant, to the pressure of what is lightest, . . . to the contact of that which is never arrives, it is by friendship that I can respond, . . . the response of passivity to the non-presence of unknown [stranger]".

- **Lévinas distance to Other**
  We call the **Lévinas distance to Other** a primary distance between the individuals in their face-to-face encounter, which the French philosopher and Talmudic scholar Emmanuel Lévinas (1906–95) discusses in his book *Totality and Infinity*, 1961.
  Lévinas considers the precognitive relation with the Other: the Other, appearing as the Face, gives itself priority, its first demand even before I react to, love or kill it, is: "thou shalt not kill me". This Face is not an object but pure expression affecting me before I start meditating on it and passively resisting the desire that is my freedom. In this asymmetrical relationship—being silently summoned by the exposed Face of the Other ("widow, orphan, or stranger") and responding by responsibility for the Other without knowing that he will reciprocate—Lévinas (in line of *Misnagdim Judaism* ethics) finds the meaning of being human and concerned about justice. For him, this ethical duty is prior to pursuit of knowledge and ontology of nature.
  According to him, before covering the distance separating the *existent* (the lone subject) from the Other, one must first go from anonymous existence to the existent, from "there is" (swarming of points) to the Being (lucidity of consciousness localized here-below). Lévinas's ethics spans the distance between the foundational chaos of "there is" and the objective or intersubjective world. Ethics marks the primary situation of the face-to-face whereas morality comes later as a set of rules emerging in the social situation if there are more than two people face-to-face. And, for Lévinas, the scriptural/traditional God is the Infinite Other.
- **Distant suffering**
  Normally, physical distance is inversely related to charitable inclinations. But the traditional morality of "universal" proximity (geographic, age, character, habits, or familial) and pity looks inadequate in our contemporary life. In fact, most important actions happen on distance and the *mediation* (capacity of the media to involve us emotionally and culturally) address our concern for the "other".
  The nonuniversal quality of humanity should be constructed. So, mass media, NGO's, aid agencies, live blogs, and celebrity advocacy use imagery in order to encourage audiences to acknowledge, care and act for far away vulnerable others. But, for Chouliaraki, 2006, the current mediation replaced earlier claims to our "common humanity" by artful stories that promise to make us better people. As suffering becomes a spectacle of sublime artistic expression, the inactive spectator can merely gaze in disbelief. Arising voyeuristic altruism is motivated by self-empowerment: to realize our own humanity while keeping the humanity of the sufferer outside the remit of our judgement and imagination,

i.e., keeping **moral distance**. Chouliaraki calls it *narcissistic self-distance* or *improper distance*.

Silverstone's (2002) **proper distance in mediation** refers to the degree of proximity required in our mediated interrelationships if we are to create a sense of the other sufficient not just for reciprocity but for a duty of care, obligation and understanding. It should be neither too close to the particularities or the emotionalities of specific instances of suffering, nor too far to get a sense of common humanity as well as intrinsic difference. Cf. **Lévinas distance** and **antinomy of distance**.

Silverstone and Chouliaraki call us to represent sufferers as active, autonomous and empowered individuals. They advocate *agonistic solidarity*, treating the vulnerable other as other with her/his own humanity. It requires "an intellectual and aesthetic openness towards divergent cultural experiences, a search for contrasts rather than uniformity" (Hannerz, 1990). For Arendt, 1978, the imagination enables us to create the distance which is necessary for an impartial judgment,

But for Dayan, 2007, a climactic Lévinasian encounter with Other is not dualistic: there are many others awaiting my response at any given moment. So, proper distance should define the point from which I am capable of equitably hearing their respective claims, and it involves the reintroduction of actual distance.

- **Moral distance**

  The **moral distance** is a measure of moral indifference or empathy toward a person, group of people, or events. Abelson, 2005, refers to moral distance as the emotional closeness between agent and beneficiary.

  But Aguiar, Brañas-Garza and Miller, 2008, define it as the degree of moral obligation that the agent has towards the recipient. So, for them the social distance is only a case of moral distance in which anonymity plays a crucial, negative role.

  The **ethical distance** is a distance between an act and its ethical consequences, or between the moral agent and the state of affairs which has occurred.

  The (moral) *distancing* is a separation in time or space that reduces the empathy that a person may have for the suffering of others, i.e., that increases moral distance. In particular, **distantiation** is a tendency to distance oneself (physically or socially, by segregation or congregation) from those that one does not esteem. Cf. **distanciation**. On the other hand, the *good distancing* (Sartre, 1943, and Ricoeur, 1995) means the process of deciding how long a given ethical link should be.

- **Simone Weil distance**

  We call **Simone Weil distance** a kind of moral God-cross radius of the Universe which the French philosopher, Christian mystic, social activist and self-hatred Jewess, Simone Weil (1909–1943) introduced in "The Distance", one of the philosophico-theological essays comprising her *Waiting for God* (Putnam, New York, 1951).

  She connects God's love to the distance; so, his absence can be interpreted as a presence: "every separation is a link" (Plato's *metaxu*). She wrote: "God did not create anything except love itself, and the means to love... Because no other could do it, he himself went the greatest possible distance, the infinite distance. This distance between God and God, this supreme tearing apart, this agony beyond all others, this marvel of love, is the crucifixion."

In her peculiar Christian theodicy, "evil is the form which God's mercy takes in this world", and the crucifixion of Christ (the greatest love/distance) was necessary "in order that we should realize the distance between ourselves and God ... for we do not realize distance except in the downward direction". Weil's *God-cross* (or *creator-creature*) distance recalls the old question: can we equate distance from God with proximity to Evil? Her main drive, purity, consisted of maximizing **moral distance** to Evil, embodied for her by "the social, Rome and Israel".

Cf. Irenaeus (second century) God-humans **epistemic distance**, which must be far enough that belief in God remains a free choice. In Irenaean teodicy, God created both, evil/suffering and free will, allowing us moral choices and development.

Cf. Pascal's (1669) *God-man-nothing* distances in *Pensées*, note 72: "... what is man in Nature? A nothing in relation to infinity, all in relation to nothing, a central point between nothing and all and infinitely far from understanding either".

Cf. Montaigne's (1580) *nothing-smallest* and *smallest-largest* distances in *Essais, III:11 On the lame*: "Yet the distance is greater from nothing to the minutest thing in the world than it is from the minutest thing to the biggest."

Cf. Tipler's (2007) *Big Bang—Omega Point* time/distance with Initial and Final singularities seen as God-Father and God-Son. Tipler's *Omega point* (technological singularity) is a variation of prior use of the term (Teilhard de Chardin, 1950) as the supreme point of complexity and consciousness: the Logos, or Christ.

Calvin's (1537) *Eucharistic theology* (doctrine on the meaning of bread and vine that Christ offered to his disciples during the last supper before his arrest) also relies on spatial distance as a metaphor that best conveys the separation of the world from Christ and of the earthly, human from the heavenly, divine.

Weil's approach reminds that of the Lurian (about 1570) kabbalistic notions: *tzimtzum* (God's concealment, withdrawal of a part, creation by self-delimitation) and *shattering of the vessels* (evil as impure vitality of husks, produced whenever the force of separation loses its distancing function and giving man the opportunity to choose between good and evil). The purpose is to bridge the distance between Infiniteness of God (or Good) and the diversity of existence, without falling into the facility of dualism (as manicheanism and gnosticism). It is done by postulating intermediate levels of being (and purity) during emanation (unfolding) within the divine and allowing humans to participate in the redemption of the Creation.

So, a possible individual response to the Creator is purification and *ascent*, i.e., the spiritual movement through the levels of emanation in which the coverings of impurity, that create distance from God, are removed progressively.

Besides, the song "From a Distance", written by Julie Gold in 1985 , is about how God is watching us and how, despite the distance (physical and emotional) distorting perceptions, there is still a little peace and love in this world.

- **Golgotha distance**
  The exact locations of the Praetorium, where Pilate judged Jesus, and Golgotha, where he was crucified, as well as of the path that Jesus walked, are not known. At present, the Via Dolorosa (600 m from the Antonia Fortress west to the Church of the Holy Sepulchre) in the Old City of Jerusalem, held to be this path.
  The first century Jerusalem was about 500 m east to west and 1,200 m north to south. Herod's palace (including Praetorium) was about 600 m from Golgotha and 400 m from the Temple. The **Golgotha distance** (total distance from Gethsemane, where Jesus was arrested, to the Crucifixion) was about 1,500 m.
  Another New Testament's distance is mentioned in Apocalypse: "And the angel thrust in his sickle into the earth, and gathered the vine of the earth, and cast it into the great winepress of the wrath of God. And the winepress was trodden without the city [Jerusalem], and blood came out of the winepress, even unto the horse bridles, by the space of 1,600 furlongs [200 miles]" (Revelation 14:19–20). It can hint to the whole length of the land of Israel, computable as 1,600 studia.
- **Distance to Heaven**
  Below are given examples of distances and lengths which old traditions related (sometimes as a metaphor) to such notions as God and Heaven.
  In the Hebrew text *Shi'ur Qomah* (*The Measure of the Body*), the height of the *Holy Blessed One* is 236×10^7 *parasangs*, i.e., 14×10^{10} (divine) *spans*. In the Biblical verse "Who has measured the waters in the hollow of his hand and marked off the heaven with a span" (Isaiah 40:12), the size of the Universe is one such span.
  The age/radius of the Universe is 13.82 billion ly. *Sefer HaTemunah* (by Nehunia ben Hakane, first century) and *Otzar HaChaim* (by Yitzchok deMin Acco, thirteenth century) deduced that the world was created *in thought* 42,000 divine years, i.e., $42,000 \times 365,250 \approx 15.3$ billion human years, ago. It counts, using the 42-letter name of the God at the start of Genesis, that now we are in the 6th of the 7 cosmic *sh'mitah* cycles, each one being 7,000 divine years long. *Tohu va-bohu* (formless and empty) followed and 6,000 years ago the creation of the world *in deed* is posited.
  In the Talmud (Pesahim, 94), the Holy Spirit points out to "impious Nebuchadnezzar" (planning "to ascend above the heights of the clouds like the Most High"): "The distance from earth to heaven is 500 year's journey alone, the thickness of the heaven again 500 years...". This heaven is the *firmament* plate, and the journey is by walking. Seven other heavens, each 500 years thick, follow and the feet of the holy Creatures are equal to the whole... Their ankles, wings, necks, heads and horns are each consecutively equal to the whole." Finally, "upon them is the Throne of Glory which is equal to the whole". The resulting journey of 4,096,000 years amounts, at the rate of 80 miles ($\approx$129 km) per day, to $\approx$2,600 AU, i.e., $\approx\frac{1}{100}$ of the actual distance to Proxima Centauri, the nearest other star. Also, in Talmud, the width of *Jacob's Ladder* (bridge to Heaven that

Jacob dreams about, described in the Book of Genesis) is computed as 8,000 parasangs.

On the other hand, *Baraita de Massechet Gehinom* affirms in Section VII.2 that Hell consists of 7 cubic regions of side 300 year's journey each; so, 6,300 years altogether. According to the Christian Bible (Chap. 21 of the Book of Revelation), New Heavenly Jerusalem (a city that is or will be the dwelling place of the Saints) is a cube of side 12,000 furlongs ($\approx$2,225 km), or a similar pyramid or spheroid.

Islamic tradition (Dawood, Book 40, Nr. 470) also attributes a journey of 71–500 years (by horse, camel or foot) between each *samaa'a* (the ceiling containing one of the seven luminaries: Moon, Mercury, Venus, Sun, Mars, Jupiter, Saturn). Besides 7 heavens (as in Judaism and Hinduism), Shia Islam and Sufism have 7 depths of esoteric meanings of Quran, with only God knowing the 4th meaning.

The Vedic text (Pancavimsab Brahmana, c. 2000 BC) states that the distance to Heaven is 1,000 Earth diameters and the Sun (the middle one among seven luminaries) is halfway at 500 diameters. A similar ratio 500–600 was expected till the first scientific measurement of 1 AU (mean Earth-Sun distance) by Cassini and Richter, 1672. The actual ratio is $\approx$11,728.

The sacred Hindu number 108 ($= 6^2 + 6^2 + 6^2 = \prod_{1 \le i \le 3} i^i$), connected to the Golden Ratio as the interior angle 108° of a regular pentagon, is traced to the following Vedic values: 108 Sun's diameters for the Earth-Sun distance and 108 Moon's diameters for the Earth–Moon distance. The actual values are (slightly increasing) $\approx$107.6 and $\approx$110.6; they could be computed during an eclipse, since the angular size of the Moon and Sun, viewed from the Earth, is almost identical. Also, the ratio between the Sun and Earth diameters is $\approx$108.6, but it is unlikely that Vedic sages knew this. In Ayurveda, the devotee's distance to his "inner sun" (God within) consists of 108 steps; it corresponds to 108 beads of *mala* (rosary): by saying beads, the devotee does a symbolic journey from his body to Heaven.

- **Swedenborg heaven distances**

  The Swedish scientist and visionary Emanuel Swedenborg (1688–1772), in Section 22 (Nos. 191–199, *Space in Heaven*) of his main work *Heaven and Hell* (1952, first edition in Latin, London, 1758), posits: "distances and so, space, depend completely on interior state of angels". A move in heaven is just a change of such a state, the length of a way corresponds to the will of a walker, approaching reflects similarity of states. In the spiritual realm and afterlife, for him, "instead of distances and space, there exist only states and their changes".

- **Safir distance**

  According to Islamic law, a traveler may shorten the prayers, combine them, and be permitted to break the fast of Ramadan if the travel (*safir*) exceeds some minimum distance. Hanafi, the largest Sunni school of jurisprudence, define such **safir distance** as 3 days of continuous journey (in the great part of the day and at a moderate speed) or 15 *farsakh* (ancient unit of length, called also *parasang*). Three other main schools define it as 2 days of such journey or as 16 farsakh, computed differently. This distance is usually approximated as 80 or 83 km and applied for travel by camel, car, plane or ship. Another strong opinion, by Ibn

Taymiyya, claims that safir is not merely a distance but also a state of mind, an exposure to the wilderness; so, any distance customarily considered traveling is safir.

- **Sabbath distance**
  The **Sabbath distance** (or *rabbinical mile*) is a **range distance**: 2,000 Talmudic cubits (960–1,1152 m, cf. **cubit** in Chap. 27) which an observant Jew should not exceed in a public thoroughfare from any given private place on the Sabbath day. It is about the distance covered by an average man in 18 min.
  Other Israelite/Talmudic length units are: a day's march, parsa, stadium (40, 4, $\frac{2}{15}$ of the rabbinical mile, respectively), and span, hasit, palm, thumb, middle finger, little finger ($\frac{1}{2}$, $\frac{1}{3}$, $\frac{1}{6}$, $\frac{1}{24}$, $\frac{1}{30}$, $\frac{1}{36}$ of the Talmudic cubit, respectively).
- **Bible code distance**
  Witztum, Rips and Rosenberg, 1994, claimed to have discovered a meaningful subtext of the Book of Genesis, formed by uniformly spaced letters. The text was seen as written on a cylinder on which it spirals down in one long line. Many reactions followed, including criticism by McKay, Bar-Natan, Bar-Hillel and Kalai, 1999, in the same journal *Statistical Science*.
  The following **Bible code distance** d_t between two letters, that are t positions apart in the text, was used. Let h be the circumference of the cylinder, and let q and r be the quotient and reminder, respectively, when t is divided by h, i.e., $t = qh + r$ with $0 \leq r \leq h - 1$. Then $d_t = \sqrt{q^2 + r^2}$ if $2q \leq h$, and $d_t = \sqrt{(q+1)^2 + (r-h)^2}$, otherwise. It is, approximatively, the shortest distance between those letters along the cylinder surface; cf. **cylindrical distance** (Chap. 20).
- **Distance numbers**
  On Maya monuments usually only one anchor event is dated absolutely, in the linear *Mesoamerican Long Count calendar* by the number of days passed since the mythical creation on August 11, 3114 BC of the fourth world, which completed a *Great Period* of 13 b'ak'tuns (≈5,125 years) on December 21, 2012. The other events were dated by adding to or subtracting from the anchor date some **distance numbers**, i.e., periods from the cyclical 52-year *Calendar Round*.
- **Antinomy of distance**
  The **antinomy of distance**, as introduced in [Bull12] for aesthetic experiences by the beholder and artist, is that both should find the right amount of **emotional distance** (neither too involved, nor too detached), in order to create or appreciate art. The fine line between objectivity and subjectivity can be crossed easily, and the amount of distance can fluctuate in time.
  The **aesthetic distance** is a degree of emotional involvement of the individual, who undergoes experiences and objective reality of the art, in a work of art. It refers to the gap between the individual's conscious reality and the fictional reality presented in a work of art. It means also the frame of reference that an artist creates, by the use of technical devices in and around the work of art, to differentiate it psychologically from reality; cf. **distanciation**.

Some examples are: the perspective of a member of the audience in relation to the performance, the psychological/emotional distance between the text and the reader, the *actor-character distance* in the Stanislavsky system of acting.
Antinomy between inspiration and technique (embracement and estrangement) in performance theory is called the *Ion hook* since Ion of Ephesus (a reciter of rhapsodic poetry, in a Platon's dialog) employed a double-consciousness, being ecstatic and rational. The acting models of Stanislavsky and Brecht are, respectively, incarnating the role truthfully and standing artfully distanced from it. Cf. **role distance**.
[Morg76] defines pastoral ecstasy as the experience of *role-distancing*, or the authentic self's supra-role suspension, i.e., the capacity of an individual to stand outside or above himself for purposes of critical reflexion. Morgan concludes: "The authentic self is an *ontological possibility*, the social self is an *operational inevitability*, and awareness of both selves and the creative coordination of both is the gift of ecstasy". Cf. **Lévinas distance to Other**.
The **historical distance**, in terms of [Tail04], is the position the historian adopts *vis-à-vis* his objects—whether far-removed, up-close, or somewhere in between; it is the fantasy through which the living mind of the historian, encountering the inert and unrecoverable, positions itself to make the material look alive. The antinomy of distance appears because historians engage the past not just intellectually but morally/emotionally. The formal properties of historical accounts are influenced by the affective, ideological and cognitive commitments of their authors.
A variation of the antinomy of distance appears in critical thinking: the need to put some emotional and **epistemic** (intellectual) distance between oneself and ideas, in order to better evaluate their validity and avoid *illusion of explanatory depth* (to fail see the trees for the forest). A related problem is how much distance people must put between themselves and their pasts in order to remain psychologically viable; Freud showed that often there is no such distance with childhoods.

- **Role distance**
In Sociology, Goffman, 1961, using a dramaturgical metaphor, defined **role distance** (or *role distancing*) as actions which effectively convey some disdainful detachment of the (real life) performer from a role he is performing. An example of social role distancing is when a teacher explains to students that his disciplinary actions are due only to his role as a teacher.
Goffman observed that children are able to merge doing and being, i.e., *embracement of the performer's role*, only from 3 to 4 years. Starting from about 5, their role distance (distinguishing being from doing) appears and expands, especially, at age 8, 11 and adult years.
Besides role embracement and role distance, one can play a role cynically in order to manage the outcomes of the situation (impression management). The most likely cause of role distancing is *role conflict*, i.e., the pressure exerted from another role to act inconsistently from the expectations of the first role.

A *frame* is a type of role (participant, observer, authority, critic, artist, etc.) given to a person in relation to a given event. The **frame distance**, introduced by Heathcote, 1980, in teaching drama, refer to a specific (implied by the frame) responsibility, interest, attitude and behavior of a person/student in this event/drama.

- **Distanciation**
  In scenic art and literature, **distanciation** (Althusser, 1968, on Brecht's *alienation effect*) consists of methods to disturb purposely (in order to challenge basic codes and conventions of spectator/reader) the *narrative contract* with him, i.e., implicit clauses defining logic behavior in a story. The purpose is to differentiate art psychologically from reality, i.e., to create some **aesthetic distance**.
  For Bakhtin, the mandate to "be outside" that which you create is a matter of subject-subject (as opposed to subject-object) relations. For Shklovsky the distancing of an object sharpens our perception and stimulates senses, thereby arousing us to artistic (as opposed to dull everyday) experience.
  One of the distanciation devices is *breaking of the fourth wall*, when the actor/author addresses the spectators/readers directly through an imaginary screen separating them. The *fourth wall* is the conventional boundary between the fiction and the audience. It is a part of the *suspension of disbelief* between them: the audience tacitly agrees to provisionally suspend their judgment in exchange for the promise of entertainment. Cf. **distancing** and **distantiation**.
- **Narrative distance**
  The author creates a persona of *narrator*, who tell the story, usually, from the point of view of first- or third-person. **Narrative distance** is (Genette, 1980) the distance between the narrator and the story's characters, setting, events and objects.
  The closest possible distance—the narrator reports on the thoughts and feeling (even unconscious ones) of the characters, while the farthest distance—reporting only actions and situations. The author can vary this distance; say, the third-person omniscient narrator can zoom in and out of character's perspectives.
- **Ironic distance**
  Rhetorical writer or speaker does not allow audience to maintain an objective or fixed distance from the story. He intrude to distance himself from characters in a story or from his own remarks. **Ironic distance** refers to the narrative irony: distance of knowledge between author/narrator/character/reader.
  As a literary device, irony implies a distance between what is said and what is meant. Irony is also the art of juxtaposing incongruous parts; so, an *ironic distance* also mean the closeness between two things that never meet.
- **Epistemic distance**
  Epistemic (or *intellectual*) **distance** from something refers to the degree of difficulty involved in knowing it. For example, conditional rhetorical constructions are used in discourse, in order to indicate this distance. Mejias (2005) *epistemological distance* between things is the difference in degree of knowledge justification.

- **Representation of distance in Painting**
In Western Visual Arts, the *distance* is the part of a picture representing objects which are the farthest away, such as a landscape; it is the illusion of 3D depth on a flat picture plane. The *middle distance* is the central part of a scene between the foreground and the background (implied horizon).
Perspective projection draws distant objects as smaller to provide additional realism by matching the decrease of their visual angle; cf. Chap, 6. A *vanishing point* (or *point of distance*) is a point at which parallel lines receding from an observer seem to converge. (For a meteor shower, the *radiant* is the point in the sky, from which meteors appear to originate.) *Linear perspective* is a drawing with 1–3 vanishing points; usually, they are placed on the horizon and equipartition it.
In a *curvilinear perspective*, there are ≥ 4 vanishing points; usually, they are mapped into and equipartition a *distance circle*. 0-*point perspective* occurs if the vanishing points are placed outside the painting, or if the scene (say, a mountain range) does not contain any parallel lines. Such perspective can still create a sense of *depth* (3D distance) as in a photograph of a mountain range.
In a *parallel projection*, all sets of parallel lines in 3D object are mapped to parallel lines in 2D drawing. This corresponds to a perspective projection with an infinite *focal length* (the distance from the image plane to the projection point).
Axonometric projection is parallel projection which is *orthographic* (i.e., the projection rays are perpendicular to the projection plane) and such that the object is rotated along one or more of its axes relative to this plane. The main case of it, used in Engineering Drawing, is **isometric projection** in which the angles between three projection axes are the same, or $\frac{2\pi}{3}$.
In Chinese Painting, the *high-distance*, *deep-distance* or *level-distance* views correspond to picture planes dominated, respectively, by vertical, horizontal elements or their combination. Instead of the perspective projection of a "subject", assuming a fixed position by a viewer, Chinese classic hand scrolls (up to 10 m in length) used axonometric projection. It permitted them to move along a continuous/seamless visual scenario and to view elements from different angles. It was less faithful to appearance and allowed them to present only three (instead of five) of six surfaces of a normal interior. But in Chinese Painting, the focus is rather on symbolic and expressionist representation.
- **Scale in art**
In drawing, the **scale** refers to the proportion or ratio that defines the size relationships. It is used to create the illusion of correct size relationships between objects and figures. The *relative scale* is a method used to create and determine the spatial position of a figure or object in the 3D picture plane: objects that are more distant to the viewer are drawn smaller in size. In this way, the relative size of an object/figure creates the illusion of space on a flat 2D picture.
In an architectural composition, the **scale** is the two-term relationship of the parts to the whole which is harmonized with a third term—the observer. For example, besides the proportions of a door and their relation to those of a wall, an observer measures them against his own dimensions.

The **scale** of an outdoor sculpture, when it is one element in a larger complex such as the facade of a building, must be considered in relation to the scale of its surroundings. In *flower arrangement* (floral decoration), the **scale** indicates relationships: the sizes of plant materials must be suitably related to the size of the container and to each other.

The *hierarchical scale* in art is the manipulation of size and space in a picture to emphasize the importance of a specific object. Manipulating the scales was the theme of *Measure for Measure*, an art/science exhibition at the Los Angeles Art Association in 2010. Examples of the interplay of the small and the large in literature are Swift's *Gulliver's Travels* and Carrol's *Through the Looking Glass*. In the cinema, the spectator can easily be deceived about the size of objects, since scale constantly changes from shot to shot.

In Advertising and Packaging, the size changes the meaning or value of an object. The idea that "bigger is better" is validated by the sales of sport utility vehicles, super-sized soft drinks and bulk food at Wal-Mart.

In reverse, the principle "small is beautiful" is often used to champion small, appropriate objects and technologies that are believed to empower people more. For example, small-sized models sell the benefits of diet programs and fitness regimes designed to scale back people's proportions. Examples of Japanese miniaturization culture are bonsai and many small/thin portable devices.

- **Distances in Interior Design**

  In Interior Design, the *scale* refer to how an item relates to the size of the room or the owner, and the *proportion* refer to the shape of an item and how it relates to other objects in the room. The vertical, horizontal, diagonal and curved lines give a room a feeling of formalness, casualness, transition and sensuality, respectively. Other required space relationships are *balance* (equal weight between objects on either side of a room) and *rhythm* (repetition of patterns, color, or line).

  Workplane is the height at which an activity takes place; it is about 90, 75–90 and 75 cm for a kitchen, bath and a dining room or desk. In a kitchen, the perimeter of the *work triangle* formed by sink, cooking surface and refrigerator ideally should be 3.5–7.5 m. In a living room, the triangle of *focal points* to emphasize is formed usually at the door or fireplace, TV, big window, sofa. Other examples of recommended distances are: 35–45 cm between the sofa (or chairs) and coffee table, 60 cm between dining chairs and at least 90 cm for traffic lanes.

  Used in lighting calculations, the **room cavity ratio** (or RCR) is $\frac{5hP}{2A}$, where h, P, A are the ceiling height, perimeter and area of the room. So, $RCR = \frac{5h(l+w)}{lw}$ for a rectangular room of length l and width w.

- **Spatialism**

  Spatialism (or *Spazialismo*) is an art movement founded by Lucio Fontana in Milan in 1947, intended to synthesize space, color, sound, movement and time into a new "art for the Space Age". Instead of the illusory virtual space of traditional *easel* (i.e., of a size and on a material suitable for framing) painting, he proposed to unite art and science to project color and form into real space by the use of up-to-date techniques, say, TV and neon lighting. His *Spatial*

Concept series consisted of holes or slashes, by a razor blade, on the surface of monochrome paintings.

- **Spatial music**
  Spatial music refers to music and sound art (especially, electroacoustic), in which the location and movement of sound sources, in physical or virtual space, is a primary compositional parameter and a central feature for the listener.
  Space music is gentle, harmonious sound that facilitates the experience of contemplative spaciousness. Engaging the imagination and generating serenity, it is particularly associated with ambient, New Age, and electronic music.
- **Distance-named cultural products**
  Far Near Distance is the name of the program of the House of World Cultures in Berlin which presents contemporary positions of Iranian artists. Examples of similar use of distance terms in modern popular culture follow.
  "Some near distance" and "Zero/Distance" are the titles of art exhibitions of Mark Lewis (Bilbao, 2003) and Jim Shrosbree (Des Moines, Iowa, 2007). "A Near Distance" is a paper collage by Perle Fine (New York, 1961); "Quiet Distance" is a fine art print by Ed Mell. "Distance" is a Windows/Mac/Linux survival racing game; "Dream Drop Distance" is a video game for Nintendo.
  "Distance" is a Japanese film directed by Hirokazu Koreeda (2001) and an album of Utada Hikaru (her famous ballad is called "Final Distance"). It is also a song by Christina Perry, the stage name of a musician Greg Sanders and the name of the rock/funk band led by Bernard Edwards. "The Distance" is a US film directed by Benjamin Busch (2000), an album by the band "Silver Bullet" and a song by the band "Cake". "Near Distance" is a musical composition by Chen Yi (New York, 1988) and lyrics by the quartet "Puressence".
  "Distance to Fault", "Distance from Shelter", "Long Distance Calling" are the rock bands. Among popular albums are "The Tyranny of Distance", "The Great Cold Distance", "Close the Distance", "The Distance to Here", "Love and Distance", "Long Distance Voyager" and "The Crawling Distance", "This Magnificent Distance" by the bands "Washington, D.C.", "Katatonia", "Go Radio", "Live", "The Helio Sequence", "The Moody Blues" and Robert Pollard, Chris Robinson.
  The terms *near distance* and *far distance* are also used in Ophthalmology and for settings in some sensor devices.
- **Distance-related quotes**
  - "Respect the gods and the devils but keep them at a distance." (Confucius)
  - "Sight not what's near through aiming at what's far." (Euripides)
  - "It is when suffering seems near to them that men have pity." (Aristotle)
  - "Distance in space or time weakened all feelings and all sorts of guilty conscience." "Distance is a great promoter of admiration." (Denis Diderot)
  - "Our main business is not to see what lies dimly at a distance, but to do what lies clearly at hand." (Thomas Carlyle)
  - "We can only see a short distance ahead, but we can see plenty there that needs to be done." (Alan Turing)

- "The foolish man seeks happiness in the distance; the wise grows it under his feet."(Julius Robert Oppenheimer)
- "The very least you can do in your life is to figure out what you hope for. And the most you can do is live inside that hope. Not admire it from a distance but live right in it, under its roof." (Barbara Kingsolver)
- "Better is a nearby neighbor, than a far off brother." (Proverbs 27:10, Bible)
- "These [patriarchs] all died in faith without receiving the things promised [Canaan, Messiah, Gospel], but they saw them and welcomed them from a distance, admitting that they were strangers and pilgrims on the earth." (Hebrews 11:13, Bible)
- "By what road", I asked a little boy, sitting at a cross-road, "do we go to the town?"—"This one", he replied, "is short but long and that one is long but short". I proceeded along the "short but long road". When I approached the town, I discovered that it was hedged in by gardens and orchards. Turning back I said to him, "My son, did you not tell me that this road was short?"—"And", he replied, "Did I not also tell you: "But long"?" (Erubin 53b, Talmud)
- "The Prophet Muhammad was heard saying: "The smallest reward for the people of paradise is an abode where there are 80,000 servants and 72 wives, over which stands a dome decorated with pearls, aquamarine, and ruby, as wide as the distance from Al-Jabiyyah [a Damascus suburb] to Sana'a [Yemen]." (Hadith 2687, Islamic Tradition)
- "The closer the look one takes at the world, the greater distance from which it looks back." (Karl Kraus)
- "Telescopes and microscopes are designed to get us closer to the object of our studies. That's all well and good. But it's as well to remember that insight can also come from taking a step back." (New Scientist, March 31, 2012)
- "Where the telescope ends, the microscope begins. Which of the two has the grander view?" (Victor Hugo)
- "Nature uses only the longest threads to weave her patterns." (Richard Feynman)
- "We're about eight Einsteins away from getting any kind of handle on the universe." (Martin Amis)
- "It is true that when we travel we are in search of distance. But distance is not to be found. It melts away. And escape has never led anywhere ... What are we worth when motionless, is the question." (Antoine de Saint-Exupéry)
- "If you want to build a ship, don't drum up people to collect wood and don't assign them tasks and work, but rather teach them to long for the endless immensity of the sea." (Antoine de Saint-Exupéry)
- "Ships at a distance have every man's wish on board." (Zora Neale Hurston)
- "If you've never stared off in the distance, then your life is a shame." (Adam Duritz)
- "Every once in a while, people need to be in the presence of things that are really far away." (Ian Frazier)
- "Only those who will risk going too far can possibly find out how far one can go." (Thomas Stearns Eliot)

- "Distance is to love like wind is to fire ... it extinguishes the small and kindles the great." (source unknown)
- "I could never take a chance of losing love to find romance
  In the mysterious distance between a man and a woman." (Performed by U2)
- "In true love the smallest distance is too great, and the greatest distance can be bridged." (Hans Nouwens)
- "Love is like a landscape which doth stand
  Smooth at a distance, rough at hand." (Robert Hegge)
- "Life is like a landscape. You live in the midst of it but can describe it only from the vantage point of distance." (Charles Lindbergh)
- "Distance between two people is only as one allows it to be." (source unknown)
- "It is only the mountains which never meet." (french proverb)
- "Nothing makes Earth seem so spacious as to have friends at a distance; they make the latitudes and longitudes." (Henri David Thoreau)
- "Distance can endear friendship, and absence sweeteneth it." (James Howell)
- "The word is distance within non-distance, that is, the width of a gap that every letter stresses while bridging it. What is said is always said in relation to what will never be expressed. At these limits we recognize ourselves." (Edmond Jabès)
- "Sad things are beautiful only from a distance ... From a distance of 130 years i'm going to distance myself until the world is beautiful..." (Tao Lin)
- "Dying away into the distance, prose turns into poetry, speech into vocalise, language into music." (Berthold Hoeckner)
- "Everything becomes romantic and poetic, if one removes it to a distance ... Distant philosophy sounds like poetry – for each call into the distance becomes a vowel ... Everything becomes poetry – poem from afar." (Novalis)
- "The appropriated way to determine whether a painting is melodious is to look at it from a distance so as to be unable to comprehend its subject or its lines." (Charles Baudelaire)
- "There is no object so large ... that at great distance from the eye it does not appear smaller than a smaller object near." (Leonardo da Vinci)
- "Distance lends enchantment to the view,
  And robes the mountain in its azure hue." (Thomas Campbell)
- "There are charms made only for distant admiration."
  "Distance has the same effect on the mind as on the eye." (Samuel Johnson)
- "Age, like distance, lends a double charm." (Oliver Wendell Holmes)
- "Distance not only gives nostalgia, but perspective, and maybe objectivity." (Robert Morgan)
- "It is the just distance between partners who confront one another, too closely in cases of conflict and too distantly in those of ignorance, hate and scorn, that sums up rather well, I believe, the two aspects of the act of judging. On the one hand, to decide, to put an end to uncertainty, to separate the parties; on the other, to make each party recognize the share the other has in the same society." (Paul Ricoeur)

- "Authority doesn't work without prestige, or prestige without distance." (Charles de Gaulle)
- "The human voice can never reach the distance that is covered by the still small voice of conscience." (Mohandas Gandhi)
- "A smile is the shortest distance between two people." (Victor Borge)
- "The shortest distance between two points is under construction." (Leo Aikman)
- "A straight line may be the shortest distance between two points, but it is by no means the most interesting." (Third Doctor from BBC TV series *Doctor Who*)
- "In politics a straight line is the shortest distance to disaster." (John P. Roche)
- "Fret not where the road will take you. Instead concentrate on the first step. That is the hardest part and that is what you are responsible for. Once you take that step let everything do what it naturally does and the rest will follow. Do not go with the flow. Be the flow." (Shams Tabrisi)
- "The distance is nothing; only the first step that is difficult." (Marie du Deffand)
- "A perfect run has nothing to do with distance. It's when your stride feels comfortable." (Sean Astin)
- "Fill the unforgiving minute with sixty seconds worth of distance run." (Rudyard Kipling)
- "The distance between dreams and reality is called Discipline." (Albert Wright)
- "Everywhere is within walking distance if you have the time." (Steven Wright)
- "Time is the longest distance between two places." (Tennessee Williams)
- "There is an immeasurable distance between late and too late." (Og Mandino)
- "They couldn't hit an elephant at this distance." (last words of John Sedgwick, seconds before he was mortally wounded)
- "The distance that the dead have gone does not at first appear;
  Their coming back seems possible for many an ardent year." (Emily Dickinson)
- "A vast similitude interlocks all ... All distances of place however wide,
  All distances of time, all inanimate forms, all souls ... " (Walt Whitman)

Chapter 29
Other Distances

In this chapter we group together distances and distance paradigms which do not fit in the previous chapters, being either too practical (as in equipment), or too general, or simply hard to classify.

29.1 Distances in Medicine, Anthropometry and Sport

- **Distances in Medicine**

  Some examples from this vast family of physical distances follow.

  In Dentistry, the **interocclusal distance**: the distance between the occluding surfaces of the maxillary and mandibular teeth when the mandible is in a physiologic rest position. The **interarch** and **interridge distances**: the vertical distances between the maxillary and mandibular arches, or, respectively, ridges.

  The **intercanine distance**: the distance between the distal surfaces of the maxillary canines on the curve (the circumference of 6 maxillary anterior teeth).. **interincisor distance**: the distance between the upper and lower incisors.

  The **interproximal distance**: the **spacing distance** between adjacent teeth; *mesial drift* is the movement of the teeth slowly toward the front of the mouth with the decrease of the interproximal distance by wear.

  The **biologic width**: the distance between the deepest point of the gingival sulcus and the alveolar bone crest. The *crown-to-root-ratio*: the ratio of the length of the part of a tooth that appears above the alveolar bone versus what lies below it.

  The **interbrow distance**: the distance between the eyebrows.

  The **interaural** (or *biauricular*) **distance**: the distance between the ears.

  The **rectosacral distance**: the shortest distance from the rectum to the *sacrum* (triangular bone at the base of the spine, inserted between the two hip bones) between the 3rd and 5th sacral vertebra. It is at most 10 mm in adults.

M.M. Deza, E. Deza, *Encyclopedia of Distances*,
DOI 10.1007/978-3-662-44342-2_29

The **anogenital distance** (or AGD): the length of the *perineum*, i.e., the region between the anus and genital area (the anterior base of the penis for a male). For a male it is 5 cm in average (twice what it is for a female). ARD is a measure of physical masculinity and, for a male, lower ARD correlates with lower fertility.

The **internipple distance**: the distance between nipples. "Ideally", the nipples and sternal notch form an equilateral triangle with a side of 21 cm, and the nipples are at the middle of the humeral shoulder-elbow distance. The average *areolar diameter* is 38 mm for a mature woman and 25 mm for a male.

The **intercornual distance**: the distance between uterine horns (2–4 cm). The **C-V distance**: clitoris-vagina distance (2.3–3 cm); $<$ 2.5 cm tend to yield reliable orgasms from intercourse alone, while >3 cm almost exclude it. The *clitoral index* (CI): product of the crosswise (3–4 mm) and lengthwise (4–5 mm) widths of the external portion of the clitoris; CI is a measure of virilization in women. The mean length of an erect penis is 13–15 cm or, counting its root, $\approx$22 cm.

A **pelvic diameter** is any measurement that expresses the diameter of the birth canal in the female. For example, the *diagonal conjugate* (13 cm) joins the posterior surface of the pubis to the tip of the sacral promontory, and the *true* (or *obstetric, internal) conjugate* (11.5 cm) is the anteroposterior diameter of the pelvic inlet.

In Obstetrics, the **fundal height** is the size the mother's uterus (the distance between the tops of uterus and pubic bone) used to assess fetal growth and development during pregnancy. The **crown-rump length** is the length of human embryos/fetuses (the distance, determined from ultrasound imagery, from the top of the head to the bottom of the buttocks) used to estimate gestational age.

Metra and uterus are (Greek and Latin) medical terms for the womb. *Metropathy* is any disease of the uterus, say, *metritis* (inflammation), *metratonia* (atony), *metrofibroma*. *Metrometer* is an instrument measuring the womb's size. *Metrography* (or *hysterography*) is a radiographic examination of the uterine cavity filled with a contrasting medium. Cf. **meter-related terms** in Chap. 27.

In Radiography, the **teardrop distance**: the distance from the lateral margin of the pelvic teardrop to the most medial aspect of the femoral head; a widening of $\geq$1 mm indicates excess hip joint fluid and so inflammation. The **intertrochanteric distance**: the distance between femurs. The **interpediculate distance**: the distance between the vertebral pedicles. The **source-skin distance**: the distance from the focal spot on the target of the X-ray tube to the subject's skin.

In *Intubation* (insertion of a tube into a body canal or hollow organ, to maintain an opening or passageway), the **insertion distance**: the distance from the body aperture at which the tubing is advanced. The *French size* of a catheter with external diameter D is $\pi D \approx 3D$; so, 20 F means $D = 6.4$ mm.

In Anesthesia, the **thyromental distance** (or TMD): the distance from the upper edge of the thyroid cartilage (laryngeal notch) to the *menton* (tip of the chin). The **sternomental distance**: the distance from the upper border of the manubrium sterni to the menton. The **mandibulo-hyoid distance**: mandibular length from menton to hyoid. When the above distances are less than 6–6.5, 12–12.5 and 4 cm, respectively, a difficult intubation is indicated.

The *depth of anesthesia* is a number expressing the likelihood of awareness by the degree of slowing and irregularity in electroencephalogram (EEG) signals. Also, at loss of consciousness, high frequency (12–35 Hz) brain waves are replaced by two (low, <1 Hz, and alpha, 8–12 Hz) superimposed waves. Even beyond a flat line EEG, some neuronal spikes come to the cortex from the hippocampus.

The **sedimentation distance** (or ESR, *erythrocyte sedimentation rate*): the distance red blood cells travel in 1 h in a sample of blood as they settle to the test tube's bottom. ESR indicates inflammation and increases in many diseases.

The **stroke distance**: the distance a column of blood moves during each heart beat, from the aortic valve to a point on the arch of the aorta.

The **distance between the lesion and aortic valve** being <6 mm, is an important predictor, available before surgical resection of DSS (discrete subaortic stenosis), or reoperation for recurrent DSS. The **aortomesenteric distance** (between aorta and superior mesenteric artery) correlates with the *body mass index*.

The **aortic diameter**: the maximum diameter of the outer contour of the aorta. It, as well as the cross-sectional diameter of the left ventricle, varies between the ends of the *systole* (the time of ventricular contraction) and *diastole* (the time between contractions). The smallest and largest cardiac dimensions are LVE (left ventricle end-) *systolic* and *diastolic diameters*; the *strain* is the ratio between them.

The **dorsoventral interlead distance** of an implanted pacemaker or defibrillator: the horizontal separation of the right and left ventricular lead tips on the lateral chest radiograph, divided by the *cardiothoracic ratio* (ratio of the cardiac width to the thoracic width on the posteroanterior film).

The **distance factor** is a crude measure $\frac{l}{d} - 1$ of arterial tortuosity, where l is the vessel length and d is the Euclidean distance between its endpoints.

In Nerve Regeneration by transplantation of cultured stem cells, the **regeneration distance** is the distance between the point of insertion of the proximal stump and the tip of the most distal regenerating axon.

The *small-for-size syndrome* (SFSS) is acute liver failure resulting from the transplantation of a too small (usually $<0.8\,\%$ of recipient weight) *graft* (donor liver).

A **distant flap** is a procedure moving tissue (skin, muscle, bone, or some combination) from one part of the body, where it is dispensable, to another part.

The length of the alimentary (mouth-to-anus) tract is $\approx$9 m in a dead and, due to muscle tone, 5–6 m in a leaving human. Transit takes 30–50 h.

In Laser Treatments, the **extinction length** and **absorption length** of the vaporizing beam are the distances into the tissue along the ray path over which 90 % (or 99 %) and 63 %, respectively, of its radiant energy is absorbed.

In Ophthalmic Plastic Surgery, the **marginal reflex distances** MRD_1 and MRD_2 are the distances from the center of the pupil (identified by the corneal reflex created by shining a light on the pupil) to the margin of the upper or lower eyelid, while the **vertical palperbal fissure** is the distance between these eyelids.

The main distances used in Ultrasound Biomicroscopy (for glaucoma treatment) are the **angle-opening distance** (from the corneal endothelium to the anterior iris) and the **trabecular ciliary process distance** (from a particular point on the *trabecular mesh-work* to the *ciliary process*).
In Medical Statistics, *length bias* is a selection bias that can occur when the lengths of intervals are analyzed by selecting random intervals in space or time. This process favors longer intervals, thus skewing the data. For example, screening over-represents less aggressive disease, say, slower-growing tumors.

- **Distances in Oncology**
In Oncology, the **tumor radius** is the mean radial distance R from the tumor origin (or its center of mass) to the tumor–host interface (the tumor/cell colony border). The cell proliferation along $[0, R]$ is ≈ 0 up to some r_0, then increases only linearly up to some r_1, and it happens mainly within $[r_1, R]$.
The **tumor diameter** is the greatest vertical diameter of any section; the *tumor growth* is the geometric mean of its three perpendicular diameters. The *average diameter* is $\frac{L+W+H}{3}$ where L, W, H are the longest length, width and height.
Tubiana, 1986, claims that for each tumor type a critical tumor diameter and mass for metastatic spread exists and this threshold may be reached before the primary tumor is detectable. For breast cancer, metastases were found in 50 % of the women whose primary tumor had a diameter of 3.5 cm, i.e., a mass $\approx$22 g.
In the *tumor, node, metastasis* (TNM) classification, describing the stage of cancer in a patient's body, the parameter T is the **tumor size** (direct extent of the primary tumor) by the categories T-1, T-2, T-3, T-4. In breast cancer, T-1, T-2, T-3 are $< 2, 2-5, > 5$ cm and T-4 is a tumor of any size that has broken through the skin, or is attached to the chest wall. A *clinical size* is 10^9–10^{11} cells.
In Oncological Surgery, the **margin distance** (or *margins of resection*) is the distance between a tumor and the ink-marked edge of *tumor bed*, i.e., normal-appearing tissue surrounding tumor that is removed along with it in order to prevent local recurrence. If the margins, as checked by a pathologist under microscope, are *positive* (cancer cells are found in the ink), then more surgery is needed. The margins are *negative* (or *clear*, *clean*) if no cancer cells are found "close" to the ink.
The **perfusion distance** is the shortest distance between the infusion outlet and the surface of the electrodes during radio-frequency tumor ablation.
In Radiation Oncology, the **maximum heart distance** MHD is the maximum distance of the heart contour (as seen in the beam's eye view of the medial tangential field) to the medial field edge, and the **central lung distance** CLD is the distance from the dorsal field edge to the thoracic wall. An "L-bar" armrest, used to position the arm during breast cancer irradiation, decreases these distances.
A **distant cancer** (or relapse, metastasis) is a cancer that has spread from the original (primary) tumor to distant organs or distant lymph nodes. It can happen by **long-distance dispersal** (cf. Chap. 23) and by dividing of *cancer stem cell*. DDFS (Distant Disease-free Survival) is the time until such an event.

According to Hanahan–Weinberg, 2000, tumor progress via evolution-like process of genetic changes which can be grouped into six hallmarks. Tumorigenesis requires a mutation pathway of four to six events among them to occur in the lineage of one cell. Spencer et al., 2006, define *tumor heterogeneity* as

$$\frac{2}{n(n-1)} \sum_{1 \le i < j \le p} n_i n_j d_{ij},$$

where $n = \sum_{1 \le i \le p} n_i$, p, n_i are the numbers of cells in a tumor, of distinct pathways, of cells with i-th pathway, and d_{ij} is the **ancestral path distance** (cf. **pedigree-based distances** in Chap. 23) between i-th and j-th pathways.
Similarly, the *distance to flu pandemic* is, say, the length of mutation pathway for a virus strain to become airborne transmissible among humans.

- **Distances in Rheumatology**
The main such distances (measured in cm to the nearest 0.1 cm) follow.
Occiput wall distance: the distance from the patient's occiput to the wall during maximal effort to touch the head to the wall, without raising the chin above its usually carrying level (when heels and, if possible, the back are against the wall).
Modified Schober test: the distance between two marked points (a point over the spinous process of $L5$ and the point 10 cm above) measured when the patient is extending his lumbar spine in a neutral position and then when he flexes forward as far as possible. Normally, the 10 cm distance increases to $\geq$16 cm.
Lateral spinal flexion: the distance from the middle fingertip to the floor in full lateral flexion without flexing forward or bending the knees or lifting the heels and attempting to keep the shoulders in the same place.
Chest expansion: the difference between full expiration and full inspiration, measured at the nipples.
Intermalleolar distance: the distance between the medial malleoli when the patient (supine, the knees straight and the feet pointing straight up) is asked to separate the legs as far as possible.
- **Distance healing**
Distance (or *distant*, *remote*) **healing** is defined (Sicher–Targ, 1998) as a conscious, dedicated act of mentation attempting to benefit another person's physical or emotional well-being at a distance. Cf. **action at a distance** in Chap. 24.
It includes prayer (intercessory, supplicative and nondirected), spiritual/mental healing and strategies purporting to channel some supra-physical energy (*non-contact therapeutic touch*, *Reiki healing*, *external qigong*).
Distant healing is part of popular alternative/complementary medicine but it is highly controversial: some positive results are attributed to a placebo effect. Still, such rejection (as well as for homeopathy) is also a matter of belief.
In Chinese Medicine, the life-energy, *qi*, flow through 20 *meridians* connecting about 400 acupuncture points. In Ayurveda, the life force, *prana*, flow

through >72,000 *nadis* connecting *chakras* (intensity points); it also lists 13 internal *srotas* (physical body channels) and, by the number of orifices, 9 external ones for males and 12 for females. Many *trigger points* (hyperirritable muscle spots) and *pressure points* in martial arts are related to above points. Many meridians are located along connective-tissue planes between muscles or muscle and bone.

Distance medicine technologies are used to transmit/treat patient information, to submit prescriptions, to create distributive patient care and distributive learning. Examples of telephonic communication with patients are in: follow-up care, reminders, interactive systems, screening and access in hospital use.

- **Brain distances**

  Diffusion MRI (Magnetic Resonance Imaging) noninvasively produces *in vivo* images of brain tissues weighted by their water diffusivity. The image intensities at each position are attenuated proportionally to the strength of diffusion in the direction of its gradient. Diffusion in tissues is described by a diffusivity tensor. Tensor data are displayed, for each voxel, by ellipsoids; their length in any direction is the diffusion distance molecules cover in a given time in this direction. The **diffusion tensor distance** is the length from the center to the surface of the diffusion tensor.

  In brain MRI, the distances considered for *cortical maps* (i.e., outer layer regions of cerebral hemispheres representing sensory inputs or motor outputs) are: MRI **distance map** from the GW (gray/white matter) interface, **cortical distance** (say, between activation locations of spatially adjacent stimuli), **cortical thickness** (the shortest distance between the GW the boundary and the innermost surface of *pia mater* enveloping the brain) and *lateralization metrics*. In fact, language, for example, tends to be on the left, attention more on the right side of the brain. The cortical thickness of Einstein's brain is 2.1 mm, while the average one is 2.6 mm; the resulting closer packing of neurons may speed up communication between them. This brain had a relatively larger (and more intricately folded) prefrontal cortex and an unusually high glia-to-neuron ratio. Also, the *corpus callosum* is thicker in many areas, indicating greater connectivity between the two hemispheres.

  Stereotaxic coordinates of a point p in the human brain are given by a triple (x_p, y_p, z_p) in mm (Talairach–Tournoux, 1988), where the *anterior commissure* is the origin $(0, 0, 0)$. The dimensions x, y, z refer to the left–right (LR), posterior–anterior (PA) and ventro-dorsal (VD or inferior–superior) orthogonal axes with positive values for the right hemisphere, anterior part and dorsal part. The **Talairach distance** of a point p is its Euclidean distance $\sqrt{x_p^2 + y_p^2 + z_p^2}$ from the origin.

  Among proto-humans, only Neanderthals had a (11.6 %) larger brain than *Homo sapiens*; we got from them the brain-size increasing gene 0.037 Ma ago. However over the last 0.03 Ma our brains have been shrinking, while our craniums have been increasing. The human brain reaches its full size by age 20 and then shrinks (faster for men) by about 1 % per year. It accounts for $\frac{1}{5}$ of the total ($\approx$100 W) body energy consumption.

- **Dysmetria**
  Dysmetria is a symptom of a cerebellar disorder or syndrome, expressed in a lack of coordination of movement typified by the undershoot (*hypometria*) or overshoot (*hypermetria*) of the intended position with the hand, arm, leg, or eye. More generally, *dysmetria* can refer to an inability to judge distance or scale, which is also one of symptoms of dyscalculia. The **distance constancy** (cf. Chap. 28) is poor in schizophrenics; so, their visual perception is lacking in depth.
  Alice in Wonderland syndrome, affecting mainly children, is that objects appear either much smaller (*micropsia*) or larger (*macropsia*) than they are. Micropsia appear also in *Charles Bonnet syndrome*, affecting mainly vision-impaired elderly.
- **Space-related phobias**
  Several **space-related phobias** have been identified: *agoraphobia*, *astrophobia*, *claustrophobia*, *cenophobia*, and *acrophobia*, *bathophobia*, *gephyrophobia*, *megalophobia* which are, respectively, fear of open, celestial, enclosed, empty spaces, and heights, depths, bridges, large/oversized objects. *Autoscopy* (or *out-of-body experience*) is the hallucination of seeing one's own body at a distance.
  Among neuropsychological spatial disorders are: *Balint's syndrome* (inability to localize objects in space), *hemispatial neglect* (bias of attention to and awareness of the side of the hemispheric lesion) and *allochiria* (left-right disorientation).
  Topographical disorientation is the inability to orient in the surrounding as a result of focal brain damage. Such agnosia with respect to self, to landmarks, to external environment, to new environments is called *egocentric*, *landmark*, *heading*, *anterograde*, respectively. *Dromosagnosia* is the loss of direction while driving.
  In Chap. 28, among applications of **spatial language** is mentioned Grove's *clean space*: a neuro-linguistic psychotherapy based on the spatial metaphors produced by the client on his present and desired "space" (state).
- **Neurons with spatial firing properties**
  Known types of **neurons with spatial firing properties** are listed below; cf. also **spike train distances** in Chap. 23.
  Many mammals have in several brain areas *head direction cells*: neurons which fire only when the animal's head points in a specific direction.
  Place cells are principal neurons in the hippocampus that fire strongly whenever an animal is in a specific location (the cell's *place field*) in an environment.
  Grid cells are neurons in the entorhinal cortex that fire periodically and at very regular distances as an animal walks. Grid cells measure distance while place cells indicate location. But only place cells are sensitive (albeit weakly) to height.
  Spatial view cells are neurons in the hippocampus which fire when the animal views a specific part of an environment. They differ from head direction cells since they represent not a global orientation, but the direction towards a specific object. They also differ from place cells, since they are not localized in space.

Border cells are neurons in the entorhinal cortex that fire when a border is present in the proximal environment.
Mirror cells are neurons that fire both when an animal acts and when it observes the same action performed by another.
Head direction cells of rats are fully developed before pups open their eyes and become mobile. Next to mature are place cells followed by grid cells. All navigational cell types mature before rat adolescence (about 30 days of age).
The smallest processing module of cortical neurons is a *minicolumn*—a vertical column (of diameter 28–40 μm) through the cortical layers of the brain, comprising 80–120 neurons that seem to work as a team. There are about 2×10^8 minicolumns in humans. Smaller minicolumns (as observed in scientists and in people with autism) mean that there are more processing units within any given cortical area; it may allow for better signal detection and more focused attention.

- **Vision distances**
The *interocular distance* is the distance between the centers of rotation of the eyeballs of an individual or between the oculars of optical instruments.
The **interpupillary distance** (or *binocular pupillary distance*) is the distance (50–75 mm) between the centers of the pupils of the two eyes. The *monocular pupillary distance* is the distance from the center of the nose to the pupil. *Stereoacuity* is the smallest detectable depth difference that can be seen in binocular vision.
The *near acuity* is the eye's ability to distinguish an object's shape and details at a near distance such as 40 cm; the **distance acuity** is the ability to do it at a far distance such as 6 m. The **distance vision** is a vision for objects that are at least 6 m from the viewer. *Optical near devices* are designed for magnifying close objects and print; **distance devices** are for magnifying things in the distance.
The *near distance* is the distance between the object and *spectacle* (eyeglasses) planes. The *vertex distance*: the distance between a person's glasses (spectacles planes) and their eyes (the corneal). The *infinite distance*: a distance of at least 6 m; so called because rays entering the eye from an object at that distance are practically as parallel as if they came from an infinitely far point.
The *default accommodation distance* (or *resting point of accommodation, RPA distance*) is the distance at which the eyes focus if there is nothing to focus on.
The *RPV distance* (or *resting point of vergence*) is the distance at which the eyes are set to *converge* (turn inward toward the nose) when there is no close object to converge on. It averages about 1.15 m when looking straight ahead and in to about 0.9 m with a 30° downward gaze angle. Ergonomists recommend the RPV distance as the eye-screen distance in sustained viewing, in order to minimize eyestrain.
The *least distance of distinct vision* (or *reference seeing distance*) is the minimum comfortable distance (usually, 25 cm) between the eye and a visible object. Ideal focus distances for reading and writing are within 37–62 cm from the eyes.
The **Harmon distance** (or *elbow distance*) is the optimal visual distance for reading and other near work. It is the distance from the elbow on the desk to the first *knuckle* (prominence of a joint connecting the finger to the hand).

The ideal *TV viewing distance* is 1.9 times the screen width, since then this width occupies a 30° angle from the viewing position. For multiple-row seating in the home theater, a viewing angle 26–36° is recommended.

The **Lechner distance** is the optimal viewing distance at which the human eye can best process the details given by High Definition TV resolution. For example, it is about 1.7 or 2.7 m for a 1080 HD TV with a screen size of 42 or 69 inches.

Lateral masking (or *crowding*) is impairment of peripheral object identification by *flankers* (nearby objects). *Critical spacing* (or *crowding distance*) is the minimum target-flanker distance that does not produce crowding of a target of fixed size.

The *throw distance* is the distance that the projector needs to be from the screen to project the optimum image. The *viewing-distance factor* is a ratio of the width of a projected image to the maximum acceptable viewer's distance from it.

The **laser hazard distance** is the safe viewing distance for direct exposure to visible laser beams.

- **Gait distances**

  Gait stride is the distance traveled between successive footfalls of the same foot. It is the double of the *step length* (distance traveled while a foot is on the ground). *Stride width* (or *walking base*) is the side-to-side distance between the line of the two feet. Normally, it is 3–8 cm for adult but it increases with gait instability.

  The *Gait Deviation* (from normality) *Index* GDI is (Schwartz–Rosumalski, 2008) the **standardized Euclidean distance** (cf. Chap. 17) in the 15D gait feature space between the abnormal state vector of a patient and the closest matching normal (mean of controls) state vector.

  The length of cane, when it is needed, should extend the distance from the distal wrist crease to the ground, when the person is placing arms at the sides.

  The average walking speed is 1–1.5 m/s; above 2 m/s, it is more efficient to run. Cadence for normal adults is 100–117 steps/min at preferred speed. As the body moves forward, its center of gravity moves vertically and laterally, with average displacement 5 and 6 cm, in a smooth sinusoidal pattern.

  Wearing high heels by women exaggerate some sex specific elements of female gait: greater pelvic rotation, increased lateral pelvic tilt, shorter strides and higher cadence. Millipedes (in fact, no species with more than 750 legs is known) have smooth wave-like polypedal gait.

  Most insects have a *tripod gait*, with front and back legs on one side moving in sync with the middle leg on the other side. But some dung beetles can gallop.

  Humans, birds and (occasionally) apes walk bipedally. Humans, birds, many lizards and (at their highest speeds) cockroaches run bipedally. But (Alexander, 2004) no animal walks or runs as we do: the trunk erect, almost straight knees at mid-stance, striking the ground with the heel alone and two-peaked force pattern in fast walking. Our walking, but not running, is relatively economical metabolically.

- **Biodistances for nonmetric traits**
In Physical Anthropology and Human Osteology (including Forensic Anthropology and Paleoanthropology), the *biodistances* (or *biological distances*) are the measures of relatedness between and within human groups, living or past, based on human cranial, skeletal or dental variation.
Nonmetric traits are skeletal **nonmetric data** (binary, nominal or ordinal, cf. Chap. 17). The main distance statistics used to compare them between populations x and y are *Mahalanobis* D^2 *statistics*, i.e., square**Mahalanobis distance** (Chap. 17) and, when the data are incomplete, the **mean measure of divergence**:

$$MMD = \frac{1}{n}\sum_{i=1}^{n}((\phi_{xi} - \phi_{yi})^2 - 4\frac{N_{xi} + N_{yi} + 1}{(2N_{xi} + 1)(2N_{yi} + 1)}).$$

Here n is the number of traits used in the comparison, ϕ_{xi} and ϕ_{yi} are the transformed frequencies in radians of the i-th trait in the groups x, y, and N_{xi} and N_{yi} are the numbers of individuals scored for the i-th trait in the two groups. The frequencies ϕ are obtained (in radians) from observed trait frequencies $\frac{k}{n}$ by the *Freeman–Tukey arcsine transformation*. The MMD can be negative. The *standardized MMD* (SMMD) is obtained by dividing MMD by its standard deviation.

- **Body distances in Anthropometry**
Besides weight and circumference, the main metric (i.e., linear continuous, cf. Chap. 17) measurements in Anthropometry are between some body landmark points or planes. The main vertical distances from a standing surface are:

  - *stature* (to the top of the head);
  - *C7 level height* (to the first palpable vertebra from the hairline down, C7);
  - *acromial height* (to the *acromion*, i.e., the lateral tip of the shoulder);
  - *L5 level height* (to the first palpable vertebra from the tailbone up, L5);
  - *knee height* (to the *patella*, i.e., kneecap plane).

  The genotype gives 60 % of the phenotypic variation of human height (stature). It was about 1.63 and 1.83 m for *Neanderthal* 0.07 Ma ago and *Homo erectus* 1.8 Ma ago. The height of the average modern man ranges from 1.37 (Mbuti people of the Democratic Republic of the Congo) to 1.84 m (the Dutch). There is small (0.15–0.20) correlation between IQ and height within national populations.
Examples of other body distances are:

  - *sitting height*: the distance from the top of the head to the sitting surface;
  - *popliteal* (or *stool*) height (seated): the distance between the underside of the foot to the underside of the thigh at the knee;
  - *hip breadth* (seated): the lateral distance at the widest part of the hips;
  - *biacromial breadth*: the distance between the acromions;
  - *buttock-knee length* : the distance from the buttocks to the patella;
  - *total foot length*: the maximum length of the right foot;

In the thigh, there are the longest ones in the human body: bone (*femur*), muscle (*sartorius*) and nerve (*sciatic*).

- **Head and face measurement distances**

The main linear dimensions of the cranium in Archeology are: lengths (of temporal bone, of tympanic plate, glabella-opistocranion), breadths (maximum cranial, minimum frontal, biauricular, mastoid), heights (of temporal bone, basion-bregma), thickness of the tympanic plate, and bifrontomolar-temporal distance.

Main viscerocranium measurements in Craniofacial Anthropometry are the *head width*, i.e., the (horizontal) maximum breadth of the head above the ears, and the *head length* (or *head depth*): the horizontal distance from the *nasion* (the top of the nose between the eyes) to the *opistocranion* (the most prominent point on the back of the head). The *cephalic index* of a skull is the percentage of width to length.

The **face length** FL: the distance between the *trichion* (midpoint on the forehead) and the *gnathion* (the lowest point of the midline of the lower jaw). It is divided by nasion and *subnasale* lines into three ("ideally", equal) parts.

The **intercanthal distance** (*medial* MC or *lateral* LC): the distance between (inner or outer) *canthi* (corners of eyes). The **face width** FW (or *bizygomatic width*) is the maximum distance between lateral surfaces of the *zygomatic arches* (cheeks). Let EW be the eye width and NW be the nose width (or **interalar distance**). "Ideally", it holds $NW = EW = MC$ and $FW = 5\,EW$.

The *upper face height* UFH is the distance between the nasion and *the prosthion* (midpoint on the alveolar arch between the median upper incisor teeth). The *superior facial index* is $\frac{UFH}{FW}$; its closeness to the *Fibonacci number* $\frac{1+\sqrt{5}}{2} \approx 1{,}618$ is one of proposed cues of female's beauty. According to Lefevre et al., 2013, the ratios fWHR$= \frac{FW}{UFH}$ and $\frac{FW}{LFH}$, where LFH is the *lower face height*, correlate with "maleness" (testosterone in mating context, aggression, status-striving etc.). On average, men have much larger faces (below the pupils), lips and chins; wider cheekbones, jaws and nostrils; and longer lower faces, but much lower eyebrows.

In Face Recognition, the sets of (vertical and horizontal) *cephalofacial dimensions*, i.e., distances between *fiducial* (standard of reference for measurement) facial points, are used. For example, the following five independent facial dimensions are derived in [Fell97] for facial gender recognition: LC, NW, FW and (vertical ones) eye-to-eyebrow distance EB and distance EM between eye midpoint and horizontal line of mouth. "Femaleness" relies on large LC, EB and small NW, FW, EM. In general, a face with larger EB is perceived as baby-like and less dominant.

Humans have the innate ability to recognize and distinguish (friend from foe) between faces from a distance. Facial attractiveness is a cultural construct found in all extant societies, and males strongly prefer neotenous facial features in females. Pallett–Link–Lee, 2009, claim that Caucasian females with $EM \approx 36\,\%$ of FL and the **interpupillary distance** $\approx 46\,\%$ of FW, have the both, most

attractive and average, faces. Cunningham et al., 1995, claim that the ideal attractive female face tends to feature: $3\,FW = 5\,EW$, chin length $\frac{h}{5}$ (where h is the height of the face), middle of eye to bottom of the eyebrow $\frac{h}{10}$, height of the visible eyeball $\frac{h}{14}$, pupil width $\frac{1}{14}$ the distance between cheekbones, nose area $< 5\,\%$ the total area of the face. But those standards could be too Western-oriented.
For example, Japanese standards of beautiful eyes changed with Westernization (comparing Meiji and modern portraits): the mean ratios to *corneal diameter* (horizontal white-to-white distance) of eye height and upper lid-to-eyebrow distance are moved from 0.62 and 2.21 to 0.82 and 1.36.
Modifying traditional canons of Facial Plastic Surgery (based on horizontal and vertical planes in 2D), Young, 2008, asserts that the iris, nasal tip and lower lip are the most prominent structures within the eye, nose and mouth. All distances which he proposed as elements of facial beauty are multiples of the diameter of the iris.
Comparing 3D facial scans with their mirror images, Djordjevic et al., 2011, found that on average, males and females have 53.5 and 58.5 % symmetry of the whole face. Cf. **distances from symmetry** in Chap. 21. Alare and pogonio were the most and the least symmetric landmark.

- **Gender-related body distance measures**
The main gender-specific body configuration features are:
for females, WHR (*waist-to-hip ratio*), LBR (*leg-to-body ratio*) and BMI (*body mass index*), i.e., the ratio of the weight in kg and squared height in m^2;
for males, height, SHR (*shoulder-to-hip ratio*) and WCR (*waist-to-chest ratio*);
androgen equation (three times the shoulder width minus one times the pelvic width) which is higher for males;
second-to-fourth digit (index to ring finger) *ratio 2D–4D* which is lower (as well as prenatal testosterone is higher) for males in the same population;
anogenital distance (cf. **distances in Medicine**) which is larger for males;
person's *center of mass* (slightly below the belly button) which is higher for males.
The female pelvis is more rounded. The sciatic notches are broader, the greater pelvis is shallower, the lesser pelvis is wider, the pelvic inlet and outlet are larger.
The mean footprint ridge density is higher among females.
The main predictor for developmental instability, increasing with age, is FA (*fluctuating asymmetry*), i.e., the degree to which the size of bilateral body parts deviates from the population mean, aggregated across several traits. At age 79–83 men (but not women) with lower facial FA have better cognitive ability and reaction time. Women prefer the odors, faces and voices of men with lower FA.
BMI and WHR indicate the percentage of body fat and fat distribution, respectively; they are used in medicine to assess risk factors. A WHR of 0.7 for women and 0.9 for men correlates with general health and fertility. As a cue to female body attractiveness for men, the ideal WHR varies from 0.6 in China to 0.85 in Africa. But Rilling et al., 2009, claim that *abdominal depth* (the depth of

the lower torso at the umbilicus) and WC (waist circumference) are stronger predictors.

In Fan et al., 2005, the main visual cue to male body attractiveness is VHI (*volume-to-height index*), i.e., the ratio of the volume in liters and squared height in m^2. Mautz et al., 2013, claim that women prefer taller men with higher SHR and FPL (flaccid penis length), but attractiveness increased quickly until FPL reached 7.6 cm and then began to slow down. Stulp et al., 2013, found that on average among speed-daters, women choose 25 cm taller men, while men choose only 7 cm shorter women, resulting in suboptimal (19 cm) pair formation.

In terms of somatotype, women prefer mesomorphs (muscular men) followed by ectomorphs (lean men) and endomorphs (heavily-set men).

In terms of the vital statistics BWH (bust–waist–hips), the average Playboy centrifold 1955–1968 has $(90.8, 58.6, 89.3)$ cm, close to the ideal hourglass figures $(90, 60, 90)$ cm and $(36, 24, 36)$ inch. The British Association of Model Agents prefers model around $(86, 60, 86)$ cm and at least 1.73 m tall. But dietitians found these proportions unhealthy and advocate waistline 80–85 cm and at most half-height.

In conversation, women are better at detecting mismatch between meaning and *prosody* (intonation and rhythm of speech), but worse at vocabulary's variety. Men's vocal cords are larger and their vocal tracts are longer than women's; so, they speak about an octave lower. In English, women use less nonstandard forms and often use different color terms and descriptive phrases from men. Pirahã (Amazon's tribe) men use larger articulatory space and, say, only men use "s".

Used as obesity indices, WC, ICO=WC/height and (proposed by Krakauer–Krakauer, 2012) ABSI=$WC/(BMI^{\frac{2}{3}} height^{\frac{1}{2}})$ are better predictors of mortality than BMI.

- **Sagittal abdominal diameter**

  Sagittal abdominal diameter (SAD) is the distance between the back and the highest point of the abdomen, measured while standing. It is a measure of visceral obesity. Normally, SAD should be under 25 cm. SAD$>$ 30 cm correlates to insulin resistance and increased risk of cardiovascular and Alzheimer's diseases. A related measurement is SAH, the abdominal height as measured in the supine position. *Inter-recti distance* (IRD) is the width of the *linea alba* (a fibrous structure that runs down the midline of the abdomen).

- **Body distances for clothes**

  Humans lost body hair around 1 Ma ago and began wearing clothes $\approx$0.17 Ma ago.

  The European standard EN 13402 "Size designation of clothes" defined, in part EN 13402-1, a standard list of 13 body dimensions (measured in cm) together with a method for measuring each one on a person. These are: body mass, height, foot length, arm length, inside leg length, and girth for head, neck, chest, bust, under-bust, waist, hip, hand. Examples of these definitions follow.

Foot length: horizontal distance between perpendiculars in contact with the end of the most prominent ones, toe and part of the heel, measured with the subject standing barefoot and the weight of the body equally distributed on both legs.
Arm length: distance from the armscye/shoulder line intersection (acromion), over the elbow, to the far end of the prominent wrist bone (ulna), with the subject's right fist clenched and placed on the hip, and with the arm bent at 90°.
Inside leg length: distance between the crotch and the soles of the feet, measured in a straight vertical line with the subject erect, feet slightly apart, and the weight of the body equally distributed on both feet.
For clothes where a larger step size is sufficient to select the right product, the standard also defines a letter code: XXS, XS, S, M, L, XL, XXL, 3XL, 4XL or 5XL. This code represents the bust girth for women and the chest girth for men.
Vanity sizing (or *size deflation*) is the marketing phenomenon of ready-to-wear clothing of the same nominal size becoming bigger in physical size over time.

- **Distance handling**
  Distance handling refers to the training of gun dogs (to assist hunters in finding and retrieving game) or sport dogs (for canine agility courses) where a dog should be able to work away from the handler.
  In agility training, the *lateral distance* is the distance that the dog maintains parallel to the handler, and the *send distance* is the distance that the dog can be sent straight away from the handler.
- **Racing distances**
  In Racing, **length** is an informal unit of distance to measure the distance between competitors; for example, in boat-racing it is the average length of a boat.
  The **horse-racing distances** and winning margins are measured in terms of the **lengths** of a horse, i.e., $\approx$8 feet (2.44 m), ranging from half the length to the **distance**, i.e., more than 20 lengths. The *length* is often interpreted as a unit of time equal to $\frac{1}{5}$ second. Smaller margins are: *short-head*, *head*, or *neck*. A *distance flag* is a flag held at a distance pole in a racecourse.
  The distances a horse travels without stops (15–25 km) and it travels in a day (40–50 km) or hour (6 km) were used as Tatar and Persian units of length.
- **Triathlon race distances**
  The **Ironman distance** (or **Ultra distance**) started in Hawaii, 1978, is a 3.86 km swim followed by a 180 km bike and a 42.195 km (*marathon distance*) run.
  The international **Olympic distance** started in Sydney, 2000, is 1.5 km (*metric mile*), 40 km and 10 km of swim, cycle and run, respectively.
  Next to it are the **Sprint distance** 0.75, 20, 5 km, the **Long Course** (or *Half Ironman*) 1.9, 90, 21.1 km and the *ITU long distance* 3, 80, 20 km.
- **Running distances**
  In Running, usually, *sprinting* is divided into 100, 200, 400 m, *middle distance* into 800, 1,500, 3,000 m and *long distance* into 5, 10 km.
  LSD (long slow distance) is a is a form of aerobic endurance training in running and cycling, in which distances longer, than those of races, are covered, but at a slower pace.

Fartlek (or *speed play*) is an approach to distance-running training involving variations of pace and aimed at enhancing the psychological aspects of conditioning. *Race-walking* is divided into 10, 20, 50 km, and *relay races* into $4 \times 100, 4 \times 200, 4 \times 300, 4 \times 400$ m. *Distance medley relay* is made up of 1200, 400, 800, 1,600 m legs.
Besides track running, runners can compete on a measured course, over an established road (*road running*), or over open or rough terrain (*cross-country running*).
Roughly, 4 units of running distance are time-equivalent to 1 unit of swimming distance. Also, one have to walk about twice the distance to burn the same amount of calories as running it. Running workout times should be multiplied by 3.5 when aiming for a similar training effect from cycling. A multiple of 0.75–1 should be used for an indoor rowing-to-running ratio.

- **Distance swimming**
  Distance swimming is any swimming race >1.5 km; usually, within 24–59 km. *DPS* (distance per swim stroke) is a metric of swimming efficiency used in training. In Rowing, *run* is the distance the boat moves after a stroke.
- **Distance jumping**
  The four Olympic jumping events are: *long jump* (to leap horizontally as far as possible), *triple jump* (the same but in a series of three jumps), *high jump* (to reach the highest vertical distance over a horizontal bar), and *pole vault* (the same but using a long, flexible pole).
  The world's records, as in 2013: 8.95, 18.29, 2.45, and 6.14 m, respectively.
- **Distance throwing**
  The four Olympic throwing events are: shot put, discus, hammer, and javelin.
  The world's records, as in 2013: 23.12, 74.08, 86.74 m, and 98.48 m, respectively.
  As in 2013, the longest throws of an object without any velocity-aiding feature are 427.2 m with a boomerang and 406.3 m with a flying ring *Aerobie*.
  Distance casting is the sport of throwing a fishing line with an attached sinker (usually, on land) as far as possible.
  Darts is a sport and a pub game in which darts are thrown at a *dartboard* (circular target) fixed to a wall so that the bullseye is 172.72 cm from the floor. The *oche* (line behind which the throwing player must stand) is 236.86 cm from the dartboard.
- **Archery target distances**
  FITA (Federation of International Target Archery, organizing world championships) **target distances** are 90, 70, 50, 30 m for men and 70, 60, 50, 30 m for women, with 36 arrows shot at each distance. Farthest accurate shot is 200 m.
- **Bat-and-ball game distances**
  The best known bat-and-ball games are bowling (cricket) and baseball. In cricket, the field position of a player is named roughly according to its polar coordinates: one word (*leg, cover, mid-wicket*) specifies the angle from the batsman, and this word is preceded by an adjective describing the distance from the batsman.

The *length* of a delivery is how far down the *pitch* (central strip of the cricket field) towards the batsman the ball bounces.
This distance is called *deep* (or *long*), *short* and *silly distance* if it is, respectively, farther away, closer and very close to the batsman. The distance further or closer to an extension of an imaginary line along the middle of the pitch bisecting the stumps, is called *wide* or *fine distance*, respectively.
In baseball, a *pitch* is the act of throwing a baseball toward home plate to start a play. The standard professional *pitching distance*, i.e., the distance between the front (near) side of the pitching rubber, where a pitcher start his delivery, and home plate is 60 feet 6 inches ($\approx$18.4 m). The distance between bases is 90 feet.

- **Three-point shot distance**
  In basketball, the *three-point line* is an arc at a set radius, called **three-point shot distance**, from the basket. A field goal made from beyond this line is worth three points. In international basketball, this distance is 6.25 m.
  Goals in *indoor soccer* are worth $1, 2$ or 3 points depending upon distance.
- **Football distances**
  In *association football* (or *soccer*), the average distance covered by a player in a men's professional game is 9–10 km. It consists of about 36 % jogging, 24 % walking, 20 % cruising submaximally, 11 % sprinting, 7 % moving backwards and 2 % moving in possession of the ball. The ratio of low- to high-intensity exercise is about 2.2:1 in terms of distance, and 7:1 in terms of time.
  In *American football*, one *yard* means usual yard (0.9144 m) of the distance in the direction of one of the two goals. A field is 120 yards long by 53.3 yards wide. A team possessing the ball should advance at least the *distance* (10 yards) in order to get a new set of (4 or 3) *downs*, i.e., periods from the time the ball is put into play to the time the play is whistled over by the officials. *Yardage* is the amount of yards gained or lost during a play, game, season, or career.
- **Golf distances**
  In golf, *carry* and *run* are the distances the ball travels in the air and once it lands. The golfer chooses a golf club, grip, and stroke appropriate to the distance. The *drive* is the first shot of each hole made from the area of *tees* (peg markers) to long distances. The *approach* is used in long- to mid-distance shots.
  The *chip* and *putt* are used for short-distance shots around and, respectively, on or near the green. The maximum distance a typical golfer can hit a ball with a particular club is the club's *hitting distance*.
  A typical *par* (standard score) $3, 4, 5$ holes measure $229, 230$–$430, \geq 431$ m. The greatest recorded drive distance, carry, shot with one hand are $471, 419, 257$ m.
  Some manufacturers stress the large range of a device in the product name, say, *Ultimate Distance* golf balls (or softball bates, spinning reels, etc.).
- **Fencing distances**
  In combative sports and arts, *distancing* is the appropriate selection of the distance between oneself and a combatant throughout an encounter.
  For example, in fencing, the *distance* is the space separating two fencers, while the distance between them is the *fencing measure*.

A *lunge* is a long step forward with the front foot. A *backward spring* is a leap backwards, out of distance, from the lunge position.
The following five distances are distinguished: *open distance* (farther than advance-lunge distance), *advance-lunge distance*, *lunging distance*, *thrusting distance* and *close quarters* (closer than thrusting distance).
In Japanese martial arts, **maai** is the *engagement distance*, i.e., the exact position from which one opponent can strike the other, after factoring in the time it will take to cross their distance, angle and rhythm of attack. In kendo, there are three maai distances: *to-ma* (long distance), *chika-ma* (short distance) and, in between, *itto-ma* $\approx$2 m, from which only one step is needed in order to strike.

- **Distance in boxing**
  The distance is boxing slang for a match that lasts the maximum number (10 or 12) of scheduled rounds. The longest boxing match (with gloves) was on April 6–7, 1893, in New Orleans, US: Bowen and Burke fought 110 rounds for 7.3 h.
- **Soaring distances**
  Soaring is an air sport in which pilots fly unpowered aircraft called *gliders* (or *sailplanes*) using currents of rising air in the atmosphere to remain airborne.
  The *Silver Distance* is a 50 km unassisted straight line flight. The *Gold* and *Diamond Distance* are cross-country flights of 300 km and over 500 km, respectively.
  Possible courses—*Straight*, *Out-and-Return*, *Triangle* and 3 *Turnpoints Distance*—correspond to 0, 1, 2 and 3 turnpoints, respectively.
  Using open class gliders, the world records in free distance, in absolute altitude and in gain of height are: 3,008.8 km (by Olhmann and Rabeder, 2003), 15,460 m (by Fossett and Enevoldson, 2006) and 12,894 m (by Bikle, 1961). The distance record with a paraglider is 501.1 km (by Hulliet, 2008).
  Baumgartner jumped in 2012 from a balloon at 39.04 km, opening his parachute at 2.52 km. He set records in altitude and unassisted speed 373 m/s = 1.24 Mach, but his free-fall was 17 s shorter than 4 min 36 s by Kittinger, 1960. The longest genuine, i.e., without the use of a drogue chute, free fall record is by Andreev, 1962: 24,500 m from an altitude of 25,458 m. A stewardess Vesna Vulović survived in 1972 a fall of 10,000 m, when JAT Flight 367 was brought down by explosives.
- **Aviation distance records**
  Absolute general aviation world records in flight distance without refueling and in altitude are: 41,467.5 km by Fossett, 2006, and 37,650 m by Fedotov, 1977.
  Distance and altitude records for free manned balloons are, respectively: 40,814 km (by Piccard and Jones, 1999) and 39,068 m (by Baumgartner, 2012).
  The general flight altitude record is 112,010 m by Binnie, 2004, on a rocket plane.
  The longest (15,343 km during 18.5 h) nonstop scheduled passenger route is Singapore Airline's flight 21 from Newark to Singapore.
  The *Sikorsky prize* (US$250,000) will be awarded for the first flight of a human-powered helicopter which will reach an altitude of 3 m, stay airborne for at least 1 min remaining within 10 m × 10 m. In 2012, a craft (32.2 kg) by a team at the University of Maryland flew 50 s at 61 cm up.

- **Amazing greatest distances**
  Examples of such distances among Guinness world records are the greatest distances:

  - goal scored in football (91.9 m),
  - being fired from a cannon (59 m),
  - walked unsupported on tightrope (130 m),
  - run on a static cycle in 1 min (2.04 km),
  - moon-walked (as Michael Jackson) in 1 h (5.125 km),
  - covered three-legged (the left leg of one runner strapped to the right leg of another runner) in 24 h (33 km),
  - jumped with a pogo stick (37.18 km),
  - walked with a milk bottle balanced on the head (130.3 km),
  - covered by a car driven on its side on two wheels (371.06 km),
  - hitchhiked with a fridge (1,650 km).

  Amazing race *The* 2904 is to drive the 2,904 miles from New York City to San Francisco for $2,904 including the vehicle, fuel, food, tolls, repairs and tickets.
- **Isometric muscle action**
  An **isometric muscle action** refers to exerting muscle strength and tension without producing an actual movement or a change in muscle length.
  Isometric action training is used mainly by weightlifters and bodybuilders. Examples of such *isometric exercises*: holding a weight at a certain position in the range of motion and pushing or pulling against an immovable external resistance.

29.2 Equipment Distances

- **Motor vehicle distances**
  The **safe following distance**: the reglementary distance from the vehicle ahead of the driver. For reglementary perception-reaction time of at least 2 s (the *two-second rule*), this distance (in m) should be $0.56 \times v$, where v is the speed (in km/h). Sometimes the *three-second rule* is applied. The stricter rules are used for heavy vehicles (say, at least 50 m) and in tunnels (say, at least 150 m).
  The **perception-reaction distance** (or **thinking distance**): the distance a vehicle travels from the moment the driver sees a hazard until he applies the brakes (corresponding to human perception time plus reaction time). Physiologically, it takes 1.3–1.5 s, and the brake action begins 0.5 s after application.
  The **braking distance**: the distance a motor vehicle travels from the moment the brakes are applied until the vehicle completely stops.
  The (total) **stopping distance**: the distance a motor vehicle travels from where the driver perceives the need to stop to the actual stopping point (corresponding to the vehicle reaction time plus the vehicle braking capability).
  The **crash distance**: (or *crushable length*): the distance between the driver and the front end of a vehicle in a frontal impact (or, say, between the pilot and the first part of an airplane to impact the ground).

The **skidding distance** (or *length of the skid mark*): the distance a motor vehicle *skidded*, i.e., slid on the surface of the road (from the moment of the accident, when a wheel stops rolling) leaving a rubber mark on the road.
The **cab-to-frame** (or **cab-to-end**, *CF*, *CE*): the distance from back of a truck's cab to the end of its frame.
The **distance to empty** (or *DTE*) displays the estimated distance the vehicle can travel before it runs out of fuel. The warning lamp start blinking at 80 km.
The **acceleration-deceleration distance** of a vehicle, say, a car or aircraft, is (Drezner–Drezner–Vesolowsky, 2009) the cruising speed v times the travel time. For a large origin-destination distances d, it is $d + \frac{v^2}{2}(\frac{1}{a} + \frac{1}{b})$, where a is the acceleration at the beginning and $-b$ is the deceleration at the end.

- **Aircraft distances**
The maximum distance the aircraft can fly without refueling is called the **maximum range** if it fly with its maximum cargo weight and the **ferry range** if it fly with minimum equipment.
For a warplane, the *combat range* is the maximum distance it can fly when carrying ordnance, and the *combat radius* is a the maximum distance it can travel from its base, accomplish some objective, and return with minimal reserves.
The FAA **lowest safe altitude**: 1,000 feet (305 m) above the highest obstacle within a horizontal distance of 2,000 feet.
A **ceiling** is the maximum *density altitude* (height measured in terms of air density) an aircraft can reach under a set of conditions.
A *flight level* (FL) is specific barometric pressure, expressed as a nominal altitude in hundreds of feet, assuming standard sea-level pressure datum of 1013.25 hPa.
The *transition altitude* is the altitude above sea level at which aircraft change from the use of altitude to the use of FL's; in US and Canada, it is 18,000 feet (5,500 m).
The **gust-gradient distance**: the horizontal distance along an aircraft flight path from the edge of the *gust* (sudden, brief increase in the speed of the wind) to the point at which the gust reaches its maximum speed.
The **distance-of-turn anticipation**: the distance, measured parallel to the anticipated course and from the earliest position at which the turn will begin, to the point of route change.
The **landing distance available** (LDA): the length of runway which is declared available and suitable for the ground run of an airplane landing. The **landing roll**: the distance from the point of touchdown to the point where the aircraft can be brought to a stop or exit the runway. The **actual landing distance** (ALD): the distance used in landing and braking to a complete stop (on a dry runway) after crossing the runway threshold at 50 feet (15.24 m); it can be affected by various operational factors. The FAA **required landing distance** (used for dispatch purposes): a factor of 1.67 of ALD for a dry runway and 1.92 for a wet runway.
The **takeoff run available** (TORA): the *runway distance* (length of runway) declared suitable for the ground run of an airplane takeoff. The **takeoff distance**

available (TODA): TORA plus the length of the clearway, if provided. The **emergency distance** (ED or *accelerate-stop distance*): the runway plus *stopway length* (able to support the airplane during an aborted takeoff) declared suitable for the acceleration and deceleration of an airplane aborting a takeoff.

The **arm's distance**: the horizontal distance that an item of equipment is located from the *datum* (imaginary vertical plane, from which all horizontal measurements are taken for balance purposes, with the aircraft in level flight attitude).

In the parachute deployment process, the **parachute opening distance** is the distance the parachute system dropped from pulling to full inflation of the canopy, while the **inflation distance** is measured from *line stretch* (when the suspension lines are fully extended) to full inflation.

Wing's aspect ratio (of an aircraft or bird) is the ratio $AR = \frac{b^2}{S}$ of the square of its span to the area of its planform. If the length of the *chord* (straight line joining the leading and trailing edges of an airfoil) is constant, then AR is length-to-breadth **aspect ratio**; cf. Chap. 1. A better measure of the aerodynamic efficiency is the *wetted aspect ratio* $\frac{b^2}{S_w}$, where S_w is the entire surface area exposed to airflow.

- **Ship distances**

  Endurance distance: the total distance that a ship or ground vehicle can be self-propelled at any specified endurance speed.

  Distance made good: the distance traveled by the boat after correction for current, *leeway* (the sideways movement of the boat away from the wind) and other errors that may be missed in the original distance measurement.

  Log: a device to measure the distance traveled which is further corrected to a distance made good. Hitherto, sea distances were measured in units of a day's sail.

  Leg (nautical): the distance traveled by a sailing vessel on a single tack.

  Berth: a safety margin to be kept from another vessel or from an obstruction.

  Length overall (LOA): the maximum length of a vessels's hull along the waterline.

  Length between perpendiculars (LPP): the length of a vessel along the waterline from the main bow perpendicular member to the main stern perpendicular member.

  Freeboard: the height of a ship's hull above the waterline. *Draft* (or *draught*): the vertical distance between the waterline and the keel (bottom of the hull).

  GM-distance (or *metacyclic height*) of a ship: the distance between its center of gravity G and the *metacenter*, i.e., the projection of the *center of buoyancy* (the center of gravity of the volume of water which the hull displaces) on the centerline of the ship as it heels. This distance, 1–2 m, determines ship's stability.

  Distance line (in Diving): a marker (say, 50 m of thin polypropylene line) of the shortest route between two points. It is used, as Ariadne's thread, to navigate back to the start in conditions of low visibility, water currents or *penetration diving* into a space (cave, wreck, ice) without vertical ascent back.

- **Distance-to-fault**
  In Cabling, DTF (**distance-to-fault**) is a test using time or frequency domain reflectometers to locate a fault, i.e., discontinuity caused by, say, a damaged cable, water ingress or improperly installed/mated connectors.
  The amount of time a pulse (output by the tester into the cable) takes for the signal (reflected by a discontinuity) to return can be converted to distance along the line and provides an approximate location of the reflection point.
  Protective *distance relays* respond to the voltage and current. The *impedance* (their ratio) per km being constant, these relays respond to the relay-fault distance.
- **Distances in Forestry**
  In Forestry, the **diameter at breast height** (d.b.h.) is a standard measurement of a standing tree's diameter taken at 4.5 feet ($\approx$1.37 m) above the ground. The *diameter at ground line* (d.g.l.) is the diameter at the estimated cutting height. The *diameter outside bark* (d.o.b.) is a measurement in which the thickness of the bark is included, and d.i.b. is a measurement in which it is excluded.
  The *crown height* is the vertical distance of a tree from ground level to the lowest live branch of the crown. The *merchantable height* is the point on a tree to which it is salable. A *log* is a length of tree suitable for processing into a wood product. *Optimum road spacing* is the distance between parallel roads that gives the lowest combined cost of *skidding* (log dragging) and road construction costs per unit of log volume. The *skid distance* is the distance logs are dragged.
  A *yarder* is a piece of equipment used to lift and pull logs by cable from the felling site to a landing area or to the road's side. The **yarding distance** is the distance from which the yarder takes logs. The **average yarding distance** is the total yarding distance for all turns divided by the total number of turns.
  A *spar tree* is a tree used as the highest anchor point in a cable logging setup. A *skyline* is a cableway stretched between two spar trees and used as a track for a log carriage. The distance spanned by a skyline is called its *reach*.
  Understory is the area of a forest which grows at the lowest height level between the forest *floor* and the *canopy* (layer formed by mature tree crowns and including other organisms). Perhaps, a half of all life on Earth could be found in canopy. The *emergent layer* contains a small number of trees which grow above the canopy.
- **Distance in Military**
  In the Military, the term **distance** usually has one the following meanings:
  the space between adjacent individual ships or boats measured in any direction between foremasts;
  the space between adjacent men, animals, vehicles, or units in a formation measured from front to rear;
  the space between known reference points or a ground observer and a target, measured in m (artillery), or in units specified by the observer. This distance along an imaginary straight line from the spotter is called *observer-target distance*.

In amphibious operations, the *distant retirement area* is the sea area located to seaward of the landing area, and the *distant support area* is the area located in the vicinity of the landing area but at considerable distance seaward of it.
Strategic depth refers to the distances between the front lines and the combatants' industrial and population core areas.
In military service, a **bad distance** of the troop means a temporary intention to extract itself from war service. This passing was usually heavily punished and equated with that of *desertion* (an intention to extract itself durably).
In US military slang, BFE (Big Fucking Empty) is an extremely distant or isolated deployment or location; used mostly about the disgust at the distance or remoteness. Also, a *klick* means a distance of 1 km.

- **Interline distance**
  In Engineering, the **interline distance** is the minimum distance permitted between any two buildings within an explosives operating line, in order to protect buildings from propagation of explosions due to the blast effect.
- **Scaled distance**
  The **scaled distance** (*SD*) is the parameter used to measure the level of vibration from a blast, when effects of the frequency characteristics are discounted.
  The minimum safe distance from a blast to a monitoring location is $SD \times \sqrt{W}$, where W denotes the maximum per delay (instantaneous) charge weight.
- **Standoff distance**
  The **standoff distance** is the distance of an object from the source of an explosion (in Warfare), or from the delivery point of a laser beam (in laser material processing). Also, in Mechanics and Electronics, it is the distance separating one part from another; for example, for insulating (cf. **clearance distance**), or the distance from a noncontact length gauge to a measured material surface.
- **Buffer distance**
  In Nuclear Warfare, the **horizontal buffer distance** is the distance which should be added to the radius of safety in order to be sure that the specified degree of risk will not be exceeded. The **vertical buffer distance** is the distance which should be added to the fallout safe-height of a burst, in order to determine a desired height of burst so that militari significant fallout will not occur.
  The term *buffer distance* is also used more generally as, for example, the buffer distance required between sister stores or from a high-voltage line.
  Cf. **clearance distance** and, in Chap. 25, **setback distance**.
- **Offset distance**
  In Nuclear Warfare, the **offset distance** is the distance the desired (or actual) ground zero is offset from the center of the area (or point) target.
  In Computation, the *offset* is the distance from the beginning of a string to the end of the segment on that string. For a vehicle, the **offset** of a wheel is the distance from its hub mounting surface to the centerline of the wheel.
  The term *offset* is also used for the **displacement** vector (cf. Chap. 24) specifying the position of a point or particle in reference to an origin or to a previous position.

- **Range of ballistic missile**
  Main **ranges of ballistic missiles** are *short* (at most 1,000 km), *medium* (1,000–3,500 km), *long* (3,500–5,500 km) and *intercontinental* (at least 5,500 km). *Tactical* and *theatre* ballistic missiles have ranges 150–300 and 300–3,500 km.
- **Proximity fuse**
  The **proximity fuse** is a fuse that is designed to detonate an explosive automatically when close enough to the target.
- **Sensor network distances**
  The **stealth distance** (or *first contact distance*): the distance traveled by a moving object (or intruder) until detection by an active sensor of the network (cf. **contact quasi-distances** in Chap. 19); the *stealth time* is the corresponding time.
  The **first sink contact distance**: the distance traveled by a moving object (or intruder) until the monitoring entity can be notified via a sensor network.
  The **miss distance**: the distance between the lines of sight representing estimates from two sensor sites to the target (cf. the **line-line distance** in Chap. 4).
  The **sensor tolerance distance**: a **range distance** within which a localization error is acceptable to the application (cf. the **tolerance distance** in Chap. 25).
  The actual distances between some pairs of sensors can be estimated by the time needed for a two-way communication. The positions of sensors in space can be deduced (cf. **Distance Geometry Problem** in Chap. 15) from those distances.
- **Proximity sensors**
  Proximity (or *distance*) **sensors** are varieties of ultrasonic, laser, photoelectric and fiber optic sensors designed to measure the distance from itself to a target. For such laser range-finders, a special *distance filter* removes measurements which are shorter than expected, and which are therefore caused by an unmodeled object. The *blanking distance* is the minimum range of an ultrasonic proximity sensor.
  The *detection distance* is the distance from the detecting surface of a sensor head to the point where a target approaching it is first detected. The *maximum operating distance* is its maximum detection distance from a standard modeled target, disregarding accuracy. The *stable detection range* is the detectable distance range in which a standard detected object can be stably detected with respect to variations in the operating ambient temperature and power supply.
  The *resolution* is the smallest change in distance that a sensor can detect. The *span* is the working distance between measurement range endpoints over which the sensor will reliably measure displacement. The *target standoff* is the distance from the face of the sensor to the middle of the span.
  Distance constant of a metereological sensor is the length of fluid flow past required to cause it to respond to 63.2 % (i.e., $1 - \frac{1}{e}$) of a step change in speed.
- **Precise distance measurement**
  The resolution of a TEM (transmission electronic microscope) is about 0.2 nm (2×10^{-10} m). This resolution is 1,000 times greater than a light microscope and about 500,000 times greater than that of a human eye which is 576 mega pixel. However, only nanoparticles can fit in the vision field of an electronic microscope.

The methods, based on measuring the wavelength of laser light, are used to measure macroscopic distances nontreatable by an electronic microscope. But the uncertainty of such methods is at least the wavelength of light, say, 633 nm. The recent adaptation of *Fabry–Perot metrology* (measuring the frequency of light stored between two highly reflective mirrors) to laser light permits the measuring of relatively long (up to 5 cm) distances with an uncertainty of only 0.01 nm.
The main devices used for low accuracy distance measurement are the rulers, engineer's scales, calipers and surveyor's wheels.

- **Laser distance measurement**
  Lasers measure distances without physical contact. They allow for the most sensitive and precise length measurements, for extremely fast recording and for the largest measurement ranges. The main techniques used are as follows.
  Triangulation (cf. **laterations**) is useful for distances from 1 mm to many km. *Pulse measurements*, used for large distances, measure the time of flight of a laser pulse from the device to some target and back. The *phase shift method* uses an intensity-modulated laser beam. *Frequency modulation methods* involve frequency-modulated laser beams. *Interferometers* allow for distance measurements with an accuracy which is far better than the wavelength of the light used.
  The main advantage of laser distance measurement is that laser light has a very small wavelength, allowing one to send out a much more concentrated probe beam and thus to achieve a higher transverse spatial resolution.
- **Radio distance measurement**
  DME **distance measuring equipment**) is an air navigation technology that measures distances by timing the propagation delay of UHF signals to a *transponder* (a receiver-transmitter that will generate a reply signal upon proper interrogation) and back. DME will be phased out by global satellite-based systems: GPS (US), GLONASS (Russia), BeiDou (China) and Galileo (EU).
  The GPS (Global Positioning System) is a radio navigation system which permits one to get her/his position on the globe with accuracy of 10 m. It consists of 32 satellites and a monitoring system operated by the US Department of Defense. The nonmilitary part of GPS can be used by the purchase of an adequate receiver. The **GPS pseudo-distance** (or *pseudo-range*) is an approximation (since the receiver clock is not so perfect as the clock of a satellite) of the distance between a satellite and a GPS receiver by the travel time of a satellite time signal to a receiver multiplied by the propagation time of the radio signal.
  The receiver uses **trilatelation** in order to calculate its position (latitude, longitude, altitude) and speed by solving a system of equations using its pseudo-distances from 4 to 7 satellites and their positions. Cf. **radio distances** in Chap. 25.
- **Laterations**
  Lateration (or **ranging**) is the determination of the distance from one location or position to another one. Usually, the term *ranging* is used for moving objects, while *surveying* is used for static terrestrial objects. Active ranging systems

operate with unilateral transmission and passive reflections, such as SONAR (SOund Navigation And Ranging), RADAR (RAdio Detection) and LIDAR (Light Detection).

A *rangefinder* is a device for measuring distance from the observer to a target. Among applications are surveying, navigation, ballistics and photography.

Triangulation is the process of locating a point P as the third point of a triangle with one known side (say, $[A, B]$ of length l) and two known angles (say, $\angle PAB = \alpha$ and $\angle PBA = \beta$). In $\mathbb{R}^2$, the perpendicular distance between P (say, a ship) and $[A, B]$ (say, a shore) is $\frac{l \sin\alpha \sin\beta}{\sin(\alpha+\beta)}$. Cf. point-line distance in Chap. 4.

Technically more complicated, **trilateration** is locating a (possibly, moving) object P, using only its distances to known locations A_1, A_2 and A_3 (for example, to stations, beacons or satellites), as the overlap of $2D$ or $3D$ spheres, centered on them and having radii $d(P, A_1), d(P, A_2), d(P, A_3)$, respectively. Using additional stations, as in GPS, permits double-checking of the measurements. Cf. the **metric basis** in Chap. 1.

More accurate generally, **multilateration** is locating a moving object P, using only two pairs (A_1, A_2), (B_1, B_2) of known locations, as the intersection of two curves defined by the relative distances $d(P, A_1) - d(P, A_2)$ and $d(P, B_1) - d(P, B_2)$, respectively.

- **Transmission distance**

  The **transmission distance** is a **range distance**: for a given signal transmission system (fiber optic cable, wireless, etc.), it is the maximal distance the system can support within an acceptable path loss level.

  For a given network of contact that can transmit an infection (or, say, an idea with the belief system considered as the immune system), the **transmission distance** is the path metric of the graph, in which edges correspond to events of infection and vertices are infected individuals. Cf. **forward quasi-distance** in Chap. 22.

- **Delay distance**

  The **delay distance** is a general term for the distance resulting from a given delay. For example, in a meteorological sensor, the *delay distance* is the length of a column of air passing a wind vane, such that the vane will respond to 50 % of a sudden angular change in wind direction. When the energy of a neutron is measured by the delay (say, t) between its creation and detection, the *delay distance* is $vt - D$, where v is its velocity and D is the source-detector distance.

  In evaluations of visuospatial working memory (when the subjects saw a dot, following a 10-, 20-, or 30-s delay, and then drew it on a blank sheet of paper), the *delay distance* is the distance between the stimulus and the drawn dot.

- **Master-slave distance**

  Given a design (say, remote manipulation, surveillance, or data transmission system) in which one device (the *master*) fully controls one or more other devices (the *slaves*), the **master-slave distance** is a measure of distance between the master and slave devices. Cf. also Sect. 18.2.

- **Flow distance**

  In a manufacturing system, a group of machines for processing a set of jobs is often located in a serial line along a path of a transporter system.

The **flow distance** from machine i to machine j is the total flow of jobs from i to j times the physical distance between machines i and j.

- **Single row facility layout**
  The **SRFLP** (or **single row facility layout problem**) is the problem of arranging (finding a permutation of) n departments (disjoint intervals) with given lengths l_i on a straight line so as to minimize the total weighted distance $\sum_{i=1}^{n-1}\sum_{i=j+1}^{n} w_{ij}d_{ij}$ between all department pairs. Here w_{ij} is the average daily traffic between two departments i and j, and d_{ij} is their **centroid distance**.
  Among applications of SRFLP, there are arranging machines in a manufacturing system, rooms on a corridor and books on a shelf.
- **Distance hart**
  In Technical Drawing, the **distance hart** means the distance from the center (the heart) of an object, as, for example, the distance hart of the toilet seat to the wall.
  The **center-to-center distance** (or **on-center**, *O.C.*) is the distance between the centers of two adjacent members (say, columns or pillars). Cf. **centroid linkage**, **centroid distance** in Chaps. 17, 19 and *center* **gear distance**.
- **Push distance**
  Precise machining of bearing rings should be preceded by part centering. In such a centering system, the **push distance** is the distance the slide must move towards the part in order to push it from its off-center position to the center of rotation.
- **Engine compression distance**
  Piston motors convert the compressed air energy to mechanical work through motion. **Engine compression distance** (or *compression height*) is the distance from the centerline of the wrist pin to the top deck of the piston.
- **Shift distance**
  A *penetrometer* is a device to test the strength of a material, say, soil. The penetrometer (usually cone-shaped) is pressed against material and the depth of the resulting hole is measured. The **shift distance** (or *friction-bearing offset*) is the distance between the cone's base and the mid-height of the friction sleeve above it.
- **Throat distance**
  The *swing* (size) of a drill/boring press is twice the **throat distance**, the distance from the center of the spindle to the column's edge.
- **Collar distance**
  In Mining, the **collar distance** is the distance from the top of the powder column to the collar of the blasthole, usually filled with stemming.
- **Quenching distance**
  Quenching is the rapid cooling of a workpiece; the **quenching distance** is the diameter of smallest hole a flame can travel through.
  The *run-up length* is the distance between initiation of a flame and onset of detonation (supersonic combustion wave). *Markstein number* is the *Markstein length*, measuring the effect of curvature on a flame, divided by the flame thickness.

- **Feeding distance**
  Carbon steel shrinks during solidification and cooling. In order to avoid resulting porosity, a **riser** (a cylindric liquid metal reservoir) provides liquid feed metal until the end of the solidification process.
  A riser is evaluated by its **feeding distance** which is the maximum distance over which a riser can supply feed metal to produce a radiographically sound (i.e., relatively free of internal porosity) casting. The **feeding length** is the distance between the riser and the furthest point in the casting fed by it.
- **Etch depth**
  Laser etching into a metal substrate produces craters. The **etch depth** is the central crater depth averaged over the apparent roughness of the metal surface.
- **Approach distance**
  In metal cutting, the **approach distance** is the linear distance in the direction of feed between the point of initial cutter contact and the point of full cutter contact.
- **Input and output distances**
  The **input distance** d_{in} of a machine M is a distance machine is moved by the *input* (applied on it) force F_{in}. The **output distance** d_{out} is a distance the *output* (exerted by it) force F_{out} results in. The *mechanical advantage* of M is $\frac{F_{out}}{F_{in}} = \frac{d_{in}}{d_{out}}$.
  For example, the *effort* (or *resistance*) *distance* and *load distance* on a lever are the distances from the fulcrum to the resistance and load, respectively.
- **Instrument distances**
  Examples of such distances follow.
  The *K-distance*: the distance from the outside fiber of a rolled steel beam to the web toe of the fillet of a rolled shape.
  The *end distance* and *edge distance* are the distances from a fastener (say, bolt, screw, rivet, nail) to the end and, respectively, edge of treated material.
  The *calibration distance*: the standard distance used in the process of adjusting the output or indication on a measuring instrument.
- **Gear distances**
  Given two meshed gears, the distance between their centers is called the *center distance*. Examples of other distances used in basic gear formulas follow.
  Pitch diameter: the diameter of the *pitch circle* (the circle whose radius is equal to the distance from the center of the gear to the pitch point).
  Addendum: the radial distance between the pitch circle and the top of the teeth.
  Dedendum: the depth of the tooth space below the pitch line. It should be greater than the addendum of the mating gear to provide *clearance*.
  Whole depth: the total depth of a tooth space, equal to addendum plus dedendum.
  Working depth: the depth of engagement (i.e., the sum of addendums) of two gears.
  Backlash: the play between mating teeth.
- **Threaded fastener distances**
  Examples of distances applied to nuts, screws and other threaded fasteners, follow.

Pitch: the nominal distance between two adjacent thread roots or crests.
Ply: a single thickness of steel forming part of a structural joint.
Grip length: the total distance between the underside of the nut to the bearing face of the bolt head.
Effective nut radius: the radius from the center of the nut to the point where the contact forces, generated when the nut is turned, can be considered to act.
Effective diameter (or *pitch diameter*): the diameter of an imaginary cylinder coaxial with the thread which has equal metal and space widths.
Virtual effective diameter: the effective diameter of a thread, allowing for errors in pitch and flank angles. *Nominal diameter*: the external diameter of the threads. *Major* and *minor diameters* are the diameters of imaginary cylinders parallel with the crests of the thread (i.e., the distance, crest-to-crest for an external or root-to-root for an internal thread), or, respectively, just touching the roots of an external (or the crests of an internal) thread.
Thread height: the distance between thread's minor and major diameters measured radially. *Thread length*: the length of the portion of the fastener with threads.

- **Distance spacer**
  A **distance spacer** is a device for holding two objects at a given distance from each other. Examples of such components are: male-female *distance bolt*, *distance bush*, *distance ring*, *distance plate*, *distance sleeve*, *distance finger*, *distance gauge*.
- **Sagging distance**
  The brazeability of brazing sheet materials is evaluated by their **sagging distance**, i.e., the **deflection** of the free end of the specimen sheet after brazing.
- **Deflection**
  In Engineering, **deflection** is the degree, in distance or angle, to which a structural part is displaced under a load/stress.
  In general, it can be a specified amount of deviation, say, the distance an elastic body or spring moves when subjected to a force, the amount by which a propagating wave or a projectile's trajectory is bent, and so on.
- **Haul distance**
  In Engineering, the **haul distance** is the average distance material is transported from where it originates to where it is deposited.
- **Distances in Structural Engineering**
  Examples of such distances related to superstructures (mainly, bridges and buildings) follow; cf. also **bar-and-joint framework** in Chap. 15.
  For a building, its *length* is the distance between out ends of wall steel lines, *width* is the distance from outside of *eave strut* (piece spanning columns at roof's edge) of one sidewall to outside of eave strut of the opposite sidewall, *height* is the distance from finished floor level to the top outer point of the eave strut.
  A *bay* refers the space between architectural elements. *Bay depth* is the distance from the building's corridor wall to the outside window. *End bay length* is the distance from outside of the outer *flange* (longitudinal part of a beam) of endwall

columns to centerline of the first interior frame column. *Interior bay length* is the distance between the centerlines of two adjacent interior main frame columns.
Clear height (or *head room*) is the vertical distance from the floor to the bottom of the lowest hanging overhead obstruction, allowable for passage.
A *beam* is a structural element that is capable of withstanding load primarily by resisting bending; *girder* is a support beam used in construction. A *truss* is a framed or jointed structure designed to act as a beam while each of its members is primarily subjected to longitudinal stress only. Given a truss or girder, its *effective length* is the distance between the points of support, *effective depth* is the perpendicular distance between the gravity lines, and *economic depth* is the depth, which will give satisfactory results from all standpoints and involving the least expenditure of money for properly combined first cost, operation, maintenance and repairs.
Effective span is the distance between supports (centres of bearings) in any structure. The **bearing distance** is the length of a beam between its bearing supports.
For a bridge, its *effective span* is the center-to-center distance of end pins, *structural height* is the maximum vertical distance from the uppermost point down to the lowest visible point, while the *deck height* is the maximum vertical distance from the *deck* (road bed) down to the ground or water surface.
Clear headway: the vertical distance from the lowest part of the superstructure to the ground or water surface; it is the measure of height of the tallest vehicle that could pass through the bridge. *Clear waterway*: the horizontal distance over the water, measured perpendicularly to the centrelines of adjacent piers.

- **Clearance distance**
  A **clearance distance** (or *separation distance*, *clearance*) is, in Engineering and Safety, a physical distance or unobstructed space tolerance as, for example, the distance between the lowest point on the vehicle and the road (*ground clearance*). For vehicles going in a tunnel or under a bridge, the *clearance* is the difference between the *structure gauge* (minimum size of tunnel or bridge) and the vehicles' *loading gauge* (maximum size). A clearance distance can be prescribed by a code or a standard between a piece of equipment containing potentially hazardous material (say, fuel) and other objects (buildings, equipment, etc.) and the public. Or, say, no vehicle should parked nearer than 15 feet (4.6 m) from a fire hydrant. In general, *clearance* refers to the distance to the nearest "obstacle" as defined in a context. It can be either a *tolerance* (the limit of an acceptable unplanned deviation from the nominal or theoretical dimension), or an *allowance* (planned deviation). Cf. **buffer distance** and **setback distance** in Chap. 25.
- **Creepage distance**
  The **creepage distance** is the shortest path distance along the surface of an insulation material between two conductive parts.
  The shortest (straight line) distance between two conductive parts is called the *clearance distance*; cf. the general term above.
- **Spark distance**
  The simplest way of measuring high voltages is by their **spark distance** (or **maximum spark length**). It is the length d of the gap between two electrodes

in a gas, at which given voltage V becomes the *breakdown voltage*, i.e., starts a discharge or electric arc (a spark jumps over). Spark distance depends on the pressure p of gas and many other factors. The *Paschen's law* estimate V as a function of pd.

- **Humidifier absorption distance**
  The **absorption distance** of a (water centrifugal atomizing) humidifier is the list of minimum clearance dimensions needed to avoid condensation.
- **Spray distance**
  The **spray distance** is the distance maintained between the nozzle tip of a thermal spraying gun and the surface of the workpiece during spraying.
- **Protective action distance**
  The **protective action distance** is the distance downwind from an incident (say, a spill involving dangerous goods which are considered toxic by inhalation) in which persons may become incapacitated.
  The *screening distance* in a forest fire is the downwind distance which should be examined for possible smoke-sensitive human sites. The *spot fire distance* is the maximum distance between a source of firebrands (a group of burning trees) and a potential *spot fire* (a fire started by flying sparks or embers from the main fire). The *response distance* is the distance to fire traveled by fire companies.
  The notion of *mean distance between people and any hazardous event* operates also at a large scale: expanding the living area of human species (say, space colonization) will increase this distance and prevent many human extinction scenarios.
- **Fringe distance**
  Usually, the **fringe distance** is the spacing between *fringes*, for example, components into which a spectral line splits in the presence of an electric or magnetic field (*Stark* and *Zeeman effects*, respectively, in Physics) or dark and light regions in the interference pattern of light beams (cf., in Chap. 24, *Pendellösung fringes* in **dynamical diffraction distances**).
  For an interferometer, the fringe distance is the value $\frac{\lambda}{2\sin\alpha}$, where λ is the laser wavelength and α is the beam angle, while the **shear distance** is the spacing between two, due to the thickness of the plate, reflections.
  In Image Analysis, there is also the *fringe distance* (Brown, 1994) between binary images (cf. **pixel distance** in Chap. 21).
- **Shooting distance**
  The **shooting distance** (or *shot distance*) is the distance achieved by, say, a bullet or a golf ball after a shot. The range of a Taser projectile delivering an incapacitating shock is called the *shocking distance*. Longest confirmed sniper kill at 2013 was 2,475 m. The **effective weapon distance** is the actual distance (as opposed to maximal range) over which it is usually deployed. For given game and rifle type, the *effective hunting distance* (or *killing distance*) is the maximal range of a "clean kill".
  For a shooting range, *firing distance* is the distance between the firing line and the target line. In shooting incident reconstruction, *firing distance*

(or *muzzle-to-target distance*) is the distance from the muzzle of the firearm to the victim's clothing.
In photography, the *shooting distance* is the camera-subject distance.

- **Lens distances**
A convex lens is converging/magnifying; a concave one is diverging/reducing.
The **focal distance** (*effective focal length*): the distance from the optical center of a lens (or a curved mirror) to the focus (to the image). Its reciprocal measured in m is called the *diopter* and is used as a unit of measurement of the (refractive) power of a lens; roughly, the magnification power of a lens is $\frac{1}{4}$ of its diopter.
The *lens effective diameter* is twice the longest lens radius measured from its center to the apex of its edge. The *back focal length* is the distance between the rear surface of a lens and its image plane; the *front focal length* is the distance from the vertex of the first lens to the front focal point.
Depth of field (DoF): the distance in the object plane (in front of and behind the object) over which the system delivers an acceptably sharp image, i.e., the region where blurring is tolerated at a particular resolution.
The *depth of focus*: the range of distance in the image plane (the eyepiece, camera, or photographic plate) over which the system delivers an acceptably sharp image.
The *vertex depth* (or *sagitta*) is the depth of the surface curve on a lens measured over a specific diameter. Given a circle, the *apothem* is the perpendicular distance from the midpoint of a chord to the circle's center; it is the radius minus the *sagitta*.
The **working distance**: the distance from the front end of a lens system to the object when the instrument is correctly focused; it is used to modify the DoF. For a flashlight, it is the distance at which the *illuminance* (maximum light falling on a surface) would fall to 0.25 lux as, say, a full moon on a clear night.
The *register distance* (or *flange distance*): the distance between the flange (protruding rim) of the lens mount and the plane of the film image.
The *conjugate image distance* and *conjugate object distance*: the distances along the optical axis of a lens from its principal plane to the image and object plane, respectively. When a converging lens is placed between the object and the screen, the sum of the inverses of those distances is the inverse focal distance.
A *circle of confusion* (CoC) is an optical spot caused by a cone of light rays from a lens not coming to a perfect focus; in photography, it is the largest blur circle that will still be perceived as a point when viewed at a distance of 25 cm.
The *close* (or *minimum, near*) *focus distance*: the closest distance to which a lens can approach the subject and still achieve focus.
The **hyper-focal distance**: the distance from the lens to the nearest point (*hyper-focal point*) that is in focus when the lens is focused at infinity; beyond this point all objects are well defined and clear. It is the nearest distance at which the far end of the **depth of field** stretches to infinity (cf. **infinite distance**).
Eye relief: the distance an optical instrument can be held away from the eye and still present the full field-of-view. The *exit pupil width*: the width of the cone of light that is available to the viewer at the exact eye relief distance.

- **Distances in Stereoscopy**
  A method of 3D imaging is to create a pair of 2D images by a two-camera system. The *convergence distance* is the distance between the *baseine* of the camera center to the *convergence point* where the two lenses should converge for good stereoscopy. This distance should be 15–30 times the **intercamera distance**.
  The *intercamera distance* (or *baseline length*, *interocular lens spacing*) is the distance between the two cameras from which the left and right eye images are rendered.
  The *picture plane distance* is the distance at which the object will appear on the *picture plane* (the apparent surface of the image). The *window* is a masking border of the screen frame such that objects, which appear at (but not behind or outside) it, appear to be at the same distance from the viewer as this frame. In human viewing, the picture plane distance is about 30 times the **intercamera distance**.
- **Distance-related shots**
  A film *shot* is what is recorded between the time the camera starts (the director's call for "action") and the time it stops (the call to "cut").
  The main **distance-related shots** (camera setups) are:

  - *establishing shot*: a shot, at the beginning of a sequence which establishes the location of the action and/or the time of day;
  - *long shot*: a shot taken from at least 50 yards (45.7 m) from the action;
  - *medium shot*: a shot from 5 to 15 yards (4.6–13.7 m), including a small entire group, which shows group/objects in relation to the surroundings;
  - *close-up*: a shot from a close position, say, the actor from the neck upwards;
  - *two-shot*: a shot that features two persons in the foreground;
  - *insert*: an inserted shot (usually a close up) used to reveal greater detail.

29.3 Miscellany

- **Range distances**
  In Mathematics, *range* is the set of values of a function or variable; specifically, it means the difference (or interval, area) between a maximum and minimum.
  The **range distances** are practical distances emphasizing a maximum distance for effective operation such as vehicle travel without refueling, a bullet range, visibility, movement limit, home range of an animal, etc. For example, the range of a risk factor (toxicity, blast, etc.) indicates the minimal **safe distancing**.
  The **operating distance** (or *nominal sensing distance*) is the range of a device (for example, a remote control) which is specified by the manufacturer and used as a reference. The **activation distance** is the maximal distance allowed for activation of a sensor-operated switch.

- **Spacing distances**
  The following examples illustrate this large family of practical distances emphasizing a minimum distance; cf. **minimum distance**, nearest-neighbor **distance in Animal Behavior**, **first-neighbor distance** in Chaps. 16, 23, 24, respectively.
  The **miles in trail**: a specified minimum distance, in nautical miles, required to be maintained between airplanes. *Seat pitch* and *seat width* are airliner distances between, respectively, two rows of seats and the armrests of a single seat.
  The **isolation distance**: a specified minimum distance required (because of pollination) to be maintained between variations of the same species of a crop in order to keep the seed pure (for example, $\approx$3 m for rice).
  The **legal distance**: a minimum distance required by a judicial rule or decision, say, a distance a sex offender is required to live away from school.
  In a restraining order, **stay away** means to stay a certain distance (often 300 yards, i.e., 275 m) from the protected person. A general **distance restriction**: say, a minimum distance required for passengers traveling on some long distance trains in India, or a distance from a voting facility where campaigning is permitted.
  The **stop-spacing distance**: the interval between bus stops; such mean distance in US light rail systems ranges from 500 (Philadelphia) to 1,742 m (Los Angeles).
  The **character spacing**: the interval between characters in a given computer font.
  The **just noticeable difference** (JND): the smallest perceived percent change in a dimension (for distance/position, etc.); cf. **tolerance distance** in Chap. 25).
- **Cutoff distances**
  Given a range of values (usually, a length, energy, or momentum scale in Physics), *cutoff* (or *cut-off*) is the maximal or minimal value, as, for example, Planck units.
  A **cutoff distance** is a cutoff in a length scale. For example, infrared and ultraviolet cutoff (the maximal and minimal wavelength that the human eye takes into account) are *long-distance* and *short-distance cutoff*, respectively, in the visible spectrum. Cutoff distances are often used in Molecular Dynamics.
  A similar notion of a **threshold distance** refers to a limit, margin, starting point distance (usually, minimal) at which some effect happens or stops. Some examples are the threshold distance of sensory perception, neuronal reaction or, say, upon which a city or road alters the abundance patterns of the native bird species.
- **Quality metrics**
  A **quality metric** (or, simply, *metric*) is a standard unit of measure or, more generally, part of a system of parameters, or systems of measurement. This vast family of measures (or standards of measure) concerns different attributes of objects. In such a setting, our distances and similarities are rather "similarity metrics", i.e., metrics (measures) quantifying the extent of relatedness between two objects.
  Examples include academic metrics, crime statistics, corporate investment metrics, economic metrics (indicators), education metrics, environmental metrics

(indices), health metrics, market metrics, political metrics, properties of a route in computer networking, software metrics and vehicle metrics.

For example, the site http://metripedia.wikidot.com/start aims to build an Encyclopedia of IT (Information Technology) performance metrics. Some examples of nonequipment quality metrics are detailed below.

Landscape metrics evaluate, for example, greenway patches in a given landscape by *patch density* (the number of patches per km^2), *edge density* (the total length of patch boundaries per hectare), *shape index* $\frac{E}{4\sqrt{A}}$ (where A is the total area, and E is the total length of edges), connectivity, diversity, etc.

Morphometrics evaluate the forms (size and shape) related to organisms (brain, fossils, etc.). For example, the roughness of a fish school is measured by its *fractal dimension* $2\frac{\ln P - \ln 4}{\ln A}$ where P, A are its perimeter (m) and surface (m^2).

Management metrics include: surveys (say, of market share, sales increase, customer satisfactions), forecasts (say, of revenue, contingent sales, investment), R&D effectiveness, absenteeism, etc.

Risk metrics are used in Insurance and, in order to evaluate a portfolio, in Finance.

Importance metrics rank the relative influence such as, for example:

- *PageRank* of Google ranking web pages;
- ISI (now Thomson Scientific) *Impact Factor* of a journal measuring, for a given two-year period, the number of times the average article in this journal is cited by some article published in the subsequent year;
- Hirsch's *h-index* of a scholar: the largest number h such that h of his/her publications have at least h citations;
- and his/her *i*10-*index*: the number of publications with at least ten citations.

- **Heterometric and homeometric**

  The adjective **heterometric** means involving or dependent on a change in size, while **homeometric** means independent of such change.

  Those terms are used mainly in Medicine; for example, *heterometric* and *homeometric autoregulation* refer to intrinsic mechanisms controlling the strength of ventricular contractions that depend or not, respectively, on the length of myocardial fibers at the end of diastole; cf. **distances in Medicine**.

- **Distal and proximal**

  The antipodal notions near (close, nigh) and far (distant, remote) are also termed *proximity* and *distality*.

  The adjective **distal** (or *peripheral*) is an anatomical term of location (on the body, the limbs, the jaw, etc.); corresponding adverbs are: distally, distad.

  For an *appendage* (any structure that extends from the main body), **proximal** means situated towards the point of attachment, while **distal** means situated around the furthest point from this point of attachment. More generally, as opposed to *proximal* (or *central*), *distal* means: situated away from, farther from a point of reference (origin, center, point of attachment, trunk). As opposed to *mesial* it means: situated or directed away from the midline or mesial plane of the body.

Proximal and distal *demonstratives* are words indicating *place deixis*, i.e., a spatial location relative to the point of reference. Usually, they are two-way as *this/that*, these/those or *here/there*, i.e., in terms of the dichotomy near/far from the speaker. But, say, Korean, Japanese, Spanish, and Thai make a three-way distinction: proximal (near to the speaker), medial (near to the addressee) and distal (far from both). English had the third form, *yonder* (at an indicated distance within sight), still spoken in Southern US. Cf. **spatial language** in Chap. 28.

A **distal stimulus** is an real-word object or event, which, by some physical process, stimulates the body's sensory organs. Resulting raw pattern of neural activity is called the **proximal stimulus**. *Perception* is the constructing mental representations of distal stimuli using the information available in proximal stimuli.

A *proximate cause* is an event which is closest to, or immediately responsible for causing, some observed result. This exists in contrast to a higher-level *ultimate* (or *distal*) *cause* which is usually thought of as the "real" reason something occurred.

Tinbergen's (1960) *proximate* and *ultimate questions* about behavior are "how" an organism structures function? and "why" a species evolved the structures it has?

- **Distance effect**

  The **distance effect** is a general term for the change of a pattern or process with distance. Usually, it is the result of **distance decay**. For example, a static field attenuates proportionally to the inverse square of the distance from the source.

  Another example of the distance effect is a periodic variation (instead of uniform decrease) in a certain direction, when a *standing wave* occurs in a time-varying field. It is a wave that remains in a constant position because either the medium is moving in the opposite direction, or two waves, traveling in opposite directions, interfere; cf. **Pendellösung length** in Chap. 24.

  The distance effect, together with the size (source magnitude) effect determine many processes; cf. **island distance effect**, **insecticide distance effect** in Chap. 23 and **symbolic distance effect**, **distance effect on trade** in Chap. 28.

- **Distance decay**

  The **distance decay** is the attenuation of a pattern or process with distance. Cf. **distance decay (in Spatial Interaction)** in Chap. 28.

  Examples of distance-decay curves: Pareto model $\ln I_{ij} = a - b \ln d_{ij}$, and the model $\ln I_{ij} = a - bd_{ij}^p$ with $p = \frac{1}{2}$, 1, or 2 (here I_{ij} and d_{ij} are the interaction and distance between points i, j, while a and b are parameters). The *Allen curve* gives the exponential drop of frequency of all communication between engineers as the distance between their offices increases, i.e., face-to-face probability decays.

  A **mass-distance decay curve** is a plot of "mass" decay when the distance to the center of "gravity" increases. Such curves are used, say, to determine an *offender's heaven* (the point of origin; cf. **distances in Criminology**) or the galactic mass within a given radius from its center (using *rotation-distance curves*).

- **Distance factor**
A **distance factor** is a multiplier of some straight-line distance needed to account for additional data. For example, 10 % increase of aircraft weight implies 20 % increase, i.e., a distance factor of 1.2, in needed take-off distance.
- **Propagation length**
For a pattern or process attenuating with distance, the **propagation length** is the distance to decay by a factor of $\frac{1}{e}$.
Cf. **radiation length** and the **Beer–Lambert law** in Chap. 24.
A **scale height** is a distance over which a quantity decreases by a factor of e.
- **Incremental distance**
An **incremental distance** is a gradually increasing (by a fixed amount) one.
- **Distance curve**
A **distance curve** is a plot (or a graph) of a given parameter against a corresponding distance. Examples of distance curves, in terms of a process under consideration, are: **time-distance curve** (for the travel time of a wave-train, seismic signals, etc.), *height-run distance curve* (for the height of tsunami wave versus wave propagation distance from the impact point), *drawdown-distance curve*, *melting-distance curve* and *wear volume-distance curve*.
A **force-distance curve** is, in SPM (scanning probe microscopy), a plot of the vertical force that the tip of the probe applies to the sample surface, while a contact-AFM (Atomic Force Microscopy) image is being taken. Also, *frequency-distance* and *amplitude-distance* curves are used in SPM.
The term *distance curve* is also used for charting growth, for instance, a child's height or weight at each birthday. A plot of the rate of growth against age is called the **velocity-distance curve**; this term is also used for the speed of aircraft.
Example of a constant rate of growth: in a month, human (hear or body) hair grow 15 and 8.1 mm, while nails (finger and toe) grow 3.5 and 1.6 mm.
- **Distance sensitivity**
Distance sensitivity is a general term used to indicate the dependence of something on the associated distance. It could be, say, commuting distance sensitivity of households, traveling distance sensitivity of tourists, distance sensitive technology, distance sensitive products/services and so on.
- **Characteristic diameters**
Let X be an irregularly-shaped 3D object, say, Earth's spheroid or a particle. A **characteristic diameter** (or **equivalent diameter**) of X is the diameter of a sphere with the same geometric or physical property of interest. Examples follow.
The **authalic diameter** and **volumetric diameter** (*equivalent spherical diameter*) of X are the diameters of the spheres with the same surface area and volume.
The **Heywood diameter** is the diameter of a circle with the same projection area.
Cf. the **Earth radii** in Chap. 25 and the **shape parameters** in Chap. 21.
The **Stokes diameter** is the diameter of the sphere with the same gravitational velocity as X, while the **aerodynamic diameter** is the diameter of such sphere of unit density. Cf. the **hydrodynamic radius** in Chap. 24.

Equivalent *electric mobility*, *diffusion* and *light scattering* diameters of a particle X are the diameters of the spheres with the same electric mobility, penetration and intensity of light scattering, respectively, as X.

- **Characteristic length**

  A **characteristic length** (or *scale*) is a convenient reference length of a given configuration, such as the overall length of an aircraft, the maximum diameter or radius of a body of revolution, or a chord or span of a lifting surface.

  In general, it is a length that is representative of the system (or region) of interest, or the parameter which characterizes a given physical quantity in, say, heat transfer or fluid mechanics. For complex shapes, it is defined as the volume of the body divided by the surface area. For example, for a rocket engine, it is the ratio of the volume of its combustion chamber to the area of the nozzle's throat, representing the average distance that the products of burned fuel must travel to escape.

- **True length**

  In Engineering Drawing, **true length** is any distance between points that is not foreshortened by the view type. In 3D, lines with true length are parallel to the projection plane, as, for example, the base edges in a top view of a pyramid.

- **Path length**

  In general, a *path* is a line representing the course of actual, potential or abstract movement. In Topology, a *path* is a certain continuous function; cf. parametrized **metric curve** in Chap. 1.

  In Physics, **path length** is the total distance an object travels, while *displacement* is the net distance it travels from a starting point. Cf. **displacement**, **inelastic mean free path**, **optical distance** and **dislocation path length** in Chap. 24. In Chemistry, (cell) **path length** is the distance that light travels through a sample in an analytical cell.

  In Graph Theory, **path length** is a discrete notion: the number of vertices in a sequence of vertices of a graph; cf. **path metric** in Chap. 1. Cf. **Internet IP metric** in Chap. 22 for **path length** in a computer network. Also, it means the total number of machine code instructions executed on a section of a program.

- **Middle distance**

  The **middle distance** is a general term. For example, it can be a precise distance (cf. **running distances**), the halfway between the observer and the horizon (cf. **distance to horizon** in Chap. 25), implied horizon of a scene (cf. **representation of distance in Painting** in Chap. 28), or the place that you can see when you are not quite focusing on the world around you (Urban Dictionary of slang).

  The *Great Declaration* by Simon Magus (first Gnostic and Gnostic Christ) claims: "Of the universal Aeons spring two shoots, without beginning or end, stemming forth from the Root, which is the invisible Power, unknowable Silence. Of these shoots, one appears from above. It is the Great Power, Universal Mind ordering all things, male. The other appears from below. It is the Great Thought, female, producing all things. They paired, uniting and appearing in the Middle Distance, the Incomprehensible Air, without beginning or end. Here is the Father by whom all those things, having a beginning and end, are sustained and nourished..."

- **Long-distance**
  The term **long-distance** usually refers to telephone communication (long-distance call, operator) or to covering large distances by moving (long-distance trail, running, swimming, riding of motorcycles or horses, etc.) or, more abstractly: long-distance migration, commuting, supervision, relationship, etc. For example, a *long-distance relationship* (LDR) is typically an intimate relationship that takes place when the partners are separated by a considerable distance.
  For example, a *long-distance* (or *distance*) *thug* has two meanings: (1) a person that is a coward in real life, but gathers courage from behind the safety of a computer, phone, or through e-mail; and (2) a hacker, spammer, or scam artist that takes advantage of the Internet to cause harm to others from a distance.
  Cf. **long-distance dispersal**, **animal** and **plant long-distance communication**, **long range order**, **long range dependence**, **action at a distance** (in Computing, Physics, along DNA).
  DDD (or *direct distance dialing*) is any switched telecommunication service (like 1+, 0 + +, etc.) that allows a call originator to place long-distance calls directly to telephones outside the local service area without an operator.
  The term *short-distance* is rarely used. Instead, the adjective **short range** means limited to (or designed for) short distances, or relating to the near future. Finally, **touching**, for two objects, is having (or getting) a zero distance between them.
- **Long-distance intercourse**
  Long-distance intercourse (coupling at a distance) is found often in Native American folklore: Coyote, the Trickster, is said to have lengthened his penis to enable him to have intercourse with a woman on the opposite bank of a lake.
  A company Distance Labs has announced the "intimate communication over a distance", an interactive installation *Mutsugoto* which draws, using a custom computer vision and projection system, lines of light on a body of a person. Besides light, haptic technology provides a degree of touch communication between remote users. A company Lovotics created *Kissinger*, a messaging device wirelessly sending kisses. *Sports over distance* is another example of implemented computer-supported movement-based interaction between remote players.
  In Nature, the *acorn barnacle* (small sessile crustacean) have the largest penis-body size ratio (up to 10 when extended) of any animal. The squid *Onykia ingens* have largest ratio among mobile animals. The male octopus *Argonauta* use a modified arm, the *hectocotylus*, to transfer sperm to the female *at a distance*; this tentacle detaches itself from the body and swims—under its own power—to the female.
  Also, many aquatic animals (say, coral, hydra, sea urchin, bony fish) and amphibians reproduce by external fertilization: eggs and sperm are released into the water. Similar transfer of sperm at a distance is pollination (by wind

or organisms) in flowering plants. Another example is *in vitro* fertilization in humans.
The shortest range intercourse happens in anglerfish. The male, much smaller, latches onto a female with his sharp teeth, fuses inside her to the blood-vessel level and degenerates into a pair of testicles. It releases sperm when the female (with about six males inside) releases eggs. But female–male pairings of a parasitic worm *Schistosoma mansoni* is monogamous: the male's body forms a channel, in which it holds the longer and thinner female for their entire adult lives, up to 30 years. Two worms *Diplozoon paradoxum* fuse completely for lifetime of cross-fertilization.

- **Go the distance**
  Go the (full) **distance** is a general distance idiom meaning to continue to do something until it is successfully completed.
  An *unbridgeable distance* is a distance (seen as a spatial or metaphoric extent), impossible to span: a wide unbridgeable river, chasm or, in general, differences.

References

[Abel91] Abels H. *The Gallery Distance of Flags*, Order, Vol. 8, pp. 77–92, 1991.

[AAH00] Aichholzer O., Aurenhammer F. and Hurtado F. *Edge Operations on Non-crossing Spanning Trees*, Proc. 16-th European Workshop on Computational Geometry CG'2000, pp. 121–125, 2000.

[AACLMP98] Aichholzer O., Aurenhammer F., Chen D.Z., Lee D.T., Mukhopadhyay A. and Papadopoulou E. *Voronoi Diagrams for Direction-sensitive Distances*, Proc. 13th Symposium on Computational Geometry, ACM Press, New York, 1997.

[Aker97] Akerlof G.A. *Social Distance and Social Decisions*, Econometrica, Vol. 65-5, pp. 1005–1027, 1997.

[Amar85] Amari S. *Differential-geometrical Methods in Statistics*, Lecture Notes in Statistics, Springer-Verlag, 1985.

[Amba76] Ambartzumian R. *A Note on Pseudo-metrics on the Plane*, Z. Wahrsch. Verw. Gebiete, Vol. 37, pp. 145–155, 1976.

[ArWe92] Arnold R. and Wellerding A. *On the Sobolev Distance of Convex Bodies*, Aeq. Math., Vol. 44, pp. 72–83, 1992.

[Badd92] Baddeley A.J. *Errors in Binary Images and an L^p Version of the Hausdorff Metric*, Nieuw Archief voor Wiskunde, Vol. 10, pp. 157–183, 1992.

[BaFa07] Baier R. and Farkhi E. *Regularity and Integration of Set-Valued Maps Represented by Generalized Steiner Points* Set-Valued Analysis, Vol. 15, pp. 185–207, 2007.

[Bara01] Barabási A.L. *The Physics of the Web*, Physics World, July 2001.

[Barb12] Barbaresco F. *Information Geometry of Covariance Matrix: Cartan-Siegel Homogenous Bounded Domains, Mostow-Berger Fibration and Fréchet Median*, in *Matrix Information Geometry*, Bhatia R. and Nielsen F. (eds.) Springer, 2012.

[Barb35] Barbilian D. *Einordnung von Lobayschewskys Massenbestimmung in either Gewissen Allgemeinen Metrik der Jordansche Bereiche*, Casopis Mathematiky a Fysiky, Vol. 64, pp. 182–183, 1935.

[BLV05] Barceló C., Liberati S. and Visser M. *Analogue Gravity*, Living Rev. Rel. Vol. 8, 2005; **arXiv: gr-qc/0505065**, 2005.

[BLMN05] Bartal Y., Linial N., Mendel M. and Naor A. *Some Low Distortion Metric Ramsey Problems*, Discrete and Computational Geometry, Vol. 33, pp. 27–41, 2005.

[Bass89] Basseville M. *Distances measures for signal processing and pattern recognition*, Signal Processing, Vol. 18, pp. 349–369, 1989.

[Bass13] Basseville M. *Distances measures for statistical data processing – An annotated bibliography*, Signal Processing, Vol. 93, pp. 621–633, 2013.

M.M. Deza, E. Deza, *Encyclopedia of Distances*,
DOI 10.1007/978-3-662-44342-2

[Bata95] Batagelj V. *Norms and Distances over Finite Groups*, J. of Combinatorics, Information and System Sci., Vol. 20, pp. 243–252, 1995.

[Beer99] Beer G. *On Metric Boundedness Structures*, Set-Valued Analysis, Vol. 7, pp. 195–208, 1999.

[BGLVZ98] Bennet C.H., Gács P., Li M., Vitánai P.M.B. and Zurek W. *Information Distance*, IEEE Transactions on Information Theory, Vol. 44-4, pp. 1407–1423, 1998.

[BGT93] Berrou C., Glavieux A. and Thitimajshima P. *Near Shannon Limit Error-correcting Coding and Decoding: Turbo-codes*, Proc. of IEEE Int. Conf. on Communication, pp. 1064–1070, 1993.

[BFK99] Blanchard F., Formenti E. and Kurka P. *Cellular Automata in the Cantor, Besicovitch and Weyl Topological Spaces*, Complex Systems, Vol. 11, pp. 107–123, 1999.

[Bloc99] Bloch I. *On fuzzy distances and their use in image processing under unprecision*, Pattern Recognition, Vol. 32, pp. 1873–1895, 1999.

[BCFS97] Block H.W., Chhetry D., Fang Z. and Sampson A.R. *Metrics on Permutations Useful for Positive Dependence*, J. of Statistical Planning and Inference, Vol. 62, pp. 219–234, 1997.

[Blum70] Blumenthal L.M. *Theory and Applications of Distance Geometry*, Chelsea Publ., New York, 1970.

[Borg86] Borgefors G. *Distance Transformations in Digital Images*, Comp. Vision, Graphic and Image Processing, Vol. 34, pp. 344–371, 1986.

[BrLi04] Bramble D.M. and Lieberman D.E. *Endurance Running and the Evolution of Homo*, Nature, Vol. 432, pp. 345–352, 2004.

[O'Bri03] O'Brien C. *Minimization via the Subway metric*, Honor Thesis, Dept. of Math., Ithaca College, New York, 2003.

[BKMR00] Broder A.Z., Kumar S. R., Maaghoul F., Raghavan P., Rajagopalan S., Stata R., Tomkins A. and Wiener G. *Graph Structure in the Web: Experiments and Models*, Proc. 9-th WWW Conf., Amsterdam, 2000.

[BGL95] Brualdi R.A., Graves J.S. and Lawrence K.M. *Codes with a Poset Metric*, Discrete Math., Vol. 147, pp. 57–72, 1995.

[Brya85] Bryant V. *Metric Spaces: Iteration and Application*, Cambridge Univ. Press, 1985.

[BuHa90] Buckley F. and Harary F. *Distance in Graphs*, Redwood City, CA: Addison-Wesley, 1990.

[Bull12] Bullough E. *"Psychical Distance" as a Factor in Art and as an Aesthetic Principle*, British J. of Psychology, Vol. 5, pp. 87–117, 1912.

[BBI01] Burago D., Burago Y. and Ivanov S. *A Course in Metric Geometry*, Amer. Math. Soc., Graduate Studies in Math., Vol. 33, 2001.

[BuKe53] Busemann H. and Kelly P.J. *Projective Geometry and Projective Metrics*, Academic Press, New York, 1953.

[Buse55] Busemann H. *The Geometry of Geodesics*, Academic Press, New York, 1955.

[BuPh87] Busemann H. and Phadke B.B. *Spaces with Distinguished Geodesics*, Marcel Dekker, New York, 1987.

[Cair01] Cairncross F. *The Death of Distance 2.0: How the Communication Revolution will Change our Lives*, Harvard Business School Press, second edition, 2001.

[CSY01] Calude C.S., Salomaa K. and Yu S. *Metric Lexical Analysis*, Springer-Verlag, 2001.

[CaTa08] Cameron P.J. and Tarzi S. *Limits of cubes*, Topology and its Appl., Vol. 155, pp. 1454–1461, 2008.

[CHKSS07] Carmi S., Havlin S., Kirkpatrick S., Shavitt Y. and Shir E. *A model of internet topology using k-shell decomposition*, Proc. Nat. Acad. Sci., Vol. 104, pp. 11150–11154, 2007.

[Cha08] Cha S.-H. *Taxonomy of nominal type histogram distance measures*, Proc. American Conf. on Appl, Math., World Scientific and Engineering Academy and Society (WREAS) Stevens Point, Wisconsin, US, pp. 325–330, 2008.

[ChLu85] Cheng Y.C. and Lu S.Y. *Waveform Correlation by Tree Matching*, IEEE Trans. Pattern Anal. Machine Intell., Vol. 7, pp. 299–305, 1985.

[Chen72] Chentsov N.N. *Statistical Decision Rules and Optimal Inferences*, Nauka, Moscow, 1972.

[ChFi98] Chepoi V. and Fichet B. *A Note on Circular Decomposable Metrics*, Geom. Dedicata, Vol. 69, pp. 237–240, 1998.

[ChSe00] Choi S.W. and Seidel H.-P. *Hyperbolic Hausdorff Distance for Medial Axis Transform*, Research Report MPI-I-2000-4-003 of Max-Planck-Institute für Informatik, 2000.

[CLMNWZ05] Coifman R.R., Lafon S., A.B., Maggioni M., Nadler B., Warner F., Zucker S.W. *Geometric diffusions as a tool for harmonic analysis and structure definition of data: Diffusion maps*, Proc. of the National Academy of Sciences, Vol. 102, No. 21, pp. 7426–7431, 2005.

[COR05] Collado M.D., Ortuno-Ortin I. and Romeu A. *Vertical Transmission of Consumption Behavior and the Distribution of Surnames*, mimeo, Universidad de Alicante, 2005.

[Cops68] Copson E.T. *Metric Spaces*, Cambridge Univ. Press, 1968.

[Cora99] Corazza P. *Introduction to metric-preserving functions*, Amer. Math. Monthly, Vo. 104, pp. 309–323, 1999.

[Corm03] Cormode G. *Sequence Distance Embedding*, PhD Thesis, Univ. of Warwick, 2003.

[CPQ96] Critchlow D.E., Pearl D.K. and Qian C. *The Triples Distance for Rooted Bifurcating Phylogenetic Trees*, Syst. Biology, Vol. 45, pp. 323–334, 1996.

[CCL01] Croft W. B., Cronon-Townsend S. and Lavrenko V. *Relevance Feedback and Personalization: A Language Modeling Perspective*, in DELOS-NSF Workshop on Personalization and Recommender Systems in Digital Libraries, pp. 49–54, 2001.

[CKK03] Cuijpers R.H., Kappers A.M.L and Koenderink J.J. *The metrics of visual and haptic space based on parallelity judgements*, J. Math. Psychology, Vol. 47, pp. 278–291, 2003.

[DaCh88] Das P.P. and Chatterji B.N. *Knight's Distance in Digital Geometry*, Pattern Recognition Letters, Vol. 7, pp. 215–226, 1988.

[Das90] Das P.P. *Lattice of Octagonal Distances in Digital Geometry*, Pattern Recognition Letters, Vol. 11, pp. 663–667, 1990.

[DaMu90] Das P.P. and Mukherjee J. *Metricity of Super-knight's Distance in Digital Geometry*, Pattern Recognition Letters, Vol. 11, pp. 601–604, 1990.

[Dau05] Dauphas N. *The U/Th Production Ratio and the Age of the Milky Way from Meteorites and Galactic Halo Stars*, Nature, Vol. 435, pp. 1203–1205, 2005.

[Day81] Day W.H.E. *The Complexity of Computing Metric Distances between Partitions*, Math. Social Sci., Vol. 1, pp. 269–287, 1981.

[DeDu13] Deza M.M. and Dutour M. *Voronoi Polytopes for Polyhedral Norms on Lattices*, arXiv:1401.0040 [math.MG], 2013.

[DeDu03] Deza M.M. and Dutour M. *Cones of Metrics, Hemi-metrics and Super-metrics*, Ann. of European Academy of Sci., pp. 141–162, 2003.

[DeHu98] Deza M. and Huang T. *Metrics on Permutations, a Survey*, J. of Combinatorics, Information and System Sci., Vol. 23, Nrs. 1–4, pp. 173–185, 1998.

[DeLa97] Deza M.M. and Laurent M. *Geometry of Cuts and Metrics*, Springer, 1997.

[DPM12] Deza M.M., Petitjean M. and Matkov K. (eds) *Mathematics of Distances and Applications*, ITHEA, Sofia, 2012.

[DiGa07] Ding L. and Gao S. *Graev metric groups and Polishable subgroups*, Advances in Mathematics, Vol. 213, pp. 887–901, 2007.

[EhHa88] Ehrenfeucht A. and Haussler D. *A New Distance Metric on Strings Computable in Linear Time*, Discrete Appl. Math., Vol. 20, pp. 191–203, 1988.

[EM98] *Encyclopedia of Math.*, Hazewinkel M. (ed.), Kluwer Academic Publ., 1998. Online edition: http://eom.springer.de/default.htm

[Ernv85] Ernvall S. *On the Modular Distance*, IEEE Trans. Inf. Theory, Vol. 31-4, pp. 521–522, 1985.

[EMM85] Estabrook G.F., McMorris F.R. and Meacham C.A. *Comparison of Undirected Phylogenetic Trees Based on Subtrees of Four Evolutionary Units*, Syst. Zool, Vol. 34, pp. 193–200, 1985.

[FaMu03] Farrán J.N. and Munuera C. *Goppa-like Bounds for the Generalized Feng-Rao Distances*, Discrete Appl. Math., Vol. 128, pp. 145–156, 2003.

[Faze99] Fazekas A. *Lattice of Distances Based on $3D$-neighborhood Sequences*, Acta Math. Academiae Paedagogicae Nyiregyháziensis, Vol. 15, pp. 55–60, 1999.

[FeWa08] Feng J. and Wang T.M. *Characterization of protein primary sequences based on partial ordering*, J. Theor. Biology, Vol. 254, pp. 752–755, 2008.

[Fell97] Fellous J-M. *Gender Discrimination and Prediction on the Basis of Facial Metric Information*, Vision Research, Vol. 37, pp. 1961–1973, 1997.

[Ferg03] Ferguson N. *Empire: The Rise and Demise of the British World Order and Lessons for Global Power*, Basic Books, 2003.

[FoSc06] Foertsch T. and Schroeder V. *Hyperbolicity, CAT(-1)-spaces and the Ptolemy Inequality*, Math. Ann., Vol. 350, pp. 339–356, 2011.

[FrSa07] Frankild A. and Sather-Wagstaff S. *The set of semidualizing complexes is a nontrivial metric space*, J. Algebra, Vol. 308, pp. 124–143, 2007.

[Frie98] Frieden B.R. *Physics from Fisher information*, Cambridge Univ. Press, 1998.

[GaSi98] Gabidulin E.M. and Simonis J. *Metrics Generated by Families of Subspaces*, IEEE Transactions on Information Theory, Vol. 44-3, pp. 1136–1141, 1998.

[Gile87] Giles J.R. *Introduction to the Analysis of Metric Spaces*, Australian Math. Soc. Lecture Series, Cambridge Univ. Press, 1987.

[GoMc80] Godsil C.D. and McKay B.D. *The Dimension of a Graph*, Quart. J. Math. Oxford Series (2), Vol. 31, pp. 423–427, 1980.

[GOJKK02] Goh K.I., Oh E.S., Jeong H., Kahng B. and Kim D. *Classification of Scale Free Networks*, Proc. Nat. Acad. Sci. US, Vol. 99, pp. 12583–12588, 2002.

[Gopp71] Goppa V.D. *Rational Representation of Codes and (L, g)-codes*, Probl. Peredachi Inform., Vol. 7-3, pp. 41–49, 1971.

[Goto82] Gotoh O. *An Improved Algorithm for Matching Biological Sequences*, J. of Molecular Biology, Vol. 162, pp. 705–708, 1982.

[GKC04] Grabowski R., Khosa P. and Choset H. *Development and Deployment of a Line of Sight Virtual Sensor for Heterogeneous Teams*, Proc. IEEE Int. Conf. on Robotics and Automation, New Orleans, 2004.

[Grub93] Gruber P.M. *The space of Convex Bodies* in *Handbook of Convex Geometry*, Gruber P.M. and Wills J.M. (eds.), Elsevier Sci. Publ., 1993.

[HSEFN95] Hafner J., Sawhney H.S., Equitz W., Flickner M. and Niblack W. *Efficient Color Histogram Indexing for Quadratic Form Distance Functions*, IEEE Transactions on Pattern Analysis and Machine Intelligence, Vol. 17-7, pp. 729–736, 1995.

[Hall69] Hall E.T. *The Hidden Dimension*, Anchor Books, New York, 1969.

[Hami66] Hamilton W.R. *Elements of Quaternions*, second edition 1899–1901 enlarged by C.J. Joly, reprinted by Chelsea Publ., New York, 1969.

[HRJM13] Harispe S., Ranwez S., Janaqi S. and Montmain J. *Semantic Measures for the Comparison of Units of Language, Concepts or Instances from Text and Knowledge Base Analysis*, **arXiv:1310.1285[cs.CL]**, 2013.

[HeMa02] Head K. and Mayer T. *Illusory Border Effects: Distance mismeasurement inflates estimates of home bias in trade*, CEPII Working Paper No 2002-01, 2002.

[Hemm02] Hemmerling A. *Effective Metric Spaces and Representations of the Reals*, Theoretical Comp. Sci., Vol. 284-2, pp. 347–372, 2002.

[High89] Higham N.J. *Matrix Nearness Problems and Applications*, in *Applications of Matrix Theory*, Gover M.J.C. and Barnett S. (eds.), pp. 1–27. Oxford University Press, 1989.

[Hofs80] Hofstede G. *Culture's Consequences: International Differences in Work-related Values*, Sage Publ., California, 1980.
[Hube94a] Huber K. *Codes over Gaussian Integers*, IEEE Trans. Inf. Theory, Vol. 40-1, pp. 207–216, 1994.
[Hube94b] Huber K. *Codes over Eisenstein-Jacobi Integers*, Contemporary Math., Vol. 168, pp. 165–179, 1994.
[HFPMC02] Huffaker B., Fomenkov M., Plummer D.J., Moore D. and Claffy K., *Distance Metrics in the Internet*, Proc. IEEE Int. Telecomm. Symp. (ITS-2002), 2002.
[InVe00] Indyk P. and Venkatasubramanian S. *Approximate Congruence in Nearly Linear Time*, Proc. 11th ACM-SIAM symposium on Discrete Algorithms, pp. 354–260, San Francisco, 2000.
[Isbe64] Isbell J. *Six Theorems about Metric Spaces*, Comment. Math. Helv., Vol. 39, pp. 65–74, 1964.
[IKP90] Isham C.J., Kubyshin Y. and Penteln P. *Quantum Norm Theory and the Quantization of Metric Topology*, Class. Quantum Gravity, Vol. 7, pp. 1053–1074, 1990.
[IvSt95] Ivanova R. and Stanilov G. *A Skew-symmetric Curvature Operator in Riemannian Geometry*, in Symposia Gaussiana, Conf. A, Behara M., Fritsch R. and Lintz R. (eds.), pp. 391–395, 1995.
[JWZ94] Jiang T., Wang L. and Zhang K. *Alignment of Trees – an Alternative to Tree Edit*, in *Combinatorial Pattern Matching, Lecture Notes in Comp. Science*, Vol. 807, Crochemore M. and Gusfield D. (eds.), Springer-Verlag, 1994.
[Klei88] Klein R. *Voronoi Diagrams in the Moscow Metric*, Graphtheoretic Concepts in Comp. Sci., Vol. 6, pp. 434–441, 1988.
[Klei89] Klein R. *Concrete and Abstract Voronoi Diagrams*, Lecture Notes in Comp. Sci., Springer-Verlag, 1989.
[KlRa93] Klein D.J. and Randic M. *Resistance distance*, J. of Math. Chemistry, Vol. 12, pp. 81–95, 1993.
[Koel00] Koella J.C. *The Spatial Spread of Altruism Versus the Evolutionary Response of Egoists*, Proc. Royal Soc. London, Series B, Vol. 267, pp. 1979–1985, 2000.
[KoSi88] Kogut B. and Singh H. *The Effect of National Culture on the Choice of Entry Mode*, J. of Int. Business Studies, Vol. 19-3, pp. 411–432, 1988.
[KKN02] Kosheleva O., Kreinovich V. and Nguyen H.T. *On the Optimal Choice of Quality Metric in Image Compression*, Fifth IEEE Southwest Symposium on Image Analysis and Interpretation, 7–9 April 2002, Santa Fe, IEEE Comp. Soc. Digital Library, Electronic edition, pp. 116–120, 2002.
[LaLi81] Larson R.C. and Li V.O.K. *Finding Minimum Rectilinear Distance Paths in the Presence of Barriers*, Networks, Vol. 11, pp. 285–304, 1981.
[LCLMV04] Li M., Chen X., Li X., Ma B. and Vitányi P. *The Similarity Metric*, IEEE Trans. Inf. Theory, Vol. 50-12, pp. 3250–3264, 2004.
[LuRo76] Luczak E. and Rosenfeld A. *Distance on a Hexagonal Grid*, IEEE Trans. on Comp., Vol. 25-5, pp. 532–533, 1976.
[MaMo95] Mak King-Tim and Morton A.J. *Distances between Traveling Salesman Tours*, Discrete Appl. Math., Vol. 58, pp. 281–291, 1995.
[Mart00] Martin K. *A foundation for computation*, Ph.D. Thesis, Tulane University, Department of Math., 2000.
[MaSt99] Martin W.J. and Stinson D.R. *Association Schemes for Ordered Orthogonal Arrays and (T, M, S)-nets*, Can. J. Math., Vol. 51, pp. 326–346, 1999.
[Masc04] Mascioni V. *Equilateral Triangles in Finite Metric Spaces*, The Electronic J. Combinatorics, Vol. 11, 2004, R18.
[Matt92] S.G. Matthews, *Partial metric topology*, Research Report 212, Dept. of Comp. Science, University of Warwick, 1992.
[McCa97] McCanna J.E. *Multiply-sure Distances in Graphs*, Congressus Numerantium, Vol. 97, pp. 71–81, 1997.
[Melt91] Melter R.A. *A Survey of Digital Metrics*, Contemporary Math., Vol. 119, 1991.

[Monj98] Monjardet B. *On the Comparison of the Spearman and Kendall Metrics between Linear Orders*, Discrete Math., Vol. 192, pp. 281–292, 1998.
[Morg76] Morgan J.H. *Pastoral ecstasy and the authentic self: Theological meanings in symbolic distance*, Pastoral Psychology, Vol. 25-2, pp. 128–137, 1976.
[MLLM13] Mucherino A., Lavor C., Liberti L. and Maculan N. (eds.) *Distance Geometry*, Springer, 2013.
[Mura85] Murakami H. *Some Metrics on Classical Knots*, Math. Ann., Vol. 270, pp. 35–45, 1985.
[NeWu70] Needleman S.B. and Wunsh S.D. *A general Method Applicable to the Search of the Similarities in the Amino Acids Sequences of Two Proteins*, J. of Molecular Biology, Vol. 48, pp. 443–453, 1970.
[NiSu03] Nishida T. and Sugihara K. *FEM-like Fast Marching Method for the Computation of the Boat-Sail Distance and the Associated Voronoi Diagram*, Technical Reports, METR 2003-45, Dept. Math. Informatics, The University of Tokyo, 2003.
[OBS92] Okabe A., Boots B. and Sugihara K. *Spatial Tessellation: Concepts and Applications of Voronoi Diagrams*, Wiley, 1992.
[OkBi08] Okada D. and M. Bingham P.M. *Human uniqueness-self-interest and social cooperation*, J. Theor. Biology, Vol. 253-2, pp. 261–270, 2008.
[OSLM04] Oliva D., Samengo I., Leutgeb S. and Mizumori S. *A Subjective Distance between Stimuli: Quantifying the Metric Structure of Representations*, Neural Computation, Vol. 17-4, pp. 969–990, 2005.
[OnGi96] Ong C.J. and Gilbert E.G. *Growth distances: new measures for object separation and penetration*, IEEE Transactions in Robotics and Automation, Vol. 12-6, pp. 888–903, 1996.
[Ophi14] Ophir A. and Pinchasi R. *Nearly equal distances in metric spaces*, Discrete Appl. Math., Vol. 174, pp. 122–127, 2014.
[Orli32] Orlicz W. *Über eine Gewisse Klasse von Raumen vom Typus B'*, Bull. Int. Acad. Pol. Series A, Vol. 8–9, pp. 207–220, 1932.
[OASM03] Ozer H., Avcibas I., Sankur B. and Memon N.D. *Steganalysis of Audio Based on Audio Quality Metrics*, Security and Watermarking of Multimedia Contents V (Proc. of SPIEIS and T), Vol. 5020, pp. 55–66, 2003.
[Page65] Page E.S. *On Monte-Carlo Methods in Congestion Problem. 1. Searching for an Optimum in Discrete Situations*, J. Oper. Res., Vol. 13-2, pp. 291–299, 1965.
[Petz96] Petz D. *Monotone Metrics on Matrix Spaces*, Linear Algebra Appl., Vol. 244, 1996.
[PM] PlanetMath.org, http://planetmath.org/encyclopedia/
[Rach91] Rachev S.T. *Probability Metrics and the Stability of Stochastic Models*, Wiley, New York, 1991.
[ReRo01] Requardt M. and Roy S. *Quantum Spacetime as a Statistical Geometry of Fuzzy Lumps and the Connection with Random Metric Spaces*, Class. Quantum Gravity, Vol. 18, pp. 3039–3057, 2001.
[Resn74] Resnikoff H.I. *On the geometry of color perception*, AMS Lectures on Math. in the Life Sciences, Vol. 7, pp. 217–232, 1974.
[RiYi98] Ristad E. and Yianilos P. *Learning String Edit Distance*, IEEE Transactions on Pattern Recognition and Machine Intelligence, Vol. 20-5, pp. 522–532, 1998.
[RRHD10] Rocher T., Robine M., Hanna P. and Desainte-Catherine M. *A Survey of Chord Distances With Comparison for Chord Analysis*, Proc. Int. Comp. Music Conf., pp. 187–190, New York, 2010.
[RoPf68] Rosenfeld A. and Pfaltz J. *Distance Functions on Digital Pictures*, Pattern Recognition, Vol. 1, pp. 33–61, 1968.
[RTG00] Rubner Y., Tomasi C. and Guibas L.J. *The Earth Mover's Distance as a Metric for Image Retrieval*, Int. J. of Comp. Vision, Vol. 40-2, pp. 99–121, 2000.
[Rumm76] Rummel R.J. *Understanding Conflict and War*, Sage Publ., California, 1976.
[ScSk83] Schweizer B. and Sklar A. *Probabilistic Metric Spaces*, North-Holland, 1983.

[Selk77] Selkow S.M. *The Tree-to-tree Editing Problem*, Inform. Process. Lett., Vol. 6-6, pp. 184–186, 1977.

[ShKa79] Sharma B.D. and Kaushik M.L. *Limits intensity random and burst error codes with class weight considerations*, Elektron. Inform.-verarb. Kybernetik, Vol. 15, pp. 315–321, 1979.

[Tai79] Tai K.-C. *The Tree-to-tree Correction Problem*, J. of the Association for Comp. Machinery, Vol. 26, pp. 422–433, 1979.

[Tail04] Tailor B. *Introduction: How Far, How Near: Distance and Proximity in the Historical Imagination*, History Workshop J., Vol. 57, pp. 117–122, 2004.

[Tymo06] Tymoczko D. *The Geometry of Musical Chords*, Science, Vol. 313, Nr. 5783, pp. 72–74, 2006.

[ToSa73] Tomimatsu A. and Sato H. *New Exact Solution for the Gravitational Field of a Spinning Mass*, Phys. Rev. Letters, Vol. 29, pp. 1344–1345, 1972.

[Vard04] Vardi Y. *Metrics Useful in Network Tomography Studies*, Signal Processing Letters, Vol. 11-3, pp. 353–355, 2004.

[VeHa01] Veltkamp R.C. and Hagendoorn M. *State-of-the-Art in Shape Matching*, in *Principles of Visual Information Retrieval*, Lew M. (ed.), pp. 87–119, Springer-Verlag, 2001.

[Watt99] Watts D.J. *Small Worlds: The Dynamics of Networks between Order and Randomness*, Princeton Univ. Press, 1999.

[Wein72] Weinberg S. *Gravitation and Cosmology: Principles and Applications of the General Theory of Relativity*, Wiley, New York, 1972.

[Weis99] Weisstein E.W. *CRC Concise Encyclopedia of Math.*, CRC Press, 1999.

[Weis12] Weiss I. *Metric 1-spaces*, **arXiv:1201.3980[math.MG]**, 2012.

[Well86] Wellens R.A. *Use of a Psychological Model to Assess Differences in Telecommunication Media*, in *Teleconferencing and Electronic Communication*, Parker L.A. and Olgren O.H. (eds.), pp. 347–361, Univ. of Wisconsin Extension, 1986.

[WFE] Wikipedia, the Free Encyclopedia, http://en.wikipedia.org

[WiMa97] Wilson D.R. and Martinez T.R. *Improved Heterogeneous Distance Functions*, J. of Artificial Intelligence Research, Vol. 6, p. 134, 1997.

[WoPi99] Wolf S. and Pinson M.H. *Spatial-Temporal Distortion Metrics for In-Service Quality Monitoring of Any Digital Video System*, Proc. of SPIE Int. Symp. on Voice, Video, and Data Commun., September 1999.

[Yian91] Yianilos P.N. *Normalized Forms for Two Common Metrics*, NEC Research Institute, Report 91-082-9027-1, 1991.

[Youn98] Young N. *Some Function-Theoretic Issues in Feedback Stabilisation*, Holomorphic Spaces, MSRI Publication, Vol. 33, 1998.

[YOI03] Yutaka M., Ohsawa Y. and Ishizuka M. *Average-Clicks: A New Measure of Distance on the World Wide Web*, J. Intelligent Information Systems, Vol. 20-1, pp. 51–62, 2003.

[Zeli75] Zelinka B. *On a Certain Distance between Isomorphism Classes of Graphs*, Casopus. Pest. Mat., Vol. 100, pp. 371–373, 1975.

Index

Note: For the sake of clarity and convenience, items explained in the book with bullet points are capitalized, while regular index entries are listed in lower case letters.

M.M. Deza, E. Deza, *Encyclopedia of Distances*,
DOI 10.1007/978-3-662-44342-2

B

M

N

MIX
Papier | Fördert gute Waldnutzung
FSC® C083411

Kolibri 360 GmbH
Ferdinand-Jühlke-Straße 7
99095 Erfurt, Deutschland
produktsicherheit@kolibri360.de